Real-Time Electromagnetic Transient Simulation of AC–DC Networks

IEEE Press
445 Hoes Lane
Piscataway, NJ 08854

Real-Time Electromagnetic Transient Simulation of AC–DC Networks

Venkata Dinavahi
Ning Lin

IEEE PRESS SERIES ON POWER ENGINEERING

IEEE PRESS
WILEY

Published by John Wiley & Sons, Inc., Hoboken, New Jersey.
Published simultaneously in Canada.

Library of Congress Cataloging-in-Publication Data

ISBN: 9781119695448

Cover design by Wiley
Cover image: © Audrius J/Shutterstock

Set in 9.5/12.5pt STIXTwoText by Straive, Chennai, India

10 9 8 7 6 5 4 3 2 1

MIX
Paper from responsible sources
FSC
www.fsc.org
FSC® C013604

Ṛgvéda 3.62.10:

ॐ भूर्भुवः स्वः तत्सवितुर्वरेण्यं ।
भर्गो देवस्य धीमहि धियो यो नः प्रचोदयात् ॥

Om bhūrbhuvaḥ svaḥ tatsáviturvaréṇyaṃ;
bhargó de̱vasya dhīmahi dhiyo̱ yo náḥ praco̱dayāt.

We meditate on that most adored Supreme Soul, the Creator, whose effulgence illumines all realms. May this divine light illumine our intellect.

Dedicated to our parents:
Late Smt. Dinavahi Sasirekha and Shri. Dinavahi Ramarao
Liyue Xie and Quanyao Lin

Contents

About the Authors *xix*
Preface *xxi*
Acknowledgments *xxv*
List of Acronyms *xxvii*

1 Field Programmable Gate Arrays *1*
1.1 Overview *1*
1.1.1 FPGA Hardware Architecture *2*
1.1.2 Configurable Logic Block *3*
1.1.3 Block RAM *4*
1.1.4 Digital Signal Processing Slice *4*
1.2 Multiprocessing System-on-Chip Architecture *6*
1.3 Communication *7*
1.4 HIL Emulation *9*
1.4.1 Vivado® High-Level Synthesis Tool *9*
1.4.2 Vivado® Top-Level Design *11*
1.4.3 Number Representation and Operations *13*
1.4.4 FPGA Design Schemes *14*
1.4.4.1 Pipeline Design Architecture *14*
1.4.4.2 Parallel Design Architecture *14*
1.4.5 FPGA Experiment *15*
1.5 Summary *16*

2 Hardware Emulation Building Blocks for Power System Components *17*
2.1 Overview *17*
2.2 Concept of HEBB *18*
2.3 Numerical Integration *18*
2.4 Linear Lumped Passive Elements *20*

2.4.1 Model Formulation *20*
2.4.1.1 Resistance R *20*
2.4.1.2 Inductance L *20*
2.4.1.3 Capacitance C *22*
2.4.1.4 *RL* Branch *23*
2.4.1.5 *LC* Branch *23*
2.4.1.6 *RLCG* Branch *24*
2.4.2 Hardware Emulation of Linear Lumped Passive Elements *26*
2.5 Sources *27*
2.5.1 Hardware Emulation of Sources *28*
2.6 Switches *30*
2.6.1 Hardware Emulation of Switches *30*
2.7 Transmission Lines *32*
2.7.1 Traveling Waves *32*
2.7.2 Traveling Wave Model *35*
2.7.2.1 Modal Transformation *36*
2.7.3 Hardware Emulation of the TWM *39*
2.7.3.1 *Transformation Unit* *39*
2.7.3.2 *Update Unit* *39*
2.7.4 Frequency Dependent Line Model *41*
2.7.5 Hardware Emulation of FDLM *46*
2.7.5.1 *Convolution* Unit *46*
2.7.5.2 *Update* Unit *47*
2.7.6 Universal Line Model *48*
2.7.6.1 Frequency-Domain Formulation *48*
2.7.6.2 Time-Domain Formulation *49*
2.7.7 Hardware Emulation of the ULM *51*
2.7.7.1 *Update* $\boldsymbol{x}$ Unit *52*
2.7.7.2 *Convolution* Unit *52*
2.7.7.3 *Interpolation* Unit *54*
2.8 Network Solver *54*
2.8.1 Hardware Emulation of Network Solver *55*
2.8.2 Paralleled EMT Solution Algorithm *55*
2.8.3 *MainControl* Module *58*
2.8.4 Real-Time Emulation Case Study *59*
2.9 Nonlinear Elements: Iterative Real-Time EMT Solver *63*
2.9.1 Compensation Method *64*
2.9.2 Newton–Raphson Method *65*
2.9.3 Hardware Emulation of Nonlinear Solver *67*
2.9.3.1 Nonlinear Function Evaluation *68*
2.9.3.2 Parallel Calculation of $\boldsymbol{J}$ and $F(i_{km})$ *68*

2.9.3.3 Parallel Gauss–Jordan Elimination *71*
2.9.3.4 Computing v_c *71*
2.9.4 Case Studies *71*
2.10 Summary *77*

3 Power Transformers *79*
3.1 Overview *79*
3.2 Nonlinear Admittance-Based Real-Time Transformer Model *80*
3.2.1 Linear Model Formulation *80*
3.2.2 Linear Module Hardware Design *82*
3.2.3 Inode Unit Module *84*
3.2.4 Nonlinear Model Solution *85*
3.2.4.1 Preisach Hysteresis Model *88*
3.2.4.2 Nonlinear Module Hardware Design *89*
3.2.5 Frequency-Dependent Eddy Current Model *90*
3.2.6 Hardware Emulation of Power Transformer *91*
3.2.7 Real-Time Emulation Case Studies *94*
3.2.7.1 Case I *94*
3.2.7.2 Case II *99*
3.3 Nonlinear Magnetic Equivalent Circuit Based Real-time Multi-Winding Transformer Model *100*
3.3.1 Topological ST EMT Model *102*
3.3.1.1 ST Operating Principle *102*
3.3.1.2 Tap-selection Algorithm *102*
3.3.1.3 High-Fidelity Nonlinear MEC-Based ST Model *102*
3.3.1.4 Iron Core Hysteresis and Eddy Currents *107*
3.3.2 High-Fidelity Nonlinear MEC-Based ST Hardware Emulation *109*
3.3.2.1 Network Transient Emulation with Embedded ST *109*
3.3.3 Real-Time Emulation Case Studies *112*
3.3.3.1 Finite Element Modeling and Validation *112*
3.3.3.2 Case Studies *112*
3.4 Real-Time Finite-Element Model of Power Transformer *123*
3.4.1 Magnetodynamic Problem Formulation *123*
3.4.1.1 Refined TLM Solution *126*
3.4.1.2 Field-Circuit Coupling *130*
3.4.2 Hardware Emulation of Finite Element Model *132*
3.4.3 Case Studies *136*
3.4.3.1 Results and Validation *137*
3.4.3.2 Speed-up and Scalability *140*
3.5 Summary *141*

4 Rotating Machines *143*
4.1 Overview *143*
4.2 Lumped Universal Machine (UM) Model *144*
4.2.1 UM Model Formulation *144*
4.2.2 Interfacing UM Model with Network *146*
4.2.3 UM HEBB *148*
4.2.3.1 *Speed & Angle* Unit *149*
4.2.3.2 *FrmTran* Unit *150*
4.2.3.3 $Compi_{dq0}$ Unit *151*
4.2.3.4 *Flux & Torque* Unit *151*
4.2.3.5 *Update & CompVc* Unit *151*
4.2.4 Real-Time Emulation Case Study *152*
4.2.5 Overall Power System HEBB for Real-Time EMT Emulation *154*
4.3 General Framework for State-Space Electrical Machine Emulation *158*
4.3.1 FPGA Design Approaches for Electrical Machine Emulation *159*
4.3.2 State-Space Representation of Machine Models *160*
4.3.3 System Configuration on FPGA *161*
4.3.3.1 Number Representation *161*
4.3.3.2 Floating-Point Implementation by VHDL *162*
4.3.3.3 Fixed-Point Implementation by Schematic *167*
4.3.4 Evaluation of Designed Architectures *170*
4.3.4.1 Real-Time Emulation Accuracy Assessment *170*
4.3.4.2 Off-line Validation *171*
4.3.4.3 Hardware Resource Utilization *172*
4.3.5 Real-Time Emulation Case Studies *174*
4.3.5.1 Case I: Induction Motor Transients *174*
4.3.5.2 Case II: Synchronous Generator Transients *174*
4.3.5.3 Case III: Line Start-Permanent Magnet Synchronous Motor Transients *176*
4.3.5.4 Case IV: DC Motor Transients *177*
4.4 Nonlinear Magnetic Equivalent Circuit Based Induction Machine Model *178*
4.4.1 Magnetic Circuit *179*
4.4.2 Interfacing of Magnetic and Electric Circuits *181*
4.4.3 Electric Circuit *182*
4.4.4 Nonlinear Solution of Detailed MEC *182*
4.4.5 Hardware Emulation of Nonlinear MEC *183*
4.4.5.1 Parallel Gauss–Jordan Elimination Unit *185*
4.4.5.2 Parallel Computational Unit for Residual Vector *187*
4.4.5.3 Nonlinear Evaluation Unit *187*

4.4.6 Evaluation of Real-Time Emulation of Induction Machine *187*
4.5 Summary *190*

5 Protective Relays *193*
5.1 Overview *193*
5.2 Hardware Emulation of Multifunction Protection System *195*
5.2.1 Signal Processing HEBB *196*
5.2.1.1 CORDIC HEBB *196*
5.2.1.2 Symmetrical Components HEBB *198*
5.2.1.3 DFT HEBB *198*
5.2.1.4 Zero-Crossing Detection HEBB *199*
5.2.2 Multifunction Protective System HEBB *203*
5.2.2.1 Fault Detection HEBB *203*
5.2.2.2 Directional Overcurrent Protection HEBB *205*
5.2.2.3 Over/Under Voltage Protection HEBB *205*
5.2.2.4 Distance Protection HEBB *205*
5.2.2.5 Under/Over Frequency Protection HEBB *209*
5.3 Test Setup and Real-Time Results *209*
5.3.1 Case I *210*
5.3.2 Case II *213*
5.4 Summary *214*

6 Adaptive Time-Stepping Based Real-Time EMT Emulation *217*
6.1 Overview *217*
6.2 Nonlinear Solution and Adaptive Time-Stepping Schemes *219*
6.2.1 Nonlinear Element Solution Methods *219*
6.2.1.1 Newton–Raphson Method *219*
6.2.1.2 Piecewise Linearization (PWL) Method *219*
6.2.1.3 Piecewise N-R Method *220*
6.2.2 Adaptive Time-Stepping Schemes *220*
6.2.2.1 Local Truncation Error Method *220*
6.2.2.2 Iteration Count Method *221*
6.2.2.3 DVDT or DIDT Method *221*
6.2.3 Combinations of Adaptive Time-Stepping Schemes *222*
6.2.3.1 Measurements and Restrictions for Real-Time Emulation *222*
6.2.4 Case Studies *223*
6.2.4.1 Diode Full-Bridge Circuit *224*
6.2.4.2 Power Transmission System *225*
6.2.4.3 FPGA Implementation *229*
6.2.4.4 Real-Time Emulation Results *234*

6.3 Adaptive Time-Stepping Universal Line Model and Universal Machine Model for Real-Time Hardware Emulation *236*
6.3.1 Subsystem-Based Adaptive Time-Stepping Scheme *237*
6.3.2 Adaptive Time-Stepping ULM and UM Models *238*
6.3.2.1 ULM Computation *238*
6.3.2.2 Universal Machine Model Computation *242*
6.3.3 Real-Time Emulation Case Study *243*
6.3.3.1 Hardware Implementation *243*
6.3.3.2 Latency and Hardware Resource Utilization *246*
6.3.4 Results and Validation *247*
6.3.4.1 Validation of the ULM Model *247*
6.3.4.2 Real-Time Emulation Results *248*
6.4 Summary *252*

7 Power Electronic Switches *253*
7.1 Overview *253*
7.2 IGBT/Diode Nonlinear Behavioral Model *255*
7.2.1 Power Diode *256*
7.2.1.1 Mathematical Model *256*
7.2.1.2 Hardware Module Architecture *257*
7.2.2 IGBT *259*
7.2.2.1 Model Formulation *259*
7.2.2.2 Hardware Module Architecture *263*
7.2.2.3 Multiple Parallel Devices *265*
7.2.3 Electro-Thermal Network *267*
7.2.4 Hardware Emulation Results *268*
7.3 Physics-Based Nonlinear IGBT/Diode Model *270*
7.3.1 Physics-Based Nonlinear p–i–n Diode Model *271*
7.3.1.1 Model Formulation *271*
7.3.1.2 Model Discretization and Linearization *272*
7.3.1.3 Hardware Emulation on FPGA *274*
7.3.2 Physics-Based Nonlinear IGBT Model *276*
7.3.2.1 Model Formulation *276*
7.3.2.2 Model Discretization and Linearization *279*
7.3.2.3 Hardware Emulation on FPGA *281*
7.3.3 Hardware Emulation Results *285*
7.3.3.1 Test circuit *285*
7.3.3.2 Results and comparison *286*
7.4 IGBT/Diode Curve-Fitting Model *292*
7.4.1 Linear Static Curve-fitting Model *293*
7.4.1.1 Static Characteristics *293*

7.4.1.2 Switching Transients *293*
7.4.2 Nonlinear Dynamic Curve-fitting Model *296*
7.4.3 Hardware Emulation Results *298*
7.5 Summary *300*

8 AC–DC Converters *301*
8.1 Overview *301*
8.2 Detailed Model *303*
8.2.1 Detailed Equivalent Circuit Model *304*
8.3 Equivalenced Device-Level Model *305*
8.3.1 Power Loss Calculation *307*
8.3.2 Thermal Network Calculation *309*
8.3.3 Hardware Emulation of SM Model on FPGA *311*
8.3.4 MMC System Hardware Emulation *314*
8.3.5 Real-Time Emulation Results *316*
8.3.5.1 Test Circuit and Hardware Resource Utilization *316*
8.3.5.2 Results and Comparison for Single-Phase Five-Level MMC *318*
8.3.5.3 Results for Three-Phase Nine-Level MMC *324*
8.4 Virtual-Line-Partitioned Device-Level Models *324*
8.4.1 TLM-Link Partitioning *326*
8.4.2 Hardware Design on FPGA *328*
8.4.2.1 Hardware Platform *329*
8.4.2.2 Controller Emulation *329*
8.4.2.3 MMC Emulation on FPGA *330*
8.4.3 Real-Time Emulation Results *335*
8.4.3.1 MMC *335*
8.4.3.2 Induction Machine Driven by Five-Level MMC *342*
8.5 MMC Partitioned by Coupled Voltage–Current Sources *344*
8.5.1 V–I Coupling *344*
8.5.2 Hardware Emulation Case of NBM-Based MMC *346*
8.5.2.1 Power Converter HIL Emulation *346*
8.5.2.2 HIL Emulation Results and Validation *347*
8.5.2.3 Islanded MMC Performance *348*
8.5.2.4 MMC–MVDC Performance *355*
8.6 Clamped Double Submodule MMC *355*
8.6.1 Operation Principles of CDSM *357*
8.6.2 Device-Level Modeling Scheme *359*
8.6.2.1 Temperature-Dependent Electrical Interface Parameter Calculation *359*
8.6.2.2 Device-Level Linearized Transient Waveform Calculation *361*
8.6.3 SM-Level Modeling Scheme *362*

8.6.4 Converter-Level Modeling Scheme *362*
8.6.5 Case Study and Hardware Implementation *363*
8.6.5.1 Design Partition *365*
8.6.5.2 Latency and Resource Consumption *367*
8.6.6 Real-Time Emulation Results and Analysis *368*
8.6.6.1 Steady-State Results *368*
8.6.6.2 DC Power Flow Control *368*
8.6.6.3 DC Fault Transient Results *371*
8.7 Summary *374*

9 DC-DC Converters *377*
9.1 Overview *377*
9.2 Buck–Boost Converter *379*
9.2.1 System-Level Modeling *379*
9.2.2 Hardware Implementation *380*
9.3 Solid-State Transformer Modeling *381*
9.3.1 MMC Arm Models *382*
9.3.1.1 TLM-Stub Model (TLM-S) *382*
9.3.1.2 Nonlinear Switch-Based Model (NSM) *383*
9.3.1.3 Hybrid Arm Model *384*
9.3.2 Three-Phase Saturable Transformer Model *385*
9.3.3 SST EMT Model *385*
9.3.4 SST HIL Emulation *386*
9.3.5 SST Real-Time HIL Emulation Results *390*
9.3.5.1 Device-Level Behavior *390*
9.3.5.2 Converter Performance *391*
9.3.5.3 System Tests *392*
9.4 Summary *394*

10 DC Circuit Breakers *397*
10.1 Overview *397*
10.2 HHB in MTDC System *399*
10.2.1 MTDC Test System Schematic *399*
10.2.2 DC Line Protection *401*
10.2.2.1 Voltage Derivative Protection *401*
10.2.2.2 Over Current Protection *401*
10.3 Proactive Hybrid HVDC Breaker *402*
10.3.1 HHB EMT Model *403*
10.3.2 Varistor Model *404*
10.3.3 General HHB Unit Model *406*
10.3.4 Two-Node IGBT Models *407*

10.3.5 IGBT Low-Order Nonlinear Behavioral Model *409*
10.3.5.1 IGBT Fourth-Order Behavioral Model *409*
10.3.5.2 Parameters Extraction *409*
10.3.5.3 Sensitivity Analysis *410*
10.3.5.4 Model Parallelization *411*
10.3.6 Electro-Thermal Network *412*
10.3.7 HHB Hardware Implementation on FPGA *412*
10.3.8 HHB HIL Emulation Results *416*
10.3.8.1 Device-Level Performance *416*
10.3.8.2 System-Level Performance *424*
10.4 Ultrafast Mechatronic Circuit Breaker *426*
10.4.1 Nonlinear Device-Level Thyristor Model *426*
10.4.1.1 Basic Device Characteristics *426*
10.4.1.2 Scalable Cascaded Thyristor Model *428*
10.4.2 UFMCB Modeling *431*
10.4.3 Relaxed Scalar Newton–Raphson (RSNR) *433*
10.4.4 UFMCB Hardware Design *435*
10.4.5 UFMCB Real-Time Tests and Validation *438*
10.4.5.1 Four-Terminal DC Grid Test Case *438*
10.4.5.2 UFMCB Design Evaluation by HIL System *438*
10.4.5.3 UFMCB in HVDC Grid *442*
10.5 Summary *444*

11 Large-Scale AC and DC Networks *447*
11.1 Overview *447*
11.2 Spatial Decomposition and Parallelism *449*
11.2.1 Functional Decomposition for Large-Scale Real-Time Emulation *449*
11.2.2 Hardware Module Parallelism *451*
11.3 Multi-FPGA Hardware Design for Real-Time EMT Emulation *453*
11.3.1 Case I: 3-FPGA Hardware Design *454*
11.3.2 Case II: 10-FPGA Hardware Design *457*
11.3.3 Performance and Scalability of the Real-Time EMT Emulator *460*
11.4 CIGRÉ DC Grid Hybrid Modeling Methodology *465*
11.4.1 Network Topology *467*
11.4.2 Control Scheme *467*
11.4.3 Hybrid Modeling Methodology *468*
11.4.3.1 Device-Level Electrothermal Model *469*
11.4.3.2 Equivalent Circuit Model *469*
11.4.3.3 Average Value Model *471*
11.4.3.4 Transmission Line Model *471*
11.4.4 Real-Time MPSoC-FPGA Based DC Grid Emulator *471*

11.4.4.1 System Decomposition *471*
11.4.4.2 Hardware Resource Allocation and Task Partitioning *472*
11.4.4.3 Design and Implementation *474*
11.4.5 Real-Time Emulation Results and Validation *475*
11.4.5.1 Steady-State Operation *475*
11.4.5.2 Power Flow Command Change *477*
11.4.5.3 DC Fault *477*
11.5 Real-Time Co-Emulation Framework for Cyber-Physical Systems *479*
11.5.1 Communication Network Simulation and Co-Simulation *481*
11.5.2 Real-Time Co-Emulation Framework *484*
11.5.2.1 RTCE Hardware Architecture *484*
11.5.3 Hardware Implementation of RTCE *487*
11.5.3.1 Multi-Board EMT Emulation *488*
11.5.3.2 Communication Protocol and Implementation *489*
11.5.4 Real-Time Emulation Results and Verification *491*
11.5.4.1 Processing Delay and Hardware Resource Cost *491*
11.5.4.2 Case Study 1: Over-Current Fault *492*
11.5.4.3 Case Study 2: Communication Link Failure *493*
11.6 Faster-Than-Real-Time Hybrid Dynamic-EMT Emulation of AC–DC Grids *495*
11.6.1 Flexible Time-Stepping Algorithm for Dynamic Emulation *496*
11.6.1.1 Transient Stability Emulation Methodology *496*
11.6.1.2 Local Equipment Based Flexible Time-stepping *497*
11.6.2 AC–DC Grid Component Modeling *498*
11.6.2.1 AC–DC Grid Interface *498*
11.6.2.2 AC Grid Modeling *499*
11.6.2.3 DC Grid Modeling *501*
11.6.3 FTRT Emulation on FPGAs *503*
11.6.4 FTRT Emulation Results and Validation *505*
11.6.4.1 Three-Phase-to-Ground Fault *506*
11.6.4.2 Generator Outage and Sudden Load Change *507*
11.7 Summary *510*

Bibliography *513*

Appendix A Parameters for Case Studies *531*
A.1 Chapter 2 *531*
A.1.1 Case in Section 2.7 *531*
A.1.2 Cases in Section 2.8 *531*
A.2 Chapter 3 *531*
A.2.1 Cases in Section 3.2 *531*

A.2.1.1 Cases Study I 531
A.2.1.2 Cases Study II 532
A.2.2 Cases in Section 3.3 532
A.2.2.1 Transformer 532
A.2.2.2 System 532
A.2.3 Cases in Section 3.4 532
A.3 Chapter 4 533
A.3.1 UM Case in Section 4.2 533
A.3.2 Cases in Section 4.3 534
A.3.2.1 State-Space Matrices of Rotating Machines 534
A.3.2.2 Parameters of Rotating Machines 538
A.3.3 MEC Case in Section 4.4 538
A.4 Chapter 5 538
A.5 Chapter 6 539
A.5.1 Cases in Section 6.2 539
A.5.2 Cases in Section 6.3 540
A.6 Chapter 7 540
A.7 Chapter 8 541
A.7.1 Equivalenced Device-Level Model in Section 8.3 541
A.7.2 MMC-IM Case in Section 8.4 541
A.7.3 MVDC Case in Section 8.5 541
A.7.4 MTDC Case in Section 8.6 541
A.8 Chapter 9 541
A.9 Chapter 10 542
A.9.1 HHB Case 542
A.9.2 UFMCB Case 542
A.10 Chapter 11 543
A.10.1 CIGRÉ B4 DC Grid Test System 543

Index 545

About the Authors

Venkata Dinavahi received the B.Eng. degree in Electrical Engineering from the Visveswaraya National Institute of Technology (VNIT), Nagpur, India, in 1993, the M.Tech. degree in Electrical Engineering from the Indian Institute of Technology (IIT), Kanpur, India, in 1996, and the Ph.D. degree in Electrical and Computer Engineering from the University of Toronto, Ontario, Canada, in 2000. Currently he is a Professor with the Department of Electrical and Computer Engineering, University of Alberta, Edmonton, Alberta, Canada. His research interests include real-time simulation of power systems and power electronic systems, electromagnetic transients, device-level modeling, large-scale systems, and parallel and distributed computing. He was the founding chair of the IEEE Power and Energy Society (PES) Task Force on Interfacing Techniques for Simulation Tools from 2006 to 2014. He actively participates in several IEEE PES committees notably in the Working Group on Modeling and Analysis of System Transients. He has published over 150 papers in peer-reviewed international journals and over 60 articles in international conference proceedings. He is a Fellow of IEEE. He was the recipient of the 2018 Outstanding Engineer Award from the IEEE PES/IAS Northern Canada Chapter. He is a member of CIGRÉ (International Council on Large Electric Systems) and a Professional Engineer in the Province of Alberta.

Ning Lin received the B.Sc. and M.Sc. degrees in Electrical Engineering from Zhejiang University, China, in 2008 and 2011, respectively, and the Ph.D. degree in Electrical and Computer Engineering from the University of Alberta, Canada, in 2018. From 2011 to 2014, he worked as an engineer on power system automation, flexible AC transmission system (FACTS), and high-voltage direct current (HVDC). Currently, he is a postdoc assistant researcher at the University of Alberta. His research interests include power electronics, real-time simulation of power systems, and high-performance computing of integrated AC/DC grids.

Preface

Promising advances in high-power electronics and the challenges facing modern power system operation in terms of integration of large-scale renewable generation and energy storage is necessitating the construction of high-voltage direct current (HVDC) transmission worldwide. Meshed multi-terminal DC grids are transporting bulk energy over longer distances, interconnecting far-flung asynchronous AC network, and are being operated with higher redundancy, efficiency, and reliability. Since the conception of real-time analog simulators (known as transient network analyzers (TNAs)) and subsequently real-time digital power system simulators in the 1970s through to the 1990s, one of their principal applications has been for the testing of control and protection functions in a hardware-in-the-loop (HIL) configuration for HVDC systems prior to their commissioning. Currently, real-time digital electromagnetic transient (EMT) simulators are used in every sector of electrical power chain: generation, transmission, distribution, and consumption. Moreover, real-time EMT simulators are also frequently employed in the transportation (automotive and railway traction), aviation, and marine industries, wherein the electrical systems share many structural and functional commonalities with terrestrial AC–DC networks. While the design, testing, and commissioning of local control and protection functions of system equipment is the primary objective of real-time EMT simulation, it is also paramount for global dynamic and interactive studies of large-scale power systems for wide-area control, protection, and system operator training. EMT simulation of large AC–DC grids in real time is a significant challenge due to the need for detailed device-level modeling of system components while simultaneously reproducing the system-level interactions accurately by accommodating large system sizes. These contrasting requirements place an enormous burden on the simulator latency constraints and the hardware selection for implementing the real-time simulation.

EMTs are the consequence of the interplay between the electric and magnetic fields in power system equipment when the system is perturbed from its steady-state operation by an event such as the switching of a line, a fault, or a lightning strike. A widely accepted classification of EMT is into four categories: low-frequency oscillations (0.1 Hz–3 kHz) e.g. temporary overvoltages, slow-front surges (50 Hz–20 kHz) e.g. switching overvoltages, fast-front surges (10 kHz–3 MHz) e.g. lightning overvotages, and very fast-front surges (100 kHz–50 MHz) e.g. circuit breaker restrike overvoltages. The switching on and off of power semiconductor devices such as insulated gate bipolar transistors (IGBTs) also generates transients albeit at the device-level, and they are typically in the 1–20 MHz range. The transients propagate near to the speed of light from their inception point and spread throughout the system causing extensive damage if not interrupted or absorbed. Of all the studies done in power systems, the modeling required to simulate EMT is the most complex. The higher the frequency of the transient, the more complex the modeling of any equipment becomes and the smaller the size of the network it spreads to. Conversely, low-frequency transients entail simpler modeling of equipment and propagate over a wider scale requiring the analysis of a larger network. For a particular EMT case study, the factors that influence the modeling complexity of any equipment are the distributed nature of parameters, frequency-dependency, and the inherent nonlinearities in the device. Accordingly, the extent to which each of these phenomena is accurately modeled for every equipment determines the computational burden experienced by the simulator. The selection of model decomposition strategy, numerical solution algorithm, and time-step also depends on these same factors. All of the above considerations need to be observed to successfully design and implement a real-time EMT simulation.

Of the many digital processors currently available on the market, field-programmable gate arrays (FPGAs) are the sole processor technology capable of meeting the rigorous real-time constraints and large hardware resource capacity required for AC–DC grid emulation with detailed equipment modeling. FPGAs contain numerous configurable logic resources, distributed memory, and inputs/outputs that can provide massive hardware parallelism for emulating detailed power system component models simultaneously. Currently available commercial FPGAs offer millions of logic gates, high-bandwidth on-chip memory, and ultrafast transceivers. The recent development of FPGA technology is not only in terms of improving their timing performance and enlarging their hardware resource capacity but also in terms of streamlining an integrated design tool ecosystem. The high-level synthesis (HLS) design methodology using C/C++ programming language instead of the conventional hardware description language (HDL) is explored as a new means of modularizing the AC–DC grid to reduce the hardware design cycle while maintaining a high synthesis efficiency. The synthesized hardware modules can then be pipelined efficiently with optimized performance regarding the resource usage and latency requirements.

This book intends to provide a detailed exposition of FPGA hardware based real-time EMT emulation for the fundamental components used in AC–DC power systems. Specific focus is afforded to detailed device-level models for their hardware realization in a massively parallel and deeply pipelined manner, and decomposition techniques for emulating large systems. In the various chapters of the book, while the hardware emulation may have been done on FPGAs from different generations or vendors, the underlying principles of model decomposition and parallel solution algorithms are generally applicable on any device. This book is intended for two groups of readers: graduate students in a university and professional research engineers and scientists in the industry. University students pursuing masters and doctoral degrees will find state-of-the-art presentation of the material on real-time EMT simulation of AC–DC grids. Such material can inspire and motivate advanced research ideas for their projects, dissertations, and publications. For industry experts, the book provides relevant academic research developments and implementation knowledge that they can incorporate in their respective product development process. The book is organized as follows: after giving a brief introduction of FPGA architecture and design flow in Chapter 1, Chapter 2 introduces the concepts of hardware emulation building blocks for fundamental power system components necessary for real-time EMT simulation: linear lumped passive elements, transmission lines, and nonlinear elements. Chapters 3 and 4 cover power transformers and rotating electrical machines building from simpler lumped linear models to detailed nonlinear magnetic equivalent circuit based models and finite element models. Chapter 5 describes the hardware emulation techniques for digital protective relays. Chapter 6 addresses the emerging and challenging topic of adaptive time-stepping based real-time EMT simulation. Chapter 7 discusses power semiconductor component models and their hardware emulation varying in complexity from simple system-level switch models to detailed device-level nonlinear behavioral and physics-based electrothermal models. Chapters 8–10 cover the modeling and emulation of various building blocks of DC grids: AC/DC converters, DC/DC converters, and DC circuit breakers. Finally, Chapter 11 culminates in examining the challenge of real-time EMT simulation of large-scale AC and DC grids.

Venkata Dinavahi
Edmonton, Alberta, Canada

Acknowledgments

We would like to express our sincere thanks to the former graduate students and members of the Real-Time Experimental Laboratory (RTX-LAB) at the University of Alberta for allowing us to use material from their publications and theses in this book: Dr. Yuan Chen, Dr. Jiadai Liu, Wentao Wang, Dr. Nariman Roshandel Tavana, Yifan Wang, Dr. Zhuoxuan Shen, Dr. Peng Liu, Dr. Tong Duan, and Shiqi Cao.

We would also like to express our profound gratitude to the Wiley Senior Publishing Team, especially Ms. Mary Hatcher, Ms. Victoria Bradshaw, Ms. Teresa Netzler, and Ms. Amudhapriya Sivamurthy for their timely feedback and for facilitating an efficient and seamless publishing process.

List of Acronyms

AC	alternating current
APU	application processing unit
ASIC	application specific integrated circuit
ATP	alternative transients program
AVM	average value model
AVR	automatic voltage regulator
AXI	advanced eXtensible interface
BJT	bipolar junction transistor
BRAM	block random access memory
CDSM	clamp double submodule
CFM	curve-fitting model
CLB	configurable logic block
CPS	cyber physical system
CPU	central processing unit
DAC	digital to analog converter
DC	direct current
DCFM	dynamic curve-fitting model
DDR4	double data rate fourth-generation
DEM	detailed equivalent model
DMA	direct memory access
DSP	digital signal processing (processor)
DUT	device-under-test
EMT	electro-magnetic transient
FACTS	flexible alternating current transmission system
FBSM	full-bridge submodule
FDLM	frequency dependent line model
FEM	finite element method
FF	flip-flop
FFT	fast Fourier transform

FIFO	first-in first-out
FMC	FPGA mezzanine card
FPGA	field programmable gate array
FPU	floating point unit
FSM	finite state machine
GPIO	general purpose IO
GPU	graphical processing unit
GT	gigabit transceiver
GTH	gigabit transceiver H
GTX	gigabit transceiver X
GTY	gigabit transceiver Y
HBSM	half-bridge submodule
HDL	hardware description language
HEBB	hardware emulation building block
HHB	hybrid HVDC breaker
HIL	hardware-in-the-loop
HLS	high-level synthesis
HVDC	high voltage direct current
I/O	input/output
IGBT	insulated gate bipolar transistor
JTAG	joint test action group
KCL	Kirchhoff's current law
KVL	Kirchhoff's voltage law
LCS	load commutation switch
LTE	local truncation error
LUT	look-up table
LVDS	low-voltage differential signaling
lwIP	light weight Internet protocol
MAC	media access layer
MFT	medium frequency transformer
MMC	modular multi-level converter
MMF	magnetomotive force
MOSFET	metal oxide semiconductor field-effect transistor
MOV	metal oxide varistor
MPSoC	multi-processor system-on-chip
MTDC	multi-terminal direct current
MVDC	medium voltage direct current
NBM	nonlinear behavioral model
NLD	nonlinear diode
NPC	neutral-point-clamped
N–R	Newton–Raphson

PCS	physical coding sublayer
PI	proportional integral
PL	programmable logic
PLL	phase locked loop
PMU	phasor measurement unit
PNR	piecewise Newton–Raphson
PS	processing system
PSC	phase-shifted carrier
PSS	power system stabilizer
PWL	piecewise linearization
PWM	pulse width modulation
QSFP	quad small form-factor pluggable
QSPI	quad serial peripheral interface
RAM	random access memory
RMS	root mean square
ROM	read only memory
RTCE	real-time co-emulation
RTL	register-transfer level
SCADA	supervisory control and data acquisition
SCR	silicon controlled rectifier
SDK	software development kit
SFP	small form-factor pluggable
SM	submodule
SST	solid-state transformer
STDC	super transient data concentrator
TCP/IP	transmission control protocol/Internet protocol
TDC	transient data concentrator
TLM	transmission line modeling
TLN	transmission level networking
TMU	transient measurement unit
TSSM	two state switch model
TWM	traveling wave model
UDP	user datagram protocol
UFD	ultrafast disconnector
UFMCB	ultrafast mechatronic circuit breaker
ULM	universal line model
UM	universal machine
USB	universal serial bus
VHDL	very high-speed integrated circuit hardware description language
VSC	voltage source converter

1

Field Programmable Gate Arrays

1.1 Overview

Field programmable gate arrays (FPGAs) were invented in the 1980s at Xilinx® as programmable array logic (PAL) devices [1, 2] and underwent three distinct stages of evolution each lasting a span of approximately 10 years [3]: Age of Invention, Age of Expansion, and the Age of Accumulation. Over this 30 year time-span, FPGAs have increased in logic capacity by a factor of over 10 000 and increased in speed by over 100; over the same period, their cost and power consumption per operation decreased by a factor of 1000. These impressive developments in processor technology led to their expansion into myriad applications [4]. The fine-grained resolution of logic resources and memory on the FPGA allows reconfiguring the hardware for a specific user application.

Reconfiguration was made possible due to static random-access memory (SRAM) based FPGAs, which can be reprogrammed using a configuration bitstream. The traditional FPGA design flow creates a register transfer level (RTL) abstraction using a hardware description language (HDL). The HDL code is translated by a synthesis tool into netlist files that are textual representation of the logic design. Then place-and-route tool is used to map the netlists to a specific FPGA device, and bit-generation tool creates a configuration bitstream. At each stage of the design, simulation can be used to verify the accuracy, timing, and functionality. While users can emulate any digital logic design on the FPGA, practical constraints such as the amount of hardware resources consumed, clock speed, and inputs/outputs (I/Os) need to be observed to achieve a successful emulation.

Although many-core graphics processing units (GPUs) may be more attractive for large-scale applications where floating-point computations predominate, algorithms implemented on GPUs must confirm to the single instruction multiple data (SIMD) paradigm to achieve the most acceleration. In comparison with GPUs, FPGAs have much lower latencies and higher computational density per watt

Real-Time Electromagnetic Transient Simulation of AC-DC Networks, First Edition.
Venkata Dinavahi and Ning Lin.

which is becoming increasingly significant as energy and cooling expenditures form the major portion of operating cost of high-performance compute (HPC) facilities. Currently available FPGAs contain millions of programmable logic gates, high-bandwidth memory, dedicated multiplier blocks, gigabit per second communication transceivers, and on-chip processors, making the platform very flexible and extensible for any specific application. FPGA vendors provide families of devices of various speeds, sizes, and functionality, each employing slightly different proprietary technologies, to target groups of application domains. In addition to hardware, vendors provide highly integrated computer-aided design (CAD) tools such as Vivado® from Xilinx® and Quartus Prime® from Intel®.

The future of the FPGA looks bright, and industry trends point to on-chip heterogeneous integration of programmable logic with multi-core CPU and many-core GPU resources to create multi-processing system-on-chip (MPSoC) [5] and adaptive computing acceleration platform (ACAP) [6] architectures to address the needs of evolving and diverse applications. FPGAs are currently being used in a wide range of application domains including consumer electronics, medical devices, robotics, industrial instrumentation and controllers, wireless communications, automotive electronics, and aerospace and defense equipment. Major cloud service providers are now integrating FPGAs [7] into their compute servers that are used for big data analytics. This chapter briefly introduces some of the terms and definitions related to FPGA architecture and design flow for hardware-in-the-loop (HIL) emulation.

1.1.1 FPGA Hardware Architecture

The FPGA is an integrated circuit containing an array of 2-D configurable logic blocks (CLBs) which are interconnected through wires and programmable switch matrices. A fundamental CLB is able to implement both combinational and sequential logic functions, and the programmable switch matrices also help to achieve hardware reconfigurability. Two typical FPGA hardware architectures are given in Figure 1.1 [8–10], which shows the I/O blocks connecting the CLBs and programmable switch matrices are arranged at the periphery of the logic array. The column-based advanced silicon modular block (ASMBL) architecture created by Xilinx® offers users a greater convenience in choosing an FPGA device with proper features for their design. This structure is adopted for the 7-series and Ultrascale+® FPGA devices. According to Figure 1.1b, the FPGA is composed of 6-input look-up tables (LUTs) based CLBs, on-chip block memory, digital signal processing (DSP) slices, precise clocking resources, enhanced PCIe® interface blocks, and the programmable switches interconnected via wires.

The Xilinx® Virtex®-7 FPGA VC707 Evaluation Kit [10] and the Xilinx® Virtex® Ultrascale+ FPGA VCU118 Evaluation Kit [11] are the two boards used in HIL

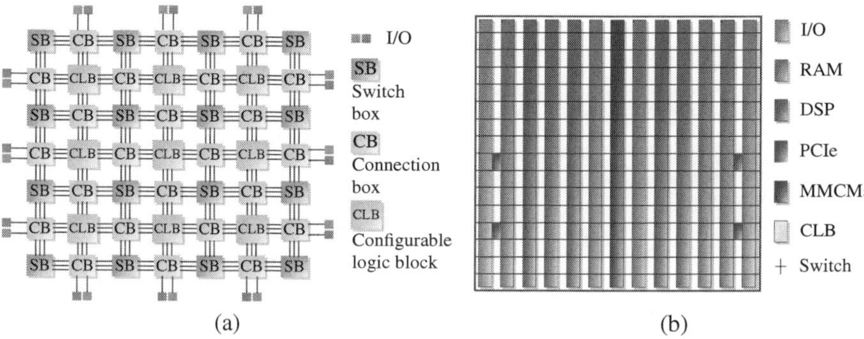

Figure 1.1 FPGA hardware: (a) mesh architecture and (b) Virtex®-7 ASMBL architecture. Source: Kilts [8], Farooq et al. [9]

emulation in many of the chapters of this book. The Virtex®-7 XC7VX485T FPGA was manufactured with 28 nm process technology, while the Virtex® Ultrascale+ XCVU9P FPGA was manufactured on 16 nm 3D FinFET process technology. A comparison of the main logic resources on the XC7VX485T and the XCVU9P devices is presented in Table 1.2 [12, 13].

While the two generations of FPGA boards share many types of resources, the latest UltraScale+ XCVU9P FPGA is more resource-abundant and efficient in data exchange than the Virtex®-7 FPGAs. The availability of some of the resources is significant as it affects the scale of the emulated system for HIL application. Thus, they are introduced in the following subsections.

1.1.2 Configurable Logic Block

The CLB is the fundamental component in the FPGA for providing basic logic and arithmetic functions as well as data storage. In the Xilinx® Virtex®-7 series FPGAs, the CLB contains two side-by-side slices, each of which is composed of four 6-input LUTs, which has two flip-flops [14, 15]. In addition, a CLB also has three wide-function multiplexers and the carry chain to perform arithmetic adding and subtracting operands in its slices.

Figure 1.2 shows the architecture of a CLB. The slices, organized as two individual columns, are not directly connected to each other. *Slice0* is at the bottom of the CLB and place in the left column, while *Slice1* locates at the top and in the right column of the die.

There are two types of CLB slices: those support data storage using distributed RAM and data shift with 32-bit registers are categorized as SLICEM, while the rest are named as SLICEL. Then, a CLB can contain either two SLICEL or one SLICEL and SLICEM. The LUT in the Virtex®-7 FPGA can be implemented as one 6-input 1-output LUT for 64-bit ROM or two 5-input LUTs with individual outputs

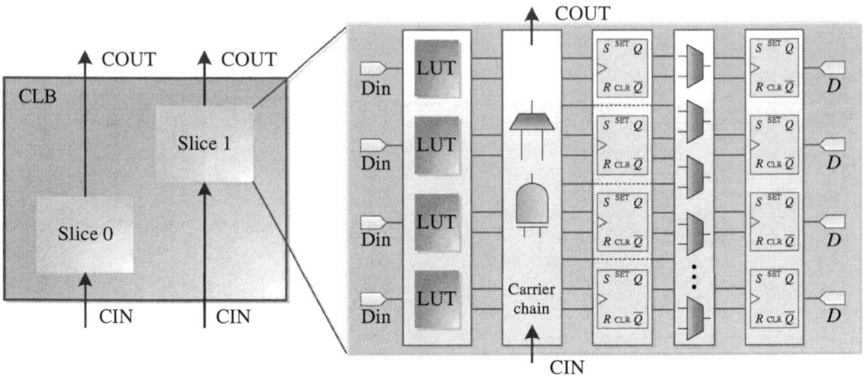

Figure 1.2 Configurable logic block architecture.

for 32-bit ROMs. The carrier chain contains multiplexers and an XOR logic gate for the addition or subtraction operation. As can be seen, the inputs and outputs of a slice are also its ports.

1.1.3 Block RAM

In Xilinx® Virtex®-7 series and Ultrascale+ FPGAs, the block RAM (BRAM) has up to 36 kbits data storage capability, and it can be implemented as either 1 RAM or 2 separate RAMs with each having 18 kbits data [16, 17]. In addition, it also has the cascaded manner when an adjacent 36 kbits BRAM is implemented, i.e. 1×64 kbits, and under simple dual-port mode, there are a variety of configurations, e.g. 1×32 kbits, 2×16 kbits, or even 72×512bits. Similar configurations are also available to the two separated 18 kbits RAMs.

Under simple dual-port mode, there is only one read-only port and write-only port, which has a high degree of independence, e.g. they are controlled by two clocks, and the data width can also be different, and independent read/write actions can take place simultaneously. Correspondingly, another BRAM type is the true dual-port RAM, whose symmetrical configuration is given in Figure 1.3. It ensures a flexible data access to either or both ports by enabling them to have an individual address, input/output data, a clock signal, write enable, etc. The description of those port names is provided in Table 1.1.

1.1.4 Digital Signal Processing Slice

Programmable logic devices are efficient carriers for DSP applications, which use many binary multipliers and accumulators. Both the 7-series and UltraScale+ FPGAs have a number of dedicated low-power DSP slices, integrating high

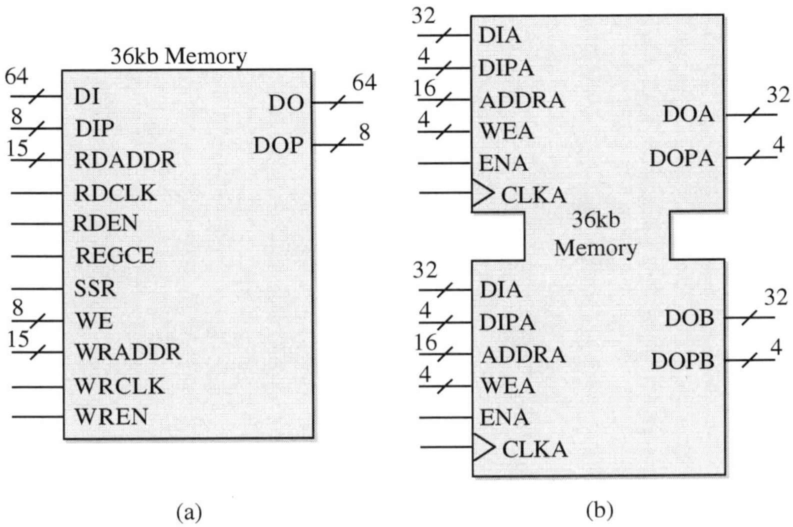

Figure 1.3 Virtex®-7 series FPGA block memory: (a) simple dual-port RAM and (b) true dual-port RAM.

Table 1.1 Dual-port RAM description.

Port	Direction	Description
DI	In	Data input bus
DIP	In	Data input parity bus
ADDR	In	Address bus
WE	In	Byte-wide write enable
EN	In	BRAM write enable
CLK	In	Clock input
DO	Out	Data output bus
DOP	Out	Data output parity bus

speed with compact size while at the same time the system design flexibility is maintained. In addition to DSP, the DSP slices also enable wide dynamic bus shifters, memory address generators, memory-mapped I/O registers, etc. On the 7-series FPGA boards, DSP48E1 slice is adopted [18], as shown in Figure 1.4, while its UltraScale+ counterpart is defined using more advanced DSP48E2 [19]. As shown in its slice architecture, the DSP48E1 slice includes 25×18 two's-complement multiplier, a 48-bit accumulator, 25-bit power-saving pre-adder, a pattern detector, etc.

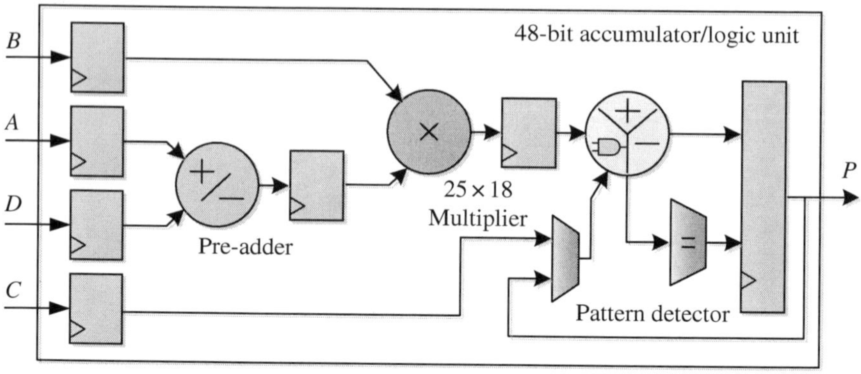

Figure 1.4 Basic DSP48E1 slice functionality.

1.2 Multiprocessing System-on-Chip Architecture

Conventionally, an MPSoC is a device that contains multiple processing elements (microprocessors) that are heterogenous in nature and address the computing needs of an application area. In the context of a MPSoC from Xilinx®, the device consists of a multi-core CPU, many-core GPU, programmable logic (FPGA), memory, and I/O peripherals. As shown in Figure 1.5, these hardware resources are divided into two processing domains: processing system (PS) and processing logic (PL). On the Xilinx® UltraScale+ XCZU9EG MPSoC, the PS consists of ARM® Cortex-A53 quad-core application processing units (APUs), ARM® Cortex-R5 dual-core real-time processing units (RPUs), the ARM® Mali-400 MP2 GPU, memory, and high-speed connectivity, whereas the PL consists of system logic resources, memory, high-speed, and general purpose I/O [5, 21]. Within each APU, various components are available to accelerate computation, such as NEON (an advanced SIMD architecture), floating point unit (FPU), etc.

The PS communicates with PL using high bandwidth and low-latency Advanced eXtensible Interface (AXI) channels. Such architecture provides high flexibility to merge the advantages of the fast sequential calculation and the hardwired parallelism to meet the requirements of high-performance computing. However, the logic resources of MPSoC are typically lower than the FPGA device with the same manufacturing technology. Due to the complex routing and additional combinational and sequential logics inferred for reconfiguration on FPGA, the clock frequency of such a device is lower than that of the PS. It is critical to analyze the characteristics of the application before determining the implementation platform. The usage of FPGA resources can be low if the implemented functions are not frequently used. The resources can be a major limitation for the simulation of complex and large-scale systems on FPGA. A combination of sequential CPU

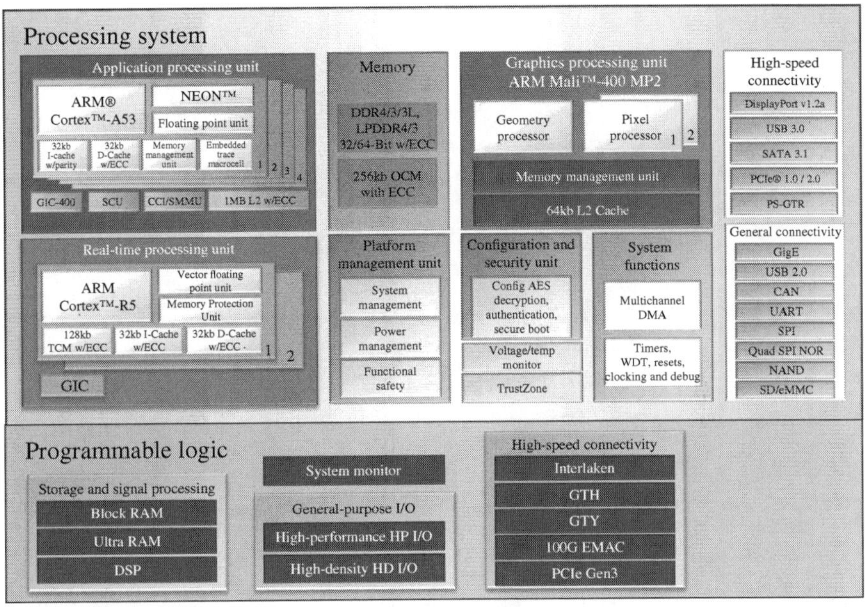

Figure 1.5 Zynq® Ultrascale+™ MPSoC block diagram. Source: Xilinx [20]. ©2018, Xilinx, Inc.

and parallel hardware implementation on the MPSoC can be an efficient and cost-effective compromise in such cases. A summary of the main resources of the devices used in this work are shown in Table 1.2.

Xilinx® provides the Software Development Kit (SDK) tools for the software design of the MPSoC device. SDK imports the hardware design from Xilinx® Vivado® tool and provide the board support package which includes various fundamental drivers for the resources of both PS and PL.

1.3 Communication

Both Xilinx® VCU118 and ZCU102 boards have plenty of components and communication interfaces, such as double data rate fourth-generation (DDR4) memory, quad serial peripheral interface (QSPI) flash, general purpose IO (GPIO), FPGA mezzanine (FMC) interface, and small form-factor pluggable (SFP) interface. Utilizing these interfaces and communication protocols, data transfer can be accomplished between multiple boards. For example, the SFP interface of ZCU102 and quad-SFP (QSFP) interface of VCU118 can be interconnected with cable and the Xilinx® Aurora IP cores can be used to accomplish the communication between the two boards as shown in Figure 1.6a. The Aurora 64B/66B core is

Table 1.2 Hardware resource comparison.

Resource	Virtex®-7 XC7VX485T FPGA	UltraScale+ XCVU9P FPGA	Zynq UltraScale+ XCZU9EG MPSoC
Logic cells	485 760	2 586 150	600 000
CLB FFs	607 200	2 364 480	548 160
CLB LUT	303 600	1 182 240	274 080
Block RAM (kbits)	37 080	75 900	32 100
Clocking (CMTs)	14	30	4
DSP slices	2 800	6 840	2 520
PCIe®	4 Gen2	6 Gen3 × 16/Gen4 × 8	—
Transceivers	GTX (12.5 Gb/s) 56	GTY (32.75 Gb/s) 120	GTH (16.3 Gb/s) 24

Source: Xilinx [12].

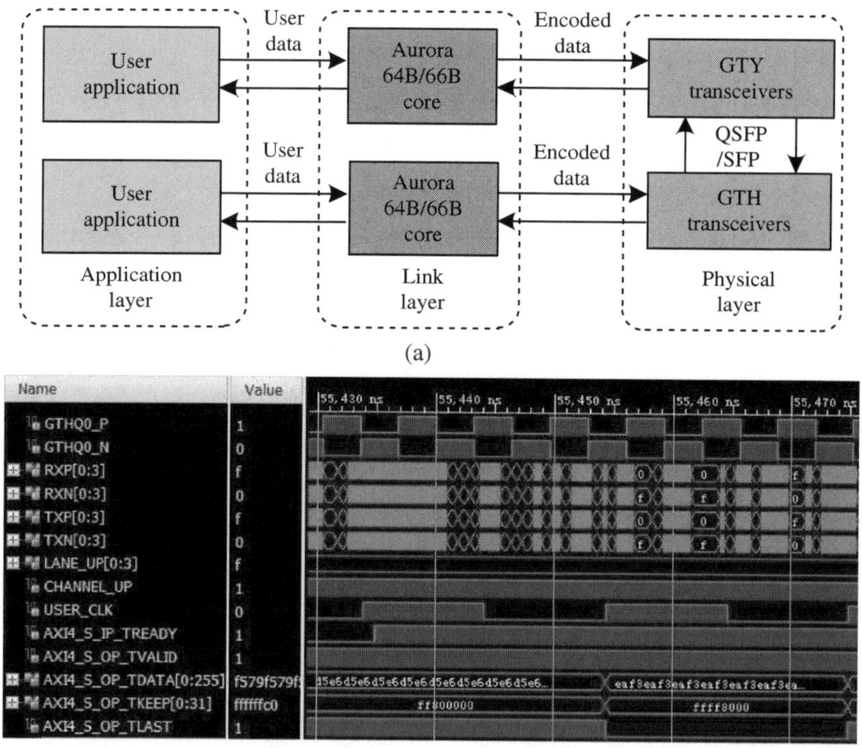

(b)

Figure 1.6 Communication process: (a) block diagram and (b) digital signal waveforms. Source: Xilinx, Inc.

a scalable, lightweight, high-data rate, link-layer protocol for high-speed serial communication, supporting bi-directional transfer of data between devices using consecutive bonded gigabit transceiver Y (GTY) on VCU118 board and gigabit transceiver H (GTH) on ZCU102 board [22]. Four transceivers located in ZCU102 SFP and VCU118 QSFP interface can be used to construct four lanes with 64-bit AXI-4 user data stream transmitting in each lane, which can achieve a throughput from 500 Mb/s to over 254 Gb/s. Figure 1.6b shows the waveforms of major signals during the communication process. When CHANNEL_UP signal is asserted, the Aurora cores have initialized and established four channel lanes for user applications to pass frames of data. User data are loaded on AXI4_TDATA[0:255] bus (64 bits × 4 lanes) at each edge of USER_CLK when AXI_TREADY is asserted. Then user data are transferred into encoded differential serial data (RXP/N[0:3] and TXP/N[0:3]) and transmitted through the four GTH or GTY transceivers and the QSFP/SFP cable.

1.4 HIL Emulation

The entire hardware design procedure for HIL emulation is depicted in Figure 1.7. The whole process can be completed in three stages, summarized as follows:

- Data entry by high-level synthesis.
- Top-level design and simulation with Vivado®.
- Bit file generation and experimental test on FPGA.

Thus, in the following each stage is specified, including tools necessary for the design, the programming language, and prototype setting.

1.4.1 Vivado® High-Level Synthesis Tool

The Xilinx® high-level synthesis software Vivado HLS® is able to transform C/C++ functions into an RTL implementation which synthesizes into the vendor's FPGAs. During this stage of design, the user can develop a hardware module using the programming language C/C++, rather than VHDL at the logic gate level, which greatly facilitates the hardware design. For example, to realize a complex multiple-input-multiple-output algebraic module, the C/C++ function can be written as:

```
void func (float ai, float bi, float *ao, float *bo){
  algebraic functions here;
}
```

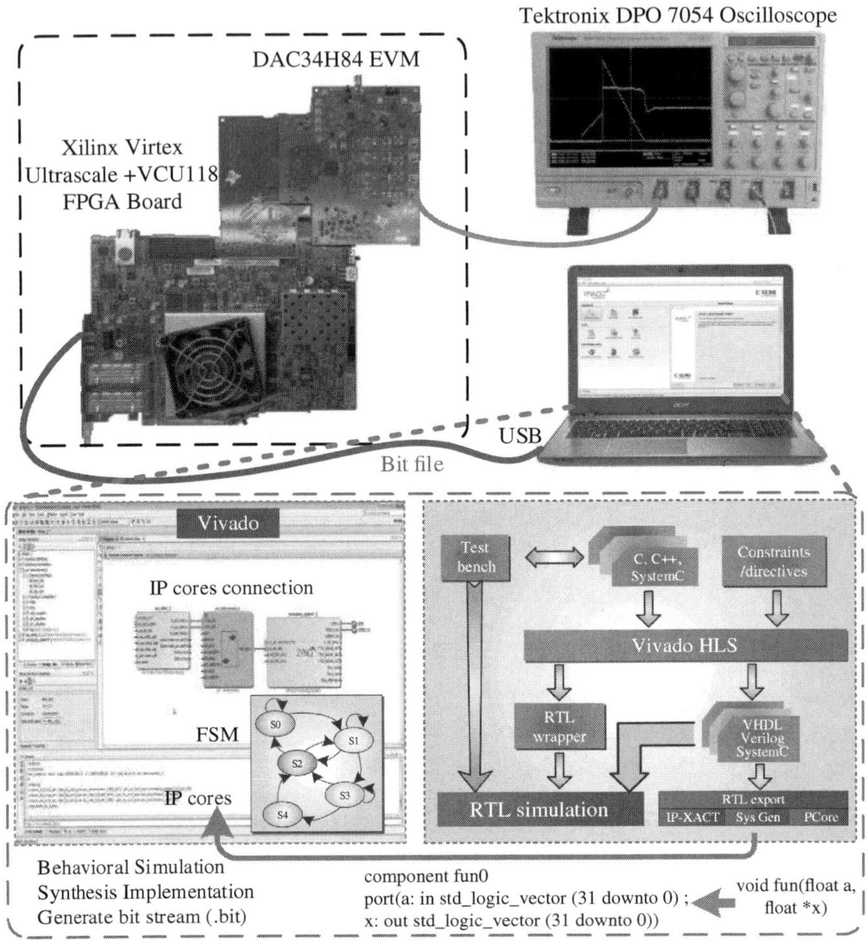

Figure 1.7 Hardware design procedure and experimental setup. Source: Xilinx, Inc.

Note that the variables are defined as a floating point which corresponds to 32 bits because it is more efficient for computation than 64-bit digits. The algebraic function description could contain potentially parallel operations, and therefore the design tool offers pipeline structure option in its directives, which greatly facilitates programming. Then, the C synthesis function provided by that tool creates the RTL design of the written function automatically. The syntax is also checked during this process: an erroneous function would lead to immature termination of C synthesis. The option *Export RTL* enables the RTL design to be exported as an intellectual property (IP), which has corresponding input/output ports in VHDL format:

```
COMPONENT func
  PORT (
  ao_ap_vld: OUT STD_LOGIC;
  bo_ap_vld: OUT STD_LOGIC;
  ap_clk: IN STD_LOGIC;
  ap_rst: IN STD_LOGIC;
  ap_start: IN STD_LOGIC;
  ap_done: OUT STD_LOGIC;
  ap_idle: OUT STD_LOGIC;
  ap_ready: OUT STD_LOGIC;
  ai: IN STD_LOGIC_VECTOR(31 DOWNTO 0);
  bi: IN STD_LOGIC_VECTOR(31 DOWNTO 0);
  ao: OUT STD_LOGIC_VECTOR(31 DOWNTO 0);
  bo: OUT STD_LOGIC_VECTOR(31 DOWNTO 0);
)
```

Since RTL design could be completed by Vivado HLS®, C/RTL co-simulation is available in the software after writing a corresponding test-bench. The co-simulation is deemed equivalent to the hardware behavioral simulation, and the results are a preliminary validation of the hardware design even though it is C-based.

1.4.2 Vivado® Top-Level Design

A user application or system model may contain a number of subsystems, normally classified according to their functionality. For HIL emulation, all those subsystems are first written as separate C/C++ functions under Vivado HLS® environment, and after IP generation and export they can be identified by Vivado® as VHDL components. Those user-defined IPs are in fact treated as the same as the default hardware modules in the IP catalog.

In a typical real-time application, signals between VHDL components are exchanged at the end of each time-step, e.g. a controller sends command signals to the modeled plant, which returns its sampled voltages and currents. In some cases, the outputs of a component should be fed into its input ports. All those data exchanges are not included in the hardware modules designed by C/C++ functions in HLS®. Instead, their connection is achieved in Vivado® using the programming language VHDL, and the typical syntax is identical for the two scenarios. Take the above hardware module *func* for example, when the outputs need to be sent to its input ports, it can be written as

```
if clk='1' and clk'event then
if ao_ap_vld='1' then ai$<$=ao; end if;
if bo_ap_vld='1' then bi$<$=bo; end if;
end if;
```

which can be synthesized into two separate latches, and Figure 1.8 shows the self-connection of the hardware module *func*. Once a valid output is generated, the corresponding data valid signal *ao_ap_vld* or *bo_ap_vld* becomes binary 1, which is taken as the clock signal of the D latch. Therefore, the outputs *ao* and *bo* will be respectively fed to the input ports *ai* and *bi* of the same module when the clock signal *ap_clk* arrives.

The *ap_ctrl* port includes four binary ports, among which the start control port is controlled by the finite state machine (FSM) along with the reset port *ap_rst*. Thus, unnecessary calculation by the module which leads to incorrect results can be avoided with an appropriate operation sequence in the FSM, and in case a rerun of the emulation is needed for observation of particular power system phenomena, giving a reset order is sufficient. On the other hand, the other three signals indicating the operation status of the module are taken by the FSM as feedbacks for state shift judgment, e.g. the emulation remains at the current state till the module with the largest hardware latency completes calculation and set its binary signal *ap_done* to 1.

Since the above design is carried out manually while the RTL design for Vivado HLS® co-simulation is conducted automatically, the results from the artificial design are not guaranteed to be correct. Thus, the behavioral simulation offered by Vivado® is a further validation approach of the top-level hardware design.

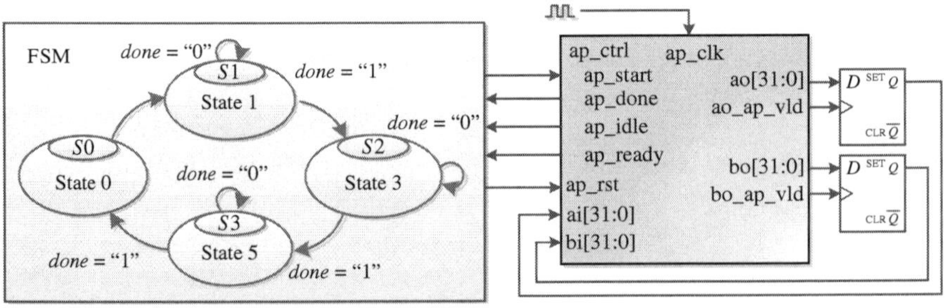

Figure 1.8 Demonstration of top-level hardware design.

1.4.3 Number Representation and Operations

For computational programming, the first step is to define the data format and arithmetic operations. In spite of the dependency of data format and the accuracy of computation, number representation also affects the hardware resource utilization in the FPGA design and programming. Basically, there are two types of number systems:

1. **Fixed-point number:**
   A fixed-point number is characterized by their word size in bits, binary point, and their sign. A common representation of a binary fixed-point number, either signed or unsigned, is shown in Figure 1.9a.
   As can be seen, a_is are the binary digits, n is the word length in bits, a_{n-1} is the most significant bit (MSB), and a_0 is the least significant bit (LSB), and the binary point is depicted three bits to the left of the LSB.
   It is worth mentioning that the preferred representation of signed fixed-point numbers is Two's complement method.
2. **Floating-point number:**
   According to *IEEE* standard 754, floating-point numbers may be represented in either single- or double-precision format.
   - *Single-precision floating-point number*:
     Any value stored in this representation has 32 bits. It is formatted as depicted in Figure 1.9b.
     A 32-bit floating-point number consists of 1 *sign* bits, 8 *exponent* bits, and 23 *fraction* bits. The single-precision floating-point number is expressed as follows:

$$d = (-1)^{sign} \times 2^{(exponent-127)} \times (1.fraction). \tag{1.1}$$

The value in this format is approximately in the range of -10^{38} to 10^{38}.

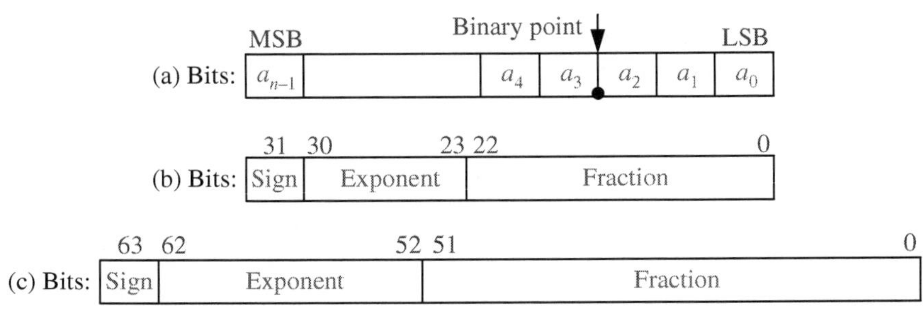

Figure 1.9 (a) Floating-point number representation, (b) single-precision, and (c) double-precision.

- *Double-precision floating-point number*:
  Any value stored in this representation has 64 bits. It is formatted as depicted in Figure 1.9c.
  A 64-bit floating-point number consists of 1 *sign* bits, 11 *exponent* bits, and 52 *fraction* bits. The double-precision floating-point number is expressed as follows:

$$d = (-1)^{sign} \times 2^{(exponent-1023)} \times (1.fraction). \tag{1.2}$$

  The value in this format is approximately in the range of -10^{308} to 10^{308}.

In this thesis, single-precision floating-point is employed for computational programming of FPGA. It actually offers high-speed computation, dynamic range of data, and acceptable accuracy. These are the main reasons for choosing single-precision format for real-time emulation.

1.4.4 FPGA Design Schemes

Parallel and pipeline paradigms are two interesting features for FPGA programming to improve computational speed and throughput. In this section, these two FPGA design techniques will be explained.

1.4.4.1 Pipeline Design Architecture

In the pipeline technique, a function is divided into the several stages, and the registers are inserted between the stages in the data path. Thus, data can march through the registers at every clock cycle. Although pipelining increases the number of clock cycles per operation, corresponding to the number of stages in a pipeline path, it improves the computational throughput by increasing the number of operations per unit of time and also conserves FPGA hardware resources. As can be seen from Figure 1.10, the pipeline path of FPGA design has three stages. The first valid results become available after three clock cycles, and the next results will be ready only after one clock cycle instead of waiting for another three clock cycles, consequently, the data throughput is one result per clock cycle.

1.4.4.2 Parallel Design Architecture

Unlike CPUs or DSPs that are sequential computational engines, an FPGA enjoys its intrinsic parallel architecture for high-speed operations and computations. Basically, an algorithm can be partitioned into several independent circuits and computed simultaneously on an FPGA. As can be seen from Figure 1.10, the operations in the first and second paths are executed concurrently on FPGA, whereas these two operations must be performed sequentially on CPU.

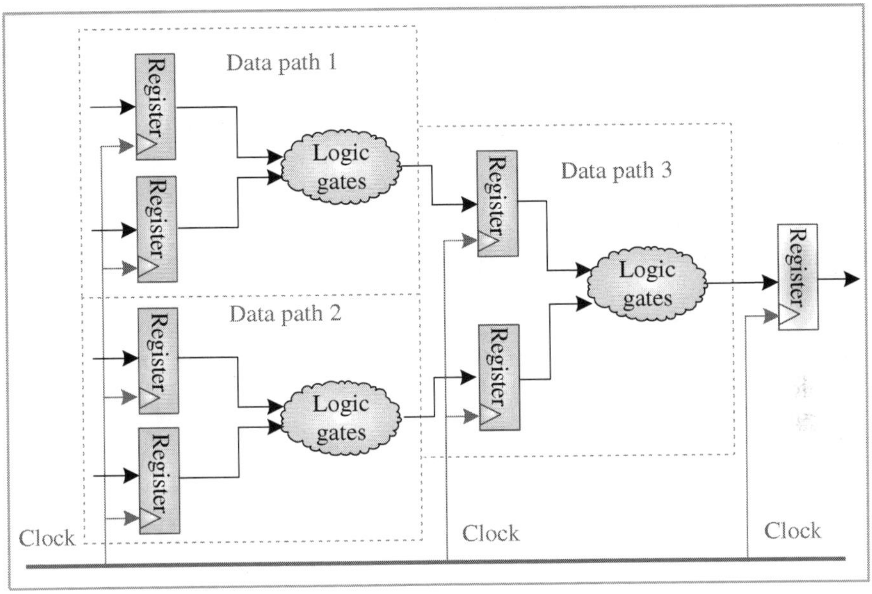

Figure 1.10 Pipeline and parallel paths of FPGA design.

1.4.5 FPGA Experiment

The HIL emulation results are ultimately expected to be observed on the oscilloscope or interfaced with an external device under real-time HIL conditions. To achieve that goal, the designed top-level needs to be implemented on the FPGA after following steps in Vivado®:

- *Run synthesis*: This is a process of transforming an RTL design into a gate-level representation.
- *Run implementation*: This includes all necessary stages to place and route the netlist onto FPGA resources, under various logical, physical, and timing constraints.
- *Generate bit stream*: This implements the embedded design and creates a bit file that can be downloaded into the targeted FPGA board.

As shown in Figure 1.7, a digital-to-analog conversion medium is mandatory since the oscilloscope channels receive analog signals. The Texas Instruments® DAC34H84 quad-channel, 16-bit, digital-to-analog converter (DAC) with a sample rate as high as 1.25 GSPS is connected to the FPGA or MPSoC board and the Tektronix® DPO7054 oscilloscope so that the hardware design results can be displayed as real-time waveforms.

1.5 Summary

This chapter described the fundamental aspects of FPGA architecture, design flow, tools, and programming technologies. Due to its intrinsic massively parallel architecture, hardware pipelining computation and custom configuration FPGAs outperform the general purpose CPU for real-time applications. In addition, the FPGA manufacturers provide all the necessary environmental support design software, and well-designed fully-tested IP cores allow the designer to implement a FPGA-based digital hardware emulator efficiently. Nevertheless, future HPC technologies are increasingly moving toward MPSoC like co-processing wherein heterogeneous compute architectures are seamlessly integrated to achieve the most efficient software and hardware workload distribution.

2

Hardware Emulation Building Blocks for Power System Components

2.1 Overview

Modern electrical power grids are evolving into large and complex interconnected systems containing myriad equipment. Historically, power systems were planned and operated in a monolithic fashion from generation to utilization and contained only a minimum number of sensors and controllers. The bulk integration of disparate energy sources at multiple levels of the grid is creating system complexities and operational challenges that can only be resolved by the pervasive use of real-time digital computation and intelligent control. The design and implementation of modern grid technologies on a wide area requires the development of real-time, parallel, reconfigurable, and high-bandwidth simulators. In particular, when a new controller, protective device, smart meter, or a security protocol needs to be tested and tuned in the HIL scenario before commissioning in the field, there is no alternative to a real-time electromagnetic transient (EMT) simulator.

Transient network analyzers (TNAs) were the predecessors of fully digital real-time simulators; however, realization of large-scale power systems with a high level of complexity and fidelity using TNAs is practically impossible. Due to the limited computational power of sequential processors, frequency-dependent network equivalents were an interim solution to accommodate realistic-sized networks in real-time simulation [23]. There is a need for real-time simulators for full-scale representation of the system for the planning, design, implementation, and testing of new grid technologies using detailed simulation models for grid components. Such simulators also need to be able to model the grid dynamics on a multi-scale basis. The interoperability of power systems, control, and communication also brings to the fore another challenge: multi-domain simulation where different modeling philosophies, solution algorithms, and numerical methods coalesce to provide a high-fidelity simulation of the smart power systems. Using FPGAs as the core computational engine, this chapter proposes the concept of hardware emulation building blocks (HEBBs) for the real-time simulation of

Real-Time Electromagnetic Transient Simulation of AC-DC Networks, First Edition.
Venkata Dinavahi and Ning Lin.

power grids [24]. The process of implementing a real-time EMT simulator on a single FPGA is described in detail. The FPGA is used to emulate the fundamental power system components including transmission lines, linear lumped RLCG elements, nonlinear elements, supply sources, and circuit breakers. The modeling of each component is described before giving the FPGA implementation details. Fully iterative Newton-based solution using sparse matrix technique is carried out for the nonlinear elements. The EMT simulation algorithm is analyzed, and the inherent parallel computations are realized in the parallel hardware architecture of FPGA. All developed hardware computational units are fully pipelined to achieve the highest throughput. IEEE 32-bit floating-point precision is used in the hardware emulation for high accuracy. Relevant case studies are used to illustrate the concepts and implementation, and comparison with off-line transient simulation is provided.

2.2 Concept of HEBB

The HEBB is a modular parallel hardwired implementation of a specific type of power system component on an FPGA using the principles of parallel processing, deep pipelining, and distributed memory management. Each HEBB is a fully self-contained standalone digital hardware component with requisite interfaces that can be quickly replicated to build large systems with several hundreds or thousands of power system components. The HEBB consumes a finite amount of digital hardware resources in terms of logic blocks, DSP blocks, and memory space and has a finite input–output latency. Therefore, by extension, the resource consumption for a large power system and the overall latency or the simulation time-step can be explicitly calculated. Depending on the FPGA resources available and the required accuracy of the real-time simulation, multiple components of the same type can be pipelined through each HEBB.

2.3 Numerical Integration

Numerically, EMT simulation is an initial value problem (IVP) where the system variables are computed at successive time points starting from a set of initial conditions. A numerical integration method provides an approximate solution to the IVP. Although the Trapezoidal Rule is popular and is widely used in off-line EMT-type programs, is it really a good choice for real-time simulation and if so why? This section seeks an answer to this question.

A general discussion on the various numerical methods for IVPs can be found in [25]. The three main criteria for selecting an integration algorithm are

Table 2.1 Numerical integration methods for real-time simulation.

Class	Name	Attributes
	Forward Euler (FE)	Low Order, Explicit, Unstable, Slow convergence
One-Step Methods		Explicit, Accurate, Orders>2 computer intensive,
	Runge–Kutta (RK)	Unstable, Employ variable step-size
	Adams-Bashforth (AB)	Explicit, Unstable, Starting problem for orders>2
Multi-Step Methods	Adams-Moulton (AM)	Implicit, Good stability, Starting problem for orders>2
	Backward Euler (BE)	AM order 1, Higher damping, Stable
	Trapezoidal Rule (TR)	AM order 2, Accurate, Conditionally Stable
	Gear's method	Implicit, Good stability, Starting problem for orders>2
	Combination of	
Predictor-Corrector Methods	AB and AM	Accurate, Variable-order, Variable step-size,
	Gear's Backward	Stable, Computer intensive
	Difference Formula (BDF)	

accuracy, speed, and stability. These criteria often contradict each other in terms of step-size and the order of the method selected which makes the selection process complicated. However, it is possible to eliminate many of the methods by careful examination of their properties. Table 2.1 summarizes the attributes of several numerical integration methods. The following guidelines can be helpful in choosing an appropriate integration method for real-time EMT simulation:

1. In general, for a fixed order and time-step, explicit numerical integration methods exhibit lower accuracy and stability than implicit methods. Therefore, they require a smaller time-step (sometimes an order of magnitude smaller) to achieve similar accuracy and stable convergent results as those of implicit methods. Consequently, their computational burden is also proportionately higher.
2. Multi-step methods that use information at several previous time steps t_{n-1}, $t_{n-2}, \ldots$ have better accuracy than single step methods that use information only at the previous time instant t_{n-1} to calculate the state at t_n.
3. Multistep methods of orders greater than 2 have a starting problem due to the lack of enough initial conditions.

4. Methods having orders higher than 2 offer better accuracy but have shrinking stability regions and involve greater computation. Thus, if computational efficiency is a requirement, as in real-time simulation, the orders have to be restricted to 2.
5. Computationally straight forward recursive schemes are better than schemes that involve iterations such as the Runge–Kutta fourth order method using Merson's error criterion to vary step-size. In RK-4, the function f is evaluated four times at each time-step, and these values are not used for any subsequent calculations.
6. For similar reasons as in (5), predictor–corrector schemes involving iterations are passed over in favor of the straight-forward Trapezoidal Rule.
7. For nonlinear system solution, in conjunction with a solution algorithm such as Newton–Raphson (N-R), there can be various criteria for adapting the time-step such as local truncation error (LTE), iteration count, and dv/dt or di/dt values. Further details can be found in Chapter 5.

In summary, while implicit methods are preferred over explicit methods for their superior stability properties, higher order explicit methods can provide comparable accuracy at smaller time-step, however, at the cost of higher computational resources. Modern FPGA devices have sufficiently enough logic resources to successfully implement either discretization methods.

2.4 Linear Lumped Passive Elements

2.4.1 Model Formulation

In the EMTP [26, 27], the linear lumped elements R, L, C, and G are represented as their discrete-time models using Trapezoidal Rule of integration. Meanwhile, the combinations of basic elements such as RL, LC, and $RLCG$ are treated as single branch element in order to reduce the number of nodes of network simulated.

2.4.1.1 Resistance *R*

The discrete-time model of a resistance R shown in Figure 2.1 is the same as its continuous-time model. The voltage and current relationship is given as

$$v_{km}(t) = Ri_{km}(t). \tag{2.1}$$

2.4.1.2 Inductance *L*

The differential equation for an inductance L shown in Figure 2.2a is

$$v_{km}(t) = L\frac{di_{km}(t)}{dt}. \tag{2.2}$$

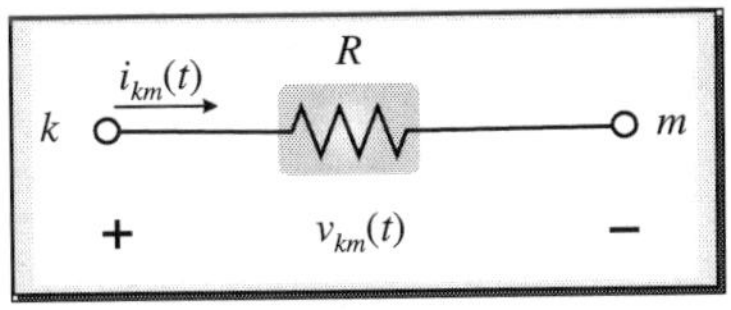

Figure 2.1 Linear resistive element and discrete-time model.

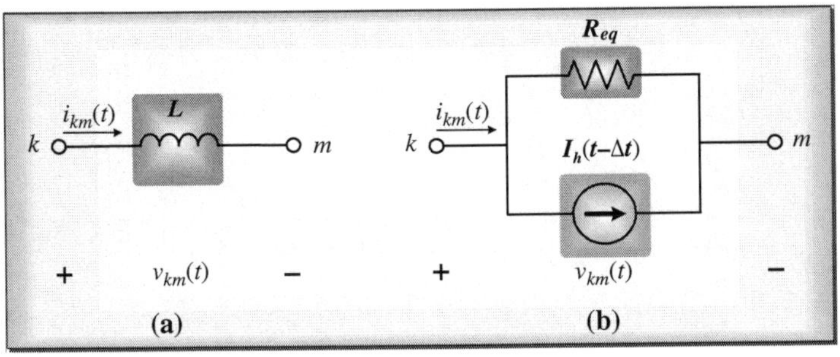

Figure 2.2 (a) Linear inductive element and (b) its discrete-time Norton equivalent.

Integrating this equation from time $(t - \Delta t)$ to t gives

$$i_{km}(t) - i_{km}(t - \Delta t) = \frac{1}{L} \int_{t-\Delta t}^{t} v_{km}(t)dt. \tag{2.3}$$

Applying the Trapezoidal Rule of integration yields

$$\begin{aligned} i_{km}(t) &= i_{km}(t - \Delta t) + \frac{\Delta t}{2L}[v_{km}(t) + v_{km}(t - \Delta t)] \\ &= \frac{\Delta t}{2L} v_{km}(t) + \left[i_{km}(t - \Delta t) + \frac{\Delta t}{2L} v_{km}(t - \Delta t) \right] \\ &= \frac{v_{km}(t)}{R_{eq}} + I_h(t - \Delta t). \end{aligned} \tag{2.4}$$

Equation (2.4) can be represented by an equivalent resistance R_{eq} in parallel with a current source (history term) $I_h(t - \Delta t)$ as shown in Figure 2.2b, where the equivalent resistance R_{eq} is defined as

$$R_{eq} = \frac{2L}{\Delta t}, \tag{2.5}$$

and the history term $I_h(t - \Delta t)$ is known from the solution at the preceding time-step and updated for use in the next time-step using

$$I_h(t - \Delta t) = \left[i_{km}(t - \Delta t) + \frac{\Delta t}{2L} v_{km}(t - \Delta t) \right], \tag{2.6}$$

or recursively using

$$\begin{aligned} I_h(t - \Delta t) &= I_h(t - 2\Delta t) + \frac{\Delta t}{L} v_{km}(t - \Delta t) \\ &= I_h(t - 2\Delta t) + \frac{2}{R_{eq}} v_{km}(t - \Delta t). \end{aligned} \tag{2.7}$$

2.4.1.3 Capacitance C

The differential equation for a capacitance C shown in Figure 2.3a is given as

$$i_{km}(t) = C\frac{dv_{km}(t)}{dt}. \tag{2.8}$$

Integrating this equation from time $(t - \Delta t)$ to t gives

$$v_{km}(t) - v_{km}(t - \Delta t) = \frac{1}{C}\int_{t-\Delta t}^{t} i_{km}(t)dt. \tag{2.9}$$

Applying the Trapezoidal Rule of integration yields

$$\begin{aligned} i_{km}(t) &= -i_{km}(t - \Delta t) + \frac{2C}{\Delta t}[v_{km}(t) - v_{km}(t - \Delta t)] \\ &= \frac{2C}{\Delta t}v_{km}(t) + \left[-i_{km}(t - \Delta t) - \frac{2C}{\Delta t}v_{km}(t - \Delta t)\right] \\ &= \frac{2C}{\Delta t}v_{km}(t) + I_h(t - \Delta t). \end{aligned} \tag{2.10}$$

Equation (2.10) can be represented by an equivalent resistance R_{eq} in parallel with a current source (history term) $I_h(t - \Delta t)$ as shown in Figure 2.3b, where the equivalent resistance R_{eq} is defined as

$$R_{eq} = \frac{\Delta t}{2C}, \tag{2.11}$$

and the history term $I_h(t - \Delta t)$ is known from the solution at the preceding time-step and updated for use in the next time-step using

$$I_h(t - \Delta t) = -i_{km}(t - \Delta t) - \frac{2C}{\Delta t}v_{km}(t - \Delta t), \tag{2.12}$$

or recursively using

$$\begin{aligned} I_h(t - \Delta t) &= -I_h(t - 2\Delta t) - \frac{4C}{\Delta t}v_{km}(t - \Delta t) \\ &= -I_h(t - 2\Delta t) - \frac{2}{R_{eq}}v_{km}(t - \Delta t). \end{aligned} \tag{2.13}$$

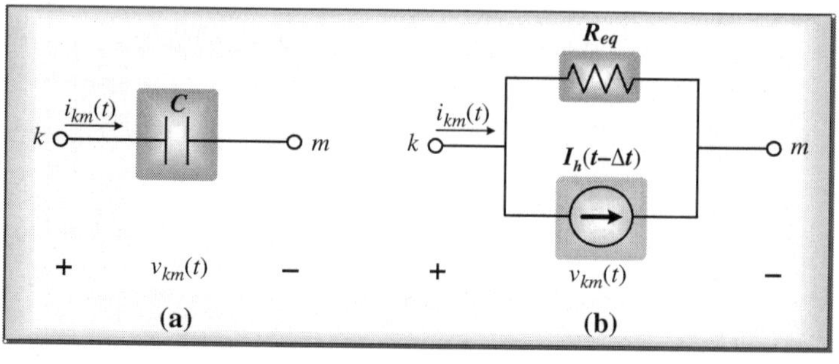

Figure 2.3 (a) Linear capacitive element and (b) its discrete-time Norton equivalent.

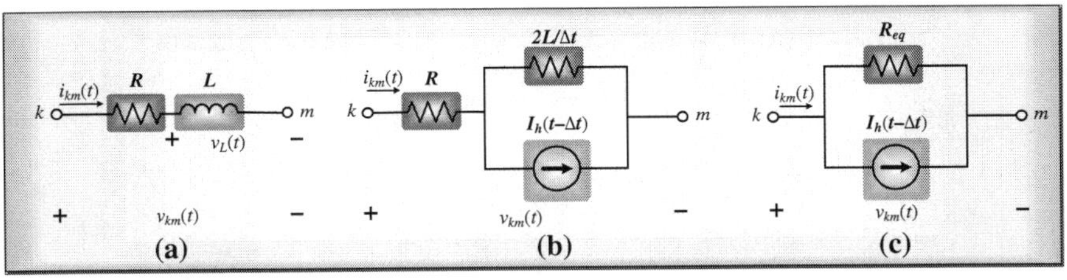

Figure 2.4 (a) Series *RL* branch, (b) combined *R* and *L* discrete-time models, and (c) Norton equivalent.

2.4.1.4 *RL* Branch

The series *RL* branch element shown in Figure 2.4a is first replaced by the discrete-time model of R and L shown in Figure 2.4b.

Combining (2.1) and (2.4) gives

$$v_{km}(t) = Ri_{km}(t) + \left[i_{km}(t) - i_{km}(t-\Delta t) - \frac{\Delta t}{2L}v_L(t-\Delta t)\right]\frac{2L}{\Delta t}. \tag{2.14}$$

Substituting $v_L(t-\Delta t)$ and rearranging yields

$$i_{km}(t) = \frac{1}{R+\frac{2L}{\Delta t}}v_{km}(t) + \frac{1}{R+\frac{2L}{\Delta t}}\left[v_{km}(t-\Delta t) - \left(R - \frac{2L}{\Delta t}\right)i_{km}(t-\Delta t)\right]. \tag{2.15}$$

Equation (2.15) can be represented by an equivalent resistance R_{eq} in parallel with a current source (history term) $I_h(t-\Delta t)$ as shown in Figure 2.4c, where the equivalent resistance R_{eq} is defined as

$$R_{eq} = R + \frac{2L}{\Delta t}, \tag{2.16}$$

and the history term $I_h(t-\Delta t)$ is known from the solution at the preceding time-step and updated for use in the next time-step using

$$I_h(t-\Delta t) = \frac{v_{km}(t-\Delta t)}{R+\frac{2L}{\Delta t}} - \frac{\left(R-\frac{2L}{\Delta t}\right)i_{km}(t-\Delta t)}{R+\frac{2L}{\Delta t}}, \tag{2.17}$$

or recursively using

$$I_h(t-\Delta t) = -\frac{R-\frac{2L}{\Delta t}}{R_{eq}}I_h(t-2\Delta t) + \frac{\frac{4L}{\Delta t}}{R_{eq}^2}v_{km}(t-\Delta t). \tag{2.18}$$

2.4.1.5 *LC* Branch

Combining the discrete-time model of L and C, a series *LC* branch element shown in Figure 2.5a can be represented as its equivalent network shown in Figure 2.5b,

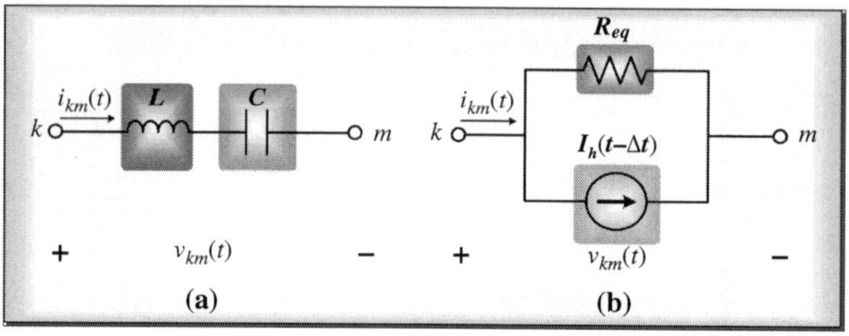

Figure 2.5 (a) Series *LC* branch and (b) its discrete-time Norton equivalent.

where the equivalent resistance R_{eq} is defined as

$$R_{eq} = \frac{2L}{\Delta t} + \frac{\Delta t}{2C}, \tag{2.19}$$

and the history term $I_h(t-\Delta t)$ is known from the solution at the preceding time-step and updated recursively for use in the next time-step using

$$\begin{cases} I_h(t-\Delta t) = \frac{1}{R_{eq}}[V_{h,L}(t-\Delta t) + V_{h,C}(t-\Delta t)], \\ V_{h,L}(t-\Delta t) = -V_{h,L}(t-2\Delta t) + \frac{4L}{\Delta t} i_{km}(t-\Delta t), \\ V_{h,C}(t-\Delta t) = V_{h,C}(t-2\Delta t) - \frac{\Delta t}{C} i_{km}(t-\Delta t), \\ i_{km}(t-\Delta t) = \frac{1}{R_{eq}} v_{km}(t-\Delta t) + I_h(t-\Delta t). \end{cases} \tag{2.20}$$

2.4.1.6 *RLCG* Branch

Combining the discrete-time model of R, L, C, and G, the *RLCG* element shown in branch Figure 2.6a can be represented as its equivalent network shown in Figure 2.6b, where the equivalent resistance R_{eq} is defined as

$$R_{eq} = R + \frac{2L}{\Delta t} + \frac{G\frac{\Delta t}{2C}}{G + \frac{\Delta t}{2C}}, \tag{2.21}$$

with G in ohms and the history term $I_h(t-\Delta t)$ is given as

$$\begin{cases} I_h(t-\Delta t) = \frac{1}{R_{eq}}[V_{h,L}(t-\Delta t) + V_{h,CG}(t-\Delta t)], \\ V_{h,L}(t-\Delta t) = -V_{h,L}(t-2\Delta t) + \frac{4L}{\Delta t} i_{km}(t-\Delta t), \\ V_{h,CG}(t-\Delta t) = \frac{G-\frac{\Delta t}{2C}}{G+\frac{\Delta t}{2C}} V_{h,CG}(t-2\Delta t) - \frac{G^2\frac{\Delta t}{C}}{\left(G+\frac{\Delta t}{2C}\right)^2} i_{km}(t-\Delta t), \\ i_{km}(t-\Delta t) = \frac{1}{R_{eq}} v_{km}(t-\Delta t) + I_h(t-\Delta t). \end{cases} \tag{2.22}$$

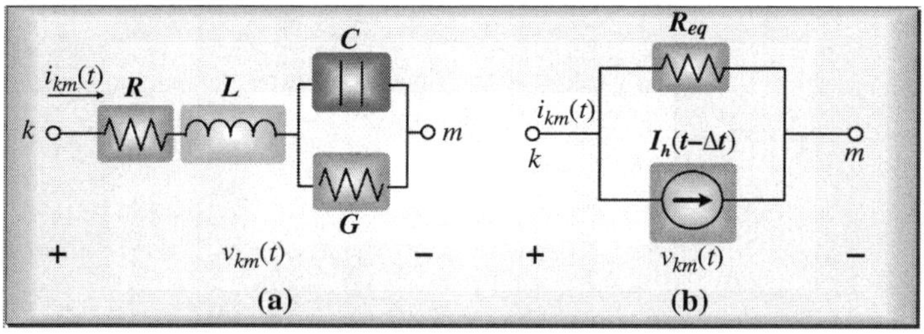

Figure 2.6 (a) *RLCG* branch and (b) its discrete-time Norton equivalent.

It is clear from the above discussion that all the RLCG elements share the same discrete-time equivalent network which is an equivalent conductance G_{eq} in parallel with a history current $I_h(t-\Delta t)$. Moreover, the history current update Eqs. (2.7), (2.13), and (2.18) have the same format, while Eqs. (2.20) and (2.22) have the same format. Thus, all the RLCG elements can be divided into two types: *RLCG-type1* which includes *R*, *L*, *C*, *RL*, and *RC* elements and *RLCG-type2* which includes *LC*, *RLC*, and *RLCG* elements. Although it is possible to have a general formula to include all types of RLCG elements, the number of *RLCG-type1* RLCG elements in power systems is higher than the number of *RLCG-type2* elements. The formula of updating history current of *RLCG-type1* is simpler than that of *RLCG-type2*, which will improve the computational efficiency significantly.

The generic equation for updating *RLCG-type1* elements history currents is

$$I_{hpe1}(t-\Delta t) = P_1 I_{hpe1}(t-2\Delta t) + P_2 v_{km}(t-\Delta t), \tag{2.23}$$

where P_1, P_2 are coefficients with respect to different elements which are listed in Table 2.2. Note for resistance R, P_1, P_2 are zeros since there is no equivalent current source.

The generic equations for updating *RLCG-type2* elements history terms are given as

$$\begin{cases} I_{hpe2}(t-\Delta t) = \frac{1}{R_{eq}}[V_{h,L}(t-\Delta t) + V_{h,c}(t-\Delta t)], \\ V_{h,L}(t-\Delta t) = P_1 V_{h,L}(t-2\Delta t) + P_2 i_{km}(t-\Delta t), \\ V_{h,C}(t-\Delta t) = P_3 V_{h,C}(t-2\Delta t) + P_4 i_{km}(t-\Delta t), \end{cases} \tag{2.24}$$

where $R_{eq}(1/G_{eq})$, P_1, P_2, P_3, and P_4 are coefficients with respect to different elements which are listed in Table 2.3.

The branch current for all RLCG elements is

$$i_{km}(t) = \frac{1}{R_{eq}} v_{km}(t) + I_{hpe}(t-\Delta t). \tag{2.25}$$

Table 2.2 Coefficients for updating *RLCG-type1* elements history currents (2.23).

RLCG-type1 elements	$R_{eq}(1/G_{eq})$	P_1	P_2
R	R	0	0
L	$\frac{2L}{\Delta t}$	1	$\frac{\Delta t}{L}$
C	$\frac{\Delta t}{2C}$	-1	$-\frac{4C}{\Delta t}$
RL	$R+\frac{2L}{\Delta t}$	$-\frac{R-\frac{2L}{\Delta t}}{R+\frac{2L}{\Delta t}}$	$\frac{\frac{4L}{\Delta t}}{(R+\frac{2L}{\Delta t})^2}$
RC	$R+\frac{\Delta t}{2C}$	$-\frac{R-\frac{\Delta t}{2C}}{R+\frac{\Delta t}{2C}}$	$-\frac{2R}{(R+\frac{\Delta t}{2C})^2}$

Table 2.3 Coefficients for updating *RLCG-type2* elements history currents (2.24).

RLCG-type2 elements	$R_{eq}(1/G_{eq})$	P_1	P_2	P_3	P_4
LC	$\frac{2L}{\Delta t}+\frac{\Delta t}{2C}$	-1	$\frac{4L}{\Delta t}$	1	$-\frac{\Delta t}{C}$
RLC	$R+\frac{2L}{\Delta t}+\frac{\Delta t}{2C}$	-1	$\frac{4L}{\Delta t}$	1	$-\frac{\Delta t}{C}$
RLCG	$R+\frac{2L}{\Delta t}+\frac{G\frac{\Delta t}{2C}}{G+\frac{\Delta t}{2C}}$	-1	$\frac{4L}{\Delta t}$	$\frac{G-\frac{\Delta t}{2C}}{G+\frac{\Delta t}{2C}}$	$-\frac{G^2\frac{\Delta t}{C}}{\left(G+\frac{\Delta t}{2C}\right)^2}$

2.4.2 Hardware Emulation of Linear Lumped Passive Elements

A hardware module *RLCG* was designed to emulate the linear lumped RLCG elements in the FPGA. Figure 2.7a shows the symbol of the *RLCG* HEBB and its input/output signals. The main function of the *RLCG* HEBB is to update history currents $I_{hpe}(t-\Delta t)$. With the node voltages v_k and v_m as inputs, the HEBB outputs the computed history current value I_{hpe} for the next time-step. The signals *PEcmd* and *SW_now* enable to computation or switching of the element to begin when the *Clock* is high, and *PEdone* signal informs the *MainControl* when the computation is finished. The pipelined hardware computation schemes for updating the history currents for *RLCG-type1* (2.23) and *RLCG-type2* (2.24) RLCG elements are shown in Figures 2.7b and 2.8, respectively.

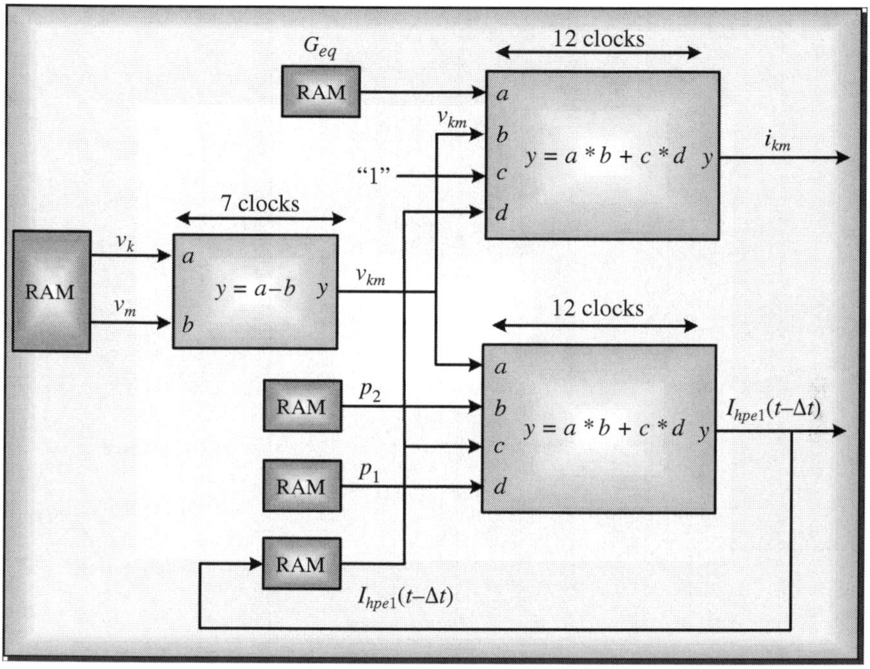

Figure 2.7 (a) RLCG HEBB module and its input/output signals. (b) Pipelined computation scheme for calculating history current of *RLCG-type1* elements.

2.5 Sources

The ideal voltage source and current source are represented using sinusoidal function (cos function) in the EMTP. The LUT is the most commonly used method to evaluate this nonlinear function. Since cos is a periodic function, as shown in Figure 2.9a, only half cycle of cos function values need to be stored in the LUT in order to save the memory space of the LUT. The accuracy of the cos value is determined by the length of the LUT. In this design, 4096 (2^{12}) cos values for half cycle are stored in the LUT; thus, the interval of the LUT is $d\theta=\pi/4096$. Figure 2.9b shows the structure of the LUT.

A voltage or current source is expressed as

$$
\begin{aligned}
v(t), i(t) &= mag \cdot \cos(\omega t + ph) \\
&= mag \cdot \cos(\omega n\Delta t + ph),
\end{aligned}
\tag{2.26}
$$

where mag, ω, and ph (0 to π) are the source magnitude, frequency, and phase angle, respectively. To retrieve the cos value from the LUT, the phase angle ph is converted to the address (0–4095) of the LUT.

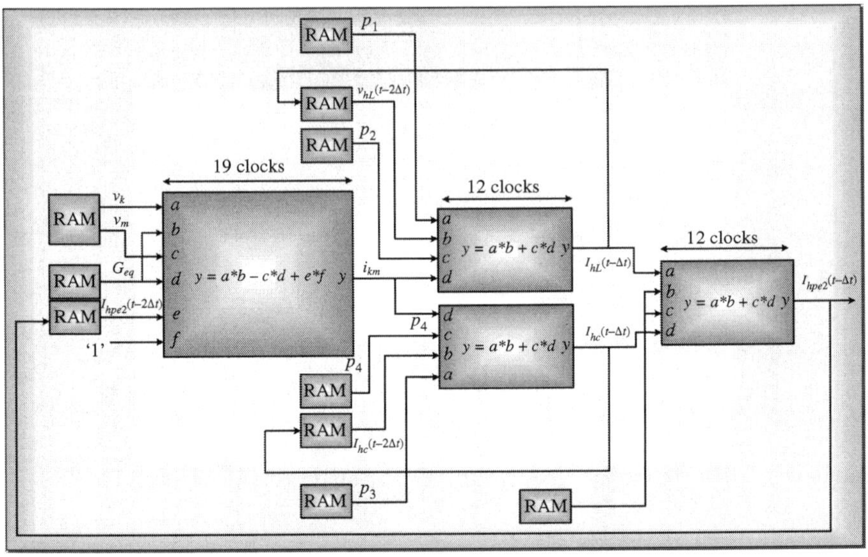

Figure 2.8 Pipelined computation scheme for calculating history current of *RLCG-type2* elements.

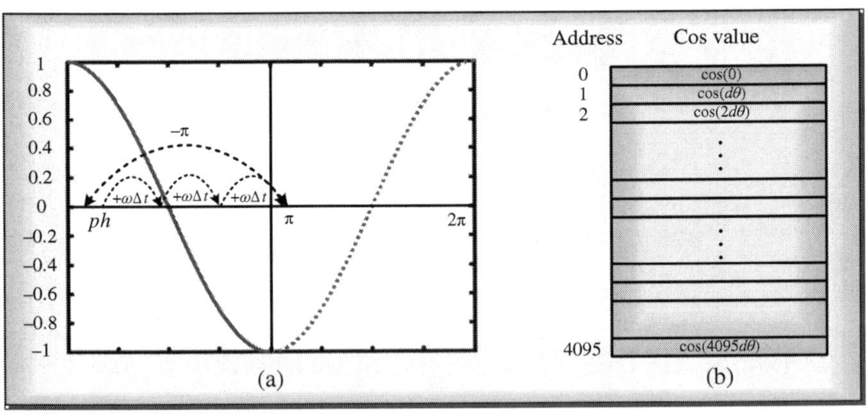

Figure 2.9 (a) Cosine function and (b) structure of cos function look-up table.

2.5.1 Hardware Emulation of Sources

A hardware module *Source* is designed to emulate the source components in the FPGA. Figure 2.10a shows the symbol of the *Source* module and its input/output signals. The main function of the *Source* module is to calculate the source values at each simulation time-step.

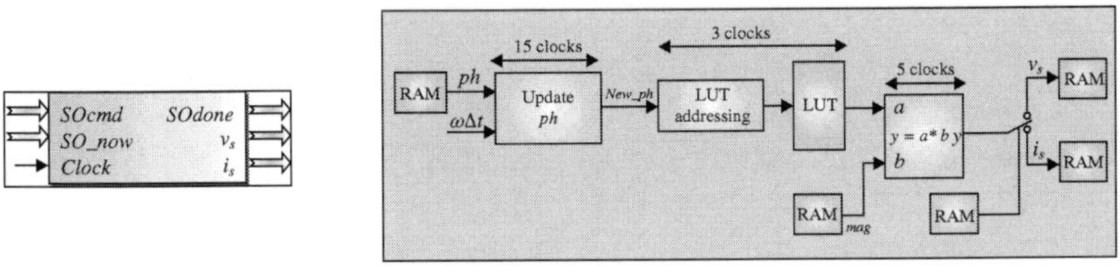

Figure 2.10 (a) *Source* HEBB and its input/output signals. (b) Pipelined configuration for calculating source values.

Address	x_i	*exp*	*dexp*	1.*mantissa*
0	0	0	0	0.000...000
1	0.0625	−4	0	1.000...000
2	0.125	−3	1	1.000...000
3	0.1875	−3	1	1.100...000
4	0.25	−2	2	1.000...000
5	0.325	−2	2	1.010...000
6	0.375	−2	2	1.100...000
7	0.4375	−2	2	1.110...000
⋮	⋮	⋮	⋮	⋮
49	3.0625	1	5	1.10001...0
50	3.125	1	5	1.10010...0
⋮	⋮	⋮	⋮	⋮

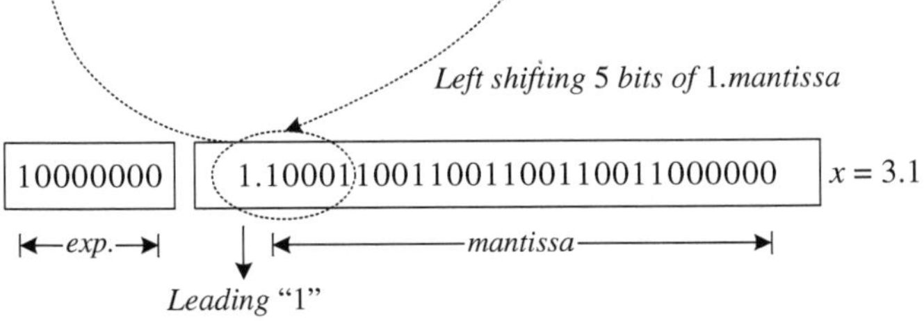

Figure 2.11 Example showing the scheme of *LUT addressing* unit.

As shown in Figure 2.10b, the calculation of the source values begins with the updating of the phase angle *ph*. The new *pha* is obtained by adding the previous *ph* by $\omega\Delta t$. Since the LUT has only half cycle of cos functions values, the calculated *ph* needs to be checked with π; if greater than π, it is subtracted by π and the sign is inverted. Then, the new *pha* is converted to the address of the LUT. Finally, the retrieved cos value is multiplied by the magnitude *mag*. The type of the source (current or voltage) is checked before the calculated source values being saved as v_s or i_s.

One challenge of implementing a floating-point LUT is that the LUT is addressed by integer number whereas the input is in floating-point format. The straightforward method is to convert the floating-point input into integer format at the expense of extra latency for the conversion. In this design, this conversion is omitted. The exponent and mantissa of the input floating-point number are used directly to access the LUT provided the interval of the LUT is always a power of 2. Assume the floating-point input is x, the *LUT addressing* unit in Figure 2.10b outputs the addresses of the point x_i making $x - x_i < step$, where *step* is the interval of the LUT. This is done by left shifting the leading "1" and mantissa of x by *dexp* bits, where the *dexp* is the difference of the exponent of input x and *step*. An example is shown in Figure 2.11. In this example, *step* is 0.0625 (2^{-4}), and input x is 3.10 whose exponent is 1 (without bias). So *dexp* is 5 (=1 − (−4)). Left shifting the leading "1" and mantissa 5 bits gives 49 which is the address of x_i (3.0625).

2.6 Switches

In the EMTP, circuit breakers are commonly represented as ideal time-controlled switches shown in Figure 2.12a. It opens at t_{op} with infinite resistance ($R = \infty$) and closes at t_{cl} with zero resistance ($R = 0$). The operation of switch causes the change of the system topology leading to the change of network admittance matrix $\boldsymbol{Y}$.

2.6.1 Hardware Emulation of Switches

A hardware module *Switch* is designed to emulate circuit breakers on the FPGA. Figure 2.12b shows the symbol of this module and its input/output signals. As shown in Figure 2.12c, the input system clock frequency (80 MHz) is divided by 800 to generate a 10 μs-period clock signal. This clock signal is counted and compared with the switch operation time *swtime* (integer of 10 μs) saved in a RAM. Once the switch's *swtime* is reached, the corresponding switch state bit in *Sw_now* register is inverted using "0"/"1" for switch open/closed.

Another important function of the *Switch* module is to generate the *dTover* signal which indicates the end of the real simulation time-step Δt. When an emulation step is finished, the *dTover* signal is checked. If it is not "1", the emulation step is finished within Δt, and thus, the real-time emulation is achieved. Otherwise, the emulation step takes longer time than Δt, and the real-time constraint is not met.

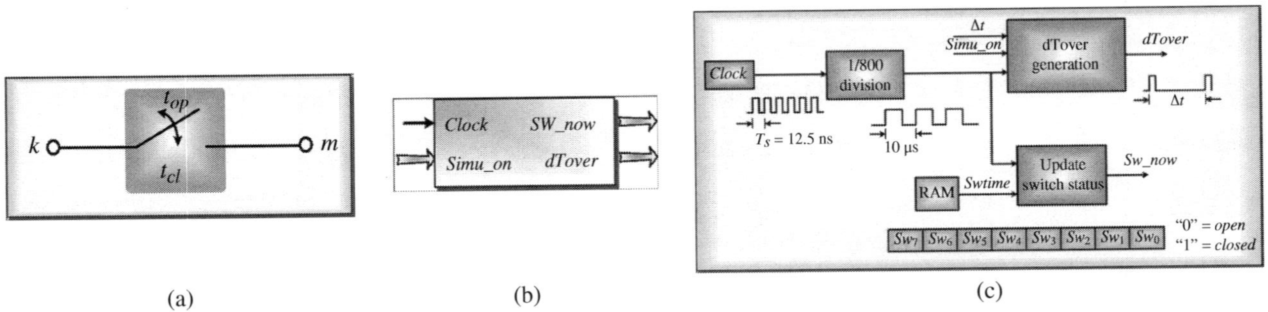

Figure 2.12 (a) Ideal time-controlled switch. (b) *Switch* HEBB and its input/output signals. (c) Functions realized in the *Switch* HEBB.

2.7 Transmission Lines

Transmission lines are characterized by four parameters: series resistance, series inductance, shunt capacitance, and shunt conductance. The resistance and conductance are lossy elements of the line. The inductance and capacitance originate due to the magnetic fields and electric fields, respectively, on the line conductors and are functions of the transmission tower geometry. These four parameters are distributed throughout the length of the line and are the reason why traveling voltage and current waves arise when a transient even such as a line switching occurs [28, 29]. The electromagnetic waves move close to the speed of light and get reflected and refracted at the terminations or junctions where other lines are connected. The traveling wave model (TWM) is the simplest model that can be used for transient simulation to represent the distributed nature of line parameters.

The propagation of EMTs on a transmission line is also greatly influenced by the frequency dependence [30–33] of its parameters namely its series impedance and shunt admittance. Traditionally, the most accurate transmission line model is the frequency-dependent line model (FDLM) built in the modal-domain. The transformation matrix between the modal-domain and the phase-domain is real and constant. This model works well for symmetrical and transposed lines; however, for asymmetrical and untransposed line configurations, the use of a constant modal transformation matrix causes error and numerical instability, especially for cable models. Although this deficiency can be overcome by taking the frequency-dependence of the transformation matrix into account [34], the practical implementation of the model is complicated. Phase-domain modeling avoids the transformation matrix by formulating the transmission line equations directly in the phase-domain [35–37]. Among these phase-domain line models, the universal line model (ULM) is considered numerically efficient and robust for both overhead lines and underground cables.

2.7.1 Traveling Waves

As shown in Figure 2.13, consider a single-phase loss-less ($R = G = 0$) transmission line section of δx m. For this line section, the series inductance is $L\,\delta x$ and shunt line capacitance is $C\,\delta x$, where L and C are the per unit length line inductance and per unit length line capacitance in H/m and F/m, respectively. Kirchhoff's voltage law (KVL) and Kirchhoff's current law (KCL) equations for this line section provide

$$v(x + \delta x, t) - v(x, t) = -L\delta x\frac{\partial i(x, t)}{\partial t}, \tag{2.27}$$

$$i(x + \delta x, t) - i(x, t) = -C\delta x\frac{\partial v(x, t)}{\partial t}. \tag{2.28}$$

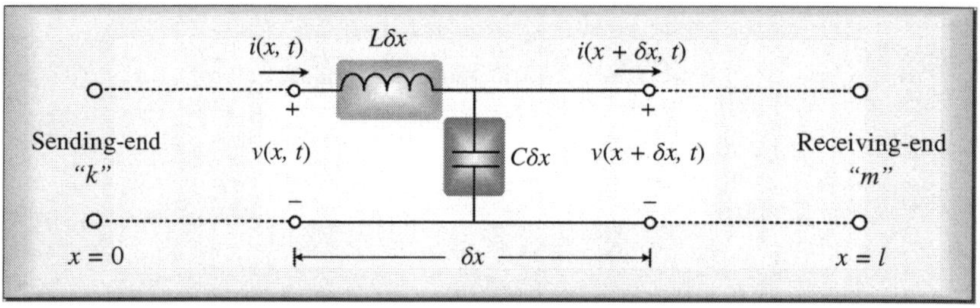

Figure 2.13 Single-phase distributed parameter loss-less line.

Dividing (2.27) and (2.28) and taking the limit as $\delta x \rightarrow 0$ gives

$$\frac{\partial v(x,t)}{\partial x} = -L\frac{\partial i(x,t)}{\partial t}, \tag{2.29}$$

$$\frac{\partial i(x,t)}{\partial x} = -C\frac{\partial v(x,t)}{\partial t}. \tag{2.30}$$

Expressing (2.29) and (2.30) in frequency-domain using Laplace transform yields

$$\frac{dV(x,s)}{dx} = -sL\, I(x,s), \tag{2.31}$$

$$\frac{dI(x,s)}{dx} = -sC\, V(x,s). \tag{2.32}$$

Differentiating (2.31) and (2.32) with respect to x and substituting for the derivatives of $I(x,s)$ and $V(x,s)$ on the right-hand side results in

$$\frac{d^2V(x,s)}{dx^2} - s^2LC\, V(x,s) = 0, \tag{2.33}$$

$$\frac{d^2I(x,s)}{dx^2} - s^2LC\, I(x,s) = 0. \tag{2.34}$$

Equations (2.33) and (2.34) are second-order linear homogeneous differential equations whose solutions can be written as

$$V(x,s) = V^+(s)e^{-\frac{sx}{v}} + V^-(s)e^{+\frac{sx}{v}}, \tag{2.35}$$

$$I(x,s) = I^+(s)e^{-\frac{sx}{v}} + I^-(s)e^{+\frac{sx}{v}}, \tag{2.36}$$

where $v = 1/\sqrt{LC}$ m/s is the velocity of wave propagation. $V^+(s)$ and $I^+(s)$ are forward voltage and current waves traveling in the positive x direction, while $V^-(s)$ and $I^-(s)$ are backward voltage and current waves traveling in the negative x direction. Converting to time-domain by taking the inverse Laplace transform of (2.35) and (2.36) and using the time-shift property $L[f(t-\tau)] = F(s)\, e^{-s\tau}$ yields

$$v(x,t) = v^+\left(t - \frac{x}{v}\right) + v^-\left(t + \frac{x}{v}\right), \tag{2.37}$$

$$i(x,t) = i^+\left(t - \frac{x}{v}\right) + i^-\left(t + \frac{x}{v}\right). \tag{2.38}$$

Using (2.35) and (2.36) in (2.32) results in

$$\frac{s}{v}\left[-I^{+}(s)e^{-\frac{sx}{v}}+I^{-}(s)e^{+\frac{sx}{v}}\right]=-sC\left[V^{+}(s)e^{-\frac{sx}{v}}+V^{-}(s)e^{+\frac{sx}{v}}\right]. \tag{2.39}$$

Now, equating the coefficients of $e^{-\frac{sx}{v}}$ and $e^{+\frac{sx}{v}}$ from both sides of (2.39) yields

$$I^{+}(s)=\frac{V^{+}(s)}{Z_c}, \tag{2.40}$$

$$I^{-}(s)=-\frac{V^{-}(s)}{Z_c}, \tag{2.41}$$

where $Z_c=\sqrt{L/C}\ \Omega$ is the characteristic or surge impedance of the line. Thus, (2.36) and (2.38) can be rewritten as

$$I(x,s)=\frac{1}{Z_c}\left[V^{+}(s)e^{-\frac{sx}{v}}-V^{-}(s)e^{+\frac{sx}{v}}\right], \tag{2.42}$$

$$i(x,t)=\frac{1}{Z_c}\left[v^{+}\left(t-\frac{x}{v}\right)-v^{-}\left(t+\frac{x}{v}\right)\right], \tag{2.43}$$

Setting $x=l$ in (2.35) and (2.42), the voltage and current at the receiving end of the line can be expressed in frequency-domain as

$$V_m(s)=V^{+}(s)e^{-\frac{sl}{v}}+V^{-}(s)e^{+\frac{sl}{v}}, \tag{2.44}$$

$$I_m(s)=\frac{1}{Z_c}\left[V^{+}(s)e^{-\frac{sl}{v}}-V^{-}(s)e^{+\frac{sl}{v}}\right]. \tag{2.45}$$

The forward and backward voltage waves can be expressed in terms of the receiving end voltage and current as

$$V^{+}(s)=\frac{V_m(s)+Z_cI_m(s)}{2}e^{+\frac{sl}{v}}, \tag{2.46}$$

$$V^{-}(s)=\frac{V_m(s)-Z_cI_m(s)}{2}e^{-\frac{sl}{v}}. \tag{2.47}$$

Setting $x=0$ in (2.35) and (2.42) and using (2.46) and (2.47), the sending end voltage and current of the line can be expressed as

$$V_k(s)=\frac{V_m(s)+Z_cI_m(s)}{2}e^{+\frac{sl}{v}}+\frac{V_m(s)-Z_cI_m(s)}{2}e^{-\frac{sl}{v}}, \tag{2.48}$$

$$I_k(s)=\frac{1}{Z_c}\left[\frac{V_m(s)+Z_cI_m(s)}{2}e^{+\frac{sl}{v}}-\frac{V_m(s)-Z_cI_m(s)}{2}e^{-\frac{sl}{v}}\right]. \tag{2.49}$$

Rewriting (2.48) and (2.49) as

$$V_k(s)=V_m(s)\left(\frac{e^{+\frac{sl}{v}}+e^{-\frac{sl}{v}}}{2}\right)+Z_c\left(\frac{e^{+\frac{sl}{v}}-e^{-\frac{sl}{v}}}{2}\right)I_m(s), \tag{2.50}$$

$$I_k(s)=\frac{1}{Z_c}V_m(s)\left(\frac{e^{+\frac{sl}{v}}-e^{-\frac{sl}{v}}}{2}\right)+\left(\frac{e^{+\frac{sl}{v}}+e^{-\frac{sl}{v}}}{2}\right)I_m(s), \tag{2.51}$$

which in turn can be expressed as

$$V_k(s) = \cos(\beta l)V_m(s) + Z_c \sin(\beta l)I_m(s), \tag{2.52}$$

$$I_k(s) = \frac{1}{Z_c}\sin(\beta l)V_m(s) + \cos(\beta l)I_m(s). \tag{2.53}$$

where $\beta = \omega\sqrt{LC}$ m^{-1} is the phase constant. Equations (2.52) and (2.53) allow the single-phase loss-less line to be represented as a two-port network for circuit analysis with the parameters $A = D = \cos(\beta l)$, $B = Z_c \sin(\beta l)$, and $C = \frac{1}{Z_c}\sin(\beta l)$.

2.7.2 Traveling Wave Model

The TWM is a discrete-time model developed by *L.* Bergeron [38] to represent the loss-less transmission line using distributed L and C parameters. Although, the high-frequency behavior of the line is not well represented, this model accurately represents the real-world propagation delay on the lines. From the solution (2.37) and (2.43) of the loss-less line equations, the following two equations can be obtained:

$$v(x,t) + Z_c i(x,t) = 2v^+\left(t - \frac{x}{v}\right), \tag{2.54}$$

$$v(x,t) - Z_c i(x,t) = 2v^-\left(t + \frac{x}{v}\right). \tag{2.55}$$

For a fictitious observer traveling at velocity v in the positive x direction $(v + Z_c i)$ remains constant when $(t - x/v)$ is constant. With τ being the travel time from terminal k to terminal m, the value of $(v + Z_c i)$ observed at time $(t - \tau)$ at terminal k must equal the value at time t at terminal m. Thus,

$$v_k(t-\tau) + Z_c i_k(t-\tau) = v_m(t) + Z_c i_m(t). \tag{2.56}$$

Similarly, for a fictitious observer traveling at velocity v in the negative x direction $(v - Z_c i)$ remains constant when $(t + x/v)$ is constant. Therefore, the value of $(v - Z_c i)$ observed at time $(t - \tau)$ at terminal m must equal the value at time t at terminal k. Thus,

$$v_m(t-\tau) - Z_c i_m(t-\tau) = v_k(t) - Z_c i_k(t). \tag{2.57}$$

From (2.56) and (2.57), the governing equations for the TWM can be written as

$$i_k(t) = I_k(t-\tau) + \frac{v_k(t)}{Z_c}, \tag{2.58}$$

$$i_m(t) = I_m(t-\tau) - \frac{v_m(t)}{Z_c}, \tag{2.59}$$

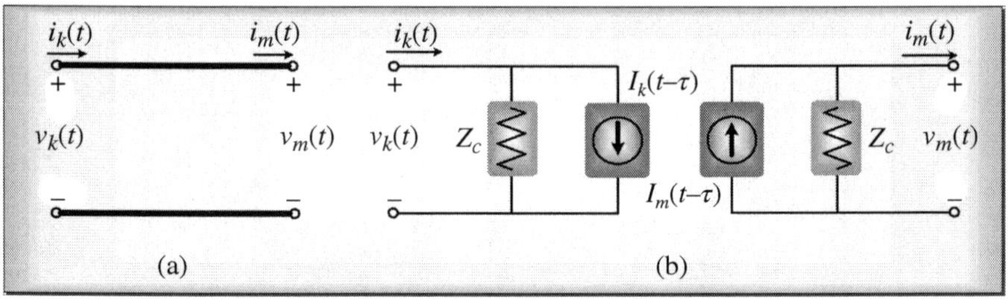

Figure 2.14 Discrete-time Norton equivalent circuit of the TWM.

where the history current sources $I_{hk}(t-\tau)$ and $I_{hm}(t-\tau)$ at sending-end and receiving-end, respectively, are given as

$$I_{hk}(t-\tau) = i_m(t-\tau) - \frac{v_m(t-\tau)}{Z_c}, \tag{2.60}$$

$$I_{hm}(t-\tau) = i_k(t-\tau) + \frac{v_k(t-\tau)}{Z_c}. \tag{2.61}$$

Also, using (2.59) in (2.60) and using (2.58) in (2.61), the history current sources can be rewritten as

$$I_{hk}(t-\tau) = I_{hm}(t-2\tau) - \frac{2v_m(t-\tau)}{Z_c}, \tag{2.62}$$

$$I_{hm}(t-\tau) = I_{hk}(t-2\tau) + \frac{2v_k(t-\tau)}{Z_c}. \tag{2.63}$$

The discrete-time Norton equivalent circuit for the TWM is shown in Figure 2.14. The inputs for this discretized model are the sending-end and receiving-end currents ($i_k(t)$ and $i_m(t)$), injection history terms ($I_{hk}(t-\tau)$ and $I_{hm}(t-\tau)$) and the surge impedance, while the outputs are the voltages on the sending-end and receiving-end ($v_k(t)$ and $v_m(t)$).

2.7.2.1 Modal Transformation

Traveling waves on transmission lines with multiple conductors travel in modes. On a line with n conductors above ground waves travel in n modes. Each wave mode is characterized by its own propagation velocity and surge impedance [39]. For example, waves on a three-phase loss-less completely transposed line with three conductors above ground travel in three modes: zero ("0"), positive ("+"), and negative ("−") modes. As a direct extension of (2.31) and (2.32), the frequency-domain equations for this line can be written as

$$\frac{d\mathbf{V}(x,s)}{dx} = -s\mathbf{L}\ \mathbf{I}(x,s), \tag{2.64}$$

$$\frac{d\mathbf{I}(x,s)}{dx} = -s\mathbf{C}\ \mathbf{V}(x,s). \tag{2.65}$$

where $\mathbf{V}(x,s)$ and $\mathbf{I}(x,s)$ are vectors of phase voltages and phase currents, respectively, and $\mathbf{L}$ and $\mathbf{C}$ are the inductance and capacitance matrices of the fully transposed line, given as

$$\mathbf{V}(x,s)=\begin{bmatrix} V_a(x,s)\\ V_b(x,s)\\ V_c(x,s)\end{bmatrix},\quad \mathbf{I}(x,s)=\begin{bmatrix} I_a(x,s)\\ I_b(x,s)\\ I_c(x,s)\end{bmatrix}, \tag{2.66}$$

$$\mathbf{L}=\begin{bmatrix} L_s & L_m & L_m\\ L_m & L_s & L_m\\ L_m & L_m & L_s\end{bmatrix}\text{H/m},\quad \mathbf{C}=\begin{bmatrix} C_s & C_m & C_m\\ C_m & C_s & C_m\\ C_m & C_m & C_s\end{bmatrix}\text{F/m}. \tag{2.67}$$

The phase voltages and currents can be transformed into their modal counterparts using

$$\mathbf{V}(x,s)=\mathbf{T}_V\mathbf{V}^M(x,s), \tag{2.68}$$

$$\mathbf{I}(x,s)=\mathbf{T}_I\mathbf{I}^M(x,s), \tag{2.69}$$

where the modal vectors are given as

$$\mathbf{V}^M(x,s)=\begin{bmatrix} V^0(x,s)\\ V^+(x,s)\\ V^-(x,s)\end{bmatrix},\quad \mathbf{I}^M(x,s)=\begin{bmatrix} I^0(x,s)\\ I^+(x,s)\\ I^-(x,s)\end{bmatrix}, \tag{2.70}$$

Substituting (2.68) and (2.69) in (2.64) and (2.65), the following can be obtained

$$\frac{d\mathbf{V}^M(x,s)}{dx}=-s(\mathbf{T}_V^{-1}\mathbf{L}\mathbf{T}_I)\,\mathbf{I}^M(x,s), \tag{2.71}$$

$$\frac{d\mathbf{I}^M(x,s)}{dx}=-s(\mathbf{T}_I^{-1}\mathbf{C}\mathbf{T}_V)\,\mathbf{V}^M(x,s). \tag{2.72}$$

In general, $\mathbf{T}_V$ and $\mathbf{T}_I$ matrices are different, however, for a fully transposed three-phase line they are the same, and their inverses are also the same, given as

$$\mathbf{T}_V=\mathbf{T}_I=\begin{bmatrix} 1 & 1 & 1\\ 1 & -2 & 1\\ 1 & 1 & -2\end{bmatrix}\quad\text{and}\quad \mathbf{T}_V^{-1}=\mathbf{T}_I^{-1}=\frac{1}{3}\begin{bmatrix} 1 & 1 & 1\\ 1 & -1 & 0\\ 1 & 0 & -1\end{bmatrix}. \tag{2.73}$$

Using (2.67) and (2.73) in (2.71) and (2.72) yields

$$\frac{d}{dx}\begin{bmatrix} V^0(x,s)\\ V^+(x,s)\\ V^-(x,s)\end{bmatrix}=\begin{bmatrix} (L_s+2L_m) & 0 & 0\\ 0 & -(L_s-L_m) & 0\\ 0 & 0 & -(L_s-L_m)\end{bmatrix}\begin{bmatrix} I^0(x,s)\\ I^+(x,s)\\ I^-(x,s)\end{bmatrix}, \tag{2.74}$$

$$\frac{d}{dx}\begin{bmatrix} I^0(x,s)\\ I^+(x,s)\\ I^-(x,s)\end{bmatrix}=\begin{bmatrix} (C_s+2C_m) & 0 & 0\\ 0 & -(C_s-C_m) & 0\\ 0 & 0 & -(C_s-C_m)\end{bmatrix}\begin{bmatrix} V^0(x,s)\\ V^+(x,s)\\ V^-(x,s)\end{bmatrix}. \tag{2.75}$$

Thus, the modal transformation results in diagonalized matrices and independent modal line equations that are identical to those of a single-phase loss-less line (2.31)–(2.32). For example, the zero mode line equations are given as

$$\frac{dV^0(x,s)}{dx} = -s(L_s + 2L_m)\, I^0(x,s), \tag{2.76}$$

$$\frac{dI^0(x,s)}{dx} = -s(C_s + 2C_m)\, V^0(x,s). \tag{2.77}$$

where

$$v^0 = \frac{1}{\sqrt{(L_s + 2L_m)(C_s + 2C_m)}}\ \text{m/s and}\ Z_c^0 = \sqrt{\frac{(L_s + 2L_m)}{(C_s + 2C_m)}}\ \Omega, \tag{2.78}$$

are the zero mode velocity and surge impedance, respectively. The velocities and surge impedances of the "+" and "−" modes are the same, given as

$$v^+ = v^+ = \frac{1}{\sqrt{(L_s - L_m)(C_s - C_m)}}\ \text{m/s}\ \ Z_c^+ = Z_c^- = \sqrt{\frac{(L_s - L_m)}{(C_s - C_m)}}\ \Omega. \tag{2.79}$$

For time-domain implementation of the three-phase loss-less TWM, the modal domain impedance matrix, injection history vectors, and voltage vectors can be written as

$$\mathbf{Z}_c^M = \begin{bmatrix} Z_c^0 & 0 & 0 \\ 0 & Z_c^+ & 0 \\ 0 & 0 & Z_c^- \end{bmatrix}, \tag{2.80}$$

$$\mathbf{I}_{hk}^M(t-\tau) = \begin{bmatrix} I_{hk}^0(t-\tau) \\ I_{hk}^+(t-\tau) \\ I_{hk}^-(t-\tau) \end{bmatrix}, \quad \mathbf{I}_{hm}^M(t-\tau) = \begin{bmatrix} I_{hm}^0(t-\tau) \\ I_{hm}^+(t-\tau) \\ I_{hm}^-(t-\tau) \end{bmatrix}, \tag{2.81}$$

$$\mathbf{V}_k^M(t) = \begin{bmatrix} V_k^0(t) \\ V_k^+(t) \\ V_k^-(t) \end{bmatrix}, \quad \mathbf{V}_m^M(t) = \begin{bmatrix} V_m^0(t) \\ V_m^+(t) \\ V_m^-(t) \end{bmatrix}. \tag{2.82}$$

The phase-domain transmission line impedance matrix, $\mathbf{Z}_c$, and injection history vectors, $\mathbf{I}_{hk}(t-\tau)$ and $\mathbf{I}_{hm}(t-\tau)$, are transferred from the modal domain as

$$\mathbf{Z}_c = \mathbf{T}_I \mathbf{Z}_c^M \mathbf{T}_V^{-1}, \tag{2.83}$$

$$\mathbf{I}_{hk}(t-\tau) = \mathbf{T}_I \mathbf{I}_{hk}^M(t-\tau), \tag{2.84}$$

$$\mathbf{I}_{hm}(t-\tau) = \mathbf{T}_I \mathbf{I}_{hm}^M(t-\tau). \tag{2.85}$$

The phase-domain transmission line sending-end and receiving-end voltage vectors are decomposed into the modal components as

$$\mathbf{v}_k(t) = \mathbf{T}_V \mathbf{V}_k^M(t), \tag{2.86}$$

$$\mathbf{v}_m(t) = \mathbf{T}_V \mathbf{V}_m^M(t). \tag{2.87}$$

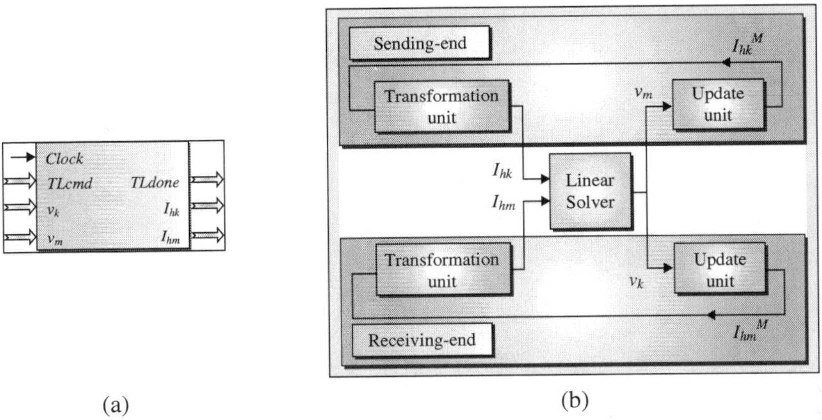

Figure 2.15 (a) *TWM* HEBB and its input/output signals. (b) Paralleled computation in the *TWM*.

2.7.3 Hardware Emulation of the TWM

Figure 2.15a shows the *TWM* HEBB module with the input and output signals. This implementation is used to perform the modal transformation by (2.83)–(2.85) and update the history terms by (2.60)–(2.61).

As can been seen from (2.58) and (2.59), the calculations at sending-end and receiving-end can be computed separately, which means in hardware they can be processed fully in parallel to achieve minimum latency. As shown in Figure 2.15b within the *TWM* module, two identical hardware parts are implemented, and each part contains one *Transformation unit* and one *Update unit*.

2.7.3.1 *Transformation Unit*

The modal transformation is performed in this unit to calculate the injection history vectors under phase-domain, $\mathbf{I}_{hk}(t-\tau)$ and $\mathbf{I}_{hm}(t-\tau)$, by reading the modal domain history terms, $\mathbf{I}_{hk}^{M}(t-\tau)$ and $\mathbf{I}_{hm}^{M}(t-\tau)$. The transformed phase-domain history vectors are then sent to the Linear Solver module to perform the nodal analysis method. A floating-point 6-input multiply-add unit is implemented for the transformation purpose. The RAM is used to write in and read out the modal domain history vectors, and ROM is used to store the matrix $\mathbf{T}_I$.

2.7.3.2 *Update Unit*

After obtaining the line sending-end and receiving-end voltages from the Linear Solver module, the *Update unit* is executed to update the history vectors in modal-domain. The first step is to transform the voltages from phase-domain to modal-domain which is achieved by a floating-point 6-input multiply-add unit. The second step is to update the history vectors in modal-domain which is

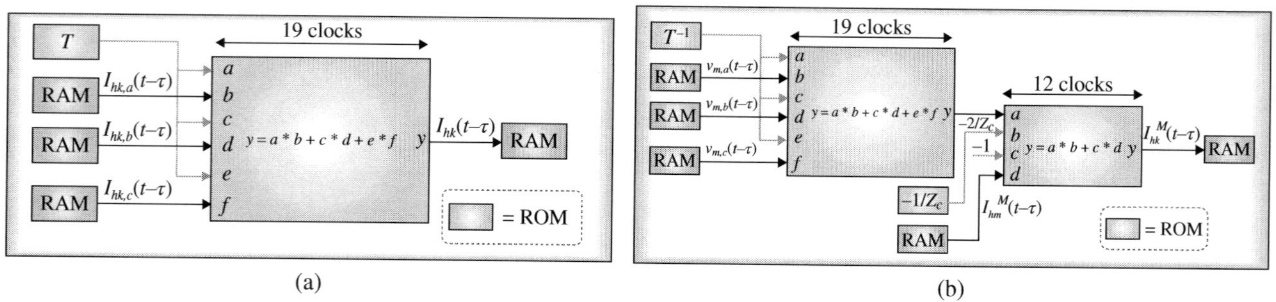

Figure 2.16 Pipelined computation scheme in (a) *Transformation unit* and (b) *Update unit*.

accomplished by a floating-point 4-input multiply-add unit. The RAM is used to write in and read out the phase-domain voltages or modal-domain history vectors, and ROM is used to store the entries of matrix $\mathbf{T}_V^{-1}$.

2.7.4 Frequency Dependent Line Model

Unlike the TWM, the parameters of a realistic transmission line are frequency dependent, i.e. $(R(\omega), L(\omega), C(\omega), G(\omega))$. This results in attenuation and phase distortion of the waves and power losses on the line. For accurate transient reproduction, this frequency dependency must be included in the line model [29]. Taking into account the frequency dependency of parameters and the distributed nature of losses in the transmission lines, the FDLM [32, 33, 40] is widely used. Similar to the Eqs. (2.52) and (2.53) for the single-phase loss-less line, the two-port network equations for a single-phase lossy frequency-dependent line can be derived in the frequency-domain as follows:

$$\begin{aligned} V_k(\omega) &= \cosh[\gamma(\omega)\ell]V_m(\omega) + Z_c(\omega)\sinh[\gamma(\omega)\ell]I_m(\omega), \\ I_k(\omega) &= \frac{1}{Z_c(\omega)}\sinh[\gamma(\omega)\ell]V_m(\omega) + \cosh[\gamma(\omega)\ell]I_m(\omega), \end{aligned} \tag{2.88}$$

where $Z_c(\omega)$ and $\gamma(\omega)$ are frequency-dependent characteristic impedance and propagation function defined as

$$Z_c(\omega) = \sqrt{\frac{R(\omega) + j\omega L(\omega)}{G(\omega) + j\omega C(\omega)}},$$

$$\gamma(\omega) = \sqrt{(R(\omega) + j\omega L(\omega))(G(\omega) + j\omega C(\omega))} = \alpha(\omega) + j\beta(\omega). \tag{2.89}$$

For the FDLM, both the $Z_c(\omega)$ and $\gamma(\omega)$ functions are complex. $\alpha(\omega)$ and $\beta(\omega)$ are the frequency-dependent attenuation and phase functions, respectively. Equations that relate currents and voltages in the frequency-domain are rewritten in terms of:

Forward traveling functions

$$\begin{aligned} F_k(\omega) &= V_k(\omega) + Z_c(\omega)I_k(\omega), \\ F_m(\omega) &= V_m(\omega) - Z_c(\omega)I_m(\omega). \end{aligned} \tag{2.90}$$

Backward traveling functions

$$\begin{aligned} B_k(\omega) &= -V_k(\omega) + Z_c(\omega)I_k(\omega), \\ B_m(\omega) &= V_m(\omega) + Z_c(\omega)I_m(\omega). \end{aligned} \tag{2.91}$$

Eliminating $V_k(\omega)$, $V_m(\omega)$, $I_k(\omega)$ and $I_m(\omega)$ from (2.88), and (2.91) yields

$$\begin{aligned} B_k(\omega) &= -A_1(\omega)F_m(\omega), \\ B_m(\omega) &= A_1(\omega)F_k(\omega), \end{aligned} \tag{2.92}$$

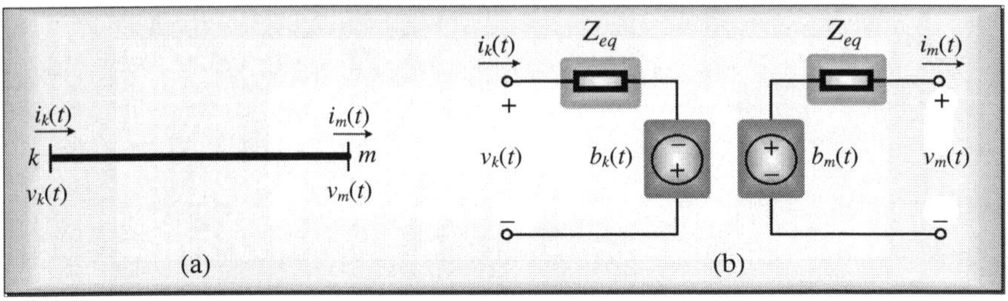

Figure 2.17 (a) Single-phase line and (b) its frequency-dependent model.

where

$$A_1(\omega) = e^{-\gamma(\omega)\ell} = \frac{1}{\cosh[\gamma(\omega)\ell] + \sinh[\gamma(\omega)\ell]}. \tag{2.93}$$

Equation (2.91) give the Thévenin equivalent network shown in Figure 2.17. The voltage sources $b_k(t)$ and $b_m(t)$ are the time-domain counterparts of $B_k(\omega)$ and $B_m(\omega)$, which are convolution integrals of (2.92) given as

$$b_k(t) = -\int_{\tau}^{\infty} f_m(t-u)a_1(u)du,$$
$$b_m(t) = \int_{\tau}^{\infty} f_k(t-u)a_1(u)du, \tag{2.94}$$

where $a_1(t)$, defined as the *weighting function*, is the time-domain form of $A_1(\omega)$, and

$$f_k(t) = 2v_k(t) + b_k(t),$$
$$f_m(t) = 2v_m(t) - b_m(t). \tag{2.95}$$

Accuracy of the FDLM greatly depends on the fitting of $Z_c(\omega)$ and $A_1(\omega)$. The vector fitting method [41] can be used for this purpose. $Z_c(\omega)$ is fitted by an N_{zc}-order rational function of the form

$$Z_{eq}(s) = k_0 + \sum_{i=1}^{N_{zc}} \frac{k_i}{s+p_i}, \tag{2.96}$$

which is realized by a series of *RC* parallel blocks as shown in Figure 2.18a. The parameters are calculated as follows

$$R_0 = k_0, \quad R_i = k_i/p_i, \quad C_i = 1/k_i. \tag{2.97}$$

Applying Trapezoidal Rule to each *RC* parallel block, its discrete-time model is obtained as shown in Figure 2.18b, where

$$R_{eqi} = \frac{R_i \frac{\Delta t}{2C_i}}{R_i + \frac{\Delta t}{2C_i}} \tag{2.98}$$

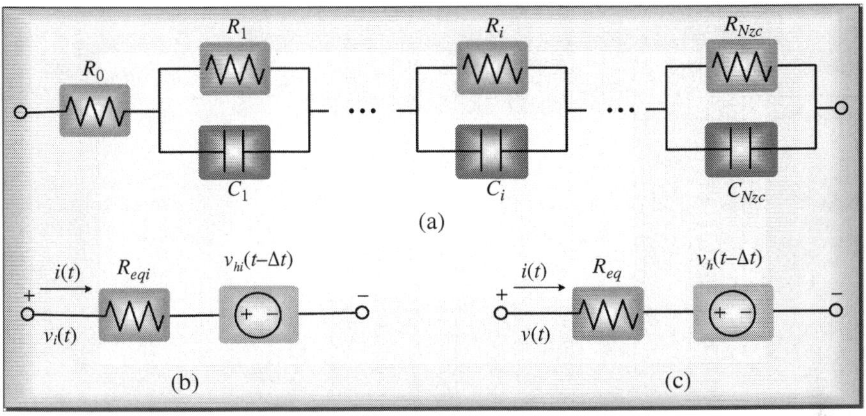

Figure 2.18 (a) *RC* network realization of $Z_{eq}(\omega)$ approximating $Z_c(\omega)$, (b) discrete-time model of *i*th *RC* block, and (c) overall Thévenin equivalent network of $Z_{eq}(\omega)$.

and

$$\begin{aligned} v_{hi}(t-\Delta t) &= \frac{R_i^2\frac{\Delta t}{C_i}}{\left(R_i+\frac{\Delta t}{2C_i}\right)^2} i(t-\Delta t) + \frac{R_i-\frac{\Delta t}{2C_i}}{R_i+\frac{\Delta t}{2C_i}} v_{hi}(t-2\Delta t) \\ &= D_{1i} i(t-\Delta t) + D_{2i} v_{hi}(t-2\Delta t). \end{aligned} \tag{2.99}$$

Combining all *RC* blocks discrete-time models together gives the overall Thévenin equivalent network shown in Figure 2.18c, where

$$R_{eq} = R_0 + \sum_{i=1}^{N_{zc}} R_{eqi} \tag{2.100}$$

and

$$v_h(t-\Delta t) = \sum_{i=1}^{N_{zc}} v_{hi}(t-\Delta t). \tag{2.101}$$

Specifically, for sending-end and receiving-end, (2.101) becomes

$$\begin{aligned} v_{hk}(t-\Delta t) &= \sum_{i=1}^{N_{zc}} v_{hki}(t-\Delta t) = \sum_{i=1}^{N_{zc}} D_{1i} i_k(t-\Delta t) + D_{2i} v_{hki}(t-2\Delta t), \\ v_{hm}(t-\Delta t) &= \sum_{i=1}^{N_{zc}} v_{hmi}(t-\Delta t) = \sum_{i=1}^{N_{zc}} D_{1i} i_m(t-\Delta t) + D_{2i} v_{hmi}(t-2\Delta t). \end{aligned} \tag{2.102}$$

Again, N_{zc} is total number of *RC* parallel blocks or the order of the approximation of $Z_c(\omega)$; i_k and i_m are line currents at both ends. Computational efficiency of the convolutions of (2.94) may be greatly increased if the weighting function $a_1(t)$

has the form of sum of exponential terms [32]. To do so, the same fitting techniques for $Z_c(\omega)$ are applied in the approximation of weighting function $A_1(\omega)$. However, rather than a proper form of $Z_{eq}(s)$, the $A_1(\omega)$ is first back-winded by the propagation delay τ to produce

$$P(\omega) = A_1(\omega)e^{j\omega\tau}, \tag{2.103}$$

and then approximated in a strictly proper manner by a N_p-order rational function

$$P_a(s) = \sum_{i=1}^{N_p} \frac{k_i}{s + p_i}. \tag{2.104}$$

Thus, a sum of exponentials is obtained from the inverse Fourier transformation as

$$a_{1a}(t) = u(t-\tau)\sum_{i=1}^{N_p} k_i e^{-p_i(t-\tau)}, \tag{2.105}$$

where $u(t-\tau)$ is step response with τ delay.

For the ith term $k_i e^{-p_i(t-\tau)}u(t-\tau)$ in approximated function $a_{1a}(t)$, the convolution integral is

$$b_i(t) = \int_{\tau}^{\infty} f(t-u)k_i e^{-p_i(u-\tau)}du. \tag{2.106}$$

$b_i(t)$ can then be directly obtained from recursive convolution [32] by

$$b_i(t) = M_i \cdot b_i(t-\Delta t) + P_i \cdot f(t-\tau) + Q_i \cdot f(t-\tau-\Delta t), \tag{2.107}$$

where

$$\begin{cases} M_i = e^{-p_i\Delta t} = \alpha, \\ P_i = \dfrac{k_i}{p_i}\left(1 - \dfrac{1-\alpha}{p_i\Delta t}\right), \\ Q_i = \dfrac{k_i}{p_i}\left(\dfrac{1-\alpha}{p_i\Delta t} - \alpha\right). \end{cases} \tag{2.108}$$

Specifically, $b_k(t)$ for sending-end and $b_m(t)$ for receiving-end are given as

$$\begin{aligned} b_k(t) &= \sum_{i=1}^{N_p} b_{ki}(t) = \sum_{i=1}^{N_p}[M_i \cdot b_{ki}(t-\Delta t) + P_i \cdot f_m(t-\tau) + Q_i \cdot f_m(t-\tau-\Delta t)], \\ b_m(t) &= \sum_{i=1}^{N_p} b_{mi}(t) = \sum_{i=1}^{N_p}[M_i \cdot b_{mi}(t-\Delta t) + P_i \cdot f_k(t-\tau) + Q_i \cdot f_k(t-\tau-\Delta t)]. \end{aligned} \tag{2.109}$$

To calculate (2.109), the past values of $f_m(t)$ and $f_k(t)$ have to be stored. The number of stored past value N_{tau} depends on the propagation delay τ and time-step Δt as follows

$$N_{tau} = \tau/\Delta t. \tag{2.110}$$

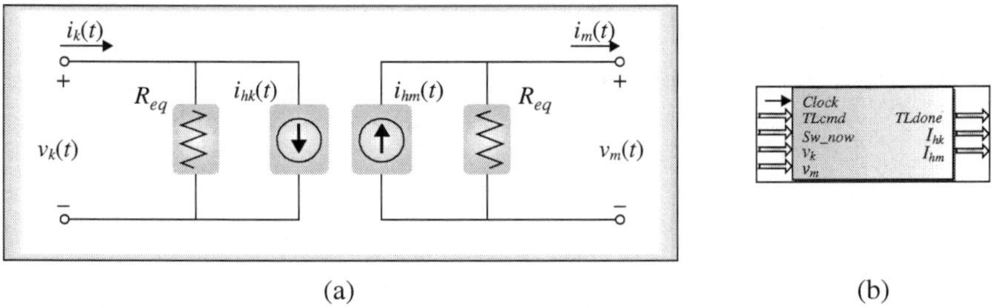

Figure 2.19 (a) Discrete-time equivalent network for FDLM. (b) *FDLM* HEBB and its input/output signals.

Finally, by combining (2.102) with (2.109), the equivalent time-domain network of FDLM is obtained as shown in Figure 2.19a, where the equivalent current sources are given by

$$\begin{aligned} i_{hk}(t) &= [-b_k(t) + v_{hk}(t-\Delta t)]/R_{eq}, \\ i_{hm}(t) &= [b_m(t) + v_{hm}(t-\Delta t)]/R_{eq}, \end{aligned} \tag{2.111}$$

and the equivalent resistance is given by (2.100). The line currents for sending-end and receiving-end are calculated as

$$\begin{aligned} i_k(t) &= v_k(t)/R_{eq} + i_{hk}(t), \\ i_m(t) &= -v_m(t)/R_{eq} + i_{hm}(t). \end{aligned} \tag{2.112}$$

For a three-phase FDLM, the modal transformation matrices of (2.68) $\mathbf{T}_V$ and $\mathbf{T}_I$ become complex and frequency-dependent. With a frequency-dependent transformation, the modes can only be accurately defined at the specific frequency of the transformation. Therefore, to decouple the three-phase line equations into three separate modal line equations, a real and constant Clarke's transformation [27] is employed. Two aerial modes (α, β) and one ground mode (0) are the result and the transformation is performed as

$$\boldsymbol{i}_{k,m} = \begin{bmatrix} i_a \\ i_b \\ i_c \end{bmatrix} = \boldsymbol{T}\boldsymbol{i}^M_{k,m} = \boldsymbol{T}\begin{bmatrix} i_0 \\ i_\alpha \\ i_\beta \end{bmatrix}, \tag{2.113}$$

$$\boldsymbol{v}_{k,m} = \begin{bmatrix} v_a \\ v_b \\ v_c \end{bmatrix} = \boldsymbol{T}\boldsymbol{v}^M_{k,m} = \boldsymbol{T}\begin{bmatrix} v_0 \\ v_\alpha \\ v_\beta \end{bmatrix}, \tag{2.114}$$

where the transformation matrix $\mathbf{T}$ is given by

$$\boldsymbol{T} = \frac{1}{\sqrt{3}}\begin{bmatrix} 1 & \sqrt{2} & 0 \\ 1 & -1/\sqrt{2} & \sqrt{3}/\sqrt{2} \\ 1 & -1/\sqrt{2} & -\sqrt{3}/\sqrt{2} \end{bmatrix}. \tag{2.115}$$

2.7.5 Hardware Emulation of FDLM

Figure 2.19b shows the symbol of the *FDLM* HEBB module and its input/output signals. The main operations carried out in this module include convolution for $b_k(t)$, $b_m(t)$ (2.109), calculating $i_{hlnk}(t)$, $i_{hlnm}(t)$ (2.111), and updating f_k, f_m (2.95) and $v_{hk}(t-\Delta t)$, $v_{hm}(t-\Delta t)$ (2.102).

Due to the traveling time delay of FDLM, the calculations at sending-end and receiving-end are naturally decoupled, which means they can be processed fully in parallel in hardware [40]. Moreover, for three-phase systems, the calculations in α mode, β mode, and 0 mode can also be carried out simultaneously. As can be seen in Figure 2.20, four identical hardware parts are implemented in the FPGA, and each part contains a *Convolution* unit and a *Update* unit whose pipelined computation schemes are shown in Figure 2.16.

2.7.5.1 *Convolution* Unit

Depending on the order of fitted rational function N_p, the recursive convolutions of (2.109) are very time-consuming in the transient solution. The *Convolution* unit is fully pipelined to improve the data throughput. Figure 2.21a shows the pipelined computation scheme in this unit. A floating-point multiply-add ($y = a \times b + c \times d + e \times f$) unit is first used to calculate $b_{ki}(t)$, followed by a floating-point accumulator ($y = \sum a_i$) unit to calculate $b_k(t)$. Then, the equivalent current $i_{hlnk}(t)$ (2.111) is calculated using another floating-point multiply-add ($y = a \times b + c \times d$) unit.

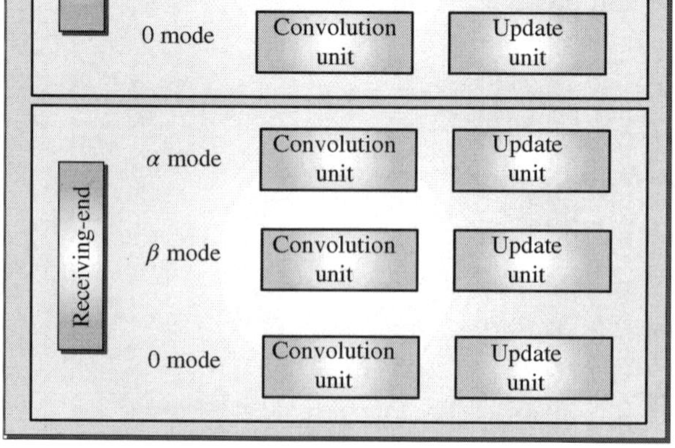

Figure 2.20 Functional units in the *FDLM* HEBB executed in parallel.

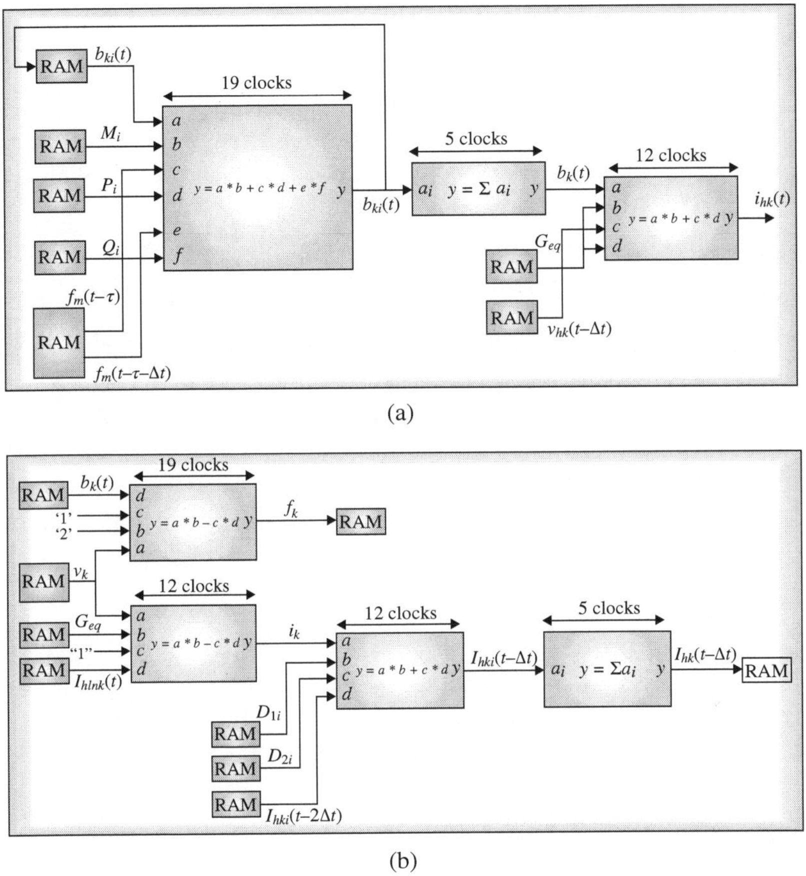

Figure 2.21 Pipelined computation scheme in (a) the *Convolution* unit and (b) in the *Update* unit.

RAM is utilized to save the parameters such as M_i, P_i, and Q_i. The RAM used to save b_{ki} is a dual-port RAM which can be read out and written in at the same time.

2.7.5.2 *Update* Unit

After the network is solved, the history terms are updated in the *Update* unit shown in Figure 2.21b. The updated terms include $f_k(t)$, $f_m(t)$ (2.95) and $v_{hk}(t-\Delta t)$, $v_{hm}(t-\Delta t)$ (2.102). As can be seen in Figure 2.21b, a floating-point multiply-add ($y = a \times b - c \times d$) unit is employed to calculate f_k, while the other floating-point multiply-add ($y = a \times b - c \times d$) unit is employed to calculate line current i_k (2.112) concurrently. Then, $v_{hk}(t-\Delta t)$ is calculated in a floating-point multiply-add ($y = a \times b + c \times d$) unit and a floating-point accumulator ($y = \sum a_i$) unit.

2.7.6 Universal Line Model

2.7.6.1 Frequency-Domain Formulation

The ULM is a direct phase-domain frequency-dependent formulation of line equations. For a n-phase transmission line as shown in Figure 2.22a, the solution of the traveling wave equations can be expressed in frequency-domain at the sending-end ("k") and the receiving-end ("m") as

$$\begin{aligned} \boldsymbol{I}_k &= \boldsymbol{Y}_c\boldsymbol{V}_k - 2\boldsymbol{I}_{ki} = \boldsymbol{Y}_c\boldsymbol{V}_k - 2\boldsymbol{H}\boldsymbol{I}_{mr}, \\ \boldsymbol{I}_m &= -\boldsymbol{Y}_c\boldsymbol{V}_m + 2\boldsymbol{I}_{mi} = -\boldsymbol{Y}_c\boldsymbol{V}_m + 2\boldsymbol{H}\boldsymbol{I}_{kr}, \end{aligned} \tag{2.116}$$

where $\boldsymbol{I}_k$, $\boldsymbol{V}_k$ and $\boldsymbol{I}_m$, $\boldsymbol{V}_m$ are n dimensional current and voltage vectors at both line ends, respectively. $\boldsymbol{I}_{ki}$ and $\boldsymbol{I}_{mi}$ are the incident currents, whereas $\boldsymbol{I}_{kr}$ and $\boldsymbol{I}_{mr}$ are the reflected currents. The two $n \times n$ matrix transfer functions which characterize a transmission line are the characteristic admittance matrix $\boldsymbol{Y}_c$ and the propagation matrix $\boldsymbol{H}$ expressed as

$$\boldsymbol{Y}_c = \sqrt{\boldsymbol{Y}/\boldsymbol{Z}} \tag{2.117}$$

and

$$\boldsymbol{H} = e^{-\sqrt{(\boldsymbol{YZ})}l}, \tag{2.118}$$

where the $\boldsymbol{Y}$ and $\boldsymbol{Z}$ are $n \times n$ shunt admittance and series impedance matrices per unit length, respectively; l is the line length.

To implement the model in time-domain, the elements of $\boldsymbol{Y}_c$ and $\boldsymbol{H}$ are approximated with rational functions. The method used here is vector fitting [41]. The elements of $\boldsymbol{Y}_c$ are in general smooth and can be easily fitted in the phase-domain. The (i, j) element of $\boldsymbol{Y}_c$ is expressed as

$$\boldsymbol{Y}_{c,(i,j)}(s) = \sum_{m=1}^{N_p} \frac{\boldsymbol{r}_{Y_c,(i,j)}(m)}{s - \boldsymbol{p}_{Y_c}(m)} + \boldsymbol{d}_{(i,j)}, \tag{2.119}$$

where N_p is the number of poles; $\boldsymbol{r}_{Y_c}$, $\boldsymbol{p}_{Y_c}$, and $\boldsymbol{d}$ are residues, poles, and proportional terms, respectively. Note that all elements of $\boldsymbol{Y}_c$ have identical poles $\boldsymbol{p}_{Y_c}$.

The fitting of the propagation matrix $\boldsymbol{H}$ is somewhat different. It is first fitted in the modal-domain with poles and time delays in each mode, followed by final fitting in the phase-domain. The (i, j) element of $\boldsymbol{H}$ is expressed as

$$\boldsymbol{H}_{(i,j)}(s) = \sum_{k=1}^{N_g} \left(\sum_{n=1}^{N_{p,k}} \frac{\boldsymbol{r}_{H,(i,j),k}(n)}{s - \boldsymbol{p}_{H,k}(n)} \right) e^{-s\tau_k}, \tag{2.120}$$

where N_g denotes the number of modes; $N_{p,k}$ and τ_k are the number of poles and time delay used for fitting the kth mode. $\boldsymbol{r}_{H,k}$ and $\boldsymbol{p}_{H,k}$ are residues and poles for kth mode. Again the poles are identical for all elements in each mode.

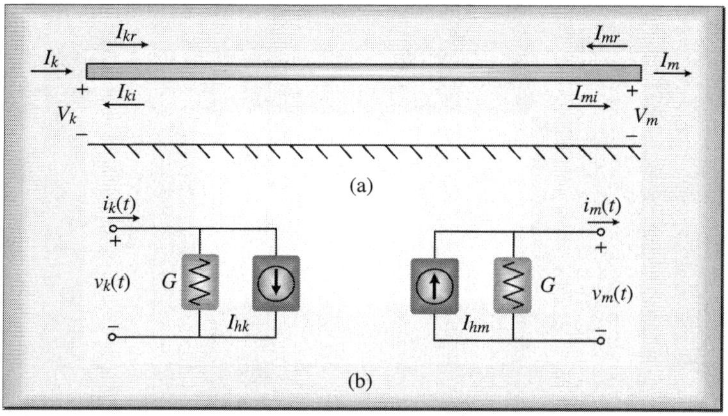

Figure 2.22 (a) An n-phase transmission line and (b) its time-domain representation.

2.7.6.2 Time-Domain Formulation

By transforming (2.116) into the time-domain using inverse Fourier Transformation, the Norton equivalent circuit for the ULM is obtained, as shown in Figure 2.22b. The history current $\boldsymbol{I}_{hln_k}, \boldsymbol{I}_{hln_m}$ and equivalent impedance matrix $\boldsymbol{G}$ are expressed as

$$\begin{aligned} \boldsymbol{I}_{hln_k} &= -\boldsymbol{Y}_c * \boldsymbol{v}_k(t) + 2\boldsymbol{H} * \boldsymbol{i}_{mr}(t-\tau), \\ \boldsymbol{I}_{hln_m} &= \boldsymbol{Y}_c * \boldsymbol{v}_m(t) - 2\boldsymbol{H} * \boldsymbol{i}_{kr}(t-\tau), \end{aligned} \tag{2.121}$$

and

$$\boldsymbol{G} = \boldsymbol{d} + \boldsymbol{r}_{Y_c} \lambda_{Y_c}, \tag{2.122}$$

where the symbol "*" denotes the matrix–vector convolution. The coefficient λ_{Y_c} is defined as

$$\lambda_{Y_c} = \left(\frac{\Delta t}{2}\right) / \left(1 - \boldsymbol{p}_{Y_c}\frac{\Delta t}{2}\right). \tag{2.123}$$

To perform the convolution of $\boldsymbol{Y}_c * \boldsymbol{v}_k(t)$, a state variable $\boldsymbol{x}_{Y_c}$ is defined as

$$\boldsymbol{x}_{Y_c}(t) = \boldsymbol{\alpha}_{Y_c}\boldsymbol{x}_{Y_c}(t-\Delta t) + \boldsymbol{v}_k(t-\Delta t). \tag{2.124}$$

The convolution is then computed using

$$\boldsymbol{Y}_c * \boldsymbol{v}_k(t) = \boldsymbol{c}_{Y_c}\boldsymbol{x}_{Y_c}(t), \tag{2.125}$$

where the coefficients $\boldsymbol{\alpha}_{Y_c}$ and $\boldsymbol{c}_{Y_c}$ are expressed as

$$\boldsymbol{\alpha}_{Y_c} = \left(1 + \boldsymbol{p}_{Y_c}\frac{\Delta t}{2}\right) / \left(1 - \boldsymbol{p}_{Y_c}\frac{\Delta t}{2}\right), \tag{2.126}$$

and

$$\boldsymbol{c}_{Y_c} = \boldsymbol{r}_{Y_c}(\boldsymbol{\alpha}_{Y_c} + 1)\lambda_{Y_c}. \tag{2.127}$$

To perform the convolution of $\boldsymbol{H} * \boldsymbol{i}_{mr}(t-\tau)$, a state variable $\boldsymbol{x}_H$ is defined as

$$\boldsymbol{x}_H(t) = \boldsymbol{\alpha}_H \boldsymbol{x}_H(t-\Delta t) + \boldsymbol{i}_{mr}(t-\tau-\Delta t). \tag{2.128}$$

The convolution is then computed using

$$\boldsymbol{H} * \boldsymbol{i}_{mr}(t-\tau) = \boldsymbol{c}_H \boldsymbol{x}_H(t) + \boldsymbol{G}_H \boldsymbol{i}_{mr}(t-\tau), \tag{2.129}$$

where the coefficient $\boldsymbol{\alpha}_H$, $\boldsymbol{c}_H$, and $\boldsymbol{G}_H$ are expressed as

$$\boldsymbol{\alpha}_H = \left(1 + \boldsymbol{p}_H \frac{\Delta t}{2}\right) / \left(1 - \boldsymbol{p}_H \frac{\Delta t}{2}\right), \tag{2.130}$$

$$\boldsymbol{c}_H = \boldsymbol{r}_H(\boldsymbol{\alpha}_H + 1)\boldsymbol{\lambda}_H, \tag{2.131}$$

and

$$\boldsymbol{G}_H = \boldsymbol{r}_H \boldsymbol{\lambda}_H, \tag{2.132}$$

with

$$\boldsymbol{\lambda}_H = \left(\frac{\Delta t}{2}\right) / \left(1 - \boldsymbol{p}_H \frac{\Delta t}{2}\right). \tag{2.133}$$

As seen from (2.128) and (2.129), $\boldsymbol{i}_{mr}(t-\tau-\Delta t)$ and $\boldsymbol{i}_{mr}(t-\tau)$ are required to perform the convolution of $\boldsymbol{H} * \boldsymbol{i}_{mr}(t-\tau)$. Linear interpolation is used to accurately calculate $\boldsymbol{i}_{mr}(t-\tau-\Delta t)$ and $\boldsymbol{i}_{mr}(t-\tau)$ since the travel time τ is normally not an integer multiple of the time-step Δt. Assume that τ is expressed as:

$$\tau = (k+\varepsilon)\Delta t, \tag{2.134}$$

where k is an integer and ε is a number between 0 and 1.

As illustrated in Figure 2.23a, $\boldsymbol{i}_{mr}(t-\tau-\Delta t)$ and $\boldsymbol{i}_{mr}(t-\tau)$ can be calculated using linear interpolation as

$$\begin{aligned}\boldsymbol{i}_{mr}(t-\tau-\Delta t) &= \boldsymbol{i}_{mr}(t-(k+2)\Delta t) \\ &\quad + (\boldsymbol{i}_{mr}(t-(k+1)\Delta t) - \boldsymbol{i}_{mr}(t-(k+2)\Delta t))(1+\varepsilon)\end{aligned} \tag{2.135}$$

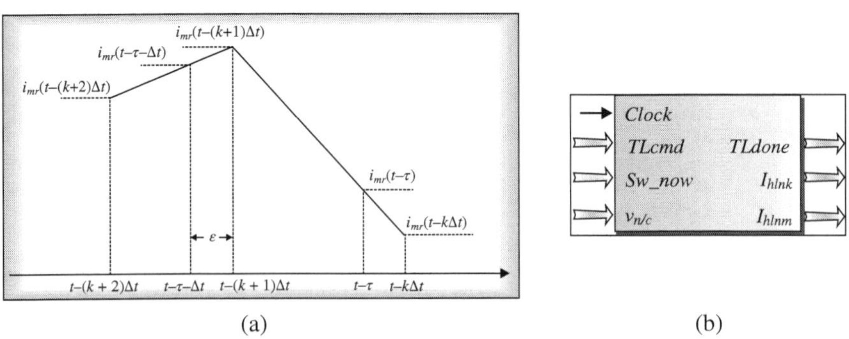

Figure 2.23 (a) Linear interpolation for calculating $\boldsymbol{i}_{mr}(t-\tau)$ and $\boldsymbol{i}_{mr}(t-\tau-\Delta t)$. (b) *ULM* HEBB and its input/output signals.

and

$$\boldsymbol{i}_{mr}(t-\tau) = \boldsymbol{i}_{mr}(t-(k+1)\Delta t) + (\boldsymbol{i}_{mr}(t-k\Delta t) - \boldsymbol{i}_{mr}(t-(k+1)\Delta t))(1+\varepsilon). \tag{2.136}$$

In order to carry out these two interpolations, $(k+2)$ time-steps history values of $\boldsymbol{i}_{mr}(t)$ need to be saved. For example, if $\tau = 9.26\Delta t$, 11 time-steps history values of $\boldsymbol{i}_{mr}(t)$ must be saved.

2.7.7 Hardware Emulation of the ULM

Figure 2.23b shows the symbol of the *ULM* HEBB module and its input/output signals. As seen from the (2.121), the main computations involved in the ULM are two convolutions at both ends of the transmission line. Depending on the number of phases of the transmission line and the number of poles of fitted rational functions, the convolution is in general quite expensive computationally. Parallel computation is essential to improve the computational efficiency. First, due to the traveling time delay of the ULM, the sending-end and receiving-end of a transmission line are decoupled. Thus, the calculations at both ends can be conducted fully in parallel. Furthermore, the two convolutions can be performed at different stage of a simulation time-step. The convolution of $\boldsymbol{H} * \boldsymbol{i}_{mr}(t-\tau)$ is independent of the node voltages, so that it can be carried out before the network is solved, whereas the convolution of $\boldsymbol{Y}_c * \boldsymbol{v}_k(t)$ must be processed after network is solved. In this way, the computation time can be reduced significantly. The computational efficiency can be improved even further for multi-phase transmission lines where the calculation in each phase can be executed simultaneously. Based on these three considerations, the detailed hardware computation units of the *ULM* HEBB module are designed, as shown in Figure 2.24. Two identical parts are implemented, for sending-end and receiving-end of transmission line, respectively. To explain these computation units easily, we rearrange (2.121), (2.125), and (2.129) to obtain

$$\boldsymbol{I}_{hln_k} = \boldsymbol{I}_{hln_{k1}} - 2(\boldsymbol{I}_{hln_{k2}} + \boldsymbol{I}_{hln_{k3}}), \tag{2.137}$$

where

$$\boldsymbol{I}_{hln_{k1}} = -\boldsymbol{c}_{Y_c}\boldsymbol{x}_{Y_c}(t), \tag{2.138}$$

$$\boldsymbol{I}_{hln_{k2}} = -\boldsymbol{c}_H\boldsymbol{x}_H(t), \tag{2.139}$$

$$\boldsymbol{I}_{hln_{k3}} = -\boldsymbol{G}_H i_{mr}(t-\tau). \tag{2.140}$$

The *Interpolation* unit is responsible for calculating $\boldsymbol{i}_{mr}(t-\tau-\Delta t)$ (2.135) and $\boldsymbol{i}_{mr}(t-\tau)$ (2.136). The *Update* $\boldsymbol{x}$ unit is used to update two state variables $\boldsymbol{x}_{Y_c}(t)$ (2.124) and $\boldsymbol{x}_H(t)$ (2.128). The *Convolution* unit is responsible for calculating $\boldsymbol{I}_{hln_{k1}}$ (2.138) and $\boldsymbol{I}_{hln_{k2}}$ (2.139). A fast floating-point multiply-accumulator (*FFPMAC*,

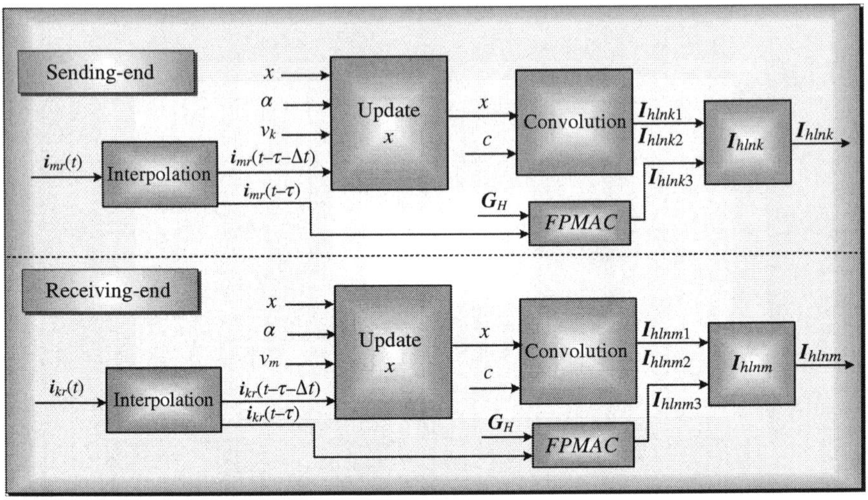

Figure 2.24 Main functional units in the *ULM* HEBB executed in parallel.

$y = \sum a_i b_i$) unit is used to calculate $\boldsymbol{I}_{hln_{k3}}$ (2.140). Finally, the line history term $\boldsymbol{I}_{hln_k}$ is obtained (2.137) from the I_{hlnk} unit.

2.7.7.1 *Update x* Unit

Figure 2.25a illustrates the parallel computation scheme of (2.124) or (2.128) in the *Update* $\boldsymbol{x}$ unit for a three-phase transmission line. The signals shown in this figure are for (2.124) only. Here the state variable $\boldsymbol{x}_{Y_c}$ and coefficient $\boldsymbol{\alpha}_{Y_c}$ are $N_p \times 3$ matrices, whereas the line voltage $\boldsymbol{v}_k$ is a 1×3 vector. Each column corresponds to one phase denoted by subscript $\{a, b, c\}$. All matrices are stored in three individual RAMs in column-wise format. The $\boldsymbol{x}_{Y_c}$ is calculated using three floating-point multiply-add ($y = a \times b + c$) units for three phases in parallel. The updated $\boldsymbol{x}_{Y_c}$ is sent back to the RAMs which are dual-port that support the "read" and "write" functions simultaneously.

2.7.7.2 *Convolution* Unit

Once $\boldsymbol{x}_{Y_c}$ or $\boldsymbol{x}_H$ is updated, the convolution of (2.138) or (2.139) can be carried out. Figure 2.25b illustrates the parallel computation scheme in the *Convolution* unit for a three-phase transmission line. The signals shown in this figure are for (2.138) only. Here the coefficient $\boldsymbol{c}_{Y_c}$ is a $3N_p \times 3$ matrix and stored in three RAMs in column-wise format. Three *FFPMAC* ($y = \sum a_i b_i$) units are used for calculations in the three phases in parallel, when a connected floating-point add ($y = a + b + c$) unit combines them to get final result of convolution. Note that this final result is not available for three phases at the same time. It comes sequentially as phase a, phase b, and phase c. Full parallelism is possible if more arithmetic units are designed subject to FPGA resource utilization.

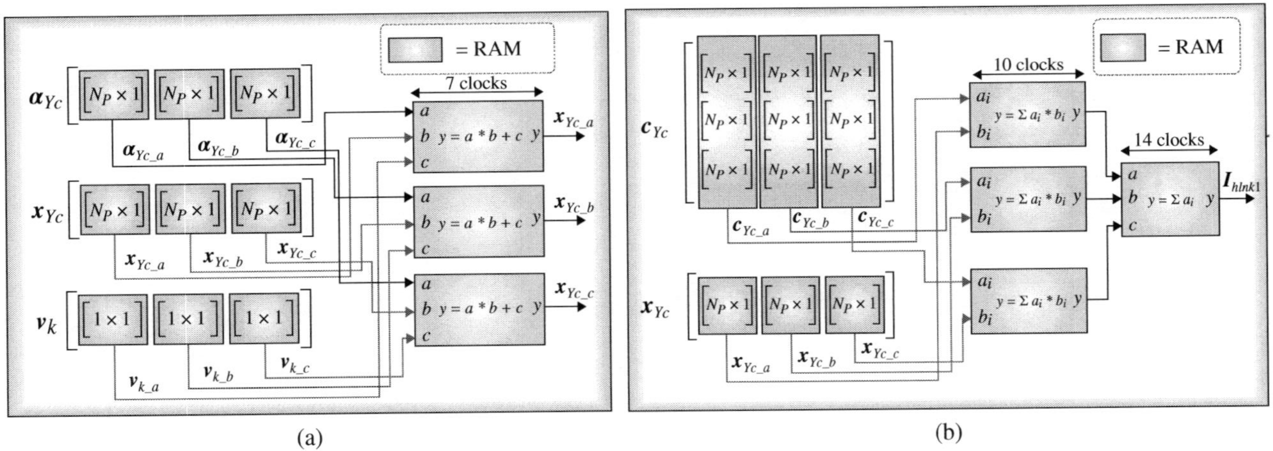

Figure 2.25 (a) Parallel computation scheme in (a) the *Update* $\boldsymbol{x}$ unit and (b) the *Convolution* unit.

2.7.7.3 *Interpolation* Unit

In the *Interpolation* unit, $\boldsymbol{i}_{mr}(t-\tau-\Delta t)$ (2.135) and $\boldsymbol{i}_{mr}(t-\tau)$ (2.136) are calculated. The hardware design of this unit is shown in Section 2.9.3.1.

2.8 Network Solver

After the discrete-time models of transmission lines and linear lumped RLCG elements are obtained, the system admittance matrix $\boldsymbol{Y}$ is assembled using the nodal analysis method. $\boldsymbol{Y}$ is symmetrical and unchanged if there is no switching operation causing the change of system topology. In general, for any type of network with n nodes, the nodal equation can be formulated as:

$$\boldsymbol{Y}\boldsymbol{v}(t) = \boldsymbol{i}(t), \tag{2.141}$$

where $\boldsymbol{v}(t)$ is vector of n node voltages and $\boldsymbol{i}(t)$ is vector of n current sources. If there are nodes with known voltages, (2.141) can be partitioned [26] into a set A of nodes with unknown voltages and a set B of nodes with known voltages. Thus (2.141) becomes

$$\begin{bmatrix} \boldsymbol{Y}_{AA} & \boldsymbol{Y}_{AB} \\ \boldsymbol{Y}_{BA} & \boldsymbol{Y}_{BB} \end{bmatrix} \begin{bmatrix} \boldsymbol{v}_A \\ \boldsymbol{v}_B \end{bmatrix} = \begin{bmatrix} \boldsymbol{i}_A \\ \boldsymbol{i}_B \end{bmatrix}. \tag{2.142}$$

The unknown voltages $\boldsymbol{v}_A$ are then found by solving the system of linear, algebraic equations

$$\boldsymbol{Y}_{AA}\boldsymbol{v}_A = \boldsymbol{i}_A - \boldsymbol{i}_{A1}, \tag{2.143}$$

where

$$\boldsymbol{i}_{A1} = \boldsymbol{Y}_{AB}\boldsymbol{v}_B, \tag{2.144}$$

where $\boldsymbol{i}_A$ is the current sources of unknown nodes. It is assembled by transmission line history currents $\boldsymbol{I}_{hln}$, lumped RLCG element history currents $\boldsymbol{I}_{hpe}$, and known current sources $\boldsymbol{i}_s$ as follows

$$\boldsymbol{i}_A = \boldsymbol{I}_{hln} + \boldsymbol{I}_{hpe} + \boldsymbol{i}_s. \tag{2.145}$$

In the off-line EMT solution, (2.143) is solved first by triangularizing the $\boldsymbol{Y}_{AA}$ matrix through a Gaussian elimination procedure and then backsubstituting for the updated values. In the real-time EMT emulator, (2.143) is solved by multiplying the precalculated inverse system admittance matrix $\boldsymbol{Y}_{AA}^{-1}$ by $(\boldsymbol{i}_A - \boldsymbol{i}_{A1})$. To account for switches in the network for a specific transient study, $\boldsymbol{Y}^{-1}$ corresponding to those switch combinations can be precalculated and stored in the memory. Meanwhile, since $\boldsymbol{Y}$ and $\boldsymbol{Y}^{-1}$ matrices are very sparse for power systems, sparse matrix technique is employed for the matrix–vector multiplication, which reduces memory utilization and increases the calculation efficiency significantly.

2.8.1 Hardware Emulation of Network Solver

Figure 2.26a shows the symbol of the *Network Solver* hardware HEBB module and its input/output signals. The main functions of this module are to calculate $\boldsymbol{i}_{A1}$ (2.144), $\boldsymbol{i}_A$ (2.145), and $\boldsymbol{v}_A$ (2.143).

As shown in Figure 2.27a, the calculation of $\boldsymbol{i}_{A1}$ and $\boldsymbol{i}_A$ are carried out concurrently. Meanwhile, $\boldsymbol{i} = \boldsymbol{i}_A - \boldsymbol{i}_{A1}$ is also calculated in order to obtain $\boldsymbol{v}_A$ shown in Figure 2.27b. In these two figures, the core unit is a *FFPMAC* unit along with a compact sparse matrix storage format.

To carry out fast sparse matrix–vector multiplication, first a very compact sparse matrix storage format which uses only one vector is defined. Each entry in this format has (i) a 32-bit *val* to store the subsequent nonzero value of matrix in row order; (ii) a 10-bit *cid* to identify column index of this nonzero value; and (iii) a 1-bit *rlb* to label all nonzero values in the same row with "0" or "1". Figure 2.26b shows an example sparse matrix and its storage format.

In the *FFPMAC* unit shown in Figure 2.28a, the accumulation is done in fixed-point format to save resources and minimize latency. Thus, the *FFPMAC* unit contains one floating-point multiplier (*FPMUL*), one floating-to-fixed point converter (*FLP2FIX*), two fixed-point adders, and one fixed-to-floating point converter (*FIX2FLP*). To calculate $\boldsymbol{Y}_{AA}^{-1}\boldsymbol{i}$ shown in Figure 2.27b, for example, the elements of sparse matrix $\boldsymbol{Y}_{AA}^{-1}$ are retrieved from its RAM, while the *cid* is used to access the $\boldsymbol{i}$ stored in its RAM. Registers are inserted for synchronizing the pipeline. The realized matrix–vector multiplication is fully pipelined and fast because there is no stall between two consecutive matrix row-vector multiplications. This is achieved by two parallel fixed-point adders (accumulators) acc_0 and acc_1 with opposite reset inputs rst_0 and rst_1 controlled by the row label bit *rlb*. Figure 2.28b shows the logic timing diagram for the FFPMC unit based on the example in Figure 2.26b. As can be seen, the accumulation of the first matrix row-vector multiplication is processed in acc_0, while the acc_1 is reset to zero which makes it ready for accumulation of the next matrix row-vector multiplication.

2.8.2 Paralleled EMT Solution Algorithm

To fully exploit the parallel architecture of the FPGA, the inherent parallelism in the EMT solution algorithm needs to be analyzed. The algorithm consists of three stages:

- *Stage 1*: Evaluating the nodal currents,
- *Stage 2*: Solving the nodal equation, and
- *Stage 3*: Updating the history terms.

These three stages form an unavoidable sequentiality in the EMT solution algorithm. For example, to solve the nodal equation, the nodal currents have to be

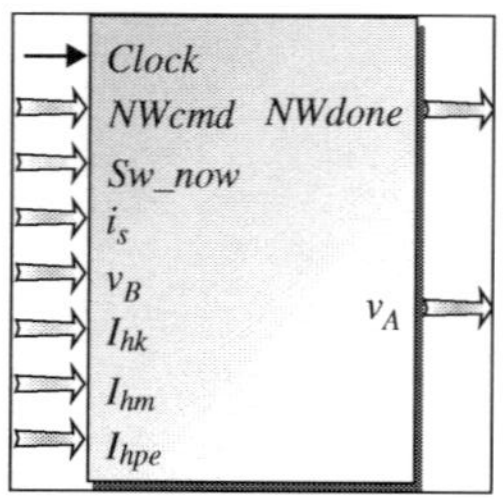

(a)

$$\begin{bmatrix} y_{00} & y_{01} & 0 & 0 & y_{04} \\ 0 & y_{11} & 0 & y_{13} & 0 \\ 0 & 0 & y_{22} & 0 & y_{24} \\ 0 & 0 & y_{32} & 0 & 0 \\ y_{40} & 0 & 0 & 0 & 0 \end{bmatrix}$$

rlb	*cid*	*val*
0	0	y_{00}
0	1	y_{01}
0	4	y_{04}
1	1	y_{11}
1	3	y_{13}
0	2	y_{22}
0	4	y_{24}
1	2	y_{32}
0	0	y_{40}

(b)

Figure 2.26 (a) *Network Solver* HEBB. (b) An example sparse matrix, and its storage format.

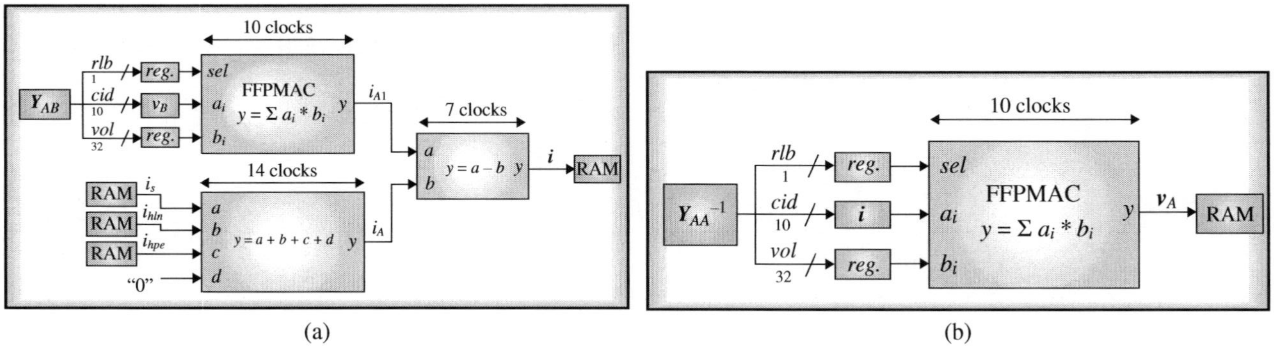

Figure 2.27 (a) Pipelined and paralleled calculation scheme for $\boldsymbol{i}_{A1}$ and $\boldsymbol{i}_A$. (b) Pipelined calculation scheme for $\boldsymbol{v}_A$.

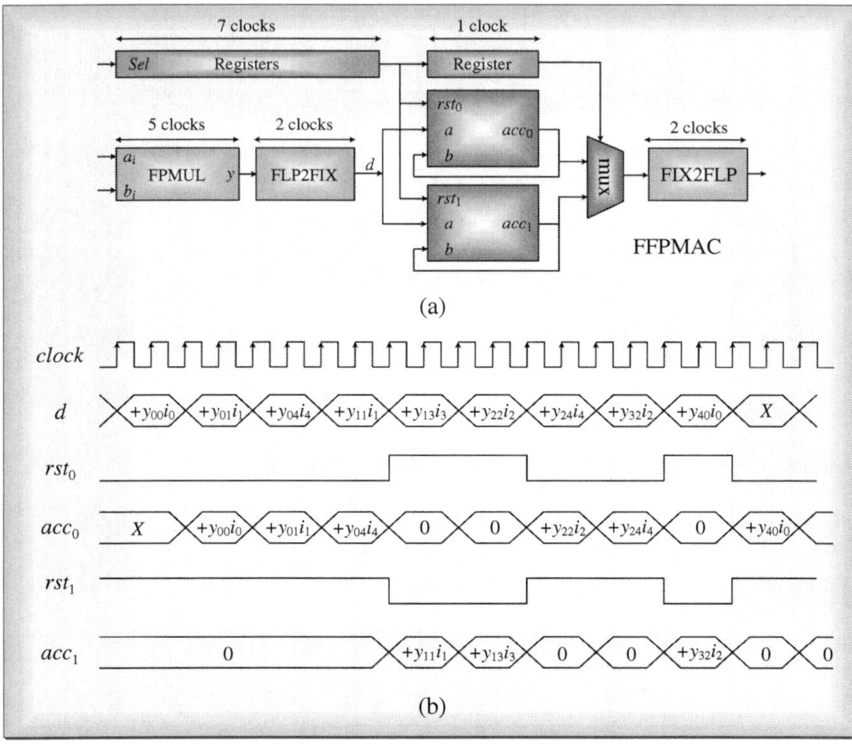

Figure 2.28 Fast floating-point multiply-accumulator unit (*FFPMAC*): (a) hardware design and (b) timing diagram.

known from *Stage 1*. Similarly, to update the history terms in *Stage 3*, we need the node voltages from *Stage 2*. Nevertheless, the operations within each of these stages can be parallelized. Figure 2.29 shows the paralleled real-time algorithm flowchart. In *Stage 1*, four operations are performed simultaneously: the calculation of the supply sources, calculation of transmission lines equivalent to current sources at both sending and receiving ends, retrieving the RLCG elements' history current sources from memory, and checking switch states. Similarly, in *Stage 3*, the history terms for the transmission lines and RLCG elements are updated simultaneously.

2.8.3 *MainControl* Module

The *MainControl* HEBB module, as shown in Figure 2.30a, coordinates the operation of the whole FPGA-based real-time EMT emulator to carry out the paralleled EMT solution algorithm. It sends out control signals (*TLcmd*, *PEcmd*, *SOcmd*,

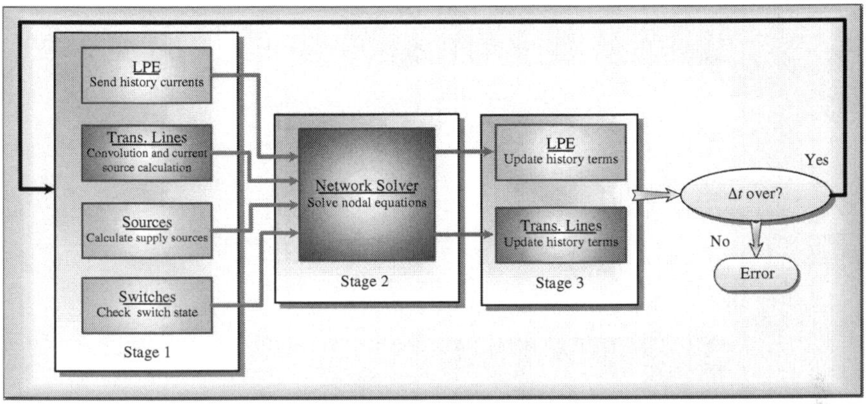

Figure 2.29 Paralleled real-time EMT solution algorithm for hardware emulation.

NWcmd) to each module to perform the required functions. Meanwhile, it receives the acknowledge signals (*TLdone*, *PEdone*, *SOdone*, *NWdone*) to judge if the function is done. Figure 2.30b shows the finite-state machine diagram of the *MainControl* HEBB.

2.8.4 Real-Time Emulation Case Study

A test power system shown in Figure 2.31 was emulated on the Altera Stratix FPGA. The system consists of 15 transmission lines, 4 generators, and 8 loads. The system data is listed in Appendix A.1. The transformers are modeled as linear two-port equivalents. The lines are modeled using the FDLM. Each of the system components is allocated to the appropriate FPGA HEBB module to process its calculation. For example, the 15 transmission line models are pipelined through the *FDLM* HEBB module.

The Altera® Stratix™ S80 Development Board with the EP1S80 FPGA was used to emulate the test system. The FPGA has the following features: 79 040 logic elements (LEs); 7 427 520 total RAM bits; 176 DSP blocks based on 9 × 9 multiplier; and 679 maximum user I/O pins. The external clock is from the 80 MHz oscillator on the development board.

An *Interface* unit was implemented to interact with the real-time emulator for uploading system input data, system reset, execution, and result recording. The user functions include changing the output voltage node number, output current number, setting the time-step, etc. The output three-phase voltages and currents were displayed on the oscilloscope via a 16-channel DAC board.

Table 2.4 lists resources utilization by each HEBB of the FPGA-based real-time EMT emulator. As can be seen, the LEs and DSP Blocks are almost fully utilized

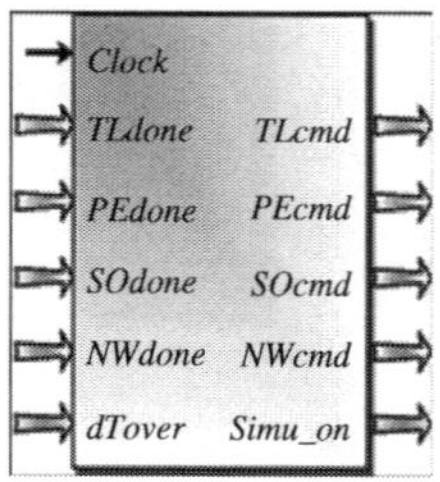

(a)

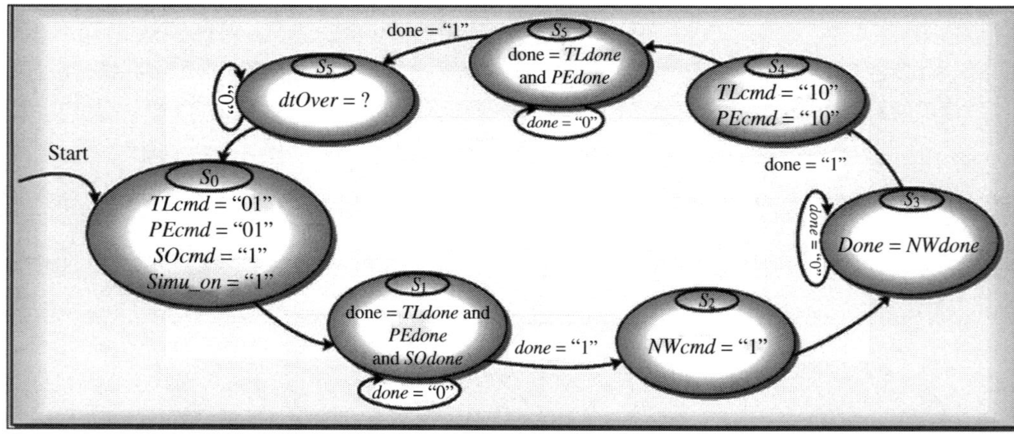

(b)

Figure 2.30 (a) *MainControl* HEBB. (b) Finite-state machine for paralleled EMT solution.

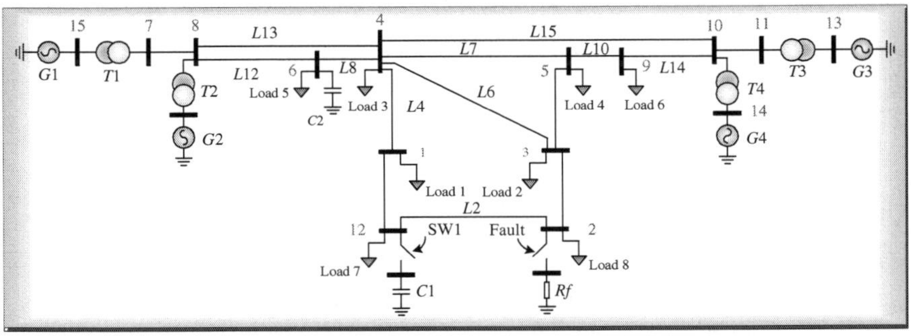

Figure 2.31 Single-line diagram of the power system used in the Case Study.

Table 2.4 Hardware resources utilized

HEBB module	Logic elements	DSP blocks	Memory bits
FDLM	37 245 (47.1%)	96 (54.5%)	733 632 (9.8%)
PElement	10 909 (13.8%)	40 (22.7%)	64 896 (0.87%)
Source	2231 (2.8%)	16 (9.0%)	162 278 (2.18%)
Switch	366 (0.4%)	0	768 (0.01%)
FormInode	6725 (8.5%)	0	8767 (0.12%)
Network Solver	18 588 (23.5%)	16 (9.0%)	49 023 (1.04%)
MainControl	163 (0.2%)	0	0
Interface	2100 (2.6%)	0	1 130 496 (15.2%)
Total	69 502 (87.8%)	168 (95.2%)	1 010 597 (13.90%)

with the *FDLM* HEBB module consuming the most logic resources. The second most resource intensive unit is the *Network Solver* HEBB module.

The execution times for each stage of the paralleled EMTP algorithm (Figure 2.29) for emulating this system are 5.5, 3.375, and 2.338 μs for *Stage*1, *Stage*2, and *Stage*3, respectively. The total execution time is 11.213 μs, while the actual time-step is 12 μs. Based on the 12.5 *ns* clock period used for the FPGA, this implies that it took 960 clock cycles to complete one loop of simulation for this case study. The achieved time-step $\Delta t = 12 \mu s$ is at least four times smaller than the acceptable time-step of 50 μs for transient simulation. As such it is possible to emulate a system that is at least four times as large i.e. a system with 60 lines, 16 generators, and 32 loads can be simulated in real time with a time-step of $\Delta t = 50 \mu s$ on this FPGA. It can also be seen that the *Stage1* utilizes the most execution time. This is due to the convolution operation of FDLM as mentioned earlier. The time-step could be decreased significantly if more parallel *FDLM* HEBB modules were implemented on a larger FPGA.

Two transient events are simulated. The first transient event is the capacitor $C1$ switched at Bus 12 at time $t = 0.05$ second. Figure 2.32a,c show the three-phase voltages and currents waveforms at Bus 12 captured in real time. As can be seen, during the capacitor transient, voltage v_a drops from 110 kV to almost 0 kV immediately. A negative peak current of 1.5 kA happens on i_b at $t = 0.05$ second and a positive peak current of 1.3 kA appears on i_a at $t = 0.053$ second. The transient lasts for about one cycle. Identical behavior can be observed from Figure 2.32b,d which show the off-line ATP simulation results with a time-step of 12 μs. The second transient event is the Bus 2 three-phase-to-ground fault with 2 Ω resistance to ground which occurs at $t = 0.05$ second. Figure 2.33a,c show the real-time three-phase voltages and fault currents waveforms at Bus 12. The voltages drop from 180 kV peak to 60 kV peak, and high-frequency transient currents occur during the fault.

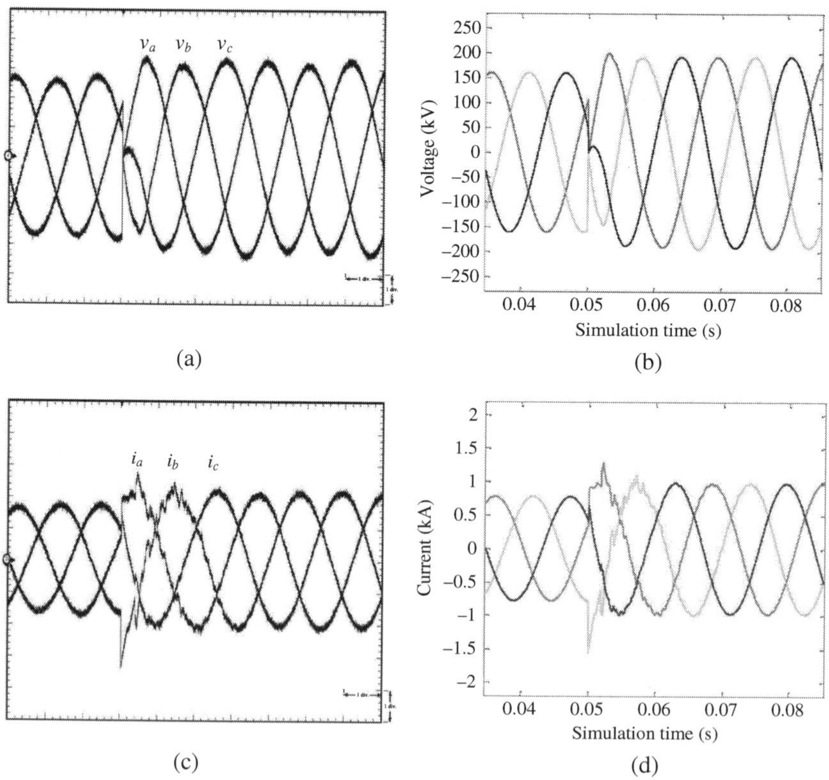

Figure 2.32 Real-time oscilloscope traces (a, c) and off-line simulation results (b, d) for a capacitor $C1$ switching transient at Bus 12. (a, b) Bus 12 voltages, (c, d) Bus 12 load currents. Scale: x-axis: 1 div. = 5 ms, y-axis: (a) 1 div. = 58 kV, (c) 1 div. = 0.44 kA.

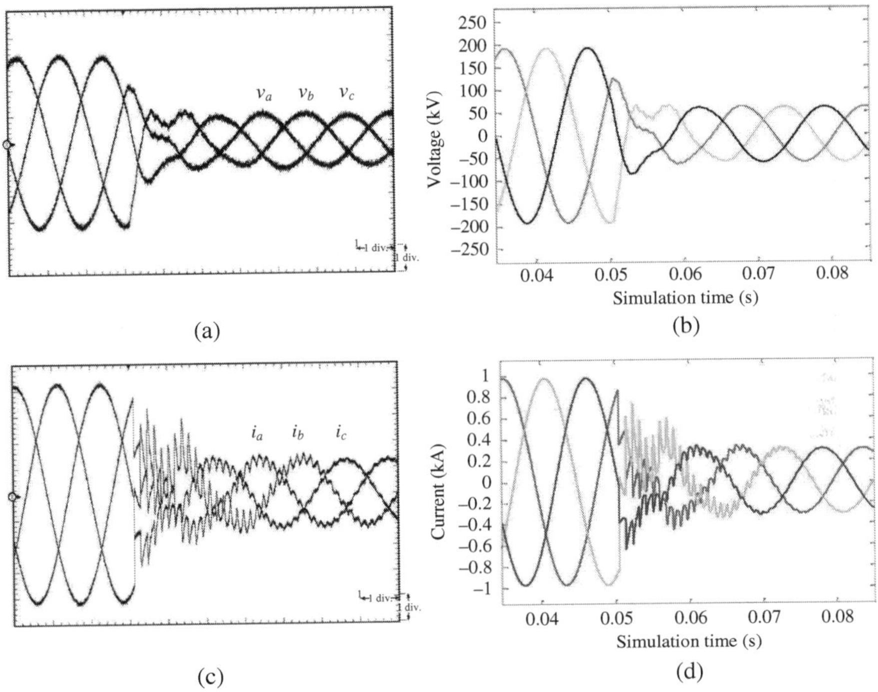

Figure 2.33 Real-time oscilloscope traces (a, c) and off-line simulation results (b, d) for a three-phase to ground fault transient at Bus 2. (a, b) Bus 12 voltages, (c, d) Bus 12 fault currents. Scale: *x*-axis: 1 div. = 5 ms, *y*-axis: (a) 1 div. = 58 kV, (c) 1 div. = 0.22 kA.

The transient lasts for about 2 cycles. Again, detailed agreement between ATP off-line simulation shown in Figure 2.33b,d and the FPGA-based real-time emulator results can be observed.

2.9 Nonlinear Elements: Iterative Real-Time EMT Solver

Nonlinear elements play a significant role in the inception and propagation of transient overvoltages and overcurrents in electrical power systems. The commonly occurring nonlinear phenomena in power systems include magnetic saturation in transformers, ferroresonance, switching surges, and lightning strikes. Accurate simulation of nonlinear phenomena is vital from the perspective of such studies as insulation coordination, protection system design, power quality, and in general for maintaining the transmission and distribution infrastructure in a reliable condition.

In off-line EMT-type software programs, the representation of nonlinear elements is either through a piecewise linear approximation method (also known as pseudo nonlinear method) or an iterative method [26, 27, 42, 43]. In the former, the nonlinearity is represented by several linear segments. At each simulation time-step, it is actually a specific linear element, such as a constant inductance, several of which can be used to represent, for example, the saturation characteristic of a transformer. However, the computational bottleneck occurs as the operating point switches from one linear segment to the next necessitating a recalculation of the system admittance matrix which results in a substantially longer run time for a large network. The frequent switching between linear segments is also known to cause numerical oscillations. For a more accurate simulation of nonlinear elements, an iterative method based on N-R can be used. Within each simulation time-step, several N-R iterations are required for convergence, and each iteration entails a recalculation of the Jacobian matrix and solution of a set of linear equations. This process is very time-consuming. In an off-line simulator, the computational burden can perhaps be overlooked; however, it becomes quite cumbersome in a real-time simulator. In a deterministic hard real-time system built using CPU- or DSP-based sequential hardware, an iterative nonlinear solution may not even be implementable using a requisite time-step mainly due to the uncertainty of convergence of the iterations within the time-step, and the limited computational power of such hardware. With FPGA being used, accurate real-time emulation of nonlinear elements using iterative method is possible.

2.9.1 Compensation Method

With the inclusion of nonlinear elements, the power system network becomes a nonlinear system. For transient solution instead of solving the entire nonlinear

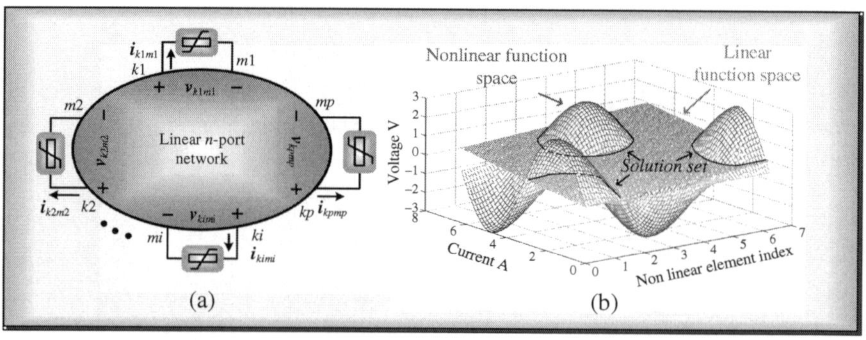

Figure 2.34 (a) Linear network interfaced with external nonlinear elements. (b) Compensation method demonstration for multiple nonlinearities.

network, the compensation method is commonly used to reduce computational burden. In this method, the nonlinear elements are first separated from the linear network, as shown in Figure 2.34a, where p nonlinear elements connected to nodes ("ki") and ("mi") $\{i = 1, 2, \ldots, p\}$ have been extracted from a n-node linear network. Seen from these nodes, the linear network can be represented as

$$\boldsymbol{v}_{km} = \boldsymbol{v}_{kmo} - \boldsymbol{R}_{eq}\boldsymbol{i}_{km}, \tag{2.146}$$

where $\boldsymbol{v}_{km}$ and $\boldsymbol{i}_{km}$ are vectors of p voltages across and currents through nonlinear elements, and $\boldsymbol{v}_{kmo}$ is a vector of p open-circuit voltages (without the nonlinear branches) between nodes ("ki") and ("mi"). $\boldsymbol{R}_{eq}$ is a $p \times p$ Thévenin equivalent resistance matrix of the linear network. To obtain $\boldsymbol{R}_{eq}$, vectors r_{thev_i} $(n \times 1)$ $\{i = 1, 2, \ldots, p\}$ which are the differences of the kith and mith columns of inverse admittance of the linear network $\boldsymbol{Y}^{-1}$ are first calculated. Then, the (i, j) element of $\boldsymbol{R}_{eq}$ is given as

$$\boldsymbol{R}_{eq}(i, j) = [\boldsymbol{r}_{thev\_j}]_{ki} - [\boldsymbol{r}_{thev\_j}]_{mi}. \tag{2.147}$$

The $\boldsymbol{v}$ and $\boldsymbol{i}$ characteristics of the nonlinear elements are expressed as

$$\boldsymbol{v}_{km} = \boldsymbol{f}(\boldsymbol{i}_{km}), \tag{2.148}$$

where $\boldsymbol{f}$ is the nonlinear functions.

When (2.146) and (2.148) are solved simultaneously as illustrated graphically in Figure 2.34b, the current $\boldsymbol{i}_{km}$ has been obtained. Then, $\boldsymbol{i}_{km}$ is superimposed on the linear network as a current source, thus the entire network is solved as follows

$$\boldsymbol{v}_c = \boldsymbol{v}_o - \boldsymbol{R}_{thev}\boldsymbol{i}_{km}, \tag{2.149}$$

where $\boldsymbol{v}_c$ is the compensated final node voltages, and $\boldsymbol{v}_o$ is the node voltages without nonlinear elements. $\boldsymbol{R}_{thev}$ is a $n \times p$ resistance matrix expressed as

$$\boldsymbol{R}_{thev} = [\ \boldsymbol{r}_{thev\_1} \quad \boldsymbol{r}_{thev\_2} \quad \cdots \quad \boldsymbol{r}_{thev\_p}\ ]. \tag{2.150}$$

The merits of compensation method are to reduce the set of nonlinear equations. The size is determined by the number of nonlinear elements p. In the three-phase power system, if there are M three-phase nonlinear elements, the total number of nonlinear equations is $p = 3 \times M$. Fortunately, the three-phase nonlinear elements are commonly separated by transmission lines. In such a case, due to the traveling time delay of transmission lines, the M three-phase nonlinear elements are decoupled with each other, which implies that M sets of three nonlinear functions can be solved simultaneously in a parallel computational environment.

2.9.2 Newton–Raphson Method

The N-R method is widely used to solve the nonlinear equations due to its quadratic convergence. Substituting (2.148) into (2.146), the nonlinear equations are obtained as

$$\boldsymbol{F}(\boldsymbol{i}_{km}) \equiv \boldsymbol{f}(\boldsymbol{i}_{km}) - \boldsymbol{v}_{kmo} + \boldsymbol{R}_{eq}\boldsymbol{i}_{km} = \boldsymbol{0}. \tag{2.151}$$

The objective is to find the solution $\boldsymbol{i}_{km}$. By applying the first-order Taylor series expansion for the nonlinear equations (2.151), the updated solution is obtained by solving the following system of linear equations:

$$\boldsymbol{J}\left(\boldsymbol{i}_{km}^{k+1} - \boldsymbol{i}_{km}^{k}\right) = -\boldsymbol{F}\left(\boldsymbol{i}_{km}^{k}\right), \tag{2.152}$$

where $\boldsymbol{J}$ is Jacobian matrix, $\boldsymbol{i}_{km}^{k+1}$ and $\boldsymbol{i}_{km}^{k}$ are the current vectors at the $(k+1)$th and kth iterations, respectively. This method is referred to as the Continuous Newton–Raphson (CNR) method since the nonlinear function is a continuous analytical equation. The Jacobian matrix and nonlinear equations are computed at each iteration using

$$\begin{cases} \boldsymbol{J} = \dfrac{\partial \boldsymbol{F}(\boldsymbol{i}_{km})}{\partial \boldsymbol{i}_{km}} = \boldsymbol{R}_{eq} + \dfrac{\partial \boldsymbol{f}(\boldsymbol{i}_{km})}{\partial \boldsymbol{i}_{km}}, \\ -\boldsymbol{F}(\boldsymbol{i}_{km}^{k}) = \boldsymbol{v}_{kmo} - \boldsymbol{R}_{eq}\boldsymbol{i}_{km}^{k} - \boldsymbol{f}(\boldsymbol{i}_{km}^{k}). \end{cases} \tag{2.153}$$

The convergence criteria is defined as

$$||\boldsymbol{i}_{km}^{k+1} - \boldsymbol{i}_{km}^{k}|| < \varepsilon_1 \quad \text{and} \quad ||\boldsymbol{F}(\boldsymbol{i}_{km}^{k+1})|| < \varepsilon_2, \tag{2.154}$$

with ε_1 and ε_2 set to sufficiently small values.

If the nonlinear function is given by a piecewise linear curve, the method becomes Piecewise Newton–Raphson (PNR) [42]. As shown in Figure 2.35a, each linear segment of the piecewise curve can be defined by

$$\boldsymbol{v}_{km} = \boldsymbol{e}_{kmj} + \boldsymbol{R}_j\boldsymbol{i}_{km}, \boldsymbol{i}_{km} \in [\boldsymbol{I}_{kmj-}, \boldsymbol{I}_{kmj+}], \tag{2.155}$$

where $\boldsymbol{e}_{kmj}$ and $\boldsymbol{R}_j$ are the intercept voltage and resistance of the jth segment of the curve, which falls within the interval $\boldsymbol{I}_{kmj-}$ and $\boldsymbol{I}_{kmj+}$.

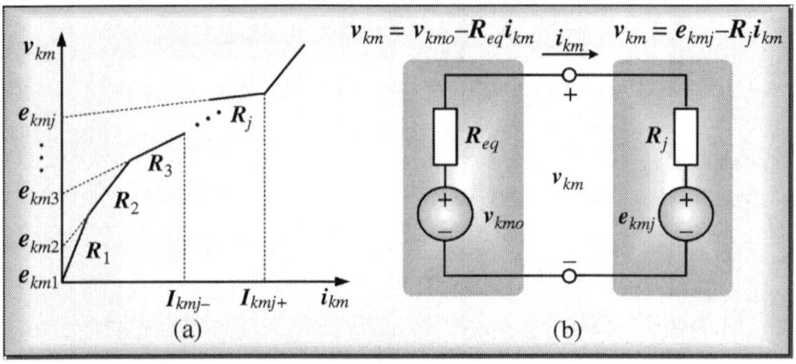

Figure 2.35 (a) Piecewise linear function and (b) its implementation in PNR.

Substituting (2.155) into (2.146), the nonlinear equations become

$$\boldsymbol{F}(\boldsymbol{i}_{km}) \equiv (\boldsymbol{v}_{kmo} - \boldsymbol{R}_{eq}\boldsymbol{i}_{km}) - (\boldsymbol{e}_{kmj} + \boldsymbol{R}_j\boldsymbol{i}_{km}) = \boldsymbol{0}. \tag{2.156}$$

Using (2.152) and (2.153) directly, we obtain

$$(-\boldsymbol{R}_{eq} - \boldsymbol{R}_j)\left(\boldsymbol{i}_{km}^{k+1} - \boldsymbol{i}_{km}^{k}\right) = \left(\boldsymbol{e}_{kmj} + \boldsymbol{R}_j\boldsymbol{i}_{km}^{k}\right) - \left(\boldsymbol{v}_{kmo} - \boldsymbol{R}_{eq}\boldsymbol{i}_{km}^{k}\right),$$

$$(\boldsymbol{R}_{eq} + \boldsymbol{R}_j)\boldsymbol{i}_{km}^{k+1} = \boldsymbol{v}_{kmo} - \boldsymbol{e}_{kmj}. \tag{2.157}$$

Note that the term $\boldsymbol{i}_{km}^{k}$ is canceled out in (2.157), which means that the calculation of Jacobian matrix $\boldsymbol{J}$ is not required in the PNR method. This is because $\boldsymbol{R}_j$ which is the derivatives of the piecewise linear curve have been precalculated. Thus, the computational burden of PNR is lower than that of CNR at the cost of reduced accuracy. Since (2.155) is for the jth segment of the piece-wise curve, the solution is then checked to see if the solution satisfies $[\boldsymbol{I}_{kmj-}, \boldsymbol{I}_{kmj+}]$. If it does, a valid solution has been found; otherwise, the next iteration begins.

2.9.3 Hardware Emulation of Nonlinear Solver

A hardware module *N-R* HEBB is designed to emulate nonlinear elements using iterative N-R method on the FPGA [44]. The symbol of this module and its input/output signals is shown in Figure 2.36a. The signal *Vn* receives open circuit voltages $\boldsymbol{v}_0$ form *Network Solver* module (Section 2.8.1). The calculated compensated voltage $\boldsymbol{v}_c$ is sent through signal *Vc*. *NRcmd* and *NRdone* are control and acknowledge signals.

As shown in Figure 2.36b, the hardware architecture of *NR* module consists of four main hardware submodules: (i) *NLFunc* (evaluating nonlinear function $\boldsymbol{f}(\boldsymbol{i}_{km})$ and $\partial \boldsymbol{f}(\boldsymbol{i}_{km})/\partial \boldsymbol{i}_{km}$; (ii) *CompJF* (computing $\boldsymbol{J}$ and $-\boldsymbol{F}(\boldsymbol{i}_{km})$ (2.153)); (iii) *GJE* (Gauss–Jordan elimination) for solving (2.152) or (2.157) for $\boldsymbol{i}_{km}$; and (iv) *CompVc* (computing the compensated node voltage $\boldsymbol{v}_c$ (2.149)).

The overall algorithm of N-R solution is in sequential; that is, one N-R iteration starts with calculating $\boldsymbol{f}(\boldsymbol{i}_{km})$ and $\partial \boldsymbol{f}(\boldsymbol{i}_{km})/\partial \boldsymbol{i}_{km}$; then the $\boldsymbol{J}$ and $-\boldsymbol{F}(\boldsymbol{i}_{km})$ can

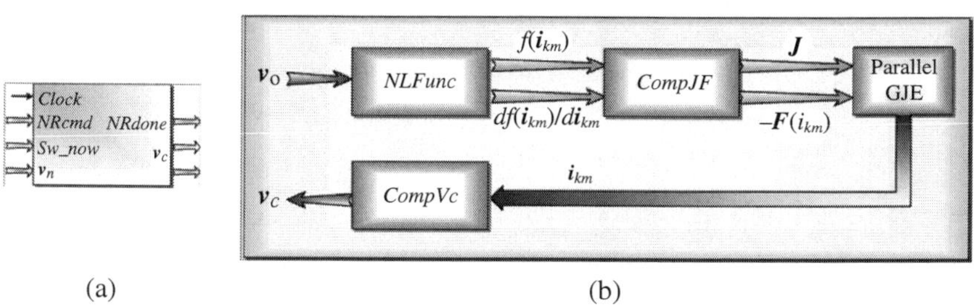

Figure 2.36 (a) *NR* HEBB and its input/output signals. (b) Overall hardware architecture of the nonlinear solver.

be calculated. After that the $\boldsymbol{i}_{km}$ is solved using GJE method. Finally, the compensated node voltages $\boldsymbol{v}_c$ is calculated. The convergence check determines if the iteration is finished or not using (2.154). However, the parallel operations still exist in each step of iteration. Figure 2.37a shows the FSM diagram for *NR* module. The sequential operations are clearly shown by state transitions $S_0 \rightarrow S_1 \rightarrow S_2 \rightarrow S_3 \rightarrow S_4 \rightarrow S_0$. The possible parallel operations, which take advantage of the hardware parallelism of the FPGA, are also shown in Figure 2.37a. For example, in state S_0, the nonlinear function evaluations $\boldsymbol{f}(\boldsymbol{i}_{km})$ and $\partial\boldsymbol{f}(\boldsymbol{i}_{km})/\partial\boldsymbol{i}_{km}$ are performed simultaneously. The computation of $\boldsymbol{J}$ and $-\boldsymbol{F}(\boldsymbol{i}_{km})$ in (2.153) are also processed concurrently in state S_1. Meanwhile, the GJE procedure is paralleled in state S_2 as discussed in more detail later.

2.9.3.1 Nonlinear Function Evaluation

At each N-R iteration, nonlinear function evaluation is required. The computational burden of this evaluation depends on the number of nonlinear elements in the original system and the nature of these nonlinearities. The commonly used methods for nonlinear function evaluation on the FPGA include the LUT method and other approximation methods such as COordinate Rotation DIgital Computer (CORDIC), series expansion, and the regular N-R iteration (an inner iterative loop inside the outer iteration) [45, 46]. Many of them have been implemented in hardware, however, based on fixed-point format mostly. The floating-point implementation is still cumbersome due to its large latency and logic resource utilization. Among the above methods, the LUT is still widely used due to its speed and convenience of implementation. To reduce the size of LUT and improve the accuracy, linear interpolation has been used in this design to compute intermediate values between two locations of the LUT. Figure 2.37b shows the hardware design of the *NLFunc* submodule. It consists of an *Address generation* unit, a dual-port RAM unit serving as a *LUT*, and a *Linear interpolation* unit. The *Address generation* unit has been utilized in *Source* module in Section 2.5.1, it is not discussed here again.

After $f(x_i)$ and $f(x_{i+1})$ are retrieved from the LUT, the linear interpolation is employed to calculate $f(x)$. The hardware design of *Linear interpolation* unit is shown in Figure 2.38a. It utilizes two floating-point subtractors ($y = a - b$) and multiplier-adder ($y = a \times b + c \times d$), while the calculation is quite straightforward.

2.9.3.2 Parallel Calculation of $\boldsymbol{J}$ and $\boldsymbol{F}(\boldsymbol{i}_{km})$

Since $\boldsymbol{f}(\boldsymbol{i}_{km})$ and $\partial\boldsymbol{f}(\boldsymbol{i}_{km})/\partial\boldsymbol{i}_{km}$ being calculated, the computations of $\boldsymbol{J}$ and $-\boldsymbol{F}(\boldsymbol{i}_{km})$ (2.153) are carried out in parallel as shown in Figure 2.38b. The computations are quite straightforward. $\boldsymbol{J}$ is calculated by an adder. $-\boldsymbol{F}(\boldsymbol{i}_{km})$ is calculated by an arithmetic unit ($y = a - b - c \times d$) after $\boldsymbol{v}_{km0}$ being computed by a subtractor.

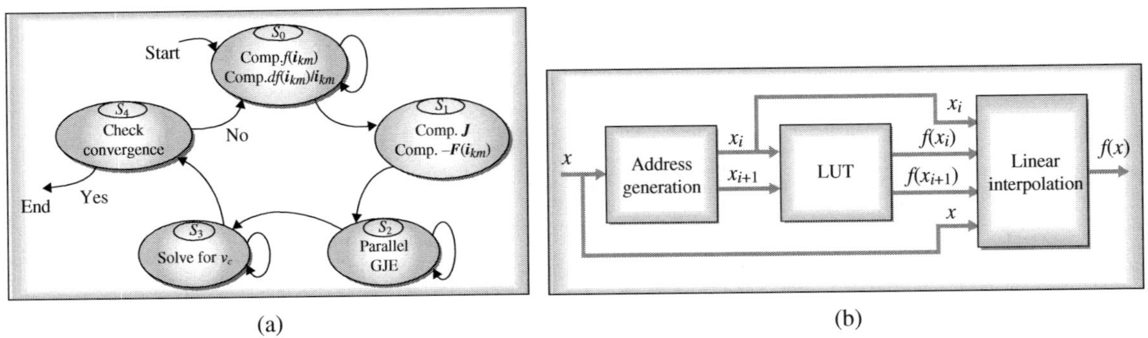

Figure 2.37 (a) Finite state machine of *NR* module. (b) Floating-point nonlinear function computation using LUT and linear interpolation.

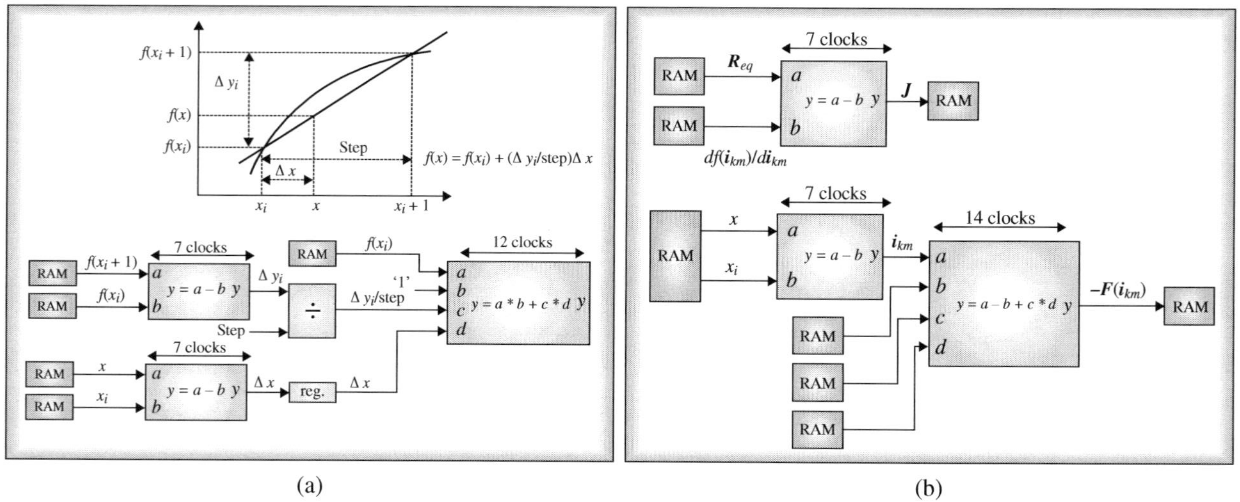

Figure 2.38 (a) Linear interpolation of $f(x)$, and its pipelined computation scheme. (b) Parallel computational scheme for calculating $\boldsymbol{J}$ and $-\boldsymbol{F}(\boldsymbol{i}_{km})$.

2.9.3.3 Parallel Gauss–Jordan Elimination

The set of linear algebraic equations in (2.152) or (2.157) are solved by GJE. Compared to other elimination methods such as Gaussian elimination with backward substitution and LU decomposition, the GJE requires more floating operations; however, the size of linear algebraic equations is only 3 under the situation that the three-phase nonlinear elements are separated by transmission lines as stated early. Furthermore, the hardware implementation of GJE is simpler than those of other methods.

To emulate M three-phase nonlinear elements, M sets of three linear equations need to be solved within each iteration of each simulation time-step. For a large M, the sequential GJE can be very time-consuming. Hence, a paralleled and pipelined scheme has been designed to speed up this process, as shown in Figure 2.39. It consists of three elimination units (*Eli1*, *Eli2*, and *Eli3*) and a factorization unit (*Fac*). Before the GJE process begins, the 3×3 matrix $\boldsymbol{J}$ and 3×1 vector $-\boldsymbol{F}(\boldsymbol{i}_{km})$ for each three-phase nonlinear element are augmented together as a 3×4 matrix. For M three-phase nonlinear elements, M sets of 3×4 matrices are combined together by rows to get a $3 \times 4M$ matrix whose each row is stored in the RAM of three elimination units. Then, the GJE computation proceeds according to the following two steps:

1. *Factorization*: The ith ($i = 1, 2, 3$) row is retrieved from the corresponding elimination unit, and the diagonal element is identified and registered as *diag* in the *Fac* unit. Then, the remaining elements within the row are divided by the registered *diag*, and the factorized row is sent back to all elimination units.
2. *Elimination*: The elimination is done simultaneously in all elimination units. In the ith elimination, unit elimination is not needed, so the factorized row is just saved in its RAM. In other two elimination units, the ith element of corresponding row is recognized and registered as *piv*, then multiplied by the elements of factorized row. The resulting product is then added to the corresponding element of row in its RAM.

2.9.3.4 Computing v_c

Finally, the complete set of node voltages can be calculated. A hardware submodule is designed to calculate $\boldsymbol{v}_c$, as shown in Figure 2.40. Since $\boldsymbol{R}_{thev}$ ia a sparse matrix, sparse matrix technique is employed here. This is realized by a *FFPMAC* unit along with a sparse matrix storage format scheme which has been discussed in detail in *Network Solver* module in Section 2.8.1.

2.9.4 Case Studies

The iterative real-time nonlinear EMT solver was implemented on a Altera Stratix III Development Board DE3. The FPGA used on this board is a Stratix III EP3SL150

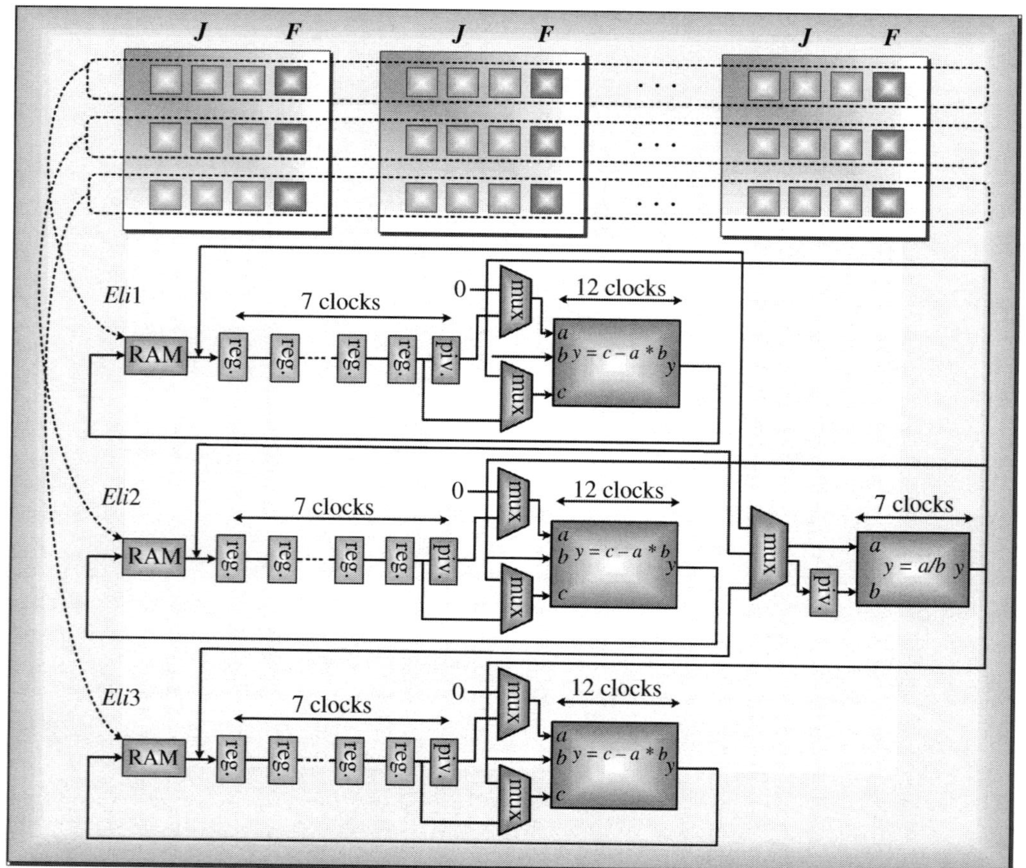

Figure 2.39 Hardware design of parallel Gauss–Jordan elimination.

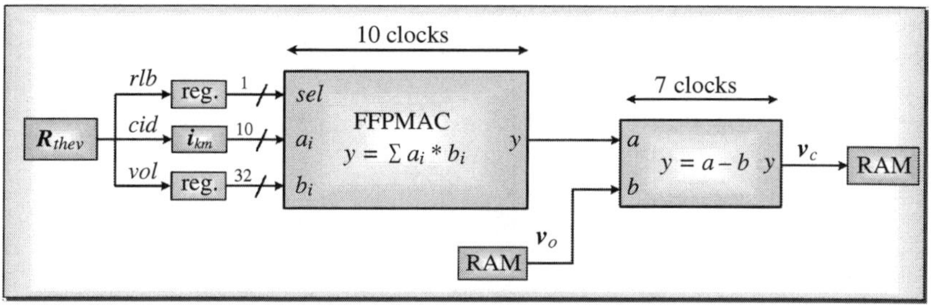

Figure 2.40 Pipelined computational scheme for calculating v_C.

Table 2.5 Hardware resources utilized by various modules

Module	Logic elements (%)	Multipliers (%)	Memory bits (%)
Global control	0.1%	0%	0%
Linear solver	4.0%	2.1%	0.5 %
Nonlinear solver	10.8%	11.5%	1.9%
Trans. lines	8.3%	18.7%	5.2%
RLCG elements	5.8%	8.3%	0.9%
Power sources	2.5%	1.1%	14.2%
Total	31.5%	41.7%	22.7%

which has the following resources: 142 500 equivalent LEs; 6390 total RAM Kbits; 384 18×18-bit multipliers; 8 phase-locked loops (PLLs), and 1152 maximum user I/O pins. The DAC card connected to the DE3 board has 14-bit resolution, 125 MSPS data rate, and $\pm$ 500 mV output voltage range. The simulation results in 32-bit floating-point format are truncated to 14-bit binary before sending to DAC, causing the loss of some information. Table 2.5 lists the FPGA resource utilization in percentage for each module of the iterative real-time nonlinear EMT solver. As can be seen from this table, the nonlinear solver utilizes the most logic resources of the FPGA. The other logic resource utilization are for the modeling of transmission lines, *RLC* elements, power sources, etc. [40].

Two case studies are used to show the effectiveness of the iterative real-time nonlinear transient solver on the FPGA. The first case study is a three-phase series-compensated transmission system whose single line diagram is shown in Figure 2.41. The compensation capacitor *C* connected in series between two lines (*line*1 and *line*2) is protected by a surge arrester. The lines are modeled using the

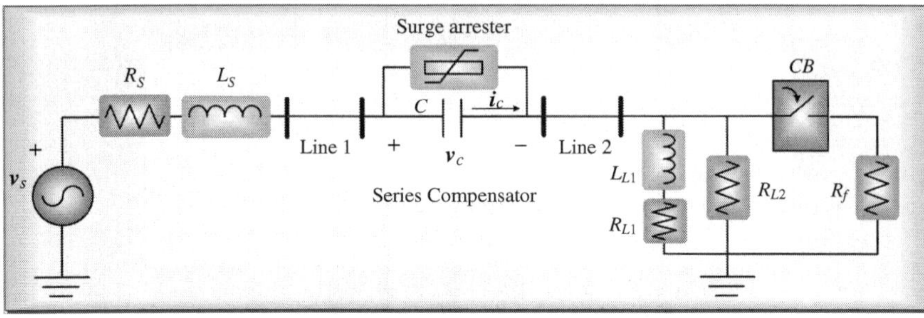

Figure 2.41 Single-line diagram for Case I.

traveling wave line (TWL) model to capture the transients on the system. The surge arresters are nonlinear resistors characterized by

$$i = p\left(\frac{v}{V_{ref}}\right)^q, \tag{2.158}$$

where q is the exponent, V_{ref} and p are arbitrary reference values. The elements R_{L1}, L_{L1}, and R_{L2} represent a composite load, and R_f is the fault resistance. The CNR method is used to solve the nonlinear equations. The parameters of the system are listed in Appendix A.1.

A three-phase fault is applied at the load terminals at $t = 0.2$ second by closing the circuit breaker *CB*. The current increases in the series capacitor and produces an overvoltage that is limited by the surge arresters. Figure 2.42a shows the three-phase voltage transients across the surge arresters captured from a real-time oscilloscope. The peak overvoltage is limited to 8 kV compared to 15 kV without

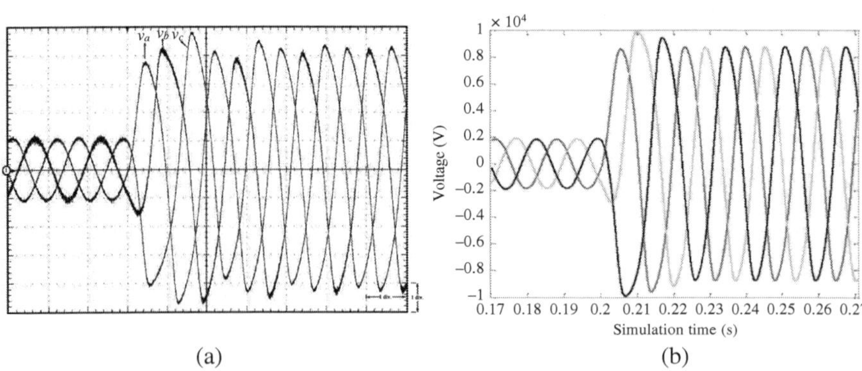

Figure 2.42 (a) Real-time oscilloscope traces and (b) off-line simulation results from ATP of the three-phase voltages across the surge arresters for a three-phase fault. Scale: x-axis: 1 div. = 10 ms, y-axis: 1 div. = 2 kV.

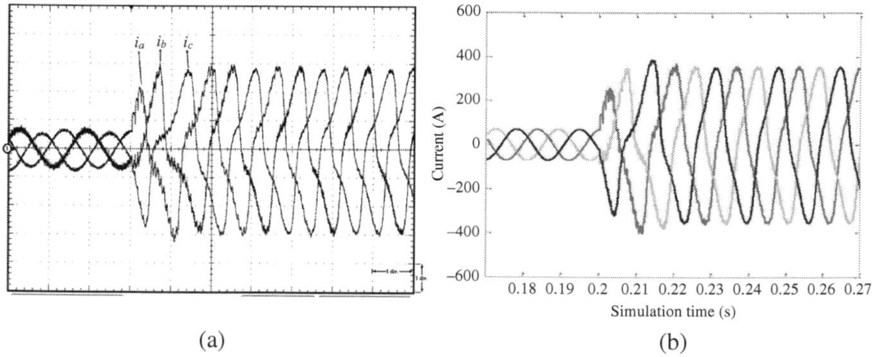

Figure 2.43 (a) Real-time oscilloscope traces and (b) off-line simulation results from ATP of the three-phase currents in the surge arresters for a three-phase fault. Scale: x-axis: 1 div. = 10ms, y-axis: 1 div. = 128A.

the arrester. Figure 2.43a shows the three-phase transient currents in the surge arresters in the real-time simulation. Large currents drawn by the arresters can be observed. Similar behavior can be observed from Figures 2.42b and 2.43b which show the off-line ATP simulation results. Small differences between the off-line and real-time results might be caused by the different initial values and switching time.

The second case study illustrates ferroresonance in a three-phase transformer. Figure 2.44a,b show the single-line and equivalent network diagrams of this case study, respectively. The transformer is connected to the system represented by a voltage source V_s through a *CB*. C_w is the circuit breaker's grading capacitance, C_s is the total phase-to-earth capacitance including transformer winding capacitance. The resister R_{fe} represents transformer core losses. The transformer current is represented by its piecewise nonlinear magnetization characteristic shown in Figure 2.44c; thus, the PNR is used in this case study with five linear segments to represent the nonlinearity. The system data is listed in the Appendix A.1. The ferroresonance response is verified by the opening the circuit breaker caused by a

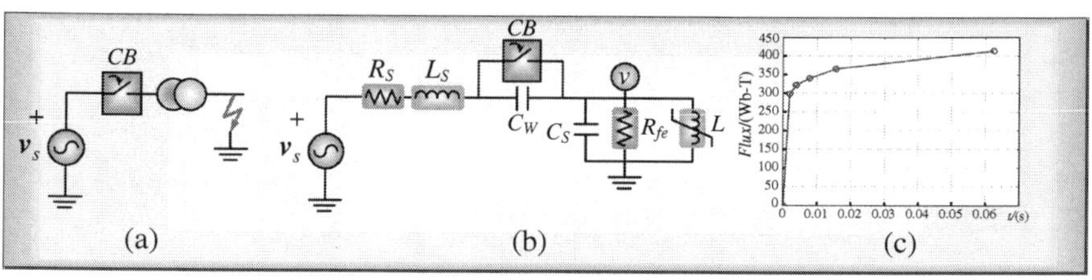

Figure 2.44 (a) Single-line diagram and (b) equivalent network diagram for Case II (ferroresonance transient). (c) Piecewise nonlinear magnetization characteristic of transformer.

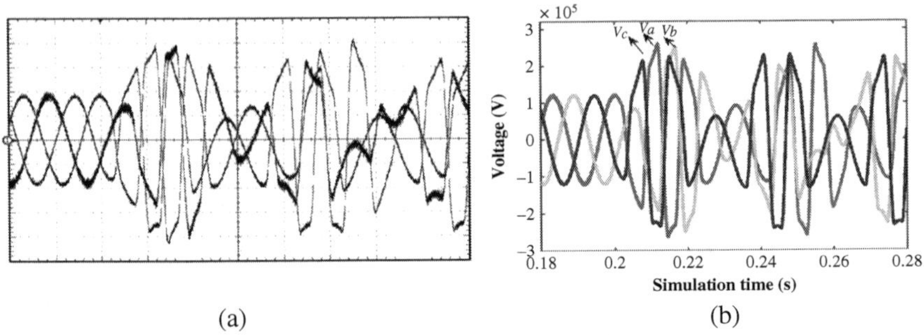

Figure 2.45 Three-phase voltages at the transformer terminals during a three-phase-to-ground fault: (a) Real-time oscilloscope traces and (b) off-line simulation results from ATP. Scale: x-axis: 1 div. = 10 ms, y-axis: 1 div. = 68 kV.

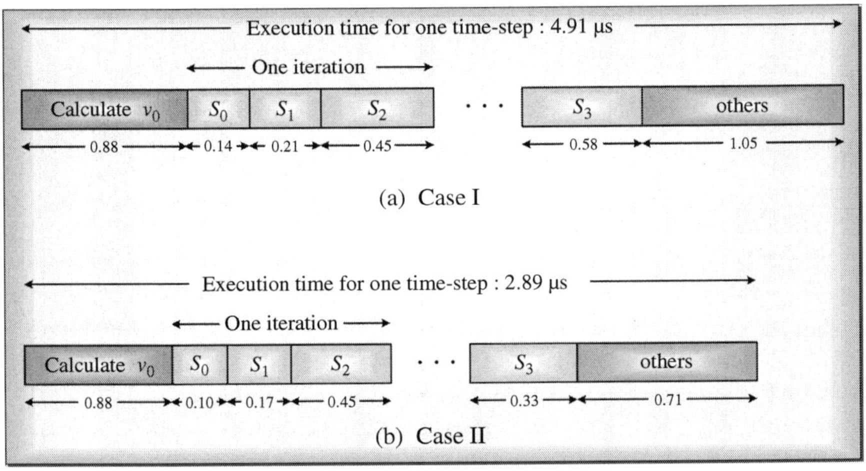

Figure 2.46 Execution time for the case studies in μs. $S_i(i = 0, \ldots, 3)$ are the states of the finite state machine of the nonlinear solver (Figure 2.37).

three-phase fault on the transformer at $t = 0.2$ second. Again similar behavior of real-time and off-line simulations can be observed from Figure 2.45.

Figure 2.46a,b show the execution time for each state in the finite state machine in Figure 2.37a for the two case studies. The execution time for "others" include time used for calculating transmission lines, *RLCG* elements, sources, etc. Based on the 60 MHz FPGA input clock of the Stratix III EP3SL150 FPGA, the total execution time for Case Study I is 4.91 μs for the average of three iterations, while the actual time-step is 5 μs, with an approximately 42% utilization of the FPGA multipliers. Thus, it is safe to say that a system of twice the size i.e. a system with at least six nonlinear elements can be emulated on this FPGA within a time-step of

5 μs which is 10 times smaller than the acceptable time-step of 50 μs for transient simulation. It is therefore possible to simulate a system with at least 60 nonlinear elements within a 50 μs time-step. For Case Study II, since there are no calculations for the Jacobian matrix the execution time is only 2.89 μs, while the actual time-step is 3 μs.

2.10 Summary

Real-time digital simulators can play a key role in modern power grid projects by enabling rapid testing of new technologies and analyzing their impact on the system. They must however be capable of detailed modeling of power grid components with a sufficiently small time-step to capture dynamic events over a wide frequency bandwidth. This chapter presented the concept of HEBBs to emulate in real-time power system components on the FPGA. The HEBB architectures for the fundamental power system components necessary for EMT simulation are presented to show how a realistic power system network can be modeled in detail on a FPGA. For nonlinear elements, a fully iterative transient solver is described for implementation on the FPGA. The nonlinear solver utilizes the compensation method with the N-R algorithm, dedicated floating-point arithmetics, sparsity techniques, and parallel GJE as its main modules. The entire design is deeply pipelined and paralleled to achieve the highest throughput and lowest latency. The case studies illustrate the accuracy of the real-time transient emulator in comparison with off-line EMT software. For emulating large-scale power systems, multiple FPGAs can be employed for hardware emulation using the natural travel time delay of a transmission line to decompose and map the smaller subsystems to individual FPGAs for transient simulation. The HEBB concept is modular and can be extended to emulate any number of multi-physical and multi-domain components to integrate them into real-time power system simulation.

3

Power Transformers

3.1 Overview

A power transformer is an essential infrastructure in power systems for transmission and distribution of electrical energy. The complicated behavior of a power transformer, such as core saturation, hysteresis, losses, frequency-dependency, and the topological aspects of core and coil structure, need to be considered when a transformer model is developed for EMT studies. These behaviors can have a great impact on the transient overvoltages and overcurrents, for example, phenomena such as inrush current, ferroresonance, and harmonics, which influence the design of protection and control systems. How to properly represent a power transformer is an important question when a power system is simulated in real time. However, due to the complex configuration and nature of the power transformer such as a large number of windings and nonlinear parameters, the modeling aspects of the transformer can be quite complicated, especially taking real-time implementations issues into consideration.

The transformer models in the literature for EMT analysis can be classified mainly into three types [47–49]: lumped admittance-based model, topology-based magnetic equivalent circuit (MEC) model, and finite element (FE) based numerical model. Lumped models are simpler and relatively easier to implement, and FE models are the most complex demanding high computational resources and time, while the MEC models provide a compromise in the amount of modeling detail and computational complexity. This chapter first describes a lumped admittance-based nonlinear hysteretic model of a single-phase two-winding transformer and its real-time implementation on the FPGA. The nonlinear solution in real-time is undertaken using full Newton iteration. Then, a high-fidelity real-time distributed nonlinear MEC-based model is presented for a three-phase transformer with multiple windings and tap changers used for power flow regulation on transmission lines and its real-time hardware emulation. Finally, real-time 2D FE simulation of the electromagnetic transients single-phase

Real-Time Electromagnetic Transient Simulation of AC-DC Networks, First Edition.
Venkata Dinavahi and Ning Lin.

power transformer coupled to external circuits is discussed. The transmission line modeling (TLM) method is deployed at the element level to decouple the nonlinear elements from the linear network. An indirect noniterative field-circuit coupling technique is presented to handle the external circuit. The self and mutual inductances of the transformer windings are calculated based on the magnetic vector potentials from the FE model, and these coupling coefficients are fed to the external circuit, which is then solved independently within each time-step.

3.2 Nonlinear Admittance-Based Real-Time Transformer Model

3.2.1 Linear Model Formulation

The admittance matrix formulation results from the mutually coupled coils concept. Under transient conditions, the governing equations for a single-phase n winding transformer (Figure 3.1) can be expressed as,

$$\frac{d}{dt}\boldsymbol{i}_T = \mathbf{L}^{-1}\boldsymbol{v}_T - \mathbf{L}^{-1}\mathbf{R}\boldsymbol{i}_T, \tag{3.1}$$

where $\boldsymbol{v}_T$ and $\boldsymbol{i}_T$ are $n \times 1$ vectors of instantaneous terminal voltages and currents, respectively, expressed as, $\boldsymbol{v}_T = \left[v_{T1}(t)\ v_{T2}(t) \dots v_{Tn}(t)\right]^T$, and $\boldsymbol{i}_T = \left[i_{T1}(t)\ i_{T2}(t) \dots i_{Tn}(t)\right]^T$. $\mathbf{R}$ is an $n \times n$ diagonal matrix of winding resistance, $\mathbf{L}$ is an $n \times n$ matrix of winding leakage inductances, given as

$$\mathbf{R} = diag\left\{R_1\ R_2\ \dots\ R_n\right\}, \mathbf{L} = \begin{bmatrix} L_{11} & L_{12} & \cdots & L_{1n} \\ L_{21} & L_{22} & \cdots & L_{2n} \\ \vdots & \vdots & \vdots & \vdots \\ L_{n1} & L_{n2} & \cdots & L_{nn} \end{bmatrix}. \tag{3.2}$$

The entries of $\mathbf{R}$ and $\mathbf{L}$ can be found directly from the short-circuit test data.

For a three-phase transformer, the governing equation is the same as (3.1), except each entry of $\mathbf{R}$ and $\mathbf{L}$ matrices is a 3×3 submatrix, $\mathbf{R}_p$ and $\mathbf{L}_{pq}$, $(p, q = 1, 2, \dots, n)$ given as

$$\mathbf{R}_p = \begin{bmatrix} R_{pa} & 0 & 0 \\ 0 & R_{pb} & 0 \\ 0 & 0 & R_{pc} \end{bmatrix}, \quad \mathbf{L}_{pq} = \begin{bmatrix} L_{paqa} & L_{paqb} & L_{paqc} \\ L_{pbqa} & L_{pbqb} & L_{pbqc} \\ L_{pcqa} & L_{pcqb} & L_{pcqc} \end{bmatrix}, \tag{3.3}$$

where a, b, and c subscripts stand for the three phases.

By multiplying dt on the both sides, the (3.1) can be written as,

$$\mathrm{d}\boldsymbol{i}_T = \mathbf{L}^{-1}\boldsymbol{v}_T\mathrm{dt} - \mathbf{L}^{-1}\mathbf{R}\boldsymbol{i}_T\mathrm{dt}. \tag{3.4}$$

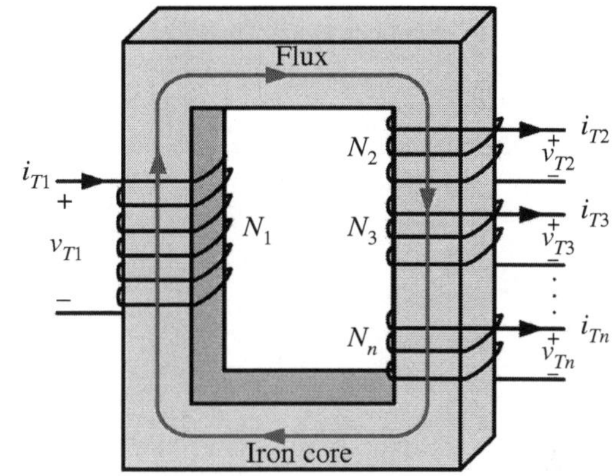

Figure 3.1 Single-phase n-winding transformer.

Using Trapezoidal rule to discretize (3.4),

$$\boldsymbol{i}_T(t) - \boldsymbol{i}_T(t-\Delta t) = \mathbf{L}^{-1}\left[\boldsymbol{v}_T(t) + \boldsymbol{v}_T(t-\Delta t)\right]\frac{\Delta t}{2} - \mathbf{L}^{-1}\mathbf{R}\left[\boldsymbol{i}_T(t) + \boldsymbol{i}_T(t-\Delta t)\right]\frac{\Delta t}{2}, \tag{3.5}$$

where Δt is the simulation time-step.

By rearranging (3.5) can be rewritten as,

$$\left[\mathbf{I} + \frac{\Delta t}{2}\mathbf{L}^{-1}\mathbf{R}\right]\boldsymbol{i}_T(t) = \frac{\Delta t}{2}\mathbf{L}^{-1}\boldsymbol{v}_T(t) + \frac{\Delta t}{2}\mathbf{L}^{-1}\boldsymbol{v}_T(t-\Delta t) - \frac{\Delta t}{2}\mathbf{L}^{-1}\mathbf{R}\boldsymbol{i}_T(t-\Delta t) + \boldsymbol{i}_T(t-\Delta t), \tag{3.6}$$

with $\mathbf{I}$ being the $n \times n$ identity matrix.

Finally, the Norton equivalent formula can be given as,

$$\boldsymbol{i}_T(t) = \mathbf{G}\boldsymbol{v}_T(t) + \boldsymbol{hist}(t-\Delta t), \tag{3.7}$$

with $\mathbf{G}$ being the $n \times n$ admittance matrix and $\boldsymbol{hist}(t-\Delta t)$ being the $n \times 1$ vector containing history terms, which are given as,

$$\mathbf{G} = \left[\mathbf{I} + \frac{\Delta t}{2}\mathbf{L}^{-1}\mathbf{R}\right]^{-1}\frac{\Delta t}{2}\mathbf{L}^{-1}, \tag{3.8}$$

$$\boldsymbol{hist}(t-\Delta t) = \mathbf{G}\boldsymbol{v}(t-\Delta t) + \left[\mathbf{I} + \frac{\Delta t}{2}\mathbf{L}^{-1}\mathbf{R}\right]^{-1}\left[\mathbf{I} - \frac{\Delta t}{2}\mathbf{L}^{-1}\mathbf{R}\right]\boldsymbol{i}_T(t-\Delta t). \tag{3.9}$$

In order to express the vector $\boldsymbol{hist}(t-\Delta t)$ in terms of $\boldsymbol{v}(t-\Delta t)$ and $\boldsymbol{hist}(t-2\Delta t)$, (3.7) is substituted into (3.9),

$$\begin{aligned} \boldsymbol{hist}(t-\Delta t) &= \mathbf{G}\boldsymbol{v}(t-\Delta t) + \mathbf{AB}\left[\mathbf{G}\boldsymbol{v}(t-\Delta t) + \boldsymbol{hist}(t-2\Delta t)\right], \qquad (3.10)\\ &= \left[\mathbf{G} + \mathbf{ABG}\right]\boldsymbol{v}(t-\Delta t) + \mathbf{AB}\boldsymbol{hist}(t-2\Delta t),\\ &= \mathbf{H}\boldsymbol{v}(t-\Delta t) + \left[\mathbf{I} - 2\mathbf{GR}\right]\boldsymbol{hist}(t-2\Delta t), \end{aligned}$$

where

$$\mathbf{A} = \left[\mathbf{I} + \frac{\Delta t}{2}\mathbf{L}^{-1}\mathbf{R}\right]^{-1}, \tag{3.11}$$

$$\mathbf{B} = \mathbf{I} - \frac{\Delta t}{2}\mathbf{L}^{-1}\mathbf{R}, \tag{3.12}$$

yielding

$$\begin{aligned}\mathbf{AB} &= \left[\mathbf{I} + \frac{\Delta t}{2}\mathbf{L}^{-1}\mathbf{R}\right]^{-1}\left[\left(\mathbf{I} + \frac{\Delta t}{2}\mathbf{L}^{-1}\mathbf{R}\right) - 2\frac{\Delta t}{2}\mathbf{L}^{-1}\mathbf{R}\right], \\ &= \mathbf{I} - 2\left[\mathbf{I} + \frac{\Delta t}{2}\mathbf{L}^{-1}\mathbf{R}\right]^{-1}\frac{\Delta t}{2}\mathbf{L}^{-1}\mathbf{R}, \\ &= \mathbf{I} - 2\mathbf{GR},\end{aligned} \tag{3.13}$$

which finally gives the formula for $\boldsymbol{hist}(t - \Delta t)$

$$\boldsymbol{hist}(t - \Delta t) = \mathbf{H}\boldsymbol{v}(t - \Delta t) + [\mathbf{I} - 2\mathbf{GR}]\,\boldsymbol{hist}(t - 2\Delta t), \tag{3.14}$$

where

$$\mathbf{H} = 2\,(\mathbf{G} - \mathbf{GRG})\,. \tag{3.15}$$

3.2.2 Linear Module Hardware Design

There are two main functions for this module: (i) calculate the history currents and send them to the linear solver and (ii) save the updated history currents within this module so as to use them in the next simulation time-step. The main control module sends a signal to this module to indicate which operation is to be executed.

As indicated by (3.14), the history current calculation can be carried out by matrix multiplication. A matrix vector multiplication (*MVM*) unit which is also part of the linear solver has been developed for this purpose. Since the entries in **R**, **L**, **G**, and **H** are the properties of transformer which are unchanged, these parameters can be saved in memory before the real-time simulation begins. Furthermore, instead of saving **R** and **L** individually, the result of $2\,(\mathbf{G} - \mathbf{GRG})$ and $(\mathbf{I} - 2\mathbf{GR})$ can be saved as **H** and **IGR** in the memory, respectively, to further reduce the computational complexity and latency. The matrix storage format (Figure 2.26b) consists of two RAMs defined to store the precalculated transformer matrices: one is a 32-bit wide RAM *val* to store the value of the matrix entities in row order and the second is a 9-bit wide RAM, consisting of the last 8-bit *cid* to store the column index of the entries, and the first bit *rlb* to label all entities in the same row with "0" or "1".

In Figure 3.2a, the RAMs (*rmHcol*, *rmH*, *rmIGRcol*, *IGR*) containing the transformer parameters are generated by the above rule, and *rmV* stores the transformer nodal voltages calculated by the nonlinear solver, while *rmIh* stores the history current terms calculated in previous time-step. When the linear module executes the updated function, the memories mentioned above are accessed

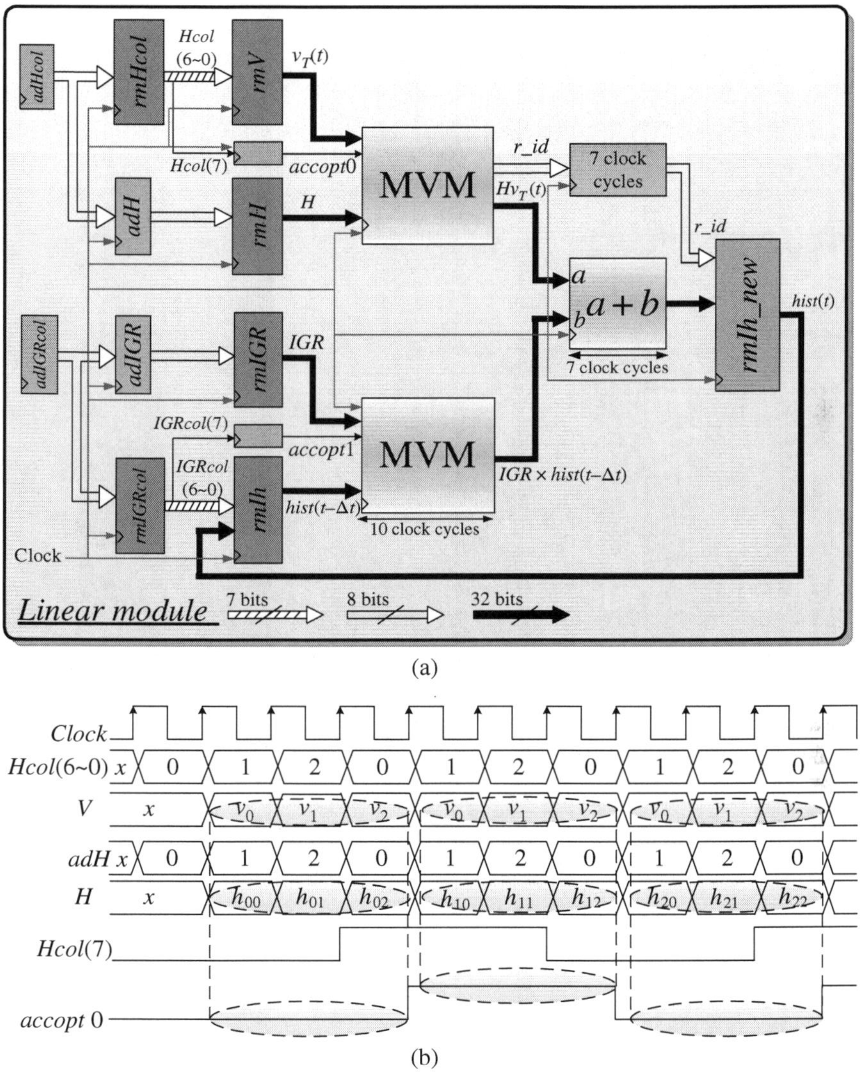

Figure 3.2 Pipelined scheme for the linear module: (a) hardware design and (b) timing diagram.

sequentially by *adHcol*, *adH*, *adIGRcol*, and *adIGR*, and the data is sent to *MVM* function unit. The timing diagram of this operation is shown in Figure 3.2b.

The *MVM* unit consists of one floating point multiplier, one floating-to-fixed point converter, one fixed-floating point converter, and two fixed point adders, as shown in Figure 3.3a. The reason for a fixed point adder in this module is that the

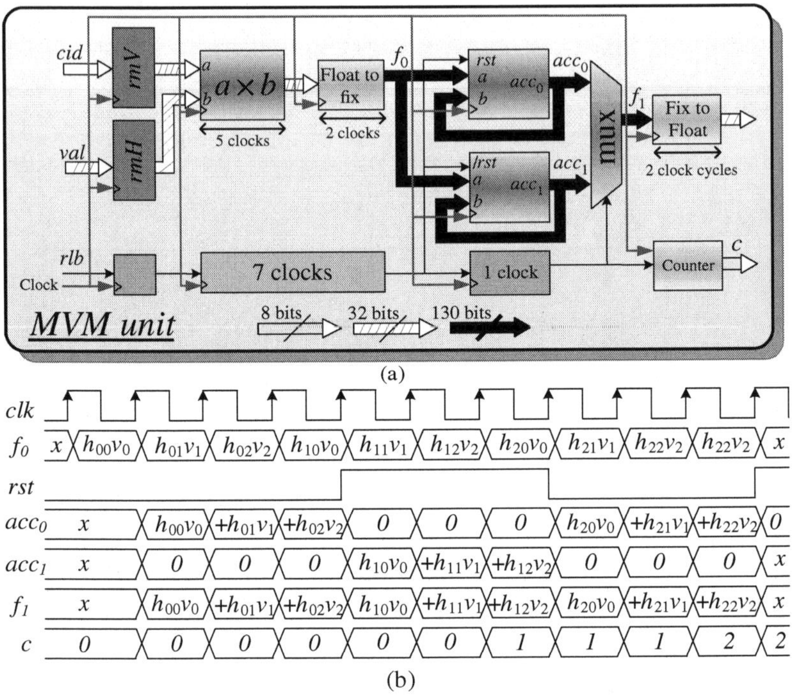

Figure 3.3 *MVM* unit: (a) hardware design and (b) timing diagram.

floating point adder needs much more logic resources and a higher clock latency. The inserted registers in this unit are used for synchronization purpose. The two adders here with opposite reset signals are for pipelining purpose, so that there is no stall between two row-vector multiplications. Figure 3.3b shows the logical timing diagram of these operations.

3.2.3 Inode Unit Module

After calculating the history currents, a *Inode Unit* module is used to form the history current vector. For a specific node, more than one history currents can be connected to this node as shown in Figure 3.4a. Therefore, the *Inode Unit* is introduced to obtain the summation history current at each node. Three vectors are defined as shown in Figure 3.4b: (i) a 12-bit *rmInodecfg* to store configuration information of each node. The bit assignments are as follows: *rmInodecfg(9-0)* indicates which history current is connected to this node, *rmInodecfg(10)* indicates its flow direction (flow into or flow out of this node) with "1" or "0", and *rmInodecfg(11)* indicates whether all history currents are connected to the same node with "1" or

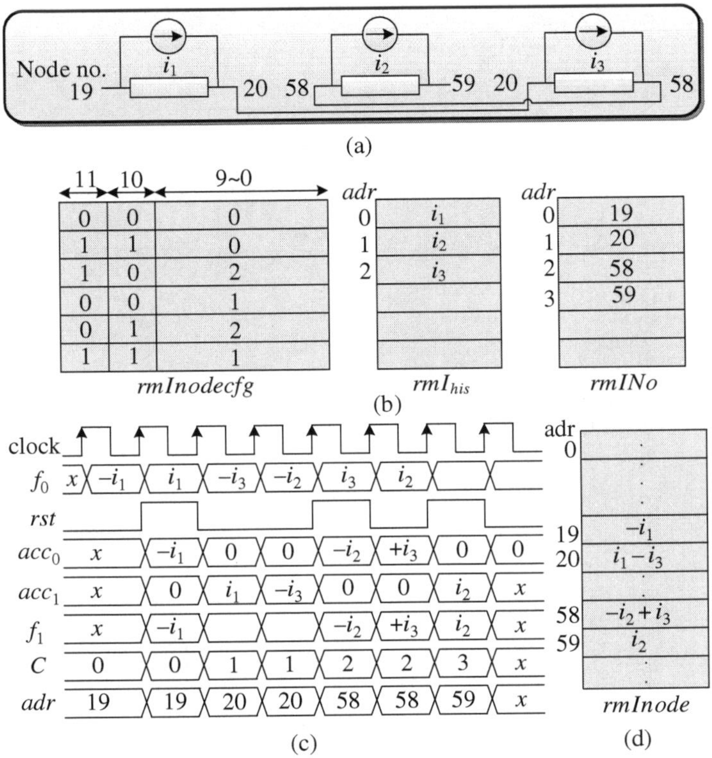

Figure 3.4 *Inode Unit*: (a) an example circuit, (b) corresponding RAM assignment, (c) timing diagram, and (d) *rmInode* vector.

"0"; (ii) a 32-bit rmI_{hist} to store the value of history current; and (iii) an 8-bit *rmINo* to store the index of each node.

Figure 3.5 shows hardware design of *Inode Unit* which is quite similar to the *MVM* unit. The *counter*, whose value increases by one if *rmInodecfg(11)* is changed, is connected to *rmINo*, so that the accumulated history current can be saved at the right address in the current vector *rmInode*. The timing diagram and the corresponding *rmInode* vector is shown in Figure 3.4c,d.

3.2.4 Nonlinear Model Solution

Inclusion of nonlinearities is necessary for accurate modeling of transformer behavior. To achieve that, nonlinear elements are attached externally at the linear model terminals. In most magnetic materials, the saturation phenomena can be seen in the continuous magnetization curve: as the current increases, the flux value increases to a maximum value asymptotically. While saturation is

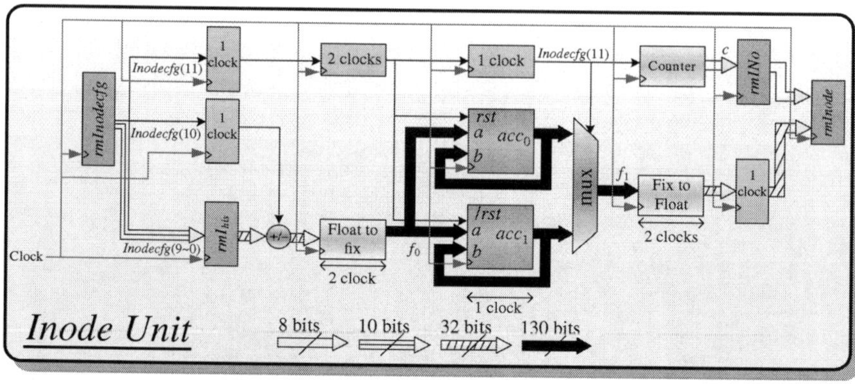

Figure 3.5 *Inode Unit*: hardware design.

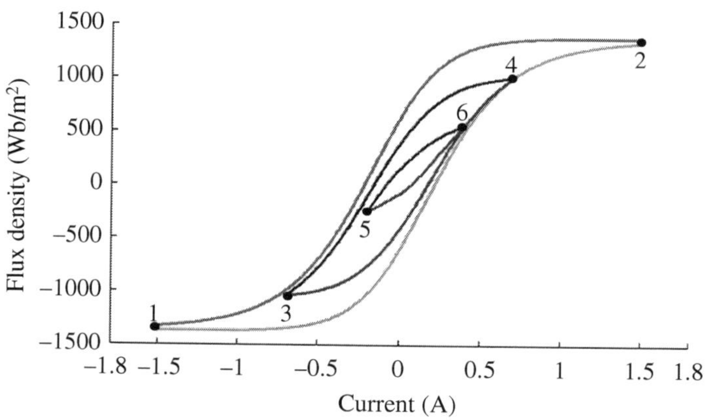

Figure 3.6 Hysteresis characteristic with major and minor loops.

independent of the history current value, the hysteresis effect is dependent on history meaning that the flux cannot be determined by instant current value; however, it can be determined given the history current value. A major hysteresis loop (Figure 3.6) is the outermost loop which enters the saturation region at the reversal points (1-2), while other closed loops are considered minor loops (3-4 and 5-6). Furthermore, saturation is dominant when current has increased to a large value, while hysteresis plays an important role when the transformer operates with a small value of current.

For the solution of the nonlinear part, the compensation method is employed. After the linear solution of a network is found, an n-node power system with p nonlinear elements is solved as illustrated in Figure 2.34b and (3.16).

$$\boldsymbol{f}(\boldsymbol{i}_{km}) = \boldsymbol{v}_{open} - \mathbf{r}_{thev}\boldsymbol{i}_{km}, \tag{3.16}$$

where $\boldsymbol{v}_{open}$ is the $p \times 1$ open circuit voltage vector (a subset of $\boldsymbol{v}_0$ the unknown node voltages) without the nonlinear elements, and $\boldsymbol{i}_{km}$ is the nonlinear elements' current injection vector with k_l and m_l $(l = 1, 2, \dots, p)$ being the node indices which the nonlinear elements are connected, expressed as:

$$\begin{aligned}\boldsymbol{v}_{open} &= [\mathrm{v}_{0k_1}(t) - \mathrm{v}_{0m_1}(t)\ \mathrm{v}_{0k_2}(t) - \mathrm{v}_{0m_2}(t) \\ &\quad \dots\ \mathrm{v}_{0k_{n_l}}(t) - \mathrm{v}_{0m_{n_l}}(t)]^T,\end{aligned} \tag{3.17}$$

$$\boldsymbol{i}_{km} = \left[\mathrm{i}_{k_1m_1}(t)\ \mathrm{i}_{k_2m_2}(t)\ \dots\ \mathrm{i}_{k_{n_l}m_{n_l}}(t)\right]^T. \tag{3.18}$$

$\mathbf{r}_{thev}$ is the $n_l \times n_l$ Thevenin equivalent resistance matrix that can be found from $\boldsymbol{Y}$, and $\boldsymbol{f}(\boldsymbol{i}_{km})$ are nonlinear functions. The process of the Compensation method and Newton–Raphson (N-R) iterations are detailed in Section 2.9.

The final network solution is found by superimposing the response of $\boldsymbol{i}_{km}$ on the linear network solution,

$$\boldsymbol{v} = \boldsymbol{v}_0 - \mathbf{R}_{Thev}\boldsymbol{i}_{km}, \tag{3.19}$$

where $\boldsymbol{v}$ is a $n \times 1$ node voltage vector with nonlinear branch, expressed as $\boldsymbol{v} = [v_1\ v_2 \dots v_n]^T$ and $\mathbf{R}_{Thev}$ is a $n \times n_l$ matrix obtained from $\boldsymbol{Y}$.

For example, for the nonlinear inductance, the characteristic function is provided in the form,

$$\lambda(t) = \boldsymbol{f}_\lambda(\boldsymbol{i}_{km}(t)), \tag{3.20}$$

where the flux $\lambda(t)$ is the integral of node voltage over a time-step:

$$\lambda(t) = \lambda(t - \Delta t) + \int_{t-\Delta t}^{t} \boldsymbol{v}'(u)du. \tag{3.21}$$

Applying Trapezoidal rule results in:

$$\lambda(t) = \frac{\Delta t}{2}\boldsymbol{v}'(t) + \lambda_{hist}(t - \Delta t), \tag{3.22}$$

where

$$\lambda_{hist}(t - \Delta t) = \lambda(t - \Delta t) + \frac{\Delta t}{2}\boldsymbol{v}'(t - \Delta t). \tag{3.23}$$

From (3.22) and (3.20), the set of nonlinear equations to determine $\boldsymbol{i}_{km}$ are,

$$\mathbf{F}_\lambda \equiv \frac{2}{\Delta t}[\boldsymbol{f}_\lambda(\boldsymbol{i}_{km}(t)) - \lambda_{hist}(t - \Delta t)] - \boldsymbol{v}_{open}(t) + \mathbf{r}_{thev}\boldsymbol{i}_{km}(t), \tag{3.24}$$

Then, the Jacobian matrix and the nonlinear equations for this case can be rewritten as:

$$\begin{cases}\mathbf{J}_\lambda = \dfrac{\partial \mathbf{F}_\lambda(\boldsymbol{i}^j_{km}(t))}{\partial \boldsymbol{i}^j_{km}(t)} = \mathbf{r}_{thev} + \dfrac{\partial \boldsymbol{f}_\lambda(\boldsymbol{i}^j_{km}(t))}{\partial \boldsymbol{i}^j_{km}(t)}\dfrac{2}{\Delta t}, \\ -\mathbf{F}_\lambda = \boldsymbol{v}_{open}(t) - \mathbf{r}_{thev}\boldsymbol{i}^j_{km}(t) - \\ \qquad [\boldsymbol{f}_\lambda(\boldsymbol{i}^j_{km}(t)) - \lambda_{hist}(t - \Delta t)]\dfrac{2}{\Delta t}.\end{cases} \tag{3.25}$$

3.2.4.1 Preisach Hysteresis Model

The Preisach model [50, 51] is employed to define the hysteresis loop. The major loop hysteresis function is geometrically represented by the following function:

$$\boldsymbol{\phi}_{p\_m}(\boldsymbol{i}_{km}) = \frac{1}{2}\sum_{u=1}^{3} a_u[\tanh(b_u \boldsymbol{i}_{km}) + c_u \text{ sec h}^2(b_u \boldsymbol{i}_{km})], \tag{3.26}$$

where a_u, b_u, and c_u are the parameters to define this hyperbolic template, $\boldsymbol{i}_{km}$ is current value vector flowing through the nonlinear inductances as before and $\boldsymbol{\phi}_{p\_m} = [\phi_{p\_m1}\ \phi_{p\_m2}\ \cdots\ \phi_{p\_mn_l}]^T$ is the flux density corresponding to this current value vector on the major loop. The subscript "p_m" stand for Preisach major loop.

Once the major loop function is fixed, all the minor loop trajectories from a reversal point follow this uniform template. A reverse point is detected, when $\phi_{p\_ml}(t + \Delta t) > \phi_{p\_ml}(t)$ on a downward trajectory, or when $\phi_{p\_ml}(t + \Delta t) < \phi_{p\_ml}(t)$ on an upward trajectory ($l = 1, 2, \dots, n_l$). In general, Preisach upward and downward trajectories from a reversal point (i_{rl}, ϕ_{rl}) can be expressed as (3.27) and (3.28), respectively [51], where $\boldsymbol{i}_r = [i_{r1}\ i_{r2}\ \cdots\ i_{rn_l}]$ and $\boldsymbol{\phi}_r = [\phi_{r1}\ \phi_{r2}\ \cdots\ \phi_{rn_l}]$. The reversal points need to be recorded in a reversal point stack (RPS), while traveling the hysteresis loop; for example when flux density starts increasing from point 5, and reaches point 6, the trajectory beyond the reversal point 6 is defined by the template beginning at point 3, as shown in Figure 3.6. Thus, if the flux density increases or decreases out of the outer loop, the last two coordinates in the stack are removed. This example is initialized by saving the two major loop reversal points (1 and 2) in the stack; these two points cannot be removed since all the trajectories cannot be beyond the major loop as illustrated in Figure 3.7. The trajectory is processed by going through the following reversal points: 1-4-3-6-5, with the corresponding stack shown in Figure 3.7a. Once the trajectory increases beyond point 6, the reversal points 5 and 6 are removed due to the reason discussed above, shown in Figure 3.7b.

The flux trajectories corresponding to the instantaneous current in a upward ϕ_{p_u} or downward ϕ_{p_d} minor loop are given as:

$$\boldsymbol{\phi}_{p\_u}(\boldsymbol{i}_{km}) = -\boldsymbol{\phi}_{p\_m}(-\boldsymbol{i}_{km}) - \boldsymbol{\phi}_{p\_m}(\boldsymbol{i}_r) + \boldsymbol{\phi}_r + 2\boldsymbol{K}(-\boldsymbol{i}_r)\boldsymbol{K}(\boldsymbol{i}_{km}), \tag{3.27}$$

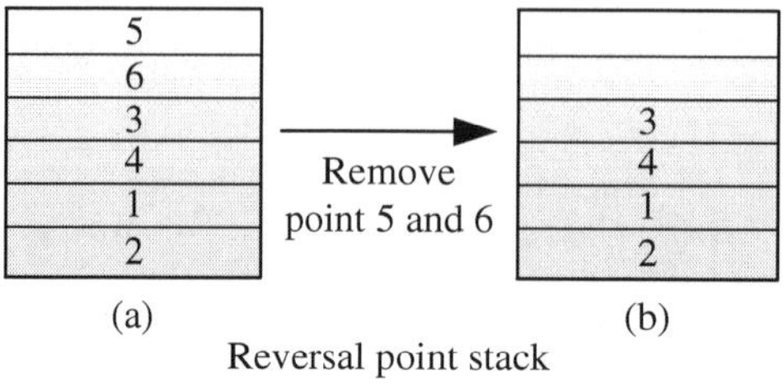

Figure 3.7 Corresponding reversal point stack before (a) and after (b) current increasing beyond the point 6.

$$\boldsymbol{\phi}_{p\_d}(\boldsymbol{i}_{km}) = \boldsymbol{\phi}_{p\_m}(\boldsymbol{i}_{km}) + \boldsymbol{\phi}_{p\_m}(-\boldsymbol{i}_r) + \boldsymbol{\phi}_r - 2\boldsymbol{K}(\boldsymbol{i}_r)\boldsymbol{K}(-\boldsymbol{i}_{km}), \tag{3.28}$$

where

$$\boldsymbol{K} = [K_1\ K_2\ \dots\ K_{n_l}],$$

$$K_l(x) = \begin{cases} \sqrt{\phi_{p\_m}(-x)}, & x < 0, \\ \dfrac{\phi_{p\_m}(-x) + \phi_{p\_m}(x)}{2\sqrt{\phi_{p\_m}(x)}}, & x > 0, \end{cases}$$

$$l = 1, 2, \dots, n_l.$$

The functions $\frac{d}{di_{km}}\boldsymbol{\phi}_{p\_u}(\boldsymbol{i}_{km})$, $\frac{d}{di_{km}}\boldsymbol{\phi}_{p\_d}(\boldsymbol{i}_{km})$ for upward and downward trajectories can be expressed as (3.29) and (3.30), respectively:

$$\frac{d}{d\boldsymbol{i}_{km}}\boldsymbol{\phi}_{p\_u}(\boldsymbol{i}_{km}) = d\boldsymbol{\phi}_{p\_m}(-\boldsymbol{i}_{km}) + 2\boldsymbol{K}(-\boldsymbol{i}_r)d\boldsymbol{K}(\boldsymbol{i}_{km}), \tag{3.29}$$

$$\frac{d}{d\boldsymbol{i}_{km}}\boldsymbol{\phi}_{p\_d}(\boldsymbol{i}_{km}) = d\boldsymbol{\phi}_{p\_m}(\boldsymbol{i}_{km}) + 2\boldsymbol{K}(\boldsymbol{i}_r)d\boldsymbol{K}(-\boldsymbol{i}_{km}). \tag{3.30}$$

Substituting (3.27) and (3.28) into (3.20), the Jacobian matrix can be computed by (3.25) using the differentials (3.29) and (3.30).

3.2.4.2 Nonlinear Module Hardware Design

In the Preisach Hysteretic Unit (PHU), look-up tables are employed instead to compute the nonlinear functions. While other methods such as COordinate Rotation DIgital Computer (CORDIC) are also available to evaluate the nonlinear functions required in the N-R iteration, the number of iterations required is unknown, and the computational burden and latency could be too high to achieve real-time execution.

As shown in Figure 3.8, taking the advantage of parallelism of logic resources, the function $f(\boldsymbol{i}_{km})$ and $\frac{d}{\boldsymbol{i}_{km}}f(\boldsymbol{i}_{km})$ can be evaluated simultaneously. The address generator (*AG*) functional unit is used to convert the 32-bit floating-point $\boldsymbol{i}_{km}$ into integer numbers to address the LUTs. (3.27)–(3.30) show that function $f(\boldsymbol{i}_{km})$ and $\frac{d}{\boldsymbol{i}_{km}}f(\boldsymbol{i}_{km})$ can be evaluated with $\boldsymbol{i}_{km}$ or $-\boldsymbol{i}_{km}$, $\boldsymbol{i}_r$ or $-\boldsymbol{i}_r$ depending on the trajectory direction. Therefore, before $\boldsymbol{i}_{km}$ is input into the *AG* unit, its trajectory direction must be known. Besides, since the function $K(\boldsymbol{i}_{km})$ is dependent on the sign of $\boldsymbol{i}_{km}$, so that two $K(\boldsymbol{i}_{km})$ LUTs ($K+$, $K-$) are precalculated and the first bit of $\boldsymbol{i}_{km}$ (or $-\boldsymbol{i}_{km}$) determines which value is involved in the calculation by using a multiplexer *mux*.

As shown in Figure 3.9, after finding the values of $f(\boldsymbol{i}_{km})$ and $\frac{d}{\boldsymbol{i}_{km}}f(\boldsymbol{i}_{km})$, the Jacobian matrix compute unit (*JMCU*) computes the values of $-\mathbf{F}$ and $\mathbf{J}$ as given in

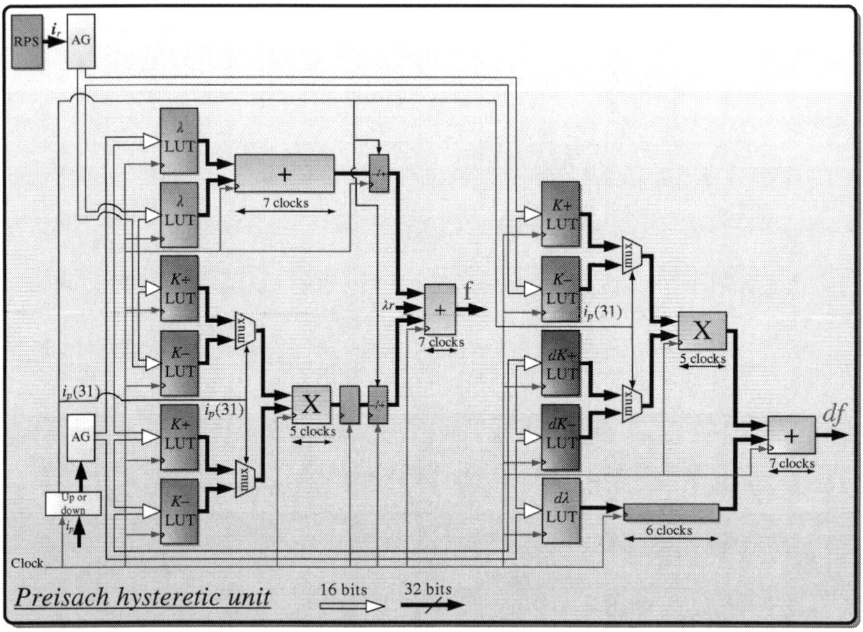

Figure 3.8 Preisach hysteretic unit (PHU) hardware design.

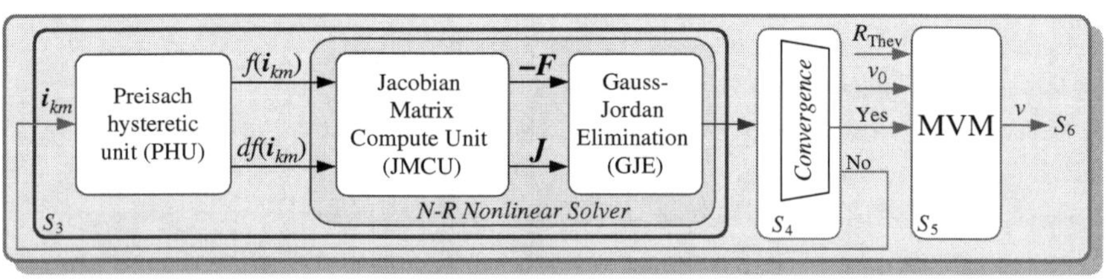

Figure 3.9 Overall nonlinear module hardware architecture.

(3.25), and then the Gauss–Jordan elimination (*GJE*) unit is used to solve the algebraic equations. If the calculated $\boldsymbol{i}_{km}$ is satisfied, then it is used to compensate the voltage $\boldsymbol{v}_0$ to find the final results $\boldsymbol{v}$ (3.19). If the $\boldsymbol{i}_{km}$ has not converged, then it is used as an input for the next iteration until convergence or the maximum number of iterations is reached.

3.2.5 Frequency-Dependent Eddy Current Model

Eddy currents, a part of the transformer core losses, are induced by the change in the magnetic field. While they can be reduced by using a laminated core, eddy current losses are frequency dependent. However, usually the eddy current model

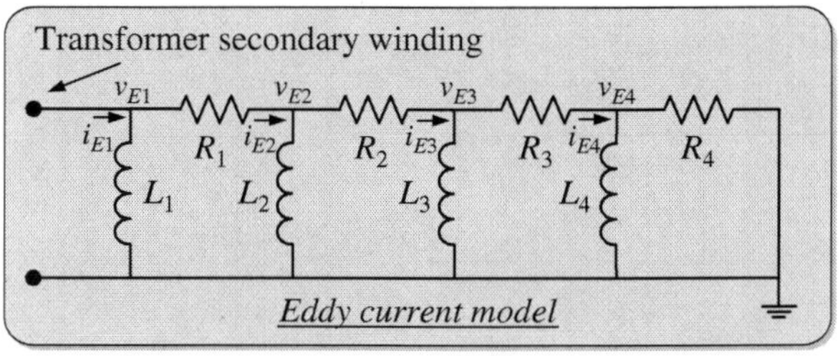

Figure 3.10 Cauer equivalent circuit for frequency-dependent eddy current model.

is normally a constant resistance in parallel with the magnetizing inductance so that this model is not valid over a wide frequency range.

A Cauer equivalent circuit is used here to represent the frequency dependency as a cascade of RL lumped sections (Figure 3.10). The accuracy of this model over a wide frequency range depends on the number of RL sections. Four sections are required to represent the frequency up to 200 kHz with an error of less than 5% [52, 53].The parameters of the Cauer model can be derived as follows:

$$\begin{aligned} L_0 &= \frac{N^2 A\mu}{l} \quad \text{and} \quad L_k = \frac{L_0}{(4k-3)}, \\ R_0 &= \frac{4N^2 A}{ld^2\gamma} \quad \text{and} \quad R_k = R_0(4k-1), \end{aligned} \tag{3.31}$$

where μ is the magnetic permeability of the steel lamination, γ is the electric conductivity of the steel lamination, A is the total cross-sectional area of all laminations, d is the thickness of the lamination, l is the length of core limb, and N is the number of coil turns.

After discretizing the cascaded R–L equivalent circuit, the Norton equivalent formula is given as (3.32), where $\boldsymbol{i}_E(t) = [i_{E1}(t)\ i_{E2}(t)\ i_{E3}(t)\ i_{E4}]^T$ and $\boldsymbol{v}_E(t) = [v_{E1}(t)\ v_{v2}(t)\ v_{E3}(t) v_{E4}]^T$ are the node current and voltage vectors, respectively, and $\boldsymbol{I}_{histE}(t)$ is the eddy current history term vector which is updated by (3.33):

$$\boldsymbol{i}_E(t) = \mathbf{Y}_E \boldsymbol{v}_E(t) + \boldsymbol{I}_{histE}(t). \tag{3.32}$$

$$\boldsymbol{I}_{histE}(t) = \mathbf{G}_E \boldsymbol{v}_E(t-\Delta t) + \boldsymbol{I}_{histE}(t-\Delta t), \tag{3.33}$$

where

$$\mathbf{G}_E = diag\left\{ \frac{\Delta t}{2L_1}\ \frac{\Delta t}{2L_2}\ \frac{\Delta t}{2L_3}\ \frac{\Delta t}{2L_4} \right\}. \tag{3.34}$$

3.2.6 Hardware Emulation of Power Transformer

The real-time transformer model hardware architecture [54] is shown in Figure 3.11. It can be divided into three parts: the linear module, the nonlinear

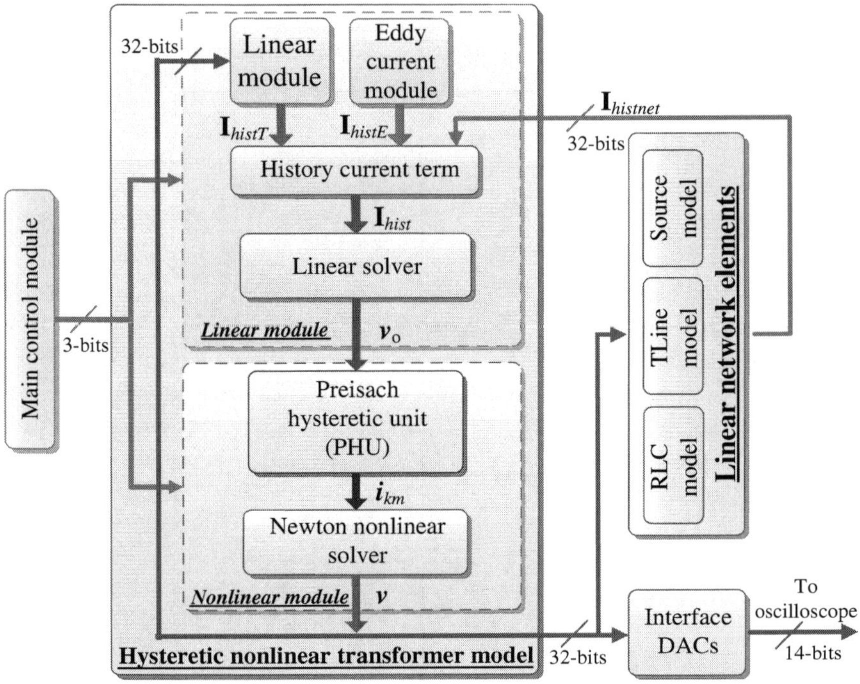

Figure 3.11 Overall architecture of the real-time nonlinear hysteretic transformer model on FPGA.

module, and the main control module. The linear module is built based on the admittance matrix approach after discretizing the transformer differential equations, and is used to update the transformer history current terms, and calculates the node voltages without the nonlinear elements according to the compensation method. The nonlinear module contains two parts: the PHU and the N-R nonlinear solver unit, for finding the solution to the nonlinear equations. Further details regarding the nonlinear module are provided in Section II(C). The control module sends command signals to other modules to control which functions will be performed by each module, and also waits for the signals sent back by each module, so that all modules can be operated either in parallel or sequentially as required by the transient algorithm. Figure 3.12 shows the detailed FSM for the main control module. S_1 calculates the summation of all the history current terms ($\mathbf{I}_{hist}$) received from the linear transformer module ($\mathbf{I}_{histT}$), eddy current module ($\mathbf{I}_{histE}$), and the network components ($\mathbf{I}_{histNET}$) which include the passive elements (RLC) module and transmission line (TWM) module, which together constitute the linear part of a power system. S_2 computes the nodal voltages ($\boldsymbol{v}_o$) without the nonlinear elements. S_3 evaluates the nonlinear

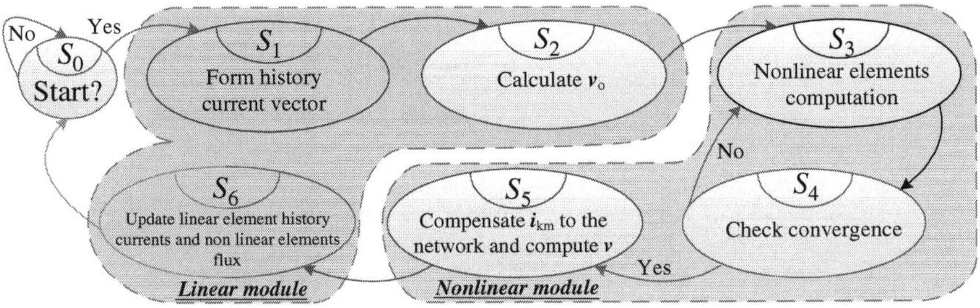

Figure 3.12 FSM of the main control module.

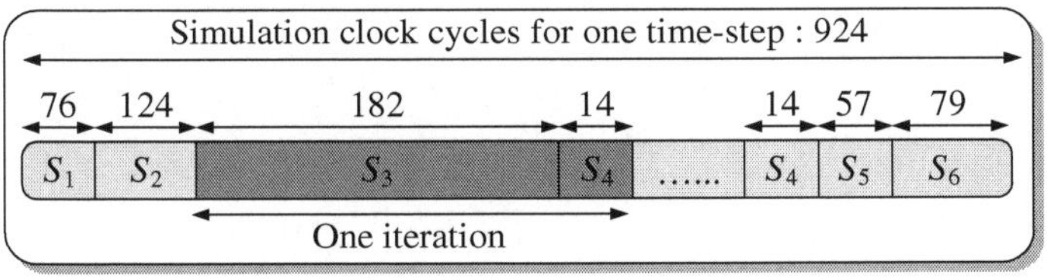

Figure 3.13 Model latencies for one time-step.

functions of $\boldsymbol{f}(\boldsymbol{i}_{com})$ and their derivative $\boldsymbol{df}(\boldsymbol{i}_{km})$. S_4 performs the N-R procedure to compute the results in each iteration; if the result is converged, the FSM goes to S_5 to calculate the nodal voltages ($\boldsymbol{v}$), otherwise it goes back to S_3. In S_6, the transformer linear module, eddy current module, and passive element and TLine modules update their history terms.

The real-time nonlinear hysteretic transformer electromagnetic transient model was implemented on the Xilinx® Virtex-7® VC707 XC7VX485T FPGA whose resources are listed in Table 1.2. The Virtex-7 FPGA outputs the real-time simulation results to a DAC board through an FMC connector integrated on the board, and a FMC-to-DAC adapter, and the real-time data are captured by an oscilloscope. The Chipscope® is a convenient tool provided by ISE software to visualize data. This tool can be used to analyze short transient periods; however, due to the limitation of the sampling number (up to 131 072), this tool cannot be used to trace the waveforms. The overall logic elements utilized by the nonlinear transformer model is 15%. The real-time simulation was carried out with a time-step of 10 μs on the FPGA input clock frequency of 100 MHz. The breakdown of model latencies in one simulation time-step is shown in Figure 3.13. The execution time for one time-step is 9.24 μs with a maximum of three Newton iterations per time-step. This means within a 50 μs time-step, which is the acceptable time-step for real-time transient simulation, this FPGA-based nonlinear transformer model can simulate five nonlinear transformers in real-time with very small hardware resource consumption. Increasing the hardware utility or the clock frequency, it is also possible to simulate a larger power system with more transformer models including the nonlinear and frequency dependent phenomena.

3.2.7 Real-Time Emulation Case Studies

Two power systems are emulated to show the effectiveness and accuracy of real-time nonlinear transformer model.

3.2.7.1 Case I

Figure 3.14 shows the first case study: single-line diagram of a three-phase system. A transformer T is connected to a voltage source G through a circuit breaker S_{W1}. C_s is the breaker's grading capacitance, and C_p is the phase-to-earth capacitance including transformer winding capacitance. The system data is listed in the Appendix A.2. The real-time simulation results are validated by off-line software ATP.

Inrush Current Transients In this inrush current transients case study, capacitors C_s and C_p are removed from Figure 3.14 and a five-segment piece-wise nonlinear function is employed to represent the saturation effect. When the breaker S_{w1} is closed at 0.1 second, the real-time captured three-phase breaker current waveforms are shown in Figure 3.15a. As can be seen from these waveforms, phases a and c have larger current amplitude, while the amplitude of phase b current is relatively small. These unbalanced current waveforms are caused due to the instantaneous voltage supply values being different in various phases when the breaker S_{w1} is closed. The peak magnitude inrush current value of phases a and c is 210% of the transformer rated current, and it decays to steady state after 0.6 second. Similar inrush current behavior can be seen in the ATP off-line simulation results in Figure 3.15b.

Hysteresis Behavior To investigate the hysteresis behavior in the magnetizing core, the PHU is used to model the saturation and hysteresis during the inrush current transient. The parameters to define the hysteresis trajectory are listed in the

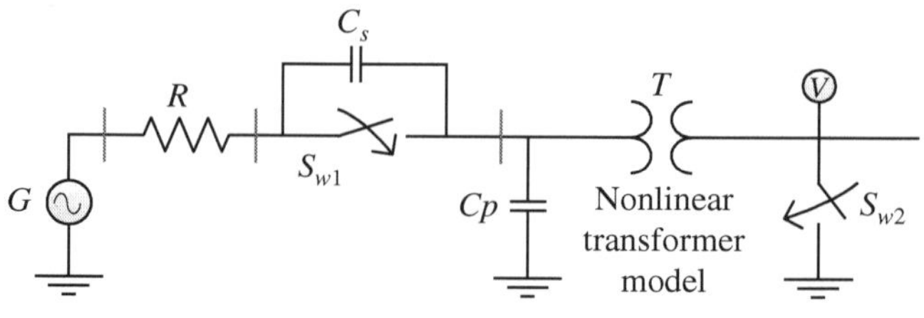

Figure 3.14 Single-line diagram for Case I.

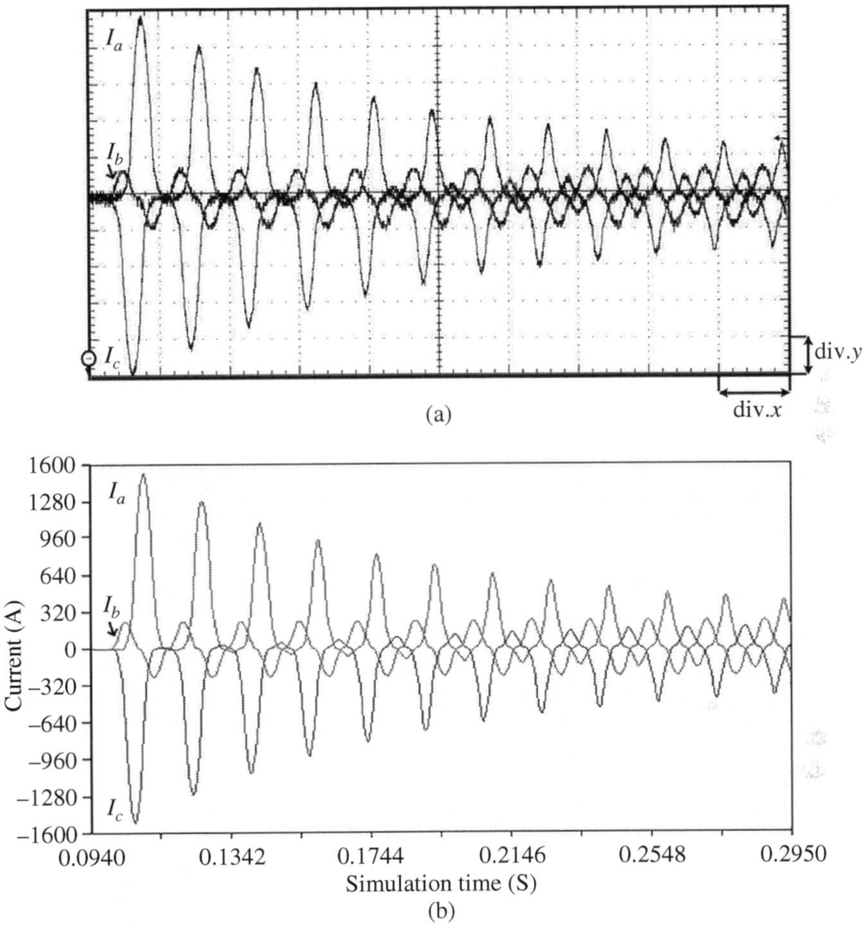

Figure 3.15 Breaker current when S_{w1} closes at 0.1 second: (a) real-time simulation results (1 div.x = 20 ms; 1 div.y = 320 A) and (b) off-line ATP simulation results.

Appendix A.2. The initial flux in the core is set on the downward major loop trajectory. Figure 3.16 shows the real-time simulation results for the flux and magnetizing current of the nonlinear transformer. From $t = 0$ to 0.106 seconds, the power system is energized from a voltage source at a value of 0.65 p.u. As shown in Figure 3.16, during the first cycle (1-2), the flux is nonsymmetrical due to the initial remanent flux and the peak flux value is 1 p.u. After the first cycle, the flux is symmetrical and travels on the minor loop since the value of voltage is not high enough to drive the transformer working in the saturation region (3-4). From $t = 0.106$ to 0.15 seconds, the breaker is opened and the flux is locked at 0.62 p.u.

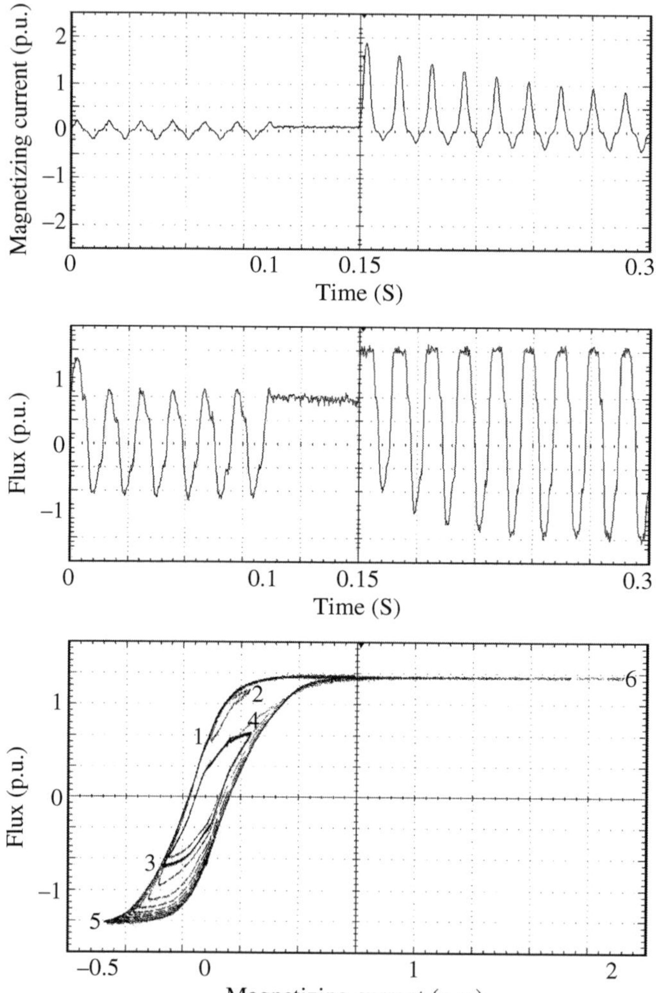

Figure 3.16 Real-time simulation results for the flux, magnetizing current, and the hysteresis characteristic of the nonlinear transformer.

From $t = 0.15$ second, the voltage rises to 1.25 p.u. and the breaker is reclosed, when the peak flux value reaches 1.25 p.u., and the high voltage value drives the flux into saturation region traveling on the major loop (5-6).

Ferroresonance Transients Ferroresonance is an electrical phenomenon which can occur when a nonlinear inductance is fed from a source. It can cause overvoltages and overcurrents in the system with subsequent risk of damage to equipment. In Figure 3.17, after opening S_{w1} while the breaker S_{w2} is closed at $t = 0.2$ second

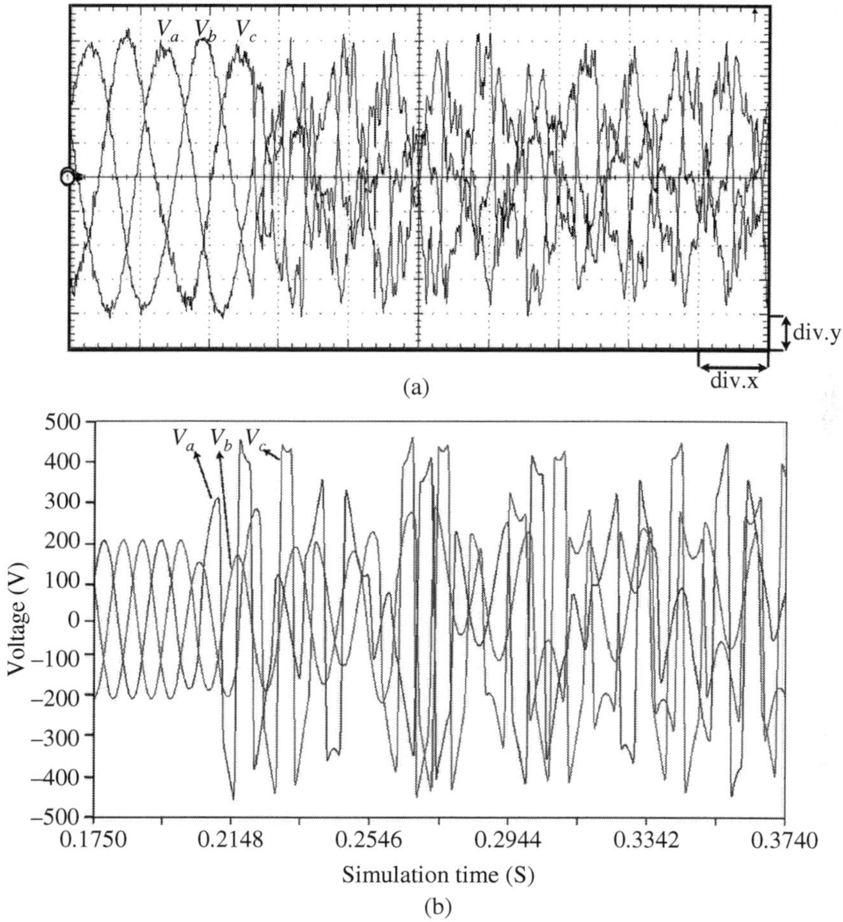

Figure 3.17 Primary side three-phase voltages when S_{w1} is opened and S_{w2} at $t = 0.2$ second: (a) real-time simulation results (1 div.$x = 20$ ms, 1 div.$y = 100$ kV) and (b) off-line (ATP) simulation results.

to simulate a three-phase ground fault, the voltage waveforms are highly distorted and have a higher peak value 220% of the steady-state voltage value. Again, similar voltage behavior can be observed in off-line simulation results in Figure 3.17b.

Eddy Current Transients Transient recovery voltages (TRV) manifest across any circuit breaker in power systems when the breaker attempts to clear a nearby fault. This case study is initialized by closing S_{w2} to cause a ground fault, then reopening S_{w2} to clear this fault. The real-time simulation results of voltage across S_{w2} by using two different models, Cauer equivalent circuit and a constant resistor, are captured. As shown in Figure 3.18, the Cauer model simulation result shows

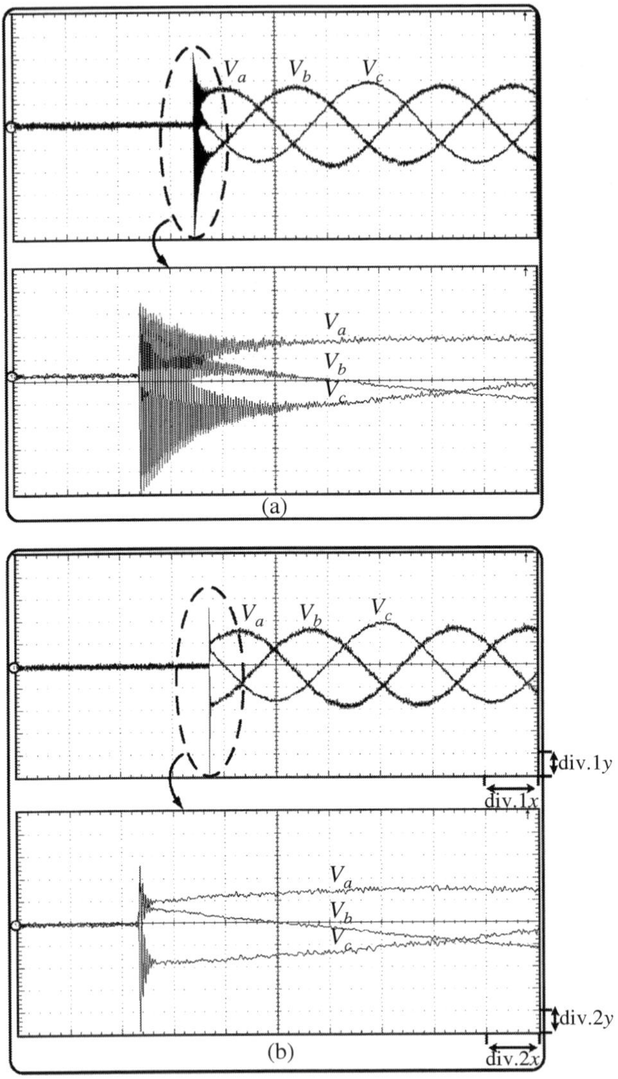

Figure 3.18 TRV real-time simulation results at $t = 0.2$ second: (a) Cauer model and (b) nonfrequency-dependent model (constant resistor) (1 div.1x = 4 ms; 1 div.11y = 84 kV, 1 div.2x = 400 μs, 1 div.2y = 84 kV).

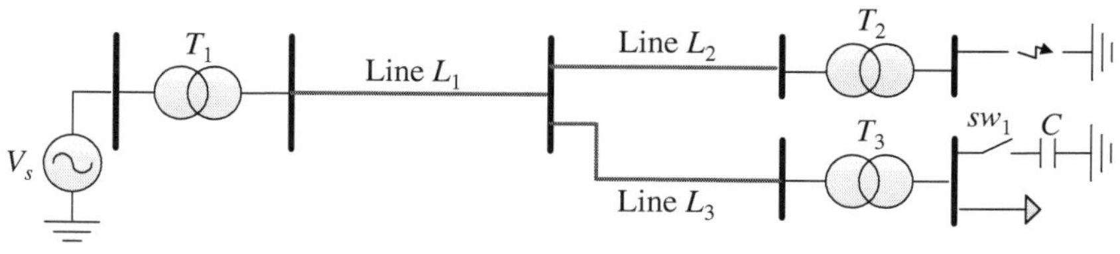

Figure 3.19 Single-line diagram for Case II.

a higher peak value and much longer damping time than that with a constant resistor model, this implies that the eddy currents increase the power losses in the transformer during the transient recovery period.

3.2.7.2 Case II

In the second case study (Figure 3.19), the power system involves: (i) three transformers (T_1, T_2, T_3), all of them employing the full nonlinear modeling; (ii) three transmission lines (L_1, L_2, L_3), modeled using distributed parameter line model (TWM); (iii) one ideal voltage source (V_s); and (iv) one switch (SW_1) and a capacitor C are used to model the capacitor switching. The parameters for this case study are also listed in the Appendix A.2.

Capacitor Switching Transients The entire power system is energized until $t = 1$ second when the capacitor switching is initiated by closing SW_1. Both real-time and off-line simulation results (Figure 3.20) are captured to show the primary winding voltages at transformer T_3. From the waveforms, after closing the switch, the voltage dropped to zero because the voltage of capacitor at the secondary winding of this transformer cannot change immediately, and then a fast voltage recovery is observed. The distorted recovery voltage is a maximum of 150% of the steady-state voltage value, and decays to steady state after three seconds.

Ground Fault Transients This case study is achieved by initiating a three-phase-to-ground fault at $t = 1$ second at the secondary winding of transformer T_2, and the voltage waveforms at the primary winding of transformer T_3 are captured both in real time and off-line, as shown in Figure 3.21. Compared with capacitor switching transients, the ground fault transients behave in a highly distorted fashion; however, there is no overshoot in voltage, and steady state is achieved after four seconds.

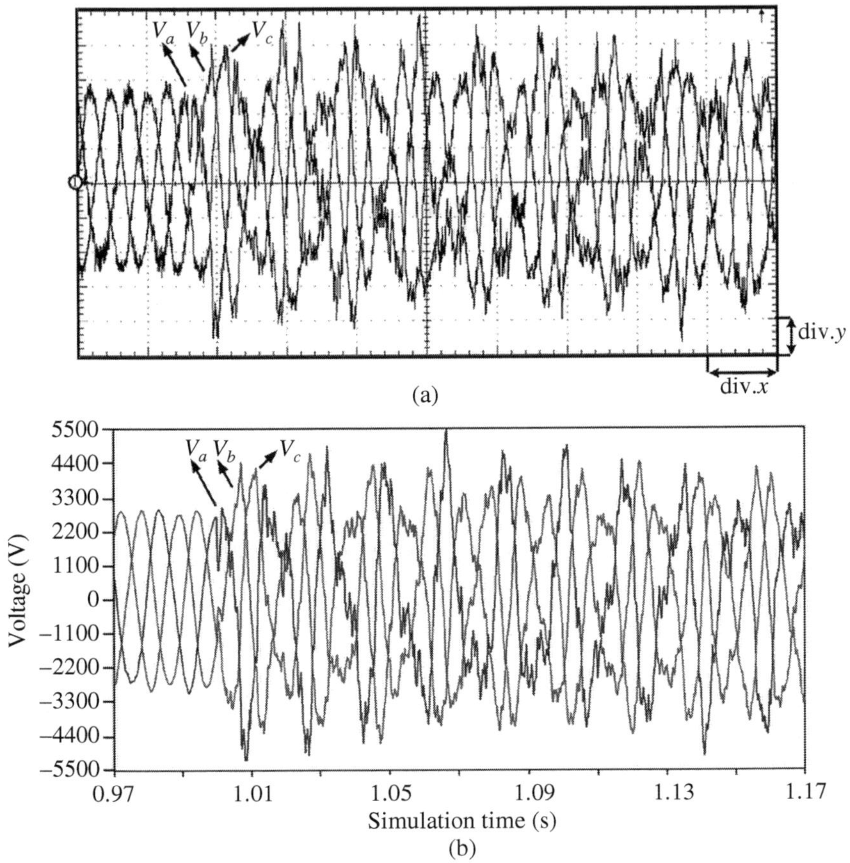

Figure 3.20 Capacitor switching at $t = 1$ second: (a) real-time simulation results (1 div.x = 10 ms, 1 div.y = 1.06 kV) and (b) off-line (ATP) simulation results.

3.3 Nonlinear Magnetic Equivalent Circuit Based Real-time Multi-Winding Transformer Model

This section presents a high-fidelity real-time nonlinear MEC based three-phase transformer EMT model emulated on the FPGA. The three-phase multi-winding transformer in question is called the Sen Transformer (ST) [55, 56]. The ST consists of a 12 winding transformer and tap changer that can regulate the power flow through a transmission line by injecting a series connected controllable voltage. The geometry-based MEC approach [49, 57] is employed to model the multi-winding transformer consisting of magnetomotive force (MMF) sources and permeance elements. The nonlinear permeance elements necessitate a

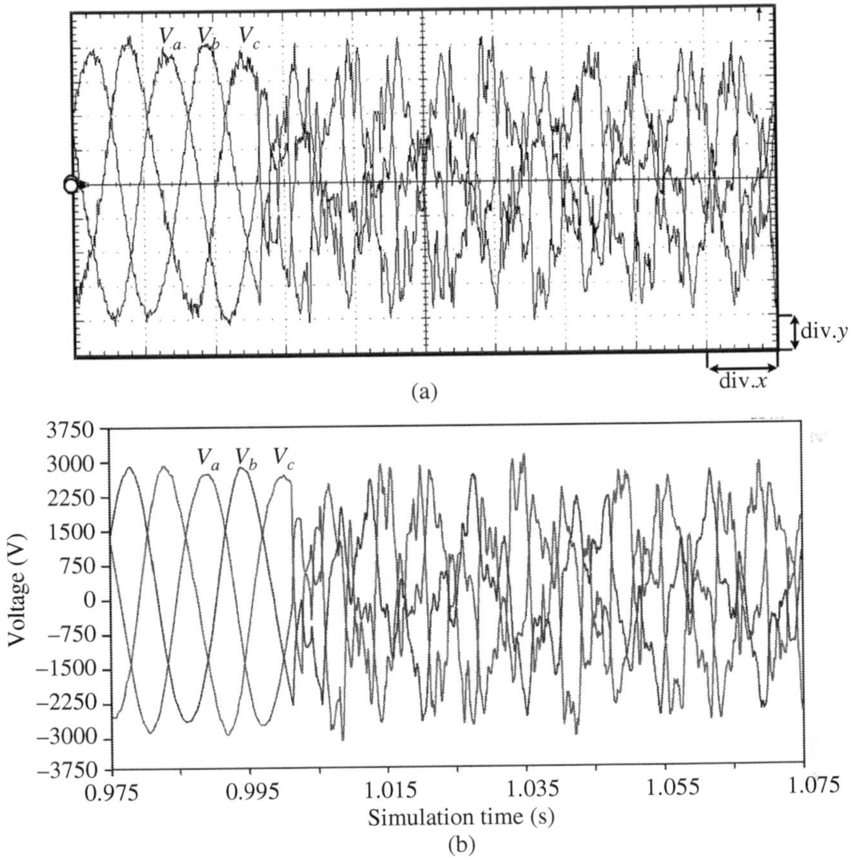

Figure 3.21 Ground fault at $t = 1$ second: (a) real-time simulation results (1 div.x = 20 ms, 1 div.y = 1.5 kV) and (b) off-line (ATP) simulation results.

time-varying transformer nodal admittance matrix which in turn require the entire power system matrix to be updated in every emulation time-step. The hysteresis and eddy current behavior are modeled based on Preisach theory [51] and frequency-dependent equivalent network [52, 53], respectively. All of these details increase the computational burden of the high-fidelity electromagnetic transient model and put upward pressure on the emulation time-step [58]. In this transformer model, there are four nonlinear permeances in each core limb carrying four different currents and the secondary winding is series connected with the transmission line; however, its formulation procedure utilizes first principles and can also be applied to other winding configurations, making the real-time model easy to modify for emulating other types of power transformers in real time.

3.3.1 Topological ST EMT Model

3.3.1.1 ST Operating Principle

The three-phase, three-limb ST circuit schematic and its interconnection to the transmission line is shown in Figure 3.22. At the sending-end, the three-phase transmission line supplies voltages, $V_{ST,pa}$, $V_{ST,pb}$, and $V_{ST,pc}$ to the ST primary windings, which are connected in Y and in shunt with the transmission line. The nine secondary windings: a_1, a_2, and a_3 are placed in the first limb; b_1, b_2, and b_3 are placed in the second limb; and c_1, c_2, and c_3 are placed in the third limb. The three secondary windings which are placed in different limbs are connected in series with the transmission line and inject a compensation voltage, e.g. windings a_1, b_1, and c_1 for the compensation voltage $V_{ST,ca}$ in phase-a; windings a_2, b_2, and c_2 for the compensation voltage $V_{ST,cb}$ in phase-b; and windings a_3, b_3, and c_3 for the compensation voltage $V_{ST,cc}$ in phase-c. By using the on-load tap changers, the active number of turns of the nine secondary windings can be configured so that the induced compensation voltage can be controlled both in magnitude and phase angle independently. If the 120° phase-shift compensation voltage is series connected with the transmission line, as shown in Figure 3.22, the previous sending-end voltage (the ST primary side voltage), $v_{ST,p}$, is modified to a new effective sending-end voltage (the ST secondary side voltage), $v_{ST,s}$, as can be seen from the top left-hand corner of Figure 3.22. The number of turns of the secondary windings can be classified into three groups: a_1, b_2, and c_3 (GP_1); b_1, c_2, and a_3 (GP_2); and c_1, a_2, and b_3 (GP_3). The windings in one group have the same tap setting; however, one group may have a different tap setting from others. The major flux paths consisting of core and leakage fluxes in the yoke and limb are also depicted in Figure 3.22.

3.3.1.2 Tap-selection Algorithm

The *Control Unit* takes the measurements of the sending end voltage, $V_{ST,p}$, and the receiving end voltage, v_r, from the transmission network as well as the required power level, P_{ref} and Q_{ref} as inputs. The calculated compensation voltage magnitude $v_{ST,c}$ and the phase angle α are further forwarded to the *Tap-setting Unit* to generate the combinations of the tap settings for each secondary winding. Further information regarding the tap-setting algorithm and its implementation can be found in [59, 60].

3.3.1.3 High-Fidelity Nonlinear MEC-Based ST Model

The three-phase, three-limb MEC based ST representation is shown in Figure 3.23. Each limb consists of four types of components: (i) MMF sources, $N_A i_{ST,a}$, $N_B i_{ST,b}$, and $N_C i_{ST,c}$, generated by the current flow through the primary windings in phases a, b, and c, respectively; $N_{a1} i_{ST,sa}$, $N_{a2} i_{ST,sb}$, $N_{a3} i_{ST,sc}$, $N_{b1} i_{ST,sa}$, $N_{b2} i_{ST,sb}$, $N_{b3} i_{ST,sc}$,

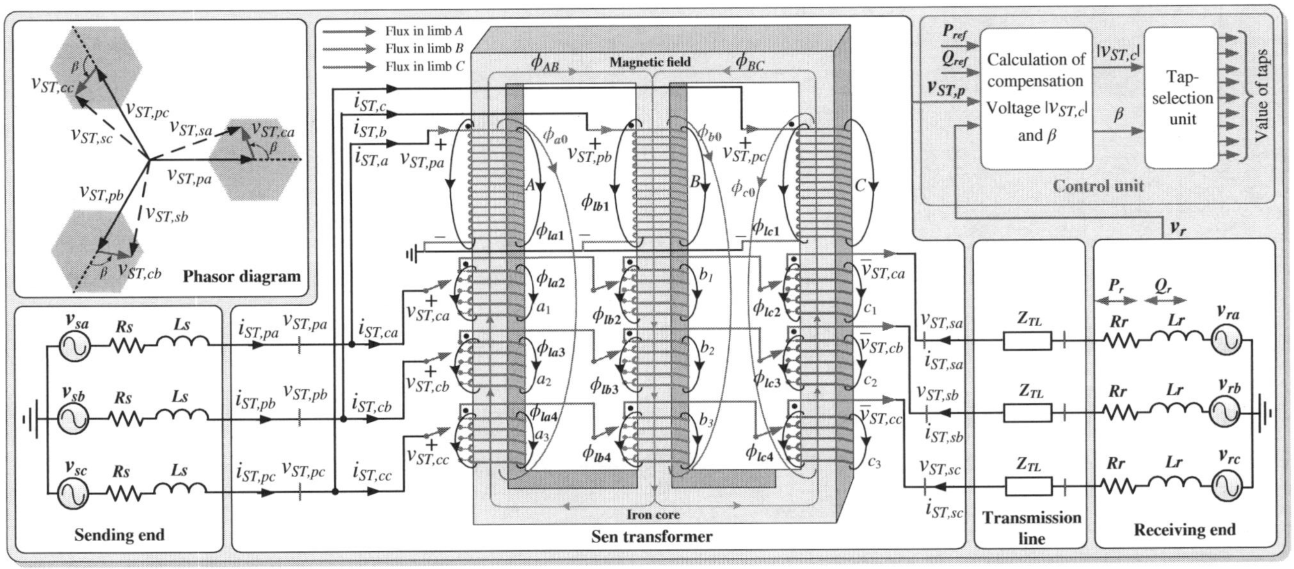

Figure 3.22 Schematic of the ST and its interconnection to the transmission network.

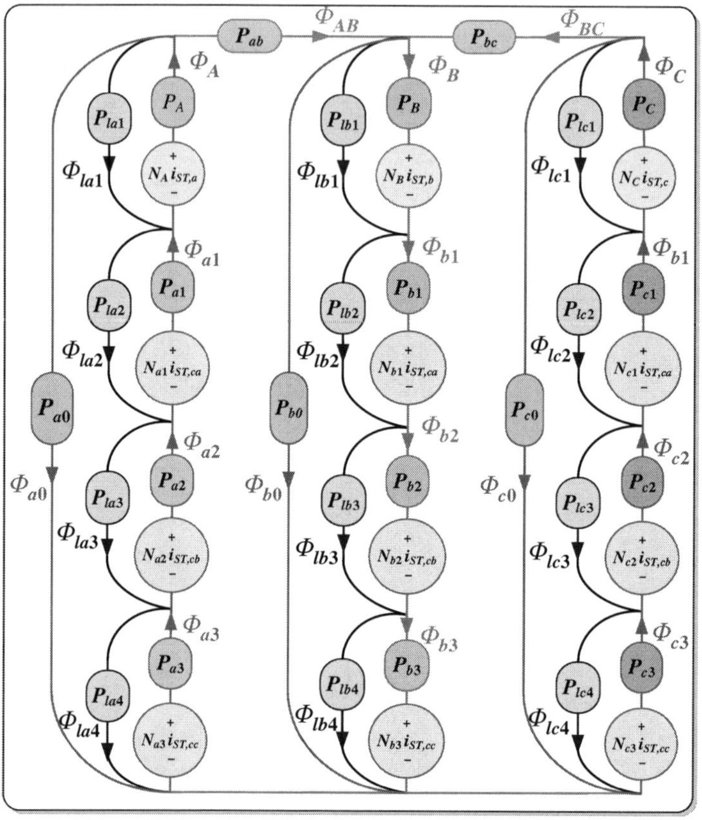

Figure 3.23 Magnetic equivalent circuit representation for the ST iron core.

$N_{c1}i_{ST,sa}$, $N_{c2}i_{ST,sb}$, and $N_{c3}i_{ST,sc}$ are generated by the current flow through the secondary windings in phases a, b, and c, respectively; (ii) nonlinear permeances, P_A, P_B, and P_C represent nonlinear iron core permeances for ST primary windings in phases a, b, and c, respectively; P_{a1}, P_{a2}, P_{a3}, P_{b1}, P_{b2}, P_{b3}, P_{c1}, P_{c2}, and P_{c3} represent nonlinear iron core permeances for the ST secondary windings in phases a, b, and c, respectively; (iii) linear permeances, P_{la1} to P_{la4}, P_{lb1} to P_{lb4} and P_{lc1} to P_{lc4} represent the ST leakage air permeances in phases a, b, and c, respectively; (iv) linear permeances P_{a0}, P_{b0}, P_{c0} represent ST zero-sequence permeances in phases a, b, and c, respectively. In addition, there are also two nonlinear iron permeances P_{ab} and P_{bc} that represent ST yoke between phase-a and phase-b, and phase-b and phase-c, respectively. Based on Ampere's and Gauss' laws [61], the relationship between the branch fluxes and the winding currents can be written as:

$$\boldsymbol{\phi} = \mathbb{P}\mathbf{P}\mathbf{N}\boldsymbol{i}, \tag{3.35}$$

where

$$\mathbb{P} = \mathbf{I} - \mathbf{PA}(\mathbf{A}^T\mathbf{PA})^{-1}\mathbf{A}^T. \tag{3.36}$$

$\boldsymbol{\phi}$ and $\boldsymbol{i}$ are $n \times 1$ vectors of branch fluxes and winding currents, $\mathbf{P}$ and $\mathbf{N}$ are $n \times n$ diagonal matrices of branch permeances and number of winding turns, $\mathbf{I}$ is the $n \times n$ identity matrix, and $\boldsymbol{A}^T$ is the node-branch connection matrix, respectively. By partitioning the permeance network into two branch sets, $\boldsymbol{\phi}^M$, for the branches with MMF sources and permeance elements, and $\boldsymbol{\phi}^P$, for branches with only permeance elements, (3.35) can be rewritten as:

$$\boldsymbol{\phi}^M = \mathbb{P}^{MM}\mathbf{P}^M\mathbf{N}^M\boldsymbol{i}^M, \tag{3.37}$$

where $\mathbb{P}^{MM}$ is a 12×12 sub-matrix of $\mathbb{P}$. $\boldsymbol{\phi}^M$ and $\boldsymbol{i}^M$ are 12×1 vectors, and $\mathbf{P}^M$ and $\mathbf{N}^M$ are diagonal 12×12 matrices, given as:

$$\begin{aligned}
\boldsymbol{\phi}^M &= \left[\phi_A(t)\ \phi_B(t)\ \phi_C(t)\ \phi_{a1}(t)\ \phi_{b1}(t)\ \phi_{c1}(t)\right.\\
&\quad \left.\phi_{a2}(t)\ \phi_{b2}(t)\ \phi_{c2}(t)\ \phi_{a3}(t)\ \phi_{b3}(t)\ \phi_{c3}(t)\right]^T,\\
\boldsymbol{i}^M &= \left[i_{ST,a}(t)\ i_{ST,b}(t)\ i_{ST,c}(t))\ i_{ST,sa}(t)\ i_{ST,sa}(t)\ i_{ST,sa}(t)\right.\\
&\quad \left.i_{ST,sb}(t)\ i_{ST,sb}(t)\ i_{ST,sb}(t)\ i_{ST,sc}(t)\ i_{ST,sc}(t)\ i_{ST,sc}(t)\right]^T,\\
\mathbf{P}^M &= diag\left\{P_A(t)\ P_B(t)\ P_C(t)\ P_{a1}(t)\ P_{b1}(t)\ P_{c1}(t)\right.\\
&\quad \left.P_{a2}(t)\ P_{b2}(t)\ P_{c2}(t)\ P_{a3}(t)\ P_{b3}(t)\ P_{c3}(t)\right\},\\
\mathbf{N}^M &= diag\left\{N_A(t)\ N_B(t)\ N_C(t)\ N_{a1}(t)\ N_{b1}(t)\ N_{c1}(t)\right.\\
&\quad \left.N_{a2}(t)\ N_{b2}(t)\ N_{c2}(t)\ N_{a3}(t)\ N_{b3}(t)\ N_{c3}(t)\right\}.
\end{aligned}$$

From Faraday's law, the winding terminal voltages can be written in matrix format as,

$$\boldsymbol{v}_{ST} = \mathbf{N}_Z\frac{d}{dt}\boldsymbol{\phi}^M(t), \tag{3.38}$$

where $\mathbf{N}_Z$ is a 6×12 winding turns matrix. $\boldsymbol{v}_{ST}$ is a 6×1 ST terminal voltage vector, given as:

$$\boldsymbol{v}_{ST} = \left[v_{ST,pa}(t)\ v_{ST,pb}(t)\ v_{ST,pc}(t)\ v_{ST,sa}(t)\ v_{ST,sb}(t)\ v_{ST,sc}(t)\right]^T,$$

and

$$\mathbf{N}_Z = \begin{bmatrix}
N_A & 0 & 0 & 0 & 0 & 0 & 0 & 0 & 0 & 0 & 0 & 0\\
0 & N_B & 0 & 0 & 0 & 0 & 0 & 0 & 0 & 0 & 0 & 0\\
0 & 0 & N_C & 0 & 0 & 0 & 0 & 0 & 0 & 0 & 0 & 0\\
N_A & 0 & 0 & N_{a1} & N_{b1} & N_{c1} & 0 & 0 & 0 & 0 & 0 & 0\\
0 & N_B & 0 & 0 & 0 & 0 & N_{a2} & N_{b2} & N_{c2} & 0 & 0 & 0\\
0 & 0 & N_C & 0 & 0 & 0 & 0 & 0 & 0 & N_{a3} & N_{b3} & N_{c3}
\end{bmatrix}.$$

The discrete-time difference equations of (3.38) can be obtained by applying Trapezoidal rule,

$$\mathbf{N}_Z\boldsymbol{\phi}^M(t) = \frac{\Delta t}{2}\boldsymbol{v}_{ST}(t) + \boldsymbol{\phi}^M_{hist}(t), \tag{3.39}$$

where Δt is the simulation time-step, with $\boldsymbol{\phi}^M_{hist}(t)$ being the 6×1 history vector term expressed as:

$$\boldsymbol{\phi}^M_{hist}(t) = \frac{\Delta t}{2}\boldsymbol{v}_{ST}(t-\Delta t) + \mathbf{N}_Z\boldsymbol{\phi}^M(t-\Delta t). \tag{3.40}$$

Substituting (3.37), the left-hand side of (3.39) can be rewritten as,

$$\mathbf{N}_Z\boldsymbol{\phi}^M(t) = \mathbf{Z}\,\boldsymbol{i}^M, \tag{3.41}$$

where $\mathbf{Z}$ is a 6×12 matrix given as:

$$\mathbf{Z} = \mathbf{N}_Z\mathbb{P}^{MM}\mathbf{P}^M\mathbf{N}^M. \tag{3.42}$$

Further (3.41) can be rewritten as:

$$\mathbf{N}_Z\boldsymbol{\phi}^M(t) = \mathbf{Z}\boldsymbol{i}^M = \mathbf{Z}'\boldsymbol{i}^{M'}, \tag{3.43}$$

with $\mathbf{Z}'$ being the 6×6 impedance matrix and $\boldsymbol{i}^{M'}(t)$ being the 6×1 current vector, given as:

$$\mathbf{Z}' = \begin{bmatrix} Z_{1,1} & \cdots & Z_{1,3} & Z_{1,4}+Z_{1,5}+Z_{1,6} & \cdots & Z_{1,10}+Z_{1,11}+Z_{1,12} \\ Z_{2,1} & \cdots & Z_{2,3} & Z_{2,4}+Z_{2,5}+Z_{2,6} & \cdots & Z_{2,10}+Z_{2,11}+Z_{2,12} \\ \vdots & \vdots & \vdots & \vdots & \vdots & \vdots \\ Z_{6,1} & \cdots & Z_{6,3} & Z_{6,4}+Z_{6,5}+Z_{6,6} & \cdots & Z_{6,10}+Z_{6,11}+Z_{6,12} \end{bmatrix},$$

and

$$\boldsymbol{i}^{M'} = \left[i_{ST,a}(t)\ i_{ST,b}(t)\ i_{ST,c}(t)\ i_{ST,sa}(t)\ i_{ST,sb}(t)\ i_{ST,sc}(t)\right]^T.$$

By noting that the ST current vector $\boldsymbol{i}_{ST}$ is expressed as,

$$\boldsymbol{i}_{ST} = \left[i_{ST,pa}(t)\ i_{ST,pb}(t)\ i_{ST,pc}(t)\ i_{ST,sa}(t)\ i_{ST,sb}(t)\ i_{ST,sc}(t)\right]^T,$$

and

$$\begin{aligned} i_{ST,pa}(t) &= i_{ST,a}(t) - i_{ST,sa}(t), \\ i_{ST,pb}(t) &= i_{ST,b}(t) - i_{ST,sb}(t), \\ i_{ST,pc}(t) &= i_{ST,c}(t) - i_{ST,sc}(t). \end{aligned}$$

Equation (3.39) can be further rewritten as

$$\mathbf{Z}_{ST}\boldsymbol{i}_T(t) = \frac{\Delta t}{2}\boldsymbol{v}_{ST}(t) + \boldsymbol{\phi}^M_{hist}(t), \tag{3.44}$$

where $\mathbf{Z}_{ST}$ is the 6×6 matrix given as,

$$\mathbf{Z}_{ST} = \begin{bmatrix} Z'_{1,1} & Z'_{1,2} & Z'_{1,3} & Z'_{1,4} + Z'_{1,1} & Z'_{1,5} + Z'_{1,2} & Z'_{1,6} + Z'_{1,3} \\ Z'_{2,1} & Z'_{2,2} & Z'_{2,3} & Z'_{2,4} + Z'_{2,1} & Z'_{2,5} + Z'_{2,2} & Z'_{2,6} + Z'_{2,3} \\ \vdots & \vdots & \vdots & \vdots & \vdots & \vdots \\ Z'_{6,1} & Z'_{6,2} & Z'_{6,3} & Z'_{6,4} + Z'_{6,1} & Z'_{6,5} + Z'_{6,2} & Z'_{6,6} + Z'_{6,3} \end{bmatrix}.$$

Then, the Norton equivalent formula for MEC-based ST is given,

$$\boldsymbol{i}_{ST}(t) = \mathbf{Y}_{ST}(t)\boldsymbol{v}_{ST}(t) + \boldsymbol{I}_{hist,ST}(t), \tag{3.45}$$

where

$$\boldsymbol{I}_{hist,ST}(t) = \mathbf{Z}_{ST}^{-1}\boldsymbol{\phi}_{hist}^{M}(t), \tag{3.46}$$

and

$$\mathbf{Y}_{ST}(t) = \frac{\Delta t}{2}\mathbf{Z}_{ST}^{-1}. \tag{3.47}$$

The time-varying matrix $\mathbf{Y}_{ST}(t)$, due to the nonlinear permeance matrices $\mathbb{P}^{MM}$ and $\mathbf{P}^{M}$ in (3.42), causes the overall power system admittance matrix $\mathbf{Y}$ to be time-varying as well. Therefore, a LU decomposition needs to be performed to invert the matrix $\mathbf{Y}$ in every simulation time-step and then to find the final network solution by using the nodal analysis method.

3.3.1.4 Iron Core Hysteresis and Eddy Currents

The permeance matrix $\mathbf{P}$ is divided into two sets $\mathbf{P}^{M}$ and $\mathbf{P}^{P}$ as described in Section II (C). The entries in the matrix $\mathbf{P}^{P}$ representing the leakage permeance are constant; however, the entries in the matrix $\mathbf{P}^{M}$ representing the nonlinear iron permeance change in every simulation time-step. The Preisach model is used to represent the nonlinearity as well as hysteresis behavior. The major loop function is represented as (3.48), once the major loop function is fixed, all the minor loop trajectories from a reversal point follow this uniform template,

$$\boldsymbol{B}_{p\_m}(\boldsymbol{H}_m) = \frac{1}{2}\sum_{u=1}^{3} a_u[\tanh(b_u\boldsymbol{H}_m) + c_u\text{sech}^2(b_u\boldsymbol{H}_m)], \tag{3.48}$$

where a_u, b_u, and c_u are the parameters to define this hyperbolic template, $\boldsymbol{H}_m = [H_{m1}\ H_{m2}\ \cdots\ H_{mn_l}]^T$ is the magnetization current vector flowing through the nonlinear elements and $\boldsymbol{B}_{p\_m} = [B_{p\_m1}\ B_{p\_m2}\ \cdots\ B_{p\_mn_l}]^T$ are the flux corresponding to this current vector on the major loop. The subscript "p_m" stands for Preisach major loop and n_l stands for the number of nonlinear permeances.

Once the major loop function is defined, all the upward and downward minor loop trajectories from a reversal point (H_{rl}, B_{rl}) follow this uniform template can be expressed as (3.49) and (3.50), respectively, with $\boldsymbol{H}_r = [H_{r1}\ H_{r2}\ \dots\ H_{rn_l}]^T$ and $\mathbf{B}_r = [B_{r1}\ B_{r2}\ \dots\ B_{rn_l}]^T$:

$$\boldsymbol{B}_{p\_u}(\boldsymbol{H}_m) = -\boldsymbol{B}_{p\_m}(-\boldsymbol{H}_m) - \boldsymbol{B}_{p\_m}(\boldsymbol{H}_r) + \boldsymbol{B}_r + 2\boldsymbol{K}(-\boldsymbol{H}_r)\boldsymbol{K}(\boldsymbol{H}_m), \tag{3.49}$$

$$\boldsymbol{B}_{p\_d}(\boldsymbol{H}_m) = \boldsymbol{B}_{p\_m}(\boldsymbol{H}_m) + \boldsymbol{B}_{p\_m}(-\boldsymbol{H}_r) + \boldsymbol{B}_r - 2\boldsymbol{K}(\boldsymbol{H}_r)\boldsymbol{K}(-\boldsymbol{H}_m), \tag{3.50}$$

where is $\boldsymbol{K}$ defined as 3.29.

A PHU shown in Figure 3.26 is designed to implement this model. As shown in Figure 3.24, a minor loop trajectory is enclosed in a major loop trajectory and a *RPS* is used to traverse the hysteresis loop. Once the nonlinear *B*–*H* curve is obtained, the nonlinear permeability μ_{nl} can be calculated as,

$$\mu_{nl} = \frac{B_{nl}}{H_{nl}}, \tag{3.51}$$

and the nonlinear permeance p_{nl} can be calculated as

$$p_{nl} = \frac{\mu_{nl}A}{0.5l}, \tag{3.52}$$

where A is cross-sectional area and l is the length of the limb.

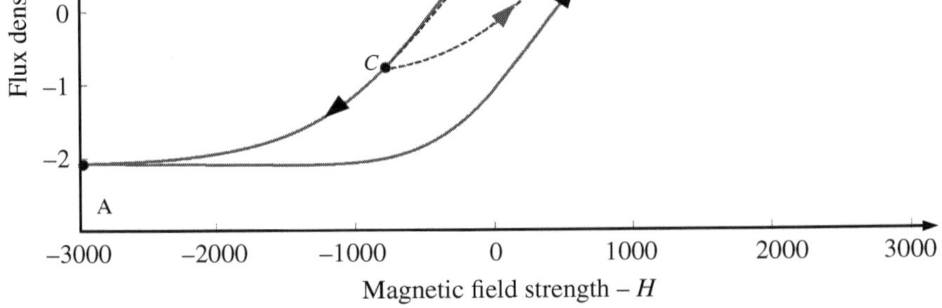

Figure 3.24 Hysteresis major and minor loops corresponding to the *RPS* within the Preisach hysteretic unit (PHU) for the ST.

To calculate the linear air permeances, the value of leakage impedance $X_{leakage}$ is necessary, which can be obtained by the short-circuit test or provided by the transformer manufacturer. By assuming the leakage inductance is equally divided among the primary and secondary windings, the leakage permeance can be obtained as

$$P_{leakage} = \frac{0.5X_{leakage}}{2\pi f N^2}, \tag{3.53}$$

where N is the number of winding turns.

Eddy currents are frequency-dependent current flows in the ST core dissipated as active power loss. In the nonlinear MEC-based ST model, the eddy current is also represented by a four section cascaded R–L lumped circuit (continued fraction model). The discretized Norton equivalent can be expressed as (3.32). To integrate the cascaded equivalent circuit into the MEC based ST model, the required parameters of the lumped R–L circuit can be calculated by (3.31). Then, the overall detailed frequency-dependent nonlinear ST model is as shown in Figure 3.25.

3.3.2 High-Fidelity Nonlinear MEC-Based ST Hardware Emulation

3.3.2.1 Network Transient Emulation with Embedded ST

The hardware emulation [62] of the complete ST model is depicted in Figure 3.26, which fully exploits pipelining and parallelism to achieve real-time emulation. Each emulation time-step can be finished in five stages, and while the stages are executed sequentially, all the modules or units within any stage are executed in parallel. All the calculations inside the hardware modules and units are fully pipelined and paralleled to achieve the lowest latency and resource consumption.

In Stage 1, the PHU is designed to evaluate the ST nonlinear iron core permeance matrix $\mathbf{P}^M$ by reading the ST core fluxes $\boldsymbol{\phi}^M$. Combining $\mathbf{P}^M$ with the linear permeance matrix $\mathbf{P}^P$ gives the overall permeance matrix $\mathbf{P}$, which is then involved in the computation of the matrix $\mathbb{P}$ using (3.36). The eddy current module, as well

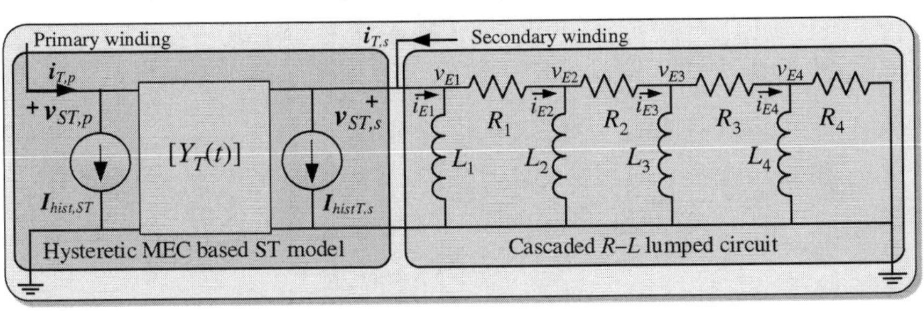

Figure 3.25 High-fidelity MEC-based ST model interfaced with cascaded equivalent circuit.

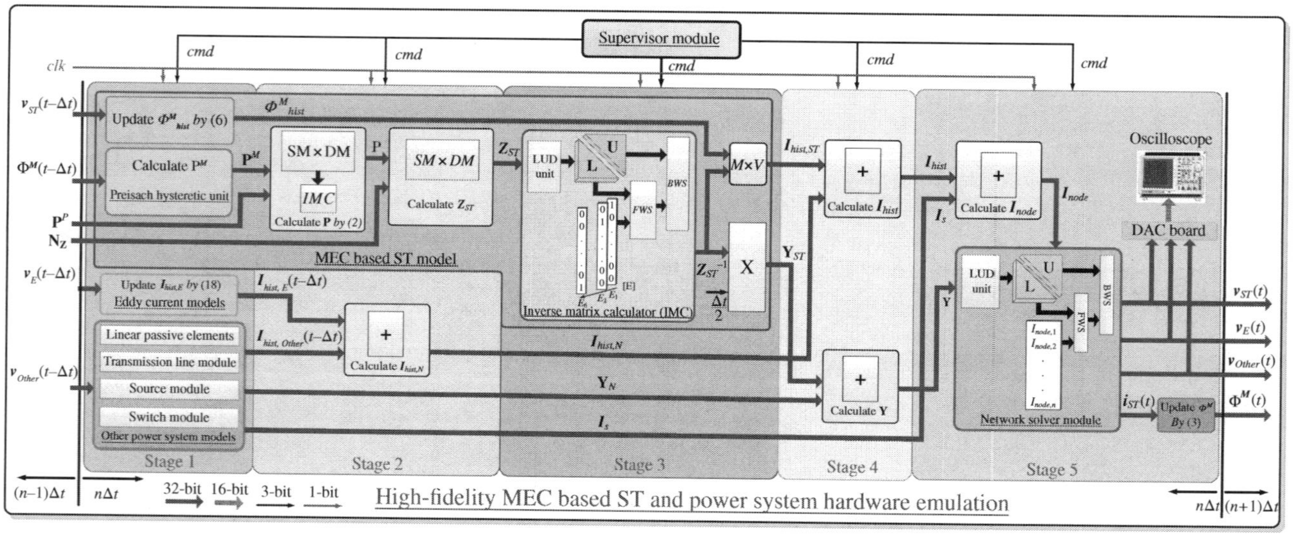

Figure 3.26 Pipelined and paralleled hardware design architecture during one time-step of emulation for the ST.

as the transmission line and passive element modules in parallel update their history current terms by reading the node voltage vectors, v_E and v_O, respectively. The source module is designed to output the source voltage by reading a sinusoidal function LUT implemented in it. The switch module sends the network admittance matrix $\mathbf{Y}_N$ to the network solver module, combined with the ST admittance matrix $\mathbf{Y}_{ST}$ to carry out the finally node voltage calculation. To efficiently perform the sparse and dense matrix–matrix and matrix–vector multiplications, such as in (3.36) and (3.46), a sparse matrix–dense matrix multiplication (*SMxDM*) unit and matrix–vector multiplication (*MxV*) unit were designed, respectively, in this work. In order to perform the matrix inversion operation in (3.36), (3.46), and (3.47), an LU based inverse matrix calculator (*IMC*) was implemented in the ST module, as described in Section III (B). The ST admittance matrix $\mathbf{Y}_{ST}$ is combined with the network admittance matrix $\mathbf{Y}_N$ to obtain the overall matrix $\mathbf{Y}$. Similarly, the ST history current vector $\boldsymbol{I}_{hist,ST}$ is combined with the network history current terms $\boldsymbol{I}_{hist,N}$ and the source current injection vector $\boldsymbol{I}_s$ resulting in the $\boldsymbol{I}_{node}$ vector. Then, the overall admittance matrix $\mathbf{Y}$ and the injection current vector $\boldsymbol{I}_{node}$ are forwarded to the network solver module to finally find the network node voltages, consisting of v_{ST}, v_E, and v_O. The LU-based decomposition (*LUD*), forward substitution (*FWS*) unit, and backward substitution (*BWS*) unit are implemented in the network solver module, and the vector $\boldsymbol{I}_{node}$ is pushed into the *FWS* unit input ports for the nodal solution.

Figure 3.27 shows the detailed FSM for the MEC-based ST and host power system hardware design. In the state S_1, the calculations in Stage 1 are executed including the transformer nonlinear iron core permeances, the eddy current module, and the passive element and transmission line modules update their history terms $\boldsymbol{I}_{hist,E}$ using (3.33) and $\boldsymbol{I}_{hist,O}$ in parallel. The Stage 2 is split into two states, S_2 and S_3. The state S_2 performs two functions in parallel: (i) calculates matrix $\mathbb{P}$ using (3.36) and (ii) calculates network history terms $\boldsymbol{I}_{hist,N}$ by combining $\boldsymbol{I}_{hist,E}$ and $\boldsymbol{I}_{hist,O}$. State S_3 calculates the ST admittance matrix $\mathbf{Z}_{ST}$ using (3.44). Stage 3 is also split into two states: S_4 and S_5. In S_4, the matrix $\mathbf{Z}_{ST}$ goes through the *IMC* unit to compute the inverse matrix $\mathbf{Z}_{ST}^{-1}$. In state S_5, the ST history term vector $\boldsymbol{I}_{hist,ST}$ and admittance matrix $\mathbf{Y}_{ST}$ are computed in parallel using (3.46) and (3.47), respectively. The Stage 4 is fully implemented in S_6, to parallel compute the summation of the all the history currents $\boldsymbol{I}_{hist}$ and construct the entire power system network admittance matrix $\mathbf{Y}$. The Stage 5 is divided into two states, S_7 and S_8. In state S_7, the current injection vector $\boldsymbol{I}_{node}$ is calculated, and the network solver module is executed to compute the network node voltages v. Finally, state S_8 updates the transformer winding and yoke fluxes using (3.37).

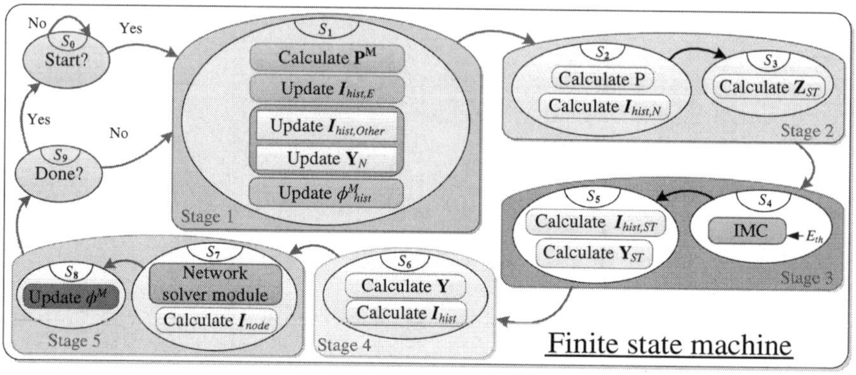

Figure 3.27 FSM of the high-fidelity nonlinear MEC-based ST model and power system.

3.3.3 Real-Time Emulation Case Studies

3.3.3.1 Finite Element Modeling and Validation

To validate hardware emulation of the real-time MEC based ST model, a 3-D three-phase, three-limb, 12 winding FE transformer model was developed in JMAG®. A half transformer core and coil model was built in JMAG Designer®, which was then converted to a full transformer model by setting boundary symmetry. The symmetry boundary target is set in the middle of the model, and its external coupled electrical circuit is shown in Figure 3.28a,b. In the coupled electrical circuit, each winding is assigned to a coil in the geometry model, and the current inflow face of each coil is as shown in Figure 3.28b. This 3D-FE ST model was connected with a voltage source on the primary side and an open-circuit on the secondary side. To build the half transformer model, 47 730 elements and 8647 nodes were used with the element size of 0.15 m. This 3D-FE ST model was connected with a voltage source on the primary side and a resistive load in the secondary side. By energizing the ST at $t = 0.05$ second, an inrush current flows through the ST, and the magnetic flux density vector and contour are plotted as shown in Figure 3.29 when the inrush current in phase-c reaches its peak at $t = 0.056$ second. The fluxes in phase-a and phase-b flow into phase-c which make the primary side of phase-c being the most saturated limb with maximum flux density close to 2.6 T. To run a $t = 0.5$ second FE simulation with the time-step of $\Delta t = 500\,\mu s$ took more than five hours on a PC featured by Intel® i7 and 8 GB memory. The geometric parameters and the rating of the transformer can be found in Appendix A.2.

3.3.3.2 Case Studies

The following case studies are analyzed in this paper to show the accuracy and efficiency of the real-time hardware high-fidelity nonlinear MEC-based ST model.

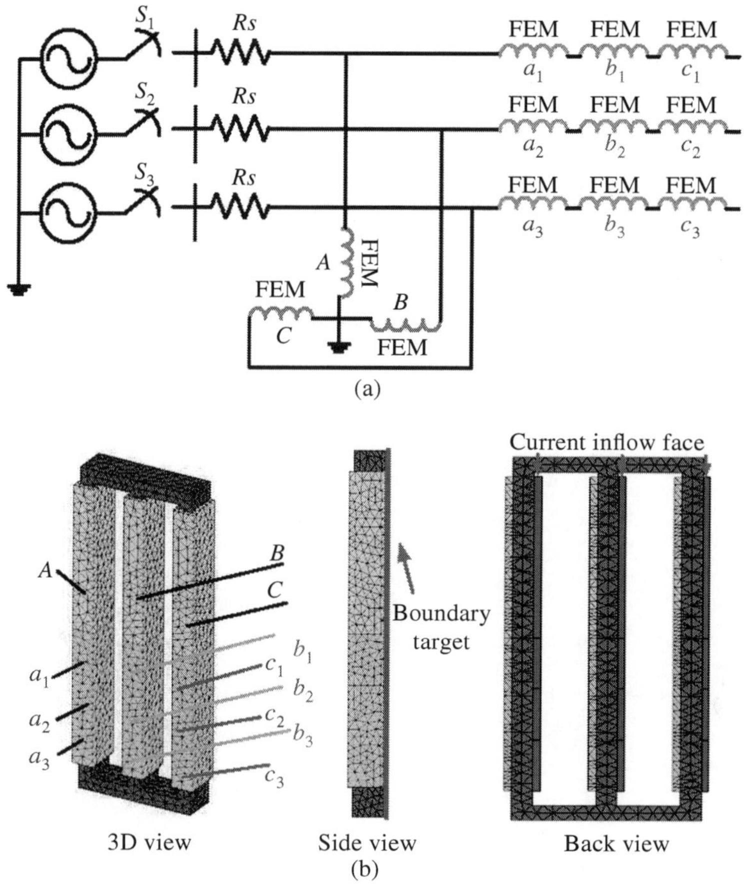

Figure 3.28 FE simulation: (a) coupled external electrical circuit and (b) 3-D FE model.

The power system shown in Figure 3.30 consists of one ST model, two voltage sources, G_{s1} and G_{s2}, and two transmission lines, L_1 and L_2, modeled by the distributed parameter line model. In the case studies, the power system network including the ST was emulated on the Xilinx® Virtex-7® VC707 XC7VX485T FPGA, and the detailed latency for each FSM in one emulation time-step is summarized in Figure 3.31. The maximum clock frequency of this design can reach to 90 MHz, with 3575 clocks latency for one time-step emulation giving the execution time of 40 μs. The hardware resource utilization is summarized in Table 3.1.

Energization Transient and Steady-State Emulation To validate against the FE simulation results, the same power system is configured in Figure 3.30 by the switching of SW_3 to position 2 and leaving other breakers open. The parameters of the power

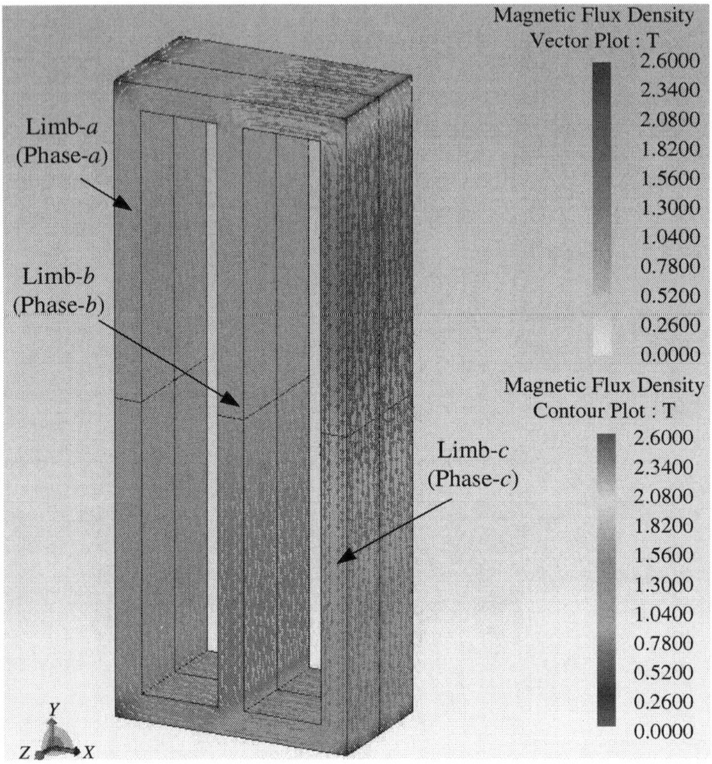

Figure 3.29 3-D FE simulation of the ST.

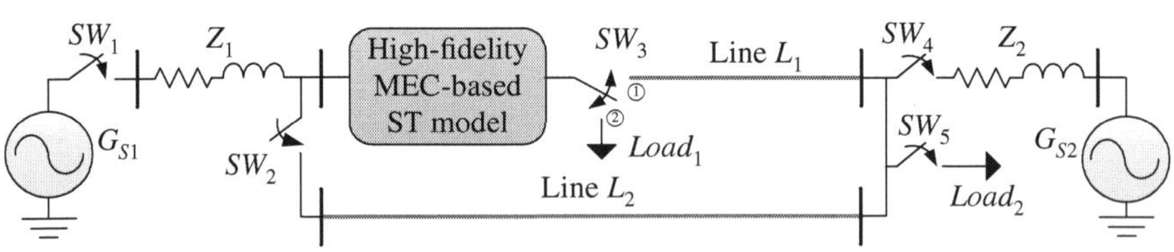

Figure 3.30 Single-line diagram of the power system with the ST model for the real-time HIL case study.

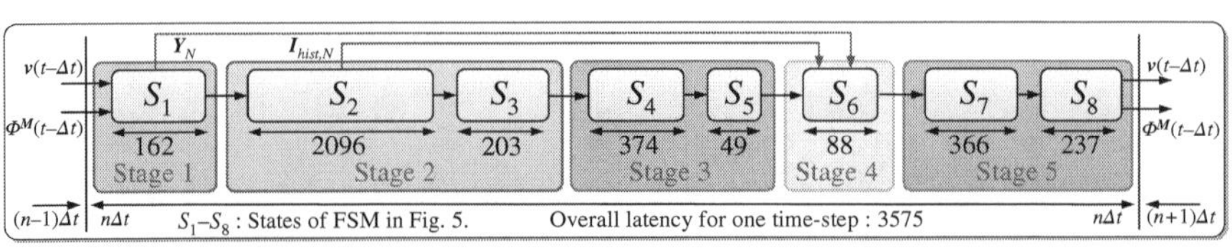

Figure 3.31 Latencies for one emulation time-step for the high-fidelity MEC-based ST model.

Table 3.1 Module level hardware resource utilization.

Resource	MEC-based ST model	Other modules
Slice Registers	113 746 (37.4%)	50 562 (16.7%)
LUTRAM	3 180 (24%)	9 636 (7.35%)
BRAM/FIFO	90 (2.9%)	59 (9.6%)
DSP48E1	746 (26.6%)	221 (7.6%)
BUFG	11 (34.4%)	4 (12.4%)

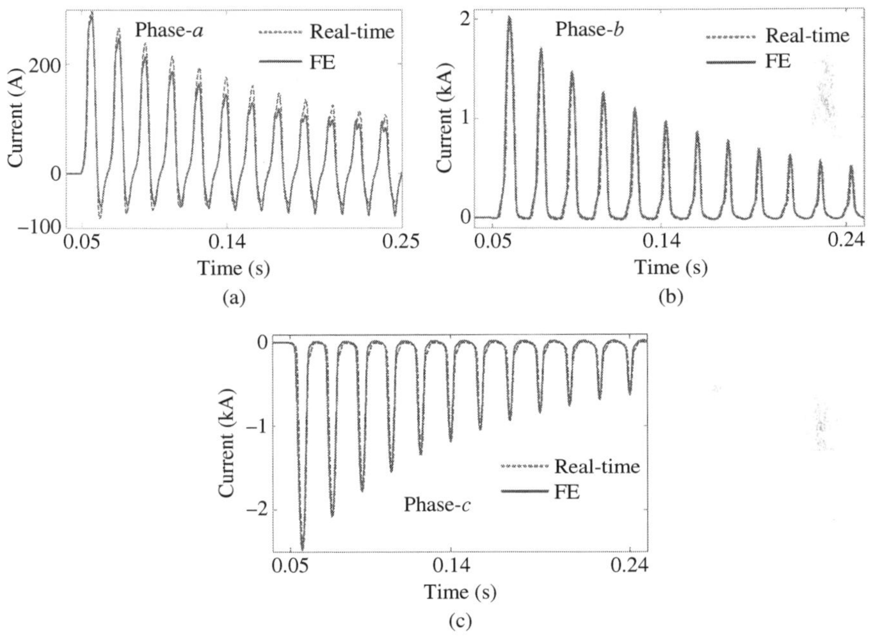

Figure 3.32 Real-time emulation and finite element simulation results for the energization transient.

system are given in Appendix A.2. The ST energization transient is initialized by closing the three-phase breaker SW_1 at $t = 0.05$ second, and the inrush transient current flows through the breaker. The real-time inrush current emulation results and the corresponding 3D-FE simulation results are plotted in Figure 3.32 while the steady-state current is shown in Figure 3.33. The magnitude of the inrush current in different phases is determined by the instantaneous magnitude of the applied voltages in different phases when the breaker SW_1 is closed. As can be seen in Figure 3.32, the magnitude of transient current in phase-*b* and phase-*c* whose peak value can reach up to 150% of the ST rated current, is much larger than that in

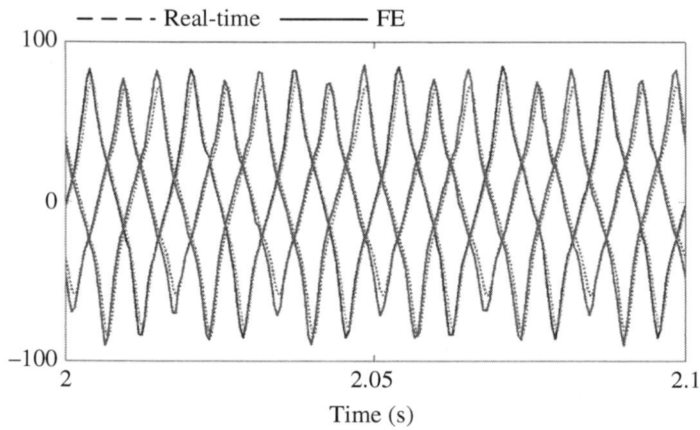

Figure 3.33 Real-time emulation and finite element simulation results for the ST steady-state current.

phase-a due to the above reason, and all the currents in each phase decay to steady state after $t = 0.3$ second. The real-time emulation results show close agreement with the 3D-FE simulation results both in transient and steady state. However, discrepancies exist in the magnitudes of the waveforms. The main reason for these differences is that even though the real-time ST model was developed based on high-fidelity MEC approach, the mesh number and node are still much smaller than those in the FEM model. The Trapezoidal discretization can also introduce numerical differences between these two simulation results. Furthermore, in the FEM model, the nonlinear effect is modeled by a piece-wise nonlinear curve; however, in the real-time MEC model, the nonlinear effect in the ST core is represented by Preisach approach. In addition, even though the transformer core nonlinearity and frequency-dependent phenomena are included in the model, it cannot predict the high-frequency effects in the windings such as skin effect or proximity effect.

The real-time primary winding-limb flux-linkage simulation results are again captured and plotted on top of the FEM simulation results both under transient and steady-state in Figure 3.34a–c,dd, respectively. As expected, the primary winding-limb three-phase flux waveforms are in phase with the three-phase breaker current waveforms, and at $t = 0.056$ second, the flux in limb-c (phase-c) reach its negative peak value and the flux in limb-a (phase-a) has the lowest peak value in the transient state. Again, close agreements can be observed between the real time and FEM results.

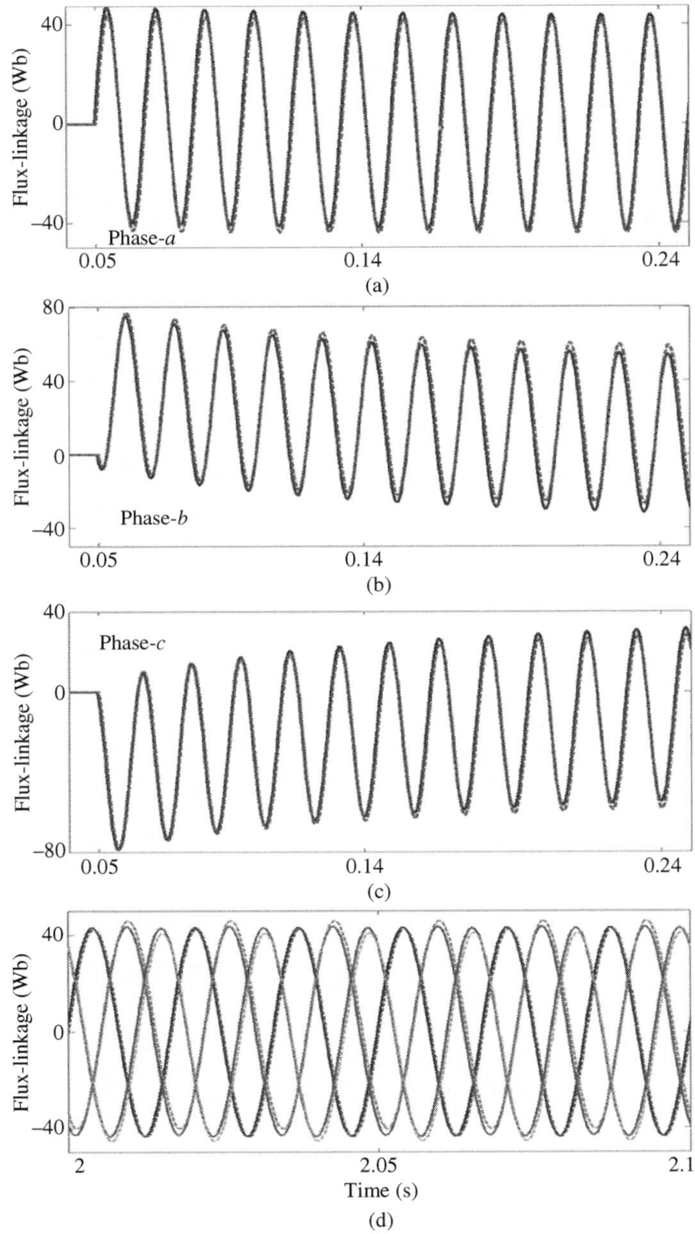

Figure 3.34 Real-time (dashed line) and finite element (solid line) simulation results: (a)–(c) flux-linkage under transient steady and (d) steady-state flux-linkage.

Hysteresis Transient Emulation The hysteresis transient is emulated initializing the transformer core flux on the downward major loop trajectory (point A). Figure 3.35 shows the real-time emulation results of the flux and the magnetizing current in the core and the hysteresis characteristic of the transformer. The parameters to define the hysteresis major loop trajectory are listed in Appendix A.2. The ST is energized by a voltage source at a value of 0.8 p.u. starting from $t = 0$ s, and the corresponding hysteresis trajectory during the first cycle (A-B) is asymmetric due to the initial flux. After the first cycle, the flux travels on the set of asymmetric minor loops (C-D) due the asymmetric magnetizing current through the transformer core. From $t = 0.087$ to $t = 0.168$ seconds, the breaker SW_1 is opened, and the flux is locked at 0.35 p.u. After $t = 0.168$ second, the breaker SW_1 is closed again and the value of voltage source increases to 1.2 p.u. The corresponding hysteresis trajectory (E-F) travels on the major loop since the voltage source is high enough to drive the transformer into the saturation region when the peak flux value reaches 1.2 p.u.

Power-Flow Regulation To investigate the power flow control capability of the ST, the case study power system is configured by closing breaker SW_1 and SW_4 and switching breaker SW_3 to position 1 to in order to obtain a constant voltage at the receiving end of line L_1. The real-time ST model works in compensation mode to make the voltage at the sending end of line L_1 variable, so that the active power and reactive power flow through this line is controllable. The power-flow variation of the active power P_r and reactive power Q_r for $V_{ST,c}$ of 0.1, 0.2, 0.3, and 0.4 p.u. at angles 0°, 60°, 120°, 180°, 240°, and 160° is given in Figure 3.36. As can be seen from this figure, the ST produce a slightly counter-clockwise hexagon profile, and for the angle of 60°, 240°, the reactive power Q_r changes much slower than the active P_r as the magnitude of injection voltage increasing.

Power-Flow Control Emulation To investigate the ST transition response, during the step-by-step regulation active and reactive powers in transmission lines, the circuit shown in Figure 3.30 was configured into a two transmission line system by opening breaker SW_4 and switch SW_3 to position 1 and leaving other breakers closed. The real-time emulation results of power regulation of the two lines, and the secondary winding current and injected compensation voltages of ST are shown in Figure 3.37. At the beginning of the emulation, as shown in these figures, the ST works in the uncompensated mode; the line L_1 active power P_{r1} and reactive power Q_{r1} is equal to 80 MW and −20 MVAr, respectively, and the line L_2 active power P_{r2} and reactive power Q_{r2} is equal to 38 MW and 20 MVAr, respectively, and the injected voltage is equal to zero. At $t_1 = 0.5$ second a compensation voltage of 0.25 p.u. at angle of 0° is requested by the controller which means that the tap-setting on the secondary winding of GP_1 should start to increase to 0.25 p.u. with step size of 0.05 p.u., which require 0.5 second for each step transition. Thus,

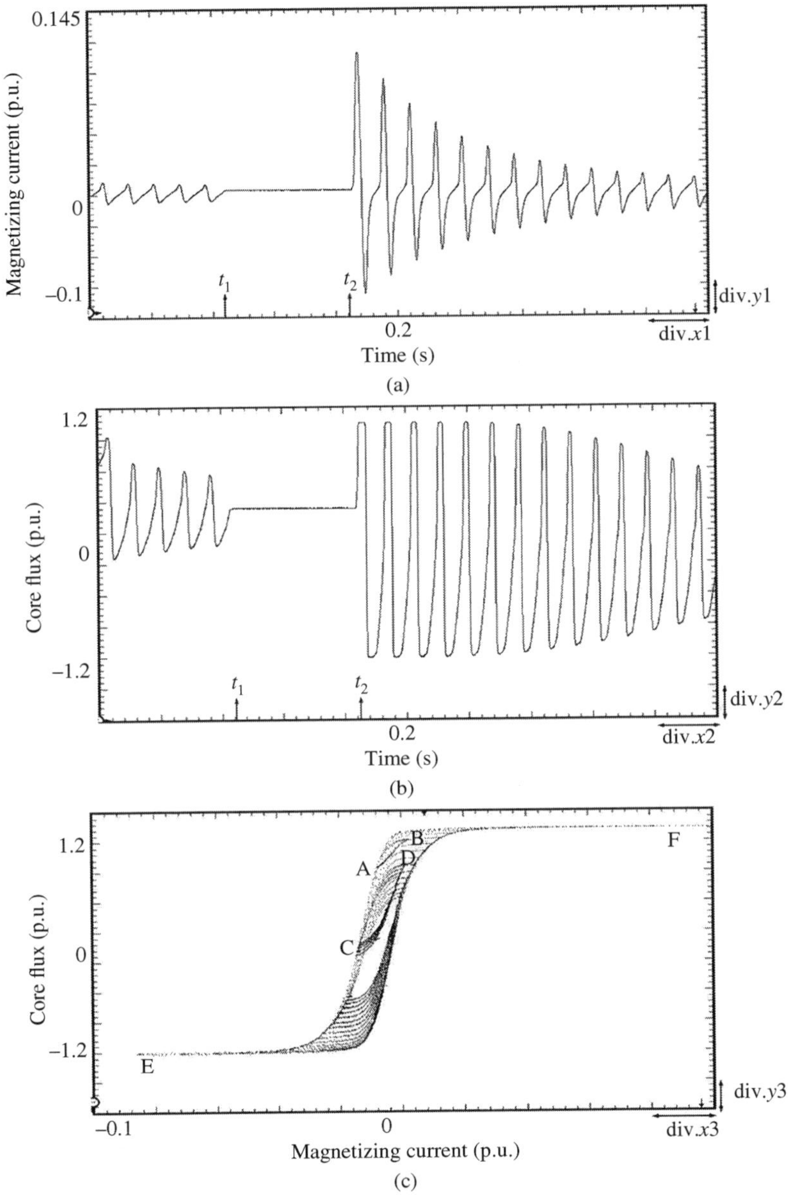

Figure 3.35 Real-time emulation results of (a) the magnetizing current (1 div.$x1$ = 0.04 second, 1 div.$y1$ = 0.025 p.u.), (b) flux in the core (1 div.$x2$ = 0.04 second, 1 div.$y2$ = 0.32 p.u.), and (c) the hysteresis characteristic of ST (1 div.$x3$ = 0.022 p.u., 1 div.$y3$ = 0.32 p.u.).

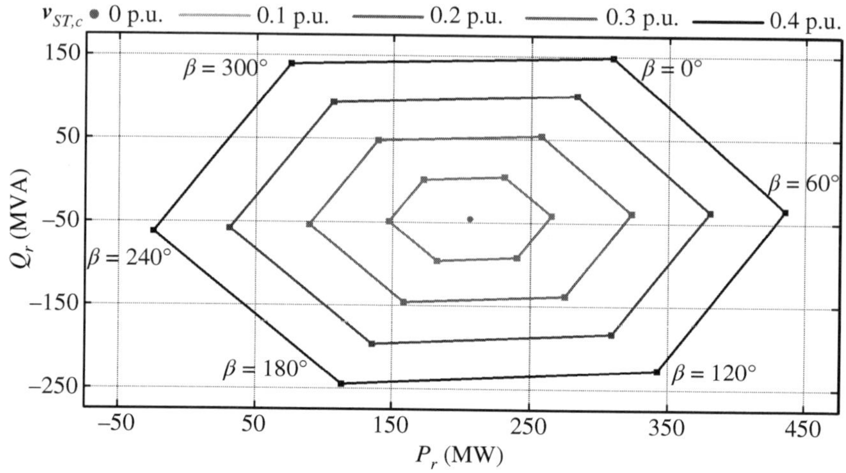

Figure 3.36 Relationship between P_r and Q_r at different injection voltage magnitude and phase angle for the ST.

from $t_1 = 0.5$ second to $t_2 = 3$ seconds, the tap-setting increased to 0.25 p.u. with the step size of 0.05 p.u. in 2.5 seconds. As can be seen in Figure 3.37 during this period, the powers P_{r1} and Q_{r1} increase while the powers P_{r2} and Q_{r2} decrease, which proves that the ST has the capability to regulate the power flow in both transmission lines. Finally, the P_{r1} and Q_{r1} increase to 134 MW and 98 MVAr, and P_{r2} and Q_{r2} decrease to 5 MW and −98 MVAr, respectively. At $t_3 = 3.5$ seconds, the controller sends a request to ST to inject a compensation voltage of 0.4 p.u. at an angle of 240°; thus, the ST secondary winding tap-setting of GP_1 needs to decrease to 0 p.u. again and GP_3 needs to increase to 0.4 p.u. Starting from $t_3 = 3.5$ seconds to $t_5 = 6$ seconds, the GP_1 windings' tap-setting decreases to 0 p.u. position, and GP_3 increases to 0.25 p.u. position, respectively. From $t_5 = 6$ seconds, the GP_3 winding increases further to 0.4 p.u. As can be seen during this period, the reactive power in both L_1 and L_2 is kept constant since the angle of the injection voltage is equal to 240°. The ST secondary winding current transition behavior during the entire period ($t = 0$–7.5 seconds) is shown in Figure 3.37c. The current magnitude decreases during the power reversal period, starting from $t_3 = 3.5$ seconds until $t_4 = 5.5$ seconds since the absolute value of active power is decreased; however at the time $t_4 = 5.5$ seconds, the current magnitude starts to increase due to the fact that the absolute active power value is increased at this time, as shown in Figure 3.37. The transition of the corresponding injected compensation voltage is shown in Figure 3.37d; the initial value is zero since the ST works in uncompensated mode, and then the step-by-step changing of the voltage magnitude can be clearly seen at the request of the controller.

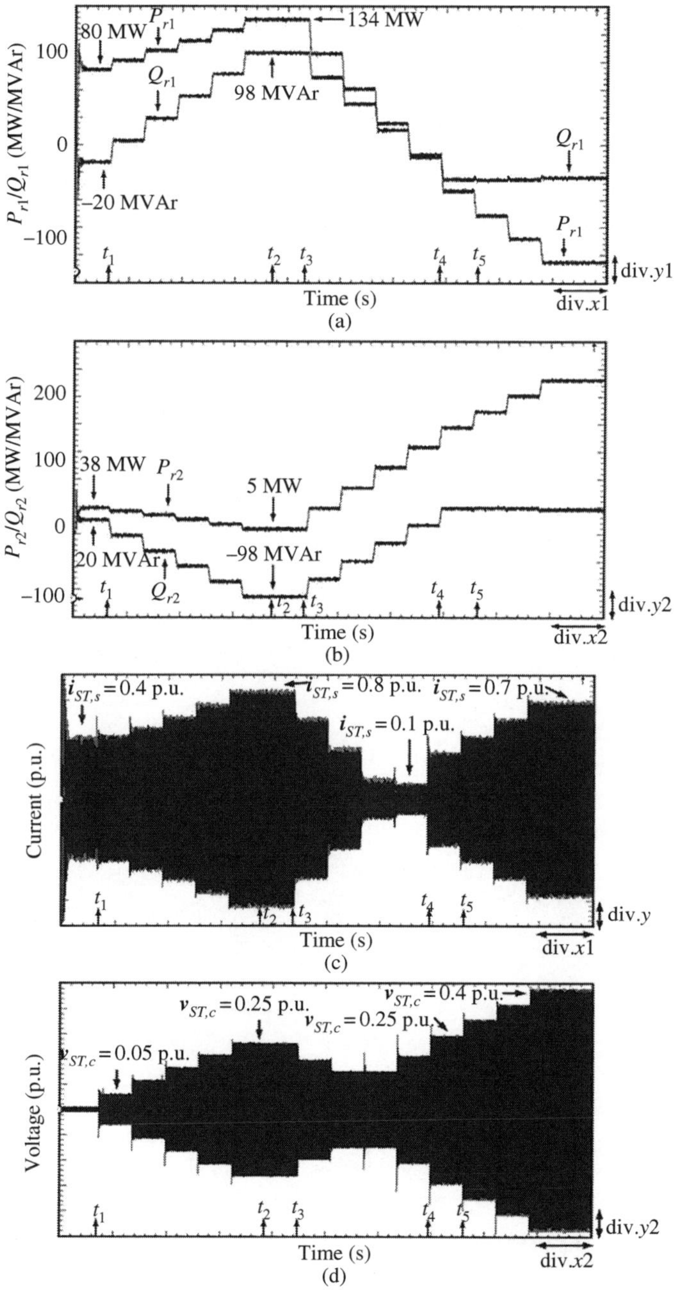

Figure 3.37 Real-time emulation of active power and reactive power regulation on the two transmission lines: (a) Line L_1 (1 div.$x1$ = 0.8 second, 1 div.$y1$ = 30 MW/MVAr), (b) Line L_2 (1 div.$y2$ = 43 MW/MVAr), (c) secondary winding current (1 div.$y1$ = 0.18 p.u.), and (d) injected compensation voltage of ST (1 div.$y2$ = 0.084 p.u.).

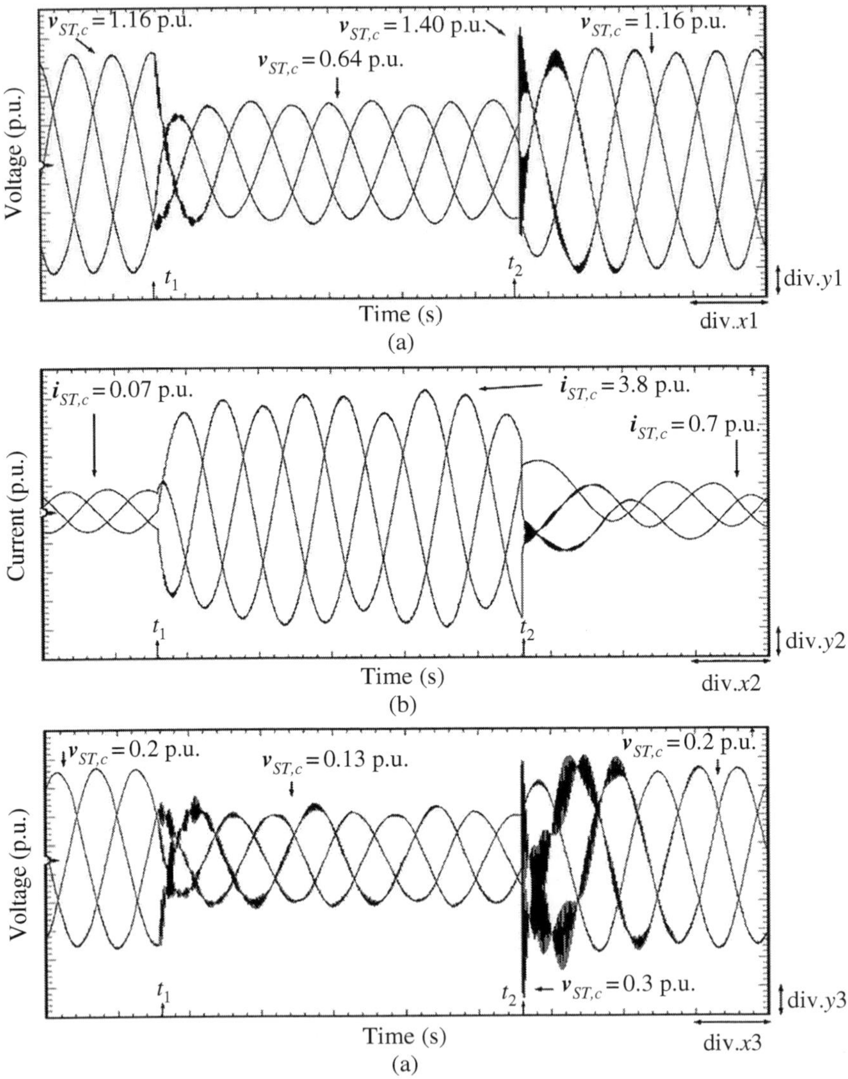

Figure 3.38 Real-time emulation results of ground fault transient (a) secondary side current (1 div.$x1$ = 0.01 second; 1 div.$y1$ = 0.95 p.u.); (b) secondary side voltage (1 div.$y2$ = 0.32 p.u.); and (c) injected compensation voltage (1 div.$y3$ = 0.067 p.u.).

Fault Transient Emulation A three-phase to ground fault was created at $t_1 = 2.1$ seconds at the receiving end of the Line L_1 and cleared at $t_2 = 2.15$ seconds. Figure 3.38 shows the real-time emulation results of the ST secondary side voltage and current, as well as the injected compensation voltage. The fault current peak magnitude reached up to 3.8 p.u. compared to the steady-state magnitude of 0.7 p.u. and decays to steady state after 2.4 seconds. The magnitude of ST secondary side fault transient voltage decreased to 0.64 p.u. from 1.16 p.u. in the steady-state, and the recovery voltage reach to 1.40 p.u. then decays to steady state in two cycles. The recovery injection compensation voltage is up to 0.3 p.u. compared to 0.2 p.u. under steady state. At the inception and clearing of the fault, the high-frequency transient can be clearly observed.

3.4 Real-Time Finite-Element Model of Power Transformer

The behavior of the electromagnetic field in a power transformer is governed by Ampere's law, and the accurate prediction of the magnetodynamic field is commonly obtained from the FEM simulation, a powerful numerical method to deal with complex geometries and material nonlinearities.

An accurate real-time electromagnetic *field-transient* simulation of a power transformer is of great benefit for the design and testing of better control and protection schemes, cost reductions, and energy efficiency improvement. However, as the permeability of the material depends on the magnetic field strength, solving the discretized nonlinear system of field equations in *space-time* requires iterative techniques such as the N-R algorithm within each simulation time-step. Thus, a new linear system has to be solved at each iteration due to the updated Jacobian matrix, and the dense computation makes the traditional FEM solver unbearably slow for real-time execution.

In this section, FE computation of electromagnetic transients of a transformer is emulated [63] in real-time on the FPGA. The TLM method is employed in the FEM to successfully decouple the nonlinear elements from the linear network so the nonlinearities could be solved individually, which is perfect for parallel processing. In addition, a noniterative field-circuit coupling is employed to interface the power transformer with the external network for EMT simulation.

3.4.1 Magnetodynamic Problem Formulation

Consider a 2-D FE model of a single-phase power transformer shown in Figure 3.39. When the windings are excited by the external circuit, the current density in the windings produces a dynamic magnetic field, which is described

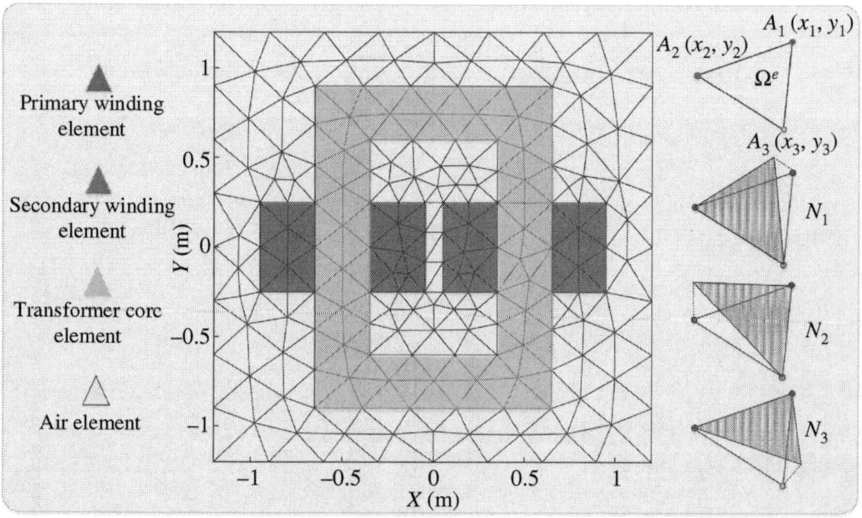

Figure 3.39 2-D FE geometry of single-phase transformer.

by the magnetic vector potential A. Since the magnetic vector potential A and the impressed current density J only have the z-components in a 2-D problem, the nonlinear magnetodynamic problem can be defined by the following diffusion equation:

$$\nabla \cdot (\upsilon \nabla A) = \sigma \frac{\partial A}{\partial t} - J_z, \tag{3.54}$$

where υ is the field-dependent reluctivity; σ is the conductivity, a constant value for a certain material; and J_z is the impressed current density, which is only nonzero in the winding zone.

The well-known Galerkin FEM [64] applied to solve the magnetodynamic problem includes the following steps: discretize the domain, form elemental equations, assemble the global matrix, and solve the system of equations.

Figure 3.39 shows the discretized domain consisting of triangular elements and nodes. For the Galerkin method, the product of the residual and the weighted function is integrated over each element, and the element equations can be formed by forcing the integral to zero. With natural boundary conditions applied, the classical weighted-integral equation of element Ω^e can be written as:

$$\iint_{\Omega^e} \upsilon^e \left(\frac{\partial A^e}{\partial x}\frac{\partial W^e}{\partial x} + \frac{\partial A^e}{\partial y}\frac{\partial W^e}{\partial y} \right) dxdy \\ + \iint_{\Omega^e} \sigma \frac{\partial A^e}{\partial t} W^e dxdy = \iint_{\Omega^e} J_z^e W^e dxdy, \tag{3.55}$$

where A^e is the magnetic vector potential over element Ω^e, and W^e the weighted function.

In Figure 3.39, the vertex values of the triangular element Ω^e are A_1, A_2, and A_3, and the magnetic vector potential over the whole element A^e can be expressed with the classical barycentric interpolation:

$$A^e = N_1A_1 + N_2A_2 + N_3A_3, \tag{3.56}$$

where N_1, N_2, and N_3 are the shape functions profiled in Figure 3.39. The values of these shape functions at any position $(x,\ y)$ are defined by

$$N_i = \frac{1}{2\Delta^e}(a_i + b_ix + c_iy) \quad (i = 1,\ 2,\ 3), \tag{3.57}$$

where Δ^e is the area of the triangular element; a_i, b_i, and c_i are related to the coordinates of the vertices:

$$\begin{aligned} a_1 &= x_2y_3 - x_3y_2,\ b_1 = y_2 - y_3,\ c_1 = x_3 - x_2,\\ a_2 &= x_3y_1 - x_1y_3,\ b_2 = y_3 - y_1,\ c_2 = x_1 - x_3,\\ a_3 &= x_1y_2 - x_2y_1,\ b_3 = y_1 - y_2,\ c_3 = x_2 - x_1. \end{aligned} \tag{3.58}$$

According to the Galerkin's method, the A^e in (3.55) is substituted by (3.56) and the weighted functions W^e are set to N_1, N_2, and N_3, respectively. Thus, three equations with three unknown vertex magnetic potential values are obtained after accomplishing the integral:

$$\frac{\upsilon^e}{4\Delta^e}\begin{bmatrix} b_1b_1 + c_1c_1 & b_1b_2 + c_1c_2 & b_1b_3 + c_1c_3 \\ b_1b_2 + c_1c_2 & b_2b_2 + c_2c_2 & b_2b_3 + c_2c_3 \\ b_1b_3 + c_1c_3 & b_2b_3 + c_2c_3 & b_3b_3 + c_3c_3 \end{bmatrix}\begin{bmatrix} A_1 \\ A_2 \\ A_3 \end{bmatrix} + \frac{\sigma^e\Delta^e}{12}\begin{bmatrix} 2 & 1 & 1 \\ 1 & 2 & 1 \\ 1 & 1 & 2 \end{bmatrix}\begin{bmatrix} \frac{\partial A_1}{\partial t} \\ \frac{\partial A_2}{\partial t} \\ \frac{\partial A_3}{\partial t} \end{bmatrix} = \frac{J_z^e\Delta^e}{3}\begin{bmatrix} 1 \\ 1 \\ 1 \end{bmatrix}. \tag{3.59}$$

Note that the reluctivity υ^e, the conductivity σ^e, and the current density J_z^e are constant over one FE. The distributions of σ^e and J_z^e are given while υ^e depends on A_1, A_2, and A_3. The dependence is not straightforward and could be quantitatively expressed by introducing the magnetic flux density B. In a 2-D problem, as A is defined by $B = \nabla \times A$, in one element B and A are related by:

$$\begin{aligned} B^2 &= \left(\frac{\partial A^e}{\partial x}\right)^2 + \left(\frac{\partial A^e}{\partial y}\right)^2 = -\frac{1}{4(\Delta^e)^2}((b_1b_2 + c_1c_2)(A_1 - A_2)^2 \\ &\quad + (b_2b_3 + c_2c_3)(A_2 - A_3)^2 + (b_1b_3 + c_1c_3)(A_1 - A_3)^2). \end{aligned} \tag{3.60}$$

The $\upsilon^e - B^2$ relationship can be found from the core material property such as the *B–H* curve representation. In this paper, the nonlinear material property of the transformer core is represented using the Preisach model [50].

The classical Preisach model regards the ferromagnetic material as an infinite set of hysterons defined by the hysteresis operator (Figure 3.40a). Thus, the magnetization moment *M* can be written as the following integral over the Preisach triangle (Figure 3.40b):

$$\begin{aligned} M(t) &= \iint_{\alpha \geq \beta} \mu(\alpha, \beta)\gamma_{\alpha\beta} H(t) d\alpha d\beta \\ &= \iint_{S^+} \mu(\alpha, \beta) H(t) d\alpha d\beta - \iint_{S^-} \mu(\alpha, \beta) H(t) d\alpha d\beta, \end{aligned} \tag{3.61}$$

where $H(t)$ and $M(t)$ are the input and output, the stair-shaped boundary of S^+ and S^- is determined by the historical extremum of $H(t)$, and $\mu(\alpha, \beta)$ is the Preisach distribution function that is usually obtained from the experimental results. Thus, the *B–H* relationship can be obtained:

$$B = \mu_0(M + H). \tag{3.62}$$

For the N-R algorithm requiring smooth nonlinearity, it is beneficial to approximate the experimental results using a mathematical distribution function. According to [65], the following 2-D Cauchy distribution function, which is analytically integrable, can achieve an accurate approximation:

$$\mu(\alpha, \beta) = c \frac{1}{1 + \left(\frac{\alpha - a_0}{b0}\right)^2} \frac{1}{1 + \left(\frac{\beta - a_1}{b1}\right)^2}, \tag{3.63}$$

where a_0, a_1, b_0, b_1, and c are coefficients given in the Appendix A.2.

Since the magnetization moment *M* depends on the past history of *H*, a reversal point vector (RPV) storing the historical extremum $(h_1, h_2, h_3, h_4, \ldots)$ is required to derive the integral. Figure 3.40c,d presents a sample input with four branches and the corresponding trajectories of the hysteresis loop. The RPV is updated or wiped out (branch $h_4 \rightarrow h_1$) based on the input sequence $H(t)$, and the *B–H* formulation in different branches are regulated by the RPV.

The problem domain is enclosed by an artificial rectangular boundary Γ (shown in Figure 3.39), and the boundary condition is $A_\Gamma = 0$. After time discretization, the above nonlinear elemental equations (3.59) can be assembled and are ready to be solved with the N-R iterative scheme. With respect to the computational burden, the TLM technique provides an efficient methodology to handle these nonlinear equations.

3.4.1.1 Refined TLM Solution

The TLM method is based on Huygens's principle and was originally developed [66] to simulate wave propagation. Then, the concepts of TLM *link* and *stub* were

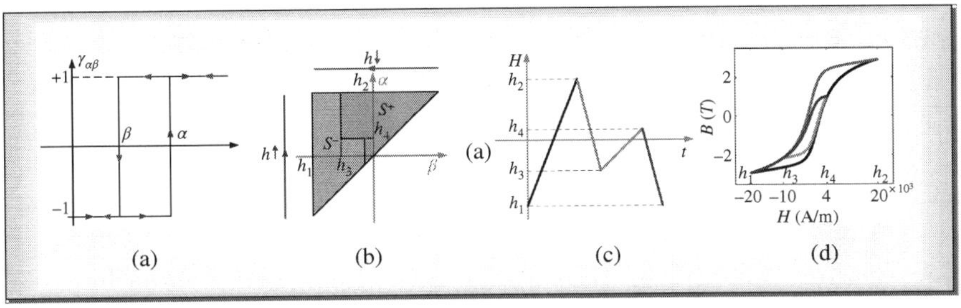

Figure 3.40 Preisach model: (a) Hysteresis operator, (b) Preisach triangle, (c) sample input of H, and (d) trajectory of the hysteresis loop.

proposed for the analysis of nonlinear networks to effectively decouple the nonlinear elements from the linear system [67–71]. Later, the analogy between the FE matrix and the nodal admittance matrix of a corresponding equivalent network was realized, and the TLM technique was successfully applied to the FEM solution [72, 73].

When the TLM technique is applied to solve an electrical circuit in the time-domain, the components are separated from the network by lossless transmission lines. The characteristic impedance of the transmission line Z_c is arbitrary for resistors, $\Delta t/2C$ for capacitors, and $2L/\Delta t$ for inductors, where Δt is the simulation time-step.

In the scattering phase, the network transmits incident impulses V_i into each element. For linear elements, the reflected pulse can be calculated by

$$V_r = K_r V_i, \tag{3.64}$$

where the reflection coefficient K_r is equal to $(R - Z_c)/(R + Z_c)$ for resistors, 1 for capacitors, and -1 for inductors. For a nonlinear element, for example, a nonlinear resistor described as $U = f(I)$, the reflected pulse V_r is obtained by solving the nonlinear equation

$$V_i + V_r = f\left(\frac{V_i - V_r}{Z_c}\right). \tag{3.65}$$

Note that if Z_c is equal to the working state value of the linear or nonlinear resistor, the reflection coefficient is 0 and only one TLM iteration is required.

In the gathering phase, all the reflected pulses return to the network and update the values of all nodes. The transmission line and the elements are replaced with their equivalent Norton circuits.

The nodal values can be updated by solving a linear network with unchanged nodal admittance matrix, then the next incident pulse can be also updated by:

$$V_i = V_d - V_r, \tag{3.66}$$

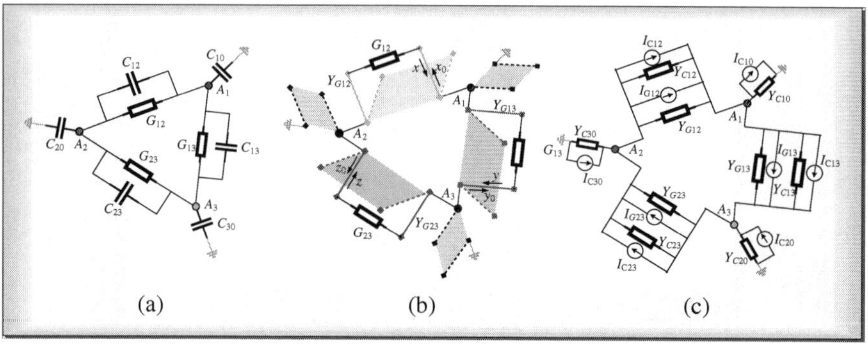

Figure 3.41 TLM technique applied to elemental equations of the FEM: (a) equivalent nonlinear network, (b) scattering phase in TLM, and (c) gathering phase in TLM.

where V_d is the nodal voltage difference. Thus, the gathering phase is complete, and the scattering phase of the next TLM iteration is ready to begin.

Equation (3.59) defines an equivalent network composed of capacitors and nonlinear resistors shown in Figure 3.41a. The magnetic vector potential is modeled as the electric potential, the right-hand side vector of (3.59) is modeled as the current sources and the Dirichlet boundary conditions are modeled as the voltage sources. The nonlinear resistors and the linear capacitors form the coefficient matrix and hold such material properties as reluctivity and conductivity. And the values of the components are given as

$$\begin{aligned}
G_{12} &= -\frac{\upsilon^e}{4\Delta^e}(b_1b_2 + c_1c_2), & Y_{G12} &= -\frac{\upsilon_g^e}{4\Delta^e}(b_1b_2 + c_1c_2),\\
G_{13} &= -\frac{\upsilon^e}{4\Delta^e}(b_1b_3 + c_1c_3), & Y_{G13} &= -\frac{\upsilon_g^e}{4\Delta^e}(b_1b_3 + c_1c_3),\\
G_{23} &= -\frac{\upsilon^e}{4\Delta^e}(b_2b_3 + c_2c_3), & Y_{G23} &= -\frac{\upsilon_g^e}{4\Delta^e}(b_2b_3 + c_2c_3),\\
C_{12} = C_{13} = C_{23} &= -\frac{\sigma^e\Delta^e}{12}, & Y_{C12} = Y_{C13} = Y_{C23} &= -\frac{\sigma^e\Delta^e}{6\Delta t},\\
C_{10} = C_{20} = C_{30} &= \frac{4\sigma^e\Delta^e}{12}, & Y_{C10} = Y_{C20} = Y_{C30} &= \frac{4\sigma^e\Delta^e}{6\Delta t}.
\end{aligned} \tag{3.67}$$

where υ^e is the prior unknown reluctivity to be solved using the 3×3 N-R scheme and υ_g^e is an initial guess value, which should be as close to υ^e as possible. With the guessed υ_g^e, the nodal admittance matrix required in the gathering phase could be obtained by assembling multiple Norton equivalent circuits (Figure 3.41c) based on the FE mesh information.

When the TLM technique is applied in a triangular network, three pulses based on the differences between nodal values will be injected into the three edges. Figure 3.41b shows the TLM links (resistors) and the TLM stubs (capacitors) in the scattering phase. The incident pulses into three edges are denoted by x_0, y_0, and z_0 and the reflected pulses to be calculated are denoted by x, y, and z, respectively. Instead of solving a single equation (3.65), the following three coupled nonlinear algebraic equations need to be solved:

$$\begin{aligned} G_{12}(x+x_0) - Y_{G12}(x_0 - x) &= 0, \\ G_{13}(y+y_0) - Y_{G13}(y_0 - y) &= 0, \\ G_{23}(z+z_0) - Y_{G23}(z_0 - z) &= 0. \end{aligned} \tag{3.68}$$

Since G_{12}, G_{13}, and G_{23} are functions of $(x+x_0)$, $(y+y_0)$, and $(z+z_0)$, and the relations can be derived from (3.60) and the Preisach model, the resulting nonlinear algebraic equations can be efficiently solved with the N-R method, and the convergence can be sped up using an appropriate relaxation factor. When calculating the 3 × 3 sensitivity matrix, the following chain rule is utilized:

$$\frac{\partial G}{\partial x} = \frac{\partial G}{\partial B^2}\frac{\partial B^2}{\partial x}. \tag{3.69}$$

In the gathering phase shown in Figure 3.41c, the reflected pulses and the TLM links or stubs are replaced with their equivalent Norton circuits. It is a linear network and the nodal values can be updated after matrix–vector multiplications if the inversion of the admittance matrix is ready at the beginning.

Note that the above descriptions of the TLM technique applied in FEM are at the triangular element level. For a problem domain composed of numerous FEs, the main computational tasks include solving (3.68) for all nonlinear elements in the scattering phase and the matrix–vector multiplications in the gathering phase. Both tasks could be perfectly parallelized on massively parallel computing architectures.

It is acknowledged that multiple TLM iterations are needed because of the mismatch of the unknown v^e and the guessed v_g^e. If v^e is equal to v_g^e, the reflected pulse is zero and only one TLM iteration is enough for convergence. Fewer TLM iterations are needed if the two values are closer. Thus, the concept of *memory* nodal admittance matrix is proposed to decrease the mismatch and to accelerate the convergence. A memory admittance matrix is featured by the impressed current density. For the transformer, once the FE model is solved with provided I_{p0} and I_{s0}, the v^e of each element can be extracted. After the assembly and inversion process, these values could be utilized to form a new admittance matrix, which contains very valuable information (memory) for the rest solution.

Figure 3.42a shows the efficiency of the memory admittance matrix in a case study (presented in Section IV) with secondary winding open-circuited ($I_s = 0$). As shown in the figure, even when the v_g^e is carefully guessed, it takes more than 30 TLM iterations for each time-step while the number is around 10 utilizing the memory admittance matrix. Note that I_p^{max} in Figure 3.42a could be set to the peak value of the rated primary winding current.

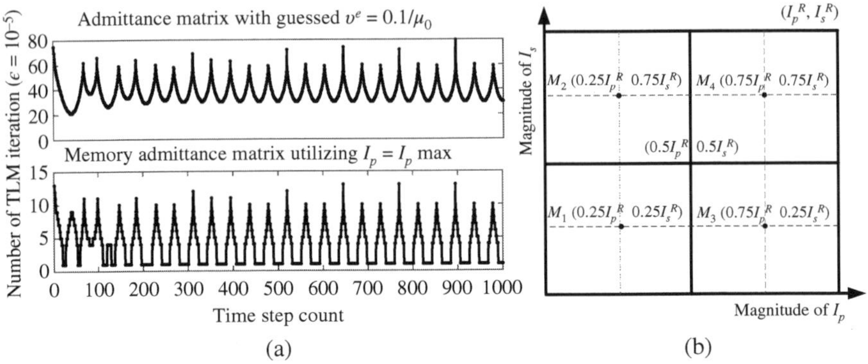

Figure 3.42 (a) TLM iteration number required utilizing guessed and memory admittance matrix. (b) Four memory matrix formation and usage.

In this work, a total of four memory admittance matrices are applied to ensure that the TLM iteration number is always less than 5 provided that I_p and I_s change arbitrarily within each time-step, as shown in Figure 3.42b, where I_p^R and I_s^R are the rated currents in the primary and secondary windings of the transformer. Different admittance matrices are used for the gathering phase based on the combination of the I_p and I_s. The decreased TLM iteration number and the parallelism within a TLM iteration make real-time simulation possible on the parallel FPGA hardware architecture.

3.4.1.2 Field-Circuit Coupling

Note that the inputs of the FE model described above are the winding currents. However, since a transformer is generally excited by voltage sources in a power system, a field-circuit coupling technique is necessary to interface with the external circuit.

Direct coupling and indirect coupling approaches have been explored to interface the FE model and the external circuit: the direct coupling approach solves the field equations and the electrical network equations simultaneously while the indirect approach solve the two subsystems separately [74]. The computational burden of the direct coupling approach is heavier than the traditional FEM solver because the circuit equations destroy the matrix symmetry [75].

In this work, an indirect coupling technique is presented by accurately extracting the self and the mutual inductances of the transformer from the FE model. Within each time-step, the field equations and the circuit equations are solved separately without having to apply any iterative scheme.

In electrical networks, a transformer model can be represented by self and mutual inductances with controlled sources, as shown in Figure 3.43. The accurate values of these nonlinear inductances (depends on the winding currents) can

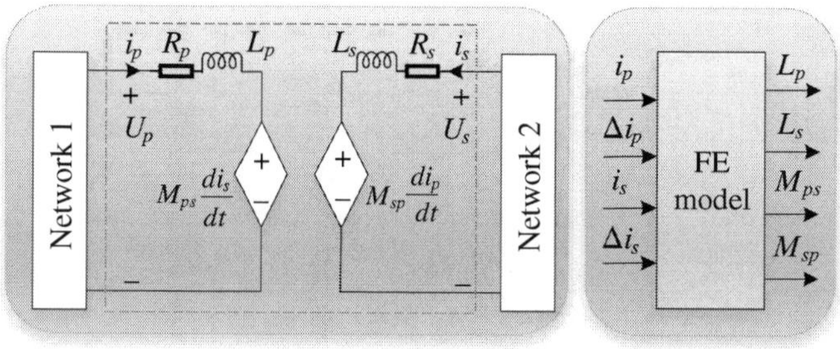

Figure 3.43 Equivalent FE transformer model coupled to external network.

be obtained with the FE model. In fact, the electric voltage U across a winding can be represented by the magnetic vector potential and the voltage drop of winding's resistance:

$$U = rI + \frac{Nl}{\Delta_S}\int_S \frac{\partial A}{\partial t} dS, \tag{3.70}$$

where I is the winding current, r the winding resistance, N the number of turns, l the axial length of each filament, S the winding zone, and Δ_S the area of the winding zone.

Since A is a function of the primary winding current i_p and the secondary winding current i_s based on the FE model, (3.70) can be rewritten as:

$$U = rI + \frac{Nl}{\Delta_S}\int_S \frac{\partial A}{\partial i_p} dS \frac{\partial i_p}{\partial t} + \frac{Nl}{\Delta_S}\int_S \frac{\partial A}{\partial i_s} dS \frac{\partial i_s}{\partial t}. \tag{3.71}$$

Applying (3.71) to the primary and the secondary winding, respectively, the self and mutual inductances of the transformer can be obtained as:

$$\begin{aligned} L_p &= \frac{N_p l_p}{\Delta_{S_p}}\int_{S_p} \frac{\partial A}{\partial i_p} dS, \quad M_{ps} = \frac{N_p l_p}{\Delta_{S_p}}\int_{S_p} \frac{\partial A}{\partial i_s} dS, \\ M_{sp} &= \frac{N_s l_s}{\Delta_{S_s}}\int_{S_s} \frac{\partial A}{\partial i_p} dS, \quad L_s = \frac{N_s l_s}{\Delta_{S_s}}\int_{S_s} \frac{\partial A}{\partial i_s} dS. \end{aligned} \tag{3.72}$$

Since

$$\begin{aligned} \frac{\partial A}{\partial i_p} &= \frac{A(i_p + \Delta i_p, i_s) - A(i_p, i_s)}{\Delta i_p}, \\ \frac{\partial A}{\partial i_s} &= \frac{A(i_p, i_s + \Delta i_s) - A(i_p, i_s)}{\Delta i_s}, \end{aligned} \tag{3.73}$$

these coupling coefficients can be calculated with the FE model by setting the Δi_p and Δi_s in (3.73) to a very small value such as 0.1 A.

It is known that by applying Trapezoidal rule, a constant inductor in the network can be replaced by a resistor $\frac{2L(t)}{\Delta t}$ in parallel with a current source I_{his}. When the inductor value $L(t)$ is time-varying, both $L(t)$ and $L(t+\Delta t)$ should be used. Since $L(t+\Delta t)$ is unknown, an error exists in the indirect coupling scheme because the time-varying inductances at time point t, calculated from $i_p(t)$ and $i_s(t)$, are utilized to solve the external networks at time point $t+\Delta t$. To reduce the asynchronization error, the inductance values at time point $t+\frac{\Delta t}{2}$ are used with a prediction scheme and the following central difference formula is utilized:

$$\begin{aligned}
\frac{\partial A}{\partial i_p}\left(t+\frac{\Delta t}{2}\right) &\approx \frac{A(i_p(t+\Delta t), i_s(t)) - A(i_p(t), i_s(t))}{i_p(t+\Delta t) - i_p(t)},\\
\frac{\partial A}{\partial i_s}\left(t+\frac{\Delta t}{2}\right) &\approx \frac{A(i_p(t), i_s(t+\Delta t)) - A(i_p(t), i_s(t))}{i_s(t+\Delta t) - i_s(t)}.
\end{aligned} \tag{3.74}$$

where $i_p(t+\Delta t)$ and $i_s(t+\Delta t)$ are unknown and can be predicted from their historical values:

$$\begin{aligned}
i_p(t+\Delta t) &= 2i_p(t) - i_p(t-\Delta t),\\
i_s(t+\Delta t) &= 2i_s(t) - i_s(t-\Delta t).
\end{aligned} \tag{3.75}$$

Simulation results show that the mean absolute relative error is less than 3% with $\Delta i_p = i_p(t) - i_p(t-\Delta t)$, $\Delta i_s = i_s(t) - i_s(t-\Delta t)$ and more than 5% using small values (0.1 A) of Δi_p and Δi_s.

Note that the calculation of $A(i_p, i_s)$, $A(i_p, i_s + \Delta i_s)$, and $A(i_p + \Delta i_p, i_s)$ can be also parallelized with enough hardware resources. Then, these parameters are fed to the external electrical network for EMT simulation [40].

3.4.2 Hardware Emulation of Finite Element Model

The hardware design was implemented on the Xilinx Virtex UltraScale+™ VCU118 board with the XCVU9P-FLG2104-E-ES1 FPGA (Table 1.2) using high-level synthesis and optimization. The sequential C and C++ functions can be compiled down to hardware IP blocks with inputs, outputs, and handshake signals for interfacing, which can be optimized based on user's preferences by adding the directives such as array partitioning, loop unrolling, and pipelining.

According to the TLM-FE solution and the field-circuit coupling technique, six hardware blocks are generated and optimized to achieve deep data pipelining on the FPGA. The detailed real-time implementation on the FPGA is shown in Figure 3.44. The block connections, pseudo codes, and the finite state machine are all illustrated. The functionalities of these blocks are presented as the following:

1. *Assemble current*: The currents generated by the reflected pulses in each triangular element (Figure 3.41) and the impressed source currents are assembled to obtain the total current injected to each node. The assembly of each node is parallelized and pipelined based on the node-element connection matrix.

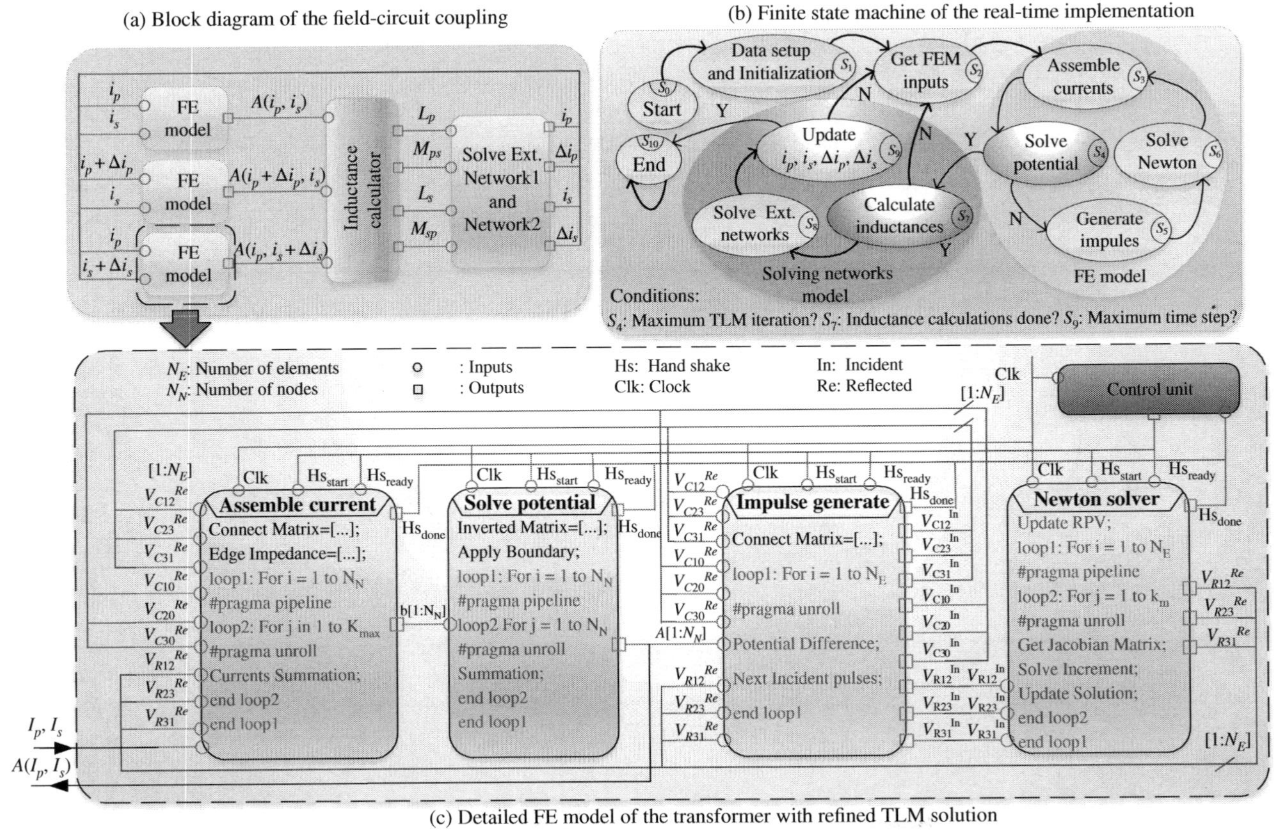

Figure 3.44 Hardware emulation of the finite-element computation for the transformer.

2. *Solve potential*: With the assembled current vector and the inverted nodal admittance matrix, the magnetic potential of each node is calculated by the matrix–vector multiplication under the Dirichlet boundary conditions. The multiplications could be totally parallelized and the summation could be accomplished using a tree adder with a latency of $O(\log_2 N)$.
3. *Impulse generator*: Based on the updated magnetic vector potential and the reflected pulses, the new incident pulses are generated for each triangular element. For each element, these operations are unrolled and executed in parallel.
4. *Newton solver*: For each triangular element, the three incident pulses are taken as inputs to obtain the three reflected pulses. The Newton solver is also unrolled and fully pipelined to treat multiple elements. In this work, by using a relaxation factor of 1.2, six iterations are enough for all elements to maintain a relative tolerance of 10^{-5}.
5. *Inductance calculator*: Once $A(i_p, i_s)$, $A(i_p, i_s + \Delta i_s)$, and $A(i_p + \Delta i_p, i_s)$ are available, the self and mutual inductances can be calculated by the integration in (3.72).
6. *Solve external networks*: The transformer is converted to the equivalent model in Figure (3.39) with all inductances known, and the external electrical networks are solved in the time-domain. For example, in Case II, when the traveling wave line is loaded, the KVL equations to be solved are given in (3.76). Note that although the execution time of the external network solution is slight compared with the FE model, the 3×3 matrix in (3.76) is also time-varying due to the time-varying transformer inductances. For larger external networks, the compensation method could be applied to handle these time-varying nonlinear transformer inductances [54] in real time.

$$\begin{bmatrix} \frac{2(L_p + L_1)}{\Delta t} + R_1 + R_p & -\frac{2M_{ps}}{\Delta t} & 0 \\ \frac{2M_{sp}}{\Delta t} & -R_s - \frac{2L_s}{\Delta t} - Z_{TL} & 0 \\ 0 & 0 & R_{L2} + \frac{2L_{L2}}{\Delta t} + Z_{TL} \end{bmatrix} \begin{bmatrix} I_p(t+\Delta t) \\ I_s(t+\Delta t) \\ I_k(t+\Delta t) \end{bmatrix}$$
$$= \begin{bmatrix} \frac{2(L_p + L_1)}{\Delta t} I_p^{his} - \frac{2M_{ps}}{\Delta t} I_s^{his} + V_{ac}(t+\Delta t) \\ \frac{2M_{sp}}{\Delta t} I_p^{his} - \frac{2L_s}{\Delta t} I_s^{his} - U_k(t) + Z_{TL} I_k(t) \\ \frac{2L_{L2}}{\Delta t} I_k^{his} + U_s(t) - Z_{TL} I_s(t) \end{bmatrix}. \tag{3.76}$$

Table 3.2 Hardware resource utilization and timing report.

FPGA device	Module	Resource utilization				Latency (clock cycles)
		BRAM	DSP	FF	LUT	
Xilinx UltraScale+ XCVU9P	Assem. curr.	84	512	130 316	253 004	393
	Solv. poten.	168	842	86 841	64 301	206
	Imp. gener.	0	24	4 418	21 316	157
	Newt. solv.	9	1 646	194 969	200 509	767
	Subtotal (FEM)	6%	44.2%	17.6%	45.6%	1 523
	Calc. induc.	0	33	4 971	5 237	71
	Solve netw.	0	5	1 611	1 758	82
	Subtotal (Netw.)	0%	1%	0%	0%	153
Xilinx UltraScale+ XCVU13P	Assem. curr.	84	512	131 490	257 507	399
	Solv. poten.	168	842	86 876	64 805	208
	Imp. gener.	0	24	5 187	21 524	159
	Newt. solv.	9	1 646	214 112	203 558	770
	Subtotal (FEM)	4.9%	24.6%	12.7%	31.7%	1 536
	Calc. induc.	0	33	4 991	5 287	72
	Solve netw.	0	5	1 811	1 958	84
	Subtotal (Netw.)	0%	1%	0%	0%	156

Table 3.2 shows the FPGA hardware utilization and the timing report of the case study with a FE model composed of 196 nodes and 358 elements, which are compiled on both Xilinx Virtex UltraScale+^{TM} XCVU9P and XCVU13P FPGAs.

Although the FPGA clock frequency of each hardware block can reach more than 115 MHz, the maximum clock frequency of the design can reach 93.1 MHz for XCVU9P and 94.2 MHz for XCVU13P after the hardware blocks are interconnected. Since five TLM iterations are required for a reasonable accuracy, according to the timing report, the execution time of the FE hardware block is $5 \times 1523 \div 93.1 = 81.8$ μs on the XCVU9P FPGA and $5 \times 1536 \div 94.2 = 81.6$ μs on XCVU13P while the latency of the circuit model is 1.7 μs. According to the hardware utilization report, two FE hardware blocks can be generated on the xcvu9p FPGA and three FE hardware blocks on the XCVU13P FPGA. Therefore, for the design on the XCVU9P FPGA, two of $A(i_p, i_s)$, $A(i_p, i_s + \Delta i_s)$, and $A(i_p + \Delta i_p, i_s)$ can be calculated in parallel and the remaining one is executed sequentially thereafter, and the total execution time of the implementation is therefore 166 μs. For the

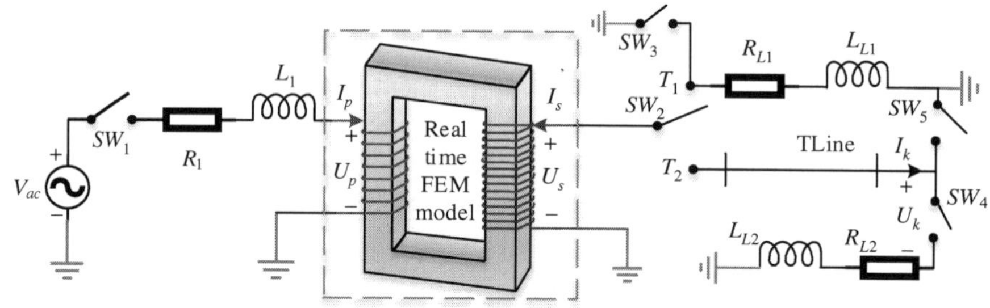

Figure 3.45 Schematic of the transformer FE model coupled with external networks for case studies.

XCVU13P FPGA, the calculations of $A(i_p, i_s)$, $A(i_p, i_s + \Delta i_s)$, and $A(i_p + \Delta i_p, i_s)$ can all be executed in parallel at the same time, and the total execution time of the implementation is 84 μs.

Although an execution time of 166 μs of the design on UltraScale+^{TM} XCVU9P FPGA in our real-time implementation is still too large for the transients requiring a time-step of 50 μs, it should be pointed out that the indirect coupling scheme can also work for multi-rate real-time simulation, which means that the external networks can utilize a much smaller time-step than that of the FE model. For example, in Case II the time-step is 180 μs for the FE transformer model and 45 μs for the external networks, implying that the transformer inductances used in the solution of the external network module are updated every four time-steps. In the future work with a larger FPGA, the applied simulation time-step for the FE model could be significantly reduced.

3.4.3 Case Studies

A single-phase power transformer rated at 37.5/202 kV was studied in this work, and the geometry and mesh (196 nodes and 358 elements) are shown in Figure 3.39. The schematic of the real-time FE model coupled to external networks is shown in Figure 3.45. The primary winding is excited by a 60 Hz AC voltage source, and a Bergeron line model is included between the secondary winding and the load. The detailed simulation parameters are provided in the Appendix A.2.

Both single-rate (Case I) and multi-rate (Case II) executions were examined with a total emulation time of 600 ms. In the single-rate case, the applied time-step is the same 180 μs for the FE transformer model and the external networks. In the multi-rate case, the applied time-step is 180 μs for the FE transformer model and 45 μs for the external networks. The following events are emulated:

1. $t = 0$, SW_1 is turned on and the transformer is energized while the secondary winding is open-circuited.
2. $t = 150$ ms, SW_2 is turned to T_1 and the transformer works with a load of R_{L1} and L_{L1}.
3. $t = 300$ ms, the second and the fourth harmonics are injected into the voltage source V_{ac}.
4. $t = 450$ ms, SW_3 is turned on and the secondary winding is short-circuited.

3.4.3.1 Results and Validation

The off-line simulation with the same case study, setup, and mesh was also performed on the commercial FEM software Comsol™. Mesh dependency test has been performed in Comsol™ and it turned out that for this 2-D problem, 196 nodes are enough to maintain reasonable accuracy compared with a finer mesh. The off-line results are regarded reliable and are thus used as the benchmark for comparison for the real-time results. The real-time simulation results of the primary winding current I_p, the primary winding voltage U_p, the secondary winding current I_s, the secondary winding voltage U_s, the transformer inductances, the time-varying winding loss, eddy current loss, hysteresis loss, and total loss are all presented in Figure 3.46 for Case I with a comparison of those results from Comsol™. Note that all the errors in the figures denote the mean absolute relative error. The zoomed-in plot of the simulation results is shown in Figure 3.47. For all practical purposes, the real-time results are identical to the Comsol™ results, and the harmonics from Fast Fourier transform of the results also match.

As can be seen from the results, when the secondary winding is open-circuited, I_p is the magnetizing current and the phase difference of I_p and U_s is always 90°. The inrush current of the primary winding occurs at $t = t_1$, whereas the inrush current of the secondary winding occurs at $t = t_5$. The inrush currents may cause some unwanted phenomenon such as saturation in the transformer, which is possible to be monitored in real time with this implementation.

The direct output of the FE model is the magnetic vector potential A, and Figure 3.48 shows the node solution of the real-time implementation compared with the Comsol™ at $t = t_1$ in Case I and at $t = t_5$ in Case II.

Other variables can also be derived after post-processing. For example, the induced electric field E and the eddy current loss L_{ec} can be calculated with

$$E = \frac{\partial A}{\partial t}, \quad L_{ec} = \iint_{\Omega_{core}} \sigma^e E^2. \tag{3.77}$$

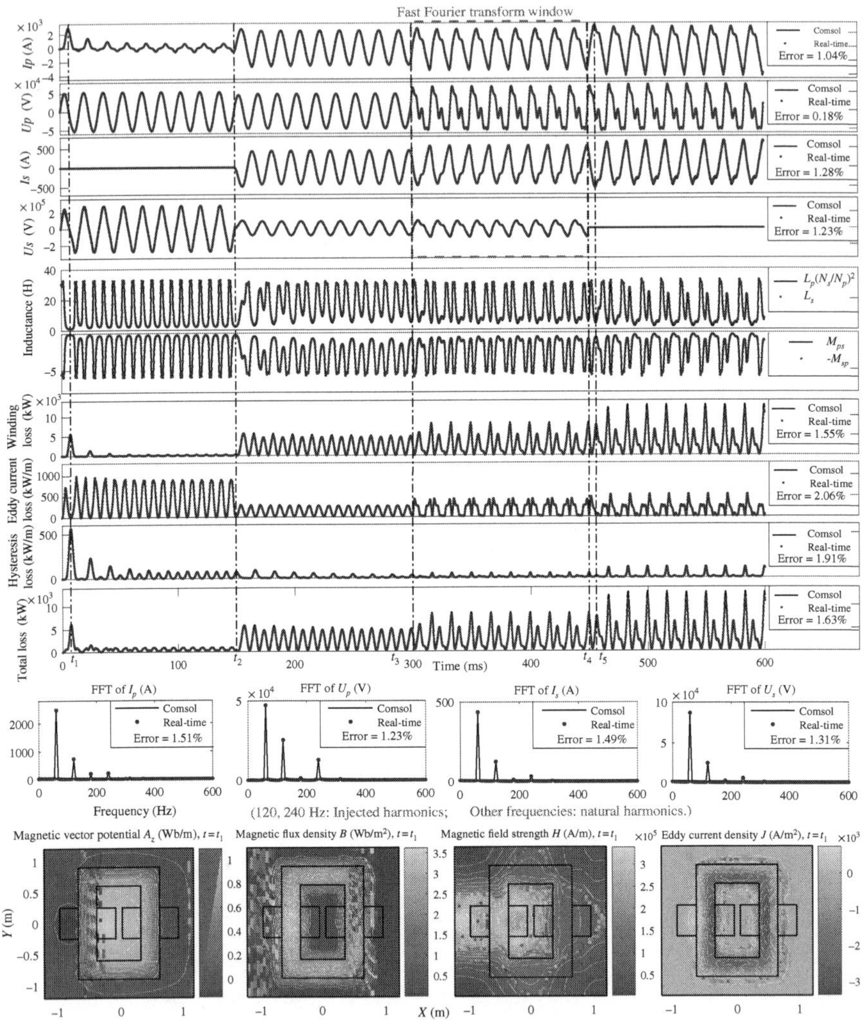

Figure 3.46 Real-time simulation results and comparison with Comsol™ off-line results: time-domain, frequency-domain, and field distributions.

Figure 3.49 shows the magnetizing current and the trajectories of the hysteresis loop in a sample element in Case I, and the hysteresis loss can be also calculated based on these loops, which are regulated by the RPV. The detailed field distributions of the magnetic vector potential A (Wb/m), the magnetic flux density B (T), the magnetic field strength H (A/m), and the eddy current density J (A/m^2) at $t = t_1$ in Case I are also presented in Figure 3.46.

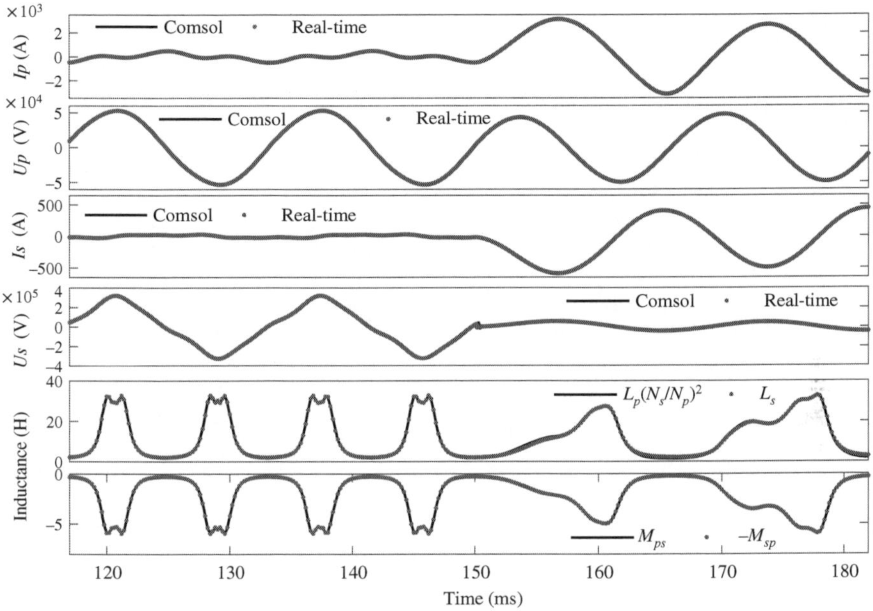

Figure 3.47 Zoomed-in time-domain result comparison in between 117 and 182 ms.

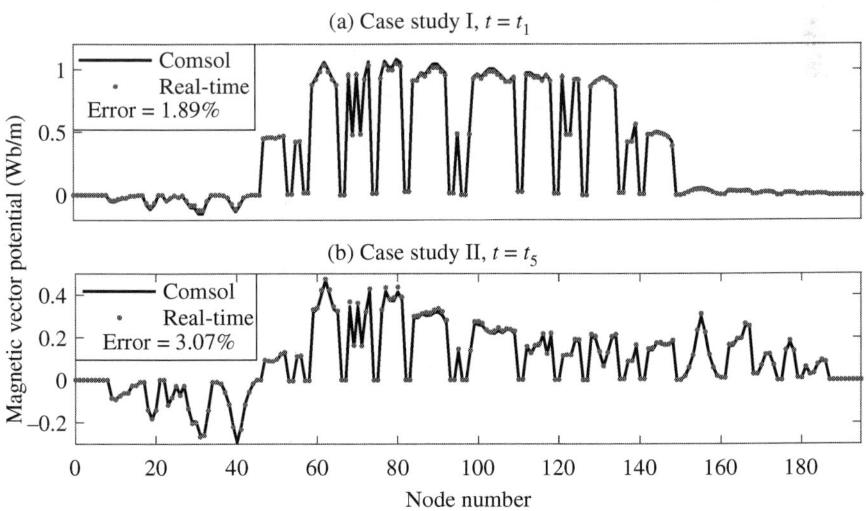

Figure 3.48 Magnetic vector potential comparison at t_1 in Case I and at t_5 in Case II.

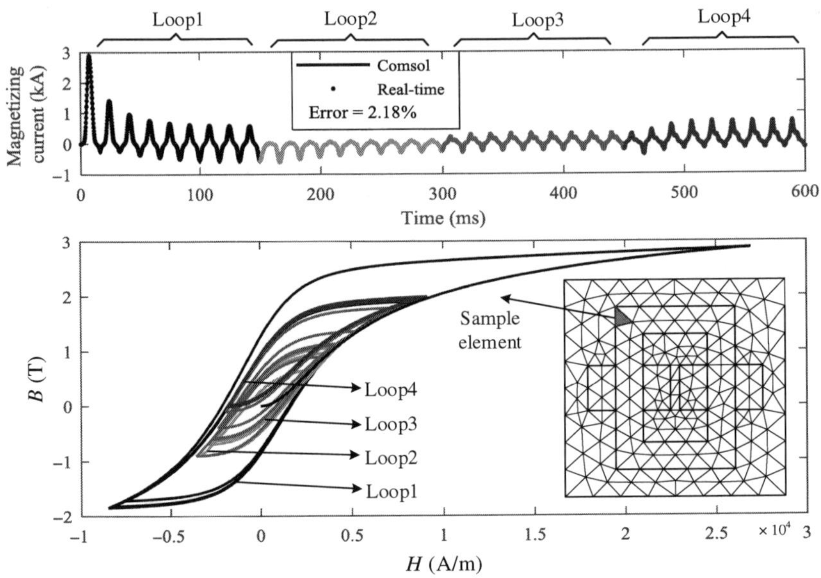

Figure 3.49 Magnetizing current and the trajectories of the hysteresis loop of a sample triangular element in Case I.

Note that the error in the case study is caused by two aspects: the FE-TLM solution and the noniterative coupling scheme. It turned out that given the same I_p and I_s, the error of the magnetic vector potential from the real-time results is less than 0.1% compared with the Comsol™ solver. Further increasing the number of TLM iterations and the Newton iterations will make this error smaller.

3.4.3.2 Speed-up and Scalability

The off-line simulation with Comsol™ was conducted on a PC with the CPU Intel Xeon E5-2620, 16 cores, 2.1 GHz clock frequency, and 32 GB RAM. The tolerance of the Comsol™ nonlinear solver is 10^{-5} and the maximum iteration number is set to 20. The order of the FE is 1. It takes 134 seconds to run 3333 simulation time-steps, and the execution time per solution step is 40 ms. By sufficiently exploring the parallelism and deep data pipelining, the speed-up of the real-time hardware implementation can reach 241 times.

The real-time model also has good scalability. Although a simple circuit is used in the case study, the real-time FE model can be easily coupled to a larger complicated external network, which can be solved independently with the accurate inductances obtained from the FE model. When applied for complex geometries that require increasing the number of nodes and elements, the resource usage of the *Assemble Current* and the *Newton Solver* hardware blocks remains unchanged and only the pipelined loop count increases, and the hardware resources usage and latency of other blocks increase linearly or logarithmically.

3.5 Summary

This chapter provided a detailed presentation of three types of power transformer EMT models and their real-time hardware emulation: a nonlinear hysteretic admittance matrix based lumped parameter model with additional frequency-dependent effects, the distributed high-fidelity nonlinear and frequency-dependent MEC based model, and the nonlinear FE model. The lumped model has the following features: linear model to update the history current terms; nonlinear phenomena such as hysteresis (Preisach model) is included with Newton iteration for solving the nonlinear equations; and a Cauer equivalent circuit was used to model eddy current behavior. Two case studies were analyzed for a single-phase transformer to show the accuracy and efficiency of this model under various transient conditions. A high-fidelity distributed topology-based model was provided for a three-phase power flow regulation transformer (the ST) with multiple windings. The real-time EMT model of the ST is based on a detailed MEC accounting for the major linear and nonlinear flux paths in the transformer core. Such a model can be used in HIL scenarios to not only implement and test new power flow control algorithms but can also be used for transformer design optimization, and for designing protection schemes. The exploitation of full parallelism and deep pipelining on the FPGA resulted in the transformer model being highly accurate with a low computational latency and resource consumption to be included in HIL simulations. The real-time model is able to accurately capture the nonlinear behavior due to core saturation, hysteresis, and eddy currents. Comparison with FE simulations proved satisfactory performance of the real-time model both under steady-state and transient conditions.

Finally, real-time FE computation of EMT in a transformer is presented. The TLM method successfully decoupled the nonlinear elements from the linear network so that the computation can be massively parallelized. The parallelism of the different phases of the TLM technique is fully explored and implemented on the FPGA with deep data pipelining. The field-circuit coupling enabled the real-time FEM to effectively interface with external circuit to capture the transient behavior of interest. Tests conducted in comparison with a commercial FEM package prove excellent accuracy and computational efficiency of the real-time FEM approach, which provided a great insight into the simulated transformer.

4

Rotating Machines

4.1 Overview

Rotating electrical machines have been an integral part of power systems for as long as the latter have existed. They are used over a wide voltage and power ratings either as generators or motors in numerous applications in power systems, and their accurate modeling is paramount for electromagnetic transient (EMT) simulation. The two commonly used circuit-oriented methods for realizing machine models are the nodal analysis and the state-space approaches. Based on these methods, there are a variety of direct and indirect methods to interface [76] a particular rotating machine model with the external power system network. The machine itself can be of many different types such as induction machine, synchronous machine, and DC machine. Each type of machine may have different electrical representation and custom interfaces between the machine and the network, and between the machine and associated mechanical system. It would be extremely cumbersome to code each of them individually into the EMT program. For this reason, the Universal Machine (UM) model was proposed [77, 78]. It is a unified, generalized model which can be used to represent up to 12 types of rotating machines. The UM model is established on the concept that all types of machines can be represented by some coupled electrical coils, and that the associated mechanical system can be replaced by the equivalent electrical analog.

To reproduce the realistic behavior of the machine for real-time HIL emulation, a detailed and accurate modeling technique is required. Just as for power transformers, the models of rotating machines can be of three kinds with varying complexity and application. Electric-equivalent lumped-parameter method such as the *qd* machine model, the magnetic equivalent circuit (MEC) based machine model, and fully distributed numerical approaches such as finite-element method (FEM) are the three options for this purpose. The *qd*-type machine models including the UM model have limited accuracy due to their inability of accounting the spatial and local nonlinear effects as well as distributed phenomena inside

Real-Time Electromagnetic Transient Simulation of AC-DC Networks, First Edition.
Venkata Dinavahi and Ning Lin.

the machine. On the other hand, although FEM offers excellent accuracy, it is computationally intensive and time-consuming even for moderately complex machine configurations, to satisfy the essential and strict execution time constraints of real-time simulation. MEC is the compromising option that can provide acceptably accurate solutions with reasonable computational effort. This chapter first focuses on the UM model emulation consistent with the nodal solution in EMTP, then a general framework for real-time emulation of state-space machine models is presented. Finally, the emulation of detailed nonlinear MEC-based machine model is demonstrated.

4.2 Lumped Universal Machine (UM) Model

4.2.1 UM Model Formulation

The UM model is established on the synchronously rotating $dq0$ reference frame. It can have at most three stator windings, any number of windings $D_1, D_2, \dots, D_m$ on the rotor direct axis (d-axis), and any number of windings $Q_1, Q_2, \dots, Q_n$ on the rotor quadrature axis (q-axis). Without loss of generality, as shown in Figure 4.1, three stator windings $\{a, b, c\}$, one field winding f, one damper winding D_1 on d-axis, and two damper windings $\{Q_1, Q_2\}$ on the q-axis are considered in this design. The voltage–current–flux relationship in these coupled windings can be described by

$$\boldsymbol{v}_{dq0} = -\boldsymbol{R}\boldsymbol{i}_{dq0} - \frac{d\boldsymbol{\lambda}_{dq0}}{dt} + \boldsymbol{u}, \tag{4.1}$$

and

$$\boldsymbol{\lambda}_{dq0} = \boldsymbol{L}\boldsymbol{i}_{dq0}, \tag{4.2}$$

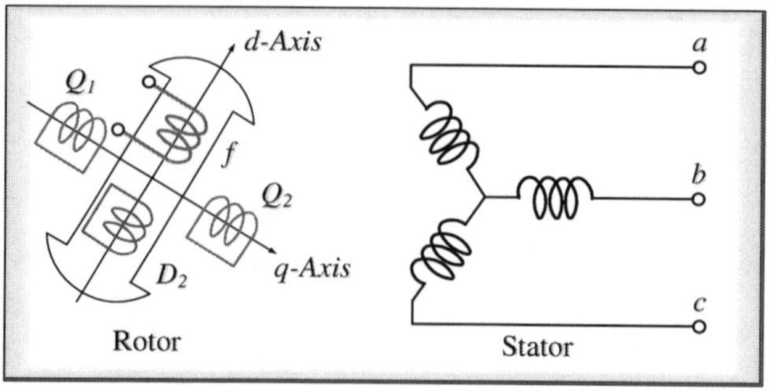

Figure 4.1 Winding representations in the UM model.

where

$$\boldsymbol{v}_{dq0} = \begin{bmatrix} v_d \\ v_q \\ v_0 \\ v_f \\ 0 \\ 0 \\ 0 \end{bmatrix}, \quad \boldsymbol{i}_{dq0} = \begin{bmatrix} i_d \\ i_q \\ i_0 \\ i_f \\ i_{D1} \\ i_{Q1} \\ i_{Q2} \end{bmatrix}, \quad \boldsymbol{\lambda}_{dq0} = \begin{bmatrix} \lambda_d \\ \lambda_q \\ \lambda_0 \\ \lambda_f \\ \lambda_{D1} \\ \lambda_{Q1} \\ \lambda_{Q2} \end{bmatrix}, \quad \boldsymbol{u} = \begin{bmatrix} -\omega\lambda_q \\ \omega\lambda_d \\ 0 \\ 0 \\ 0 \\ 0 \\ 0 \end{bmatrix}, \tag{4.3}$$

are vectors of voltages, currents, flux linkages, and speed voltages of the windings.

$$\boldsymbol{R} = \begin{bmatrix} R_d & 0 & 0 & 0 & 0 & 0 & 0 \\ 0 & R_q & 0 & 0 & 0 & 0 & 0 \\ 0 & 0 & R_0 & 0 & 0 & 0 & 0 \\ 0 & 0 & 0 & R_f & 0 & 0 & 0 \\ 0 & 0 & 0 & 0 & R_{D1} & 0 & 0 \\ 0 & 0 & 0 & 0 & 0 & R_{Q1} & 0 \\ 0 & 0 & 0 & 0 & 0 & 0 & R_{Q2} \end{bmatrix}, \tag{4.4}$$

is a diagonal matrix of winding resistances, while

$$\boldsymbol{L} = \begin{bmatrix} L_d & 0 & 0 & M_{df} & M_{dD1} & 0 & 0 \\ 0 & L_q & 0 & 0 & 0 & M_{qQ1} & M_{qQ2} \\ 0 & 0 & L_0 & 0 & 0 & 0 & 0 \\ M_{df} & 0 & 0 & L_f & M_{fD1} & 0 & 0 \\ M_{dD1} & 0 & 0 & M_{fD1} & L_{D1} & 0 & 0 \\ 0 & M_{qQ1} & 0 & 0 & 0 & L_{Q1} & M_{Q1Q2} \\ 0 & M_{qQ2} & 0 & 0 & 0 & M_{Q1Q2} & L_{Q2} \end{bmatrix},$$

is a symmetrical matrix of the winding leakage inductances with L and M denoting the self and mutual inductances, respectively.

The electromagnetic torque is calculated using

$$T_e = \lambda_d i_q - \lambda_q i_d. \tag{4.5}$$

Instead of a mechanical model of the mass-shaft system as in most machine models, the UM uses an equivalent electric network with lumped R, L, and C elements to represent the mechanical part of machine [77]. This gives the greatest flexibility to model multi-mass mechanical systems using the existing electrical models in the EMTP framework. By replacing the mechanical parameters with their analog electrical quantities as shown in Table 4.1, the differential equation (4.6) describing the dynamics of the rotor can be modeled by an electrical circuit. Figure 4.2 shows an example of a single-mass mechanical system and its electrical analog.

$$T_m = J\frac{d\omega}{dt} + D\omega + T_e. \tag{4.6}$$

Table 4.1 Equivalence between mechanical and electrical quantities for UM model.

Mechanical	Electrical
T_m: load torque (N m)	I_m: current source (A)
T_e: air gap torque (N m)	I_e: current (A)
J: inertia (kg m^2)	C: capacitance (F)
D: damping (N m s/rad)	$1/R$: conductance (S)
ω: rotor speed (rad/s)	V: voltage (V)

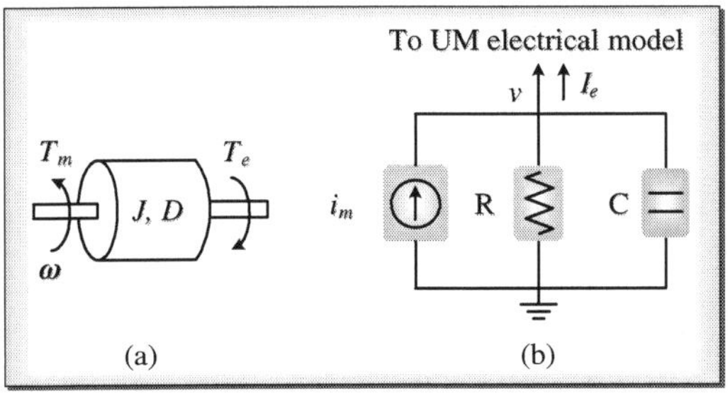

Figure 4.2 (a) Mechanical system of rotor and (b) its electrical analog.

4.2.2 Interfacing UM Model with Network

To interface the UM model to the power system network, the differential equation (4.1) is discretized using the trapezoidal rule of integration to yield:

$$\boldsymbol{v}_{dq0}(t) = -\boldsymbol{R}\boldsymbol{i}_{dq0}(t) - \frac{2}{\Delta t}\boldsymbol{\lambda}_{dq0}(t) + \boldsymbol{u}(t) + \boldsymbol{v}_{hist}, \tag{4.7}$$

where Δt is the time-step and $\boldsymbol{v}_{hist}$ is the history term expressed as

$$\begin{aligned}\boldsymbol{v}_{hist} &= -\boldsymbol{v}_{dq0}(t-\Delta t) - \boldsymbol{R}\boldsymbol{i}_{dq0}(t-\Delta t)\\ &+ \frac{2}{\Delta t}\boldsymbol{\lambda}_{dq0}(t-\Delta t) + \boldsymbol{u}(t-\Delta t).\end{aligned} \tag{4.8}$$

To incorporate the UM model into the network transient solution, the compensation method is employed. The network is first solved without the machine and represented by a Thévenin impedance $\boldsymbol{R}_{eq}$ and a voltage source $\boldsymbol{v}_{abc\_0}$ expressed as

$$\boldsymbol{v}_{abc} = \boldsymbol{v}_{abc\_0} + \boldsymbol{R}_{eq}\boldsymbol{i}_{abc}. \tag{4.9}$$

To link the $dq0$ quantities of the machine with the abc phase quantities of the rest of the network, an orthogonal Park's transformation matrix $\boldsymbol{P}$ is used, defined as

$$P = \sqrt{\frac{2}{3}} \begin{bmatrix} \cos(\beta) & \cos(\beta - 120°) & \cos(\beta + 120°) \\ \sin(\beta) & \sin(\beta - 120°) & \sin(\beta + 120°) \\ 1/\sqrt{2} & 1/\sqrt{2} & 1/\sqrt{2} \end{bmatrix}, \tag{4.10}$$

where β is the rotor angle.

Transforming (4.9) into $dq0$ frame, we obtain

$$\begin{bmatrix} v_d \\ v_q \\ v_0 \end{bmatrix} = \begin{bmatrix} v_{d\_0} \\ v_{q\_0} \\ v_{0\_0} \end{bmatrix} + \boldsymbol{R}_{eq\_dq} \begin{bmatrix} i_d \\ i_q \\ i_0 \end{bmatrix}, \tag{4.11}$$

where

$$\begin{bmatrix} v_{d\_0} \\ v_{q\_0} \\ v_{0\_0} \end{bmatrix} = \boldsymbol{P} \begin{bmatrix} v_{a\_0} \\ v_{b\_0} \\ v_{c\_0} \end{bmatrix}, \tag{4.12}$$

and

$$\boldsymbol{R}_{eq\_dq} = \boldsymbol{P}\boldsymbol{R}_{eq}\boldsymbol{P}^{-1}. \tag{4.13}$$

Substituting (4.11) into (4.7), the voltages are eliminated as

$$\boldsymbol{A}\boldsymbol{i}_{dq0} = \boldsymbol{b}, \tag{4.14}$$

where

$$\begin{aligned} \boldsymbol{A} &= \boldsymbol{R}_c + \boldsymbol{R}_{eq\_dq1} + \omega \boldsymbol{L}_2, \\ \boldsymbol{b} &= \boldsymbol{v}_{dq0\_0} + \boldsymbol{v}_{hist}, \end{aligned} \tag{4.15}$$

with

$$\boldsymbol{R}_c = \boldsymbol{R} + \frac{2}{\Delta t}\boldsymbol{L}, \tag{4.16}$$

$$\boldsymbol{R}_{eq\_dq1} = \begin{bmatrix} [\boldsymbol{R}_{eq\_dq}]_{3\times 3} & \boldsymbol{0}_{3\times 4} \\ \boldsymbol{0}_{4\times 3} & \boldsymbol{0}_{4\times 4} \end{bmatrix}, \tag{4.17}$$

$$\boldsymbol{L}_2 = \begin{bmatrix} 0 & L_q & 0 & 0 & 0 & M_{qQ1} & M_{qQ2} \\ -L_d & 0 & 0 & -M_{df} & -M_{dD1} & 0 & 0 \\ 0 & 0 & 0 & 0 & 0 & 0 & 0 \\ 0 & 0 & 0 & 0 & 0 & 0 & 0 \\ 0 & 0 & 0 & 0 & 0 & 0 & 0 \\ 0 & 0 & 0 & 0 & 0 & 0 & 0 \\ 0 & 0 & 0 & 0 & 0 & 0 & 0 \end{bmatrix}. \tag{4.18}$$

After the machine terminal currents i_d, i_q, i_0 are solved in (4.14), they are transformed back to i_a, i_b, i_c in the abc frame and superimposed into the rest of the network to calculate the compensated voltages $\boldsymbol{v}_c$. Figure 4.3 illustrates this compensation method.

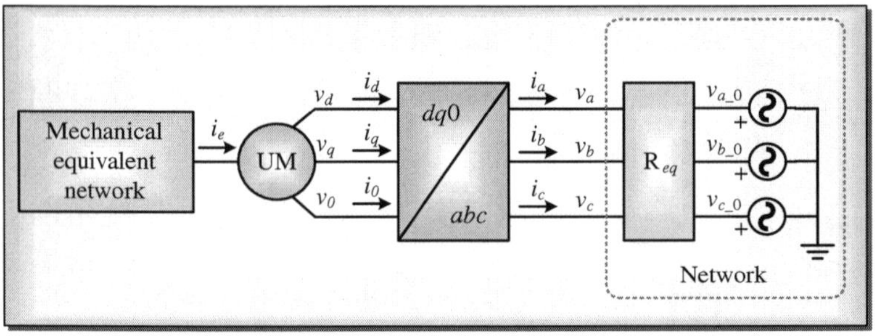

Figure 4.3 Interfacing of UM model to the network using the compensation method.

4.2.3 UM HEBB

A hardware module `UM` is designed to emulate the UM model in the FPGA. Figure 4.4a shows the symbol of this module and its input/output signals. As shown in Figure 4.4b, `UM` module mainly consists of six functional units. The `Speed & Angle` unit is responsible for predicting and calculating rotor speed ω and rotor angle β. The `FrmTran` unit transforms the UM quantities between the *abc* and *dq*0 frames. The `Parallel GJE` unit is used to calculate the machine current $\boldsymbol{i}_{dq0}$. The flux linkages $\boldsymbol{\lambda}_{dq0}$ and torque T_e are solved in the `Flux & Torque` unit. The `Update` unit updates machine history terms $\boldsymbol{v}_{hist}$ and the history terms of the equivalent mechanical network. Finally, the `ComVol` unit calculates the complete voltages of the network $\boldsymbol{v}_c$.

To solve the nonlinear equations of the machine, iterations are required. The number of iterations is generally small (1–3) due to the relatively large inertia of machine. Figure 4.5 shows the FSM diagram of the iteration process. When the network is solved without machines for applying the compensation method [43], the UM procedure starts with the prediction of rotor speed ω using linear

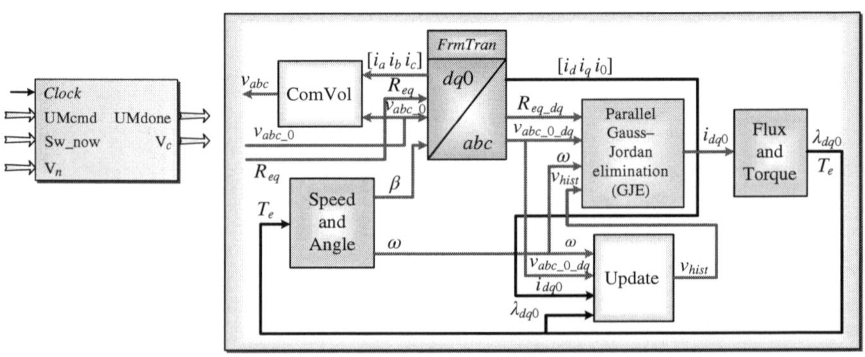

Figure 4.4 (a) *UM* HEBB and its input/output signals. (b) Main functional units in the *UM* HEBB.

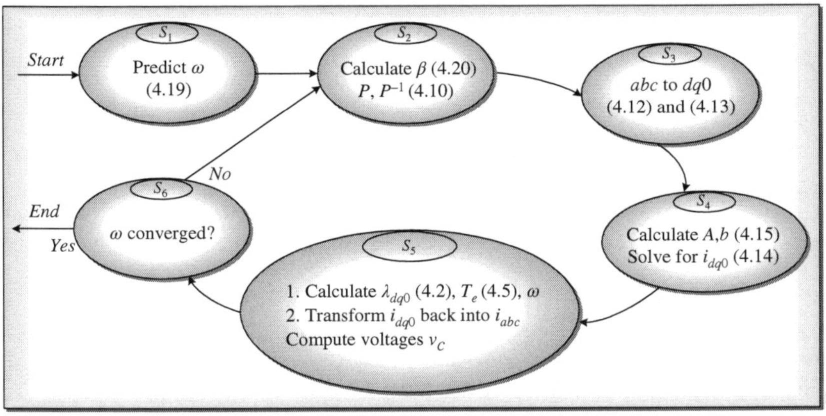

Figure 4.5 Finite-state machine for the iteration process in the *UM* HEBB.

extrapolation in state S_1. Then in state S_2, the rotor angle β is calculated, followed by Park's transformation matrices $\boldsymbol{P}$ and $\boldsymbol{P}^{-1}$ (4.10). The $\boldsymbol{P}$ and $\boldsymbol{P}^{-1}$ matrices are calculated using a sinusoidal function look-up table (LUT). The Thévenin equivalent for the rest of network is then transformed into the *abc* frame in state S_3. Then in state S_4, the current $\boldsymbol{i}_{dq0}$ is calculated using a parallel Gauss–Jordan elimination (GJE) method. Once the $\boldsymbol{i}_{dq0}$ is available, the flux linkages λ_{dq0}, electromagnetic torque T_e, and rotor speed ω are computed sequentially in *dq*0 frame in state S_5. Meanwhile, the $\boldsymbol{i}_{dq0}$ is transformed back to *abc* frame to superimpose into the rest of network to calculate the complete node voltages of the network $\boldsymbol{v}_c$ in state S_5. The calculated ω is compared with the predicted ω in state S_6. If the difference is within the given tolerance, the iteration process is terminated; otherwise, the next iteration is started.

Solving the UM model requires iterations, and each iteration involves six sequential steps. To improve the hardware computational efficiency of the UM model, parallelism has to be employed. Although the overall procedure is sequential, there exist possible parallel processing paths. For instance, in state S_3 of Figure 4.5, (4.12) and (4.13) are calculated simultaneously; in state S_4, calculating $\boldsymbol{A}$ and $\boldsymbol{b}$ (4.15) are carried out concurrently; and in state S_5, after $\boldsymbol{i}_{dq0}$ is calculated, the machine variables λ_{dq0}, T_e, and ω are calculated in *dq0* frame, while $\boldsymbol{i}_{dq0}$ is transformed back to *abc* frame thus the complete voltages $\boldsymbol{v}_c$ are calculated simultaneously. Furthermore, parallel processing can also be applied into each individual calculation as discussed in detail as follows.

4.2.3.1 *Speed & Angle* Unit

In the *Speed & Angle* unit, the rotor speed ω is predicted with linear extrapolation as

$$\omega(t) = 2\omega(t - \Delta t) - \omega(t - 2\Delta t). \tag{4.19}$$

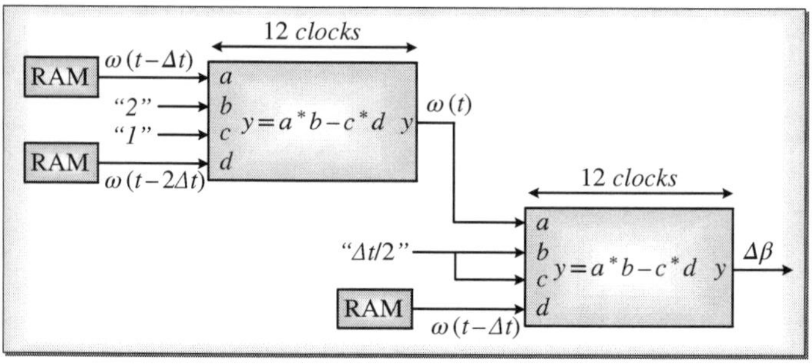

Figure 4.6 Pipelined computation scheme in the *Speed & Angle* unit.

The rotor position (incremental) $\Delta\beta$ is then calculated using the predicted speed with trapezoidal rule of integration as

$$\Delta\beta(t) = \beta(t) - \beta(t - \Delta t) = \frac{\Delta t}{2}[\omega(t - \Delta t) + \omega(t)]. \tag{4.20}$$

ω and $\Delta\beta$ are calculated in the *Speed & Angle* unit shown in Figure 4.6.

4.2.3.2 *FrmTran* Unit

In the *FrmTran* unit, the quantities in $dq0$ and abc frame are transformed. This includes transforming $\boldsymbol{v}_{abc\_0}$ into $\boldsymbol{v}_{dq0\_0}$, $\boldsymbol{R}_{eq}$ into $\boldsymbol{R}_{eq\_dq}$ in the $dq0$ frame, and $\boldsymbol{i}_{dq0}$ back to $\boldsymbol{i}_{abc}$ in the abc frame. Figure 4.7 shows its parallel computation scheme. First, the Park matrices $\boldsymbol{P}^{-1}$ and $\boldsymbol{P}$ are calculated using a LUT based

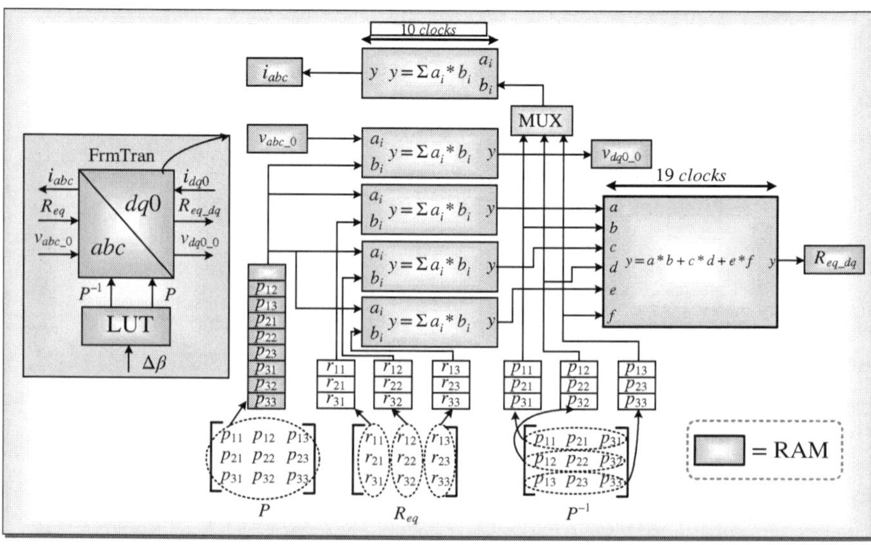

Figure 4.7 Parallel computation scheme in the *FrmTran* unit.

on the calculated $\Delta\beta$ from the *Speed & Angle* unit. The process of the LUT has been discussed in Section 2.5.1. To perform the frame transformation in parallel, $\boldsymbol{P}$ and $\boldsymbol{P}^{-1}$ are saved in various RAMs. The matrix $\boldsymbol{P}$ is stored in a single RAM, and $\boldsymbol{P}^{-1}$ is stored in three individual RAMs but in a row-wise format, while the matrix $\boldsymbol{R}_{eq}$ is also stored in three individual RAMs but in column-wise format. With this configuration, the matrix multiplication of $\boldsymbol{PR}_{eq}$ is performed by three fast floating-point multiply-accumulate (*FFPMAC*, $y = \sum a_i b_i$) units in parallel. Thus, one row of $\boldsymbol{P}$ can multiply three columns of $\boldsymbol{R}_{eq}$ resulting in three elements in one row of the product matrix simultaneously. Then these intermediate results along with three elements in one column of matrix $\boldsymbol{P}^{-1}$ are multiplied using a floating-point multiply-add ($y = a \times d + c \times d + e \times f$) unit resulting in one element of the final product matrix. Meanwhile another *FFPMAC* unit is used to calculate $\boldsymbol{v}_{dq0\_0}$ at the same time. It is worthy to note that this parallel configuration is important to reduce the execution time. For example, performing $\boldsymbol{PR}_{eq}\boldsymbol{P}^{-1}$ sequentially needs 54 clock cycles, while only 9 clock cycles are required in the presented scheme. The same configuration is used for calculating $\boldsymbol{i}_{abc}$, but a multiplexer is used to select the elements of $\boldsymbol{P}^{-1}$.

4.2.3.3 *Compi*$_{dq0}$ Unit

In the *Compi*$_{dq0}$ unit, a set of algebraic equations (4.14) are solved for $\boldsymbol{i}_{dq0}$. First, $\boldsymbol{A}$ and $\boldsymbol{b}$ are calculated (4.15). Since $\boldsymbol{A}$ is a 7×7 matrix for 7 windings representation, 49 clock cycles (not including hardware latency) are required to calculate $\boldsymbol{A}$. This is very time-consuming especially when the number of machines is large. A parallel scheme shown in Figure 4.8 is designed to improve the computational efficiency. Each row of matrix $\boldsymbol{A}$ is calculated independently. Since only first two rows of the $\boldsymbol{L}_2$ (4.18) and first three rows of the $\boldsymbol{R}_{eq\_dq1}$ (4.17) have nonzero elements, the first two rows of $\boldsymbol{A}$ are calculated by two arithmetic ($y = a + b + c \times d$) units, while the third row of $\boldsymbol{A}$ is carried out by an adder. The rest rows of $\boldsymbol{A}$ are from matrix $\boldsymbol{R}_c$ (4.16), so they are retrieved directly from corresponding RAMs. The calculation of $\boldsymbol{b}$ is performed concurrently by an adder. After $\boldsymbol{A}$ and $\boldsymbol{b}$ are available, $\boldsymbol{i}_{dq0}$ are solved in a parallel *GJE* unit which has been discussed in Section 2.9.3.3. Here, seven independent elimination units are utilized in the parallel *GJE* unit for solving a set of seven algebraic equations in parallel.

4.2.3.4 *Flux & Torque* Unit

In the *Flux & Torque* unit, the flux λ_{dq0} (4.2) and torque T_e (4.5) are calculated. As can be seen in (4.2), the matrix $\boldsymbol{L}$ is very sparse, so the sparse matrix technique is employed, as shown in Figure 4.9. After λ_{dq0} is available, T_e is calculated.

4.2.3.5 *Update & CompVc* Unit

In the *Update* unit, the history term for machine $\boldsymbol{v}_{hist}$ and for the passive elements used to represent the equivalent mechanical part of the machine are updated.

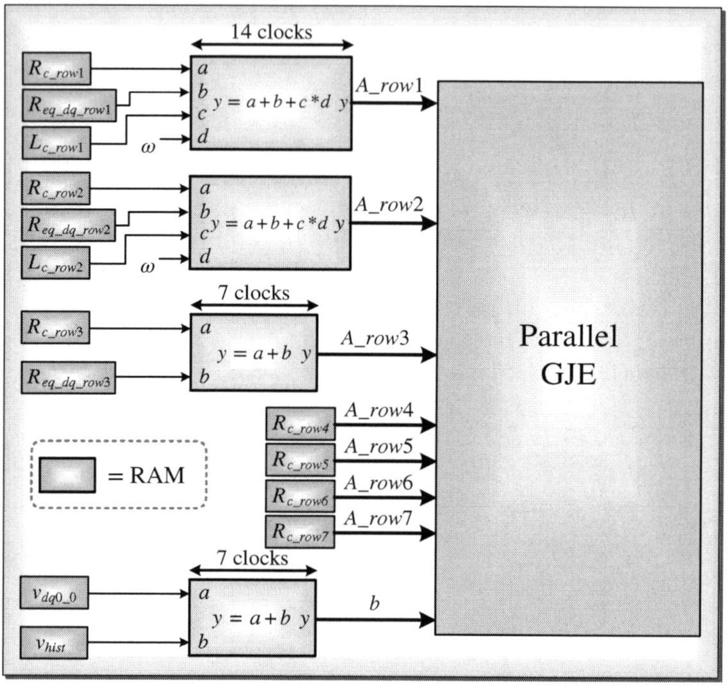

Figure 4.8 Parallel computation scheme in the *Comp*$\boldsymbol{i}_{dq0}$ unit.

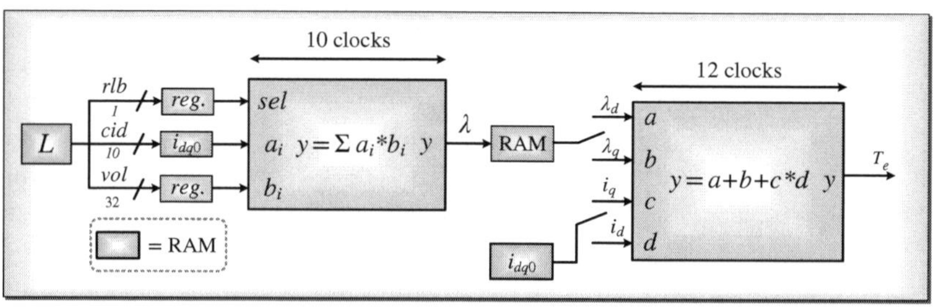

Figure 4.9 Pipelined computation scheme in the *Flux & Torque* unit.

Finally, the compensated node voltages $\boldsymbol{v}_c$ are calculated by superimposing $\boldsymbol{i}_{abc}$ into the linear network.

4.2.4 Real-Time Emulation Case Study

To show the effectiveness of UM hardware design, an example power system shown in Figure 4.10 is simulated. The system consists of three UM synchronous

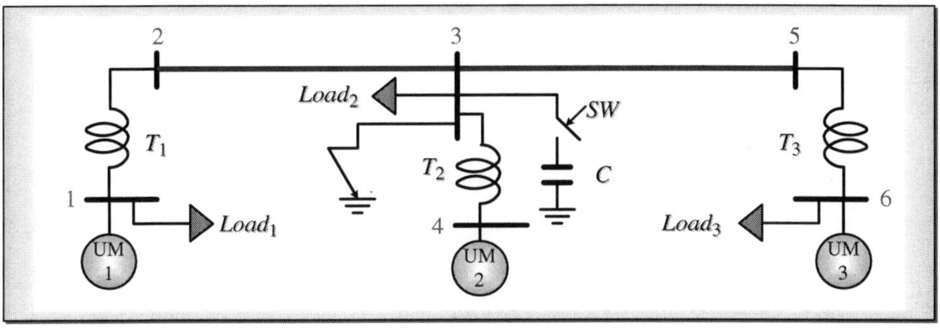

Figure 4.10 Single-line diagram of power system for the case study.

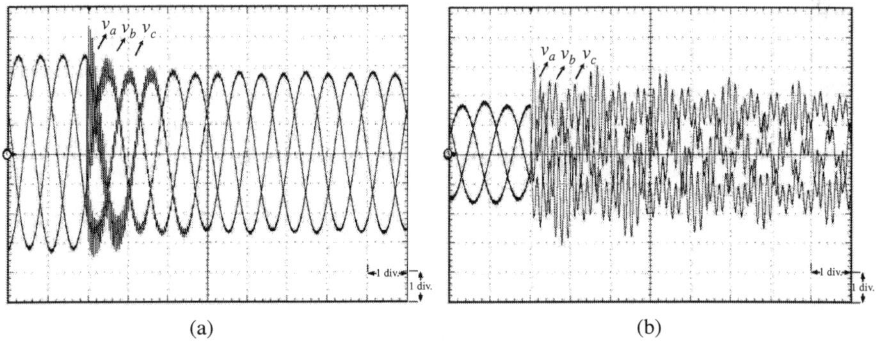

Figure 4.11 Real-time oscilloscope traces of: (a) voltages at Bus 2 during a three-phase fault at Bus 3 (scale: x-axis 1 div. = 10 ms, y-axis 1 div. = 5.2 kV) and (b) voltages at Bus 3 during a capacitor switching at Bus 3 (scale: x-axis 1 div. = 10 ms, y-axis 1 div. = 10.4 kV).

generators, two ULM lines, three transformers, and three loads [79]. The transformers are modeled as equivalent leakage reactance only. Each ULM utilizes ninth order rational functions to fit its characteristic admittance matrix and propagation matrix.

The first transient event is a three-phase-to-ground fault at Bus 3 which occurs at $t = 0.1$ second. Figure 4.11a shows the three-phase voltages waveforms at Bus 2 captured from a real-time oscilloscope. The second transient event is the capacitor C switched at Bus 3 at time $t = 0.1$ second. Figure 4.11b shows the three-phase voltages waveforms at Bus 3. These real-time simulation results have been validated by off-line transient simulation [79].

The achieved time-step in this implementation is 6.72 μs based on a FPGA clock frequency of 130 MHz. The consumed hardware resources for the UM module include 22% logic elements, 22% multipliers, and 30% memory utilization on the Altera Stratix III EP3SL150 FPGA. A detailed breakdown of this time-step

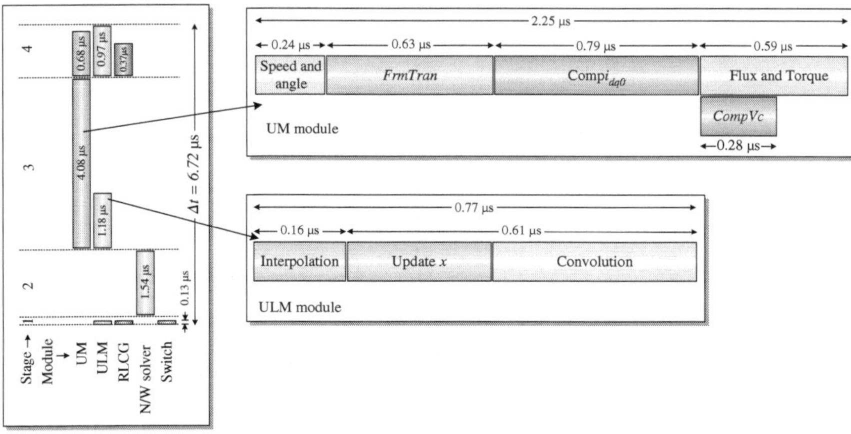

Figure 4.12 Breakdown of one time-step (Δt) in μs for various stages and modules for the case study.

for various stages of computation and modules is shown in Figure 4.12. The computation for the UM equations in *Stage 3* is most time-consuming, and the updating of history terms in the *ULM* module has the second highest latency. The detailed execution time of one UM generator and one ULM line in *Stage 3* are also shown in Figure 4.12. The computation of the three UM generators in the system is pipelined through one *UM* module, while the computation of the two ULM lines is pipelined through one *ULM* module. Depending on the size of the system and the time-step required for a certain transient, more UM and ULM modules can be replicated on a larger FPGA to either reduce the time-step or accommodate a large system.

4.2.5 Overall Power System HEBB for Real-Time EMT Emulation

To integrate all the equipment hardware emulation building blocks (HEBBs) developed in preceding chapters into a single FPGA, a power system hardware emulation building block (*PS* HEBB) was built, as shown in Figure 4.13a. Each equipment HEBB has dedicated hardware resources. The (*PS* HEBB) is extensible and flexible. Each type of equipment can have multiple HEBBs depending on the objective of the transient study, the model complexity of the component HEBBs, and the available hardware resources. A control module was designed to coordinate the operation of all HEBBs to fulfill the power system transient emulation. The finite-state machine of the *PS* HEBB is shown in Figure 4.13b. The real-time emulation time-step involves three stages: solving the network for voltages $\boldsymbol{v}_n$ without taking into account machines and nonlinear elements (S_0, S_1), solving machine equations and nonlinear equations for compensated voltages $\boldsymbol{v}_{c1}$

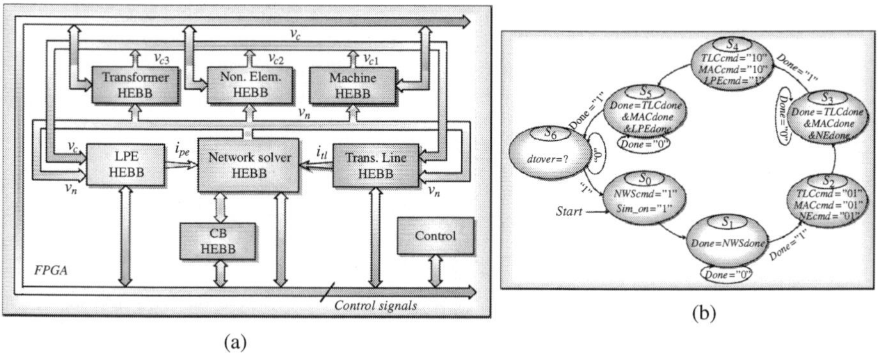

Figure 4.13 (a) Power system HEBB and (b) finite-state machine for *PS* HEBB.

and $\boldsymbol{v}_{c2}$ (S_2, S_3), and finally updating history terms (S_4, S_5). The admittance matrix $\boldsymbol{Y}$ of the system is assembled based on the discrete-time equivalents of all the system functional components. Then, the system nodal equation $\boldsymbol{Y}\boldsymbol{v}_n = \boldsymbol{i}$ is solved at every emulation time-step Δt for the node voltage vector $\boldsymbol{v}_n$ (without the machines and nonlinear elements). The current injection vector $\boldsymbol{i}$ is assembled from $\boldsymbol{i}_{tl}$ and $\boldsymbol{i}_{pe}$. Then, the unknown node voltages are obtained by a sparse matrix vector multiply of the inverse system nodal admittance matrix $\boldsymbol{Y}^{-1}$ with the current injections $\boldsymbol{i}$. While in this method the $\boldsymbol{Y}^{-1}$ has to be calculated *a priori* for all combinations of the circuit breaker switches, it has the advantage of reducing memory utilization and increasing the calculation efficiency significantly. A formal sparse LU decomposition and substitution method can also be used to solve for $\boldsymbol{v}_n$ provided there are enough hardware resources available. As shown in Figure 2.27b, the current vector $\boldsymbol{i}$ is first assembled by an adder. Then, a floating-point multiply-accumulator (*FPMAC*) unit is used for sparse matrix-vector multiplication for calculating node voltage $\boldsymbol{v}_n$.

To accommodate large network sizes or reduce the simulation time-step for high frequency transients, multiple HEBBs of various component types can be implemented depending on the available hardware resource of the FPGA. Depending on the study system requirements, the aim was to implement more parallel HEBBs of the same type to maximize the resource utilization of the FPGA, and minimize hardware latency. In this design, the Altera Stratix III EP3SL340 FPGA was used. Table 4.2 lists the main logic resources utilized by each type of HEBB. As can be seen, in one *PS* HEBB, two HEBBs are implemented for each of the *Trans. Line*, *Non. Elem.*, and *Machine*, while one HEBB each is implemented for *LPE*, *Switch*, and *NWS* components. When the utilized resources exceed the available resources, pipelining can be used to accommodate more components at the cost of a higher latency and thus a larger time-step. A larger FPGA would be able to accommodate more HEBBs of the same type.

Table 4.2 Resources utilized by *PS* HEBB.

HEBB	Combinational ALUTs	DSP 18-bit multipliers	Memory (kbits)
1 × *Linear passive element*	33 481 (12.38%)	64 (11.11%)	842.16 (5.18%)
2 × *Transmission line*	89 737 (33.19%)	192 (33.33%)	1 407.74 (8.65%)
2 × *Machine*	49 951 (18.47%)	176 (30.56%)	1 745.18 (10.73%)
2 × *Nonlinear element*	23 384 (8.65%)	32 (5.56%)	406.80 (2.50%)
1 × *Switch*	50 (0.02%)	0 (0%)	408 (2.50%)
1 × *Network solver*	35 212 (13.02%)	64 (11.11%)	808.28 (4.97%)
Total	231 814 (85.73%)	528 (91.67%)	5 617.54 (34.53%)

A test system was built using the modified IEEE New England 39-bus test system, as shown in Figure 4.14. The *PS* HEBB residing on the Stratix III EP3SL340 FPGA was able to accommodate three such systems connected with transmission lines. A time-step of $\Delta t = 45$ μ seconds was used for the real-time emulation of this interconnected system. For capturing EMTs of up to 10 kHz in real time, a time-step of $\Delta t = 50$ μ seconds is acceptable according to Nyquist criterion. The total number of three-phase components emulated in real time within a $\Delta t = 45$ μ seconds time-step for high-fidelity transient emulation include 300 generators, 1170 transmission lines, 990 linear passive elements, and 90 nonlinear surge arresters. Note that the transformers were modeled using equivalent RLC elements. All ULM lines utilized ninth order rational functions for $\boldsymbol{Y}_c$ and $\boldsymbol{H}$, and all synchronous machines utilize the eighth order UM model. On a device with higher logic resources, either the time-step can be decreased significantly for a given system size, or the number of emulated components can be scaled up for a fixed time-step. This choice is primarily determined by the needs of the specific case study being carried out, and can be achieved by emulating multiple parallel HEBBs of each component type on the FPGA comprising the *PS* HEBB.

Figure 4.15 shows a sampling of real-time transient results captured on the oscilloscope. A three-phase fault is emulated at Bus 2 (Figure 4.14). The fault is initialized at $t_1 = 0.5$ second and cleared at $t_2 = 0.6$ second. Figure 4.15a shows the three-phase voltages at Bus 3 over a period of 200 ms. The high-frequency EMTs can be seen clearly at both t_1 and t_2 at fault initiation and fault recovery respectively. Figure 4.15b shows the zoomed view of the voltage transients around fault recovery at time t_2. Figure 4.15c shows the three-phase fault currents over a period of 200 ms at Bus 2, with the maximum fault current of 4.3 kA in phases b and c. The slower electromechanical dynamics of the synchronous generators over a period of 500 ms are shown in Figure 4.15d which depicts the rotor angle (δ)

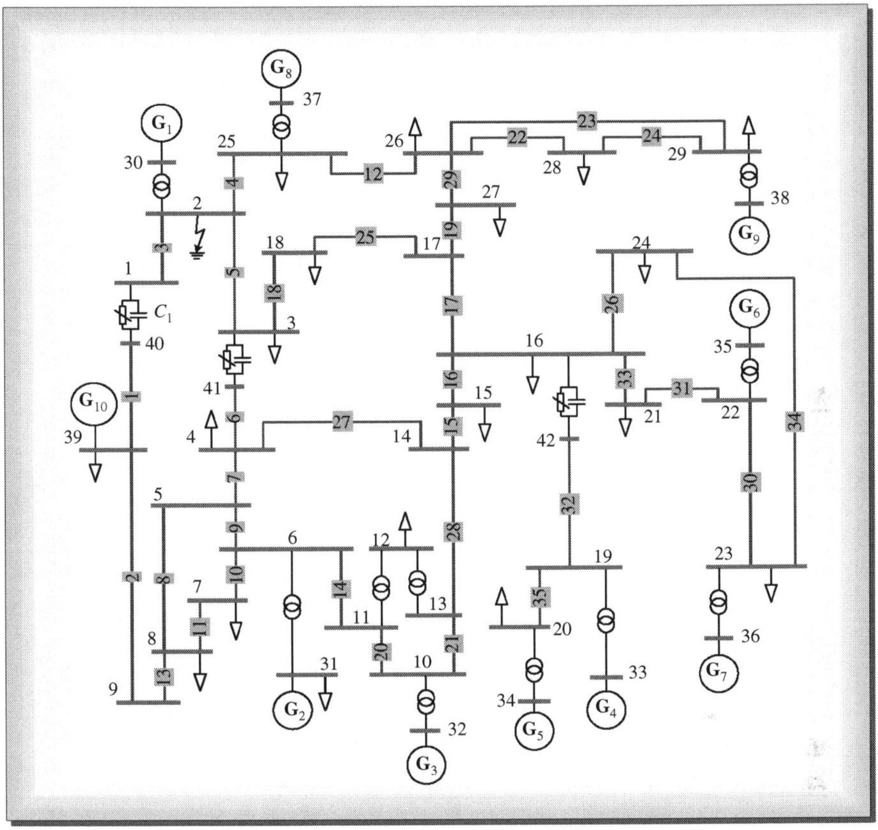

Figure 4.14 Single-line diagram of the power system modeled in case study.

oscillations of $\boldsymbol{G_1}$, $\boldsymbol{G_2}$ and $\boldsymbol{G_{10}}$. As can be seen from Figure 4.14, $\boldsymbol{G_1}$ is the nearest generator to the fault, while $\boldsymbol{G_2}$ and $\boldsymbol{G_{10}}$ are further away from the fault. During the fault, the oscillations of $\boldsymbol{G_1}$ are more pronounced compared to those of $\boldsymbol{G_2}$ and $\boldsymbol{G_{10}}$.

For real-time emulation of large-scale power grids, multiple *PS* HEBBs have to be utilized. Practically, a large power system can be partitioned into smaller subsystems interconnected using transmission lines. Due to the inherent travel time delay of the transmission line, each subnetwork can be simulated in a *PS* HEBB independently. At the end of each emulation time-step, the node voltages between the neighboring *PS* HEBBs would have to be exchanged. However, this may be difficult to achieve on a single FPGA device due to the limitation on available logic resources. Multiple FPGAs would have to be employed invariably to enable massive parallelism and pipelining for large-scale systems.

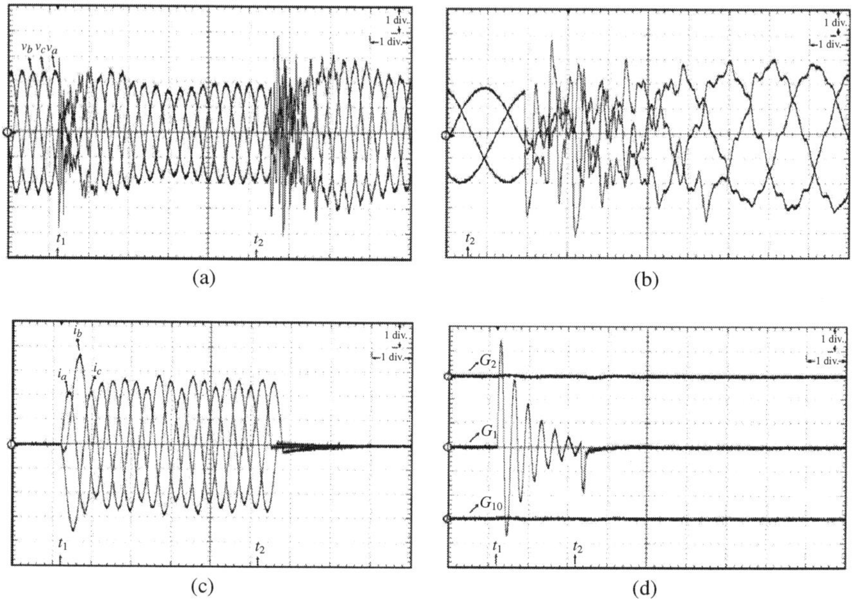

Figure 4.15 (a) Real-time transients for Bus 3 voltages during a three-phase fault at Bus 2 (x-axis 1 div. = 20 ms, y-axis 1 div. = 1.8 kV), (b) zoomed view of Bus 3 voltages (x-axis 1 div. = 5 ms, y-axis 1 div. = 1.8 kV), (c) fault currents during the three-phase fault at Bus 2 (x-axis 1 div. = 50 ms, y-axis 1 div. = 1.2 kA), and (d) power angles (δ) for generator G_1, G_2, and G_{10} (x-axis 1 div. = 50 ms, y-axis 1 div. = 0.0082 rad).

4.3 General Framework for State-Space Electrical Machine Emulation

This section presents a unified framework for FPGA-based real-time emulation of rotating machine models based on the state-space approach. Off-line transient simulations have been used as a successful method to evaluate the performance of electrical machines for a long time. Although execution time is still a matter of concern in such off-line tools, it is not as critical as in a real-time digital simulator, which has to interact with external devices in a HIL scenario. The main application of real-time emulation of electrical machines is to evaluate the behavior of newly designed machines, drive systems, controllers, and protective devices in a HIL configuration in an effective and economic approach before applying them in a real system. In this section, a comprehensive comparison between fixed-point and floating-point implementation, deeply pipelined and paralleled architecture, as well as two user-oriented realization methods is presented to evaluate FPGA-based real-time emulation of the machine models in the general framework. Finally, the effectiveness of the emulated machine models are assessed and highlighted by providing a number of case studies.

4.3.1 FPGA Design Approaches for Electrical Machine Emulation

The techniques to emulate a system on FPGA can be broadly classified into two groups: (i) textual programming language (TPL) such as hardware description language (HDL) and (ii) schematic method from vendor specific blocksets. The textual programming method by means of HDL such as VHDL or Verilog HDL is a powerful method to develop a digital hardware design without any restrictions. However, it can be very complex and cumbersome to debug. Even experts in machine modeling and simulation may find programming in HDL a daunting task. The schematic or the model-based method relies on a library of basic combinatorial and sequential building blocks offered in the Altera DSP Builder®, Xilinx System Generator® blocksets, etc. within the Matlab/Simulink® environment. This method allows users to go from system simulation using the industry-standard Mathworks simulation tools to hardware implementation in a short time. This method is user-friendly, easier to troubleshoot, easily understandable, and specifically useful for novice users of embedded digital systems. Expert users will of course be able to save development time by adopting intellectual property (IP) core blocks already available in such blocksets with no HDL hard-coding. However, this approach is still limited to applications in which no complex sequencers and deeply pipelined structures are required [80].

Furthermore, the FPGA is a space-oriented logic device that enables full hard-wired parallelism to be achieved to the extent permitted by the implemented user model and algorithm. A large number of customized parallel processing units can be easily configured. Thus, a highly parallel implementation is the best realization according to the FPGA architecture (one data per hardware module per time-step) for the algorithms where a lower amount of hardware resources for computational processing is required such as fixed-point arithmetic based designs. The integrated massive memory blocks can be partitioned into many independent types such as RAM, ROM, FIFO, single port, or dual-port user memory units through which multiple data can be accessed simultaneously. Although multiple data can be processed in parallel on an FPGA, due to resource limitation, it is difficult to achieve the ideal parallelism for large systems and not area-optimized for the algorithms that utilize massive amount of hardware resources such as floating-point calculation based designs. In such cases, the pipelining technique has to be used. In a pipelined scheme, a function is divided into several stages by inserting registers between stages, allowing multiple data to be processed at different stages at once, resulting in a high computational throughput (multiple data per hardware module per time-step).

As a consequence of the above reasons, the seamless and user-friendly schematic method is a suitable technique for a highly parallel implementation of real-time machine model based on fixed-point operations, while on the other hand, the TPL method is the most appropriate approach for the pipeline realization of floating-point calculation based real-time system, resulting in an area-optimized hardware architecture acceptable in the industrial applications.

4.3.2 State-Space Representation of Machine Models

The governing equations describing magnetically coupled stator and rotor circuits in an electrical machines, whose windings are identical, and symmetrically placed and has constant parameters, can be written as follows:

$$\begin{bmatrix}\mathbf{V}_{abc_s}\\ \mathbf{V}_{abc_r}\end{bmatrix} = \begin{bmatrix}\mathbf{r}_s & \mathbf{0}\\ \mathbf{0} & \mathbf{r}_r\end{bmatrix}\begin{bmatrix}\mathbf{I}_{abc_s}\\ \mathbf{I}_{abc_r}\end{bmatrix} + p\begin{bmatrix}\lambda_{abc_s}\\ \lambda_{abc_r}\end{bmatrix},$$
$$\begin{bmatrix}\lambda_{abc_s}\\ \lambda_{abc_r}\end{bmatrix} = \begin{bmatrix}\mathbf{L}_{abc_{ss}} & \mathbf{L}_{abc_{sr}}\\ \mathbf{L}_{abc_{rs}} & \mathbf{L}_{abc_{rr}}\end{bmatrix}\begin{bmatrix}\mathbf{I}_{abc_s}\\ \mathbf{I}_{abc_r}\end{bmatrix}, \tag{4.21}$$

where $\mathbf{V}$, $\mathbf{I}$, λ, $\mathbf{r}$, and $\mathbf{L}$ denote voltage, current, flux linkage, resistance, and inductance matrices, respectively. Additionally, subscript letters including abc, r, and s are phase domain, rotor, and stator indexes, respectively.

The equations representing the dynamic behavior of machines consist of an inductance matrix as a function of rotor position. Thus, a change of frame is used to reduce the complexity of this matrix by referring the machine variables to a reference frame that rotates at an arbitrary velocity given by [81]:

$$\mathbf{f}_{qd0} = \mathbf{K}(\theta)\,\mathbf{f}_{abc},$$
$$\mathbf{K}(\theta) = \frac{2}{3}\begin{bmatrix}\cos\theta & \cos\left(\theta - \frac{2\pi}{3}\right) & \cos\left(\theta + \frac{2\pi}{3}\right)\\ \sin\theta & \sin\left(\theta - \frac{2\pi}{3}\right) & \sin\left(\theta + \frac{2\pi}{3}\right)\\ \frac{1}{2} & \frac{1}{2} & \frac{1}{2}\end{bmatrix}, \tag{4.22}$$

where $\mathbf{f}$ represent voltage, current, or flux linkage vector of stator and rotor circuits. Also, $\mathbf{K}(\theta)$ is a transformation matrix and θ is arbitrarily selected.

Using state-space approach in the orthogonal qd axis model, the simulation of electrical side of various machines can be expressed as:

$$\begin{cases}\frac{d}{dt}\mathbf{x}(t) = \mathbf{A}\mathbf{x} + \mathbf{B}\mathbf{u},\\ \mathbf{y} = \mathbf{C}\mathbf{x} + \mathbf{D}\mathbf{u},\end{cases} \tag{4.23}$$

where $\mathbf{x} \in \mathbb{R}^h$ is the state vector, $\mathbf{u} \in \mathbb{R}^p$ is the input vector, and $\mathbf{y} \in \mathbb{R}^q$ is the output vector. Plus, $\mathbf{A}_{h\times h}$, $\mathbf{B}_{h\times p}$, $\mathbf{C}_{q\times h}$, and $\mathbf{D}_{q\times p}$ denote state, input, output, and feedback matrices in state-space equations. Flux linkages (λ_{qd}), supply voltages ($\mathbf{V}_{qd}$), and output currents ($\mathbf{I}_{qd}$) are selected as the state, input, and output variables, respectively. The matrices and vectors in (4.23) are defined in Appendix A.3 for different types of machines.

The second set of equations is for the mechanical side of machines for which electromagnetic torque (T_e), and rotor speed (ω_r) are the input and state vectors, represented by

$$T_e = \left(\frac{3}{2}\right)\left(\frac{P}{2}\right)\left(\lambda_{ds} i_{qs} - \lambda_{qs} i_{ds}\right), \tag{4.24}$$

$$\dot{\omega}_r = \frac{P}{2J} T_{effective} = \frac{P}{2J}\left(T_e - T_{mech} - T_{damp}\right), \tag{4.25}$$

where T_e, T_{mech}, and T_{damp} represent electromagnetic, mechanical, and damping torques. The rotor inertia and number of pole pairs are denoted by J, and P, respectively.

For the hardware realization, the differential equations of machine model should be discretized. The implicit techniques for discretization are more expensive due to a root finding procedure at any given time-step (T_s). The important observation regarding explicit methods are that the unknown quantities at each time-step are given in terms of history parameters. Thus, the discretized equations with explicit technique can be computed within a shorter elapsed time compared with the implicit one with the same order, resulting in a reduced local truncation error at every time-step, as well as a reduced global error, and a higher simulation accuracy. However, the drawback arises from the limitation on the time step-size in the explicit method to ensure numerical stability. Since the FPGA can provide nanosecond computation clock cycles, a very small simulation time-step, which is much smaller than electrical and evidently mechanical time constants of machines can be achieved to capture all transients of machine behavior.

The explicit Adams-Bashforth (A-B) method can be employed as a suitable choice to discretize the machine equations based on $s = \frac{z-1}{T_s}$ for first-order or $s = \frac{2}{T_s}\frac{z^2-1}{3z-1}$ for second-order Adams-Bashforth transformation [82]. In this work, the first-order A-B method is chosen for discretization of the state-space equations as follows:

$$\begin{cases} \mathbf{x}(t) = \mathbf{x}(t - T_s) + T_s \times [\mathbf{A}(t - T_s)\mathbf{x}(t - T_s) + \mathbf{B}(t - T_s)\mathbf{u}(t - T_s)], \\ \mathbf{y}(t) = \mathbf{C}(t)\mathbf{x}(t) + \mathbf{D}(t)\mathbf{u}(t). \end{cases} \tag{4.26}$$

4.3.3 System Configuration on FPGA

4.3.3.1 Number Representation

Choosing either a fixed-point or floating-point number representation is the first step for any hardware design. To provide a comparison for hardware resource utilization and achieved accuracy, a 32-bit single precision floating-point format

(*IEEE* Standard 754) by TPL method (VHDL) and 32-bit fixed-point format (binary point is located where the integer part of the maximum or minimum value of flowing data is effectively fitted in the underlying integer part of fixed-point number for the highest accuracy) by Schematic method are used for real-time emulation in this work. In the realized machine blocks for real-time emulation, TPL (HDL coding) and schematic method for different hardware platforms remain the same and the only difference is between the IP cores of one FPGA type to another designed just for the implementation of basic arithmetic operations.

4.3.3.2 Floating-Point Implementation by VHDL

The hardware modules that assist the real-time emulation to be executed include: *Main Control*, *Source*, *Timer & Switch*, and *Electrical Machine*.

Figure 4.16 shows the overall procedure in an emulation time-step in the proposed hardware design. First, *Source Module* generates input voltages for the three phase machine terminals, whereas *Switch Module* checks the switch states to apply or remove faulty conditions and load torque. Then *Electrical Machine Module* starts to solve machine's equations. It is obvious that the parallel processing exists in each stage, while preserving the necessary sequential stages in the overall simulation algorithm.

The *Main Control Module* coordinates the operation of the whole emulator to carry out the algorithm. It sends out control signals (*Simu_on*, *Socmd*, *EMcmd*) to each module to perform the required functions. Meanwhile, it receives the acknowledged signals (*Sodone*, *EMdone*, *dtOver*, *Sw*) to judge if the functions are done or the switches are on or off.

Using the *rst* function on FPGA board, the hardware emulator starts real-time execution. At first, *Main Control* sends out a command signal (*Socmd*) to the *Source Module* for generating supply voltages while it checks the switch status signals (*Sw0*, *Sw1*, *Sw2*, …) to apply or remove voltage source to electrical machine terminals or mechanical torque to the machine shaft. Once acknowledged signals are received from *Source* and *Switch Modules*, main control sends out a command signal (*EMcmd*) to the *Electrical Machine Module* to compute the state variables of the machine. Then, it waits to be given acknowledge signal (*EMdone*) from machine module and transmits real-time signals to output ports of emulator. Finally, *Main Control* checks whether or not real-time procedure is performed within the emulation time-step (*dtOver*) and decides to go to the next time-step or send the error signal to output ports and terminate the real-time emulation procedure.

The voltage source in the *Source Module*, as shown in Figure 4.17, is represented using sinusoidal function (cos function). The accuracy of the cos value is determined by the length of the LUT used to store this nonlinear function. In this design, 4096 (2^{12}) cos values for half cycle are stored in the LUT; thus, the

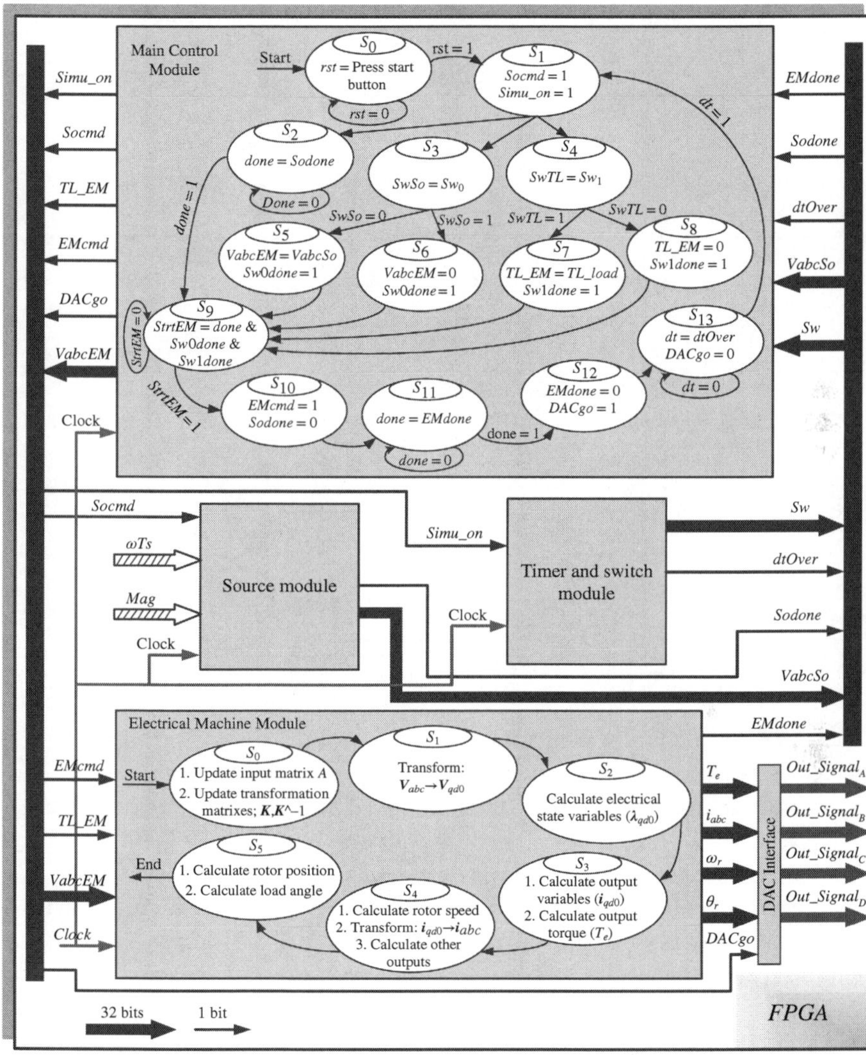

Figure 4.16 FSM for paralleled and pipelined real-time electrical machine emulation.

Resolution of the LUT is $d\theta = \pi/4096$. The calculation of the source values begins with the updating of the *phase* angle. The *new_phase* is obtained by adding the previous *phase* by ωT_s, where ω denotes angular frequency. Since the LUT has only half cycle of cos functions values, the calculated *phase* needs to be checked with π; if greater than π, it is subtracted by π and the sign of the result is inverted. Then, the *new_phase* is converted to the address of the LUT. Finally, the retrieved cos value is multiplied by the magnitude *Mag*. The exponent and mantissa of the input

θ_i	dexp	1. mantissa(θ_i/Resolution)
0.00000	0	0.0000000000000...000
0.00077	0	1.0000000000000...000
0.00153	1	1.0000000000000...000
⋮	⋮	⋮
0.74723	9	1.1110011100000...000
0.74800	9	1.1110011110000...000
⋮	⋮	⋮
1.43692	10	1.1101010001000...000
1.43769	10	1.1101010010000...000
1.43846	10	1.1101010011000...000
⋮	⋮	⋮
3.14083	11	1.1111111111000...000
π	11	1.1111111111100...000

Address	cos(θ_i)
0	1.000000
1	1.000000
2	0.999999
⋮	⋮
974	0.733573
975	0.733052
⋮	⋮
1873	0.133473
1874	0.132713
1875	0.131952
⋮	⋮
4094	−1.000000
4095	−1.000000

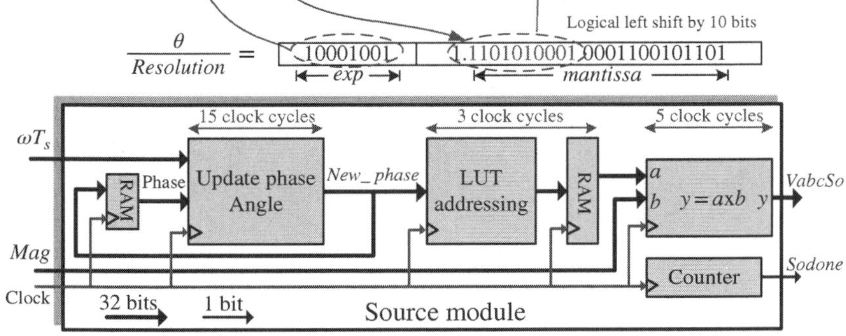

Figure 4.17 Pipelined configuration for calculating *Source Module.*

floating-point number are used directly to access the LUT when the step length is always a power of 2. Assume the floating-point input is θ, the *LUT addressing* unit in Figure 4.17 outputs the addresses of the point θ_i making $\theta - \theta_i < Resolution$, where *Resolution* is the interval of the LUT. This is done by left shifting the leading "1" and mantissa of $\theta/Resolution$ by *dexp* bits, where the *dexp* is the exponent of $\theta/Resolution$ (without bias). An example is shown in Figure 4.17. In this example, input θ is 1.437 whose *dexp* is 10 (without bias). Left shifting the leading "1" and mantissa 10 bits gives 1873 which is the address of $\theta_i = 1.436\ 92$ and $\cos(\theta_i) = 0.133\ 473$.

A hardware module of *Timer & Switch* is designed to simulate switches. Figure 4.18 shows the details of this module and its input/output signals. Since the switches are time controlled, the core of this module is a real-time clock generator. The best achievable clock frequency is generated by the input system clock frequency (200 MHz). This clock signal is counted and compared with the switch operation times saved in a RAM. Once the switch times are reached, the corresponding switch state bit in *Sw* register (*Sw0*, *Sw1*, *Sw2*, …) is inverted using "0"/"1" for switch open/closed. Another important function of the *Switch Module*

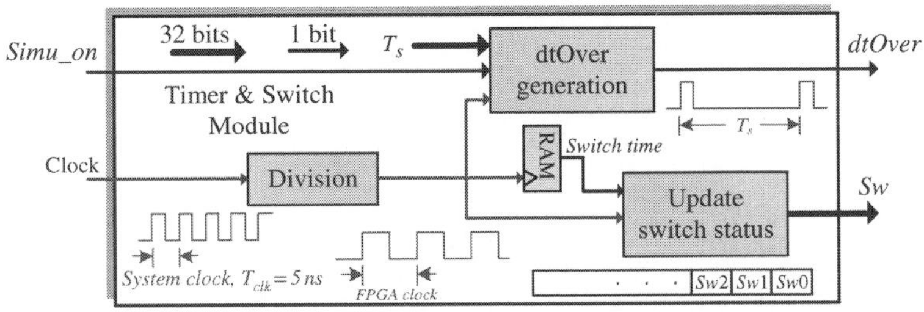

Figure 4.18 Functions realized in the *Timer & Switch Module.*

is to generate the *dtOver* signal which indicates the end of the real simulation time-step T_s. When a simulation step is finished, the T_s-over signal is checked. If it is not "1," the simulation step is finished within T_s; thus, the real-time simulation is achieved. Otherwise, the simulation step takes longer time than T_s, and the real-time constrain is not met [24].

The FSM for the hardware realization of machine models is depicted in Figure 4.16. The procedure starts with the update of input Matrix (**A**) using the calculated speed from the previous time-step in parallel with the computation of transformation matrices (**K** and $\mathbf{K}^{-1}$) using sinusoidal function LUT based on information about rotor and reference frame position from the passed time-step in state S_0. Then, in state S_1, $\mathbf{V}_{abc}$ is transformed to an arbitrary reference frame (*qdo*) to provide input vector for state-space equations. Once $\mathbf{V}_{qd}$ is available, the electrical state variables (λ_{qd}) are calculated in state S_2. The output state variables ($\mathbf{I}_{qd}$) and output torque (T_e) are computed concurrently in state S_3. The $\mathbf{I}_{qd}$ is transformed back to *abc* frame in state S_4, meanwhile, rotor speed (ω_r) and other required outputs are obtained. Finally, rotor position and load angle are calculated simultaneously in state S_5. Obviously, since the implementation involves six sequential steps, it is time-consuming. To improve the hardware computational efficiency of the deeply pipelined modules, all possible parallel processing paths have been taken into account, although the overall procedure is sequential.

The hardware realization of electrical machines is achieved by manipulation of sparse matrix multiplication (*SprMxMul*) and floating-point multiply-add/subtract (*FLPMAS*) units. These submodules can be chained to realize many functions and computations such as simple addition/subtraction, multiplication, conversion between floating- and fixed-point numbers, $a \times b \times c$, $a \times b \pm c \times d \pm e \times f$, $(a \pm b) \times c$, $\mathbf{A}_{n\times1} \pm c \times \mathbf{B}_{n\times1}$, $\mathbf{A}_{n\times n} \times \mathbf{B}_{n\times1} \pm c \times \mathbf{D}_{n\times1}$, etc.

A fast sparse matrix multiplication submodule where a compact sparse matrix storage format (Figure 4.19) which uses only one vector is defined for the realization of $\mathbf{A}_{n\times n} \times \mathbf{B}_{n\times1}$. Each entry in this format has the following: (i) A 32-bit value to store the subsequent nonzero value of matrix $\mathbf{A}_{n\times n}$ in row order; (ii) a

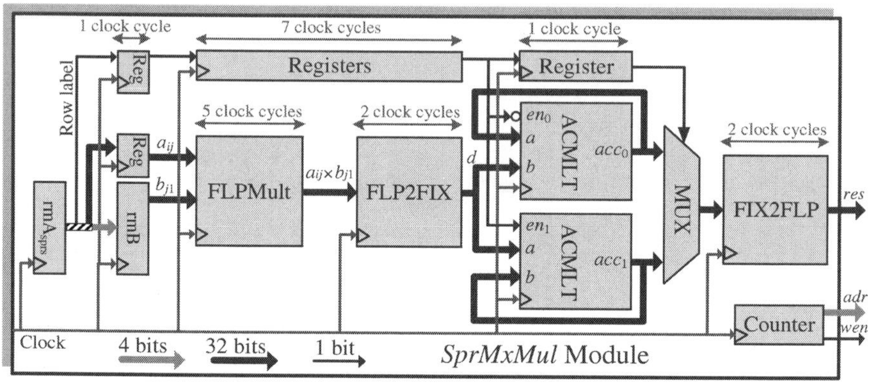

(a)

$$A = \begin{bmatrix} a_{11} & 0 & 0 & 0 \\ 0 & 0 & a_{23} & a_{24} \\ a_{31} & a_{32} & 0 & a_{34} \\ 0 & 0 & a_{43} & 0 \end{bmatrix}$$

$$B = \begin{bmatrix} b_{11} \\ b_{21} \\ b_{31} \\ b_{41} \end{bmatrix}$$

rmB	
RAM address	RAM value (31:0)
0	b_{11}
1	b_{21}
2	b_{31}
3	b_{41}

rmA_{sprs}			
RAM address	RAM value		
	Row label (36)	A Column (35:32)	A elements (31:0)
0	0	1	a_{11}
1	1	3	a_{23}
2	1	4	a_{24}
3	0	1	a_{31}
4	0	2	a_{32}
5	0	4	a_{34}
6	1	3	a_{43}

(b)

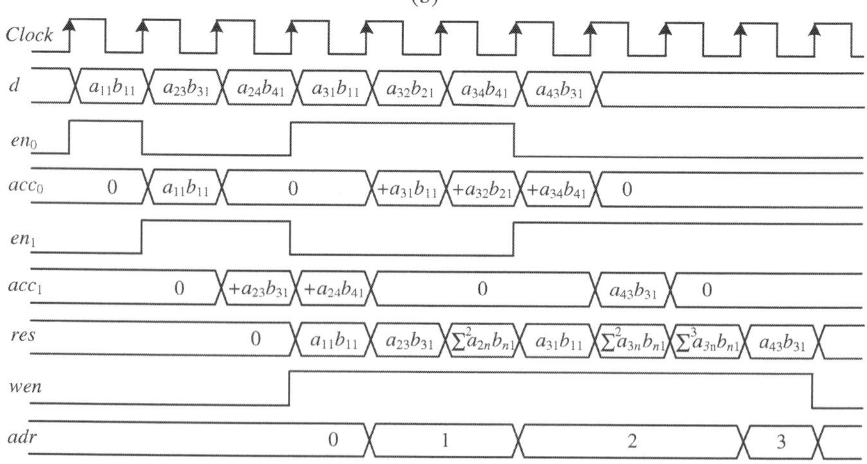

(c)

Figure 4.19 *SprMxMul* unit: (a) hardware design, (b) RAM initialization, and (c) timing diagram.

4-bit column number to identify column index of this nonzero value; and (iii) a 1-bit row index to label all nonzero values in the same row with "0" or "1." Figure 4.19b shows an example sparse matrix and its storage format. In the *SprMxMul* submodule, the accumulation is done in fixed-point format. The reason is that floating-point accumulator has a much longer latency and requires much more logic resources to implement. The fixed-point accumulator needs only one adder with one clock cycle latency. The used fixed-point number format is 40.100, which has 40 integer bits and 100 fraction bits to guarantee both the range and precision [24]. As shown in Figure 4.19a, the *SprMxMul* submodule contains one floating-point multiplier, one floating-to-fixed-point converter, two fixed-point adders, and one fixed-point converter. The elements of sparse matrix $\mathbf{A}_{n\times n}$ are retrieved from RAM A_{sprs}, while 4-bit column indexes are used to access the $\mathbf{B}_{n\times 1}$ matrix stored in RAM *B*. The registers are inserted for synchronization in the computation. The realized matrix-vector multiplication is fully pipelined and fast because there is no stall between two consecutive matrix row-vector multiplications. This is achieved by the two parallel fixed-point adders (accumulators) acc_0 and acc_1 with opposite enable inputs en_0 and en_1 controlled by the row label information. Figure 4.19c shows the logic timing diagram for the *SprMxMul* unit based on the example in this figure. As can be seen, the accumulation of the first matrix row-vector multiplication is processed in the acc_0, while the acc_1 is reset to zero, which makes it ready for the accumulation of the next matrix row-vector multiplication.

Moreover, the basic floating-point functions including addition/subtraction and multiplication are combined to build a basic arithmetic unit. Figure 4.20 presents the data flow of various outputs of this submodule. The latency of the longest path is 19 clock cycles. It can be used for floating-point arithmetic operations of both scalar and vector quantities, for example the realization of Eqs. (4.24)–(4.26). The *FLPMAS* unit is pipelined to achieve high data throughput.

4.3.3.3 Fixed-Point Implementation by Schematic

The schematic method is performed under Matlab/Simulink® software environment. The procedure starts by the development of a functional model of a system using basic Simulink continuous-time blocks or M-file coding. The behavior of the developed model is verified by the performance of the machine model from SimPowerSystem® toolbox and the fixed-point formats are then defined for all coefficients, variables, and data paths inside the model. In addition, the number of required clock cycles and suitable real-time simulation time-step size are determined to compute all governing equations of the system model.

At this stage, the emulation is realized by the development of a digital fixed-point model by using Xilinx System Generator® (XSG). The final step is linked with FPGA-based implementation. The HDL code of the design

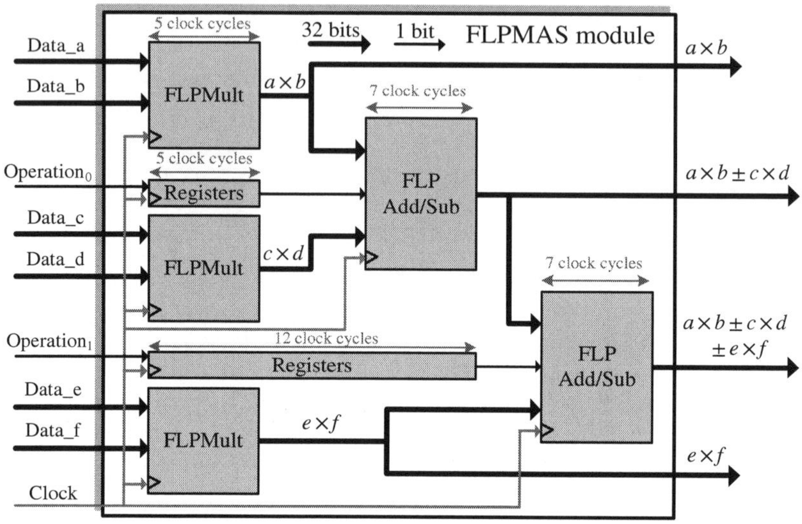

Figure 4.20 Pipelined realization of *FLPMAS module*.

architecture is automatically generated. The main control and interface files as well as pin assignments suitable for the FPGA platforms or HIL test setups are written and organized. The modifications and adjustments are performed in this step to get a balance in terms of area/time performances. Synthesis, map, place, and route processes and analysis of the static timing performances are done in the Xilinx® ISE software. The binary files produced are then transferred to FPGA via a serial interface (JTAG) for its reconfiguration.

The generated FPGA digital and analog I/O signals, which transmit low-level voltages and currents, can be interfaced with an amplifier that generates and absorbs high-level power where an actual real-time emulated machine works in the HIL configuration.

A generic digital machine model implementation is shown in Figure 4.21. It mainly consists of five functional steps. The algorithm starts from the calculation of voltage source in the three-phase domain. Phase angle is incremented over each time-step to feed the Xilinx DDS Compiler block, which implements high performance, optimized phase to sinusoidal circuits. The core sources sinusoidal waveforms and consists of a SIN/COS look up table. Moreover, the appropriate voltage magnitude can be achieved by a multiplication unit at the output port of this module.

The transformation module is in charge to change *abc* to an arbitrary reference frame. As can be seen in the associated unit for *abc* to stationary transformation, to obtain V_q, two sets of calculations ((i) Production of $2/3$ and V_a, and (ii) Production of $1/3$ with the summation of V_b and V_c) must be performed in parallel mode

Figure 4.21 Design configuration for one emulation step of induction motor by schematic method.

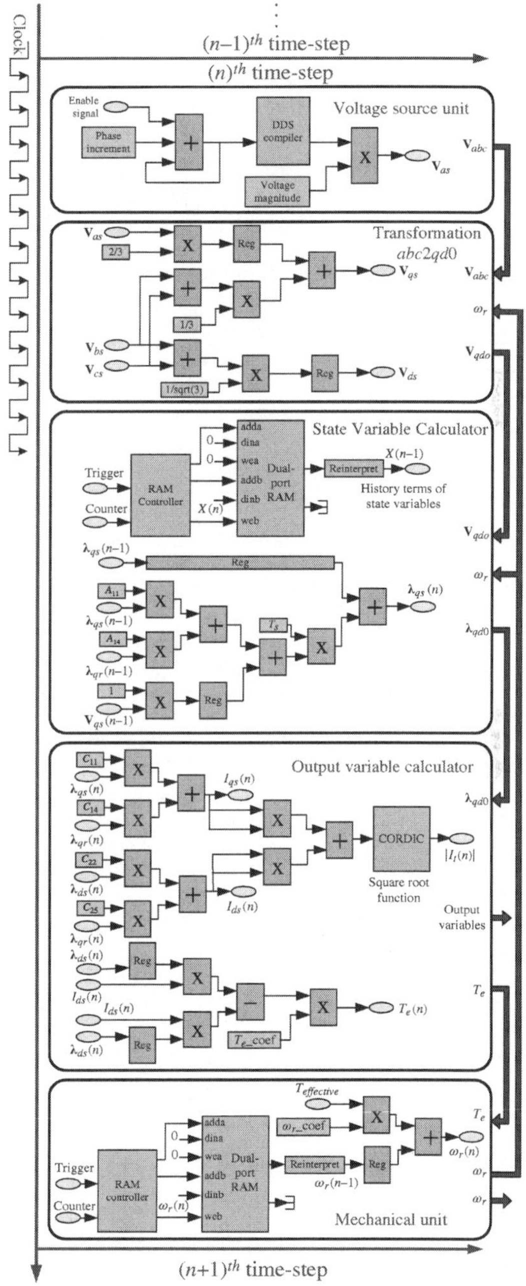

with the same latency before the final addition happens. This is why a register is inserted in the first path to synchronize the data flow with the second parallel path. Consequently, the final addition is only done when the operations at the two input ports are completed at the same time. A similar strategy is applied to compute V_d and also can be used for the transformation of other reference frames.

The state-space variable calculator, depicted in Figure 4.21, includes a memory controller that schedules the reads and writes to memory, and a highly parallel processing unit that computes state variables. To show how to implement a state-space system without losing the generality of the problem, the first row of the induction motor state equations is expanded and realized. All potential parallelism and data synchronism by registers are considered in the data flow graph of the following equation:

$$\lambda_{qs}\left((n)T_s\right) = \lambda_{qs}\left((n-1)T_s\right) + T_s \times [A_{11} \cdot \lambda_{qs}\left((n-1)T_s\right) + A_{14} \cdot \lambda_{qr}\left((n-1)T_s\right) + V_{qs}\left((n-1)T_s\right)]. \tag{4.27}$$

In this step, the dual-port RAMs are employed to store the history terms of the state variables in the $(n-1)$th time-step and call them for the integration procedure in the (n)th time-step.

Once the parallel computation of the state variables is performed, the required output variables are calculated. Most of output variables such as stator and rotor currents, electromagnetic torque can be obtained by using elementary operators including adders, multipliers, multiplexers, and registers (calculation of stator current and electromagnetic torque are provided in the corresponding module in Figure 4.21). However, to obtain some output variables, for example, the magnitude of a space vector, a square-root function is required, for which the generalized CORDIC function has been used.

Additionally, the structure of blocks arranged to calculate rotor speed is realized based on A-B integration technique. The RAM controller is responsible to manage the data flow of history information of rotor speed synchronized by the incoming current speed difference for an accumulation over each time-step. The same pattern is applied to calculate rotor position based on currently computed rotor speed information.

4.3.4 Evaluation of Designed Architectures

4.3.4.1 Real-Time Emulation Accuracy Assessment

Emulation time-step is an important criteria from practical point of view. In reality, currently available FPGA-based HIL test setups are not able to accommodate real-time models with a time-step more than a few microseconds due to practical limitations to communicate and transfer data to amplifiers and actual devices under real-time tests.

Table 4.3 Latencies and time-step sizes for machine emulator.

Real-time emulator	Fixed-point		Floating-point	
	Latency	Time-step (ns)	Latency	Time-step μs
Induction motor	46	230	234	1.755
Synchronous generator	55	275	256	1.920
Line start-PMSM	52	260	240	1.80
DC motor	29	145	126	0.945

Note that although in the off-line simulation, an increase in the order of discretization method or in the precision of numbers and operations results undoubtedly in an increase in the accuracy of final results, in the real-time emulation it does not necessarily lead to an increase of accuracy.

As noticed about all case studies, with the same discretization method, single-precision floating-point implementation of machine models expend more time, listed in Table 4.3, to finish all calculations within the time-step compared with fixed-point. It means that the executed real-time emulation of floating-point realization is carried out with longer time-step size than that of fixed-point. Therefore, although floating-point calculations mathematically offer higher level of accuracy in comparison with fixed-point, an increase in the associated simulation time-step size of digital hardware implementation increases the truncation error at each time-step. Consequently, it may reduce the accuracy of the results in the time-marching or time-stepping digital real-time simulation. That is why in real-time emulation, an increase of discretization order and precision of calculations may not guarantee more accurate results, whereas in the off-line simulation they can. The machine models were realized by 5 and 7.5 ns of FPGA clock frequency for fixed-point and floating-point implementation, respectively, resulting in an efficient implementation with smaller real-time emulation time-step sizes compared with [79].

4.3.4.2 Off-line Validation

The first validation step is an off-line simulation of design architecture according to the best case achievable frequency obtained at the end of place and route report. This step can be performed by using ISim®, ModelSim®, or Matlab® software tools. The good functionality and accuracy of the design algorithm written by TPL or schematic method can be verified in this step.

In the following transient off-line study, the first two oscillations of stator current of a 3hp induction machine, whose specifications are given in [83], by the various approaches are plotted in Figure 4.22. The induction machine is started from stall. The reference solution is obtained using the *qd* induction machine

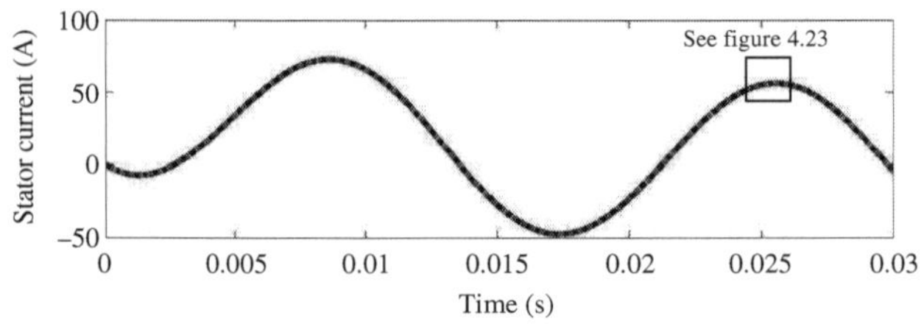

Figure 4.22 Stator current of induction machine using different methods presented in the caption of Figure 4.23.

model of SimPowerSystem toolbox and solved with Runge–Kutta fourth-order (RK4) method and double-precision (DP) operations using a small time-step of 100 ns. The simulation results obtained by Xilinx System Generator® [84], Matlab® M-file, and ISim® software are overlaid with the reference solution. As can be observed in Figure 4.22, the transient responses produced by all methods coincide and converge to the reference solution. This clearly demonstrates that all approaches predict the machine behavior with the acceptable accuracy. A magnified fragment of Figure 4.22 is also shown in Figure 4.23 for better comparison. As seen in Figure 4.23, the behavior of deeply pipelined digital hardware implementation of machine model by VHDL coding and single-precision (SP) floating-point operations obtained by ISim® (top solid line) with time-step equal to 1.755 μs gives the most deviation from the reference (solid line), while a minor improvement can be achieved by double-precision calculations with the same algorithm and time-step (dashed line) obtained by Matlab M-file coding. The result produced by a digital fixed-point (FiP) hardware implementation (32 bits) of induction motor with 230 ns time-step (dotted line) by schematic method (XSG) provides more accurate results compared with previous ones. The most noticeable difference of this waveform from others is its stair step shape. The reason is attributed to the errors that occur when a value lies outside the representable range and when the number of fractional bits is insufficient to represent the fractional portion of a value in fixed-point architecture. Moreover, double-precision calculated stator current by M-file coding (dash-dot line) with the same condition of fixed-point simulation is also provided for further comparison.

4.3.4.3 Hardware Resource Utilization

The electrical machine hardware designs were targeted to Xilinx® Virtex-7® XC7VX485T FPGA. The hardware resources of this FPGA are given in Table 1.2. The resource utilization of various types of emulated machines can be seen in Tables 4.4 and 4.5. Note that although fixed-point operators are thrifty, massively parallel fixed-point implementation of machine models consumes more hardware resources in terms of registers and LUTs compared with deeply pipelined floating-point realization containing extravagant floating-point calculations.

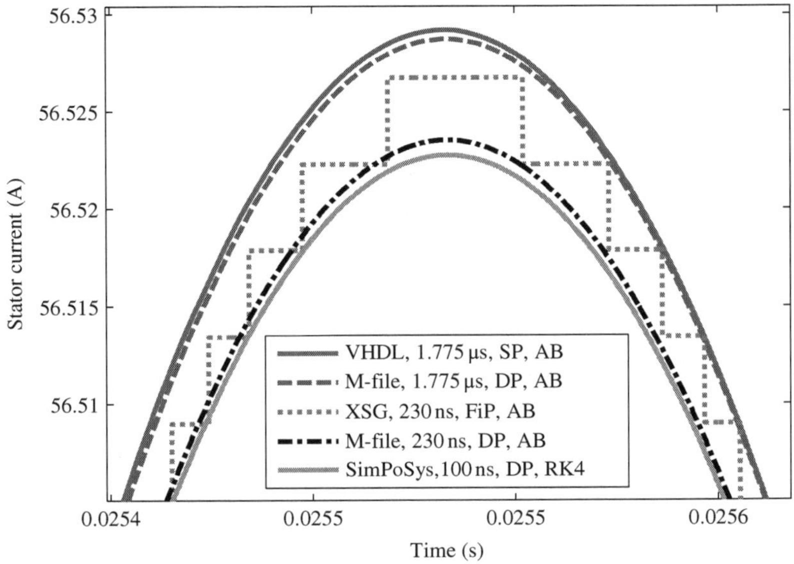

Figure 4.23 Zoomed in view of Figure 4.22 identified by a legend (realization technique, time step-size, number representation, discretization method). Top solid line: (digital hardware realization by VHDL in ISim, 1.755 μs, SP, A-B), dashed line: (off-line simulation by M-file, 1.755 μs, DP, A-B), dotted line: (digital hardware realization by XSG, 230 ns, FiP (32 bits), A-B), dash-dot line: (off-line simulation by M-file, 230 ns, DP, A-B), and solid line: (off-line simulation by SimPowerSystem tools, 100 ns, DP, RK4).

Table 4.4 Resource utilization using fixed-point arithmetic operations.

	Fixed-point by schematic method		
Real-time emulator	**Number of slice registers (607 200 available)**	**Number of slice LUTs (303 600 available)**	**Number of DSPs (2 800 available)**
Induction motor	25 127	36 567	26
Synchronous generator	42 287	61 024	43
Line start-PMSM	36 915	53 298	28
DC motor	9 886	14 551	9

Table 4.5 Resource utilization using floating-point arithmetic operations.

	Floating-point by TPL		
Real-time emulator	**Number of slice registers (607 200 available)**	**Number of slice LUTs (303 600 available)**	**Number of DSPs (2 800 available)**
Induction motor	6 689	29 115	38
Synchronous generator	7 670	30 493	67
Line start-PMSM	7 253	29 540	59
DC motor	13 102	6 585	27

4.3.5 Real-Time Emulation Case Studies

To demonstrate the usefulness of proposed hardware designs for real-time emulated electrical machines in HIL environment, the models of induction motor, synchronous generator with field windings, line start-permanent magnet synchronous motor (PMSM), and shunt-connected DC motor are tested and evaluated. The results were captured on a 500-MHz 4-channel oscilloscope that was connected to the DAC card on the Xilinx Virtex-7 VC707 FPGA development board. The results show the details of the electrical machine transients under normal and disturbed conditions. The parameters and specifications of the machines are listed in Appendix A.3.

4.3.5.1 Case I: Induction Motor Transients

The dynamic performance of the induction motor was emulated in real-time, and the associated electromagnetic torque transient is depicted in Figure 4.24. The oscilloscope traces of real-time emulator corresponding to induction motor transients during free acceleration from stall are captured in this case. Moreover, the corresponding off-line and experimental results are presented and discussed in Section 4.4.6. Good agreement between off-line, real-time, and experimental results confirms the effectiveness of real-time emulation of the induction machine.

4.3.5.2 Case II: Synchronous Generator Transients

This case focuses on the real-time emulation of dynamic performance of the synchronous generator during a three-phase fault at the machine terminals. The stability of synchronous machines in a power system following a fault is of importance to determine line loading limits. The real-time oscilloscope traces shown in Figures 4.25 and 4.26a illustrate the dynamic behavior of the synchronous generator during and following a three-phase fault. The machine is

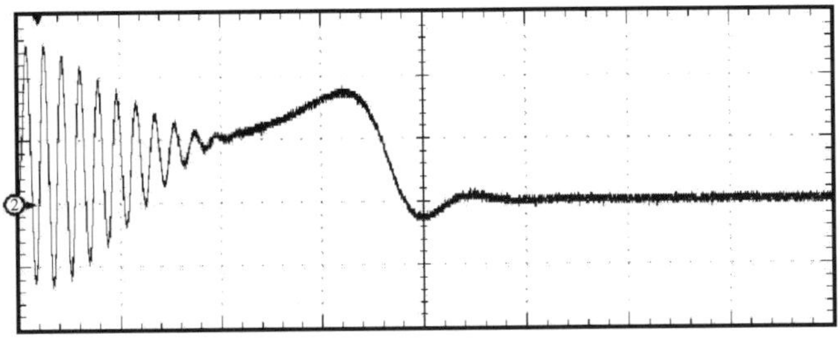

Figure 4.24 Real-time trace of torque response of induction motor. [Time: 100 ms/div., T_e: 12.8 (N m)/div.]

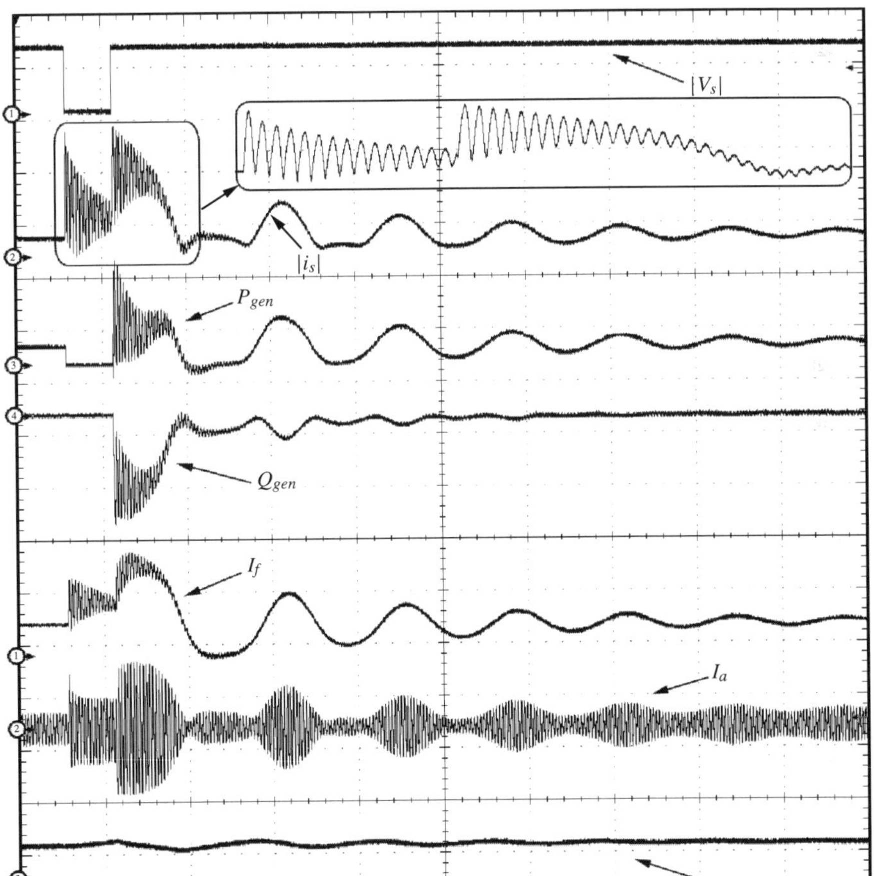

Figure 4.25 Real-time traces of synchronous generator dynamic behavior under normal and faulted condition. [Time: 450 ms/div., v_s (stator voltage phasor): 11.3 kV/div., i_s (stator current phasor): 104.4 kA/div., P_{gen}: 2660 MW/div., Q_{gen}: 2660 MVar/div., I_f: 98.5 kA/div., I_a: 166.5 kA/div., ω_r: 590 rps/div.]

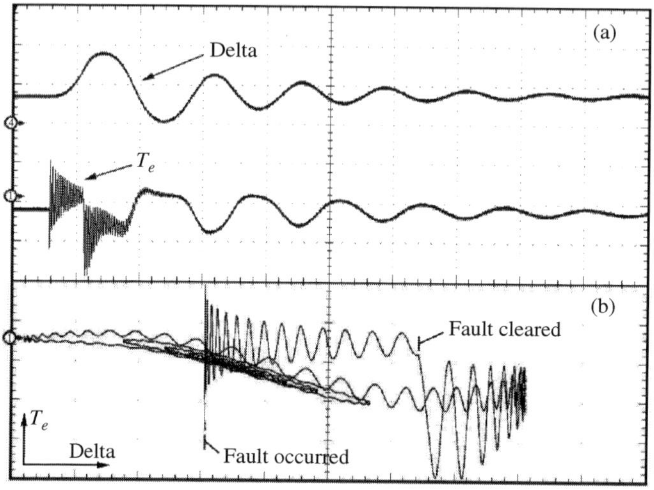

Figure 4.26 (a) Real-time traces of torque and rotor angle of synchronous generator vs. time. [Time: 450 ms/div., δ: 1.4 rad/div., T_e: 12.8M (N m)/div.] (b) Real-time $X-Y$ mode trace of torque vs. rotor angle of synchronous generator. [T_e: 7.9M (N m)/div., δ: 0.33 rad/div.]

initially connected to an infinite bus delivering rated apparent power at nominal power factor. The input torque and field voltage are held constant. With the machine operating in steady-state, a three-phase fault occurs at the machine terminals at $t = 0.25$ second. During the fault, the terminal voltage is 0, the machine is unable to transmit power to system. Hence, all of the input torque, with the exception of the ohmic losses, accelerates the rotor. The fault is cleared at $t = 0.5$ second and machine returns to its original operating condition after experiencing a transient condition. Plus, Figure 4.26b shows the impact of this disturbance on torque-rotor angle characteristics by a real-time oscilloscope Lissajous curve.

4.3.5.3 Case III: Line Start-Permanent Magnet Synchronous Motor Transients

Line start-permanent magnet synchronous motor (LSPMSM) is a high efficient alternative to replace induction motors in the constant speed operations with load variations. The dynamic performance of real-time emulated LSPMSM is depicted in Figure 4.27 by oscilloscope traces for applied three-phase supply voltages. The motor accelerates from stall. Once the steady-state operation is established, the load torque is suddenly stepped to 80% of nominal torque at $t = 1$ second. Motor increases load angle (δ) to maintain steady-state operation and then three phase fault is occurred at the motor terminal after $t = 1.5$ seconds whereupon the motor loses its synchronism and reestablishes the steady-state condition when the fault is cleared at $t = 1.65$ seconds. Induction component (T_{ind}) of electromagnetic

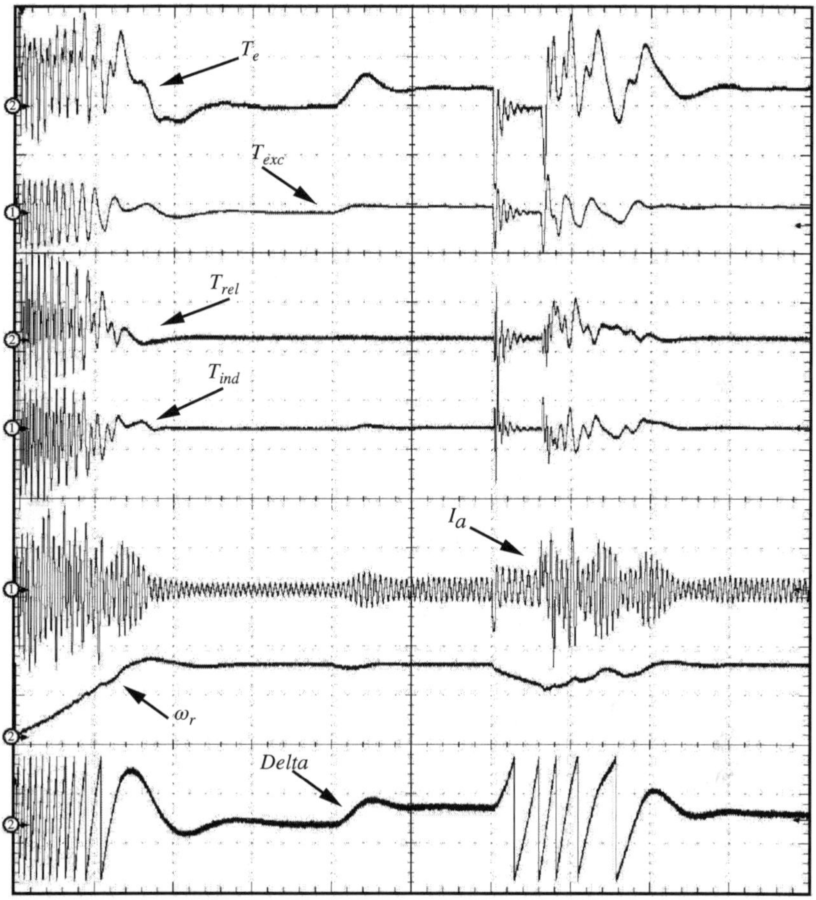

Figure 4.27 Real-time traces of LSPMSM performance. [Time: 250 ms/div., T_e: 18.2 (N m)/div., T_{exc}: 39.5 N m)/div., T_{rel}: 91.7 (N m)/div., I_a: 46.1 A/div., ω_r: 269.3 rps/div., δ: 2.6 rad/div.]

torque (T_e) significantly contributes to accelerate the rotor during starting and faulty condition while excitation component (T_{exc}) plays a considerable role in the steady-state operation where reluctance torque (T_{rel}) has a small effect in this case.

4.3.5.4 Case IV: DC Motor Transients

The results from real-time emulator of DC machine are assessed in this case study. A shunt-connected DC motor is selected for evaluation due to its desirable features and characteristics in the industrial and adjustable speed drive applications. The starting transients of the motor are represented in Figure 4.28, when it is started

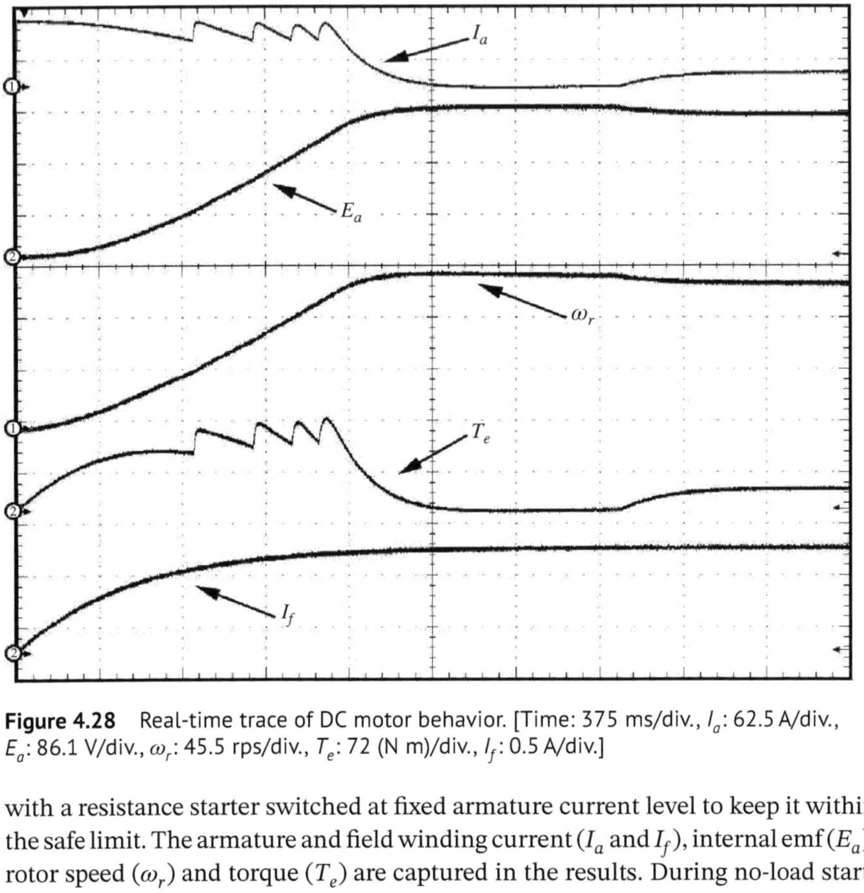

Figure 4.28 Real-time trace of DC motor behavior. [Time: 375 ms/div., I_a: 62.5 A/div., E_a: 86.1 V/div., ω_r: 45.5 rps/div., T_e: 72 (N m)/div., I_f: 0.5 A/div.]

with a resistance starter switched at fixed armature current level to keep it within the safe limit. The armature and field winding current (I_a and I_f), internal emf (E_a), rotor speed (ω_r) and torque (T_e) are captured in the results. During no-load starting, the switching of the four resistor segments is being triggered by the crossing of I_a below the threshold value. After a transient condition, steady-state operation is achieved at $t = 2$ seconds approximately. Then, the rated load torque is suddenly applied at $t = 2.7$ seconds. There is an increase in the armature current and a decrease in speed for this increase in load torque.

4.4 Nonlinear Magnetic Equivalent Circuit Based Induction Machine Model

A detailed nonlinear MEC model is presented for rotary induction machine in this section. It is intended to retain much of the salient features of extensive and complex models [85–88] and yet meet the computation time constraint of real-time simulation [89]. This section explains the necessary steps to develop the nonlinear MEC model for induction machine. The complete details of the

FPGA-based real-time hardware realization of the associated nonlinear equations are then presented. The experimental measurements and finite-element results are finally used to confirm the correctness of presented approach for real-time emulation of the induction machine.

4.4.1 Magnetic Circuit

Similar to meshing in FEM, the MEC model requires a permeance element network. A carefully selected permeance network not only improves the accuracy of final results but also minimizes the complexity and accelerates the necessary computations of corresponding nonlinear equations which must be solved within the simulation time-step in real-time.

According to a general 2-D structure of rotor and stator sectors, the permeance elements are created by a polar grid with several angles (φ_0 to φ_5) and radiuses (R_0 to R_8) outlined by the dashed lines in Figure 4.29 in such a way that the flux tubes are coincident with the main flux paths. Generally, the generated permeance network is similar to the one proposed in [85] in terms of number of nodes and structure. The reason why the minimum number of nodes (two nodes per tooth) is considered for permeance network is that any extra node leads to an extra equation in the system model, which dramatically increases the computational time. However, it enjoys finer permeance element sizes to define flux tubes in the MEC model more precisely, while the parallel hardware architecture of FPGA allows computing a large number of permeance elements simultaneously. In the presented MEC, the corresponding permeances for polar direction of flux paths in stator and rotor tooth (E_{02} to E_{04}, and E_{11} to E_{14}) are neglected, and additionally, permeance elements for the radial direction of flux paths in the stator and rotor slot closure (E_{07} to E_{10}, and E_{05} to E_{06}), stator yoke (E_{15} to E_{18}), and rotor central elements (E_{00} and E_{01}) are eliminated to reduce the real-time computational efforts significantly with only a slight reduction in the results accuracy due mainly to the fact that the main flux path does not flow anywhere except in the direction of the path of magnetic materials.

In the MEC, the elements of E_{02}, E_{04}, E_{11}, and E_{13} to E_{18} are simplified by trapezoidal flux prisms and the rest by rectangular prisms. The corresponding permeances, as shown in Figure 4.30, can be approximated as follows [87]:

$$P = \begin{cases} \mu L\left(\frac{R_3\varphi_3 - R_2\varphi_2}{R_3 - R_2}\right)\left(\ln\left(\frac{R_3\varphi_3}{R_2\varphi_2}\right)\right)^{-1}, \\ \qquad \text{Element Number: } E_{02}, E_{04}, E_{11}, E_{13}, E_{14}, \\ \mu L\left(\frac{R_3 - R_2}{R_3\varphi_3 - R_2\varphi_2}\right)\ln\left(\frac{R_3\varphi_3}{R_2\varphi_2}\right), \\ \qquad \text{Element Number: } E_{15}, E_{16}, E_{17}, E_{18}, \\ \mu L\left(\frac{(R_1 + R_0)(\varphi_1 + \varphi_0)}{4(R_1 - R_0)}\right), \quad \text{Element Number: Others,} \end{cases} \tag{4.28}$$

where μ and L are magnetic permeability and machine length, respectively.

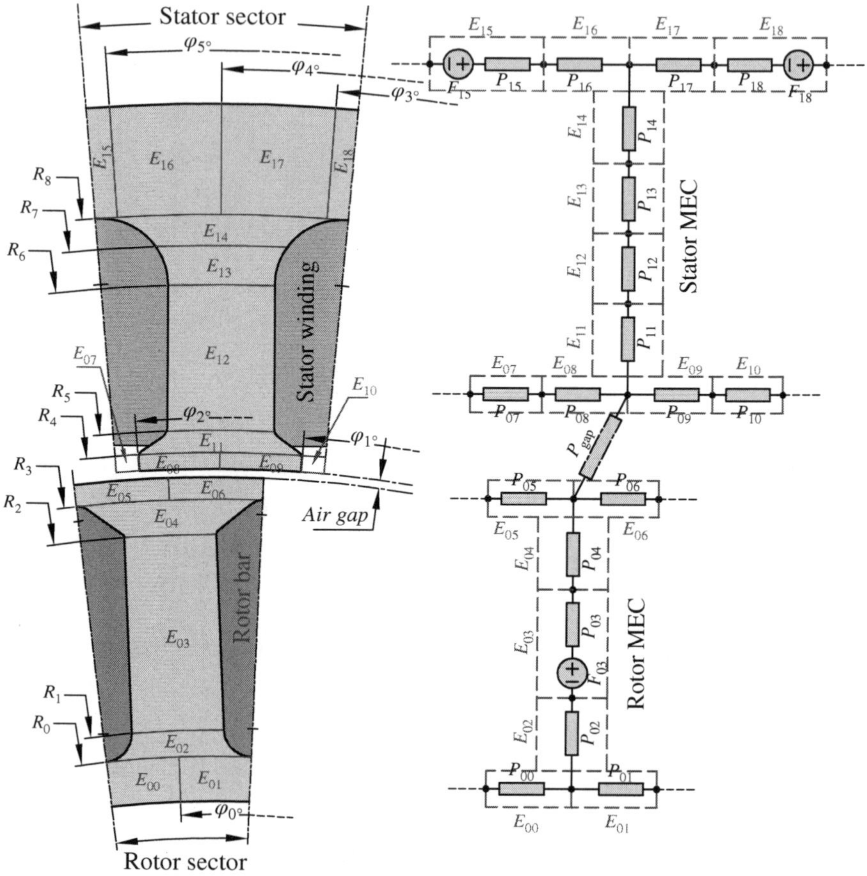

Figure 4.29 One sector of MEC of induction machine.

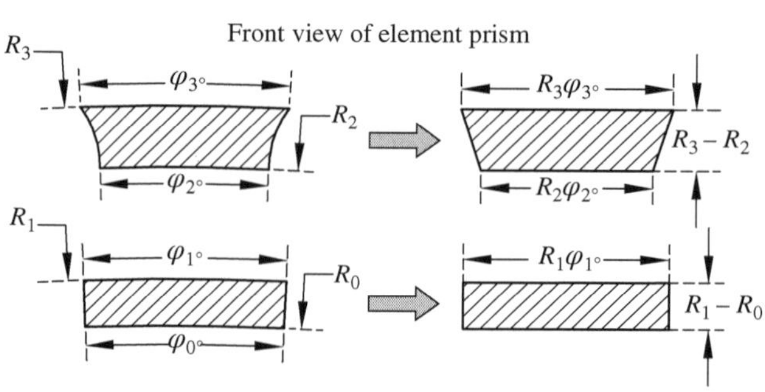

Figure 4.30 Simplification of magnetic flux prism.

The air-gap permeances are calculated based on the area of overlap between the stator and rotor teeth. The air-gap permeance between ith rotor tooth and jth stator tooth with instant polar angle difference of φ can be obtained by

$$P_{gap}(i,j) = \begin{cases} P_{max} = \frac{\mu_0 A_{st} L}{g}, \\ \qquad \text{if } |\varphi| < \varphi_{ovr}; \text{ Region I}, \\ \frac{\varphi_{linr} - \varphi}{\varphi_{linr} - \varphi_{ovr}} P_{max}, \\ \qquad \text{if } \varphi_{ovr} \leq |\varphi| \leq \varphi_{linr}; \text{ Region II}, \\ 0, \qquad \text{O.W.}, \end{cases} \tag{4.29}$$

where Region I defines the area of overlap and Region II is where the overlapped area is changing linearly when the motor is rotating. Additionally, g and A_{st} denote mechanical air-gap length and stator tooth face width, respectively.

Moreover, the stator and rotor MMF sources are determined by Ampere-turn formulae corresponding to the associated winding turns in the stator and rotor slots.

Once the permeances and MMF sources are calculated, nonlinear equations of MEC can be formulated based on standard circuit nodal analysis given by [88]:

$$\mathbf{P}(\mathbf{M}, \mathbf{i}) \cdot [\mathbf{M}, \mathbf{i}]^T = \mathbf{0}, \tag{4.30}$$

where $\mathbf{P}(\mathbf{M}, \mathbf{i})$ represents the permeance matrix which is a function of magnetic scalar potentials of nodes ($\mathbf{M}$) as well as stator and rotor loop currents ($\mathbf{i}$).

4.4.2 Interfacing of Magnetic and Electric Circuits

MMF sources are the elements by which magnetic and electric circuit are connected to each other. The magnetic flux generated by stator winding (electric circuit) traveling through the MEC can be expressed by

$$\left(\mathbf{W} \cdot \times \mathbf{P}_{st}^T\right) \cdot \left(\mathbf{M}_{st} - \mathbf{M}_{sy}\right) = \lambda_s, \tag{4.31}$$

wherein $\mathbf{W}$ is the winding function for the three-phase stator system [81], and $\cdot\times$ represents element-wise multiplication of two vectors. Equation (4.31) illustrates that adding up the number of turns times the magnetic flux entering the stator teeth with the permeance of $\mathbf{P}_{st}$ is almost equal to the stator flux linkages (λ_s). In this equation, $\left(\mathbf{M}_{st} - \mathbf{M}_{sy}\right)$ defines the difference of magnetic scalar potentials across the permeance of the stator tooth ($\mathbf{P}_{st}$). Additionally, $\mathbf{M}_{st}$ and $\mathbf{M}_{sy}$ denote magnetic scalar potentials of stator tooth and stator yoke, respectively.

Note that in Eq. (4.31), the small amount of stator leakage flux crossing inside stator slot and also fringing flux for outside and inside faces of stator core are neglected due mainly to the real-time computation while significant amount of this flux crossing from stator tooth tip to the adjacent tooth tip, and the flux traveling from stator to rotor and again back to stator without inducing rotor current are taken into account in the MEC.

Furthermore, the rotor flux linking the end ring (λ_e) and nth current loop (λ_r^n) can be given by:

$$\lambda_r^n = -P_b^{n-1} i_b^{n-1} + P_b^n i_b^n + 2P_e^n i_r^n - P_e^n i_e + \phi_r^n,$$
$$n = 1, 2, \dots, N_r, \tag{4.32}$$

$$\lambda_e = \sum_{i=1}^{N_r} P_e^n \left(i_e - i_r^n \right), \tag{4.33}$$

and

$$i_b^n = i_r^n - i_r^{n+1}, \quad n = 1, 2, \dots, N_r, \tag{4.34}$$

where P_b^n is the permeance of the magnetic flux path inside nth rotor bar, and P_e^n is the permeance of nth segment of end ring of rotor cage, which can be found in [85]. Also, the current of nth rotor loop, nth rotor bar, and end ring are denoted by i_r^n, i_b^n, and i_e, respectively.

In (4.32), ϕ_r^n represents the magnetic flux flowing through nth rotor tooth in the rotor structure with N_r slots.

4.4.3 Electric Circuit

The three-phase voltage equations of the stator electric circuit of induction machine can be defined by:

$$\frac{d}{dt}\lambda_s = \mathbf{v}_s - \mathbf{r}_s \mathbf{i}_s, \tag{4.35}$$

where $\mathbf{v}_s$, $\mathbf{r}_s$, and $\mathbf{i}_s$ represent applied voltages, resistances, and currents of stator windings, respectively.

Furthermore, the electrical equations for the rotor side are as follows:

$$\frac{d}{dt}\lambda_r^n = r_b^{n-1} i_b^{n-1} - r_b^n i_b^n + 2r_e^n i_r^n + r_e^n i_e,$$
$$n = 1, 2, \dots, N_r, \tag{4.36}$$

$$\frac{d}{dt}\lambda_e = -\sum_{i=1}^{N_r} r_e^n \left(i_e - i_r^n \right), \tag{4.37}$$

where r_b^n denotes the resistance of nth rotor bar, and r_e^n is the resistance of nth segment of end ring.

4.4.4 Nonlinear Solution of Detailed MEC

In the governing equations of the developed MEC, the permeance elements are defined as a nonlinear function of magnetic scalar potentials due to iron saturation effect. Thus, the combination of Eqs. (4.30)–(4.37) results in a system of nonlinear algebraic equations $\mathbf{f}(\mathbf{x})$ with a large dimension rewritten as

$$\mathbf{f}(\mathbf{x}) = \mathbf{A}(\mathbf{x})\mathbf{x} - \lambda = \mathbf{0}, \tag{4.38}$$

where $\mathbf{x}$ and λ are the unknown and known vectors in each simulation time-step, respectively.

The time-marching transient simulation of MEC of induction machine is performed for a 230 V, three-phase, 3hp squirrel cage Baldor induction machine, whose geometry and specifications are presented in Appendix A.3.

The motor has 36 stator slots and 28 rotor bars. The system of nonlinear algebraic equations for the presented MEC has a total of 160 unknowns in this case which is excessively time-consuming for real-time simulation.

To cope with this problem, an anti-periodicity and moving band technique, which is widely used in FEM, is applied to the MEC model to reduce the domain of study to one pole-pitch (a quarter) and increase the computational speed. Since the geometry of a squirrel-cage IM is normally composed of a repetitive section of domain and the MEC model is distributed across the geometry of the machine, the method allows to reduce the number of stator slots from 36 to 9 and the number of rotor slots from 28 to 7. Thus, the total number of equations in the augmented magnetic system decreases from 160 to 43.

The forward Euler method is chosen as a numerical integration technique to discretize the differential equations of electric circuits in each simulation time-step and N-R method is employed to solve the system of nonlinear equations. In order to improve convergence, the nonlinear iteration is decelerated by applying an under-relaxation factor (α), as follows [90]:

$$\mathbf{x}_{k+1} = \mathbf{x}_k - \alpha \mathbf{J}_k^{-1}\mathbf{f}(\mathbf{x}_k), \tag{4.39}$$

where $\mathbf{J}$ is the Jacobian matrix. Additionally, $\mathbf{x}_{k+1}$ and $\mathbf{x}_k$ are the solutions of nonlinear procedure at $(k+1)$th and (k)th iteration, respectively.

In this work, a constant relaxation factor of 0.32 is considered for all iterations, and the maximum number of N-R iterations is limited to 15. Consequently, hardware realization of proposed approach for an MEC model of induction machine on FPGA can be executed within a time-step of 400 μs for real-time emulation according to the best case achievable for FPGA clock frequency obtained at the end of place and route process of hardware design architecture.

4.4.5 Hardware Emulation of Nonlinear MEC

A 32-bit single-precision floating-point format (*IEEE* standard 754) is employed for hardware implementation. As shown in Figure 4.31, FSM for real-time emulation of MEC for the induction machine consists of four main modules: *Main Control Module*, *Source Module*, *Timer & Switch Module*, and *MEC Module*.

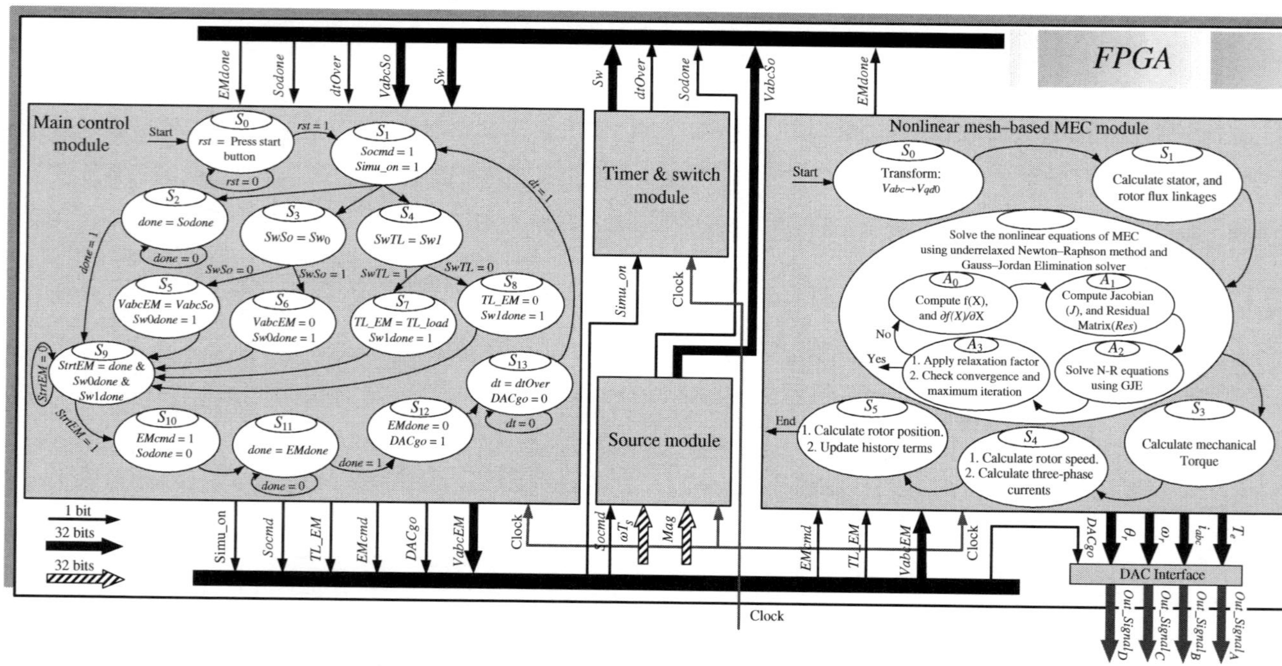

Figure 4.31 Finite-state machine for paralleled and deeply pipelined emulation of nonlinear MEC model of induction machine.

The *Main Control Module* coordinates the operation of the whole emulator to carry out the algorithm by sending control signals (*Simu_on*, *Socmd*, *EMcmd*) to each module to perform the required functions and receiving the acknowledged signals (*Sodone*, *EMdone*, *dtOver*, *Sw*) to judge if the functions are done.

The three-phase voltage source is created by the *Source Module*. Furthermore, a hardware module of *Timer & Switch* is designed to simulate switch status, and generate *dtOver* signal to check whether or not the simulation is executed within the real-time step-size [83].

The emulation procedure starts with the transformation of voltage sources from 3-phase domain to stationary reference frame by the help of floating-point matrix multiplication. The required flux linkages are then computed by dedicated floating-point arithmetic units and forward Euler integration method. The under-relaxed N-R is responsible to compute unknown variables of the MEC model, and mechanical performance of the machine is then predicted in the next states. The realization of N-R is the core of hardware design which is organized by using the three main computational submodules: (i) parallel GJE module; (ii) residual vector calculator; and (iii) nonlinear function evaluator of $\mu(H)$ and $d\mu(H)/dH$.

The global control schedules all sequential and parallel operations in the MEC model. The sequential operations can be seen in Figure 4.31 by state transitions, and all possible parallel calculations, which take the advantages of hardware parallelism in the FPGA, are taken into account for the realization. For example, the elements of $f(M, i)$, $\partial f(M, i)/\partial M$, and $\partial f(M, i)/\partial i$ are calculated in parallel in state S_0, and the evaluation of Jacobian and residual vector is carried out concurrently in state S_1. The detailed architecture of main submodules inside the MEC Module is described forthwith.

4.4.5.1 Parallel Gauss–Jordan Elimination Unit

As can be seen from Figure 4.32a, the hardware architecture for parallel GJE consists of 43 elimination units (Eli_1, Eli_2, …, Eli_{43}) and one factorization unit. First, the Jacobian matrix (**J**) and residual vector (**Res**) are combined together to make a 43×44 matrix whose rows are stored in RAMs and then GJE starts computing unknown vector. In the factorization module, the ith row is retrieved from the corresponding elimination unit, and the diagonal element is registered in the *Factorization unit* and the remaining elements within the row are divided by the registered diagonal element and returned to the elimination units. The elimination is carried out concurrently in all elimination units. For the ith row, no elimination is performed, and the factorized row is saved in RAM for the next call. However, for the jth row ($j \neq i$), the ith element is identified and registered as Eli_j. The elimination is then processed by using the elements of ith factorized row, registered Eli_j, and the corresponding elements of jth row [83].

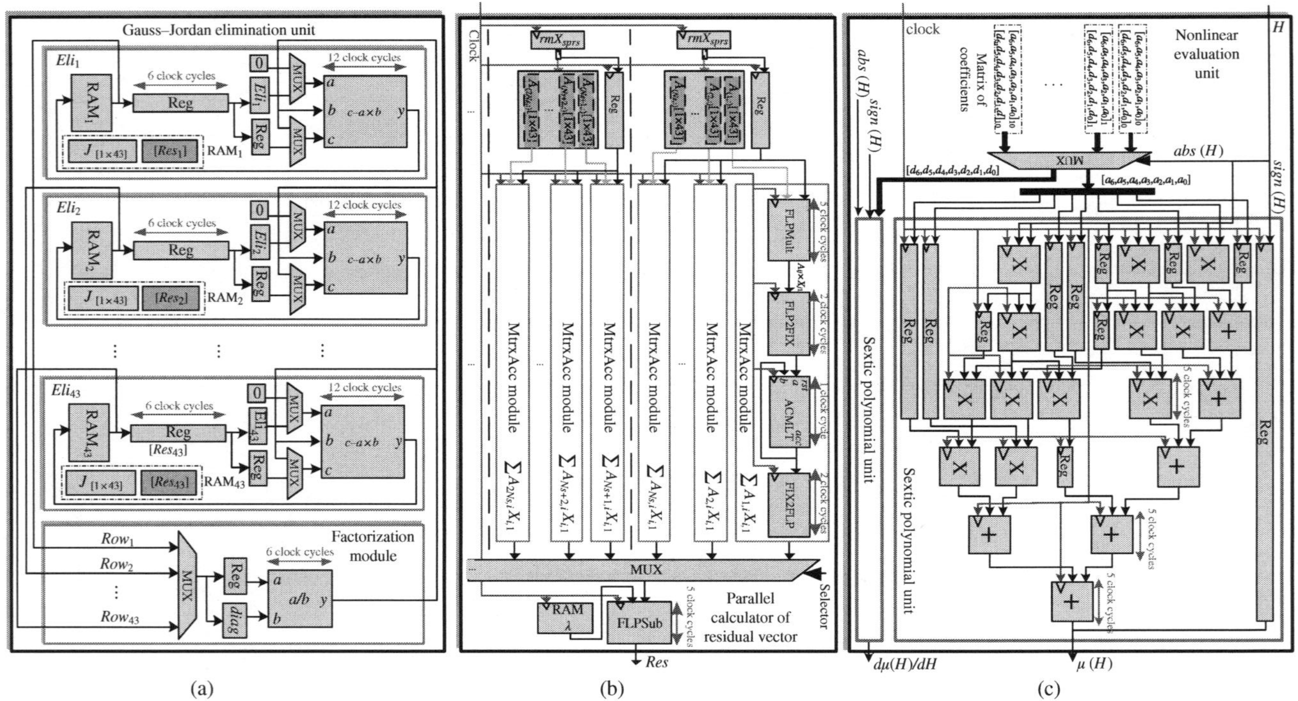

Figure 4.32 (a) Massively parallel Gauss–Jordan elimination module. Source: Based on Chen and Dinavahi [44]. ©2011, IEEE. (b) Parallel implementation of residual vector calculator. (c) Fully pipelined realization of nonlinear function evaluator.

4.4.5.2 Parallel Computational Unit for Residual Vector

As the direct multiplication of $\mathbf{A}$ matrix with a 43×43 dimensions to 43-element $\mathbf{x}$ vector takes long time in each iteration of N-R algorithm within the simulation time-step, the massively parallel hardware architecture of sparse matrix vector multiplications is designed to compute the residual vector. It can be observed from Figure 4.32b that a RAM (rmXsprs) is employed to store the contents of $\mathbf{x}_{43\times1}$ vector and the address of nonzero elements of each row of $\mathbf{A}_{43\times43}$ matrix. Such a hardware realization splits $\mathbf{A}_{43\times43}$ matrix to 43 individual rows and computes sparse row-vector multiplications simultaneously, for instance $\sum_{j=1}^{43} A_{ij}x_{j1}$ of ith row. The *MatrxAcc* submodule is in charge to provide the element-wise multiplication of two input vectors in a pipelined paradigm, convert the obtained results to fixed-point format (*FLP2FIX*), accumulate them in every single clock cycle (*ACMLT*), and again convert to floating-point representation (*FIX2FLP*). This is the fastest way to perform the matrix multiplications with large dimensions, although its digital realization consumes a large amount of hardware resources.

4.4.5.3 Nonlinear Evaluation Unit

Due to saturation effect, the permeability of iron core (μ) is a nonlinear function of magnetic field intensity (H) presented in Appendix A.3. In this work, the eleven-segment sextic polynomial functions are used to closely predict the nonlinearity of $\mu(H)$ an $d\mu(H)/dH$ for the calculation of permeances and Jacobian matrix elements in each N-R iteration. As can be seen in Figure 4.32c, the deeply pipelined structure of the unit is implemented by simple arithmetic operations such as adder, multiplier, and multiplexer. Additionally, registers are inserted on the data paths to synchronize the data flow in the hardware architecture.

Note that the shape of nonlinear function defines the convergence rate and the number of iterations in the Newton–Raphson method in each time-step regardless of how the nonlinear functions are called for the computation of residual vectors or Jacobian matrix in the iterative solution procedure. In other words, more or less detailed representation of a nonlinear function does not impact the rate of convergence. However, the only difference between less and more detailed representation of a nonlinear function is the speed of the nonlinear function call in each iteration, especially when a sequential processor (CPU) is used for computation. Unlike the CPU, using the advantage of parallel processing on FPGA allows us to represent the nonlinear function with eleven-segment sextic polynomial function with the higher accuracy and fast computational time.

4.4.6 Evaluation of Real-Time Emulation of Induction Machine

Hardware design of the induction machine was targeted to Xilinx® Virtex-7® XC7VX485T FPGA. Table 4.6 shows the FPGA hardware resource utilization

Table 4.6 Resource utilization of real-time emulator.

Real-time emulator	Number of slice registers (607 200 available)	Number of slice LUTs (303 600 available)	Number of DSPs (2 800 available)
Main control	11	10	0
Switch module	56	86	0
Source module	767	697	7
MEC module			
N-R unit	123 372	215 772	1 346
Others	6 590	12 052	74
Total	130 796 (22%)	228 617 (76%)	1 427 (51%)

of the MEC model of induction machine. It should be noted that the real-time hardware emulation of machine model in this work is realized by 10 ns FPGA clock frequency for the computational processing within the simulation time-step of 400 μs. The iterative under-relaxed N-R realization takes 387.15 μs in which one iteration spends 25.81 μs.

The detailed execution times according to the states of the nonlinear MEC model presented in Figure 4.31 for one simulation time-step and one iteration of under-relaxed N-R method for the computation of unknown variables within the simulation time-step are shown in Figure 4.33. As the MEC model for real-time emulation of induction machine is obtained by some simplifications, it is necessary to evaluate the extent of accuracy of the real-time model. Thus, a squirrel cage induction machine with closed rotor slots mechanically coupled with a DC machine is utilized as an experimental setup, as depicted in Figure 4.34b, to assess the performance of FPGA-based real-time emulation of induction machine. Moreover, a 2-D nonlinear time-stepping transient FEM by JMAG® software was employed for validation as well. Figure 4.34b shows the flux density distribution and flux lines inside the analyzed machine computed by FEM. In the FEM simulation, a relaxed N-R technique is used to obtain the nonlinear solution and the incomplete Cholesky–conjugate gradient method is chosen as the linear solver.

A 3hp induction machine, which is mechanically coupled to a DC generator is employed for experimental test. The transient performance of the real-time emulation during a free acceleration from stall of the induction machine is investigated by using a no-load test where the machine is started directly from a 208 V three-phase supply. The captured results are plotted in Figure 4.35. For further comparison, the FEM prediction of the current and rotor speed are also provided in Figure 4.35a,b.

There is good agreement between the real-time emulated, finite-element calculated, and experimentally measured performance of the motor. All currents decay

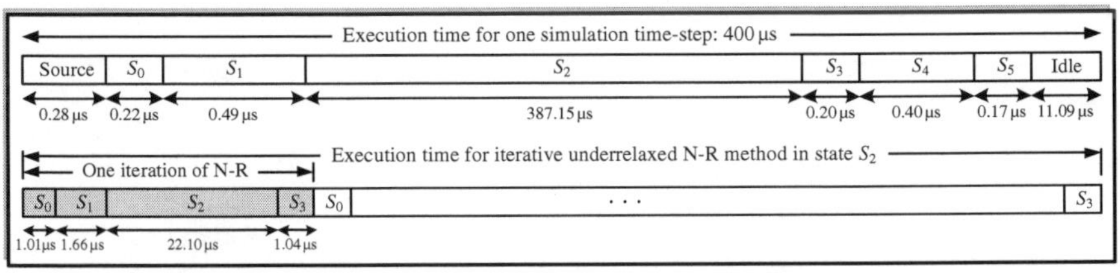

Figure 4.33 Detailed execution time for one simulation time-step and one iteration of under-relaxed N-R method.

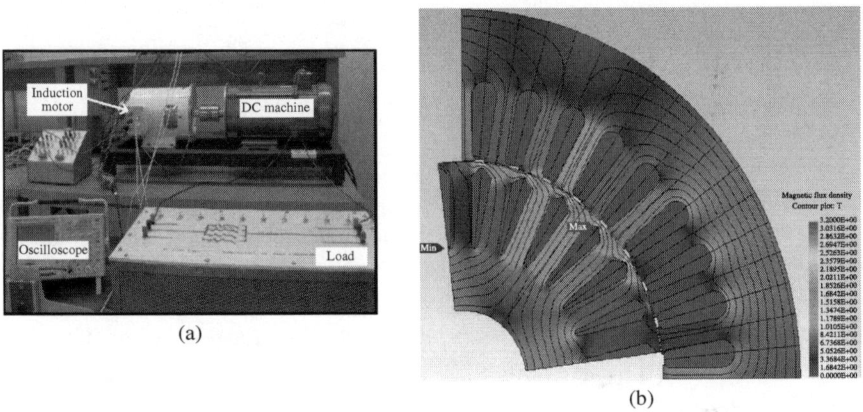

Figure 4.34 (a) Experimental setup. (b) Magnetic flux density distribution and flux lines of the analyzed machine by FEM.

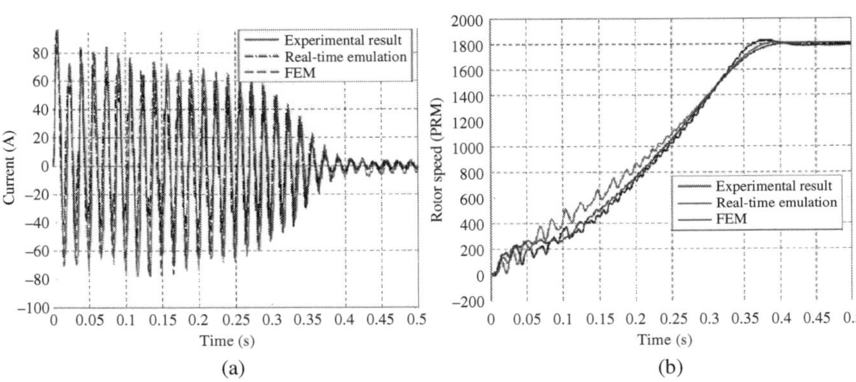

Figure 4.35 Start-up transient of: (a) stator current and (b) rotor speed.

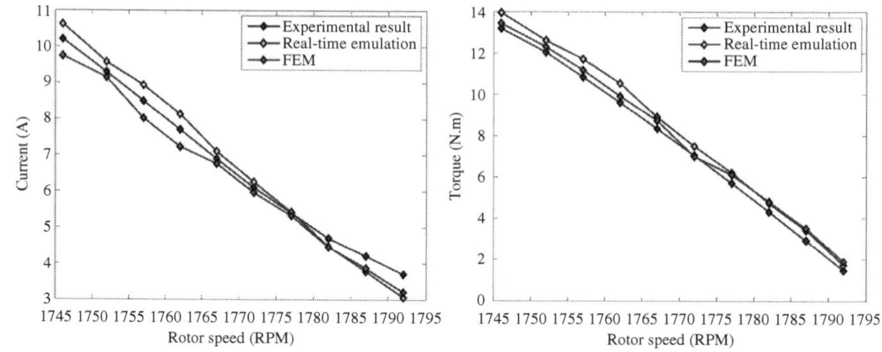

Figure 4.36 Steady-state performance of: (a) stator current and (b) electromagnetic torque at different rotor speeds.

to steady-state over almost 0.4 second and the speeds follow the similar trajectory closely. However, discrepancies are observed between the current amplitudes and speed fluctuations especially in the lower speed region or transient period due mainly to the backlash between the induction machine shaft and the DC machine which magnifies the effect of torque pulsations on the experimentally measured current and speed. Another reason is that although real-time induction machine model is based on a distributed element approach, the model is still lumped and not being able to include all distributed circuit effects perfectly for the prediction of transient behavior resulting in a limited accuracy in this region.

In the next assessment, the machine performance is investigated during steady-state operating points with different rotational speeds. Figure 4.36 illustrates the real-time emulated, finite-element calculated, and measured current and torque values. It can be seen that real-time MEC predicts the stator current of steady-state performance with acceptable accuracy. The discrepancy of the results can be attributed in part to neglecting the core loss in the real-time MEC. Another reason could be the neglecting of skin effects in the rotor parameters for real-time computation. Furthermore, simplifying the elements of machine structure with trapezoidal and rectangular prisms, neglecting the skew effect in the rotor bars can be taken into account as other reasons for the differences.

4.5 Summary

Accurate EMT simulation of power systems requires detailed and generalized models as well as solution strategies for rotating machines. The chapter began with the presentation of UM model, a sufficiently generalized machine model which can accurately represent several types of rotating machines for transient studies, its network interface, and detailed hardware implementation. Then, a unified framework for digital hardware emulation of electrical machines based on the state-space approach was presented and a comprehensive comparison was provided. Hardware designs were developed for induction motor, synchronous generator, LSPMSM, and DC machine. All parallel paths were explored for fast computational processing. Close agreement between the real-time emulated performances, off-line simulation, and experimental measurements confirmed the effectiveness of the implementation schemes. Such a real-time emulation of electrical machine can be used in HIL simulations to test new controller and drive systems against a virtual model of the machine. Moreover, in this study the results demonstrated that although floating-point arithmetic utilizes more hardware resources compared with fixed-point calculations, deeply pipelined implementation by floating-point number representation can consume less hardware resources than massively parallel realization by fixed-point operations.

Finally, a detailed geometry-based nonlinear MEC-based induction machine model and its complete real-time FPGA implementation has been presented. Fast and parallel iterative nonlinear and linear solvers are employed for the model computation. In the proposed MEC, the corresponding permeances for polar direction of flux paths in stator and rotor tooth were eliminated, and permeance elements for the radial direction of flux paths in the stator and rotor slot closure, stator yoke, and rotor central elements were neglected to reduce the real-time computational time significantly with just a slight reduction in the accuracy of the results due mainly to the fact that the magnetic flux flows mainly in the direction of the path of magnetic materials. The steady-state and transient results demonstrated that real-time emulated machine model followed the behavior of the induction machine obtained from experimental measurement and finite-element analysis closely. Consequently, such a model can be used in the iterative design optimization procedure of the machine to find the best set of machine parameters and dimensions in a large multidimensional search space to fulfill the desired objectives in a short time, and also for real-time machine fault analysis in hardware-in-the-loop applications. Research is ongoing on novel solution schemes for making nonlinear transient finite-element simulation of electrical machines feasible in real-time [91, 92].

5

Protective Relays

5.1 Overview

An electrical power system undertakes the responsibilities of generation, transmission, and distribution of electricity economically and reliably. Faults occurring on the system under abnormal conditions could lead to hazards to people, expensive electrical facilities, and the system's stable and safe operation. Therefore, protection against faults and abnormal events in the power system is crucial. A protective relay is an essential device in power systems to warn or isolate faults by detecting, locating the faults, and sending the trip signal to the corresponding circuit breaker. Protective relays make the trip decisions based on power system quantities such voltage, current, frequency, etc. In the industry, protective relays have experienced mainly three generations of evolution [93, 94]: electromechanical (EM) relays, solid-state (SS) relays, and numerical relays.

EM relays were based on moving parts to perceive abnormal changes of current or voltages to generate the mechanical torque. In EM relays, the actuating forces are created by a combination of the input signals, stored energy in springs and dashpots. Therefore, this type of relay contains an electromagnet and a moving part. When the actuating quantity exceeds a certain predetermined value, an operating torque is developed and applied on the moving part. The torque caused the moving part to finally close a contact to energize the trip coil of the circuit breaker.

SS relays based on analog electronic devices such as transistors, diodes, and other electronic components were the static replacements of EM relays which contain no moving elements. All of the functions and characteristics available with EM relays can be performed by SS devices. SS relays use low-power components with rather limited capability to tolerate extreme environmental factors or overvoltages and overcurrents. In general, SS relays are more accurate than EM relays

Real-Time Electromagnetic Transient Simulation of AC-DC Networks, First Edition.
Venkata Dinavahi and Ning Lin.

and have the advantages of reduced size of SS devices. Furthermore, absence of mechanical moving component in SS relays which may normally bring noise and cause contact problem, and less maintenance is also an improvement compared to the electromagnetic relays.

Currently, most commercial relays are digital numeric relays based on microprocessor technology [95], and sequential software programmability. This type of relay is introduced with the development of very large-scale integration (VLSI) technology and fast microprocessors. The core of this type of relay is a digital signal processor. The usual relay inputs are power system voltages and currents. Analog signals are filtered and converted to the digital form by analog-to-digital converter (ADC). Then, the relaying algorithm processes the sampled data to generate a digital output. The most obvious advantage of the microprocessor-based relay is the flexibility due to its programmable approach. However, the sequentiality of software code inherently limits the operating speed of the designs, and precludes the exploitation of parallelism in the implementation. In addition, the growing demand for computational speed to accommodate multiple protection functions, or flexible reconfigurable features cannot be easily met by microprocessor-based platforms at low cost and higher reliability.

The growing complexity and size of modern power systems demand faster and better performing protective relays. Especially, the traditional notion of power systems is mutating into the smart grid concept with the widespread use of distributed generation and smart loads, which pose significant demands on power system protection and security [96–98]. Advanced concepts such as adaption, self-healing, wide-area monitoring, agent-based, transient operation, and artificial intelligence need computationally powerful devices to be implemented in real-time applications [99–101]. There is a need for a high-capacity, high-bandwidth protective relay that can cope with the demand of signal processing, intelligence, and communication functions that such advanced concepts entail. In such applications, field programmability (FP) is a desirable feature to have in a protective relay. An FP device, whether analog or digital, is an integrated electronic circuit whose firmware can be reprogrammed in the field after manufacture.

In this chapter, a multifunction protection system is presented based on the hardware emulation building block (HEBB) concept introduced in Chapter 2 which has been successfully used to emulate detailed models for other power system equipment on the field-programmable gate array (FPGA). The entire hardware design is implemented in VHDL in 32 (64)-bit floating point format for high accuracy. In the hardware emulation, parallelism is fully exploited to achieve a low-latency realization of complex algorithms. Pipelining design is also used to increase the overall design throughput. Test cases are used to demonstrate the efficacy of the hardware multifunction protection system.

5.2 Hardware Emulation of Multifunction Protection System

The overall architecture of the proposed hardware multifunction protection system on FPGA is shown in Figure 5.1 wherein the design is constructed in the form of HEBBs. It consists of two major parts [102]: the signal processing HEBB (SPH) and the protective relay HEBB (PRH).

Within SPH, the main components include: (i) COordinate Rotation Digital Computer (CORDIC) HEBB; (ii) Discrete Fourier Transform (DFT) with DC offset removal HEBB; (iii) Symmetrical Components HEBB; and (iv) Zero-crossing detection HEBB.

The PRH consists of: (i) Fault detection HEBB; (ii) Distance protection HEBB; (iii) Directional overcurrent protection HEBB; (iv) Over/under voltage protection HEBB; and (v) Over/under frequency protection HEBB.

After the voltage and current data enter the protection system, they are processed by the SPH. For simplicity, potential and current transformer nonlinear effects on fault data are not considered in the design. The CORDIC HEBB provides trigonometric and nonlinear function values that are necessary for arithmetic computations, while the DFT HEBB extracts fundamental amplitude and phase of the required signals. The symmetrical components HEBB computes the sequence components of the three-phase system, while the zero-crossing HEBB is for power frequency estimation. Depending on the protection function

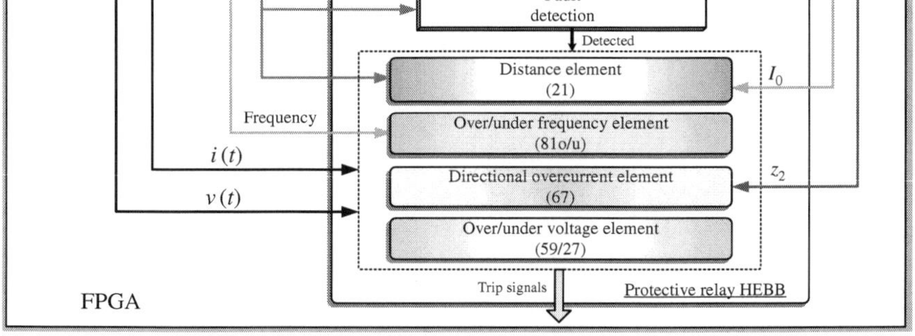

Figure 5.1 Overall architecture of the hardware multifunction protection system.

chosen, the required signal processing is carried out, and the corresponding trip signal is issued. In the following subsections, the hardware design details of each of the above HEBBs are given.

5.2.1 Signal Processing HEBB

The SPH computes the power system quantities needed by the various protection functions.

5.2.1.1 CORDIC HEBB

The iterative CORDIC method [103, 104] has the advantage of higher precision over the conventional look-up table (LUT)-based method whose precision may be limited by the volume of ROM limits. The basic trigonometric and hyperbolic functions are first computed by rotation mode and vectoring mode of the iterative CORDIC algorithm. The general form of the CORDIC iteration is as follows with FPGA implementation given in Figure 5.3a:

$$\begin{aligned} x_{i+1} &= x_i - m \cdot \sigma_i \cdot y_i \cdot 2^{-i}, \\ y_{i+1} &= y_i + \sigma_i \cdot x_i \cdot 2^{-i}, \\ z_{i+1} &= z_i - \sigma_i \cdot \alpha_{m,i}, \\ i &= i + 1, \end{aligned} \tag{5.1}$$

where x_i, y_i, and z_i are the 2-D rotating vector coordinates and angle of rotation respectively in i_{th} iteration. $\alpha_{m,i}$ is the value of $\tan^{-1}(2^{-i})$ when $m = 1$ or $\tanh^{-1}(2^{-i})$ when $m = -1$.

In each iteration, the plane vector $v_i = (x_i, y_i)^T$ rotates to $v_{i+1} = (x_{i+1}, y_{i+1})^T$, and z_i tracks the angle at each rotation. For instance, in the rotation mode (Figure 5.2), when estimating the trigonometric values of a given angle, z_i rotates toward 0 with a micro rotation angle of θ_i which has the tangent value of 2^{-i} (or hyperbolic tangent value of 2^{-i} in hyperbolic coordinates) in each iteration. These values are pre-calculated and stored in the FPGA memory. The σ_i indicates the rotation direction. At the last iteration, when the sum of rotated angles reaches the input angle, the x and y coordinates indicate the sin and cos values of the input angle. On the other hand, in the vectoring mode the coordinates of a vector are given *a priori* while the magnitude and angular argument of the original vector are computed. Therefore, after a certain number of iterations, the CORDIC HEBB can provide the trigonometric and hyperbolic functions based on its mode of operation. The parameter descriptions of (5.1) and the CORDIC mode operation summary are given in Table 5.1.

FPGA realization of CORDIC iteration requires only addition/subtraction and shifting operations (multiplication by 2^{-i} can be done by bit shifting to the right),

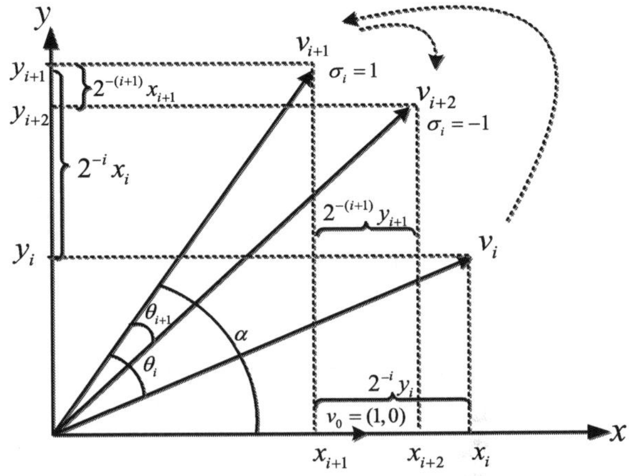

Figure 5.2 CORDIC rotation mode for the circular coordinate system.

Table 5.1 CORDIC iteration parameters and function description.

	Circular coordinates ($m = 1$)	**Hyperbolic coordinates ($m = -1$)**
Parameters	$\alpha_{1,i} = \tan^{-1}(2^{-i})$	$\alpha_{-1,i} = \tanh^{-1}(2^{-i})$
Rotation:	$\sin(x)$,	
$z_i \rightarrow 0$		$\tan^{-1}(x)$
$\sigma_i = sign(z_i)$	$\cos(x)$	
Vectoring:	$\sinh(x)$,	
$y_i \rightarrow 0$		$\tanh^{-1}(x)$
$\sigma_i = -sign(x_i y_i)$	$\cosh(x)$	

which can be easily achieved in hardware as shown in Figure 5.3a. The iterative stage n in this design is set to 20 to provide adequate precision. While the iterative CORDIC algorithm is based on fixed-point data, all the incoming data which are normally in 32-bit floating-point format are converted to fixed-point number using a float-to-fix module. The computation results are converted back to floating-point numbers by a fix-to-float module at the end of iteration.

With the computation of the elementary functions through CORDIC, the non-linear functions required in DFT such as e^x and $\ln(x)$ could be obtained from:

$$\begin{aligned} e^x &= \sinh(x) + \cosh(x), \\ \ln(x) &= 2\tanh^{-1}\left|\frac{x-1}{x+1}\right|. \end{aligned} \tag{5.2}$$

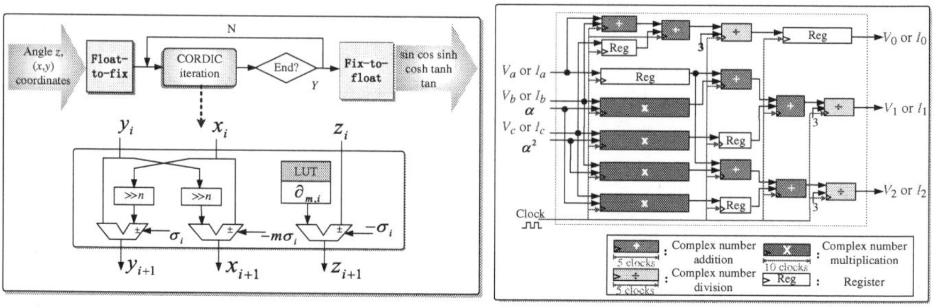

(a) (b)

Figure 5.3 Signal processing: (a) iterative CORDIC HEBB and (b) symmetrical components HEBB.

5.2.1.2 Symmetrical Components HEBB

This hardware module calculates the sequence components of a three-phase system. The positive-, negative-, and zero-sequence components are obtained by multiplying a constant sequence transform matrix with the original three-phase signals [105]. The symmetrical components of the current $I_{abc} = [I_a\ I_b\ I_c]^T$, for example, can be obtained as follows:

$$I_{012} = A^{-1} I_{abc} = \frac{1}{3}\begin{bmatrix} 1 & 1 & 1 \\ 1 & \alpha & \alpha^2 \\ 1 & \alpha^2 & \alpha \end{bmatrix}\begin{bmatrix} I_a \\ I_b \\ I_c \end{bmatrix}, \tag{5.3}$$

where $A = [1\ 1\ 1; 1\ \alpha^2\ \alpha; 1\ \alpha\ \alpha^2]$, the subscripts 0, 1, and 2 denote zero-, positive- and negative-sequence components respectively, and $\alpha = e^{j2\pi/3}$.

The hardware implementation of sequence components calculation is shown in Figure 5.3b. Inside each elementary complex number computation, the floating-point calculation is processed in the module with latencies shown in the figure. The registers inject synchronizing latencies.

5.2.1.3 DFT HEBB

This HEBB computes the fundamental frequency components of input voltage and current signals. DC offset removal is essential since the current signal will be mostly distorted by it during a fault. The full-cycle DFT (FCDFT) calculation with DC offset removal [106] that requires one cycle plus two samples is chosen.

The traditional FCDFT algorithm calculates the fundamental components as:

$$\begin{aligned} X_{(1)} &= \frac{2}{N}\sum_{n=0}^{N-1} x(n) \times (\cos \omega_1 n \Delta T - j \sin \omega_1 n \Delta T) \\ &= \frac{2}{N}\sum_{n=0}^{N-1} x(n) \times \left(\cos \frac{2\pi n}{N} - j \sin \frac{2\pi n}{N}\right), \end{aligned} \tag{5.4}$$

$$A_1 = \sqrt{X_{1real}^2 + X_{1imag}^2}, \tag{5.5}$$

$$\theta_1 = \arctan(X_{1imag}/X_{1real}), \tag{5.6}$$

where $x(n)$ is the discrete-time sinusoidal input signal, N is the number of samples in a fundamental period, with ω_1 being the fundamental angular frequency, and ΔT the sampling interval. The amplitude A_1 and phase θ_1 of X_1 can be calculated as in (5.5) and (5.6). For the DC offset removal strategy, the input signal contains a decaying DC offset $Ae^{-t/\tau}$ (time constant $\tau = q\Delta T$). After computing $X_{real(N)}, X_{real(N+1)}, X_{real(N+2)}$ which are real parts of three FCDFT fundamental phasor calculation results, the parameters of the DC offset are obtained as:

$$e^{-1/q} = \frac{(X_{real(N+2)} - X_{real(N+1)})\cos(2\pi/N)}{(X_{real(N+1)} - X_{real(N)})\cos(4\pi/N)}, \tag{5.7}$$

$$A = \frac{0.5N(X_{real(N+1)} - X_{real(N)})}{\cos(2\pi/N)e^{-1/q}(e^{-N/q} - 1)}. \tag{5.8}$$

The DC offset can be then subtracted from the sampled signal $x(k)$ as:

$$z(k) = x(k) - Ae^{-k/q}. \tag{5.9}$$

The DFT calculation is based on a moving window strategy. Once every full-cycle window (68 samples) of data calculation is completed, the window moves forward by one sampling point to the next computation cycle. The DFT HEBB gets the *set–offset* control signal from the finite state machine (FSM) controller (Figure 5.4) which has four transition states going from $S1$ to $S4$. Under the control strategy, the controller first senses the fault *detected* signal, and waits for two more sample calculation cycles to finish DC offset parameters calculation. Then, the offset could be subtracted if fault is detected and the input signal is the current. Within the DFT HEBB shown in Figure 5.4, the real and imaginary parts are calculated in parallel. The accumulation process is achieved by using the floating-point adder that has only one clock cycle latency. Figure 5.4 also shows the logic timing diagram for the module. After the very first round of calculation, which takes 112 clock cycles for amplitude and 117 clock cycles for phase output, with every fundamental window of interval the module generates a new calculation result.

5.2.1.4 Zero-Crossing Detection HEBB

This hardware module provides the algorithm necessary to estimate the fundamental frequency that is crucial for the frequency protection function. The zero-crossing method is used here as the frequency estimation algorithms [107]. The theoretical illustration and hardware implementation is given in Figure 5.5. For higher accuracy requirement, a 64-bit precision is used for implementation.

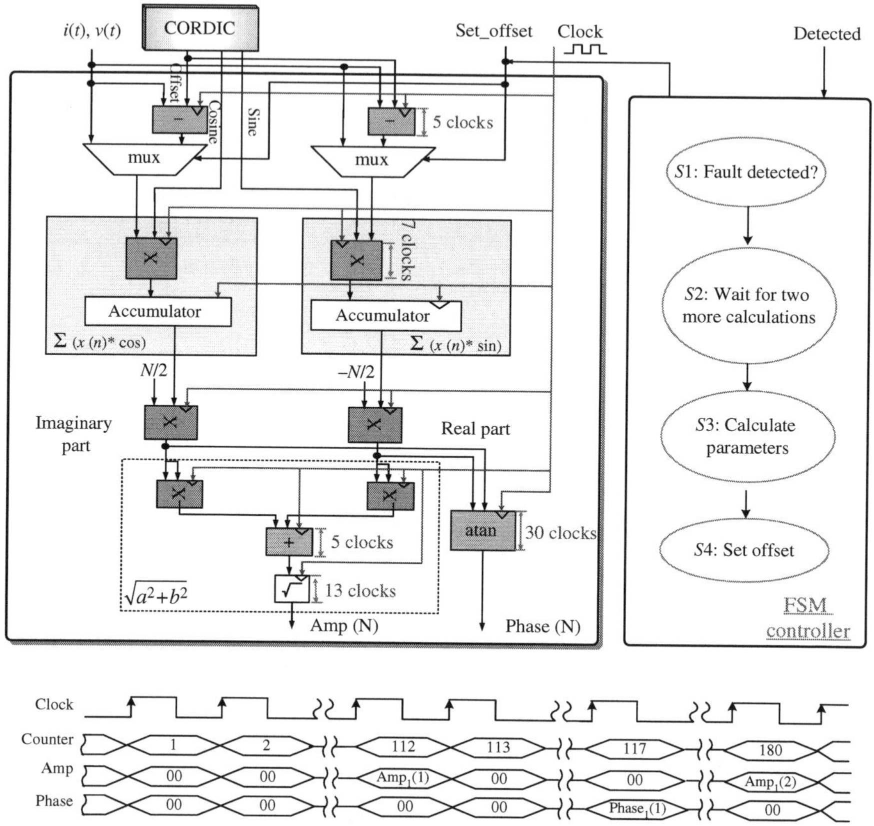

Figure 5.4 Signal processing: DFT HEBB with timing diagram.

The major step is to measure the time interval between the two successive zero crossings of the voltage signal to find the frequency using:

$$f = \frac{1}{T} = \frac{1}{2(t_2 - t_1)}. \tag{5.10}$$

The zero-crossing points are first approximated by comparing the signs of two neighboring points. Then, five samples around the zero-crossing point are used to fit a cubic interpolant. For instance, in Figure 5.5a, the five sample indices $[n - 2, n - 1, n, n + 1, n + 2]$, with n and $n + 1$ being the indices which have opposite signs, are used to determine the cubic polynomial:

$$f(x) = ax^3 + bx^2 + cx + d. \tag{5.11}$$

The coefficients of (5.11) can be determined by least squares method as:

$$A = \begin{bmatrix} a & b & c & d \end{bmatrix}^T = (K^T K)^{-1} K^T V, \tag{5.12}$$

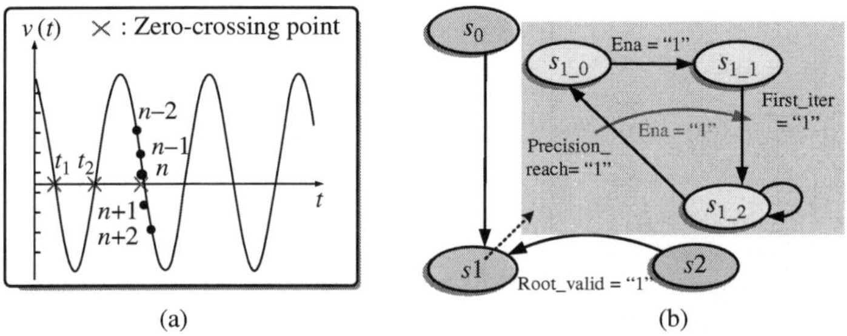

Figure 5.5 HEBB for frequency estimation using the zero-crossing method: (a) principle of operation and (b) state machine.

where K is the constant Vandermonde matrix, and V is the voltage sample vector, given as:

$$K = \begin{bmatrix} 0 & 0 & 0 & 1 \\ 1 & 1 & 1 & 1 \\ 2^3 & 2^2 & 2 & 1 \\ 3^3 & 3^2 & 3 & 1 \\ 4^3 & 4^2 & 4 & 1 \end{bmatrix}, \quad V = \begin{bmatrix} v_{n-2} \\ v_{n-1} \\ v_n \\ v_{n+1} \\ v_{n+2} \end{bmatrix}. \tag{5.13}$$

Hardware realization of (5.12) (see Figure 5.5c) is the process of matrix multiplication. The matrices are stored in the FPGA RAM. The four inner product computations are done in parallel. After the coefficients are determined, the three roots of the cubic equation can be found. One of the roots represents the zero-crossing point. The algorithm for calculating the roots of the cubic equation in hardware involves computing the eigenvalues of companion matrix of the coefficient polynomial.

For the polynomial in (5.11), the companion matrix is calculated as:

$$B = \begin{bmatrix} -\frac{b}{a} & -\frac{c}{a} & -\frac{d}{a} \\ 1 & 0 & 0 \\ 0 & 1 & 0 \end{bmatrix}. \tag{5.14}$$

Figure 5.6 gives the hardware design of companion matrix calculation. The eigenvalue calculation relies on an iterative QR decomposition. The iteration is initiated by decomposing the previously calculated B matrix into an orthogonal Q matrix and upper triangular R matrix, $B_0 = Q_0R_0$. The Gram–Schmidt process [108] is employed here. Then, multiplying the two matrices in reverse order leads to a new matrix $B_1 = R_0Q_0$. Repeating the process we can obtain the $(n+1)$th matrix $B_{n+1} = R_nQ_n$. After a certain number of iterations (12 in this case), the matrix converges to an upper triangular matrix whose diagonal entries are the eigenvalues of the matrix B. The whole iteration process is controlled by

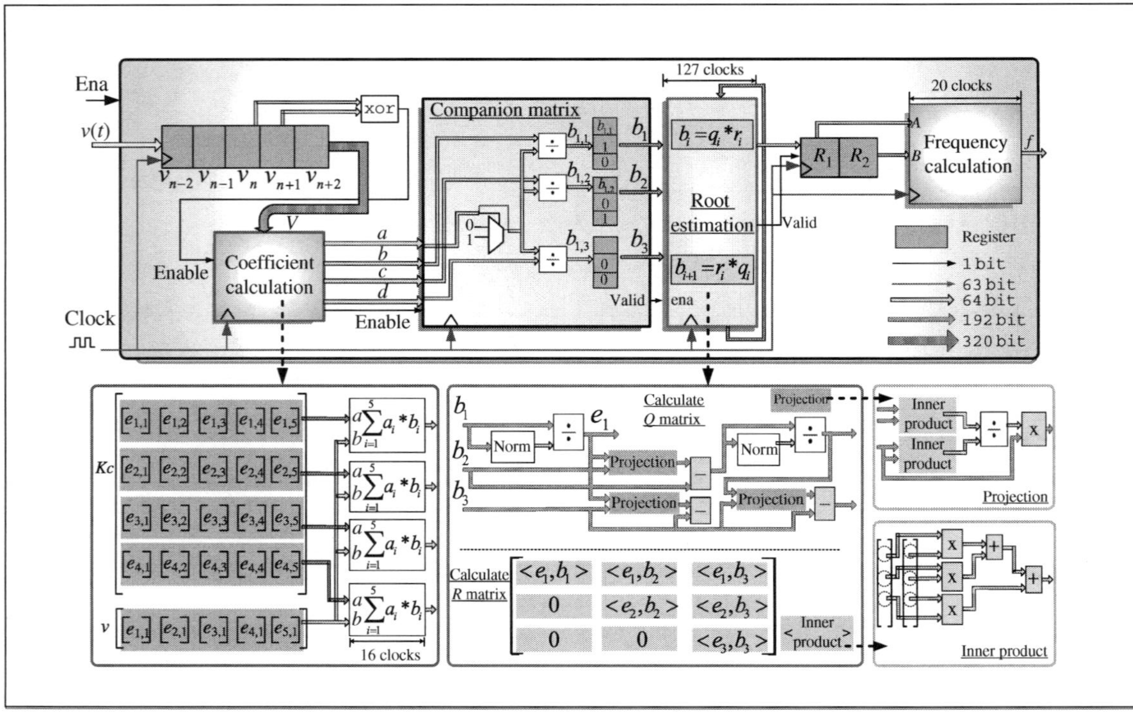

Figure 5.6 Hardware design for frequency estimation using the zero-crossing method.

a FSM shown in Figure 5.5b from S_{1-0} to S_{1-2}. When ena = “1”, the input of first iteration in S_{1-1} is the companion matrix. Then in S_{1-2}, the output of the present iteration serves as the next iteration’s input until required precision is reached. The outer FSM from S_0 to S_2 serves to record the time interval between successive zero-crossing points by using counters.

The Gram–Schmidt process for QR decomposition is described as follows with the hardware design also given in Figure 5.5c. Consider matrix B in (5.14) with three columns, $B = [b_1, b_2, b_3]$. The Q matrix can be obtained by the following:

$$\begin{aligned} e_1 &= \frac{u_1}{\|u_1\|}, \quad u_1 = b_1, \\ e_2 &= \frac{u_2}{\|u_2\|}, \quad u_2 = b_2 - proj_{e_1} b_2, \\ e_3 &= \frac{u_3}{\|u_3\|}, \quad u_3 = b_3 - proj_{e_1} b_3 - proj_{e_2} b_3, \end{aligned} \tag{5.15}$$

where $\|u_1\|$ stands for the norm operation of a vector and “proj” stands for the projection operation: $proj_e b = \frac{inner\ product\langle e,b\rangle}{inner\ product\langle e,e\rangle} e$. And Q_1 equals to $[e_1, e_2, e_3]$. With the results from (5.15), R can be calculated using:

$$R = \begin{bmatrix} \langle e_1, b_1\rangle & \langle e_1, b_2\rangle & \langle e_1, b_3\rangle \\ 0 & \langle e_2, b_2\rangle & \langle e_2, b_3\rangle \\ 0 & 0 & \langle e_3, b_3\rangle \end{bmatrix}. \tag{5.16}$$

After the eigenvalues are found, the roots of the cubic equation are determined. Two successive roots are stored in the register R_1 and R_2 for the frequency calculation by (5.10).

5.2.2 Multifunction Protective System HEBB

5.2.2.1 Fault Detection HEBB

The fault detection HEBB detects the initiation of the fault and triggers the subsequent protection processes. This detection algorithm uses the DFT current amplitude results to decide if a fault occurred. As the DFT window moves in real time, the calculated amplitude is sensed by this module. The strategy is shown in Figure 5.7a. If the amplitude of any phase current exceeds a certain threshold value three times consecutively, it indicates that the protected element is exposed to an abnormal situation, and a fault detection signal is sent out.

The fault detector HEBB emulation is shown in Figure 5.7b. The DFT amplitude results for each phase enter the HEBB and are compared with the corresponding threshold values. The *enable* signal indicates validation of each DFT computation result which permits the comparison and registration process. The register after the *enable* signal is need for synchronization, while the three single-bit registers record the comparison results. A fault detection signal is generated when all three registers record a “1.”

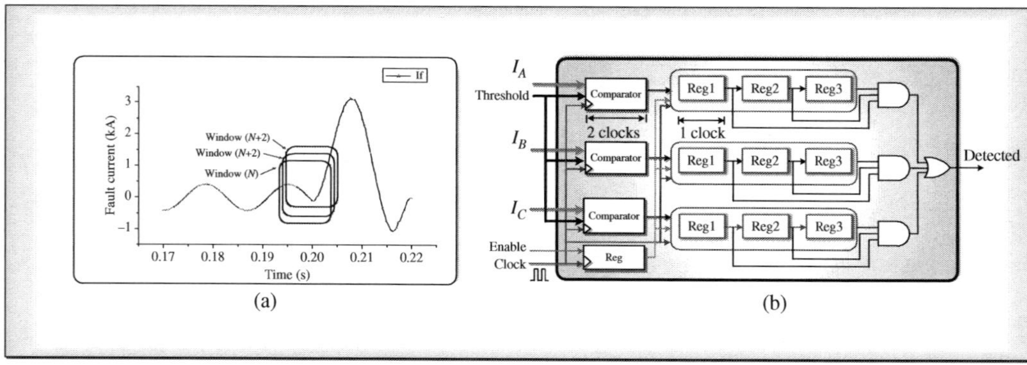

Figure 5.7 Fault detection HEBB: (a) principle of operation and (b) hardware emulation.

5.2.2.2 Directional Overcurrent Protection HEBB

Overcurrent protection is a simple and economical way to protect power system equipment. It is generally used for low-cost subtransmission lines, distribution circuits, and industrial systems [109, 110]. The overcurrent protection operates when the current exceeds the pickup value. In order to provide better selectivity, it is also necessary to determine the fault current direction to avoid unintended outages. A directional element is therefore added. It uses the negative-sequence impedance (Z_2) seen by the relay to get the direction information. Under faulted conditions, an approximate 180° of impedance angle change can be observed [111, 112] for two directional faults as shown in Figure 5.8a. For a reverse fault, $Z_2 = Z_{2sB} + Z_{2line}$, where Z_{2sB} is system-B's negative sequence impedance, and Z_{2line} is the apparent negative sequence line impedance. The process to determine the direction is to compare the angle of the measured sequence impedance with the maximum torque angle (MTA). The forward and reverse action areas of two direction faults are seen in Figure 5.8a.

From the overall FPGA design (Figure 5.8b for the directional overcurrent protection, it is apparent that only when the demands for both the forward direction element and the overcurrent element are met, the relay will initiate final tripping. The three-step zone overcurrent protection design is shown in Figure 5.8c. Each protection zone has its own setting and delay timer. This module can also be configured for the voltage protection which will be discussed later. After a fault is detected, any timer's time-up signal can activate the trip signal.

5.2.2.3 Over/Under Voltage Protection HEBB

The maintenance of normal and stable voltage magnitude is important for the health of any power system. Voltage deviations deteriorate the power quality and life-expectancy of electrical infrastructure. Over/under voltage elements in a multifunction relay can provide the phase-to-ground and phase-to-phase protection by sensing voltage deviations. The phase-to-ground undervoltage elements operate when the phase voltage is smaller than the low-setting threshold, while the overvoltage elements trips when voltage is larger than the high-setting threshold. The phase-to-phase element works under the same mechanism. If any one of the three voltages rise above or drop below a certain set value, a trip signal is generated. This element also comes with three-step zone protection scheme similar to the overcurrent protection. In Figure 5.8a, after configuring the input signal to voltage signal and the choosing of under or over options, the hardware module can be used for voltage protection.

5.2.2.4 Distance Protection HEBB

The operation of distance protection is based on the measurement and evaluation of the short-circuit impedance which is proportional to the distance between the

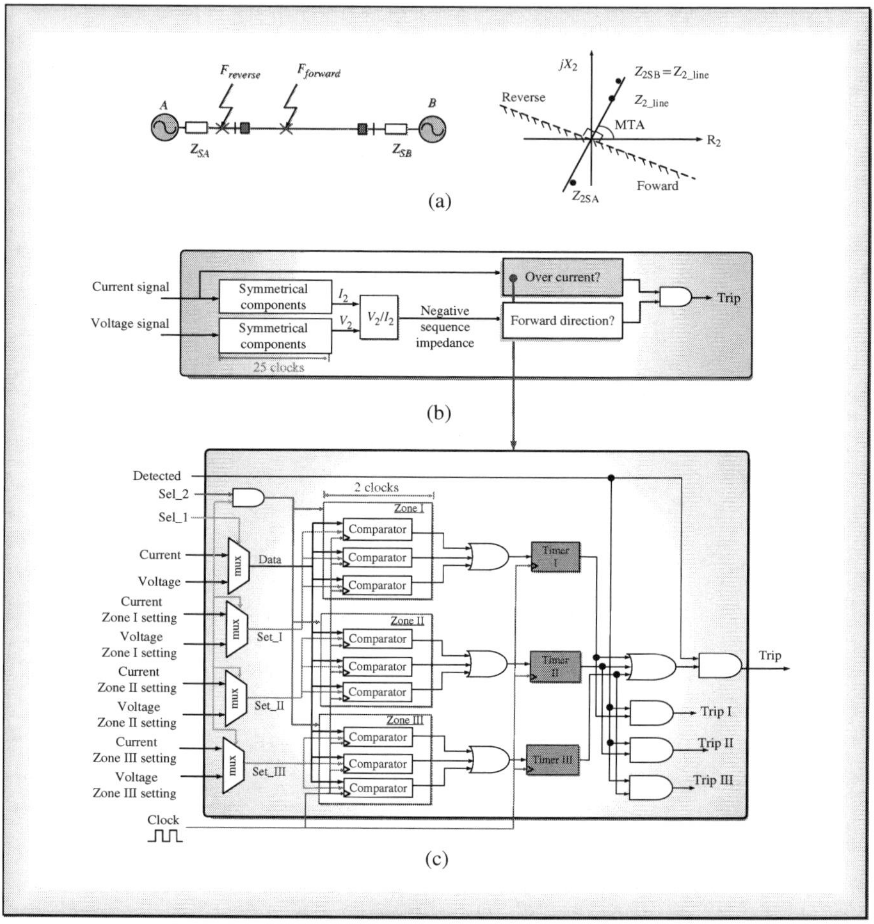

Figure 5.8 Directional overcurrent protection HEBB: (a) principle of operation, (b) overall hardware design, and (c) three-step zone protection design.

relay and the fault location. The two types of impedance calculation in the distance protection HEBB are phase-to-ground and phase-to-phase impedance given as follows:

$$
\begin{aligned}
Ground &: Z_g = V_{p-g}/(I_{p-g} + 3k_0 * I_0), \\
Phase &: Z_p = (V_{phase1} - V_{phase2})/(I_{phase1} - I_{phase2}),
\end{aligned}
\tag{5.17}
$$

where I_0 is zero sequence current calculated from $(I_A + I_B + I_C)/3$; $k_0 = (Z_0 - Z_1)/3Z_1$. Z_0 and Z_1 are the zero- and positive-sequence line impedances from the relay location to protection zone, respectively. Hardware realization of impedance calculation involves straightforward complex number computations. The calculated

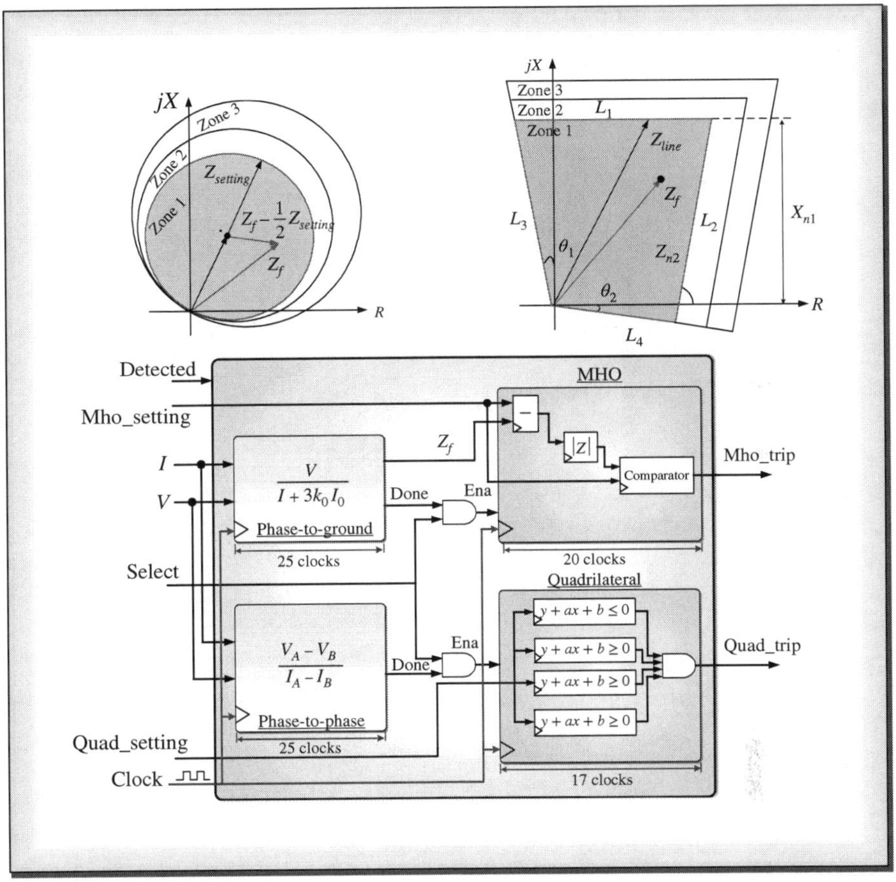

Figure 5.9 Distance protection HEBB with mho and quadrilateral characteristics.

impedance is then processed by a mho or quadrilateral characteristic [94, 113, 114] relay element to decide in-zone tripping. Figure 5.9 gives the two distance characteristics with three protection zones. If the calculated impedance locates within the tripping area which is inside the characteristic shape, a trip signal is generated. The application of zone protection can be achieved by setting different zone areas of the characteristic shapes.

Figure 5.9 also illustrates the hardware implementation [115] for the distance protection HEBB. The voltage and current DFT calculation results enter the module at first. After the apparent impedance calculation, mho or quadrilateral characteristic can be selected for tripping decision. For the mho characteristic, if the fault impedance is inside the circle, the vector $Z_f - 1/2Z_{setting}$ has a magnitude less than $1/2Z_{setting}$. For the quadrilateral characteristic, the trip area is shaped by four straight lines. $L1$ is the reactance relay setting (X_{n1}) and $L2$ is the angle-impedance relay setting (Z_{n2}). $L3$ and $L4$ are the directional relay with angle settings (θ_1, θ_2). The apparent impedance coordinates are then substituted into the four line equations to determine whether they are in the tripping region. In the quadrilateral zone, the calculated impedance is compared with the four sides of the geometry by checking the relative position of impedance point and the line. It should meet the conditions of locating below $L1$, above $L4$, on the left of $L2$, and to the right of $L3$ to get a in-the-zone decision.

Instantaneous-signal-based Distance Relay Element To provide a faster tripping decision, this work also gives another option of distance protection that is based on instantaneous signal processing. The basic trip strategy of this option is the same as the mho phase comparator; however, unlike the DFT distance relay option relying on the fundamental phasor extraction illustrated before, the instantaneous-signal-based algorithm processes instantaneous voltages and current signals. As a result, it can shorten the operating time to less than half a fundamental cycle, while the DFT method requires collecting one full cycle of data to calculate the phasors and then perform further protection logic. In this module, the input waveforms are first filtered to remove any high-frequency components that could influence the accurate decision making of the relay.

The hardware implementation of the instantaneous-signal based method is shown in Figure 5.10a. The filtered fault signal goes into the module and is compared with an offset value (here set as "0") to produce a square wave corresponding to the input sinusoidal value. The square wave carries the phase information of the sinusoidal wave. Then, the square waves of the two inputs (the polarizing vector V_p and operating vector V_o) are compared with each other using

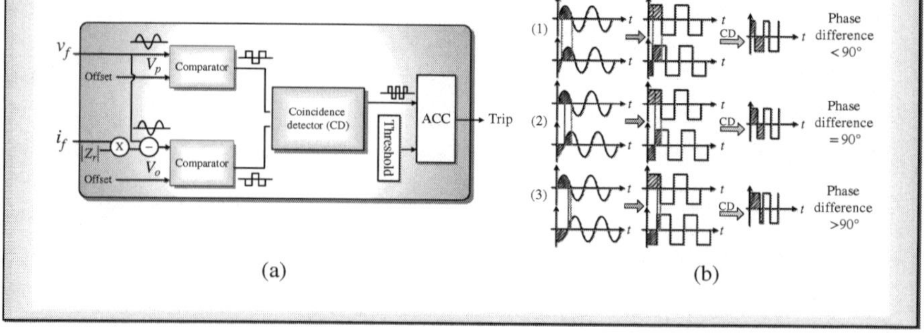

Figure 5.10 Instantaneous-signal-based distance relay: (a) functional block diagram of the hardware design and (b) principle of operation.

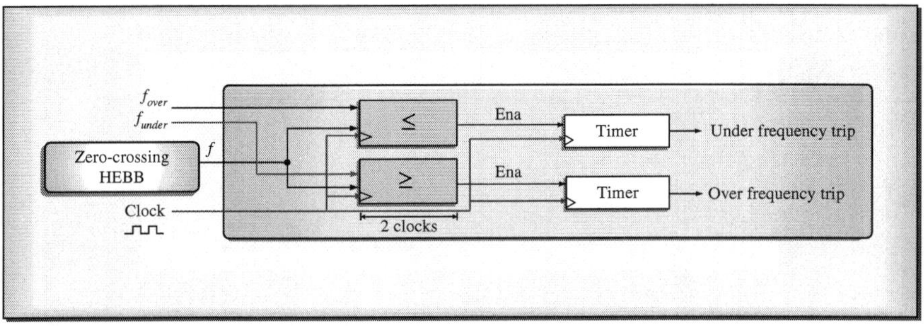

Figure 5.11 Frequency protection HEBB.

a coincidence detector. A "−1" is produced during the times both square waves agree in polarity while a "+1" is produced during times the two square waves have opposite polarity; thus, a coincidence detection is achieved resulting in a double frequency square wave. If the two input waves have a phase difference of ±90°, the double frequency square wave has equal positive and negative pulses as in Figure 5.10b (2). In other cases, for instance a fault, the coincidence is biased that can be clearly seen in Figure 5.10b (1 or 3). The *ACC* block takes in the coincidence results and accumulates the biased part points in the double frequency waveform. Once the accumulation number exceeds a set threshold, a trip output signal is sent to isolate the fault.

5.2.2.5 Under/Over Frequency Protection HEBB

Frequency is another important quantity to monitor the well-being of a power system. It reflects the balance between the generation and system load plus losses. During sudden generation-load imbalances, frequency can deviate from its nominal value. The frequency protection HEBB senses the abnormal frequency deviation to make the trip decision. It provides under/over-frequency protection with or without a time delay. The functional architecture of frequency protection HEBB is shown in Figure 5.11. Using the output of zero-crossing (see Figure 5.5) frequency calculation results, the under/over- frequency elements compare the signal with the prespecified low-setting or high-setting to determine the trip logic.

5.3 Test Setup and Real-Time Results

The complete hardware multifunction system design was targeted to the Xilinx® Virtex-7® XC7VX485T FPGA whose hardware resources are listed in Table 1.2. The design was implemented wholly in VHDL with 32 (64)-bit floating-point precision. The Xilinx ISE® environment was used to synthesize and map the design to the

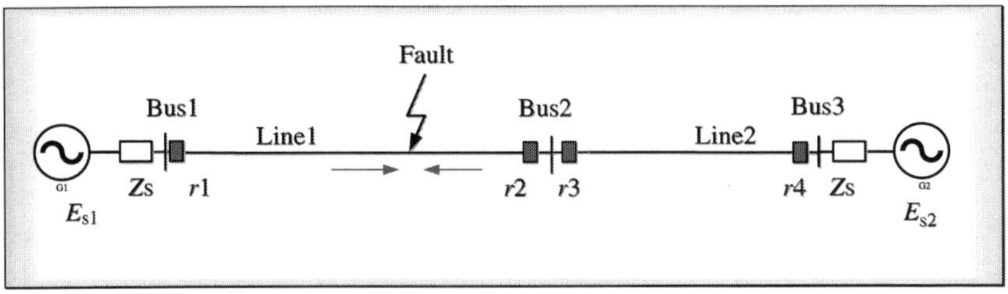

Figure 5.12 Case I test system.

Table 5.2 Hardware resource consumption and latencies of the multifunction protection system.

HEBBs	Slice registers	Slice LUTs	DSP blocks	Latency (clock cycles)
Directional overcurrent	4 614 (1%)	1 669 (0.6%)	32 (1%)	47
Distance relay	64 118 (10%)	28 576 (9%)	53 (2%)	209
Over/under voltage	2 520 (0.4%)	1 417 (0.5%)	0 (0%)	7
Over/under frequency	34 983 (6%)	88 042 (29%)	510 (18%)	1 566
Total usage	106 235 (17%)	119 704 (39%)	595 (21%)	—

physical device. The highest frequency the whole design can run at is 40 MHz due to the long critical path latency of the frequency HEBB. To observe output waveforms, a 16-bit 4-channel digital to analog converter (DAC) board was connected to the FPGA board. The Tektronix® 1 GHz DPO4104 with four analog channels and 5 GS/s mixed-signal oscilloscope was used to capture the output of the DACs as shown in Figure 5.12a. The hardware resource usage and overall latency of each protection function (the signal processing and detection functions are included in the respective protective function designs) is summarized in Table 5.2. The frequency protection consumes the most hardware resources due to the large floating point and iteration computational burden, while the voltage protection consumes the least resource. Overall, the hardware consumption remains at a reasonable level. In the following subsections, two test cases are set up to verify the functions of the protection system.

5.3.1 Case I

This case study shows the test results of directional overcurrent protection, voltage protection, and distance protection. The test power system can be seen in Figure 5.12 and consists of two synchronous generators and two transmission lines

(1 and 2) which are 110 and 100 km long, respectively. The parameters of this system are in the Appendix A.4.

The studied fault for overcurrent and distance protection is a phase-a-to-ground fault. The oscilloscope captures the current waveform (in Figure 5.13a) at location 1 from $t = 0.1$ second to $t = 0.565$ second. The fault happens on Line 1 at $t = 0.28$ second, 50 km away from the relay location 1. Figure 5.13a gives the DFT

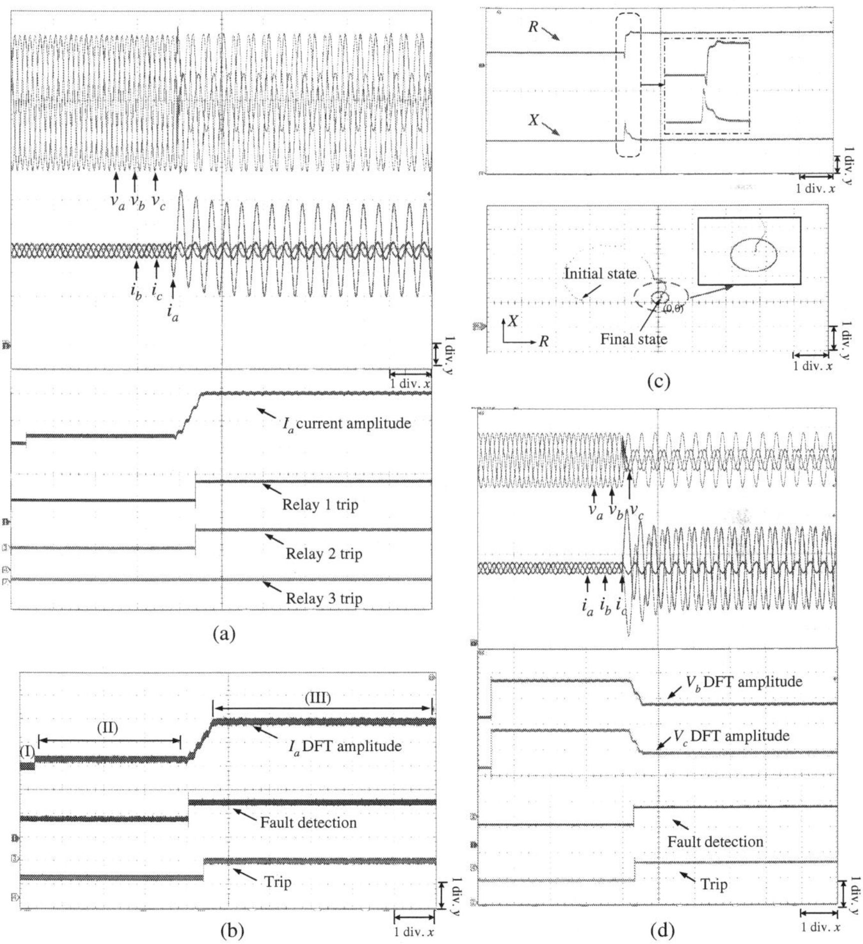

Figure 5.13 Real-time test results for Case I: (a) voltage and current waveforms of a phase-a-to-ground fault, and directional overcurrent protection signals, (b) DFT, detection and trip signals for distance protection, (c) impedance captures of distance protection, and (d) voltage and current waveforms of a phase-b-to-c fault, and voltage protection signals. Scale (time: 1 div. $x = 46.5$ ms; voltage: 1 div. $v = 70$ kV; current: 1 div. $i = 0.97$ kA; 1 div. r or $x = 500\,\Omega$).

and no-time-delay trip signals for the directional overcurrent protection case study. The relays 1 and 2 see the overcurrent fault in the forward direction and send the trip signals, while relay 3 sees the reverse direction and remains inactive. Due to the coordination between relay 4 and 2, the trip signal of relay 4 is blocked since relay at 1 and 2 on Line 1 successfully trip.

For distance protection, the real-time results can be observed in Figure 5.13b,c. Figure 5.13c gives the real-time trajectory of the apparent impedance seen by the distance protection relay. The mho Zone 1 reach is 80% of the line length, which is $(3.14 + j44.69)\,\Omega$. By coupling the time-based resistance R and reactance X coordinates, the real-time impedance locus is obtained. Before the fault at the initial state $(-628.3 + j1.7)\,\Omega$, the impedance is outside the mho circle. After the fault occurs, the impedance enters into the circle and converges to its final state of $(5.85 + j24.33)\,\Omega$, which is 56% of the zone reach and 44% of the line length. In the DFT processing results of the faulted phase shown in Figure 5.13b, the amplitude of Segment I stays zero until one full cycle of fault data collection is finished. During Segment II, the I_a amplitude value stays constant under normal operating condition until at $t = 0.28$ *s*when the ground fault occurs. Accordingly from the DFT calculation, the amplitude of I_a starts to increase after a small transient oscillation. The fault detection module senses the abnormal current increase and operates to send the fault *detected* signal. The smooth Segment III indicates the effectiveness of the DC offset removal of fault current in the faulted phase. The final trip signal gives at $t = 0.303$ *s*.

The trip time results of a double-phase-to-ground fault and a three-phase-to-ground fault are given in Table 5.3. According to these results, the proposed hardware digital distance relay has achieved its fundamental objectives. While an exact comparison of the FPGA-based distance relay with existing numerical relays might be somewhat misleading since the implemented algorithms and the design logic might not be exactly the same, to put this work into context, a few examples from existing relays in the industry are provided. ABB's REL 650 line distance protection relay has the distance measuring typical operate time of 30 ms, and 24 ms for the REL 670 platform. In SEL's 421 protection, automation,

Table 5.3 Trip times for different types of faults.

Faults	Phasor-based (ms)	Instantaneous-signal based (ms)
a–g	25.12	8.46
b–c	22.89	6.22
b–c–g	22.39	6.21
a–b–c–g	22.14	5.47

Table 5.4 Comparison based on different sampling rates: detection times (DFT-based) and trip times (instantaneous-signal based).

Faults	Detection times (DFT-based) (ms)			Trip times (Inst.-signal based) (ms)		
	Sample rates		Reduction	Sample rates		Reduction
	(68/cycle)	(334/cycle)	(ms)	(68/cycle)	(334/cycle)	(ms)
a–g	7.35	6.80	0.55	8.46	6.58	1.88
b–c	5.15	4.49	0.66	6.22	5.59	0.63
b–c–g	5.39	4.89	0.50	6.21	5.79	0.42
a–b–c–g	4.42	3.84	0.58	5.47	4.69	0.78

and control system, the high-speed elements operate around 0.5 fundamental cycle, while the standard-speed elements longest operating times is close to 1.5 fundamental cycles. These operating times are similar to the ones shown in Table 5.4 for the designed hardware distance relay. This is because, using FCDFT, one cannot initiate the Fourier algorithm until a full cycle of data has been collected. Furthermore, it is worth mentioning that the initial chosen sampling rate for a full-cycle for this hardware design is 68. However, the overall low-latency of the hardware design leads to high-speed computation on the FPGA, and thus higher sampling rates can be chosen (more recent numerical relays use sampling rates that are as high as 96 samples per period [95]). In the proposed hardware design, when the sampling rate is 334 (50 μs time interval) per cycle, results in Table 5.4 for the DFT-based and for the instantaneous-signal-based options show that the fault detection time or trip time could be shortened by over 10%.

A b–c double-phase fault applied to the same power system is studied to verify the voltage protection function. Figure 5.13d shows the three-phase voltage, current waveforms, and the DFT amplitude of faulted phase voltage. The fault detection and tripping signals are also depicted in this figure. After the fault happens, the faulted phases b and c voltage amplitudes drop dramatically. The undervoltage element senses this decrease in voltage and issues a trip signal.

5.3.2 Case II

The power system shown in Figure 5.14 whose parameters are given in Appendix A.4 is used to test the performance of frequency protection. In this system, three generators are connected to a common bus that supplies the loads. The waveforms trace the time period of $t = 0$ second to $t = 50$ seconds. The real-time tracking of the system frequency has a precision of $\pm$ 0.01 Hz. One branch that has an R–L load can emulate the frequency disturbance source. When the load is suddenly connected at $t_1 = 1$ second, the system experiences under-frequency condition as shown in Figure 5.15 and the frequency goes down gradually. As

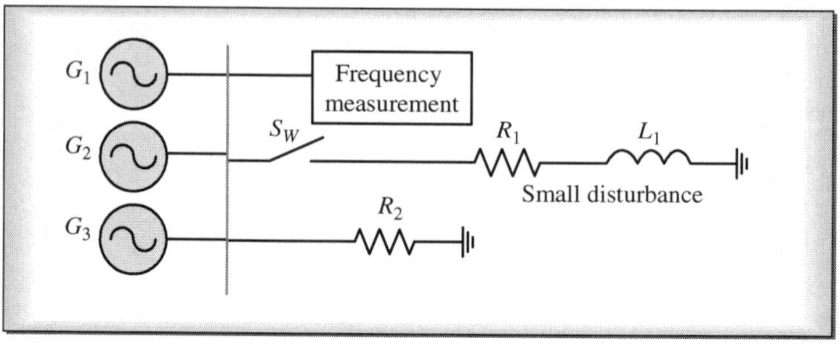

Figure 5.14 Test system for Case II.

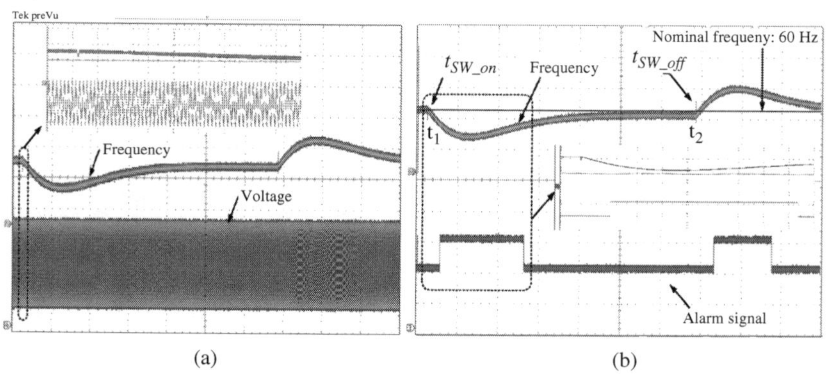

Figure 5.15 Results for Case II: (a) frequency and voltage signals and (b) frequency and alarm signals (Time: 1 div. x = 5 seconds; voltage: 1 div. v= 7.3 kV; frequency: 1 div f= 1.0 Hz).

suggested in [116], abnormal frequency can be operated continuously between 59.5 and 60.5 Hz. In this case, when frequency reaches to 59.5 Hz, a no-time-delay alarm signal is issued. When the frequency gets back to above 59.5 Hz after about 10 seconds, the alarm signal is cleared. At t_2 when the disturbance branch is switched off, this introduces an over-frequency condition. The alarm signal is issued when the frequency goes over 60.5 Hz.

5.4 Summary

This chapter presented a complete real-time multifunction protection system design in hardware. FPGA technology provides a flexible and efficient platform to develop complex protection functions in power systems. The total hardware usage of the multifunction protection system is 39% which is reasonably low to reduce power consumption and increase life cycle of the device. All the relay functions are emulated in VHDL using 32 (64)-bit floating point precision ensuring high fidelity. Parallelism and pipelining was fully exploited in both the signal processing hardware emulation blocks as well as the protection function blocks. Based on the designed hardware multifunction relay, there are several advantages related to design reliability, maintainability, power consumption, and flexibility that make FPGAs suitable for protective relay applications. Whereas in a numerical relay with a digital signal processing (DSP) software design, the executing algorithm relies on transferring information to and from memory and sequential processing, the FPGA is composed of configurable logic components, and all the user-defined functional elements can be achieved by the hard-wired architecture. Each task is allocated its own hardware resources, such as the parallel distributed memory throughout the device, and runs independently with less interference between two functional blocks. The VHDL code for the designed hardware multifunction relay has a hierarchical modular structure, and the various VHDL modules distributed on the chip were interconnected by input and output ports. The connections between the various modules are hardwired. Testing the entire design involved testing every functional circuit and then testing the top-level design connected together. This makes the VHDL program easy to troubleshoot and deploy.

The designed multifunction relay has a very low hardware latency while operating at a much lower clock frequency (100 MHz for FPGA vs. 1 GHz for DSP) thereby reducing chip power consumption. Thus, a much higher sample rate can be used for the FPGA-based relay which leads to better reproduction of the original signals and better transient analysis. Furthermore, new protection algorithms based on traveling waves and transient signals can be developed on the FPGA.

6

Adaptive Time-Stepping Based Real-Time EMT Emulation

6.1 Overview

Since transients can happen in just few tens of microseconds and the overvoltages can be multiple times the corresponding ratings, small time-steps are required for the electro-magnetic transient (EMT) simulation to accurately capture the transient behavior. However, for the simulation of normal steady-state operation, a small time-step is not necessary for observation and can be a great computational burden for optimum simulation speed. Currently, existing off-line power system EMT simulation tools employ a fixed user-defined simulation time-step. The concept of adaptive or variable time-step has been traditionally used in many device-level power electronics circuit simulation tools such as PSpice®, HSPICE®, SaberRD® [117–119]. Such simulation tools often use iterative methods, like the N-R or the Katzenelson methods combined with various adaptive time-stepping (ATS) schemes to solve the nonlinear circuit system.

However, in the context of real-time simulation, which is a powerful tool for hardware-in-the-loop testing, fixed time-step (FTS) is exclusively used in commercial EMT-type simulators due to the following reasons:

1. The real-time simulators are often required to interface with external device-under-test (DUT) such as a control system or protection system. The sampling process of the interface signals, such as assigned instantaneous voltage and current values collected by control and protection system, or the gating signals of switching devices received by the simulator, is inherently required at fixed time intervals.
2. Computation efforts are large and are unable to be predicted precisely for iterative and variable time-stepping schemes. The iterative methods used in device-level circuit simulation tools often have less stability, and the solution may not converge under certain circumstances.

Real-Time Electromagnetic Transient Simulation of AC-DC Networks, First Edition.
Venkata Dinavahi and Ning Lin.

Nevertheless, imperfectness exists in the interface process between the real-world power system and the control and protection system, which includes communication delay, analog-to-digital conversion delay, drive circuit delay, the reaction and process time of execution devices, etc. These inherent delays are considered in the design of control systems. A small amount of delay for the real-time simulator is tolerable depending on the system type and control purpose, varying from tens of microseconds to a few milliseconds. Based on this fact, variable time-stepping schemes are implementable within a fixed longer sampling period for real-time simulation.

In many academic research projects, the real-time simulator often works in stand-alone mode where the control algorithm is built inside the simulator along with the host system model. In some industrial training applications, the real-time simulator is used to train and test the response of the operators under certain circumstances. For such applications, reliable and accurate results and overall real-time performance are required over the simulation duration; however, it is not necessary to ensure the real-time performance for all time instances.

At this point, the concept of *hard* real-time, *firm* real-time, and *soft* real-time, which originally comes from the area of general real-time computing, is discussed within the scope of power systems real-time simulation [120]. The hard real-time applications refer to those where real-time constraint must be met in every simulation time-step; therefore, an FTS must be used. The firm real-time applications refer to those where the interface sampling period is significantly larger than the minimum simulation time-step; however, it is still less than a certain value, such as 1 ms, where an adaptive simulation time-step can be used limitedly. The soft real-time applications refer to those which do not require hardware interfacing or have a very large interface sampling period, where variable time-stepping schemes can be used fully or with minimum restriction. The use of firm and soft real-time does not necessarily mean the degradation of the simulation results, but the emphasis on the different functions and purposes. As later demonstrated in this chapter, with the same computation capability, fast transients can be better presented if an ATS scheme is applied.

This chapter presents methodologies for using ATS in real-time EMT simulation. Various combinations of nonlinear solution methods and ATS schemes are explored in detail revealing the challenges and corresponding characteristics. ATS universal line model (ULM) and universal machine (UM) model are presented. Relevant case studies are used to highlight the advantages of ATS for real-time simulation and the field programmable gate array (FPGA) implementation uses both HDL and HLS for the hardware design.

6.2 Nonlinear Solution and Adaptive Time-Stepping Schemes

The selection of time-stepping scheme is related to the nonlinear element solution method, which provides certain variables for the time-step control algorithm. Therefore, the combinations can be various, but not arbitrary. In this section, several widely used nonlinear element solution methods and time-stepping schemes are introduced [27, 119, 121–123]. The challenges of applying ATS for real-time simulation and the mitigation solutions are then elaborated.

6.2.1 Nonlinear Element Solution Methods

6.2.1.1 Newton–Raphson Method

Applying N-R method to solve a general nonlinear circuit using nodal analysis, the matrix equation is giving by:

$$\mathbf{G}(\mathbf{v}^k)\mathbf{v}^{k+1} = \mathbf{I}_{eq}(\mathbf{v}^k), \tag{6.1}$$

where $\mathbf{v}$ is the unknown node voltage vector; $\mathbf{I}_{eq}$ is the equivalent current source vector; k is the iteration number; and $\mathbf{G}$ is the Jacobian matrix for conductances, given by:

$$\mathbf{G}(\mathbf{v}^k) = \begin{pmatrix} \frac{\partial i_1}{\partial v_1}\Big|_{\mathbf{v}^k} & \frac{\partial i_1}{\partial v_2}\Big|_{\mathbf{v}^k} & \cdots & \frac{\partial i_1}{\partial v_n}\Big|_{\mathbf{v}^k} \\ \vdots & \vdots & & \vdots \\ \frac{\partial i_n}{\partial v_1}\Big|_{\mathbf{v}^k} & \frac{\partial i_n}{\partial v_2}\Big|_{\mathbf{v}^k} & \cdots & \frac{\partial i_n}{\partial v_n}\Big|_{\mathbf{v}^k} \end{pmatrix}. \tag{6.2}$$

The computation effort of N-R method is relatively large, because of the iteration process, which often involves the matrix refactorization. The complex process of N-R method, on the other hand, provides various convergence information, such as iteration counts and local truncation errors (LTEs), which can be used as the criteria for time-stepping schemes.

6.2.1.2 Piecewise Linearization (PWL) Method

This method, also known as pseudo-nonlinear method, uses piecewise linearized segments to fit the nonlinear curve. The segment is determined at the beginning of each time-step based on the node voltages of previous time instance, which may cause the problem of overshoot. This method is less accurate however faster compared with other iteration methods, which is often used by conventional real-time simulators.

6.2.1.3 Piecewise N-R Method

N-R method uses continuous functions to describe the nonlinear elements, while PNR method instead uses piecewise linearized functions, which simplifies the nonlinear function resolving process and lowers the convergence criteria. PNR can also been seen as the iterative version of piecewise linearization (PWL) method. During fast transients, iteration is necessary to locate the correct piecewise segment without delay, which significantly increases the result accuracy.

6.2.2 Adaptive Time-Stepping Schemes

The ATS process is determined by the properly selected discretization method and time-step control scheme, and based on which of the specific equipment models are involved in the EMT simulation.

6.2.2.1 Local Truncation Error Method

Typically, a power system containing linear or nonlinear dynamic elements can be formulated as a series of differential equations expressed as: $\dot{\boldsymbol{x}} = f\ (\boldsymbol{x}, t)$. The practical approach for solving these differential equations is to repeatedly discretize them and solve the resulting algebraic equations. The uniform discretization method is the linear multistep (LMS) method:

$$\sum_{j=0}^{k} \alpha_j \boldsymbol{x}_{n-j} = \Delta t \sum_{j=0}^{k} \beta_j \boldsymbol{f}_{n-j}, \tag{6.3}$$

where k is the step-number and if $k = 1$, it is called one-step method. If $\beta_0 = 0$, the method is explicit, otherwise it is implicit.

The time-step Δt in (6.3) is assumed to be fixed. As for the ATS method in which the time-step Δt may change, the LMS method requires to be re-derived for the case of nonequidistant data. But one-step methods require no modification to their formulas, i.e.

$$\sum_{j=0}^{1} \alpha_j \boldsymbol{x}_{n-j} = \Delta t_n \sum_{j=0}^{1} \beta_j \boldsymbol{f}_{n-j}, \tag{6.4}$$

$$\Delta t_n = t_n - t_{n-1}. \tag{6.5}$$

Higher order formulas are useful because they can be more accurate; however, it becomes more complicated to derive the LMS method with an adaptive time-step. Since the commonly used discretization methods such as the forward Euler (FE), backward Euler (BE), and Trapezoidal rule (TR) method are all one-step LMS methods. Among these methods, BE and TR are both convergent and zero-stable, which means the time-step control can be based solely on the consideration of accuracy without considering the convergence and stability. Therefore, in this work, the TR method is applied for all the system equipment, and the time-step control is based on consideration of accuracy, which is always represented by the LTE defined as follows:

$$LTE(t_n) = \boldsymbol{x}(t_n) - \tilde{\boldsymbol{x}}_n \approx C_{p+1}\Delta t_n^{p+1}\boldsymbol{x}^{(p+1)}(t_{n-1}), \tag{6.6}$$

where $\boldsymbol{x}(t_n)$ is the exact solution at t_n, and $\tilde{\boldsymbol{x}}_n$ is the solution returned by the k-step LMS method when artificially setting $\boldsymbol{x}_{n-j} = \boldsymbol{x}(t_{n-j})$ for $j = 0, 1, \dots, k-1$. C_{p+1} is the error constant of a specific LMS method; $\boldsymbol{x}^{(p+1)}(t_{n-1})$ means the $(p+1)$th derivative of $\boldsymbol{x}$ evaluated at t_{n-1}, which is usually difficult to find. Thus, alternative approaches are applied to estimate the LTE:

$$LTE(t_n) \approx C_{p+1}\Delta t_n^{p+1}(p+1)!\ \mathbf{g}[t_n, \dots, t_{n-1-p}], \tag{6.7}$$

where $\mathbf{g}[t_n, \dots, t_{n-k}]$ can be calculated step-by-step: $\mathbf{g}(t_n) = \boldsymbol{x}_n$ and

$$\mathbf{g}[t_n, \dots, t_{n-k}] = \frac{\mathbf{g}[t_n, \dots, t_{n-k+1}] - \mathbf{g}[t_{n-1}, \dots, t_{n-k}]}{t_n - t_{n-k}}. \tag{6.8}$$

Once the LTE at t_n is obtained, the next time-step is supposed to change to:

$$\Delta t_{n+1} \approx \Delta t_n min\left|\frac{\epsilon(i)}{\epsilon_n(i)}\right|^{\frac{1}{p+1}}, \tag{6.9}$$

where $\epsilon_n = LTE(t_n)$ is computed and ϵ is the desired LTE at t_{n+1}.

6.2.2.2 Iteration Count Method

This method simply records the N-R iteration count for each time instance. If many iterations are required to obtain the convergent results, the time-step will decrease for future time instances.

6.2.2.3 DVDT or DIDT Method

DVDT and DIDT are the derivatives of the node voltages and branch currents detecting the large changing rates. When a sudden large voltage or current surge appears in the circuit, small time-steps are used to capture the transients more clearly and accurately.

The three time-stepping schemes mentioned above have the major focus on the LTE induced by dynamic element modeling, the degree of convergence for nonlinear elements, and the observations needed for accurate large-amplitude high-frequency transients, respectively. The circuit type, modeling method, and the simulation purpose all affect the selection of time-stepping scheme and the corresponding threshold parameters. Another issue is the choice of the time-step changing rate, and multiple methods exist in the literature [122, 123]. In this work, a factor of 2 multiplication or division is chosen as the adaptive rate for time-step, which is straightforward and has been adopted by SaberRD® software.

6.2.3 Combinations of Adaptive Time-Stepping Schemes

Various combinations of nonlinear element solution methods and dynamic time-stepping schemes are feasible for real-time simulation. DVDT or DIDT method can be used with all three above-mentioned nonlinear solution methods with or without iterations. It is noted that the combination of PWL method and DVDT/DIDT method is compatible with off-line EMT-type software engine. Therefore, this combination could be helpful for boosting the off-line simulation for some applications. Although PNR method does require iterations for some time instances, however in other circumstances the nonlinear segment does not change and the results can be solved without iteration, which is the same as PWL method. Therefore, iteration count scheme is not available for PNR and PWL method. LTE scheme is normally used with N-R like method, where highly nonlinear functions exist. Under such circumstances, the numerical stability is relied on the appropriate choice of time-steps. Since PNR and PWL methods generally have good numerical stability, the LTE scheme is not necessary though the predictor–corrector procedures may still be applicable for some applications.

Since the computation effort for N-R method is higher, the simulated circuit scale is smaller than those using other simpler methods. The smaller circuit using N-R method can be cooperated with other larger linear circuit by using various circuit separation schemes such as introducing delays, using transmission line model for connection. The connected larger circuit can use FTS with the interface similar to the one used for the external control system.

6.2.3.1 Measurements and Restrictions for Real-Time Emulation

The relatively large computation effort for iterative schemes and the requirement of a fixed interface sampling rate are major challenges for applying ATS schemes. Some restrictions and measurements are applied to mitigate the issues, listed as follows.

Limiting the Range of Time-Steps The smaller range, especially limiting the maximum time-step, can improve the convergence property of the iteration schemes, and reduce the chance of rejecting solutions since large LTE or iteration counts are often caused by using large time-steps. To maximize the usage of the computation capability, the variable time-steps shall be changed around the maximum FTS, which ensures the real-time performance. The maximum time-step cannot exceed the interface sampling period, while the minimum time-step is restricted by the computation capability.

Applying Conservative Time-Step Criteria The time-step changing criteria is referred to the threshold values of LTE, iteration counts, or DVDT. The conservative threshold criteria choose smaller time-steps, which may increase the total time instances; however, it may avoid frequent time-step changing. However, because of the smaller time-steps, the total iteration counts tend to be smaller, and the chance of solution rejection is smaller. Combined with the limited time-step range, the computational effort is relatively stabler and easier to predict during the simulation.

Using Smaller Maximum Iteration Count Number The purpose of this restriction is to limit the computation burden for a time instance. It is particularly useful for PNR method with numerical stable applications, where solution rejection is not necessary since limiting iterations just affect the result accuracy. For such applications, the extreme case of applying this restriction is the same as using PWL method.

Precalculating Constants for Different Time-Steps When applying smaller range of time-steps and using constant changing rate for time-step, there are only several determined time-steps during the simulation. Before the simulation begins, the constants and the conductance matrices for all these time-steps are calculated in advance and partially decomposed. When the time-step is changed, these precalculated constants are loaded to the solver directly without recalculation. The loading process can be extremely fast for FPGA with merely one clock cycle delay.

Using Foreknown Transient Information Some test conditions, such as faults and surges, are often preset to the simulator and triggered at a determined time instance. This knowledge can let the simulator set the time-step to a minimum value just before the condition happens, since high-frequency transients may occur. This measurement will not affect testing results, since the device under test, which can be a controller, is not aware of any test conditions in advance.

6.2.4 Case Studies

Two case studies, diode full-bridge circuit (Case 1) and a transmission system with nonlinear surge arrestors (Case 2), are presented in this section, using different time-stepping schemes. The parameters of the case studies are given in [10]. The off-line circuit simulation programs for the case studies are written in Matlab® script, where the restrictions and measurements for real-time simulation are considered. The result accuracy is verified by comparing with SaberRD® or PSCAD/EMTDC® software. The solution combinations with the best performance are then implemented on FPGA and run in real-time.

6.2.4.1 Diode Full-Bridge Circuit

Diode Model Description The topology of diode bridge circuit is shown in Figure 6.1, where a physics-based nonlinear p–i–n diode model including accurate reverse recovery phenomenon is used [124], given as:

$$i(t) = \frac{q_E(t) - q_M(t)}{T_M}, \tag{6.10}$$

$$0 = \frac{dq_M(t)}{dt} + \frac{q_M(t)}{\tau} - \frac{q_E(t) - q_M(t)}{T_M}, \tag{6.11}$$

$$q_E(t) = I_S\tau \left[e^{\frac{v(t)}{nV_T}} - 1\right], \tag{6.12}$$

where $q_E(t)$ is the junction charge variable; $q_M(t)$ is the charge in the middle of intrinsic region; $i(t)$ and $v(t)$ is the current and voltage across the diode; T_M is the diffusion transit time; τ is the carrier lifetime; I_S is the diode saturation current constant; V_T is the thermal voltage; and n is 2 for high level injection. Figure 6.1 presents the equivalent circuit model of p–i–n diode and the reverse recovery phenomenon from T_0 to T_2 with the peak value I_{RM} at T_1. Since the diodes are modeled by highly nonlinear and relatively complex functions, N-R method is chosen to obtain the results accurately. The TR is applied to discretize and linearize the differential equation (6.11), given as:

$$q_M(t) = \frac{\Delta t}{2T_M(1 + k_1)} \cdot q_E(t) + \frac{q_{hist}(t - \Delta t)}{1 + k_1}, \tag{6.13}$$

where

$$q_{hist}(t - \Delta t) = q_M(t - \Delta t) \cdot (1 - k_1) + \frac{\Delta t}{2T_M} \cdot q_E(t - \Delta t), \tag{6.14}$$

$$k_1 = \frac{\Delta t}{2}\left(\frac{1}{\tau} + \frac{1}{T_M}\right). \tag{6.15}$$

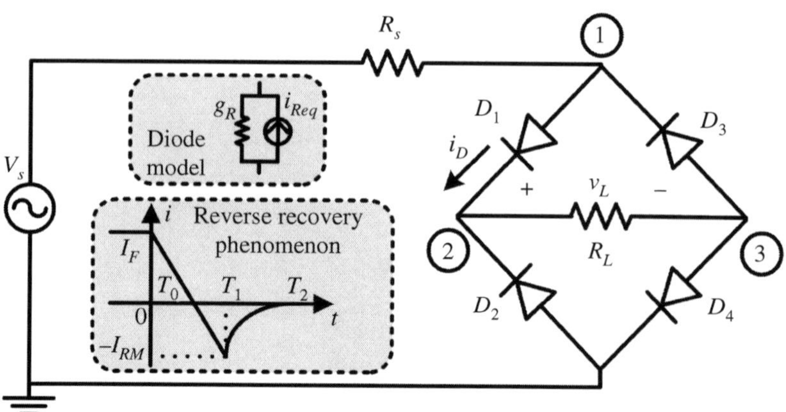

Figure 6.1 Circuit topology, diode model, and reverse recovery phenomenon for diode full-bridge circuit case study.

The time-step Δt for the calculation of $q_M(t)$ and $q_{hist}(t-\Delta t)$ can be different when applying dynamic time-stepping schemes. Using N-R method, the equivalent conductance g_R and the current source i_{Req} are given as:

$$g_R = \frac{\partial i_R}{\partial v(t)} = k_2 I_S \tau \cdot \frac{1}{nV_T} \cdot e^{\frac{v(t)}{nV_T}}, \tag{6.16}$$

$$i_{Req} = k_2 I_S \tau \left[e^{\frac{v(t)}{nV_T}} - 1\right] - k_3 q_{hist}(t-\Delta t) - g_R \cdot v(t), \tag{6.17}$$

where

$$k_2 = \frac{1}{T_M} - \frac{\Delta t}{2T_M^2(1+k_1)}, \tag{6.18}$$

$$k_3 = \frac{1}{T_M(1+k_1)}. \tag{6.19}$$

Off-line Simulation Results The off-line circuit simulation results for the diode bridge circuit using LTE, and iteration count schemes based-on N-R method are shown in Figure 6.2b,c. Since the voltage and current waveforms do not change abruptly, DVDT and DIDT methods are not effective for this application. The off-line simulation results, including the diode current i_D for D_1 and load voltage v_L, are compared with SaberRD® (Figure 6.2a), where the power diode model is employed. All the current and voltage waveforms have sufficient accuracy clearly showing the diode reverse recovery phenomenon. The time-steps for both LTE and iteration count schemes are limited within the range of 1.25–5 μs, while the minimum time-step is not limited for Saber® result. By limiting the range, the calculated points are reduced, while accuracy is preserved. Using a fixed small time-step to limit the LTE for all points to ensure sufficient accuracy would be computationally inefficient. Compared with the iteration count scheme, LTE scheme utilizes less total iteration counts; therefore, it is chosen to be implemented on the FPGA.

6.2.4.2 Power Transmission System

Circuit Structure The three-phase power transmission system shown in Figure 6.3 is simulated to observe the transients of the lightning strike, occurring in the middle of a transmission line. On the generator side, an equivalent voltage source V_s in series with a resistor R_s is connected to a Y–Y connected transformer T. Series connected inductor L_L and resistor R_L are used to represent the load, connected through a 100 km transmission line using Bergeron line model (distributed parameter traveling wave model). To determine the history current value $i(t-\tau)$ for Bergeron line model with variable time-stepping scheme, the exact time of the history current value is required to be stored. The $i(t-\tau)$ can be obtained by linear interpolation of the two history current values between $t-\tau$. The six gap-less surge arresters SA_{1-6} are connected to the both sides of the transmission line, which is separated into two segments L_1 and L_2 with the lightning directly hitting phase C represented by a current source I_{LS}.

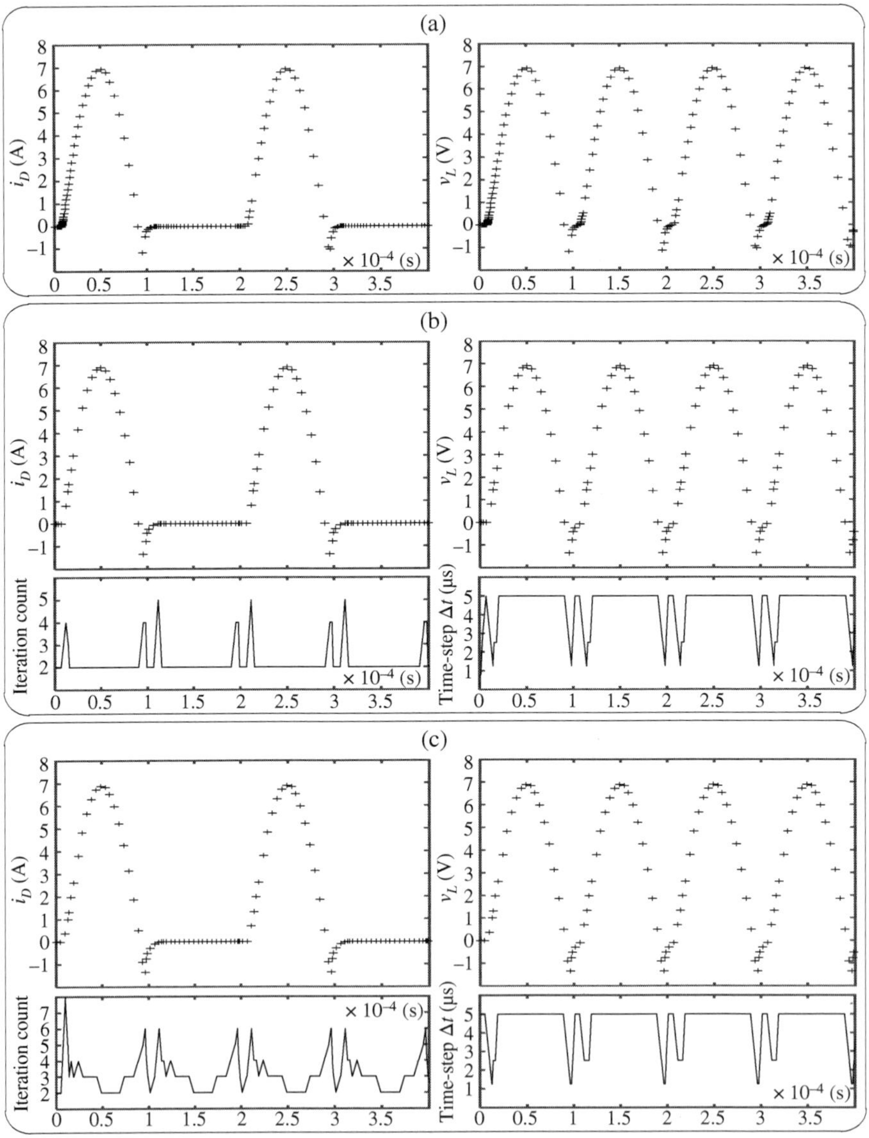

Figure 6.2 Off-line results of diode current i_D for D_1, load voltage v_L, iteration count, and time-step Δt: (a) SaberRD®, (b) LTE scheme based on N-R method, and (c) iteration count scheme based on N-R method.

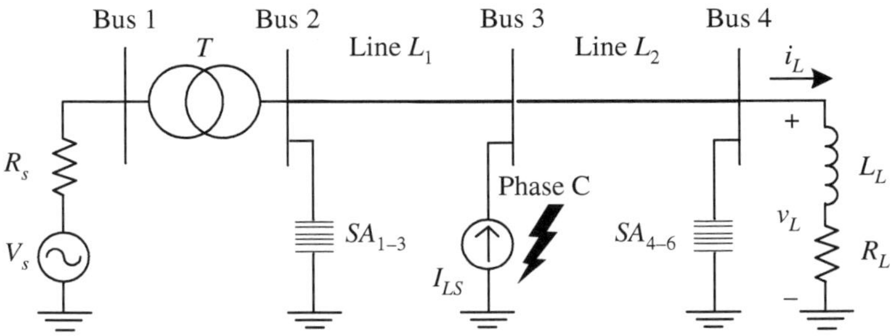

Figure 6.3 Single-line diagram for power transmission system case study.

Arrester Model and Lightening Surge Metal-oxide surge arrester is commonly used to suppress the overvoltage caused by lightning, switching, faults, etc. [125]. Piece-wise linearization method is applied to model the surge arrester, consuming less computation effort than using a continuous nonlinear function. The segment data for surge arresters are presented in Appendix A.5.

The standard 8/20 μs current surge testing waveform is used in this work, given as [126]:

$$I_{LS}(t) = AI_p t^k \exp(-t/\tau), \tag{6.20}$$

where $A = 0.012\,43\ (\mu s)^{-3}$; $k = 3$; $\tau = 3.911\ \mu s$; I_p, the maximum value of the surge current, is 5000 A.

Off-line Simulation Results PNR and PWL methods combined with DVDT scheme are utilized for the off-line power transmission system. The lightning current surge starts from exactly 10 *ms* of the simulation. The results are compared with PSCAD/EMTDC® using FTSs, shown in Figure 6.4. By changing the time-step value from 5 to 10 μs in PSCAD/EMTDC®, the result accuracy is significantly reduced, which can be observed by comparing the maximum values of overvoltages and overcurrents. When using PNR with various time-steps between 5 and 20 μs, the results are very close to the one from PSCAD/EMTDC® with 5 μs as the time-step. If PWL scheme is used, the results from variable time-stepping are no longer accurate; however, they are still better than the PSCAD/EMTDC® results with 10 μs. Since the transient frequency is very high, the change of time-steps can pose great impact on the result accuracy. In this application, by using dynamic time-stepping, the computation effort is similar to the one using 10 μs as FTS, while results with PNR method are more accurate which are close to the PSCAD/EMTDC® results using 5 μs as the FTS. Admittedly, the small-amplitude high-frequency transients are somewhat diminished, which is due to the DVDT algorithm used for the application. For hardware implementation, PNR method is chosen for its better accuracy.

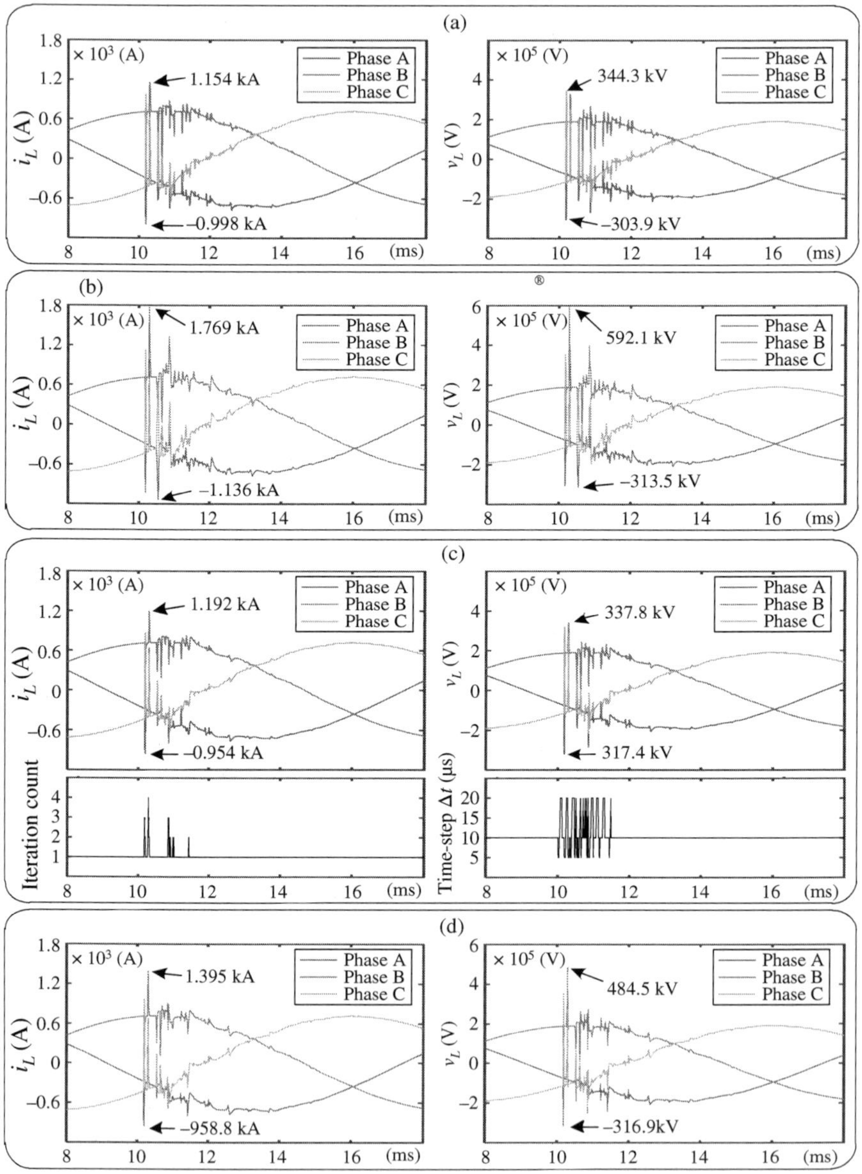

Figure 6.4 Off-line results of load current i_L, load voltage v_L, iteration count, and time-step Δt: (a) PSCAD/EMTDC® with fixed time-step 5 μs, (b) PSCAD/EMTDC® with fixed time-step 10 μs, (c) DVDT scheme based on PNR method, and (d) DVDT scheme based on PWL method.

6.2.4.3 FPGA Implementation

The hardware implementation and data flow of the two case studies are presented in Figure 6.5. The cells in dashed-line boxes are implemented for the diode full-bridge case study and the power transmission system case study, respectively. For other parts, the two case studies share the same hardware structure.

The finite-state machine of the real-time emulator controls the overall emulation flow by interfacing with other modules through control signals, and the state chart is shown in Figure 6.6. It receives the exact time t_{real} from real-time counter, and communicates with another internal FSM in real-time I/O control module, which manages the I/O data transfer between the emulator and the external devices. When the reset signal is detected, the state changes from S_0 to S_1, starting the initialization module, which calculates the constants for all time-steps. These constants are processed from user-defined circuit parameters to avoid unnecessary repetitive calculations, and some of them can be applied directly to the conductance matrix entries, such as the conductance of linear resistor. Taking the example of the nonlinear diode, the expression $k_2 I_s \tau$ is a typical Δt-dependent constant, while $1/nV_T$ is a precalculated constant without such dependence in (6.16) and (6.17). The time-step Δt dependent constants are stored in the Δt-dependent constant memory connecting to a multiplexer, while others are transferred to the corresponding modules directly.

After the initialization and loading process, the FSM then enters to the repetitive processes from S_2 to S_8, which marks the start of the real-time emulation enabling t_{real}. In S_2, the Δt-dependent constants are reloaded to the electrical source module, linear element module, and nonlinear element module through the multiplexer with Δt as the selection signal.

Once S_2 is done, all the cells in the linear element module and the electrical source module start simultaneously in S_3, which execute only once for one time-step. These cells calculate the conductance entries and the current terms of various elements, similar to the g_R and i_{Req} in (6.16) and (6.17), but for other sources and linear elements, such as the transformer and the transmission lines, using TR for discretization. Then in S_4, the equivalent conductances and current terms for nonlinear elements are calculated, which are exactly the g_R and i_{Req} of the diodes for Case 1. For nonlinear elements, iterations between S_4 and S_5 may be required, and g_R and i_{Req} need to be recalculated several times until the results converge.

The matrix solution module running in S_5 is composed of three sequential processes: matrix formation, matrix solution, and convergence judgment. Matrix formation is applied adding the conductances and current terms of multiple elements connecting to the same node, and forming the matrix $\mathbf{G}$ and vector $\mathbf{I}_{eq}$, which are then solved by Gauss–Jordan method. For the matrix solver, LU factorization, conventional Gauss elimination and Gauss–Jordan method could all be the candidates

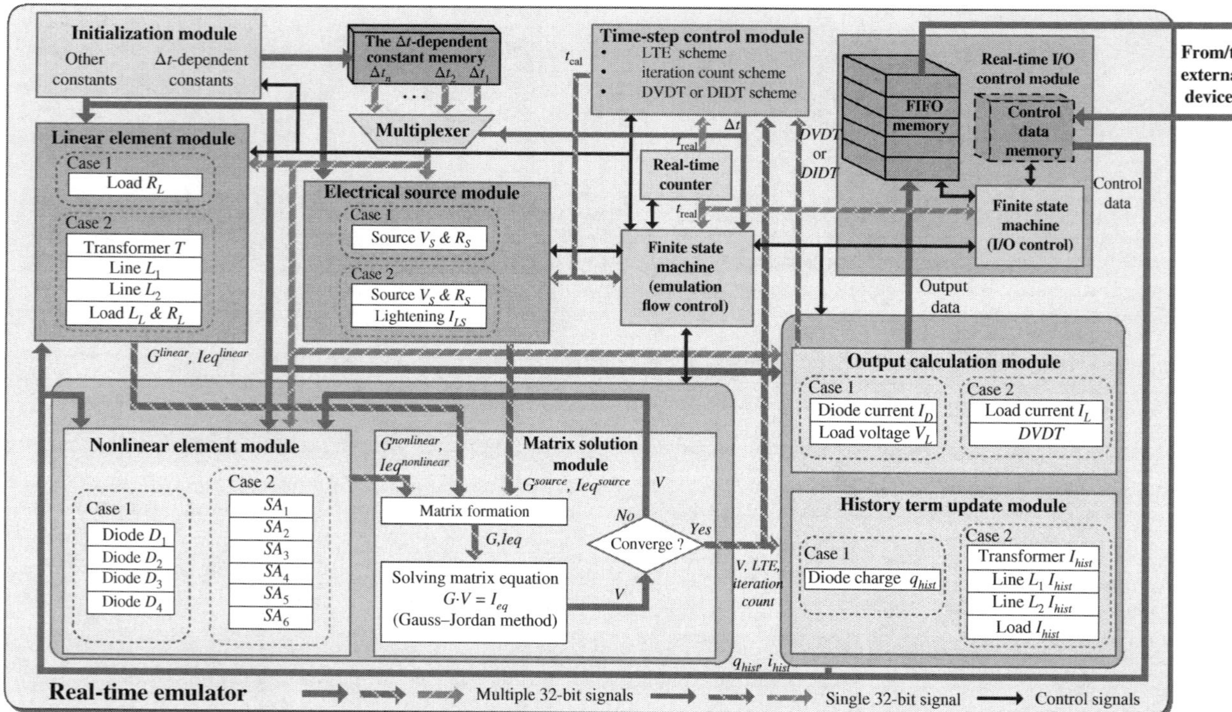

Figure 6.5 Hardware implementation and data flow of real-time emulator for the case studies.

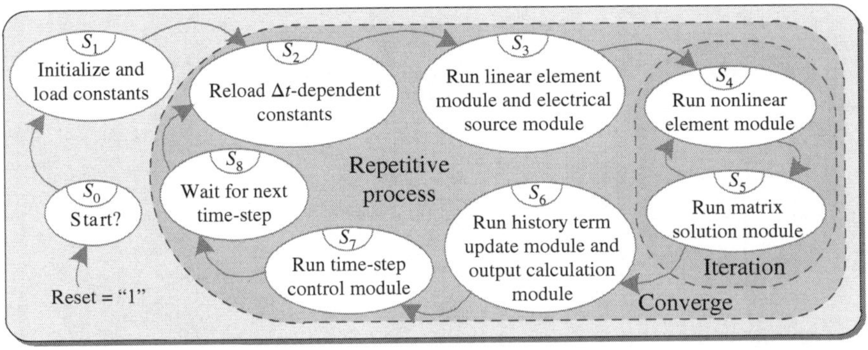

Figure 6.6 State chart for real-time emulation flow control.

for the real-time emulator [127]. LU factorization is very useful when conductance matrix G does not change over time. However when nonlinear elements are present, the factorization may need to be re-applied if any element of G changes. In such case, LU factorization is less efficient than simply using Gauss elimination, since LU method requires two sequential substitution processes (forward and backward substitutions). Both Gauss elimination and Gauss–Jordan methods include the forward elimination and backward substitution processes. The Gauss–Jordan method can effectively parallel the two processes, which always run completely for every solution calculation, though the total sequential computation effort is slightly higher. For the purpose of achieving highest parallelism and lowest latency, Gauss–Jordan method is applied for the FPGA-based implementation. Then, the matrix solution module conducts the convergence judgment by comparing the node voltages or the segments of surge arresters of the last two sets of results in Case 1 and Case 2, respectively.

Once the converged results are obtained, the history term update module and the output calculation module run simultaneously in S_6. The history term update module calculates some variables which will be used for future calculations for dynamic elements, such as the q_{hist} in (6.14) for diode. Various outputs, such as the currents of the diodes in Case 1, require further calculation, which is accomplished by the output calculation module. In S_7, the next Δt is calculated in the time-step control module.

When the next time instance comes, the state changes from S_8 to S_2, and the computation for next time-step begins. Due to the usage of variable time-stepping, S_8 may go through without actual waiting, when the delay time t_{delay}, the difference between t_{real}, and the time for the current calculated solution t_{cal}, is positive.

The algorithms of the time-step control module for the case studies are shown in Figure 6.7 in the form of pseudo-code. The time-steps can be either limited to a range between the maximum value Δt_{max} and the minimum value Δt_{min} or predetermined with several choices. For diode full-bridge circuit, the LTE

```
if  LTE > LTE_th1
    Δt ← Δt/2
    if  Δt < Δt_min
        Δt ← Δt_min
else
    if  LTE < LTE_th2
        Δt ← Δt × 2
        if  Δt > Δt_max
            Δt ← Δt_max
t_cal ← t_cal + Δt
(LTE_th1 > LTE_th2)                    (a) Case 1
```

```
t_delay ← t_real − t_cal
if  lightening surge starts
    Δt ← Δt_1
else
    if  DVDT > DVDT_th
        Δt ← Δt_1
    else
        if  t_delay > t_delay,th
            Δt ← Δt_3
        else
            Δt ← Δt_2
t_cal ← t_cal + Δt
(Δt_1 < Δt_2 < Δt_3)                   (b) Case 2
```

Figure 6.7 Pseudo-code for time-step control module: (a) diode full-bridge circuit and (b) power transmission system.

is compared with the threshold value LTE_{th1} and LTE_{th2} to determine the next time-step. While for power transmission system, the time-step is not only decided by $DVDT$ and the threshold value $DVDT_{th}$ but also the delay time t_{delay} and the threshold value $t_{delay,th}$. When the computation time exceeds the minimum time-step, a large time-step will be applied to balance the overall computation time ensuring real-time performance. Due to numerical stability consideration, this strategy is not applied for the diode full-bridge case study. For the diode full-bridge case study, the results can be calculated slightly ahead of real-time, when interfacing is not required (soft real-time condition). In other words, t_{delay} can be negative, while the results are sent out in exact real time.

The real-time I/O control module synchronizes and exchanges data between the emulation system and external devices, such as monitors, control and protection systems. For the case studies in this chapter, the external device is a four-channel oscilloscope connected by a 16-bit DAC (digital to analog converter). The FIFO memory receives the results and sends them out during the next interfacing period. The data of several time instances will be available during one interfacing period, and the external devices can receive the data of either one time instance or all time instances with the delay maximum of up to one interfacing period based on the setting in the internal FSM. A control data memory can be added to the emulator, which is not implemented for the two case studies. When the control data arrives, it will be first registered and remain the same for the interfacing period. It will then be loaded to other modules when the calculation for the next time-step starts in S_2. For the case studies, the FSM for the I/O control module only has two states: idle state and real-time output data sending state, which can be expanded for controller interface or other user-defined features. Once the results of output calculation module are ready, they are sent to the FIFO memory, which is not controlled by the FSM of real-time I/O control module.

The initialization module, the linear element module, the electrical source module, the nonlinear element module, the matrix solution module, the history term update module, and the output calculation module, shown in Figure 3.44, are synthesized by high-level synthesis (HLS), and then packaged as user designed IP cores, respectively. These modules generally include relatively complex algorithm, and the design effort can be much lowered by C programming. The real-time counter and the real-time I/O control module are designed by VHDL code for their strict timing requirement. The finite-state machine (emulation flow control), the Δt-dependent constant memory, and the multiplexer have their own fixed or classic hardware structures, and can be designed by VHDL code very efficiently and easily. IEEE 32-bit floating-point number precision is applied for the hardware design. All these modules and structures are assembled with VHDL coding, and then implemented generating the configuration bit stream.

Hardware Resource Utilization and Latency The case studies were emulated on the Xilinx® Virtex-7® XC7VX485T FPGA with complete resources listed in Table 1.2. The hardware resource consumption for the case studies is presented in Table 6.1.

The latencies of various states and the total latency for the two case studies are presented in Table 6.2. The latencies for adder/subtractor, multiplier, and divider are 4, 4, and 9 clock cycles, respectively, for the case studies with 10 ns as the clock period. The latencies may vary for different modules, due to the module scale, and are also influenced by the FPGA device, and the uncertainty of the clock period. Since S_1 only executes once before the repetitive process, and is therefore not taken account into the total latency. For both S_3 and S_6, two modules run concurrently, and the maximum latency of the modules is used as the latency of the corresponding state. The latency of linear element module in Case 2 is relatively long because it requires multiple comparisons for transmission line model to find the correct history term and then conduct the linear interpolation. The maximum iteration number is four for the both case studies, with the given parameters and test conditions. For the case studies, the average allocated computation times for one

Table 6.1 Hardware resource utilization for case studies.

Resource	Case 1	Case 2
LUT	51 043 (16.81%)	62 495 (20.58%)
LUTRAM	903 (0.69%)	539 (0.41%)
FF	32 678 (5.38%)	63 804 (10.51%)
BRAM	7 (0.68%)	102 (9.90%)
DSP	344 (12.29%)	327 (11.68%)

Table 6.2 State latency for case studies.

State	Module	Case 1 (μs)	Case 2 (μs)
S_1	Initialization module	0.36	1.54
S_2	Δt-dependent constant memory and multiplexer	0.01	0.01
	Linear element module	0.01	2.27
S_3	Electrical source module	0.16	0.43
S_4	Nonlinear element module	0.29	0.05
S_5	Matrix solution module	0.81	2.8
	History term update module	0.4	0.86
S_6	Output calculation module	0.4	0.19
S_7	Time-step control module	0.16	0.28
	Total Minimum latency (iteration number)	2.93 (two)	6.27 (one)
	Total Maximum latency (iteration number)	5.13 (four)	14.82 (four)

time-step are 5 and 10 μs for the two case studies. In such setting, the maximum delays due to the use of variable time-stepping and iterative method are 9.1 and 75 μs, respectively. The minimum average computation times for a time-step could be 3.9 and 8.2 μs, measured from the time scopes of 0.15–0.65 and 7–17 ms, for two case studies. However, the maximum delay could be much larger, if using the minimum average computation time as the allocated value with no redundancy.

6.2.4.4 Real-Time Emulation Results

Real-time emulation results for the case studies obtained from the oscilloscope are shown in Figures 6.8 and 6.9, with the same parameters and test condition as the off-line simulations. All calculated data points were captured and sent to the oscilloscope. Since the matrix solver and data precision of the real-time emulator are different from the ones from Matlab®, the results are slightly different from the off-line simulation. For both case studies, the real-time emulator used the iteration counts maximum up to four times for any time-instances, and did not reject any solutions. The results from hardware emulation have good accuracy compared with the off-line SaberRD® and PSCAD/EMTDC®, respectively.

Scalability Scalability is composed of two major parts, the increase of sequential calculation and the parallelism capability and corresponding cost of larger system.

During periodic calculation, the two major steps are matrix element update (S_3, S_4, and S_6) and matrix solution (S_5). The longest latency for matrix element update is only dependent on the most complex single element, not the element number. For EMT-type power system real-time simulation, the matrix is sparse and can be naturally decomposed to multiple smaller matrices through existing

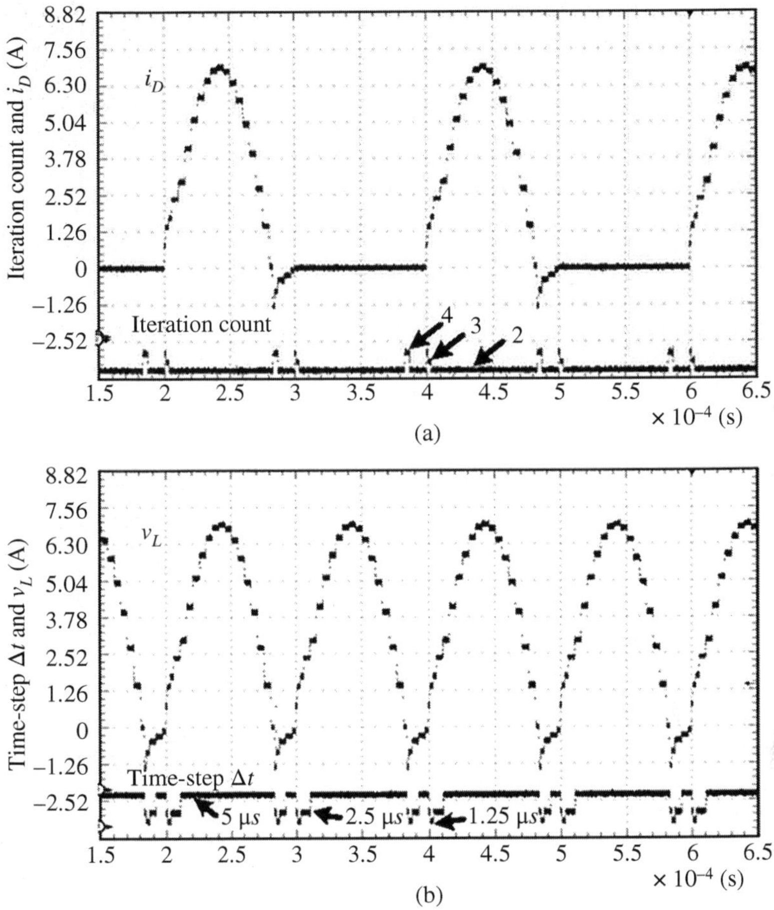

Figure 6.8 Real-time emulation results for Case 1: (a) diode current i_D for D_1 and iteration count and (b) load voltage v_L and time-step Δt.

transmission line model by properly numbering the nodes, which is typically used in commercial real-time simulators. The latency of matrix solution is therefore decided by the size of largest sub-matrix.

The iteration number is mostly affected by the nonlinearity of the element, modeling schemes, and the time-stepping schemes. The system size may also have an effect, since more nonlinear elements are involved. It can affect the stability for N-R method, while only affect the result accuracy for PNR method with a certain maximum iteration number.

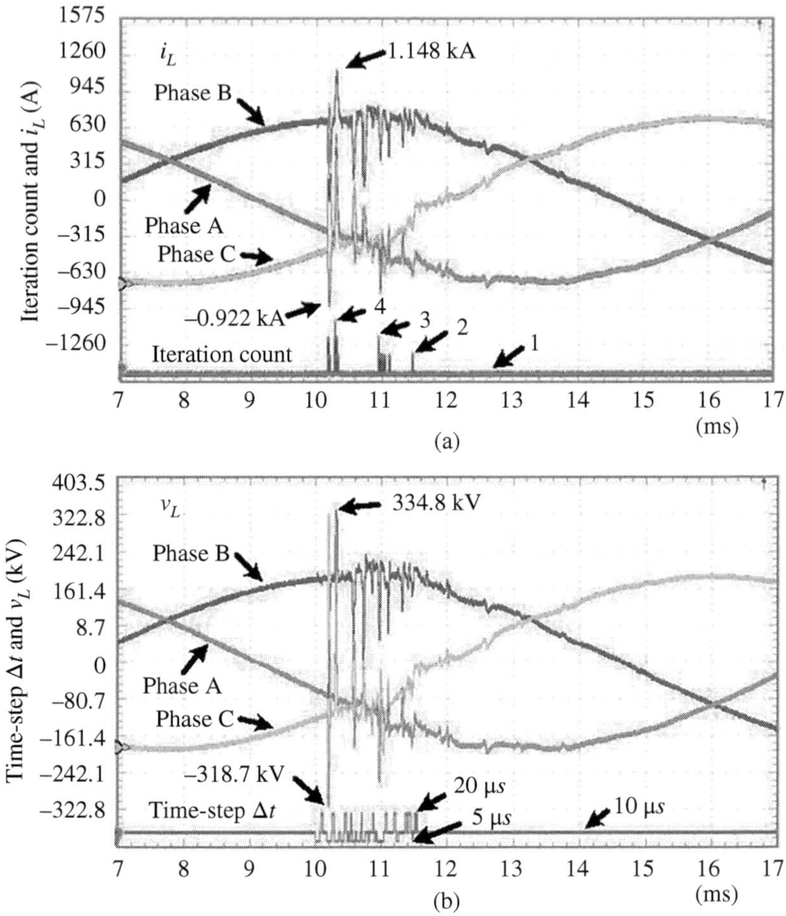

Figure 6.9 Real-time emulation results for Case 2: (a) load current i_L and iteration count and (b) load voltage v_L and time-step Δt.

6.3 Adaptive Time-Stepping Universal Line Model and Universal Machine Model for Real-Time Hardware Emulation

Transmission lines and rotating machines that widely exist in power systems should be accurately modeled in real-time EMT simulation for obtaining precise results for HIL applications. In the conventional EMT simulator, the time-step is fixed, which may lead to inefficiencies when the time constants of the system change – the ATS method can efficaciously solve this problem.

Since the ULM and UM models serve a wide range of transmission lines and rotating machines, they are required to be properly modeled for ATS-based EMT simulation. Although the system decomposition can be applied to use different time-steps for different subsystems [123], the time-step of each subsystem also changes in the ATS simulation; thus, how to deal with data exchange and synchronization between subsystems with adaptive time-steps remains to be discussed. Based on this observation, this section first derives the ATS model for ULM and UM, and then the proposed models are emulated in both real-time and "*faster-than-real-time*" (FTRT) on the FPGA with large-scale test systems. Different with the work [128] that uses dv/dt or di/dt as the time-step changing criteria, the time-step control here is based on the LTE. Since the dv/dt value will also change as a sine function in a steady-state sinusoidal waveform, there will be a big difference in dv/dt value change when the transient happens at the different places of a waveform, which may not be accurate enough to show the system perturbation. By applying the subsystem-based ATS simulation architecture, the simulation process can be accelerated to achieve an FTRT simulation.

6.3.1 Subsystem-Based Adaptive Time-Stepping Scheme

Since in this work the system is simulated in hardware and all the parameters related to the time-steps are stored in advance, it is infeasible to change the time-step arbitrarily. Thus in general, there should be several pre-determined candidate time-steps, and the time-step adaption is determined by the LTE of the previous step and the threshold. Considering that the state-variables and time-constants of different equipment (such as ULMs, UMs, and transformers) are also different, this work proposes a subsystem-based ATS scheme, as shown in Figure 6.10. Each subsystem ($S_i, 1 \leqslant i \leqslant N$) holds its own time-step set $\{\Delta t_{i,1}, \Delta t_{i,2}, \dots, \Delta t_{i,n_i}\}$ and LTE threshold $\{\xi_{i,1}, \xi_{i,2}, \dots, \xi_{i,n_i}\}$ and changes the time-step Δt_i locally while synchronizing to the simulation time of the whole system. Note that since the LTE threshold of various components within

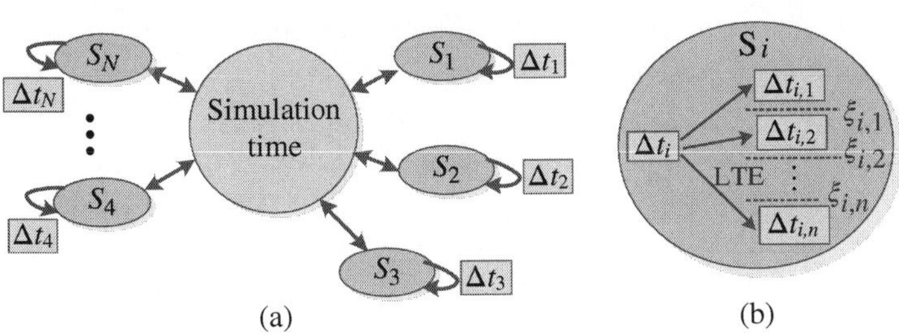

Figure 6.10 Subsystem-based adaptive time-stepping scheme: (a) global scheme and (b) local subsystem scheme.

a subsystem is also different, the time-step of a subsystem will change if any components exceed their own LTE threshold.

Although the whole system can be divided into subsystems and each subsystem can run in different time-steps, their time-step size also cannot be arbitrarily assigned due to the necessity of synchronization. For example, if a subsystem uses a 10 μs time-step and the connected subsystem uses a 13 μs time-step, then it will be extremely complicated to synchronize the two subsystems because they could not reach at the same synchronization point after each time-step. Therefore, in this work, the time-step sizes of different subsystems are always linearly proportional to each other. Then, all subsystems can proceed to the same synchronization point with different time-steps, and the time-space to the next synchronization point is the maximum time-step of these subsystems. Since their time-steps are all linearly proportional, the other subsystems with smaller time-steps are only required to execute several times to reach that synchronization point. Therefore, the adaptive multiple variable time-stepping method will be executed as follows:

1. Calculate the LTE of different equipment at t_{n-1};
2. Increase or decrease the time-step locally based on the specific threshold;
3. Implement the time-step change and calculation, by accordingly updating the numerical parameters;
4. Synchronize the results between interconnected subsystems.

6.3.2 Adaptive Time-Stepping ULM and UM Models

Although TR is one-step method, the ULM and UM model computation for ATS is not the same as that for the FTS because the model parameters may change when the time-step changes.

6.3.2.1 ULM Computation

The equivalent circuit for ULM is shown in Figure 6.11, in which the two ends (k and m) of the line are abstracted into two disconnected parts, and each part combines an equivalent conductance matrix in parallel with a compensating current source ($\boldsymbol{i}_k^{hist}$ and $\boldsymbol{i}_m^{hist}$) through which the two ends are interacting.

The electromagnetic behavior of a transmission line in frequency domain can be characterized by two matrix transfer functions: $\mathbf{Y}_c(\omega)$, the characteristic admittance matrix, and $\mathbf{H}(\omega)$, the propagation matrix. The time–domain relationship between currents and voltages at the two ends can be expressed using these two matrices:

$$\boldsymbol{i}_k^{hist} = \mathbf{G}_k \boldsymbol{v}_k(t) - \left[\mathbf{y}_c * \boldsymbol{v}_k(t) - 2\mathbf{h} * \boldsymbol{i}_{mr}(t)\right], \tag{6.21}$$

$$\boldsymbol{i}_m^{hist} = \mathbf{G}_m \boldsymbol{v}_m(t) - \left[\mathbf{y}_c * \boldsymbol{v}_m(t) - 2\mathbf{h} * \boldsymbol{i}_{kr}(t)\right], \tag{6.22}$$

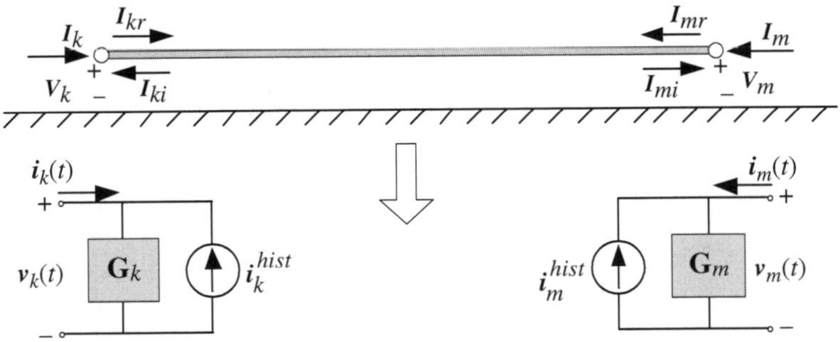

Figure 6.11 Equivalent circuit of the ULM.

where $\mathbf{y}_c$ and $\mathbf{h}$ are obtained via an inverse Fourier transform for $\mathbf{Y}_c(\omega)$ and $\mathbf{H}(\omega)$, $\boldsymbol{i}_{kr}$ and $\boldsymbol{i}_{ki}$ are the incoming current wave and reflected current wave at k point, respectively.

By applying proper fitting techniques, the time–domain elements of $\mathbf{Y}_c$ and $\mathbf{H}$ can be simplified as:

$$\mathbf{y}_c^{(i,j)}(t) = \sum_{k=1}^{N_p} \mathbf{r}_{Y_c}^{(i,j,k)} e^{\boldsymbol{p}_{Y_c}^{(k)} t} + \mathbf{d}^{(i,j)} \delta(t), \tag{6.23}$$

$$\mathbf{h}^{(i,j)}(t) = \sum_{k=1}^{N_m} \sum_{n=1}^{N_{p,k}} \mathbf{r}_H^{(i,j,k,n)} e^{\mathbf{p}_H^{(k,n)}(t-\tau_k)}, \tag{6.24}$$

where $N_p, N_{p,k}$ are the number of poles, N_m is the number of modes, the superscript of each symbol indicates the elements of a vector or matrix. Note that the italic symbols represent vectors and nonitalic symbols denote matrices. The residue matrix $\mathbf{r}$ is a three-dimension matrix (for $\mathbf{Y}_c$) because it contains a 3×3 matrix (three conductors for example) for each pole. The pole parameter $\boldsymbol{p}$ is a vector for $\mathbf{Y}_c$ but is a matrix for $\mathbf{H}$ because it has multiple modes. $\mathbf{d}$ is the proportional terms, and τ_k is the time delay for the kth mode.

In EMT simulation, the $\mathbf{g}(t) = \mathbf{y}_c * \boldsymbol{v}_k(t)$ convolution calculation can be discretized to be performed step-by-step by using a state variable $\boldsymbol{x} = [\boldsymbol{x}^{(1)}, \boldsymbol{x}^{(2)}, \ldots, \boldsymbol{x}^{(N_p)}]$ defined below:

$$\dot{\boldsymbol{x}}^{(i)}(t) = \boldsymbol{p}_{\mathbf{Y}_c}^{(i)} \boldsymbol{x}^{(i)}(t) + \boldsymbol{v}_k(t), \quad 1 \leqslant i \leqslant N_p, \tag{6.25}$$

$$\mathbf{g}(t) = \sum_{i=1}^{N_p} \mathbf{r}_{Y_c}^{(:,:,i)} \boldsymbol{x}^{(i)}(t) + \mathbf{d}\boldsymbol{v}_k(t), \tag{6.26}$$

where the superscript $(:,:,i)$ means the ith 3×3 matrix of $\mathbf{r}_{Y_c}$. By applying the TR discretization, (6.25) could be calculated as:

$$\boldsymbol{x}^{(i)}(n) = \boldsymbol{\alpha}_{Y_c}^{(i)}(n)\boldsymbol{x}^{(i)}(n-1) + \boldsymbol{\lambda}_{Y_c}^{(i)}(n)\boldsymbol{v}_k(n) + \boldsymbol{\mu}_{Y_c}^{(i)}(n)\boldsymbol{v}_k(n-1), \tag{6.27}$$

$$\mathbf{g}(n) = \sum_{i=1}^{N_p} \mathbf{r}_{Y_c}^{(:,:,i)} \boldsymbol{x}^{(i)}(n) + \mathbf{d}\boldsymbol{v}_k(n), \tag{6.28}$$

where

$$\boldsymbol{\alpha}_{Y_c}^{(i)}(n) = \left(1 + \boldsymbol{p}_{Y_c}^{(i)}\frac{\Delta t_n}{2}\right) / \left(1 - \boldsymbol{p}_{Y_c}^{(i)}\frac{\Delta t_n}{2}\right), \tag{6.29}$$

$$\lambda_{Y_c}^{(i)}(n) = \boldsymbol{\mu}_{Y_c}^{(i)}(n) = \left(\frac{\Delta t_n}{2}\right) / \left(1 - \boldsymbol{p}_{Y_c}^{(i)}\frac{\Delta t_n}{2}\right). \tag{6.30}$$

Traditional Model The traditional method that is widely used in fixed time-stepping ULM model is to use a new state variable $\boldsymbol{x}^{*(i)}(n) = \boldsymbol{x}^{(i)}(n) - \lambda_{Y_c}^{(i)}(n)\boldsymbol{v}_k(n)$, $1 \leqslant i \leqslant N_p$ instead to eliminate the $\boldsymbol{v}_k(n)$ item:

$$\begin{aligned}\boldsymbol{x}^{*(i)}(n) &= \boldsymbol{\alpha}_{Y_c}^{(i)}(n)\boldsymbol{x}^{*(i)}(n-1)\\ &\quad + \left(\boldsymbol{\alpha}_{Y_c}^{(i)}(n)\lambda_{Y_c}^{(i)}(n-1) + \boldsymbol{\mu}_{Y_c}^{(i)}(n)\right)\boldsymbol{v}_k(n-1),\end{aligned} \tag{6.31}$$

$$\mathbf{g}(n) = \sum_{i=1}^{N_p} \mathbf{r}_{Y_c}^{(:,:,i)}\boldsymbol{x}^{*(i)}(n) + \mathbf{G}_k(n)\boldsymbol{v}_k(n). \tag{6.32}$$

And $\mathbf{G}_k(n)$ is the equivalent conductance matrix:

$$\mathbf{G}_k(n) = \mathbf{G}_m(n) = \mathbf{d} + \sum_{k=1}^{N_p} \lambda_{Y_c}^{(k)}(n)\mathbf{r}_{Y_c}^{(:,:,k)}. \tag{6.33}$$

Note that $\mathbf{G}(n)\boldsymbol{v}_k(n)$ in (6.32) can be eliminated by the current items in (6.21) when calculating the equivalent current source $\boldsymbol{i}_k^{hist}$; thus, $\boldsymbol{v}_k(n)$ is not required, which makes the equivalent current source to be computed only depending on the results of previous steps. The $\mathbf{h} * \boldsymbol{i}_{mr}(t)$ convolution has the same computation flow, but only the history items $\boldsymbol{i}_{mr}(t - \tau_k), k = 1, \dots, N_m$ are needed for the calculation at time t because there is a time-delay τ_k in the $\mathbf{h}$ matrix function.

However, this traditional method for FTS may not be applicable for adaptive time-step. Because when the time-step changes, the constant λ_{Y_c} will change to a new value, which makes the state variable $\boldsymbol{x}^*$ actually change to a different new state variable from the one before time-step changes. That means, the new state-variable may need several steps to make itself stable and consistent with the former state-variable, which will cause instability when the time-step changes. This phenomenon will be verified later in the simulation section.

Process-Reversed Model To solve the instability problem of the traditional method, this work proposes a novel current-source based method that has a reverse logic of the traditional method. The rest of the system in each end of the transmission line can be equivalenced to a current source $\boldsymbol{i}_{eq}$ in parallel with a conductance $\mathbf{G}_{eq}$, as shown in Figure 6.12. At the port k, the system equation can be written as:

$$[\mathbf{G}_{eq} + \mathbf{G}_k]\boldsymbol{v}_k(t) = \boldsymbol{i}_{eq}(t) + \boldsymbol{i}_k^{hist}. \tag{6.34}$$

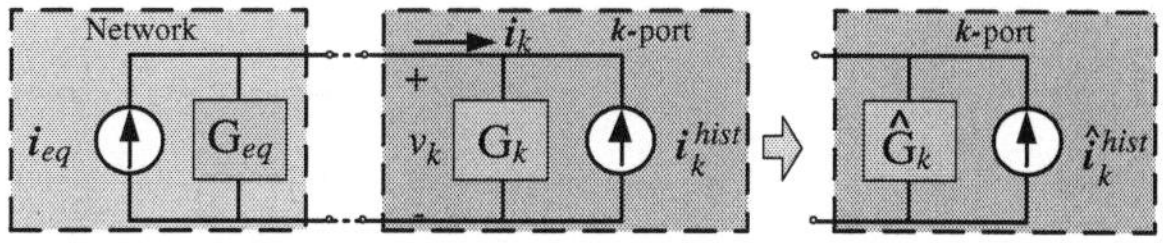

Figure 6.12 Illustration for the process-reversed model for the ULM.

Substituting $\boldsymbol{i}_k^{hist}$ by (6.21) and expressing as discrete-time:

$$[\mathbf{G}_{eq} + \mathbf{G}_k]\boldsymbol{v}_k(n) = \boldsymbol{i}_{eq}(n) + \mathbf{G}_k\boldsymbol{v}_k(n) - \left[\mathbf{y}_c * \boldsymbol{v}_k(n) - 2\mathbf{h} * \boldsymbol{i}_{mr}\right]. \tag{6.35}$$

From (6.27) and (6.28), we know:

$$\mathbf{y}_c * \boldsymbol{v}_k(n) = \sum_{i=1}^{N_p} \mathbf{r}_{Y_c}^{(:,:,i)}\boldsymbol{x}^{(i)}(n) + \mathbf{d}\boldsymbol{v}_k(n). \tag{6.36}$$

Here, $\mathbf{G}_k(n) = \mathbf{d}$. Combining (6.27), (6.35), and (6.36) together, we get:

$$[\mathbf{G}_{eq} + \mathbf{G}_k + \mathbf{G}_r]\boldsymbol{v}_k(n) = \boldsymbol{i}_{eq}(n) + 2\mathbf{h} * \boldsymbol{i}_{mr} - \boldsymbol{x}^{hist}, \tag{6.37}$$

where

$$\mathbf{G}_r = \sum_{k=1}^{N_p} \lambda_{Y_c}^{(k)}(n)\mathbf{r}_{Y_c}^{(:,:,k)}, \tag{6.38}$$

$$\boldsymbol{x}^{hist} = \sum_{i=1}^{N_p} \mathbf{r}_{Y_c}^{(:,:,i)}\left[\boldsymbol{\alpha}_{Y_c}^{(i)}(n)\boldsymbol{x}^{(i)}(n-1) + \boldsymbol{\mu}_{Y_c}^{(i)}(n)\boldsymbol{v}_k(n-1)\right]. \tag{6.39}$$

Since the convolution $\mathbf{h} * \boldsymbol{i}_{mr}$ at time t_n actually only needs the history items $\boldsymbol{i}_{mr}(t_n - \tau_k), k = 1, \dots, N_m$, the only unknown variable to be solved at t_n in (6.37) is the node voltage $\boldsymbol{v}_k(n)$. Equations (6.37) and (6.39) actually generate a new equivalent admittance matrix $\hat{\mathbf{G}}_k$ and $\hat{\boldsymbol{i}}_k^{hist}$, as shown in Figure 6.12, where

$$\hat{\mathbf{G}}_k = \mathbf{G}_k + \mathbf{G}_r, \quad \hat{\boldsymbol{i}}_k^{hist} = 2\mathbf{h} * \boldsymbol{i}_{mr} - \boldsymbol{x}^{hist}. \tag{6.40}$$

And the $\hat{\mathbf{G}}_k$ will change if the time-step changes. After the node voltages are solved, the other variables ($\boldsymbol{x}, \boldsymbol{i}_{kr}(n), \boldsymbol{i}_{mr}(n)$, etc.) at t_n can be solved based on the node voltages.

The logic of this method is totally different from the traditional method. In the traditional method, the state variables $\boldsymbol{x}^{(i)}(n)$ are calculated first to obtain $\boldsymbol{i}_k^{hist}$, and then the node voltages are solved; however, this method will cause instability when the time-step changes. In the process-reversed method, the calculation sequence is reversed: firstly, the node voltages are solved and then the state variables $\boldsymbol{x}(n)$ are updated. In the fixed time-stepping simulation, these two methods essentially have the same presentation; however, in the ATS simulation, using the process-reversed method, the state variables $\boldsymbol{x}(n)$ will remain the same no matter how the time-step changes, which greatly improves the stability.

6.3.2.2 Universal Machine Model Computation

For the UM model, there are three stator windings $\{d, q, 0\}$, several damper windings $\{kd, kq\}$ in the direct and quadrature (d and q) rotor axis, and one filed windings $\{f\}$. In this work, the machine with one kd winding and two kq windings is considered as a general model to represent the synchronous generators. The fixed time-stepping model and hardware implementation can be found in Chapter 4 wherein the relationship between voltages and currents can be expressed as follows:

$$\boldsymbol{v}_{um}(t) = -\mathbf{R}_{um}\boldsymbol{i}_{um}(t) - \frac{d}{dt}\boldsymbol{\psi}_{um}(t) + \boldsymbol{u}(t), \tag{6.41}$$

$$\boldsymbol{\psi}_{um}(t) = \mathbf{L}_{um} \cdot \boldsymbol{i}_{um}(t), \tag{6.42}$$

where $\boldsymbol{v}_{um} = [v_d,\ v_q,\ v_0,\ v_f,\ 0,\ 0,\ 0]^T$, $\boldsymbol{i}_{um} = [i_d,\ i_q,\ i_0,\ i_f,\ i_{kd},\ i_{kq1},\ i_{kq2}]^T$, $\boldsymbol{\psi}_{um} = [\psi_d,\ \psi_q,\ \psi_0,\ \psi_f,\ \psi_{kd},\ \psi_{kq1},\ \psi_{kq2}]^T$, $\boldsymbol{u} = [-\omega\psi_q,\ \omega\psi_d,\ 0,\ 0,\ 0,\ 0,\ 0]^T$, $\mathbf{R}_{um} = diag(R_d, R_q, R_0, R_f, R_{kd}, R_{kq1}, R_{kq2})$ and $\mathbf{L}_{um}$ is the leakage inductance matrix.

To solve the machine equations above in discrete time, (6.41) is discretized using TR, which leads to a Thévenin voltage source representation [129] shown in Figure 6.13:

$$\boldsymbol{v}_{um}(n) = -\left[\mathbf{R}_{um} + \frac{2}{\Delta t_n}\mathbf{L_{um}} - \omega\mathbf{L_u}\right]\boldsymbol{i}_{um}(n) + \boldsymbol{v}^{hist}, \tag{6.43}$$

where

$$\begin{aligned}\boldsymbol{v}^{hist} &= \boldsymbol{u}(n-1) - \boldsymbol{v}_{um}(n-1)\\ &\quad - \left[\mathbf{R}_{um} - \frac{2}{\Delta t_n}\mathbf{L}_{um}\right]\boldsymbol{i}_{um}(n-1),\end{aligned} \tag{6.44}$$

and $\boldsymbol{u}(n) = \omega\mathbf{L_u}\boldsymbol{i}_{um}(n)$, $\mathbf{L_u} = [-\mathbf{L_{um}}(2); \mathbf{L_{um}}(1); \mathbf{0}; \mathbf{0}; \mathbf{0}; \mathbf{0}; \mathbf{0}]$. Note that the time-step Δt_n is not a constant but may change during the simulation process, which is different from the FTS model.

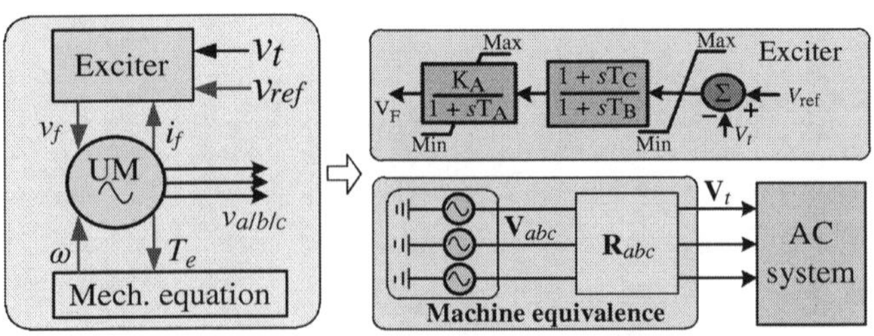

Figure 6.13 Equivalent circuit for the UM model. Source: Based on Wang and Jatskevich [129].

Let $\mathbf{R}_{um,eq} = [\mathbf{R}_{um} + \frac{2}{\Delta t}\mathbf{L_{um}} - \omega\mathbf{L_u}]$, and $\mathbf{R}_{um,eq} = [\mathbf{R}_{ss}\ \mathbf{R}_{sr};\ \mathbf{R}_{rs}\ \mathbf{R}_{rr}]$. Then, the dq0 frame can be extracted from the vector of (6.43):

$$\boldsymbol{v}_{dq0}(n) = -\mathbf{R}_{ss}\boldsymbol{i}_{dq0}(n) - \mathbf{R}_{sr}\boldsymbol{i}_r(n) + \boldsymbol{v}_{dq0}^{hist}, \tag{6.45}$$

$$\boldsymbol{v}_r(n) = -\mathbf{R}_{rs}\boldsymbol{i}_{dq0}(n) - \mathbf{R}_{rr}\boldsymbol{i}_r(n) + \boldsymbol{v}_r^{hist}, \tag{6.46}$$

where $\boldsymbol{i}_r = [i_f, i_{kd}, i_{kq1}, i_{kq2}]^T$, and $\boldsymbol{v}_r = [v_f, 0, 0, 0]^T$ only contains the field voltage of which the value at time t_n is known from the exciter module. Thus from (6.45) and (6.46), the relationship between $\boldsymbol{v}_{dq0}(n)$ and $\boldsymbol{i}_{dq0}(n)$ is derived:

$$\boldsymbol{v}_{dq0}(n) = -\mathbf{R}_{dq0}\boldsymbol{i}_{dq0}(n) + \boldsymbol{v}_{dq0,eq}^{hist}, \tag{6.47}$$

where

$$\mathbf{R}_{dq0} = \mathbf{R}_{ss} - \mathbf{R}_{sr}\mathbf{R}_{rr}^{-1}\mathbf{R}_{rs}, \tag{6.48}$$

$$\boldsymbol{v}_{dq0,eq}^{hist} = -\mathbf{R}_{sr}\mathbf{R}_{rr}^{-1}\{-\boldsymbol{v}_r(n) + \boldsymbol{v}_r^{hist}\} + \boldsymbol{v}_{dq0}^{hist}. \tag{6.49}$$

The equivalent voltage source $\boldsymbol{v}_{abc,eq}(n)$ and resistance $\mathbf{R}_{abc,eq}(n)$ can be obtained by transforming (6.47) into *abc* frame using the Park's transformation matrix $\mathbf{P}_n$:

$$\mathbf{R}_{abc,eq}(n) = \mathbf{P}_n^{-1}\mathbf{R}_{dq0}\mathbf{P}_n, \quad \boldsymbol{v}_{abc,eq}(n) = \mathbf{P}_n^{-1}\boldsymbol{v}_{dq0,eq}^{hist}. \tag{6.50}$$

The change of ω is handled by the mechanical equation, and since ω at time t_n is unknown while solving the machine equations, it is predicted first and then updated iteratively by solving the network equations until convergence.

The AC4A type exciter model in PSCAD/EMTDC® [121] is adopted to make the machine work stable. As can be seen in the diagram, it also involves the convolution operation when passing the transfer function that has the same format with transmission lines. Thus, it has the same step-by-step calculation flow of convolution as described in the ULM model.

6.3.3 Real-Time Emulation Case Study

To test and verify the proposed ATS transmission line and machine models on a realistic-sized power system, an IEEE 39-bus system composed of 34 transmission lines, 10 generators, 12 transformers, and 19 loads was selected as the case study as shown in Figure 6.14a. A lumped parameter model based on admittance matrix representation (Chapter 3) is used to represent the transformers.

6.3.3.1 Hardware Implementation

The Virtex® UltraScale+ FPGA VCU118 board [11] used in this work contains both highly programmable UltraScale+ XCVU9P device (Table 1.2) and

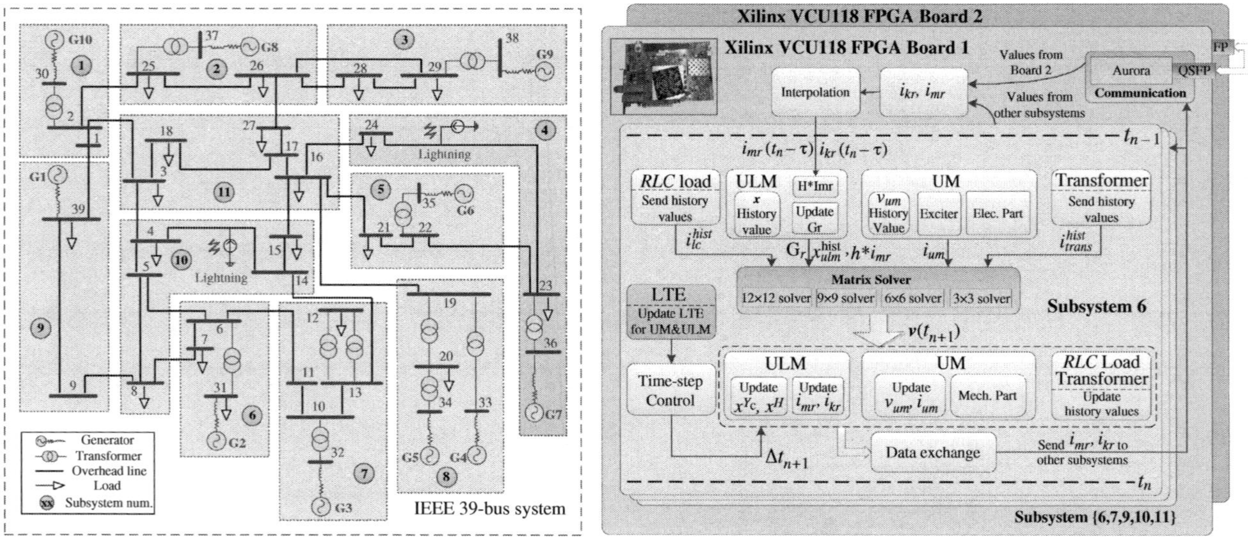

(a) (b)

Figure 6.14 Test system and the hardware emulation on two interfaced FPGA boards.

rich external resources. Considering the heavy computation task within each time-step and the large system scale, the multi-board solution is adopted [130], in which two VCU118 boards were interfaced through SFP transceivers as shown in Figure 6.14b, and each board only models a part of the system.

Subsystem Allocation As shown in Figure 6.14a, the power system is divided into 11 subsystems taking advantage of the distributed transmission lines, and the connected subsystems exchange data (i_{mr} and i_{kr}) after calculation of each time-step. Since almost every generator is connected with a transformer, it is convenient to implement if every generator–transformer pair is allocated to an individual subsystem, as marked as subsystem $S_1 \sim S_6$ in Figure 6.14a. In each subsystem, the transmission lines can again divide the subsystem into sub-subsystems, which effectively reduces the size of admittance matrix to be solved. For example, the maximum matrix size of $S_9 \sim S_{11}$ is 3×3, $S_1 \sim S_6$ is 6×6. S_7 and S_8 have the largest matrix to solve due the coupled transformers, which are 9×9 and 12×12, respectively. Since the transients happen in subsystem S_{10} and S_4, to reduce the latency of data exchange between connected subsystems, the subsystems ($S_6, S_7, S_9, S_{10}, S_{11}$) that are connected with S_{10} are all allocated at the same FPGA board, and the other six subsystems are allocated to another FPGA board. The board-to-board communication is handled by the lightweight communication protocol Aurora [131], which is supported by different type of transceivers such as GTX, GTY, and GTH, was used. The two boards are interfaced via two Aurora lanes, with 32 bits floating point data transferred in each lane.

Adaptive Time-Step Control In this work, four time-step sizes are applied: 5, 10, 20, and 40 μs. Considering the matrix to be solved in different subsystems may be different and some subsystems contain generators that will occupy more time-space to compute, the time-step set of different subsystems are different. After the minimum computing latency of each subsystem is obtained through hardware implementation, the proper time-step set can be determined: subsystem $S_1 \sim S_6$ has the same time-step set {10 μs, 20 μs, 40 μs}, subsystem $S_7 \sim S_8$ has the same time-step set {20 μs, 40 μs}, and subsystem $S_9 \sim S_{11}$ has the same time-step set {5 μs, 10 μs, 20 μs}. The time-step increase or decrease is determined by the LTE of the previous step and the threshold, wherein the threshold of various subsystems are also different. For the subsystems where transients happen, the LTE threshold is relatively smaller compared with that of the other subsystems.

Faster-Than-Real-Time (FTRT) Execution Although the system with FTS can also be implemented in real-time, the main advantage of variable time-step is to accelerate

Table 6.3 Demonstration of FTRT

Mode	Subsystem	Δt (μs)	t_p (μs)	Subsystem	Δt (μs)	t_p (μs)
RT	S_{10}	5	5	S_1	10	10
		10	10		10	10
FTRT2	S_{10}	10	5	S_1	20	10
		20	10		40	20
FTRT4	S_{10}	20	5	S_1	40	10

the simulation progress without loss of accuracy. Thus, the proposed ATS simulation could be even FTRT, which means, the actual processing time of each step is smaller than the time-step applied for the simulation. FTRT can be achieved by proper coordination between different subsystems. First, the maximum processing latency of a subsystem limits the minimum time-step of the subsystem, which implies that the large time-step is actually achieved by adding a time delay between two steps. If the time space size can be narrowed synergistically, the simulation time will be reduced when the time-step changes to be larger than the minimum time-step. Secondly, the narrowed time space should be consistent among different subsystems because the time-step of various subsystems may be different and it will easily cause inconsistency if the actual latency is not reduced proportionally.

Take subsystems S_1 and S_{10} as an example, their time-step Δt and actual processing time t_p is illustrated in Table 6.3. The minimum time-step of S_1 and S_{10} is near 10 and 5 μs, respectively; thus, the time-step of 10 and 5 μs cannot be reduced. As shown in the second row, if S_1 runs at 10 μs, the latency of S_{10} with time-step size of 10 μs also could not be reduced because the processing time of S_1 cannot be smaller than 10 μs. This indicates that no matter which subsystem was running at the minimum time-step (typically is under transient condition), the latency of other subsystems cannot be reduced. Under normal steady-state conditions, the time-step of all subsystems is usually changed to larger values; in this case, the actual latency can be reduced to make the simulation faster, as shown in the last three rows in Table 6.3. Note that FTRT2 means two times faster, and FTRT4 means four times faster.

6.3.3.2 Latency and Hardware Resource Utilization

The latency of different subsystems on hardware are recorded in Table 6.4, which indicates the minimum time-step that can be applied for different subsystems. For example, subsystem S_8 consumes the most latency because it involves the iteration for UM and has the largest matrix (12 × 12) to solve. The hardware resource consumption on the two VCU118 boards is presented in Table 6.5.

Table 6.4 Processing latency of subsystems

Subsystem/element	Latency (μs)	Subsystem/element	Latency (μs)
Subsystem 1–6	9.7	Subsystem 7	14.6
Subsystem 8	19.2	Subsystem 9	4.8
Subsystem 10–11	3.3	Trans. Line	1.05

Table 6.5 Hardware resource utilization of the case study

Resource	VCU118-Board1	VCU118-Board2
Subsystems	S_6,S_7,S_9,S_{10},S_{11}	S_1,S_2,S_3,S_4,S_5,S_8
LUT	767 174 (64.9%)	819 148 (69.3%)
FF	927 012 (39.2%)	1 014 011 (42.8%)
DSP	6 448 (94.3%)	6 714 (98.1%)
BRAM	570 (26.4%)	624 (28.9%)

6.3.4 Results and Validation

The test case was emulated on the two FPGA boards, and the results are compared with PSCAD/EMTDC® to show the effectiveness of the proposed ATS models. The clock frequency of FPGA boards is set at 100 MHz.

6.3.4.1 Validation of the ULM Model

Before the ULM model is integrated into the IEEE 39-bus system, it is necessary to verify the stability of the proposed process-reversed method. Note that the UM model in ATS is stable because the state variable remains the same when the time-step changes. To solely validate the ULM model, the subsystem S_9 in Figure 6.14a with ideal voltage source is simulated while keeping the load and connection with other subsystems as open circuit. At time 0.005 second, the time-step changes from 10 to 20 μs; and at time 0.035 second, the time-step changes from 20 to 50 μs.

The results of i_{39-9} (from Bus 39 to Bus 9) and $\mathbf{Y}_c * v_{39}$ using the traditional model are shown in Figure 6.15a, when the time-step changes, there will have a large oscillation. Since the actual state variable changes when the time-step changes, the initial value of the new state variable will be incorrect, which causes an abrupt change of the convolution results. However, when using the proposed process-reversed model (Figure 6.15b), it can be observed that the convolution results remain continuous and stable, which results in a stable current i_{39-9} when the time-step changes.

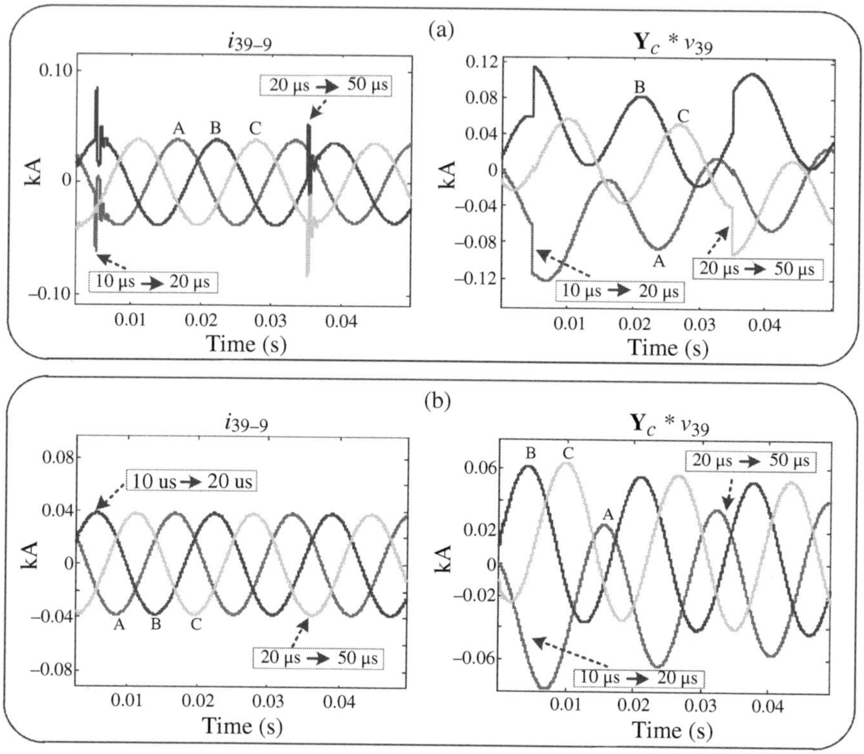

Figure 6.15 Waveforms under time-step change operation: (a) traditional ULM model and (b) proposed process-reversed ULM model.

6.3.4.2 Real-Time Emulation Results

To simulate the behavior of the system, the lightning surge at phase C of transmission line L_{4-14} (between Bus 4 and Bus 14) and L_{23-24} is chosen as the transient test. The results are evaluated by the proposed emulator and PSCAD/EMTDC® that used a FTS of 10 μs while the emulator used adaptive time-steps. The standard 10/350 μs lightning surge current [126] applied in this work is given as:

$$I_{LS}(t) = CI_m(t/\tau_1)^k e^{\frac{-t}{\tau_2}}/[1 + (t/\tau_1)^k], \tag{6.51}$$

where the coefficient $C = 1.075$, $k = 10$; the time constant $\tau_1 = 19$ μs, $\tau_2 = 485$ μs; the maximum value of the surge current $I_m = 3$ kA. The lightning surge current at L_{4-14} and L_{23-24} is applied at exactly 2 and 2.5 seconds of the simulation to demonstrate the transient behavior of ULM and UM, respectively.

First, the simulation is executed in real-time and compared with PSCAD and the hardware emulator with FTS, and the results are recorded in Figure 6.16 and Table 6.6. The FTS emulator applies the same time-step for the entire system. Since the maximum processing latency of the decomposed subsystems is near 20 μs, FTS-20 μs is actually the FTS emulator with minimum time-step size that can be achieved in real time. But through the proposed subsystem-based ATS scheme, the entire system can be simulated in real-time with smaller time-step sizes by applying different variable time-steps for different subsystems. From Figure 6.16a,b with the same time-step size it can be observed that, the peak values of the machine terminal voltage under transients are −708.8 and −659.5 kV, respectively, which indicates that the UM model implemented in this work is more stable than PSCAD.

In Figure 6.16b,c and Table 6.6, the emulator with 10 μs time-step size produces a smaller LTE, and the peak value of LTEs of ULM and UM reduces 63.1% and 73.7%, respectively, compared with that using a 20 μs time-step. That means reducing the time-step size will generate a more accurate result. The results using ATS are shown in Figure 6.16d, from which we can see the large time-step is applied under normal operation, and when the transient happens, the time-step is reduced automatically according to the LTE. For example, the time-step is 20 μs for L_{14-4} before the lightning, and when the transient happens, the LTE directly rises to an extremely large value that exceed the max threshold ξ_1 immediately, then the time-step reduces to the minimum one (5 μs) directly. When the LTE is below ξ_2, the time-step increases into a larger one (10 μs), and as the LTE is reduced below the minimum threshold ξ_3, the time-step regains to the maximum one (20 μs). Since the time-step size of ULM reduces to 5 μs under transients, the maximum LTE reduces 76.5% compared with that of FTS-20 μs. The UM has the same process although the time-step set is not the same since L_{14-4} and L_{23-24} belongs to the different subsystems.

The minimum time-step size of UM is 10 μs, resulting in a LTE 62.0% smaller than that of FTS-20 μs. The LTE of ATS is a little different with FTS-10 μs because the time-step sizes in ATS also vary between subsystems while FTS-10 μs applies the same time-step for the whole system.

Secondly, to show the acceleration of ATS simulation, the transient simulation is carried in FTRT2 mode (shown in Table 6.3) through elaborate configurations and compared with the results of real-time simulation. Figure 6.17 demonstrates the Bus 36 voltage of subsystem S_4, it can be observed the two versions have the same numerical results because the time-steps and parameters applied are the same. But due to the different actual processing time of each step, the output rate of the results is different. The FTRT version can produce data at a faster rate than the real-time simulation under normal operations, although under lightning transient they have the same output rate. In this work, only the simulation with 40 μs is accelerated by two times, because as described in Section 6.3.3.1, the minimum time-step cannot be accelerated. Since the duration of simulation with 20 μs is quite small, to reduce the complexity of simulation time control logic, it is not necessary to accelerate it. Besides, the maximum time-step is only four times of

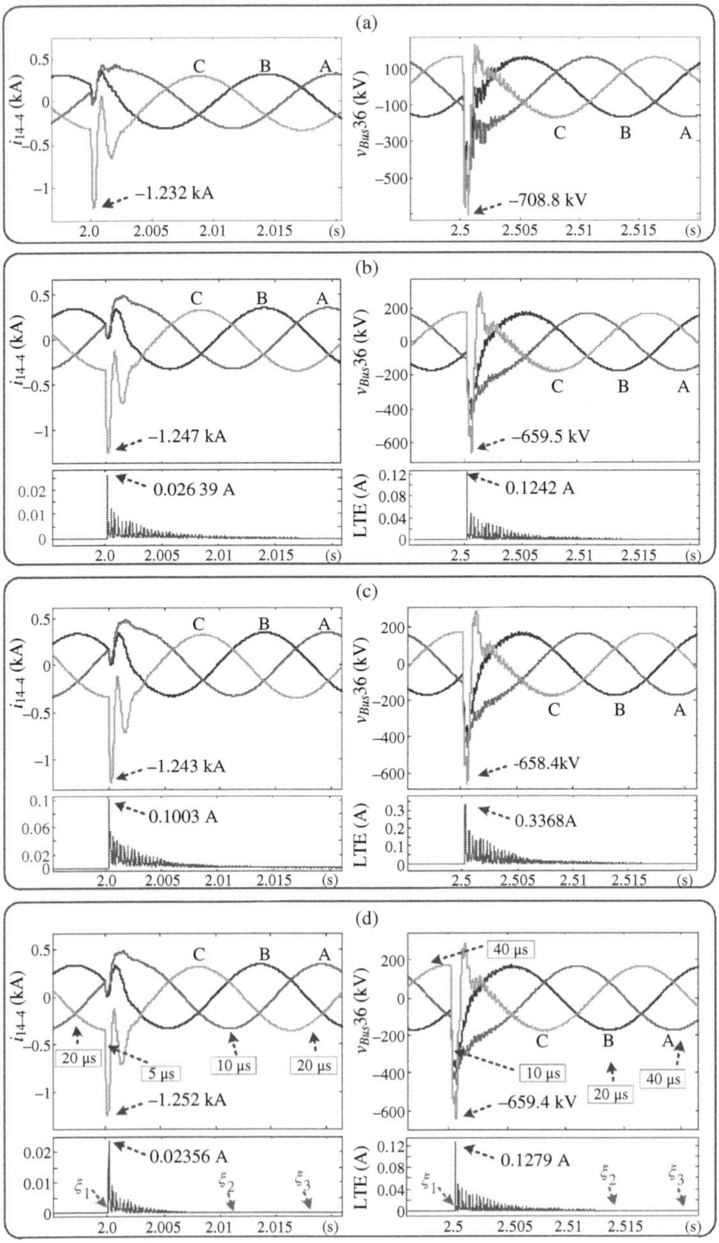

Figure 6.16 Lightning transient results of i_{14-4} and v_{Bus36}: (a) PSCAD/EMTDDC® results with 10 μs fixed time-step, (b) and (c) real-time emulator with 10 and 20 μs time-steps, and (d) real-time emulator with ATS.

Table 6.6 Minimum time-step size and maximum LTE of ULM and UM under lightning transient

Element	FTS-10 μs	FTS-20 μs	ATS
ULM	10 μs, 0.026 39 A	20 μs, 0.1003 A	5 μs, 0.023 56 A
UM	10 μs, 0.124 2 A	20 μs, 0.336 8 A	10 μs, 0.127 9 A

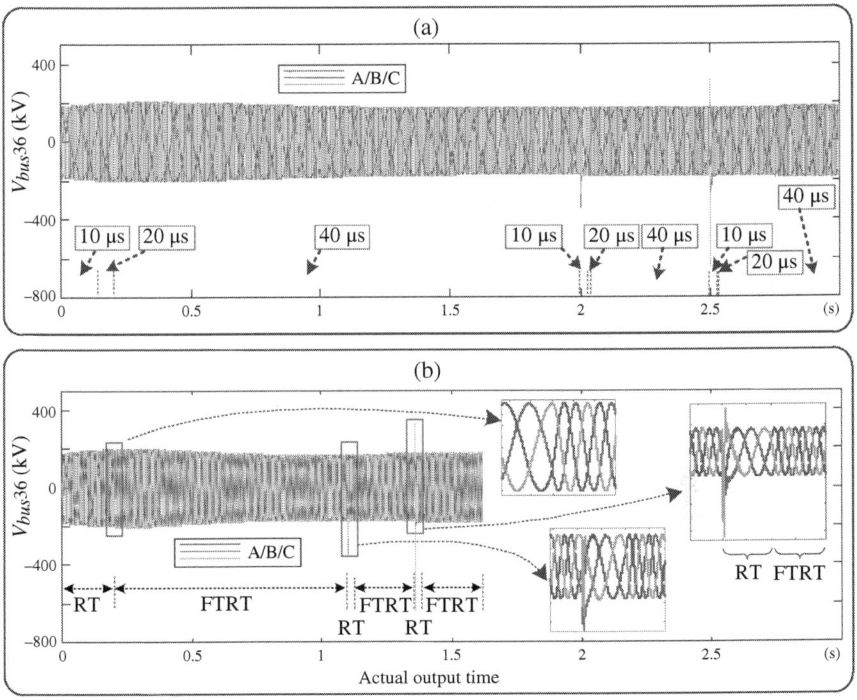

Figure 6.17 Demonstration of FTRT results: (a) real-time emulator results with ATS and (b) results of FTRT2 mode in ATS.

the minimum time-step in this work, it can be concluded that if the maximum time-step is larger, the output rate will be even faster than real time.

6.4 Summary

Real-time emulation of nonlinear transients in power systems can be greatly enhanced in both accuracy and efficiency by adopting adaptive time-steps. This chapter presented multiple dynamic ATS schemes coupled with nonlinear element solution methods and discussed the strengths and limitations for real-time emulation in detail. Two case studies are presented to illustrate the performance of the proposed ATS schemes and their efficiency in capturing nonlinear transients in both off-line and real time. For optimum realization of paralleled hardware implementation and smaller coding effort, the FPGA and HLS are chosen to emulate the proposed schemes on hardware. As verified, using ATS in both off-line simulation and real-time emulation can result in better accuracy with the same amount of computation effort especially during the transients.

Further in the chapter, an ATS universal line model and universal machine model are presented for accurate EMT simulation of power systems in real-time and FTRT mode. In the proposed ULM model, through a novel process-reversal of the traditional model, the stability is significantly improved during the time-step change operation. By using LTE as the time-step change criteria, the time-step can be adjusted properly when the transient happens. The hardware emulator for realistic-sized power system is presented, in which the system is divided into small subsystems and each subsystem maintained its own time-step set and LTE threshold. By coordination between subsystems, the FTRT mode was achieved. The hardware resource cost and processing delay of different subsystems were evaluated, which showed the practicality of ATS for transient simulation. The real-time emulation results are captured and compared with PSCAD/EMTDC®, which demonstrates the effectiveness of the proposed models. The proposed ATS-based ULM and UM model and the hardware emulation architecture can be used for FTRT EMT simulation of large power systems. ATS power electronic apparatus models can also be integrated with the proposed ATS-based ULM and UM models for a comprehensive AC–DC grid real-time emulation.

7

Power Electronic Switches

7.1 Overview

The past decades have witnessed a boom of power electronic apparatus integrated into the transmission and distribution systems, such as the static synchronous compensator (STATCOM) and the unified power flow controller (UPFC) as a family of flexible AC transmission system (FACTS), and the thyristor-based line commutated converter (LCC) or the voltage source converter (VSC) acting as the terminal in high-voltage direct current (HVDC) power transmission. The formation of integrated AC/DC system manifested the importance of electro-magnetic transient (EMT) simulation that facilitates its design, analysis, and test. Yet the credibility and the capability of those EMT-type solvers in providing all the required information are heavily dependent on the modeling of power semiconductor switches that are fundamental in the construction of all aforementioned apparatus. Categorized according to the complexity, seven types of computational models for the power converters have been identified, which are [132]:

1. *Type 1*: Full physics-based models
2. *Type 2*: Full detailed models
3. *Type 3*: Simplified switchable resistances
4. *Type 4*: Detailed equivalent circuit models
5. *Type 5*: Average value models based on switching functions
6. *Type 6*: Simplified average value models
7. *Type 7*: RMS load-flow models.

The document, drafted by the CIGRÉ working group B4.57, also classified models suitable for EMT simulators, i.e. Type 2–6, whereas Type 1 being the most detailed model was excluded due to its complexity related to the device's semiconductor physics as well as the small time-step it required, while Type 7 fell short of even system dynamics due to oversimplification. A number of models can be classified as the Type 1 [133], such as the finite element method (FEM) model,

Real-Time Electromagnetic Transient Simulation of AC-DC Networks, First Edition.
Venkata Dinavahi and Ning Lin.

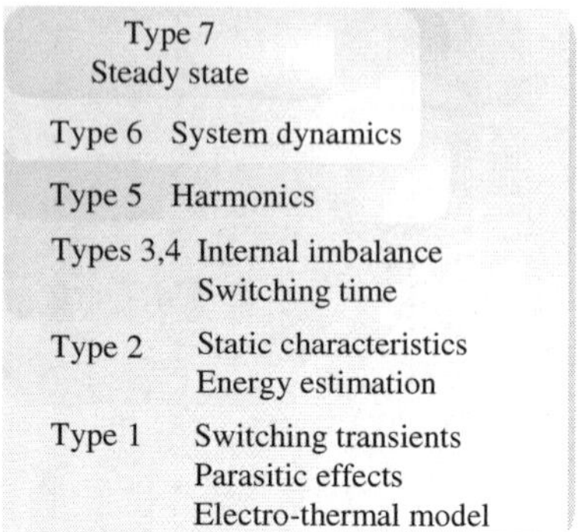

Figure 7.1 Power semiconductor switch model categorization.

the physics-based model, the behavioral model, all of which are nonlinear and have a high order. The curve-fitting model (CFM) has been seen as an alternative to the Type 1 model since it extracts key phenomena caused by the p–n junctions when the overall semiconductor is subjected to galvanization. Figure 7.1 shows the extent of information that each model type is able to reveal, where a higher-level model with a smaller number covers a lower-level model.

As low-level types have insufficient accuracy and fidelity, Type 3 model, also known as the two-state switch model (TSSM), became the prevalent model that current EMT-type solvers provide in their libraries for constructing various power electronic apparatus. Historically, when real-time digital simulators utilized DSPs, with limited computational capability, for implementing power electronic switch models, complicated correction algorithms were required for synchronizing discrete switching signals from the controller with the discrete time-step of the real-time digital simulator [134, 135]. In this case, the detailed system-level computation results were not be compromised; nevertheless, omitting the complex device physics of the power semiconductor switch or related phenomena meant that the Type 3 based on the two-state resistor concept with distinct on- and off-state values fell short of providing device-level details – and even Type 2 fell short as well which took extra static characteristics into consideration.

The VSCs in the DC grid are the major interface to the AC grid, and currently the two-level converter and the modular multilevel converter (MMC) are two prevalent forms of a converter station, the latter becoming predominant. It turned out to be quite inefficient in simulating the MMC even if Type 3 was adopted since mathematically the large scale of the converter corresponded to a huge admittance matrix that induced an enormous computational burden when solving the nodal voltages. Type 4 which has the Thévenin equivalence of Type 3 was provided to improve the circuit solution efficiency. However, as the semiconductor switch model is a mere two-state resistor, both types lack proper reverse blocking capability, resulting in erroneous results under some operation conditions. The VSCs can further be simplified into the average value models by employing the Types 5 and 6 [136], and as expected, a lower accuracy led to fewer applications.

Types 1 and 2 avoid the above issue, and in addition disclose device-level information, but the computational burden will increase dramatically once they are included in the simulation without utilizing any optimization of the program. As the operation status of a single switch in a high-power converter is increasingly concerned with ascertainment of its safe operation, Type 2 turned out to be the simplest model – in fact, Type 1 that has a much higher fidelity and versatility is preferred for testing various scenarios. For small-scale power converter simulation, the device-level models are frequently involved since the total number of nodes is acceptable; nevertheless, in MMC-based HVDC application, hundreds or even thousands of such devices are connected, making the circuit solution extremely slow if proceeded sequentially, and due to an extensive distribution of nonlinearity, the simulation often ends up with an early termination due to numerical divergence.

In this chapter, three models belonging to Type 1 are introduced for hardware-in-the-loop (HIL) simulation on the field programmable gate array (FPGA), in addition to the mainstream Types 3 and 4 models employed in computing the DC grid.

7.2 IGBT/Diode Nonlinear Behavioral Model

The nonlinear behavioral model (NBM) reveals the necessary static and dynamic device characteristics in circuit simulation while omitting excessive semiconductor physical details. Therefore, it gains computational advantages over the full physics-based model or the FEM model. There are several variants, e.g. the macro-model [137] and the Hammerstein configuration [138]. Due to the data-sheet-driven feature, the NBM exempts itself from acquiring a long list of inaccessible device-correlated parameters as in other analytical or numerical models and consequently is more widely applicable.

The major difference between a high-order nonlinear model and the linearized device-level model such as the CFM is the versatility: due to an excessive omission of nonlinearities as well as a low order, the waveform shapes produced by the latter model cannot change along with the electromagnetic environment, typically the gate driver circuit conditions, while the former virtually behaves as an actual device.

7.2.1 Power Diode

7.2.1.1 Mathematical Model

The NBM of the power diode available in SaberRD® is composed of various circuit parts to reflect all of its features including static I–V characteristics, the reverse and forward recovery phenomena, and junction capacitance [139]. However, only the static features and the reverse recovery dynamics are preserved; other components are omitted due to their insignificant impact on the overall performance, as shown in Figure 7.2a [140].

The diode static characteristics represented by the nonlinear diode (NLD) symbol reflects an exponential relationship between the static current I_d and the junction voltage V_j, as expressed by

$$I_d = I_s \cdot \left(e^{\frac{V_j}{V_b}} - 1 \right), \tag{7.1}$$

where I_s is the leakage current and V_b denotes the junction barrier potential. Its discrete-time Norton equivalent circuit, which is shown in Figure 7.2b, becomes available by taking partial derivative and subsequent linearization, as expressed by

$$G_j = \frac{\partial I_d}{\partial V_j} = \frac{I_s}{V_b} e^{\frac{V_j}{V_b}} \quad \text{and} \tag{7.2}$$

$$I_{jeq} = I_d - G_j \cdot V_j, \tag{7.3}$$

respectively, where G_j and I_{jeq} are the conductance and the equivalent current contribution of NLD.

The diode reverse recovery is attained by the R_L–L pair in addition to the voltage-controlled current source (VCCS) with a coefficient of K, with a proper

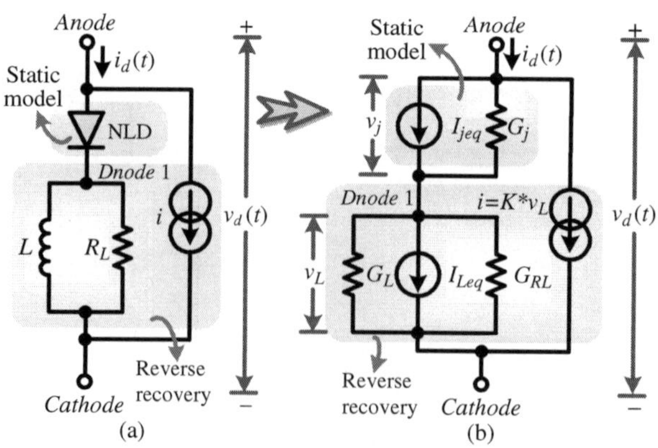

Figure 7.2 Nonlinear power diode model: (a) simplified power diode model and (b) linearized discrete-time equivalent circuit. Source: Lin et al. [140]. ©2018, IET.

selection of parameters by

$$R_L = \frac{ln10 \cdot L}{t_{rr} - I_{rrm}\left(\frac{dt}{dI_r}\right)} \quad \text{and} \tag{7.4}$$

$$K = \frac{I_{rrm}}{L}\left(\frac{dt}{dI_r}\right)\left\{(1 - \exp\left[\frac{-I_{Fo} - I_{rrm}}{L\frac{dI_r}{dt}\left(K + \frac{1}{R_L}\right)}\right]\right\}, \tag{7.5}$$

where I_{rrm} is the peak value of reverse current, $\frac{dI_r}{dt}$ indicates the current slope at turn-off, and I_{Fo} the on-state current prior to turning off.

In the device-level modeling and simulation where the time-step is always small, the Backward Euler method is preferred over the Trapezoidal rule or other implicit or explicit integration methods due to its lower latency in hardware implementation. Therefore, the Norton equivalent circuit of the linear inductor L composing of an impedance and the history current is derived by the following two equations

$$G_L = \frac{L}{\Delta t}, \tag{7.6}$$

$$I_{Leq}(t) = i_L(t - \Delta t), \tag{7.7}$$

where Δt is the simulation time-step and the iterative inductor current $i_L(t)$ takes the form of

$$i_L(t) = I_{Leq}(t) + G_L \cdot v_L(t). \tag{7.8}$$

Hence, the matrix equation of the simplified diode model can be expressed as

$$\mathbf{G}^{Diode} \cdot \mathbf{v}^{Diode} = \mathbf{I}_{eq}^{Diode}, \tag{7.9}$$

where, arranged from anode to cathode, the 3×3 admittance matrix is given by

$$\mathbf{G}^{Diode} = \begin{bmatrix} G_j & K - G_j & -K \\ -G_j & G_j + G_L + G_{RL} & -G_L - G_{RL} \\ 0 & -G_L - G_{RL} - K & G_L + G_{RL} + K \end{bmatrix}, \tag{7.10}$$

$\mathbf{v}^{Diode}$ is a vector of diode nodal voltages, and the equivalent current source contribution vector is

$$\mathbf{I}_{eq}^{Diode} = \left[\; -I_{jeq}, \;\; I_{jeq} - I_{Leq}, \;\; I_{Leq} \;\right]^T. \tag{7.11}$$

The device-level simulation tool SaberRD® provides some typical parameters of the diode which can be obtained from device datasheet, as given in Table A.4.

7.2.1.2 Hardware Module Architecture

The hardware design of the power diode model is a process that realizes computation of its basic equations by the FPGA hardware modules that conform to parallelism and pipelined structure. The introduction of Vivado HLS® greatly shortens the design cycle by enabling C/C++ programming, especially when

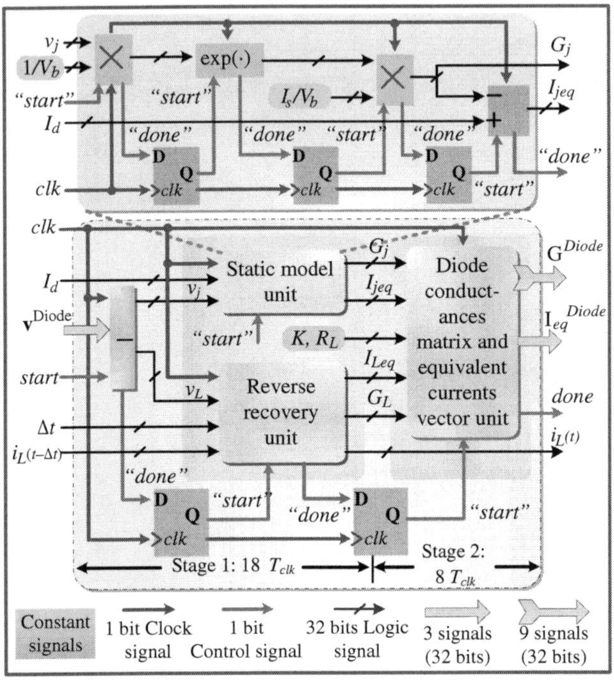

Figure 7.3 Hardware module architecture of the nonlinear power diode model. Source: Lin et al. [140]. ©2018, IET.

complex models are involved in the emulation. Nevertheless, the traditional way using VHDL is still a choice and well suits for the diode model which is considered simple, as Figure 7.3 shows.

As can be seen, the diode hardware module is mainly composed of three blocks, with each enclosing a controllable *start* input port and a *done* signal output port. The structure of static model unit is taken for instance to illustrate the way mathematical equations are being materialized as hardware modules by algebraic operators and registers on the FPGA, and most of the operators are available in the library of Vivado®. The Law of Mathematical Operations determines that in (7.2), the exponent is computed first, followed by the exponential operation and the multiplication of coefficient $\frac{I_s}{V_b}$. However, to ensure correct data transmission, the flip-flops are always needed between two sequentially connected hardware blocks, and the *done* signal of the first block is always the token of data transfer to the next block. For the entire diode module, once the outer start command is received by the subtraction circuit at Stage 1, voltages of internal components are first calculated, followed by the concurrent calculation of the static model and reverse recovery unit, as demonstrated in Figure 7.4. Then, formations of the diode's admittance

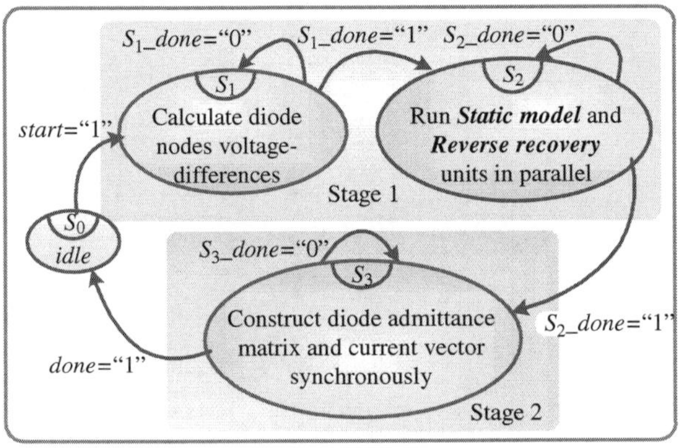

Figure 7.4 Finite state machine for power diode behavioral model hardware implementation.

matrix and current contribution vector begin simultaneously at Stage 2. Following an automatic generation of the *done* signal, the outputs are prepared for external modules that solve the overall circuit with a larger matrix equation. This strictly organized execution sequence is realized by flip-flops, which, as part of the finite state machine (FSM), translate the internal done signals to start orders in one clock cycle for subsequent hardware modules. The overall hardware latency is generally the combination of both stages, which have 18 and 8 clock cycles, respectively.

7.2.2 IGBT

7.2.2.1 Model Formulation

The insulated gate bipolar transistor (IGBT) symbol and its behavioral model are shown in Figure 7.5a, where *PWLD* denotes a piecewise linear diode, R_g is the resistance to the gate pole, and elements such as VCCSs i_{mos} and i_{tail} as well as inter-electrode capacitors C_{ce} and C_{cg} are all nonlinear. The collector to emitter voltage v_{ce} and the gate to emitter voltage v_{ge} determine the value of current source i_{mos}. The tail current i_{tail}, which only emerges during the IGBT turn-off process, is controlled by the internal parallel R_{tail}–C_{tail} pair [141].

The basic operation principle can be summarized when the collector–emitter voltage value v_{CE} is less than the threshold voltage V_{on}, *PWLD* keeps off and the collector current i_C is zero; when the value of v_{CE} is between V_{on} and the saturation voltage V_{sat}, the device behavior will be represented by i_{mos} in the quasi-linear region; then, when v_{CE} is greater than V_{sat}, i_C will mainly depend on the gate-emitter voltage v_{ge} and v_{ce}. Using the IGBT tool provided by SaberRD®, the static and dynamic parameters of a device can be acquired based on the corresponding characteristics and curves from the device datasheet.

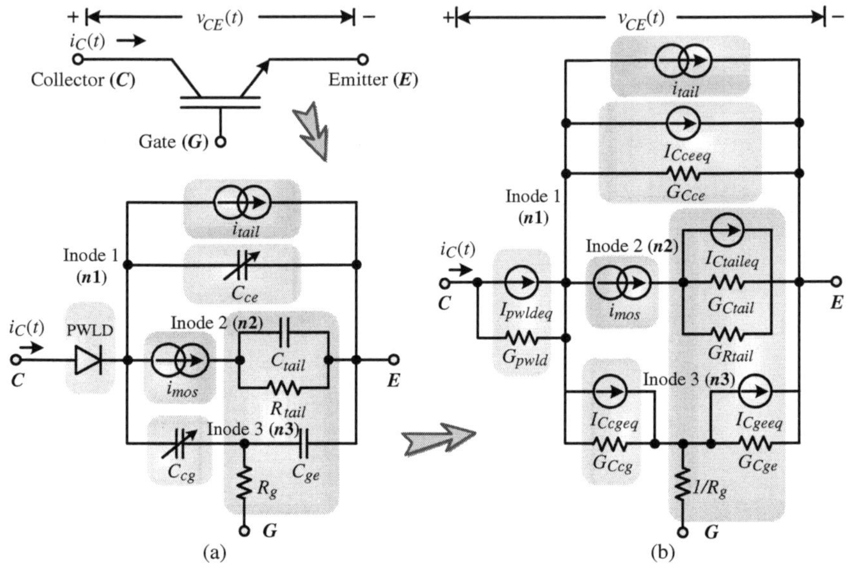

Figure 7.5 Nonlinear IGBT EMT model: (a) continuous-time behavioral model and (b) linearized discrete-time equivalent circuit. Source: Lin et al. [140]. ©2018, IET.

In the model, *PWLD* can be basically treated as a binary conductor whose on- and off-state values are denoted by g_{on} and g_{off}, respectively, and in EMT simulation, its Norton equivalent model is

$$G_{pwld} = \begin{cases} g_{on} \ (v_{pn} > V_{onpwld}), \\ g_{off} \ (v_{pn} \leq V_{onpwld}), \end{cases} \tag{7.12}$$

$$I_{pwldeq} = -G_{pwld} \cdot V_{onpwld}, \tag{7.13}$$

where v_{pn} is the voltage across the component, V_{onpwld} is the *PWLD* forward threshold voltage which could be regarded as V_{on} for the collector to emitter threshold voltage when the IGBT conducts.

Using a similar procedure illustrated in the previous diode section, internal components of the IGBT model can be discretized and linearized, and the outcome is shown in Figure 7.5b. Since the VCCSs (e.g. i_{mos} and i_{tail}) are dependent on voltages over other components which are similar to the diode reverse recovery current i unit, they could not be drawn out as several parts and shown in the figure. However, they could still be treated as a combination of conductances (e.g. G_{mosvd}) and equivalent current sources (e.g. I_{moseq}) in discretization and linearization processes. Meanwhile, circuit elements such as C_{tail} and C_{ge} which are linearly passive can be discretized as follows:

$$G_{Cx} = \frac{C_x}{\Delta t}, \tag{7.14}$$

$$i_{Cx} = G_{Cx} \cdot [v_{Cx}(t) - v_{Cx}(t-\Delta t)], \tag{7.15}$$

$$I_{Cxeq} = -G_{Cx} \cdot v_{Cx}(t-\Delta t), \tag{7.16}$$

where C_x can be referred to as either C_{tail} or C_{ge}. With regard to those that are nonlinear, such as C_{ce} and C_{cg}, they are discretized in a same fashion as taking C_{cg}, for example

$$G_{Ccg} = \begin{cases} \dfrac{\left(ccgo\cdot\left(1+\frac{v_{Ccg}}{vcgo}\right)^{-M}\right)}{\Delta t} & v_{Ccg} > 0, \\ \dfrac{ccgo}{\Delta t} & v_{Ccg} \le 0, \end{cases} \tag{7.17}$$

$$i_{Ccgeq} = \frac{q_{Ccg}(t) - q_{Ccg}(t-\Delta t)}{\Delta t} - G_{Ccg} \cdot v_{Ccg}(t), \tag{7.18}$$

where M is the Miller capacitance exponent coefficient that affects the current rise and fall times as well as the switching power losses, and it is usually set to be 0.5 by default.

The two VCCSs, i.e. i_{mos} and i_{tail}, are more complicated, because their EMT models are taken as a combination of equivalent current sources in parallel with conductance or transconductance. For example, as the essential part of the IGBT NBM, the metal oxide semiconductor field-effect transistor (MOSFET) effect takes the form of

$$i_{mos} = \begin{cases} 0, & (v_{Cge} < V_t)||(v_d \le 0), \\ a_2 \cdot v_d^{(z+1)} - b_2 \cdot v_d^{(z+2)}, & v_d < (y \cdot (v_{Cge} - V_t))^{\frac{1}{x}}, \\ \frac{(v_{Cge}-V_t)^2}{a_1+b_1\cdot(v_{Cge}-V_t)}, & \text{(others)}, \end{cases} \tag{7.19}$$

where a_1, b_1, a_2, b_2, x, y, and z are internal parameters, V_t is the channel threshold voltage, and v_d the potential difference between $Inode1$ and $Inode2$. It indicates that i_{mos} can branch off the conductance $G_{mosvcge}$, which can be derived by $\frac{\partial i_{mos}}{\partial v_{Cge}}$

$$G_{mosvcge} = \begin{cases} 0, & (v_{Cge} < V_t)||(vd \le 0), \\ \frac{\partial a_2}{\partial v_{Cge}} \cdot v_d^{(z+1)} - \frac{\partial b_2}{\partial v_{Cge}} \cdot v_d^{(z+2)}, & v_d < (y \cdot (v_{Cge} - V_t))^{\frac{1}{x}}, \\ \frac{2\cdot(v_{Cge}-V_t)}{a_1+b_1\cdot(v_{Cge}-V_t)} - \frac{b_1\cdot(v_{Cge}-V_t)^2}{(a_1+b_1\cdot(v_{Cge}-V_t))^2}, & \text{(others)}. \end{cases} \tag{7.20}$$

The current source is also determined by its own terminal voltage, and consequently, G_{mosvd} is derived by the partial derivative $\frac{\partial i_{mos}}{\partial v_d}$

$$G_{mosvd} = \begin{cases} 0, & (v_{Cge} < V_t)||(vd \le 0), \\ a_2(z+1)v_d^z - b_2(z+2)v_d^{(z+1)}, & v_d < (y \cdot (v_{Cge} - V_t))^{\frac{1}{x}}, \\ 0, & \text{(others)}. \end{cases} \tag{7.21}$$

Thus, its equivalent current I_{moseq} can be obtained by

$$I_{moseq} = i_{mos} - G_{mosvd} \cdot v_d - G_{mosvcge} \cdot v_{Cge}. \tag{7.22}$$

Similarly, the tail current is also dependent on the voltages of other components, which is expressed as

$$i_{tail} = \left(\left| \frac{v_{tail}}{R_{tail}} - i_{mos} \right| + \left(\frac{v_{tail}}{R_{tail}} - i_{mos} \right) \right) \cdot \frac{i_{rat}}{2}. \tag{7.23}$$

Therefore, the equivalent current contribution from this unit can also be found using

$$I_{taileq} = i_{tail} - G_{tailvd} \cdot v_d - G_{tailvcge} \cdot v_{Cge} - G_{tailvtail} \cdot v_{tail}, \tag{7.24}$$

where the three transconductors G_{tailvd}, $G_{tailvcge}$, and $G_{tailvtail}$ are

$$G_{tailvtail} = \begin{cases} 0, & \frac{v_{tail}}{R_{tail}} \leq i_{mos}, \\ \frac{1}{R_{tail} \cdot i_{rat}}, & \frac{v_{tail}}{R_{tail}} > i_{mos}, \end{cases} \tag{7.25}$$

$$G_{tailvd} = \begin{cases} 0, & \frac{v_{tail}}{R_{tail}} \leq i_{mos}, \\ -\frac{G_{mosvd}}{i_{rat}} & \frac{v_{tail}}{R_{tail}} > i_{mos}, \end{cases} \tag{7.26}$$

$$G_{tailvcge} = \begin{cases} 0, & \frac{v_{tail}}{R_{tail}} \leq i_{mos}, \\ -\frac{G_{mosvcge}}{i_{rat}}, & \frac{v_{tail}}{R_{tail}} > i_{mos}. \end{cases} \tag{7.27}$$

Equations (7.20)–(7.27) indicate all conductances related to i_{mos} and i_{tail} are nonlinear. They are cross-correlated to voltages over other branches, rather than the voltage between *Inode*1 and *Inode*2 for i_{mos} or between *Inode*1 and the emitter for i_{tail}. For example $G_{mosvcge}$ is derived by i_{mos} taking partial derivative with respect to v_{Cge} which is the potential difference between *Inode*3 and the emitter. Moreover, the conductances from i_{tail} are even related to the conductance from i_{mos}. Thus, these features make the model more complicated. Finally, all the subunits are combined to construct a 5 × 5 admittance matrix $\mathbf{G}^{IGBT}$ and current source contribution vector $\mathbf{I}_{eq}^{IGBT}$. Combined with IGBT nodal voltage vector $\mathbf{v}^{IGBT}$, the matrix equation for the linearized system can be obtained as follows:

$$\mathbf{G}^{IGBT} \cdot \mathbf{v}^{IGBT} = \mathbf{I}_{eq}^{IGBT}, \tag{7.28}$$

where $\mathbf{G}^{IGBT}$, $\mathbf{v}^{IGBT}$, and $\mathbf{I}_{eq}^{IGBT}$ are expressed in Eqs. (7.29), (7.30), and (7.31), respectively. It is noted that V_s is the gate voltage exerted between the gate and the emitter.

$$\mathbf{v}^{IGBT} = [v_{Collector}, v_{Inode1}, v_{Inode2}, v_{Inode3}, v_{Emitter}]^T. \tag{7.29}$$

The device-level simulation tool SaberRD® provides some typical parameters of the Siemens® BSM300GA160D IGBT module [142], as given in Table A.5.

$$\mathbf{G}^{IGBT}=\begin{bmatrix} G_{pwld} & -G_{pwld} & 0 & 0 & 0\\ -G_{pwld} & \begin{matrix}G_{Cce}+G_{Ccg}+\\ G_{mosvd}+G_{tailvd}\\ +G_{pwld}\end{matrix} & \begin{matrix}G_{tailvtail}-\\ G_{mosvd}-G_{tailvd}\end{matrix} & \begin{matrix}G_{mosvcge}+\\ G_{tailvcge}\\ -G_{Ccg}\end{matrix} & \begin{matrix}-G_{mosvcge}-\\ G_{tailvtail}-G_{Cce}\\ -G_{tailvcge}\end{matrix}\\ 0 & -G_{mosvd} & \begin{matrix}G_{mosvd}+\\ G_{tail}+G_{Ctail}\end{matrix} & -G_{mosvcge} & \begin{matrix}G_{mosvcge}-\\ G_{tail}-G_{Ctail}\end{matrix}\\ 0 & -G_{Ccg} & 0 & \begin{matrix}G_{Cge}+G_{Ccg}\\ +\frac{1}{R_g+R_{gs}}\end{matrix} & \begin{matrix}-G_{Cge}-\\ \frac{1}{R_g+R_{gs}}\end{matrix}\\ 0 & \begin{matrix}-G_{Cce}-\\ G_{tailvd}\end{matrix} & \begin{matrix}G_{tailvd}-\\ G_{tail}-G_{Ctail}\\ -G_{tailvtail}\end{matrix} & \begin{matrix}-G_{Cge}-\\ G_{tailvcge}\\ -\frac{1}{R_g+R_{gs}}\end{matrix} & \begin{matrix}G_{Cge}+G_{Cce}+G_{tail}\\ +G_{tailvtail}+G_{tailvcge}\\ +G_{Ctail}+\frac{1}{R_g+R_{gs}}\end{matrix} \end{bmatrix}. \tag{7.30}$$

$$\mathbf{I}_{eq}^{IGBT}=\begin{bmatrix} -I_{pwldeq}, & \begin{matrix}I_{pwldeq}-I_{moseq}\\ -I_{Cceeq}-I_{Ccgeq}\\ -I_{taileq},\end{matrix} & \begin{matrix}I_{moseq}\\ -I_{Ctaileq},\end{matrix} & \begin{matrix}I_{Ccgeq}-I_{Cgeeq}\\ +\frac{V_s}{R_g+R_{gs}},\end{matrix} & \begin{matrix}I_{taileq}+I_{Ctaileq}\\ +I_{Cgeeq}+I_{Cceeq}\\ -\frac{V_s}{R_g+R_{gs}}\end{matrix} \end{bmatrix}^T. \tag{7.31}$$

7.2.2.2 Hardware Module Architecture

The hardware module of IGBT is also designed in a pipelined structure with internal functional subunits implemented in parallel. As shown in Figure 7.6, where all subunits are horizontally scaled with respect to latency, the overall process is completed within three stages. At the first stage, similar to the power diode module, voltages over each internal component are first calculated, followed by the simultaneous start and concurrent implementation of the six independent subunits such as capacitor units, *PWLD* unit and i_{mos} unit. This sequential relation is coordinated by the IGBT FSM as shown in Figure 7.7: the parallel operation of subunits in state S_2 will not start unless the *done* signal of state S_1 is detected. Similarly, the release of S_2*_done* signal marks the end of *Stage 1*, or, in other words, the beginning of a new stage, where the operation of i_{tail} unit is being carried out. The reason for setting up this particular stage is that the i_{tail} unit has a sequential relationship with the i_{mos} unit. Then, the signal S_3*_done* triggers *Stage 3*, and all previously calculated conductances and current contributions by reactive components are grouped based on their positions to form the admittance matrix $\mathbf{G}^{IGBT}$ and equivalent current vector $\mathbf{I}_{eq}^{IGBT}$. After that, the *done* signal indicating the completion of one cycle is issued, and the FSM keeps idle until a new start order for the next Newton–Raphson (N-R) iteration or time-step is received.

For the twofold aim of saving hardware resource as well as reducing the latency, constant terms such as R_{tail} and several internal parameters are calculated in

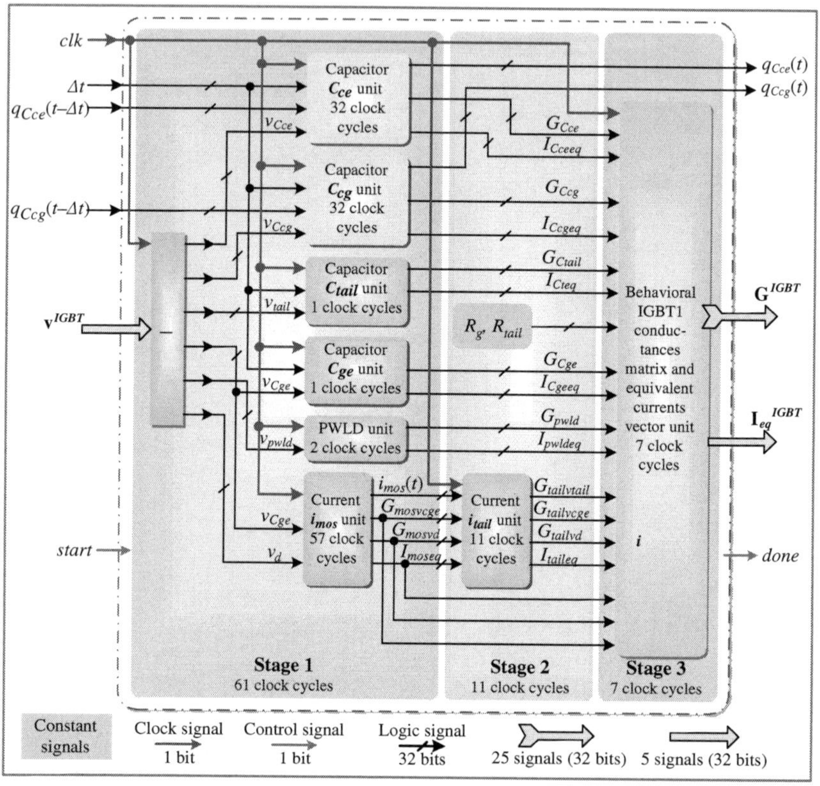

Figure 7.6 Hardware architecture of the nonlinear behavioral IGBT model. Source: Lin et al. [140]. ©2018, IET.

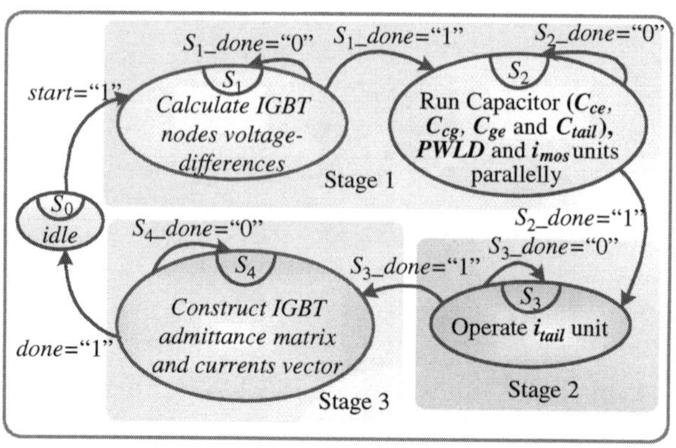

Figure 7.7 Finite state machine of the nonlinear behavioral IGBT hardware module.

Table 7.1 Hardware resources utilization of IGBT and diode hardware modules.

Resources	IGBT	Diode
LUT	29 064 (9.57%)	5 555 (1.83%)
LUTRAM	168 (0.13%)	20 (0.02%)
FF	21 070 (3.47%)	4 722 (0.78%)
BRAM	6.50 (0.63%)	2.50 (0.24%)
DSP48	518 (18.5%)	100 (3.57%)
I/O	98 (14.0%)	98 (14.0%)
BUFG	1 (3.12%)	1 (3.12%)

advance and coded in the VHDL-based hardware design as constant signals. The circuit nodal voltages are the input signals to the IGBT hardware module including the collector voltage v_C, internal nodes voltages v_{Inode1} and v_{Inode2}, gate voltage v_{Inode3}, and emitter voltage v_E. The two history signals $q_{Cce}(t-\Delta t)$and $q_{Ccg}(t-\Delta t)$ are also the inputs to the hardware module. After N-R iteration converges in the present time interval, their two output counterparts $q_{Cce}(t)$ and $q_{Ccg}(t)$ can be correctly calculated, which indicates that the necessary history values are successfully updated after one time-step Δt.

To sum up, the total latency – equivalent to the time one calculation needs or to go through one cycle of the FSM – is about 79 clock cycles according to its critical path. Like the power diode module, this IGBT hardware module fully demonstrates the concept of parallelism and pipeline. Meanwhile, it is as configurable as an actual device so that it can be conveniently fitted in the hardware design of an arbitrary power converter.

Table 7.1 lists the hardware resource utilization of the designed individual IGBT and power diode hardware modules on the Xilinx® Virtex-7® XC7VX485T FPGA with total resources listed in Table 1.2.

7.2.2.3 Multiple Parallel Devices

For small-scale power converters such as the chopper circuit given in Figure 7.8a, the conventional circuit solution approach is preferred for its efficacy, i.e. solving directly one nodal equation corresponding to the overall circuit, which has a dimension of 5 in this case, since potentials at IGBT emitter and diode cathode are known, and the internal node between R and L can be eliminated by merging them together when EMT modeling is carried out, as expressed by

$$i(t) = \frac{v(t)}{R+\frac{L}{\Delta t}} + \frac{\frac{L}{\Delta t}}{R+\frac{L}{\Delta t}} i(t-\Delta t), \tag{7.32}$$

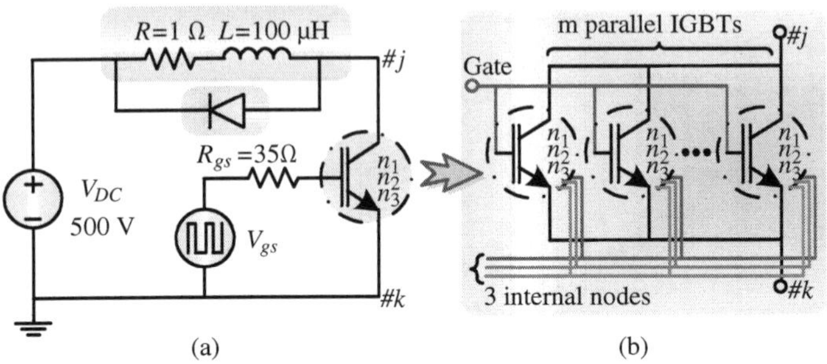

Figure 7.8 Small-scale test circuit for nonlinear behavioral model emulation: (a) chopper circuit and (b) IGBTs in parallel.

where Δt is the simulation time-step, and $i(t)$ and $v(t)$ are current and terminal voltage of the R–L pair. Nevertheless, for the convenience of illustration, a general condition under which potentials of all IGBT and diode nodes are assumed unknown is considered. Consequently, taking IGBT for example, its two terminal nodes are tagged as j and k, and three internal nodes are abbreviated by $n_{1,2,3}$. In applications where the collector current exceeds the capacity of a single IGBT, constructing an array of them in parallel is one common solution, which brings more nodes to off-line simulation tools. However, the fact that numerical solution always leads to a well-balanced condition means all these external and internal nodes can be deemed as connected, and consequently, the total number of nodes remains the same, as demonstrated in Figure 7.8b. Thus, for m parallel IGBTs, their equivalent admittance and current matrix are simply multiplied by that coefficient, i.e. $m \cdot G^{IGBT}$ and $m \cdot \mathbf{I}_{eq}^{IGBT}$, and the matrix equation corresponding to the IGBT portion is constructed as

$$\left(m \cdot \mathbf{G}^{IGBT} + \begin{bmatrix} G_{j1} & 0 & \cdots & 0 \\ 0 & 0 & \cdots & 0 \\ \vdots & & \ddots & \vdots \\ 0 & 0 & \cdots & G_{k5} \end{bmatrix}_{5\times5} \right) \cdot \mathbf{U}_{5\times5} = \left(m \cdot \mathbf{I}_{eq}^{IGBT} + \begin{bmatrix} J_{eq1} \\ 0 \\ 0 \\ 0 \\ J_{eq5} \end{bmatrix} \right), \quad (7.33)$$

where G_{j1}, G_{k5}, J_{eq1}, and J_{eq5} are conductance and current contributions from the external circuit. Similarly, this approach can also be applied to parallel diodes as well. By doing so, the hardware resource utilization for m switches is the same to a single switch, and the computational time is greatly shortened compared with solving each switching element individually with an originally much larger matrix equation.

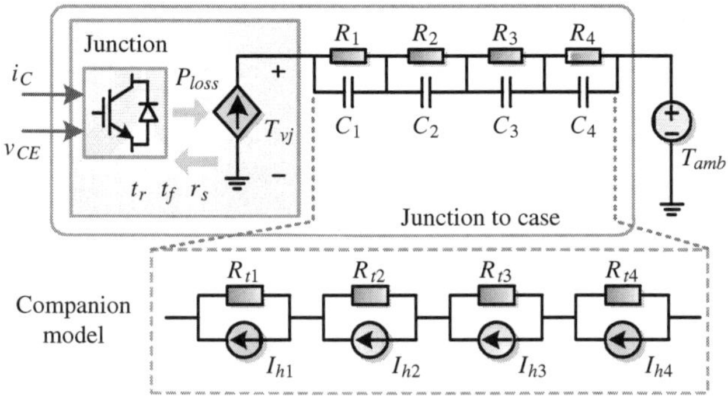

Figure 7.9 IGBT inherent electro-thermal transient network.

7.2.3 Electro-Thermal Network

The IGBT power loss due to conduction and switching produces heat which diffuses through the junction and raises its temperature that in turn affects the device characteristics and thereby the performance. Hence, establishing a dynamic electro-thermal relationship as in Figure 7.9 is essential for cooling system capacity evaluation.

In the network, the IGBT power loss P_{loss} is numerically equal to the instantaneous value of the input current source, whose terminal voltage represents the junction temperature T_{vj}. The dynamic junction to case thermal impedance has the following expression:

$$Z_{th} = \sum_{i=1}^{n} R_i \left(1 - e^{-\frac{t}{\tau_i}}\right), \tag{7.34}$$

where R_i and τ_i are constants available in manufacturer's device datasheet. For circuit simulation purpose, the thermal impedance is embodied by an R–C network, with the capacitance being calculated by

$$C_i = \frac{\tau_i}{R_i}. \tag{7.35}$$

Then, the companion circuit R_{ti}–I_{hi} can be obtained by discretization. Subsequently, the junction temperature is calculated as

$$T_{vj} = \sum_{i=1}^{4} \left[(P_{loss} + I_{hi}) \times \left(\frac{1}{R_i} + \frac{2\tau_i}{R_i \Delta t} \right)^{-1} \right] + T_{amb}, \tag{7.36}$$

where T_{amb} is the ambient temperature set at 25 °C, and I_{hi} is the current source contribution of the capacitors' transmission line modeling (TLM) stub model, written as

$$I_{hi} = 2 \cdot t^i_{Ci} \cdot \frac{2\tau_i}{R_i \Delta t}, \tag{7.37}$$

in which t^i_{Ci} is the incident pulse of capacitor's TLM stub model and is updated by

$$t^i_{Ci}(t) = \left[(P_{loss} + I_{hi}) \times \left(\frac{1}{R_i} + \frac{2\tau_i}{R_i \Delta t} \right)^{-1} \right] - t^i_{Ci}(t - \Delta t). \tag{7.38}$$

As shown by the device datasheet, the junction temperature has a significant impact on IGBT static performance. Thus, these static parameters should be expressed as functions of the temperature. Linear functions that have the form of

$$y(T_j) = k \cdot T_j + p \tag{7.39}$$

are applied to the calculation of these parameters because in datasheet only two temperature curves, at 25 and 125 °C, are provided. However, if more data are available, nonlinear functions can be employed, so as to describe the dynamic electrothermal features more precisely.

7.2.4 Hardware Emulation Results

The chopper converter given in Figure 7.8a is a typical topology chosen to test the nonlinear behavioral IGBT and diode models. The hardware emulation time-step is chosen to be 200 ns, and the latency of the converter hardware design for a full iteration, i.e. from initial inputs to final outputs, within a single time-step is 164 clock cycles. Thus, based on the fact that the FPGA board clock frequency is chosen as 100 MHz as well as a minimum three iterations are needed for convergent results, the estimated time of the FPGA emulation for an interval of 100 ms is about 2.5 seconds. On the other hand, the executive time of SaberRD® for the same interval is about 6.44 seconds using its built-in variable time scheme, indicating the emulation is around 2.6 times faster than the off-line simulation.

An overall view of current and voltage waveforms of IGBT and diode are shown in Figure 7.10a,d, respectively. As can be observed, the two switching devices conduct alternately. Consequently, an IGBT current surge emerges when it starts to conduct due to diode reverse recovery process which occurs simultaneously, as specified in Figure 7.10b,e. Then, the IGBT terminal voltage v_{CE} declines accompanied by the rising of current i_C due to the charging of the R–L branch. The time annotated in oscilloscope results has been divided by a factor of 25, and therefore, for instance, the reading value of 76 μs for t_{vf} is discounted to 3 μs. In the meantime, the diode endures a reverse voltage, which increases while the diode

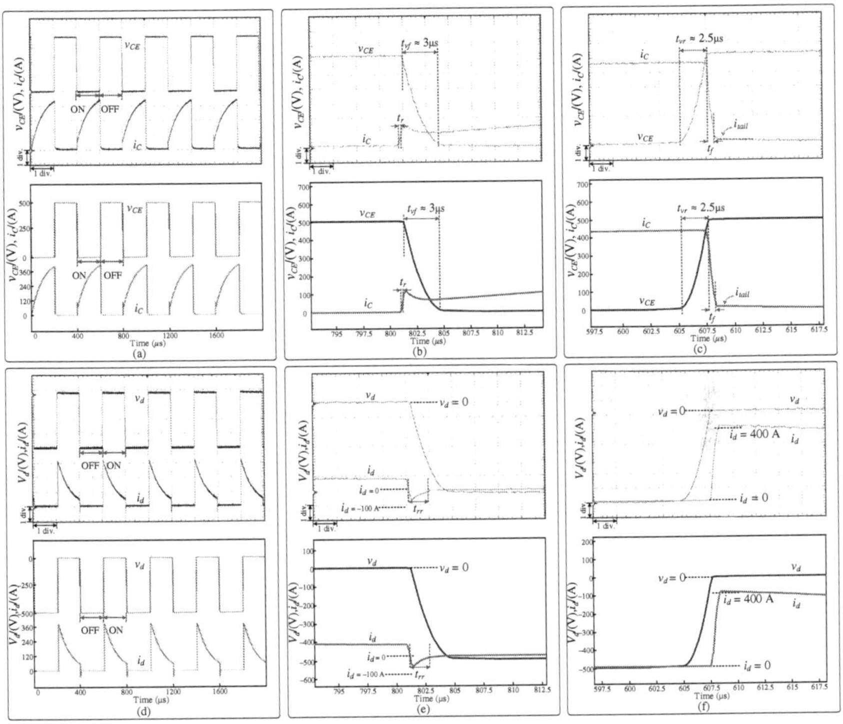

Figure 7.10 Chopper circuit hardware emulation results on oscilloscope (top) and off-line simulation results from SaberRD® (bottom). (a), (b), and (c) IGBT collector–emitter voltage and collector current; (d), (e), and (f) diode voltage and current. Oscilloscope *y*-axis: (a), (d) 132 V/div., 129 A/div., (b), (c), (e), and (f) 80 V/div., 77 A/div.; *x*-axis: (a), (d) 5 ms/div., (b), (c), (e), and (f) 50 μs/div.

is recovering to off-state. Similarly, the reverse processes, i.e. IGBT turn-off and diode conduction, are shown in Figure 7.10c,f, where a tailing current accounting for about 5% of the steady-state value can be observed in both the hardware emulation and SaberRD® simulation. In general, this comparison demonstrates that under a switching frequency of 2.5 kHz all oscilloscope waveforms show a high degree of consistency with simulation results from SaberRD®, proving that the hardware emulation of the power converter is able to show its device-level details. Some minor differences are expected as they are mainly caused by the DAC, which has a precision of 16 bits. In addition, the small fluctuations in current and voltage waveforms in the oscilloscope are caused by environmental noise.

In Table 7.2, the static and dynamic features of IGBT and power diode models are compared between the hardware emulation and SaberRD® simulation under major operation conditions. The power losses at conduction state for both devices

Table 7.2 Comparison of IGBT and power diode nonlinear models between the hardware emulation and SaberRD® simulation.

	SaberRD®	FPGA	Error
	Transient time		
t_{rr}^{Diode}	2080 ns	1970 ns	5.2%
t_{r}^{IGBT}	200 ns	205 ns	2.5%
t_{f}^{IGBT}	640 ns	600 ns	6.6%
	Power dissipation		
P_{rr}^{Diode}	3572 W	3378 W	5.4%
P_{cond}^{Diode}	1117 W	1119 W	0.2%
P_{r}^{IGBT}	16.53 kW	16.46 kW	0.4%
P_{f}^{IGBT}	94.08 kW	90.87 kW	3.4%
P_{cond}^{IGBT}	1012 W	1012 W	<0.1%

are measured at 300 A since the static currents flowing through them are either increasing or decreasing. It shows that the reverse recovery time of diode lasts up to 2 μs, much longer than IGBT's turn-on and turn-off period, defined as time interval between 10% and 90% of steady-state current, which is around 200 and 640 ns, respectively. The errors for those data in HIL emulation are around 5%. The power loss calculation is more precise, with diode reverse recovery power dissipation having the largest error of 5.4% and IGBT turn-off loss next to it, at 3.4%, the errors in the remaining data are negligible, particularly steady-state power losses distinguished by subscript *cond*, for the reason that the static I–V characteristics of both devices are much easier to model. Moreover, the numerical results explicitly illustrate that transient power losses are much higher than those under steady state, especially for the IGBT, meaning that they should be taken into consideration for the sake of safe operation of a converter and underlining the importance of device-level nonlinear switch models.

7.3 Physics-Based Nonlinear IGBT/Diode Model

Different from the NBM which sketches the IGBT/diode characteristics from a macro perspective, the physics-based model involves a lot semiconductor physics and therefore poses a higher computational challenge to the HIL simulation. This type of device-level power semiconductor switch model is the most prevalent of its kind and has been included in the libraries of commercial power electronics circuit simulators such as SaberRD® and PSpice®.

7.3.1 Physics-Based Nonlinear p–i–n Diode Model

7.3.1.1 Model Formulation

In contrast to the normal p–n diode, the p–i–n structure has a wide intrinsic region between the p-layer and n-layer, which enables it to work in fast switching and high-voltage operations, standard for most power diodes. Therefore, for power converter hardware emulation, a detailed model for p–i–n power diode is desirable. The physics-based p–i–n diode model, depicted in Figure 7.11, contains the following five major phenomenological characteristics: (i) reverse recovery, (ii) forward recovery, (iii) emitter recombination, (iv) junction capacitance, and (v) contact resistance. This model is based on the lumped charge concept [11, 143], and utilizes transport equations in the semiconductor to represent the diode physics.

The reverse recovery phenomenon is represented by the following equations:

$$i_R(t) = \frac{q_E(t) - q_M(t)}{T_M}, \tag{7.40}$$

$$0 = \frac{dq_M(t)}{dt} + \frac{q_M(t)}{\tau} - \frac{q_E(t) - q_M(t)}{T_M}, \tag{7.41}$$

$$q_E(t) = I_S\tau\left[e^{\frac{v_E(t)}{V_T}} - 1\right], \tag{7.42}$$

where $i_R(t)$ is the diffusion current from the center of the intrinsic region, $v_E(t)$ and $q_E(t)$ are the junction voltage and charge variable, respectively, $q_M(t)$ is the

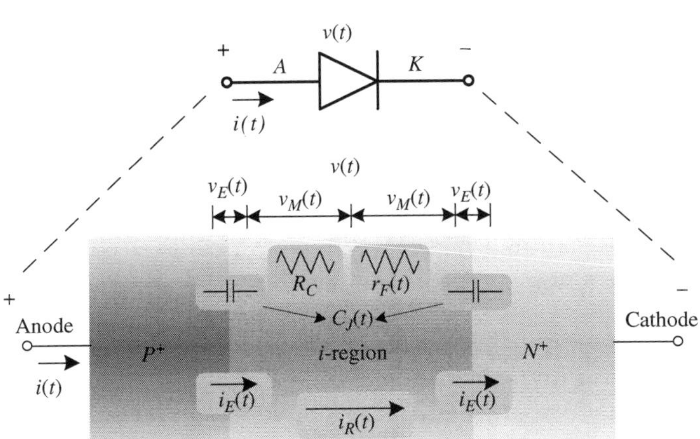

Figure 7.11 Physical structure of power p–i–n diode.

charge in the middle of intrinsic region, T_M denotes the diffusion transit time, τ is the carrier lifetime, I_S represents the diode saturation current constant, and the constant V_T is the thermal voltage. The forward recovery phenomenon is described by

$$v_M(t) = \frac{V_T T_M R_{M0} i(t)}{q_M(t) R_{M0} + V_T T_M}, \tag{7.43}$$

where $i(t)$ represents the whole diode current, $v_M(t)$ is voltage across half of the intrinsic region, and R_{M0} is the initial resistance in the intrinsic region. The emitter recombination phenomenon is formulated as

$$i_E(t) = I_{SE}\left[e^{\frac{2v_E(t)}{V_T}} - 1\right], \tag{7.44}$$

where $i_E(t)$ is the end region recombination current, and I_{SE} is the emitter saturation current constant. The contact resistance R_C is modeled by

$$v(t) = 2v_M(t) + 2v_E(t) + R_C \cdot i(t), \tag{7.45}$$

where $v(t)$ is the voltage across the diode. The charge stored in the junction capacitance $q_J(t)$ which contributes to the diode current $i(t)$ takes the form of

$$i(t) = i_R(t) + i_E(t) + \frac{dq_J(t)}{dt}. \tag{7.46}$$

The junction capacitance $C_J(t)$ and its charge $q_J(t)$ are determined as

$$C_J(t) = \begin{cases} \dfrac{C_{J0}}{\left[1 - \frac{2v_E(t)}{V_J}\right]^m}, & v_E < \frac{V_J}{4}, \\ \dfrac{m \cdot C_{J0} \cdot 2v_E(t)}{V_J \cdot 0.5^{m+1}} - (m-1)\dfrac{C_{J0}}{0.5^m}, & v_E \geq \frac{V_J}{4}, \end{cases} \tag{7.47}$$

$$q_J(t) = \int C_J(t) d[2v_E(t)], \tag{7.48}$$

where C_{J0} is the zero-biased junction capacitance, V_J is the junction built-in potential, and m is the junction grading coefficient.

7.3.1.2 Model Discretization and Linearization

As seen from above, the power diode model described by (7.40)–(7.48) is nonlinear and time-varying. Discretization and linearization are conducted toward the model to obtain a discrete-time equivalent circuit for time-domain simulation. The charge $q_M(t)$ differential equation (7.41) is first discretized using the Trapezoidal rule as

$$q_M(t) = \frac{\Delta t}{2T_M\left(1 + \frac{k_1 \Delta t}{2}\right)} \cdot q_E(t) + \frac{q_{hist}(t - \Delta t)}{1 + \frac{k_1 \Delta t}{2}}, \tag{7.49}$$

where the history charge is given as

$$q_{hist}(t) = q_M(t)\left(1 - \frac{\Delta t}{2}\cdot k_1\right) + \frac{\Delta t}{2T_M}\cdot q_E(t) \quad \text{and} \tag{7.50}$$

$$k_1 = \frac{1}{\tau} + \frac{1}{T_M}. \tag{7.51}$$

Similarly, in (7.46), the differential term $\frac{dq_J(t)}{dt}$ represented by the current contributed by junction capacitor $i_J(t)$ is discretized as

$$i_J(t) = \frac{2}{\Delta t}\cdot q_J(t) - \frac{2}{\Delta t}\cdot q_J(t-\Delta t) - i_J(t-\Delta t), \tag{7.52}$$

where the junction capacitor charge $q_J(t)$ can be deduced from

$$q_J(t) = \begin{cases} \frac{-V_J C_{J0}}{1-m}\cdot\left[1 - \frac{2v_E(t)}{V_J}\right]^{(1-m)}, & v_E(t) < \frac{V_J}{4}, \\ \frac{2^{m+2} m C_{J0} v_E^2(t)}{V_J} - \\ 2^{m+1}(m-1)C_{J0}v_E(t), & v_E(t) \geq \frac{V_J}{4}. \end{cases} \tag{7.53}$$

Meanwhile, the power diode model's nonlinear feature is obvious from (7.42), (7.43), (7.44), and (7.47). The nonlinearity coming from the reverse recovery characteristics (7.42) is linearized as a dynamic conductance g_R and a parallel current source i_{Req} given as

$$g_R = \frac{\partial i_R}{\partial(2v_E)} = k_2 I_S \tau \cdot \frac{1}{2V_T}\cdot e^{\frac{v_E(t)}{V_T}}, \tag{7.54}$$

$$i_{Req} = k_2 I_S \tau\left[e^{\frac{v_E(t)}{V_T}} - 1\right] - k_3 q_{hist}(t-\Delta t) - g_R\cdot 2v_E(t), \tag{7.55}$$

where

$$k_2 = \frac{1}{T_M} - \frac{\Delta t}{2T_M^2\left(1 + \frac{\Delta t k_1}{2}\right)} \quad \text{and} \quad k_3 = \frac{1}{T_M(1 + \frac{\Delta t k_1}{2})}. \tag{7.56}$$

Similarly, the emitter recombination (7.44) and the junction capacitance (7.47) nonlinearities are linearized to obtain the corresponding conductance and current source pairs (g_E, i_{Eeq}) and (g_J, i_{Jeq}), respectively,

$$g_E = \frac{\partial i_E}{\partial(2v_E)} = \frac{I_{SE}}{V_T}e^{\frac{2v_E(t)}{V_T}}, \tag{7.57}$$

$$i_{Eeq} = I_{SE}\left[e^{\frac{2v_E(t)}{V_T}} - 1\right] - g_E\cdot 2v_E(t), \tag{7.58}$$

$$g_J = \frac{\partial i_J}{\partial(2v_E)} = \frac{2}{\Delta t}C_J(t), \quad \text{and} \tag{7.59}$$

$$i_{Jeq} = i_J(t) - g_J\cdot 2v_E(t). \tag{7.60}$$

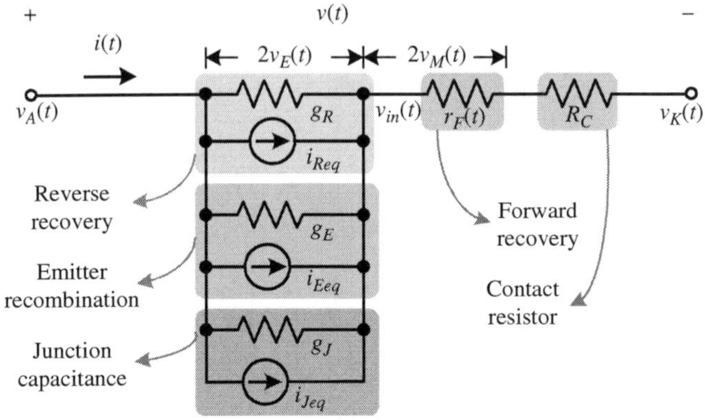

Figure 7.12 Discrete-time linearized equivalent circuit for the power diode.

The forward recovery effect (7.43) is represented as a time-varying resistor:

$$r_F(t) = \frac{2v_M(t)}{i(t)} = \frac{2V_T T_M R_{M0}}{q_M(t)R_{M0} + V_T T_M}. \tag{7.61}$$

Finally, the complete discretized and linearized power diode (Figure 7.12) can be written in the following form:

$$\mathbf{G}^{Diode} \cdot \mathbf{v}^{Diode} = \mathbf{i}_{eq}^{Diode}, \tag{7.62}$$

where $\mathbf{G}^{Diode}$ is a 3×3 conductance matrix given as

$$\begin{pmatrix} g_R + g_E + g_J & -g_R - g_E - g_J & 0 \\ -g_R - g_E - g_J & g_R + g_E + g_J + \frac{1}{r_F(t)+R_C} & -\frac{1}{r_F(t)+R_C} \\ 0 & -\frac{1}{r_F(t)+R_C} & \frac{1}{r_F(t)+R_C} \end{pmatrix},$$

and $\mathbf{v}^{Diode} = [v_A \ v_{in} \ v_K]^T$ is the node voltage vector, and $i_{eq}^{Diode} = [-i_{Req} - i_{Eeq} - i_{Jeq} \ i_{Req} + i_{Eeq} + i_{Jeq} \ 0]^T$ is the equivalent current source vector.

7.3.1.3 Hardware Emulation on FPGA

The power diode hardware module includes six units, which are implemented in three stages (Figure 7.13). The Junction Limit unit runs in Stage 1, which is used to limit the p–n junction voltage value during the Newton iterations. Specifically, the input signal v_E_old (history v_E value from the previous Newton iteration) is updated by the output signal v_E_new within the current iteration for this unit. This limit models a bounded p–n junction voltage for the diode exponential characteristic. Otherwise, the values related to the diode module will be so high that the system equations may become singular and could not be solved, and sometimes they could even hardly be represented as 32-bit floating-point numbers.

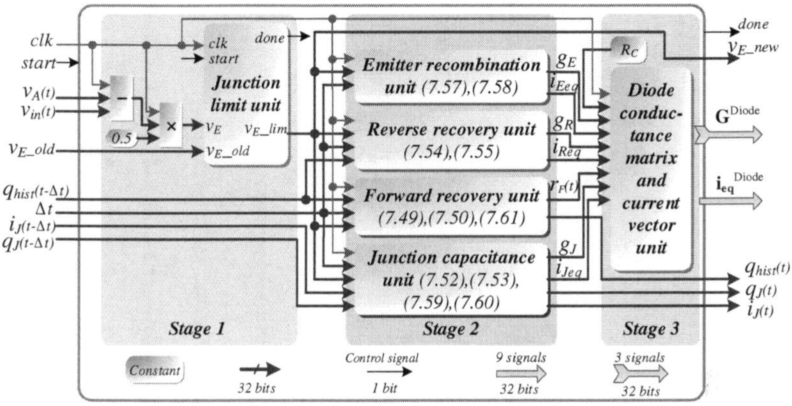

Figure 7.13 Architecture of power diode hardware module.

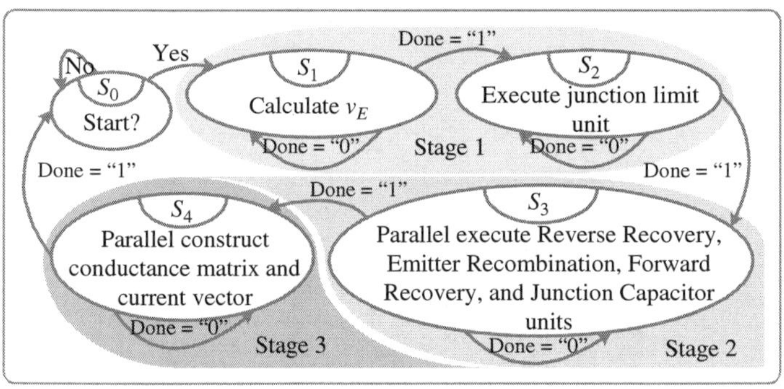

Figure 7.14 Finite state machine for the power diode module.

Stage 2 comprises of four units executed in parallel: the Emitter Recombination Unit, the Reverse Recovery Unit, the Forward Recovery Unit, and the Junction Capacitance Unit. Corresponding dynamic conductors and equivalent current sources calculated by these four units are used to form the conductance matrix and the equivalent current source vector in Stage 3. The sequence of operations for this module is also reflected in the FSE in Figure 7.14, where all the four aforementioned units execute simultaneously in State S_3, before the formation of $\mathbf{G}^{Diode}$ and $\mathbf{i}_{eq}^{Diode}$ within the next state S_4. Furthermore, the hardware structure of each of the six units is highly paralleled and pipelined. For example, within the Reverse Recovery Unit depicted in Figure 7.15, the calculations of the current i_R and conductance g_R are fully in parallel. In order to save the hardware resources, part of the resources even contribute to both the calculations of i_R

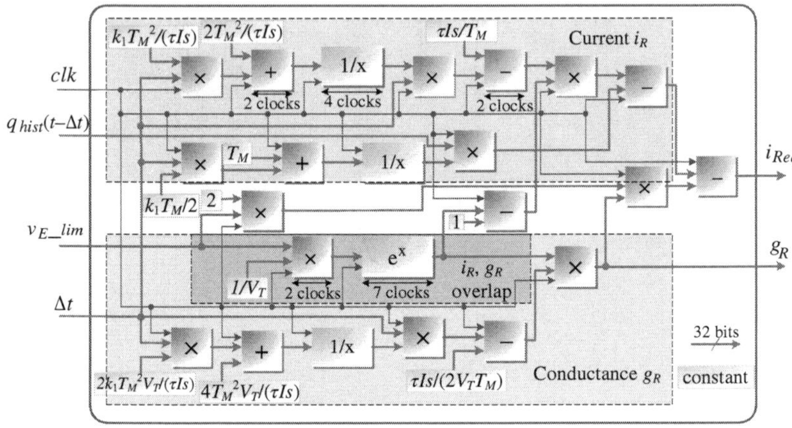

Figure 7.15 Hardware structure of the Reverse Recovery Unit.

and g_R. Accordingly, the final result of equivalent source i_{Req} is obtained. Moreover, even those computations for their subcomponents conform to such policy of deep pipeline and parallelism. The result is a highly efficient and fast hardware architecture. Terms such as $\frac{k_1 T_M^2}{\tau I_S}$ and $\frac{1}{V_T}$ are precalculated and used as constants in the hardware design, which reduce a great deal of computational complexity and latency. The nonlinear operators such as e^x and $1/x$ are coming directly from Xilinx® IP core function package.

7.3.2 Physics-Based Nonlinear IGBT Model

7.3.2.1 Model Formulation

The physics-based IGBT contains the following two main phenomenological characteristics: (i) internal MOSFET phenomena and (ii) internal BJT phenomena. The base current of the BJT is fed by the MOSFET, thereby combining the advantages of low on-state resistance plus high current capacity of the power BJT and excellent gate drive control of power MOSFET. However, both the internal conceptual BJT and MOSFET function differently from their microelectronic counterparts, because their structures differ greatly based on their design goals. Specifically, Hefner interpreted the intrinsic nonlinear physical phenomena into circuit elements (nonlinear capacitors and dependent current sources, see Figure 7.16), which enable device-level circuit simulators to implement the dynamic phenomenological model of IGBT [144].

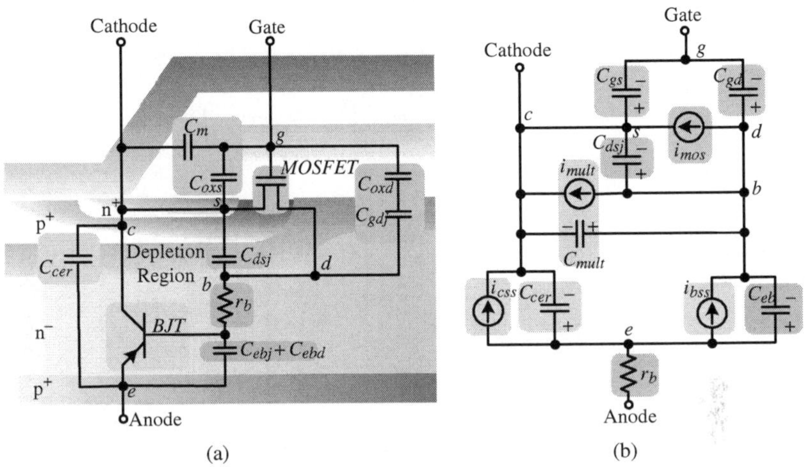

Figure 7.16 (a) Phenomenological circuit and (b) analog equivalent circuit of IGBT.

The MOSFET channel current is modeled as a current source i_{mos}, which is determined by

$$i_{mos} = \begin{cases} 0, & v_{gs} < V_{th}, \\ \dfrac{K_p K_f \left[(v_{gs} - V_{th}) v_{ds} - \frac{K_f |v_{ds}| v_{ds}}{2} \right]}{1 + \theta (v_{gs} - V_{th})}, & |v_{ds}| \leq \frac{v_{gs} - V_{th}}{K_f}, \\ \dfrac{K_p (v_{gs} - V_{th})^2}{2[1 + \theta (v_{gs} - V_{th})]}, & v_{ds} > 0, \\ \dfrac{-K_p (v_{gs} - V_{th})^2}{2[1 + \theta (v_{gs} - V_{th})]}, & v_{ds} < 0, \end{cases} \tag{7.63}$$

where v_{gs} is the gate-source voltage, v_{ds} the drain–source voltage, K_p is transconductance parameter, K_f is triode region transconductance factor, V_{th} is the channel threshold voltage, and θ the transverse field transconductance factor [145]. Its gate-source capacitance C_{gs} equals the combination of the source metallization capacitance C_m and gate-source overlap oxide capacitance C_{oxs}. The gate-drain capacitance C_{gd} is the sum of gate-drain overlap oxide capacitance C_{oxd} and gate-drain junction depletion capacitance C_{gdj}. C_{dsj} is the drain–source junction depletion capacitance.

The BJT is modeled by the steady-state base current source i_{bss} and collector current i_{css}. The former is decided by

$$i_{bss} = \begin{cases} \dfrac{Q}{\tau_{HL}} + \dfrac{4Q^2 N_B^2}{Q_B^2 n_i^2} I_{sne}, & v_{eb} > 0, \\ 0, & v_{eb} \leq 0, \end{cases} \tag{7.64}$$

where τ_{HL} stands for the base high-level lifetime, v_{eb} emitter–base voltage, Q_B background mobile carrier base charge, N_B base doping concentration, n_i intrinsic carrier concentration, and I_{sne} emitter electron saturation current. Q is the instantaneous excess carrier base charge, which is decided by

$$Q = p_0 qAL \tanh\left(\frac{W}{2L}\right), \tag{7.65}$$

where q is the electron charge, A is the active device area, L is the ambipolar diffusion length, p_0 is the carrier concentration at emitter end of base, which is decided by a nonlinear equation:

$$\left(\frac{p_0}{n_i^2} + \frac{1}{N_B}\right)(N_B + p_0)^{1-\beta} N_B^{\beta} = e^{\frac{qv_{eb}}{kT}}, \tag{7.66}$$

where β is order of the nonlinearity, k the Boltzmann constant, T is the room temperature, and W is quasi-neutral base width given as

$$W = W_B - \sqrt{\frac{2\epsilon_{si}(v_{bc} + 0.6)}{qN_B}}, \tag{7.67}$$

where W_B is the metallurgical base width, ϵ_{si} is the silicon dielectric constant, and v_{bc} which equals to v_{ds} is the base–collector voltage.

The BJT collector current is decided by

$$i_{css} = \begin{cases} \frac{i_T}{1+b} + \frac{4bD_p}{W^2(1+b)} Q, & v_{eb} > 0, \\ 0, & v_{eb} \leq 0, \end{cases} \tag{7.68}$$

where i_T represents the anode current, b ambipolar mobility ratio, and D_p hole diffusivity. Its emitter–base capacitance C_{eb} represents the joint effect of the emitter–base junction depletion capacitance C_{ebj} and the emitter–base diffusion capacitance C_{ebd}. C_{cer} is the collector–emitter redistribution capacitance. r_b is the conductivity-modulated base resistance, through which and the anode current i_T, conforms to $i_T = v_{ae}/r_b$, where v_{ae} stands for anode–emitter voltage. The avalanche multiplication current source i_{mult}, given by

$$i_{mult} = (M - 1)(i_{mos} + i_{css}) + M \cdot i_{gen}, \tag{7.69}$$

where M is the avalanche multiplication factor defined as

$$M = \frac{1}{1 - \left(\frac{v_{ds}}{BV_{cb0}}\right)^{BV_n}}, \tag{7.70}$$

where BV_n is the avalanche multiplication exponent, BV_{cb0} is collector–base breakdown voltage (emitter open), and i_{gen} is collector–base thermally generated current given by

$$i_{gen} = \frac{qn_iA}{\tau_{HL}} \sqrt{\frac{2\epsilon_{si}v_{bc}}{qN_B}}, \tag{7.71}$$

and capacitor C_{mult} supplement the base to collector current for the BJT.

Notably, all the aforementioned capacitances (except for constant C_{gs}) are charge-dependent, whose values are directly connected to the depletion region width of the p–n junctions inside the IGBT and they all conform to the following general formulation:

$$C_x = \frac{A_x \cdot \epsilon_{si}}{W_x}, \tag{7.72}$$

where A_x is the active area related to the capacitance, and W_x is the depletion width related to the capacitance.

7.3.2.2 Model Discretization and Linearization

The discrete-time linearized IGBT model in Figure 7.17 can be obtained using procedures similar to the power diode module, and it contains 11 conductances (e.g. g_{bsseb}), 11 equivalent current sources (e.g. i_{bsseq}), and 10 VCCSs (e.g. $g_{bssbc} \cdot v_{bc}$). All the capacitances other than C_{cer} are modeled as pairs of conductance and equivalent source (e.g. g_{Cgs} and i_{Cgseq}). For example, the drain–source capacitance C_{dsj} has its conductance g_{Cdsj} formulated as

$$g_{Cdsj} = \begin{cases} 0, & W = W_B, \\ \frac{2\epsilon_{si}(A-A_{gd})}{\Delta t \cdot (W_B - W)}, & W \neq W_B, \end{cases} \tag{7.73}$$

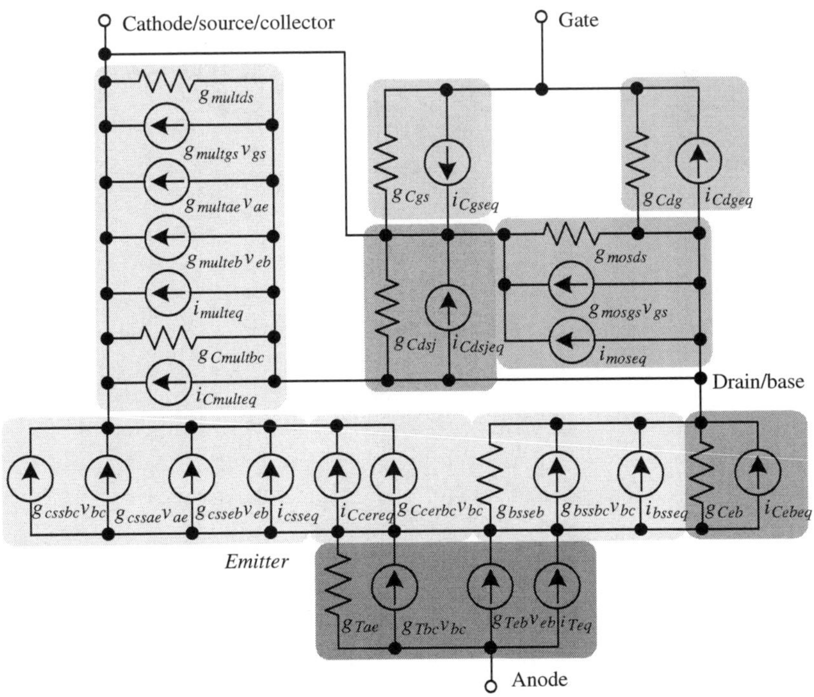

Figure 7.17 Discrete-time linearized equivalent circuit for the IGBT.

where A_{gd} is the gate-drain overlap area, and the drain–source capacitor charge q_{Cdsj} is expressed by

$$q_{Cdsj} = qN_B(A - A_{gd})(W_B - W), \tag{7.74}$$

the drain–source capacitor current i_{Cdsj} takes the form of

$$i_{Cdsj} = \frac{2}{\Delta t}[q_{Cdsj} - q_{Cdsj}(t - \Delta t)] - i_{Cdsj}(t - \Delta t), \tag{7.75}$$

and, accordingly, its equivalent source i_{Cdsjeq} is written as

$$i_{Cdsjeq} = i_{Cdsj} - g_{Cdsj}v_{ds}. \tag{7.76}$$

In contrast, the current sources (e.g. i_{mos}) are modeled as not only pairs of conductance and equivalent source (except for i_{css}) but also as VCCSs (e.g. $g_{mosgs} \cdot v_{gs}$). The MOSFET channel current i_{mos}, for example has its conductance g_{mosgs} $(\frac{\partial i_{mos}}{\partial v_{gs}})$ formulated as

$$g_{mosgs} = \begin{cases} \frac{K_pK_f v_{ds} - i_{mos}\theta}{1+\theta(v_{gs}-V_{th})}, & |v_{ds}| \leq \frac{v_{gs}-V_{th}}{K_f}, \\ \frac{K_p(v_{gs}-V_{th})-i_{mos}\theta}{1+\theta(v_{gs}-V_{th})}, & v_{ds} > 0, \\ \frac{-K_p(v_{gs}-V_{th})-i_{mos}\theta}{1+\theta(v_{gs}-V_{th})}, & v_{ds} < 0, \\ G_{min}, & v_{gs} < V_{th}, \end{cases} \tag{7.77}$$

where G_{min} is the MOSFET minimum conductance. The drain–source conductance g_{mosds} $(\frac{\partial i_{mos}}{\partial v_{ds}})$ is derived by

$$g_{mosds} = \begin{cases} \frac{K_pK_f(v_{gs}-V_{th}-K_f|v_{ds}|)}{1+\theta(v_{gs}-V_{th})}, & |v_{ds}| \leq \frac{v_{gs}-V_{th}}{K_f}, \\ 0, & v_{ds} > 0, \\ 0, & v_{ds} < 0, \\ G_{min}, & v_{gs} < V_{th} \end{cases} \tag{7.78}$$

and its equivalent current i_{moseq} can be obtained as

$$i_{moseq} = i_{mos} - g_{mosds}v_{ds} - g_{mosgs}v_{gs}. \tag{7.79}$$

All aforementioned conductances and transconductances (including those formulated in the VCCSs) and equivalent current sources are finally combined to form a 5×5 conductance matrix $\mathbf{G}^{IGBT}$ and current source vector $\mathbf{I}_{eq}^{IGBT}$, which satisfy the equation:

$$\mathbf{G}^{IGBT} \cdot \mathbf{v}^{IGBT} = \mathbf{i}_{eq}^{IGBT}, \tag{7.80}$$

where $\mathbf{v}^{IGBT} = [v_c\ v_g\ v_a\ v_d\ v_e]^T$ is the IGBT node voltage vector, and $\mathbf{G}^{IGBT}$ and $\mathbf{I}_{eq}^{IGBT}$ are given in (7.81) and (7.82), respectively.

$$\mathbf{G}^{IGBT} = \begin{bmatrix} \begin{matrix} g_{Cgs} + g_{Cdsj} + \\ g_{mosds} + g_{mosgs} + \\ g_{multds} + g_{cssbc} + \\ g_{Ccerbc} + g_{multgs} \\ +g_{Cmultbc} \end{matrix} & \begin{matrix} -g_{Cgs} \\ -g_{multgs} \\ -g_{mosgs} \end{matrix} & \begin{matrix} -g_{cssae} \\ -g_{multae} \end{matrix} & \begin{matrix} -g_{Cdsj} - g_{mosds} \\ +g_{csseb} - g_{multds} \\ -g_{Cmultbc} - g_{cssbc} \\ -g_{Ccerbc} + g_{multeb} \\ -g_{Ccerbc} + g_{multeb} \end{matrix} & \begin{matrix} g_{cssae} - g_{csseb} \\ -g_{multeb} \\ +g_{multae} \end{matrix} \\ -g_{Cgs} & \begin{matrix} g_{Cgs} \\ +g_{Cdg} \end{matrix} & 0 & -g_{Cdg} & 0 \\ -g_{Tbc} & 0 & g_{Tae} & -g_{Teb} + g_{Tbc} & g_{Teb} - g_{Tae} \\ \begin{matrix} -g_{Cdsj} - g_{mosds} \\ -g_{mosgs} - g_{multgs} \\ -g_{multds} - g_{Cmultbc} \\ +g_{bssbc} \end{matrix} & \begin{matrix} -g_{Cdg} \\ +g_{multgs} \\ +g_{mosgs} \end{matrix} & g_{multae} & \begin{matrix} g_{Cdg} + g_{Cdsj} \\ +g_{mosds} + g_{bsseb} \\ +g_{multds} - g_{multeb} \\ +g_{Ceb} \end{matrix} & \begin{matrix} -g_{Ceb} - g_{bsseb} \\ +g_{multeb} \\ -g_{multae} \end{matrix} \\ \begin{matrix} -g_{Ccerbc} - g_{bssbc} \\ -g_{cssbc} + g_{Tbc} \end{matrix} & 0 & \begin{matrix} g_{cssae} - \\ g_{multae} \end{matrix} & \begin{matrix} -g_{Ceb} - g_{bsseb} \\ -g_{csseb} + g_{Teb} + \\ g_{bssbc} - g_{Tbc} \\ +g_{Ccerbc} + g_{cssbc} \end{matrix} & \begin{matrix} g_{Ceb} + g_{bsseb} \\ -g_{Teb} + g_{csseb} \\ -g_{cssae} + g_{Tae} \end{matrix} \end{bmatrix}. \tag{7.81}$$

$$\mathbf{i}_{eq}^{IGBT} = \begin{bmatrix} \begin{matrix} i_{moseq} + i_{csseq} \\ +i_{Cgseq} + i_{Cdsjeq} \\ +i_{Ccereq} + i_{Cmulteq} \\ +i_{multeq}, \end{matrix} & \begin{matrix} -i_{Cgseq} \\ +i_{Cdgeq}, \end{matrix} & -i_{Teq}, & \begin{matrix} -i_{moseq} + i_{bsseq} \\ -i_{Cdsjeq} - i_{Cdgeq} \\ +i_{Cebeq} - i_{Cmulteq} \\ -i_{multeq}, \end{matrix} & \begin{matrix} -i_{bsseq} - i_{csseq} \\ -i_{Cebeq} \\ +i_{Teq} \\ -i_{Ccereq} \end{matrix} \end{bmatrix}^T. \tag{7.82}$$

7.3.2.3 Hardware Emulation on FPGA

The basic strategy of the IGBT model hardware design is to turn all the components of the analog equivalent circuit in Figure 7.16b into 14 hardware units given by Figure 7.18, to fully exploit the possibility of hardware parallelism. Also, it is highly efficient to code and debug all the hardware units individually. The customizable nature of the IGBT hardware module makes it extremely flexible to fit into different power converter topologies with varying module complexities. All the current units including i_{mos}, (i_T, i_{bss}, i_{css}), and i_{mult} are responsible for the calculations of the corresponding currents, dynamic conductances, and equivalent current sources. All the capacitance units including the C_{gs}, C_{gd}, C_{dsj}, (C_{cer}, C_{eb}),

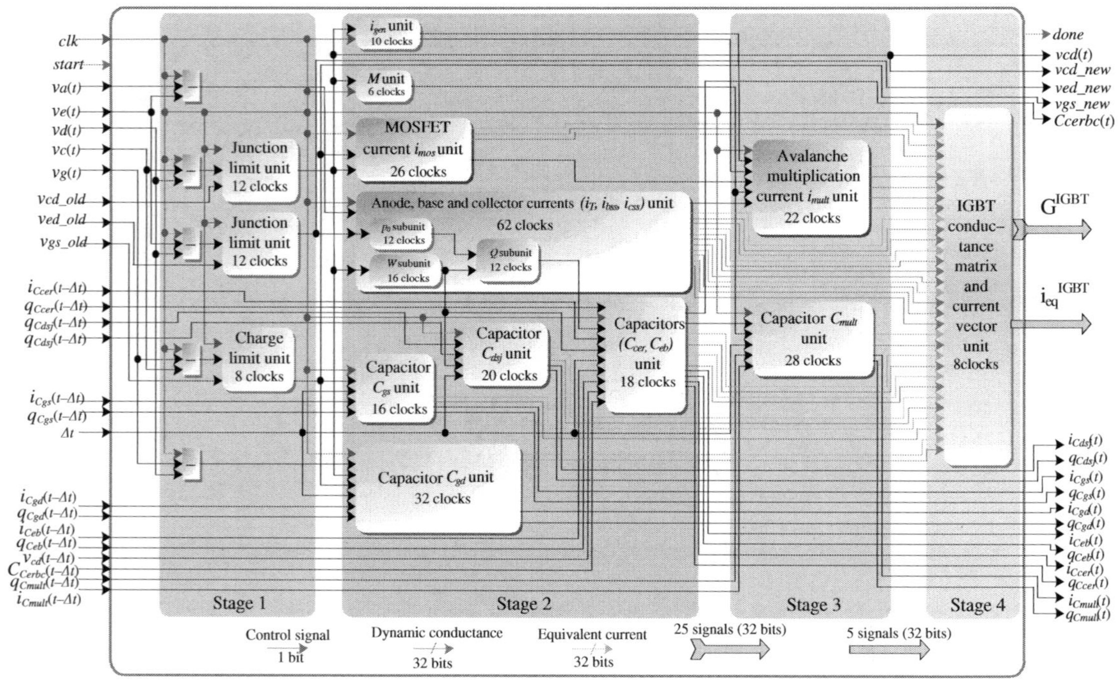

Figure 7.18 Architecture of the IGBT hardware module with all subunits horizontally scaled with respect to latency.

and C_{mult} are responsible for the calculations of not only the currents, dynamic conductances, and equivalent currents but also the charges. The Junction Limit Units and the conductance matrix and current vector unit have similar functions to those used in the power diode hardware module. Other units like the Charge Limit Unit is used to limit the charge variation range between successive Newton iterations, while the avalanche factor unit calculates M, and the i_{gen} unit calculates i_{gen} and its base–collector conductance, which is essential for the calculations in the i_{mult} unit and C_{mult} unit. The input signals to the IGBT hardware module include the circuit node voltages such as v_c (cathode/source/collector), v_g (gate), v_a (anode), v_d (drain/base), and v_e (emitter). The three input signals v_{cd}_old, v_{ed}_old, v_{gs}_old, and their three output counterparts v_{cd}_new, v_{ed}_new, and v_{gs}_new are used to transfer the necessary history values between successive Newton iterations.

As shown in Figure 7.18, the IGBT hardware module executes in four stages. In Stage 1, the two junction limit units and the charge limit unit run in parallel. In Stage 2, eight other units, including i_{gen}, M, i_{mos}, (i_T, i_{bss}, i_{css}), C_{gs}, C_{dg}, C_{dsj}, and (C_{cer}, C_{eb}), run concurrently. In Stage 3, the i_{mult} unit and C_{mult} unit run independently. In Stage 4, all the dynamic conductances and equivalent currents calculated by the above units are combined to the output conductance matrix $\mathbf{G}^{IGBT}$ and equivalent current source vector $\mathbf{I}_{eq}^{IGBT}$. Moreover, the internal structure of all the aforementioned units or subunits are deeply pipelined. The operation sequence of the IGBT hardware module can be seen in Figure 7.19, where six different states (S_3–S_8) contribute to Stage 2. The reason is that W subunit and Q subunit of the (i_T, i_{bss}, i_{css}) unit interconnect to the C_{dsj} unit and the (C_{cer}, C_{eb}) unit, respectively. So for synchronization within the (i_T, i_{bss}, i_{css}) unit, the two subunits responsible for W, p_0 computations are also executed in parallel, these units need to be triggered at the right time.

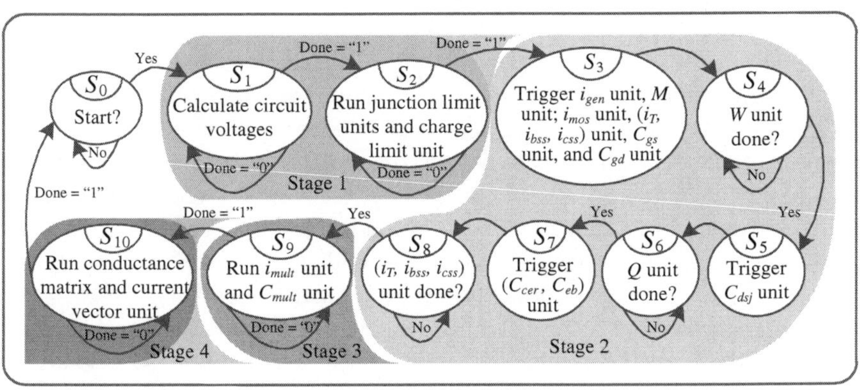

Figure 7.19 Finite state machine of the IGBT hardware module.

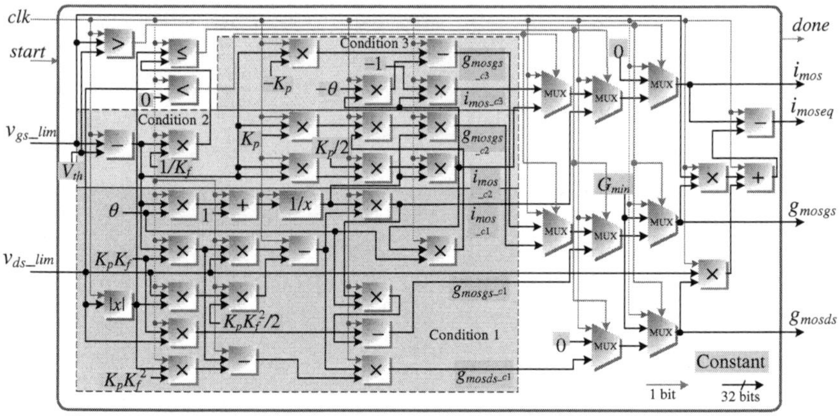

Figure 7.20 Hardware structure of the MOSFET current i_{mos} unit.

The internal hardware structure of one of the units, the MOSFET current i_{mos} unit, is shown in Figure 7.20, to illustrate the parallelism and pipelining. There are three comparators and five multiplexers which are responsible for selecting four groups of results among three nested conditions, see (7.77) and (7.78). Since one group of results (when $v_{gs} < V_{th}$) is obtained directly, the remaining three groups of temporary results for i_{mos}, g_{mosgs},and g_{mosds} are all calculated before being selected by the three layers of multiplexers under different conditions as the final results. As for i_{moseq}, its value can only be computed after i_{mos}, g_{mosgs}, and g_{mosds} are obtained.

It is highlighted that the hardware structures to calculate the results under three conditions are fully parallelized. Some of the computing units are repetitively used to calculate different results. Since under the fourth condition, the output g_{mosgs} and g_{mosds} are constant, G_{min} is sent to two multiplexers, repetitively. The nonlinear operators such as $|x|$ and $1/x$ come directly from the Xilinx® IP core function package.

With detailed hardware modules of the power diode, the IGBT and the linear circuit components, the emulation of a complete power converter can be realized. As shown in Figure 7.21, it is composed of three intervals within a simulation time-step Δt. Interval 1 is responsible for selecting the proper time-step Δt utilizing the variable time-step control module (VTCM) as well as the input voltage of the converter circuit. Interval 2 is responsible for performing the Newton iterations, which includes the calculations of the conductance matrix and equivalent current source vector for the whole converter circuit and obtaining the solution for circuit node voltages using a parallel linear solver. Although the linear component module is triggered simultaneously with the power diode and IGBT modules, their input node voltages are deliberately kept constant unlike those of the nonlinear ones. The reason to put them inside the Newton iteration loop is to increase the

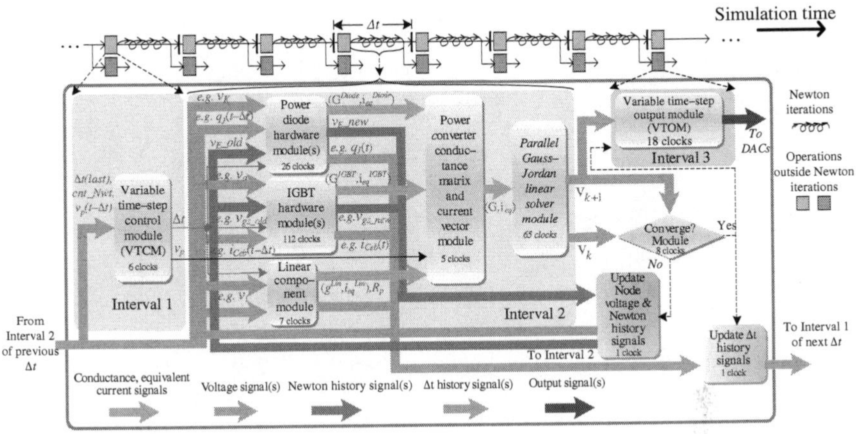

Figure 7.21 Sequence of three intervals within a time-step Δt of the power converter hardware emulation.

parallelism of the whole structure so as to reduce latency. Most of the Δt history terms and Newton history terms are related to the modules in this interval, which are updated during the calculations between successive time-steps and Newton iterations, respectively.

7.3.3 Hardware Emulation Results

7.3.3.1 Test circuit

The test circuit used for the power converter hardware emulation is a DC–DC buck converter given in Figure 7.22, and the circuit parameters are provided in the Appendix A.6. There are two nonlinear components (IGBT and Diode), four linear components, and a voltage source in the circuit. The IGBT node voltages v_c, v_g, v_a, v_d, and v_e are also the circuit node voltages v_1, v_2, v_7, v_3, and v_4, respectively.

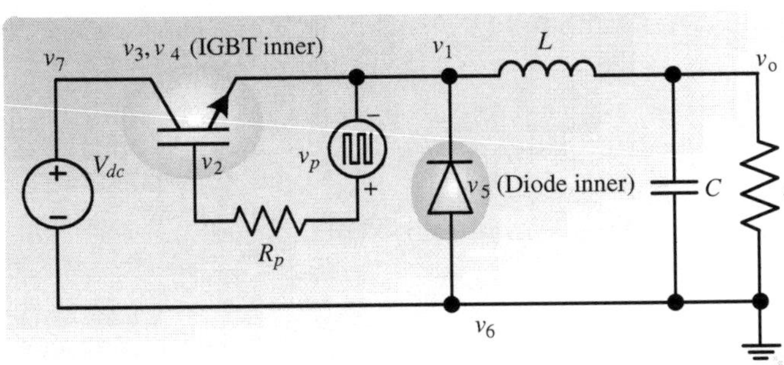

Figure 7.22 DC–DC buck converter circuit.

Table 7.3 Hardware resources utilized by nonlinear components.

Resource	Diode	IGBT	Overall circuit
FF	7 763 (1.3%)	61 384 (10%)	72 181 (12%)
LUT	10 741 (3.5%)	88 061 (29%)	107 264 (35%)
Memory LUT	86 (0.066%)	603 (0.46%)	729 (0.56%)
I/O	32 (4.6%)	32 (4.6%)	91 (13%)
BRAM	1 (0.049%)	34 (1.7%)	38 (1.8%)
DSP48	158 (5.6%)	1 419 (51%)	1 657 (59%)
BUFG	0 (0.0%)	0 (0.0%)	1 (3.1%)

The diode node voltage v_A, v_{in}, and v_K correspond to the circuit node voltages v_6, v_5, and v_1. The linear components, R, L, and C, can be collectively discretized as a conductance g^{Lin} and a parallel equivalent current source i_{eq}^{Lin}. Similarly, v_p with a serial resistor R_p can be equally changed to a current source $\frac{v_p}{R_p}$ and a parallel resistor R_p in order to reduce the dimension of the whole system [146]. The dimension of the discrete-time linearized system is originally 7, but since the circuit node voltages are $v_6 = 0$ (ground) and $v_7 = V_{dc}$, it was reduced to 5.

The DC–DC converter was emulated [147] on the Xilinx® Virtex-7® VC707 board with XC7VX485T FPGA whose resources are listed in Table 1.2, and the off-line simulation for validation was carried out on a Windows®-7 PC with Intel® Core™i7-2600K 3.4 GHz CPU and 8 GB RAM. The execution time for the off-line simulation of the converter for 100 ms of run time in SaberRD® using variable time-step strategy (with an initial time-step of 10 ns and maximum time-step of 1 μs) was 17.5 seconds, while the hardware emulation time was 0.58 seconds. Therefore, the speed-up is more than 30 times. It is conceivable that off-line simulation of converter circuits with more IGBTs and diodes employing detailed device-level models would be much slower than hardware emulation. The latency of one time-step calculation is 579 clock cycles, the highest frequency is 115 MHz, and the resource utilization is listed in Table 7.3.

7.3.3.2 Results and comparison

The steady-state results for the converter output voltage v_o, the IGBT collector–emitter voltage $v_{ce}(= v_7 - v_1)$, the IGBT collector current i_c, the diode voltage $v_d(= v_6 - v_1)$, and the diode current i_d of the DC–DC converter hardware emulation under the switching frequency of 2.5 kHz are shown in Figure 7.23a–e. Compared with SaberRD®, the steady-state results from hardware emulation are accurate. Since SaberRD® did not converge when R_{M0} is set to 50 Ω, the waveforms

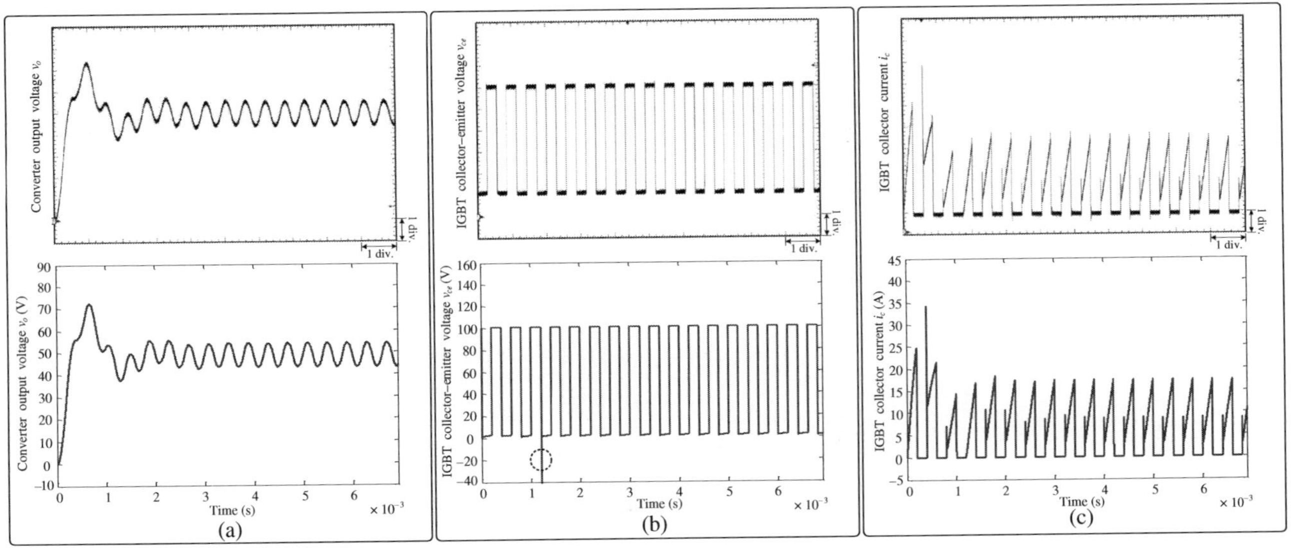

Figure 7.23 Steady state and transient results for DC–DC converter from hardware emulation (oscilloscope) and off-line simulation (SaberRD®) at 2.5 kHz switching frequency. Scale: (a) y-axis: 10 V/div., (b) y-axis: 20 V/div., (c) y-axis: 5 A/div., (d) y-axis: 20 V/div., (e) y-axis: 5 A/div., (f) y-axis: 20 V/div. (v_{ce}), 2 A/div. (i_c), (g) 20 V/div. (v_d), 2 A/div. (i_d), (h) y-axis: 20 V/div. (v_{ce}), 4 A/div. (i_c), (i) y-axis: 20 V/div. (v_d), 4 A/div. (i_d); (a)–(e) x-axis: 4.0 ms, (f) x-axis: 320 ns, (g) x-axis: 3.2 μs, (h, i) x-axis: 1 μs.

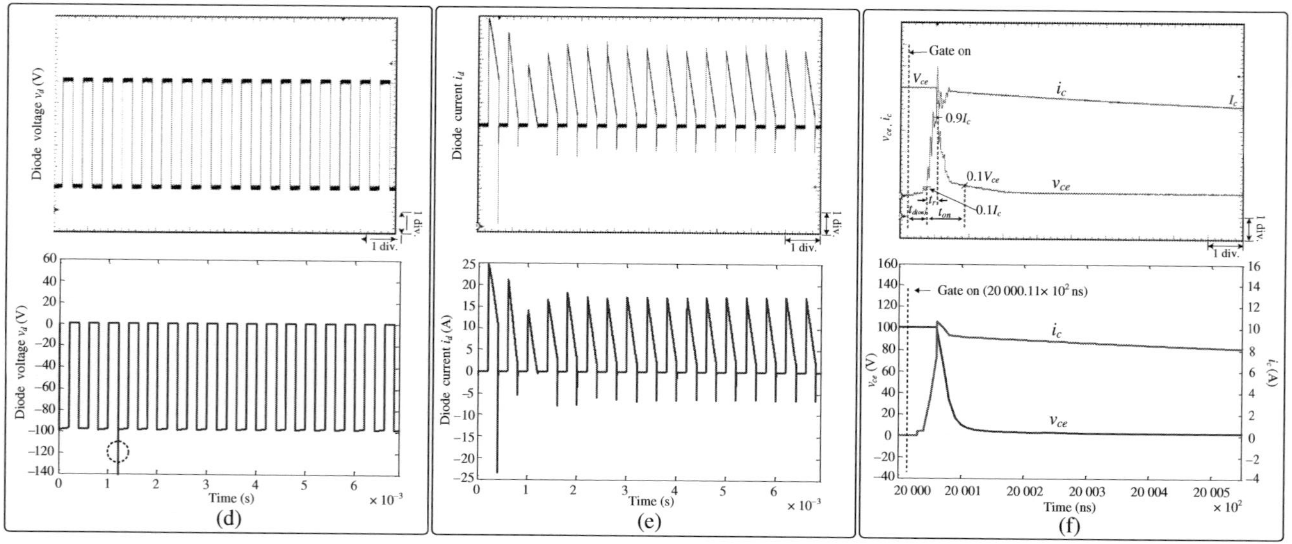

Figure 7.23 (*Continued*)

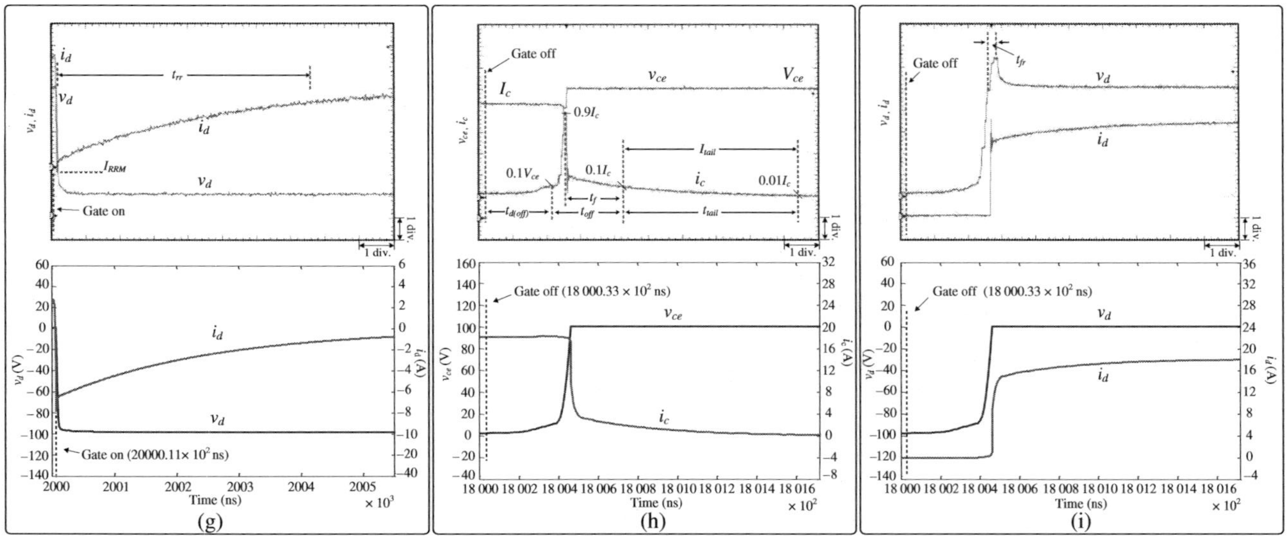

Figure 7.23 (*Continued*)

Table 7.4 Comparison of device switching times and power dissipation.

	SaberRD®	FPGA	Error
	Switching time		
$t_{d(on)}(IGBT)$	28.8 ns	29.5 ns	2.43%
$t_{on}(IGBT)$	60.0 ns	58.7 ns	2.17%
$t_r(IGBT)$	18.6 ns	17.8 ns	4.30%
$t_{d(off)}(IGBT)$	327 ns	334 ns	2.14%
$t_{off}(IGBT)$	380 ns	368 ns	3.16%
$t_{tail}(IGBT)$	860 ns	879 ns	2.21%
$t_f(IGBT)$	285 ns	293 ns	2.81%
$t_{rr}(Diode)$	4009 ns	4020 ns	0.27%
$t_{fr}(Diode)$	n/a	42.1 ns	n/a
	Dissipated power		
$P_{on}(IGBT)$	96.05 mW	93.87 mW	2.27%
$P_{off}(IGBT)$	539.6 mW	522.5 mW	3.17%
$P_{cond}(IGBT)$	11.46 W	11.05 W	3.58%
$P_{rev,recov}(Diode)$	5.434 W	5.507 W	1.34%
$P_{cond}(Diode)$	4.768 W	4.782 W	0.29%

shown in Figure 7.23a–h were obtained by setting R_{M0} to 0 Ω (no forward recovery). The dash circled areas in Figure 7.23b,d are the nonconvergence points of SaberRD®.

The device-level transient results for (v_{ce}, i_c), and (v_d, i_d) during turn-on and turn-off switching of the DC–DC converter hardware emulation under the switching frequency of 2.5 kHz are shown in Figure 7.23f–i. Compared with SaberRD®, the device-level transient results of the hardware emulation show good agreement. Specifically, the transient waveform of diode voltage from oscilloscope in Figure 7.23i was obtained by setting R_{M0} to 50 Ω, which shows the forward recovery clearly. The comparison of IGBT and diode switching times from SaberRD® and hardware emulation is shown in Table 7.4. The hardware emulation was also tested under high-switching frequency conditions, up to a maximum of 100 kHz. The results for a 40 kHz switching frequency are shown in Figure 7.24. Again, a close agreement with SaberRD® can be observed.

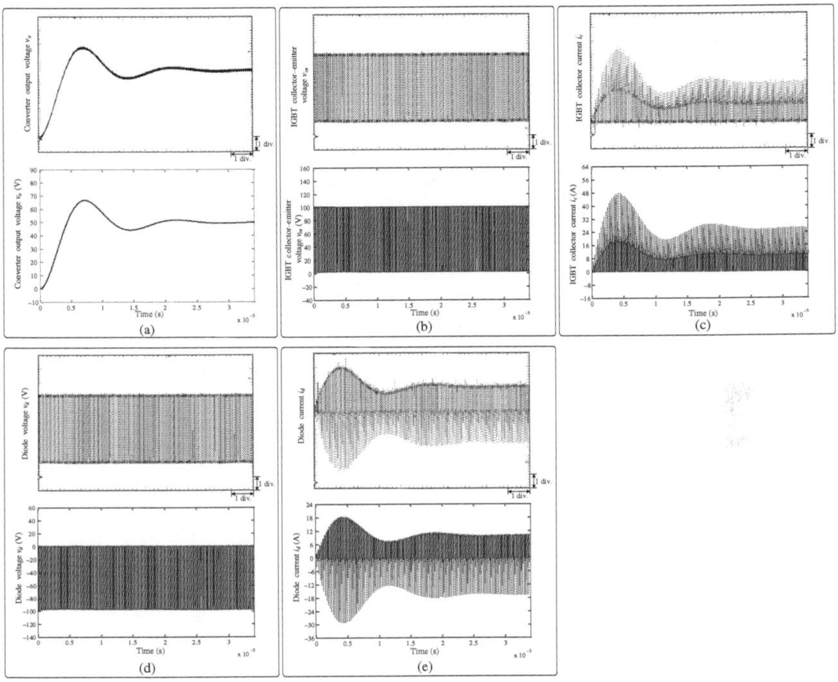

Figure 7.24 Steady-state results for DC–DC converter from hardware emulation (oscilloscope) and off-line simulation (SaberRD®) at 40 kHz switching frequency. Scale: (a) *y*-axis: 10 V/div., (b) *y*-axis: 20 V/div., (c) *y*-axis: 8 A/div., (d) *y*-axis: 20 V/div., (e) *y*-axis: 6 A/div.; (a)–(e) *x*-axis: 12.5 ms.

The power dissipation during a switching cycle of IGBT mainly comes from the switching loss and conduction loss. The instantaneous power dissipation is calculated by multiplying the IGBT collector–emitter voltage (v_{ce}) and collector current (i_c). By choosing corresponding time range from t_0 to t_1, the energy losses in one switching cycle are obtained by integrating the instantaneous power dissipation during turn-on, turn-off, and conduction period, respectively. The power loss P_{loss} is a quotient of the energy loss and switching period T_s, expressed as

$$P_{loss} = \frac{\int_{t_0}^{t_1} v_{ce}(t) \cdot i_c(t)dt}{T_s}. \tag{7.83}$$

The power dissipation of reverse recovery and conduction period for the diode are obtained similarly. Table 7.4 shows the power dissipation results under the switching frequency of 2.5 kHz from both SaberRD® and hardware emulation. The relative errors between the hardware emulation and SaberRD® have several reasons. The first one can be attributed to the modeling error. Specifically,

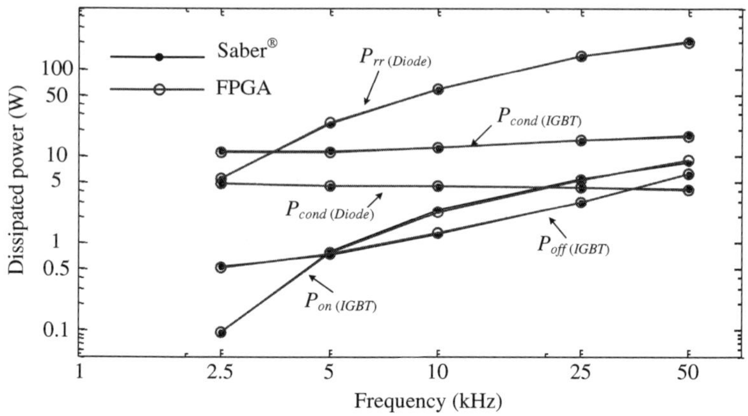

Figure 7.25 Comparison of power dissipation of IGBT (a) and diode (b) between SaberRD® and hardware emulation.

the power diode model used by SaberRD® is more sophisticated than Lauritzen's diode model used in this work. For the IGBT model, although SaberRD® and the hardware emulation both use Hefner's model, SaberRD®'s model is more refined judging from its larger parameter set. Another reason for the error is the limit of the 32-bit floating-point number used in the hardware emulation, which is very accurate under ordinary scenarios, but still restrictive when used to emulate the highly complicated physics-based IGBT and power diode model. Moreover, the linear solver working with a 32-bit floating-point number precision may also contribute to the loss of accuracy. Also, some specific methods used in the hardware emulation to deal with the nonlinearity of the model (e.g. look-up tables), although performed very fast, but at the cost of certain level of accuracy.

Figure 7.25 shows the switching power loss and conduction power loss for both IGBT and diode at the switching frequency of 2.5, 5, 10, 25, and 50 kHz. The specific switching cycles around 2 ms for different frequencies are chosen for power loss calculation to ensure that the circuit conditions are similar for all cases. The switching losses of IGBT and power diode increase notably along with a rising switching frequency, whereas the conduction power losses remain relatively stable.

7.4 IGBT/Diode Curve-Fitting Model

The CFM depicts the complex dynamic characteristics of the power semiconductor switch by multiple linear or nonlinear sections [148]. Based on the versatility, there are two categories, i.e. the static curve-fitting model (SCFM) and the dynamic curve-fitting model (DCFM). The difference is the current and voltage waveform shapes do not change in the former type, but that of the latter is affected by several factors. The rise and fall time automatically changes along with the electromagnetic environment, including the collector current, gate voltage as well as resistance, thermal impact, etc. Meanwhile, taking into consideration the linearity, there are in total four types, which are static linear and nonlinear models [149–151], and dynamic linear CFM and nonlinear CFM [152]. However, all those categories are incapable of simulating extreme scenarios since some information is omitted because of the numerical order reduction. The modeling process generally includes two steps, i.e. the static characteristics and the switching transients, as specified below in two examples.

7.4.1 Linear Static Curve-fitting Model

7.4.1.1 Static Characteristics

Different from the simple two-state resistors, the SCFM utilizes junction temperature-dependent resistors $r(T_{vj})$ and voltage sources $v(T_{vj})$ to represent the IGBT modules. The slope resistance $r_{on}(T_{vj})$ and threshold voltage $v_{on}(T_{vj})$ for the IGBT, and the diode are obtained from the static I_C–V_{CE} and the I_F–V_F curves by piecewise linearization [153]. R_{off} is the off-state resistance which can be estimated by the collector–emitter cut-off current I_{CES}. Figure 7.26 shows the output and forward characteristic curves for the IGBT and the diode at $T_1 = 25$ °C and $T_2 = 125$ °C from the data-sheet [154]. Linear interpolation is used to estimate $r_{on}(T_{vj})$ and $v_{on}(T_{vj})$ at any junction temperatures T_{vj} for either IGBT or diode, when values in T_1 and T_2 are known, given as

$$r_{on}(T_{vj}) = \frac{T_{vj} - T_2}{T_2 - T_1}\left(r_{on}^{T_2} - r_{on}^{T_1}\right) + r_{on}^{T_2}, \tag{7.84}$$

$$v_{on}(T_{vj}) = \frac{T_{vj} - T_2}{T_2 - T_1}\left(v_{on}^{T_2} - v_{on}^{T_1}\right) + v_{on}^{T_2}. \tag{7.85}$$

7.4.1.2 Switching Transients

The linearized transient waveforms are generated in this part for IGBT modules to be computed in real-time, based on the solution from system-level simulation and the transient time information from datasheet. Rise time t_r^{rated}, fall time t_f^{rated}, turn-on delay $t_{d,on}^{rated}$, and turn-off delay $t_{d,off}^{rated}$ at $T_1 = 25$ °C and $T_2 = 125$ °C in standard test conditions, with rated current i_{rated} and rated voltage v_{rated} through and across IGBT module, are provided by the datasheet. The actual and linearized voltage and current waveforms in turn-on and turn-off transients are shown in Figure 7.27. During the turn-on transition of the linearized waveform, after turn-on delay time, the IGBT collector current rises from the bottom to

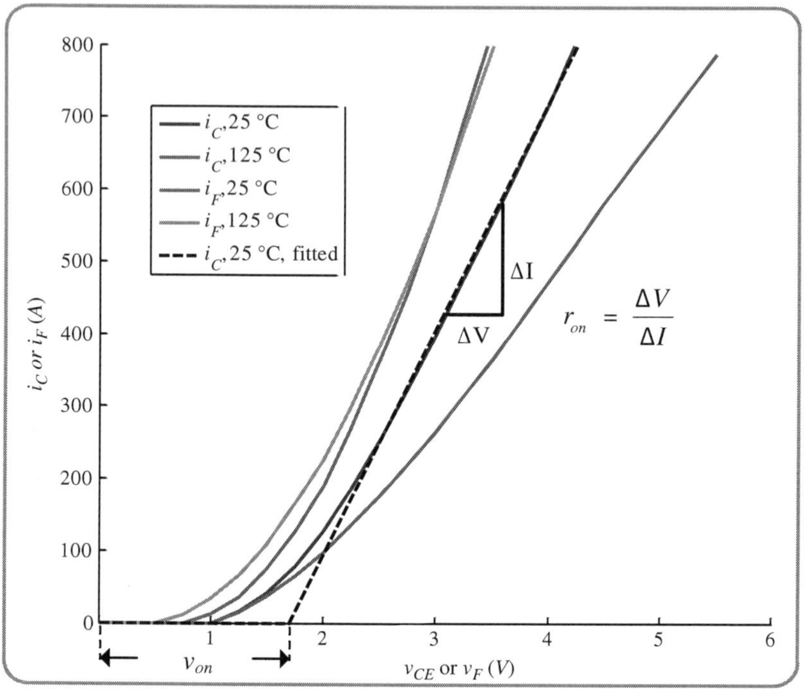

Figure 7.26 Output characteristics for IGBT and diode at $T_1 = 25$ °C and $T_2 = 125$ °C.

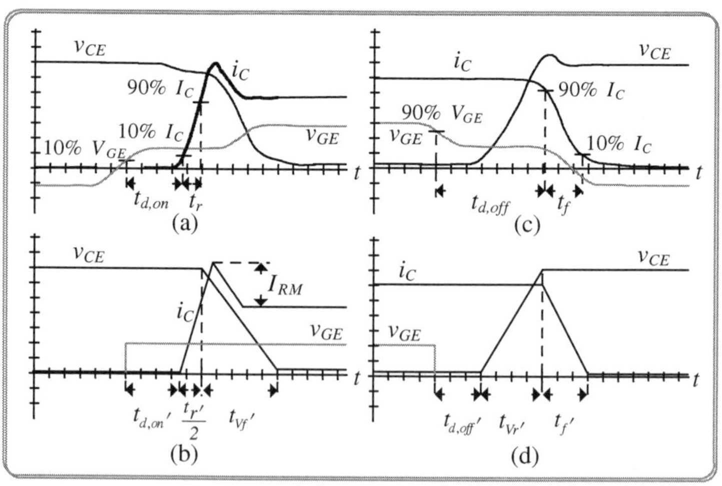

Figure 7.27 Device-level transient waveforms for IGBT module: (a) actual waveforms during turn-on transition, (b) linearized waveforms during turn-on transition, (c) actual waveforms during turn-off transition, and (d) linearized waveforms during turn-off transition.

the steady-state conduction value; after the exact point of the current reaching the steady-state value, the voltage across the IGBT starts to drop, as shown in Figure 7.27a,b. The turn-off transition in Figure 7.27c,d is similar except that voltage rises first, and then the current drops.

The revised rise time $t'_r(T_{vj}, i(t))$ and fall time $t'_f(T_{vj}, i(t))$ of current waveform as a function of junction temperature T_{vj} and conduction current through IGBT module $i(t)$ are estimated as follows:

$$t'_r(T_{vj}, i(t)) = \frac{i(t)}{i_{rated}} \cdot \frac{1}{0.8} \left(\frac{T_{vj} - T_2}{T_2 - T_1} \left(t_r^{T_2,rated} - t_r^{T_1,rated} \right) + t_r^{T_2,rated} \right), \tag{7.86}$$

$$t'_f(T_{vj}, i(t)) = \frac{i(t)}{i_{rated}} \cdot \frac{1}{0.8} \left(\frac{T_{vj} - T_2}{T_2 - T_1} \left(t_f^{T_2,rated} - t_f^{T_1,rated} \right) + t_f^{T_2,rated} \right). \tag{7.87}$$

In (7.86) and (7.87), the current changing rates are assumed to be constant for different conduction currents, and the coefficient 0.8 is used simply to expand the transient time range of current changing from 80% to 100%, since the provided current rise and fall time are between 10% and 90%.

The rise time $t'_{Vr}(T_{vj}, i(t), v(t))$ and fall time $t'_{Vf}(T_{vj}, i(t), v(t))$ for voltage waveforms are not provided by datasheet, which can be estimated as follows:

$$t'_{Vr}(T_{vj}, i(t), v(t)) = \frac{2E_{off}^{IGBT}(T_{vj}, i(t), v(t))}{v(t) \cdot i(t)} - t'_f(T_{vj}, i(t)), \tag{7.88}$$

$$t'_{Vf}(T_{vj}, i(t), v(t)) = \frac{2E_{on}^{IGBT}(T_{vj}, i(t), v(t))}{v(t) \cdot i(t)} - t'_r(T_{vj}, i(t)). \tag{7.89}$$

Finally, the revised turn-on delay $t'_{d,on}(T_{vj}, i(t))$ and turn-off delay $t'_{d,off}(T_{vj}, i(t), v(t))$ are given as

$$t'_{d,on}(T_{vj}, i(t)) = \frac{T_{vj} - T_2}{T_2 - T_1} \left(t_{d,on}^{T_2,rated} - t_{d,on}^{T_1,rated} \right) + t_{d,on}^{T_2,rated} - \frac{1}{10} t'_r(T_{vj}, i(t)), \tag{7.90}$$

$$\begin{aligned} t'_{d,off}(T_{vj}, i(t), v(t)) = {} & \frac{T_{vj} - T_2}{T_2 - T_1} \left(t_{d,off}^{T_2,rated} - t_{d,off}^{T_1,rated} \right) + t_{d,off}^{T_2,rated} - t'_{Vr}(T_{vj}, i(t), v(t)) \\ & - \frac{1}{10} t'_f(T_{vj}, i(t), v(t)). \end{aligned} \tag{7.91}$$

It is noted that when the IGBT modules are configured in half-bridge circuit topology, the diode reverse recovery current will add to the IGBT current during turn-on transient. This over current cannot be ignored, since the reverse recovery current can be even larger than the steady-state current of on-state. To present this phenomenon, the exact amount of maximum diode reverse recovery current I_{RM}, which is temperature-dependent and proportional to the steady-state current, is added to the current waveform. This addition happens when the IGBT current begins to rise as shown in Figure 7.27b. By doing so the current rise slope is doubled, which means the exact rise time of the hardware emulation will be only half of the result calculated in (7.86).

Admittedly, the linearized transient waveforms are not obtained from accurate differential equation solution for the IGBT module, and therefore cannot reflect the exact detailed physical phenomenon of the IGBT modules, which is why the DCFM is introduced in the following section for attaining real-time, along with aforementioned nonlinear high-order models. Nevertheless, they are still able to provide a fairly accurate estimate of the switching transients in the real-time emulation for MMC system based on the limited information available from the datasheet.

7.4.2 Nonlinear Dynamic Curve-fitting Model

Piecewise linearizing the IGBT static I–V curves provided in the manufacturer's datasheet or obtained based on experimental measurement into six segments, the collector current in the jth segment can be written as

$$I_C = k_j(T_{vj})V_{CE} - b_j(T_{vj}), \tag{7.92}$$

where b_j and k_j are linear functions of junction temperature T_{vj} since data at two different temperatures are available. Taking the IGBT under steady-state as a resistor, its value can then be deduced as

$$r_s = \frac{V_{CE}}{I_C} = \frac{I_C + b_j(T_{vj})}{k_j(T_{vj})I_C}. \tag{7.93}$$

It should be pointed out that the IGBT off-state accounts for one of the six segments. Meanwhile, switching transients must be included as part of the model. In addition to T_{vj}, the rise and fall times generally denoted by $t_{r,f}$ are also affected by factors such as gate resistance R_g, and collector current I_C, each of which, according to device datasheet, can be expressed by a piecewise linear function

$$t_{r,f}(x_i) = A_i x_i + B_i, \tag{7.94}$$

where x_i represents either T_{vj}, R_g, or I_C, and A_i, B_i are coefficients. However, when two or more factors are combined, the relationship is still nonlinear; therefore, the overall effect can be described by a polynomial function

$$t_{r,f}(x_1, x_2, x_3) = k_0 \cdot \prod_{i=1}^{3}(x_i) + \sum_{i,j=1\to 3}^{i\neq j} k_i x_i x_j + \sum_{i=1}^{3} b_i x_i + b_0, \tag{7.95}$$

where k_i and b_i are coefficients that are obtained in a way that sets two variables constant and forces the function to be equal to (7.94) with the remaining variable, i.e.

$$t_{r,f}(x_i) = t_{r,f}(x_i, x_j, x_k)|_{x_j, x_k = C}. \tag{7.96}$$

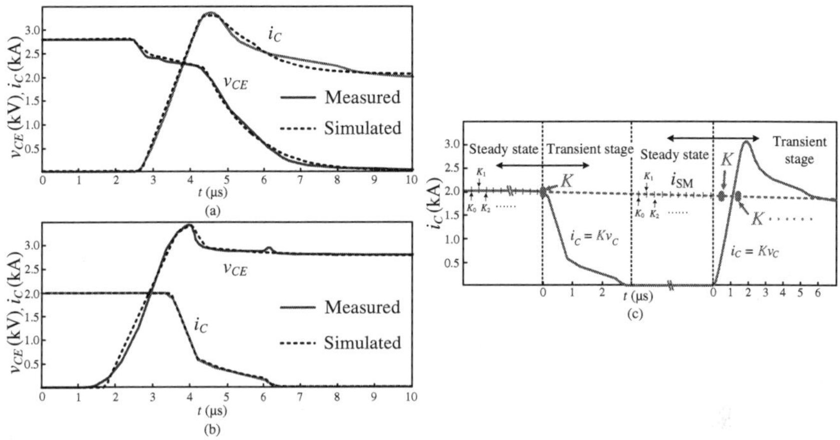

Figure 7.28 IGBT transient waveforms from a bridge-structure test circuit: (a) turn-on process, (b) turn-off process, and (c) coefficient K determination.

Note that the gate-driving voltage does not appear in (7.95) because for specific applications its amplitude is fixed. Nevertheless, it can be added to (7.95) once the $t_{r,f}$–V_g relationship is available.

The shape of IGBT transient waveforms is influenced by the test circuit. In Figure 7.28a,b, the turn-on and turn-off waveforms of 5SNA 2000K450300 Stak-Pak IGBT module are obtained from a bridge-structure test circuit which provides the same electromagnetic environment to that of an MMC submodule [155] to ensure the applicability of the fitted model. As a result, the diode's reverse recovery reflected by current surge in Figure 7.28a is automatically included in the IGBT transient waveforms. Figure 7.29a,b is the transient model for IGBT in which the output of a per-unit circuit is amplified by K times. The VCCS and current-controlled current source (CCCS) are able to reproduce simulated curves that virtually fit with those measured experimentally, as shown in Figure 7.28. The descending curves can be modeled as the capacitor voltage of a discharging RC circuit with a time constant τ. Take the collector current for instance, the fall time t_f, defined as the current dropping from 90% to 10% of the initial value along an extrapolated straight line drawn between the time instants when the current is

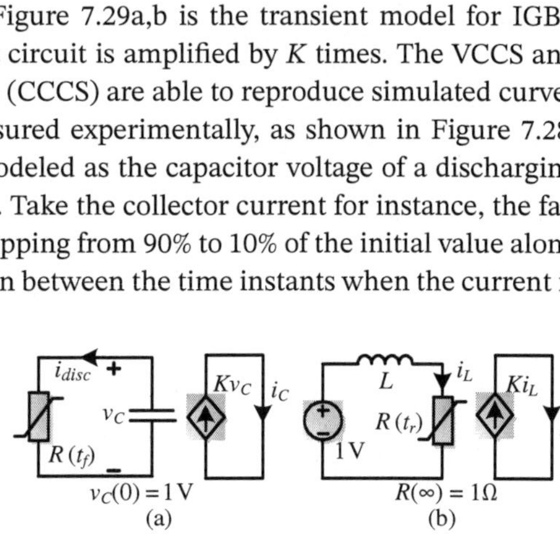

Figure 7.29 Complete IGBT dynamic electrothermal model: (a) VCCS for descending curves and (b) CCCS for rising curves.

90% and 60% of its initial value, is located on a virtually straight line. To achieve that, the initial capacitor discharging rate i_{disc} should be controlled at

$$i_{disc} = C\frac{dv_C}{dt} = C\frac{(90\% - 10\%)v_C(0)}{t_f}. \tag{7.97}$$

After v_C drops to about 33% of its initial value, the curvy tail current emerges. Then, the control object shifts to the resistance, while the capacitance is kept constant, and in the jth nonlinear segment of the curve, it is

$$i_{Cj} = K \cdot v_{Cj}(0)e^{-\frac{t}{\tau_j}}, \tag{7.98}$$

where $v_{Cj}(0)$ denotes the initial capacitor voltage of that segment, and coefficient K is the last steady-state value and instantaneous arm current for turn-off and turn-on currents, respectively, as shown in Figure 7.28c.

Similarly, the ascending curves are realized by an RL circuit. The overshoot is achieved by charging purely the inductor and introducing a time-varying resistor forces to the curve to decline with the desired slope. The rise time, defined as the time between instants when the collector current rises from 10% to 90% of the final value, decides the inductance. Since the segment where t_r locates is a straight line, the inductance can be derived as follows:

$$L = U \cdot \frac{dt}{di} = \frac{t_r(T_{vj}, R_g, I_c) \times 1(\text{V})}{(90\% - 10\%)(\text{A})}. \tag{7.99}$$

7.4.3 Hardware Emulation Results

The collector current has a significant influence on IGBT's turn-on and turn-off times, as tested in Figure 7.30a,b. When I_C climbs from 500 to 3500 A, the turn-on time increases steadily, while the turn-off time first declines from 1300 ns to around 800 ns, and then rises again. Comparison with corresponding values provided by the datasheet proves the accuracy of the IGBT model, and with more linear segments to approximate the nonlinear $t_{r,f}$–I_C curves, the accuracy of the results can be further improved. Figure 7.30c gives corresponding energy consumption, both turn-on and turn-off energies E_{on}/E_{off} closely follow datasheet values when identical test conditions are set. A high degree of agreement between simulation and experimental data indicates that the IGBT DCFM qualifies for simulation of power converters such as an MMC to give design guidance.

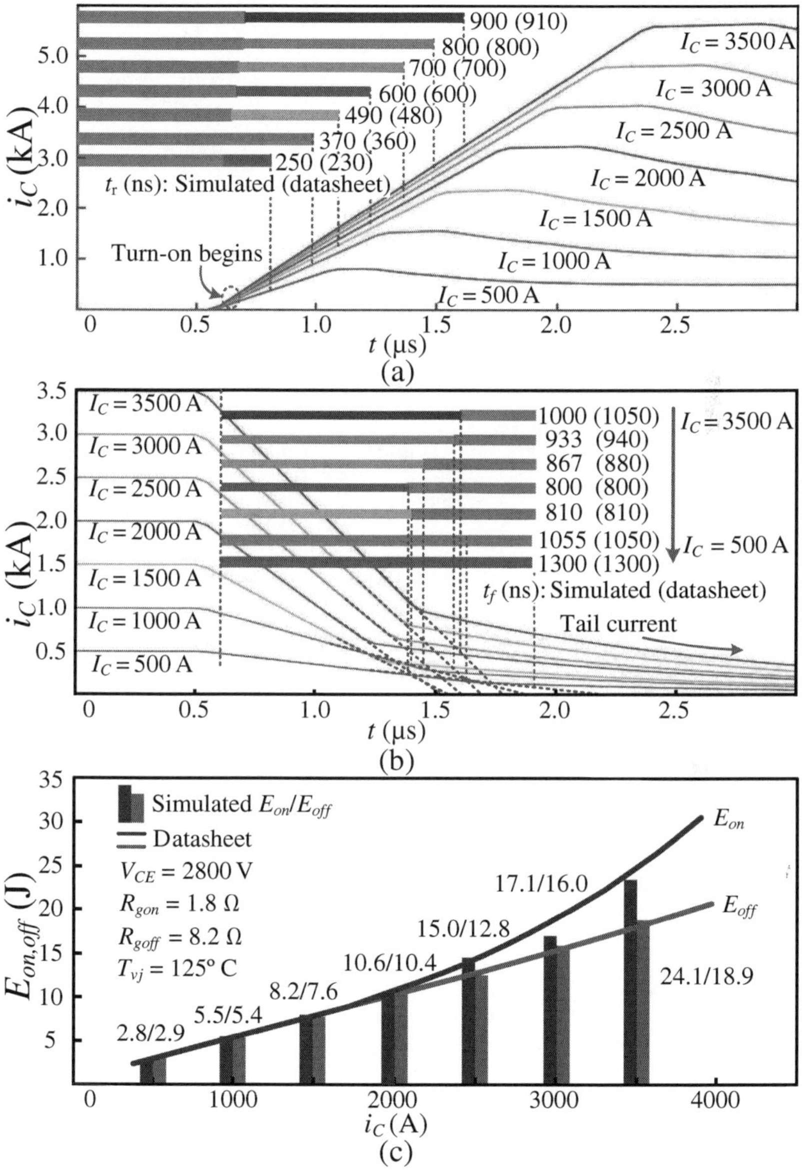

Figure 7.30 IGBT transient tests under different collector current at T_{vj}=125 °C: (a) turn-on process, (b) turn-off process, and (c) turn-on and turn-off energies.

7.5 Summary

This chapter presented device-level models of the power semiconductor IGBT and diode that constitute the power electronic converters in the DC grid. As the most complex type of its kind, this category of models is able to provide an insight into the detailed behaviors of the IGBT/diode during turn-on and turn-off transient processes in addition to the basic on-state conductance, and therefore, it enables power loss calculation and the subsequent junction temperature estimation for equipment design guidance. Details of three most prevalent models in device-level EMT simulation, i.e. the behavioral model, the physics-based model, and the CFM, as well as their hardware design methods for emulation on the FPGA are provided. While the nonlinear high-order models are subjected to the N-R iteration for numerical convergence, their first-order counterpart exempts itself from repetitive calculations in the same time-step for being inherently convergent. The hardware emulation speed on current available FPGAs determines their scope of applications: the CFM is best suited for real-time HIL testing, the physics-based model reveals the most comprehensive information, and the NBM is a trade-off between fidelity and computational efficiency. High-fidelity real-time device-level emulation of power semiconductors has significant advantages in advanced transport applications [156–159] as well such as electric ship, high-speed railway traction drives, and more electric aircraft power systems.

8

AC–DC Converters

8.1 Overview

The voltage source converters (VSCs) including the AC/DC and DC/AC types categorized as the rectifier and inverter, respectively, are the interfaces that interlink the AC grid and its DC counterpart. In this chapter, they are uniformly named as the AC–DC converter, which has mainly three configurations, i.e. the two-level VSC, the neutral-point clamped (NPC) converter, and the modular multi-level converter (MMC). Currently, the most popular topology for VSCs for both high- and medium-voltage applications is the MMC. Therefore, the focus of this chapter is primarily on the MMC topology.

The complexity of a converter model is not solely determined by the circuit topology; in fact, demand for the extent of information to be revealed for various purposes by the simulation also has a significant impact. For example, the grid-connected two-level VSC with all its six IGBT/diode pairs adopting the physics-based model or the nonlinear behavioral model is computationally more challenging to the simulator than the MMC using merely the two-state resistor model, albeit the latter has dozens or hundreds of IGBT/diode pairs. Therefore, it is reasonable to classify the VSCs according to the power semiconductor switch models they contain rather than their topology in the electromagnetic transient (EMT) simulation, and according to Chapter 7, the general categories are the following:

1. Average value model (AVM);
2. Detailed model including its equivalent;
3. Iterative nonlinear device-level model, including the hybrid model.

Sorted out by the complexity, it becomes more time-consuming for the CPU, as well as the field-programmable gate array (FPGA) or other types of processors, to simulate the converter when its modeling complexity increases from the AVM to the iterative nonlinear models not only because of a much larger node

Real-Time Electromagnetic Transient Simulation of AC-DC Networks, First Edition.
Venkata Dinavahi and Ning Lin.

number due to a sharp increase in model order but also the nonlinear nature of the insulated gate bipolar transistor (IGBT) and diode which require repetitive calculations in a single time-step in the form of Newton–Raphson iterations to ensure numerical correctness and convergence. Nevertheless, as mentioned in Chapter 7, a high-order model provides more information that is helpful to the converter design assessment in addition to system-level preview and therefore, it has aroused a wide interest in recent years following the rapid development of converter topologies applied in power system which resulted in a large amount of power loss during operation. Among all the aforementioned basic converter configurations, the MMC is the most promising topology in forming the high voltage direct current (HVDC) grid, where this VSC has been gaining momentum and is expected to overtake traditional thyristor-based line-commutated converters as the main vehicle for electrical energy conversion due to its advantages such as resilience to commutation failure and capacity of regulating reactive power. Meanwhile, it also has merits over traditional two-level or other multilevel converters such as high-power quality quasi-sinusoidal output waveforms, obviating the need for bulky filtering equipment and scalability, which allows for the number of submodules to be flexibly changed to adjust to different voltage stresses or to produce the demanded voltage levels.

In Figure 8.1, a three-phase MMC adopting the half-bridge submodule (HBSM) is shown, where the IGBT/diode models can be either device-level or use the ideal switch model based on two-state resistors. Hitherto, the detailed model including its Thévenin or Norton equivalents is the most popular type in terms of MMC

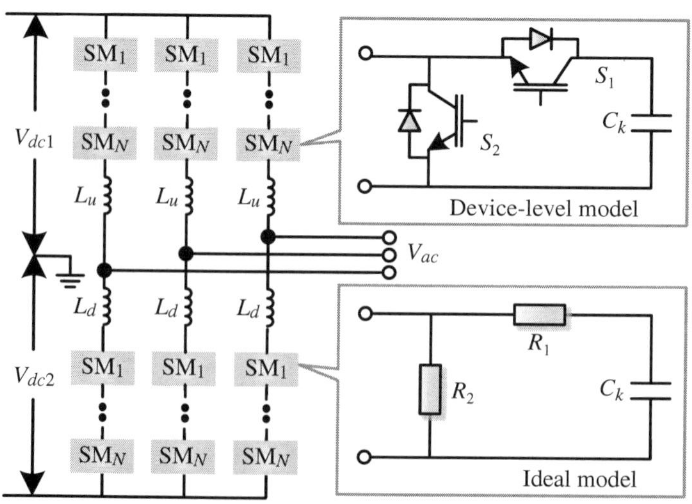

Figure 8.1 MMC configuration and its half-bridge submodule models.

modeling and EMT simulation [160]. It certainly is more detailed than the AVM, which to the utmost is merely able to provide harmonics and the submodule capacitor voltages that barely fit the requirement of EMT simulation. However, when it comes to simulating the DC line fault which is one of the common contingencies in an actual grid, the challenge shows up, i.e. the AC grid inappropriately halts feeding electricity to the low-impedance DC side after the blocking order from the secondary protection devices is executed because both models fall short of the diode freewheeling feature. Though technically, it can be solved by introducing the converter-level performance, it is less convincing since the MMC performance at this stage is based on the analysis of the overall converter and a desired operation status is always assumed, which is why we need a device-level to reveal the fault mechanism from a more specific level involving every individual power semiconductor switch.

In addition, using power semiconductor device-level models for converter EMT simulation addresses the growing emphasis on power efficiency in modern power systems by being able to provide abundant detailed information of each submodule including conduction and switching power losses, junction temperatures, and linearized or natural switching transient waveforms of IGBT modules, which are important indicators for converter design guide and assessment, including but not limited to

1. evaluating the power converter efficiency under various control methods;
2. tuning control parameters such as switching frequency based on converter efficiency and thermal requirements;
3. choosing appropriate current rating for IGBT modules and design qualified heat sink based on thermal requirements;
4. obtaining a more accurate fast Fourier transform (FFT) analysis for MMC voltage and current waveforms when considering nonideal switching characteristics.

Therefore, in this chapter, all three models of the MMC are introduced and compared. Details of the one based on device-level model are specifically given, along with the hardware architecture design on the FPGA and subsequent emulation results.

8.2 Detailed Model

The detailed model describes the converter that is topologically identical to a real configuration, as has been demonstrated in Figure 8.1, where each IGBT/diode pair is taken as a gate-pulse controlled two-state resistor, i.e. the

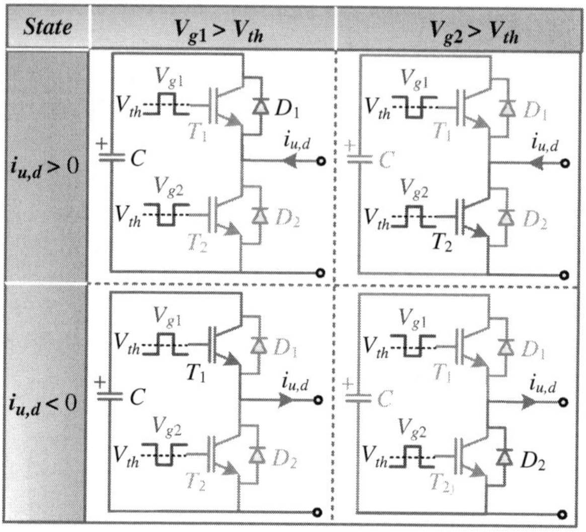

Figure 8.2 MMC submodule operation states.

power semiconductor switch is either under on-state or off-state with two distinct fixed resistances, and the transition between the two states is completed instantly.

In the HBSM, when the upper switch is under on-state, the DC capacitor is either charging or discharging, meaning the terminal voltage in this case is equal to the capacitor voltage; otherwise, the submodule is bypassed and the submodule terminal voltage is 0, as given in Figure 8.2. Therefore, summing all the submodule outputs yields the multiple voltage levels.

8.2.1 Detailed Equivalent Circuit Model

The basic one-step integration method determines that the reactive components can be converted into the Thévenin or Norton equivalences, meaning that circuit simplification by merging all parallel or series components turn out to be possible, which is exactly the case in the MMC detailed equivalent model. Using any one-step integration methods such as the Trapezoidal rule or Backward Euler method, the Thévenin equivalent circuit of a submodule can be expressed by the following set of two equations:

$$R_{eq} = \frac{R_1 R_2 + R_2 Z_{Ck}}{R_1 + R_2 + Z_{Ck}}, \tag{8.1}$$

$$V_{eq}(t) = \frac{R_2 V_{Ceqk}(t)}{R_1 + R_2 + Z_{Ck}}, \tag{8.2}$$

where Z_{Ck} and V_{Ceqk} compose the Thévenin equivalent circuit of the submodule capacitor, R_1 and R_2 are resistances of the two complementary switches. Therefore, the N submodules in an arm can be summarized, and to further eliminate internal node caused by the arm inductor, the entire arm is merged, leading to

$$R_{arm\_eq} = Z_{Lu,d} + \sum_{i=1}^{N} R_{eq(i)}, \tag{8.3}$$

$$V_{arm\_eq}(t) = V_{Leq}(t) + \sum_{i=1}^{N} V_{eq}(t), \tag{8.4}$$

where $Z_{Lu,d}$ and $V_{Leq}(t)$ constitutes the Thévenin equivalent circuit of the arm inductor. For the convenience of solving the nodal voltages as a vector in EMT computation, it is not difficult to convert the above two equations into the Norton equivalent circuit form:

$$G_{arm\_eq} = \frac{1}{Z_{Lu,d} + \sum_{i=1}^{N} R_{eq(i)}}, \tag{8.5}$$

$$J_{arm\_eq}(t) = G_{arm\_eq}\left(V_{Leq}(t) + \sum_{i=1}^{N} V_{eq}(t)\right). \tag{8.6}$$

8.3 Equivalenced Device-Level Model

Figure 8.3a shows the SM structure of a two-level half-bridge topology consisting of an upper IGBT S_1, an upper diode D_1, a lower IGBT S_2, a lower diode D_2, and an energy-storing capacitor C. As mentioned in Figure 8.2, with different combinations of gate signals $V_{g1}(t)$ and $V_{g2}(t)$, the SM has on-state, off-state, blocking, and the short-circuit mode if both power semiconductor switches are turned on, which is not allowed due to its potential damage to the devices and even the entire

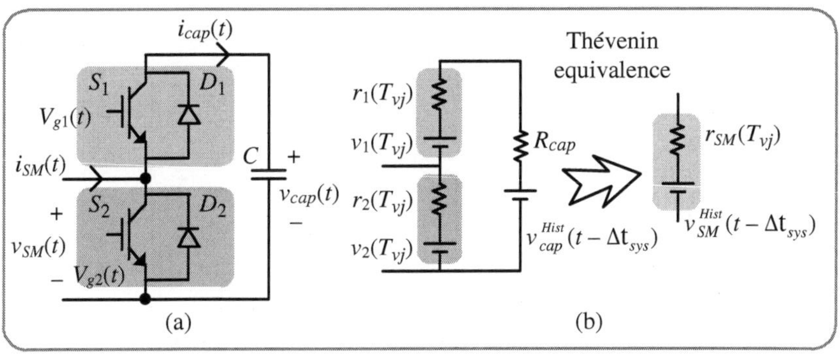

Figure 8.3 The sub-module: (a) two-level half bridge topology, (b) temperature-dependent modeling and Thévenin equivalence.

MMC system. In off-state, current goes through either S_2 or D_2 without affecting the capacitor voltage, while in on-state, the capacitor is charging or discharging according to the direction of the SM current $i_{SM}(t)$. Other than fault, blocking mode also appears in dead-time period when the gate signals are retrieved from both switches.

As there are three nodes in the HBSM, node reduction is carried out, as demonstrated in Figure 8.3b, where the linear static curve-fitting model using series r–v circuit introduced in Chapter 7 is adopted. The capacitor voltage $v_{cap}(t)$ of the SM is derived using Trapezoidal integration rule with the history value $v_{cap}^{Hist}(t-\Delta t_{sys})$ expressed as follows:

$$v_{cap}(t) = (2-\alpha)R_{cap}i_{cap}(t) + v_{cap}^{Hist}(t-\Delta t_{sys}), \tag{8.7}$$

where

$$R_{cap} = \frac{\Delta t_{sys}}{2C}, \quad \text{and} \tag{8.8}$$

$$v_{cap}^{Hist}(t-\Delta t_{sys}) = \alpha R_{cap}i_{cap}(t-\Delta t_{sys}) + v_{cap}(t-\Delta t_{sys}), \tag{8.9}$$

$$\alpha = 2\left(\frac{t_{gate} - t_{comm} - (t-\Delta t_{sys})}{\Delta t_{sys}}\right), \quad (0 \leq \alpha \leq 2). \tag{8.10}$$

C is the capacitance of the SM capacitor; $i_{cap}(t)$ is the current through the capacitor; Δt_{sys} is the system-level time-step; t_{gate} is the exact time when the emulator receives the updated gate signals; t_{comm} is the communication delay of gate signals from the controller to the real-time emulator. The coefficient α is introduced for the purpose of reducing the capacitor voltage error when gate signals are changing, while, for the rest of the time, α shall be 1 for the Trapezoidal integration rule. This coefficient is only necessary when gate signal sampling time-step is significantly smaller than the system-level time-step and the communication delay cannot be neglected. In this section, the circuit and its controller are designed on the same FPGA board, and the communication delay can be neglected since it is only one clock cycle or 10 ns. Moreover, the gate signal sampling rate is the same as that of system-level computation, and therefore, α is always set as 1.

Using Thévenin equivalence, the circuit model for the SM is represented by $r_{SM}(T_{vj})$ and $v_{SM}^{Hist}(t-\Delta t_{sys})$ in series, and the SM output voltage $v_{SM}(t)$ is given as follows:

$$v_{SM}(t) = r_{SM}(T_{vj})i_{SM}(t) + v_{SM}^{Hist}(t-\Delta t_{sys}), \tag{8.11}$$

where

$$r_{SM}(T_{vj}) = \frac{r_2(T_{vj})(r_1(T_{vj}) + R_{cap})}{r_1(T_{vj}) + r_2(T_{vj}) + R_{cap}}, \tag{8.12}$$

$$v_{SM}^{Hist}(t-\Delta t_{sys}) = \frac{r_1(T_{vj})+R_{cap}}{r_1(T_{vj})+r_2(T_{vj})+R_{cap}} v_2(T_{vj}) + \frac{r_2(T_{vj})}{r_1(T_{vj})+r_2(T_{vj})+R_{cap}} \left(v_{cap}^{Hist}(t-\Delta t_{sys}) - v_1(T_{vj})\right). \tag{8.13}$$

Similarly, Thévenin equivalence for all SMs of the MMC are processed in parallel on the FPGA. $r_{arm}(T_{vj})$ and $v_{arm}^{Hist}(t-\Delta t_{sys})$ representing all SMs in one converter arm are the interface circuit elements to the system-level calculation, which are given as follows:

$$r_{arm}(T_{vj}) = \sum_{k=1}^{n} r_{SM}^{k}(T_{vj}), \tag{8.14}$$

$$v_{arm}^{Hist}(t-\Delta t_{sys}) = \sum_{k=1}^{n} v_{SM}^{k,Hist}(t-\Delta t_{sys}), \tag{8.15}$$

where n is the number of SMs in one converter arm. The dynamics of the other linear passive elements can also be discretized by Trapezoidal rule. After solving the nodal equations for the circuit of entire MMC system, $i_{SM}(t)$, which is the same as $i_{arm}(t)$, is known. $i_{cap}(t)$ for each SM capacitor is updated as follows:

$$i_{cap}(t) = \frac{r_2(T_{vj})i_{SM}(t)+v_1(T_{vj})+v_2(T_{vj})-v_{cap}^{Hist}(t-\Delta t_{sys})}{r_1(T_{vj})+r_2(T_{vj})+R_{cap}}. \tag{8.16}$$

Finally, a recursive equation is applied for the calculation of v_{cap}^{Hist}, given as follows:

$$v_{cap}^{Hist}(t) = 2R_{cap}i_{cap}(t) + v_{cap}^{Hist}(t-\Delta t_{sys}). \tag{8.17}$$

Equation (8.17) is applicable when the coefficient α for capacitor voltage error correction is 1, otherwise (8.7)–(8.10) are used to update $v_{cap}^{Hist}(t)$. At this point, a set of electrical circuit solution equations (8.7)–(8.17) and (7.84) and (7.85) in Chapter 7) for the SM are established.

8.3.1 Power Loss Calculation

After the system-level calculation and the update of capacitor voltages, the power losses of the IGBTs and the diodes in the SMs can be calculated for each system-level time-step. The power losses calculation method used in this chapter is the same in [161]. Power losses for IGBT $P_{loss}^{IGBT}(t)$ are mainly composed of conduction power losses $P_{cond}^{IGBT}(t)$, turn-on power losses $P_{on}^{IGBT}(t)$, and turn-off power losses $P_{off}^{IGBT}(t)$, while the power losses for diode $P_{loss}^{Diode}(t)$ are mainly from conduction power losses $P_{cond}^{Diode}(t)$ and reverse recovery power losses $P_{rr}^{Diode}(t)$.

The equations of conduction power losses for the IGBT and the diode during a system-level simulation time-step Δt_{sys} are given as follows:

$$P_{cond}^{IGBT}(t) = \left(r_{on}^{IGBT}(T_{vj})i_C(t) + v_{on}^{IGBT}(T_{vj})\right) i_C(t), \tag{8.18}$$

$$P_{cond}^{Diode}(t) = \left(r_{on}^{Diode}(T_{vj})i_F(t) + v_{on}^{Diode}(T_{vj})\right) i_F(t), \tag{8.19}$$

where $i_C(t)$ is the collector current in IGBT and $i_F(t)$ is the forward current in diode, which can be determined based on $i_{SM}(t)$ and switching condition. $r_{on}^{IGBT}(T_{vj})$, $v_{on}^{IGBT}(T_{vj})$, $r_{on}^{Diode}(T_{vj})$, and $v_{on}^{Diode}(T_{vj})$ are slope resistances and threshold voltages of either upper or lower IGBT and diode pair. The IGBT switch-on energy E_{on}^{IGBT}, switch off energy E_{off}^{IGBT}, and the diode reverse recovery energy E_{rr}^{IGBT} as functions of the current $i_C(t)$ or $i_F(t)$, when junction temperature T_{vj} is 125 °C and the voltage across the switch in off-state is the rated value v_{rated} (1800 V), are provided in the datasheet [162]. Second-order polynomials are adopted here to fit these curves with the T_{vj} of 125 °C (T_2), given as follows:

$$E_{sw}^{T_2}(i(t), v(t)) = (a \cdot i^2(t) + b \cdot i(t) + c) \cdot \frac{v(t)}{v_{rated}}, \tag{8.20}$$

where $E_{sw}^{T_2}$ represents E_{on}^{IGBT,T_2}, E_{off}^{IGBT,T_2} or E_{rr}^{Diode,T_2}, $i(t)$ represents $i_C(t)$ or $i_F(t)$, $v(t)$ represents the voltage across the switch in off-state.

Based on the datasheet information, the fitting parameters are presented in Table 8.1, and the comparison between fitting curves and original curves from datasheet is shown in Figure 8.4. However, besides the loss curves at 125 °C, the datasheet only provides the switching energy losses at the rated value (400 A) and 25 °C. Based on the assumption that all points with different currents in the switching loss curves at 25 °C (T_1) follow the proportional relationship of the rated current case, and the curves at other junction temperature are linearly distributed, the energy estimation equations can be derived as follows:

$$E_{sw}(T_{vj}, i(t), v(t)) = \frac{T_2 - T_{vj}}{T_2 - T_1}\left(E_{sw}^{T_1}(i(t), v(t)) - E_{sw}^{T_2}(i(t), v(t))\right) + E_{sw}^{T_2}(i(t), v(t)), \tag{8.21}$$

Table 8.1 Extracted a, b, and c for switching energy curve fitting.

$E_{sw}^{T_2}(T_2 = 125\ °C)$	a	b	c
E_{on}^{IGBT,T_2}	0.002575	1.478	179.7
E_{off}^{IGBT,T_2}	0.0003982	1.209	58.23
E_{rr}^{Diode,T_2}	−0.00068631	1.075	177.2

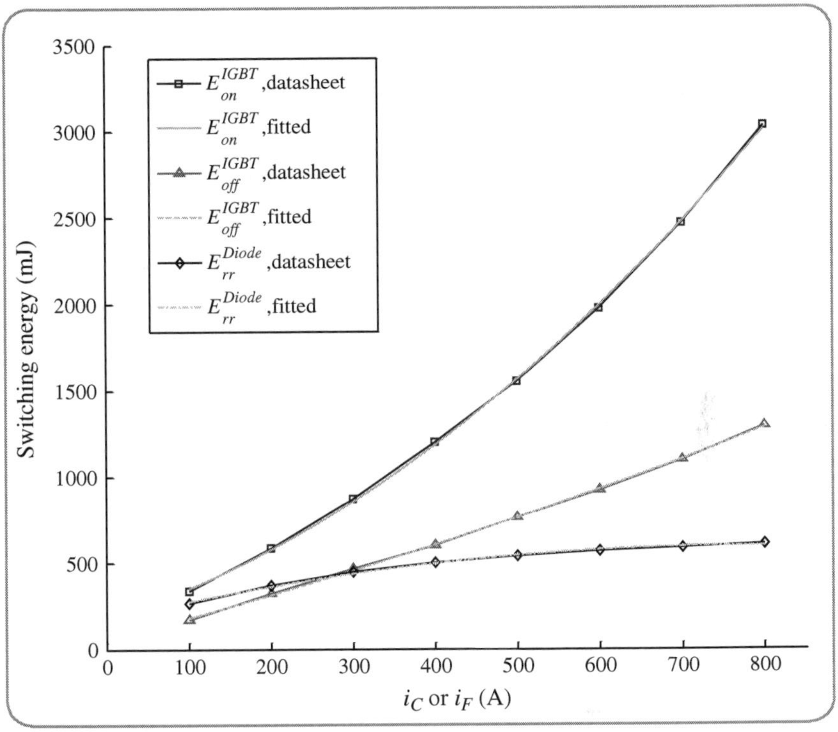

Figure 8.4 Switching energy losses at the T_{vj} of 125 °C from fitted equations and datasheet.

where

$$E_{sw}^{T_1}(i(t), v(t)) = \frac{E_{sw}^{T_1,rated}}{E_{sw}^{T_2,rated}} E_{sw}^{T_2}(i(t), v(t)). \tag{8.22}$$

$E_{sw}^{T_1,rated}$, and $E_{sw}^{T_2,rated}$ are the energy losses at rated test condition at T_1 and T_2. The averaged switching power loss in one system-level time-step is given as follows:

$$P_{sw}(T_{vj}, i(t), v(t)) = \frac{E_{sw}(T_{vj}, i(t), v(t))}{\Delta t_{sys}}. \tag{8.23}$$

8.3.2 Thermal Network Calculation

The calculated power losses now become the input of the thermal network to compute the junction temperatures for all IGBTs and diodes in the SM. Partial fraction thermal circuit model composed of multiple levels of thermal resistor and

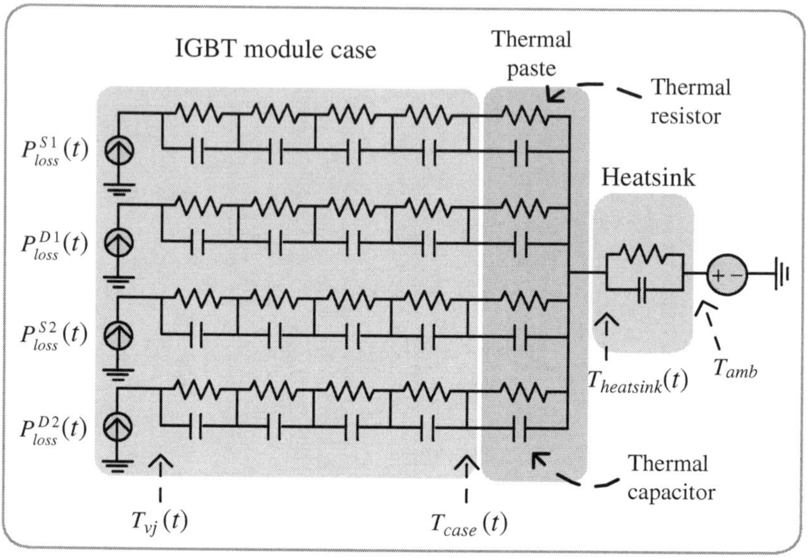

Figure 8.5 Thermal circuit structure.

capacitor pairs are employed to compute the junction temperature, as shown in Figure 8.5. The junction temperature $T_{vj}(t)$ can be calculated as follows:

$$T_{vj}(t) = P_{loss}(t) \cdot (Z_{thjc} + Z_{thch}) + P_{total}(t) \cdot Z_{thha} + T_{amb}, \tag{8.24}$$

where $P_{loss}(t)$ is the power loss for a single IGBT or diode; $P_{total}(t)$ is the power loss for all devices mounted on the same heatsink; T_{amb} is the ambient temperature which is fixed in the emulation; Z_{thjc}, Z_{thch}, and Z_{thha} are the thermal impedances from junction to case, case to heatsink, and heatsink to ambient, respectively. In this work, we consider that two IGBT modules are mounted on the same 10 K/kW water-cooled heatsink. The thermal impedance Z_{thjc} and thermal resistor R_{thch} from case to heatsink are given by the datasheet. All the thermal impedance parameters for this work are shown in Table 8.2 in the form of thermal resistances and time constants.

Applying Trapezoidal rule, the numerical equation to solve for the junction temperature for either the IGBT or the diode is given as follows:

$$\begin{aligned} T_{vj}(t) = \sum_{i=1}^{6} \Delta T_{th}^{i}(t) + T_{amb} = \sum_{i=1}^{5} (\alpha_i (P_{loss}(t) + P_{loss}(t - \Delta t_{thm})) \\ + \beta_i \Delta T_{th}^{i}(t - \Delta t_{thm})) + \alpha_6 (P_{total}(t) + P_{total}(t - \Delta t_{thm})) \\ + \beta_6 \Delta T_{th}^{6}(t - \Delta t_{thm}) + T_{amb}, \end{aligned} \tag{8.25}$$

Table 8.2 Impedances for thermal network.

	Z_{thjc}				Z_{thch}	Z_{thha}
Thermal impedance	$i = 1$	$i = 2$	$i = 3$	$i = 4$	$i = 5$	$i = 6$
$R_{th}^{i,IGBT}$ (K/kW)	11.475	6.375	1.53	6.12	24	10
$\tau_{th}^{i,IGBT}$ (s)	0.03	0.1	0.3	1	3	45
$R_{th}^{i,Diode}$ (K/kW)	22.95	12.75	3.06	12.24	48	10
$\tau_{th}^{i,Diode}$ (s)	0.03	0.1	0.3	1	3	45

where

$$\alpha_i = \frac{R_{th}^i \cdot \Delta t_{thm}}{2\tau_{th}^i + \Delta t_{thm}}, \tag{8.26}$$

$$\beta_i = \frac{2\tau_{th}^i - \Delta t_{thm}}{2\tau_{th}^i + \Delta t_{thm}}, \quad i = 1, 2, \dots, 6. \tag{8.27}$$

Δt_{thm} is the thermal time-step, which has the same value as Δt_{sys} in this work. Considering the fact that thermal time constant is much longer than the MMC system emulation time-step, the thermal network can have a separate time-step that could be set larger without a significant impact on accuracy.

8.3.3 Hardware Emulation of SM Model on FPGA

For each SM in the MMC system, a dedicated SM hardware unit such as Figure 8.6 emulated on the FPGA consists of operators, the FSM, and five hardware subunits: the electrical model hardware subunit, the power loss hardware subunit, the thermal network hardware subunit, the temperature dependent parameter update hardware subunit, and the device-level waveform generation hardware subunit. All SM hardware units run simultaneously, which means the computation time for SM hardware units of the MMC will not increase with the number of SMs.

In Figure 8.6, the signal flow routes and the execution of hardware subunits follows the algorithm described in Section 8.3.2. For instance, the power loss hardware subunit requires the gate signals $g_1(t)$, $g_2(t)$ and the direction of SM current $i_{SM}(t)$ from the MMC controller and the system network solution to figure out the SM operation mode, which determines the conditions of all switching devices (switch-on, switch-off, conduction or off). Then, $r_{on}(T_{vj})$, $v_{on}(T_{vj})$, $T_{vj}(t)$ of the corresponding devices, $v_{cap}(t)$, $i_{SM}(t)$ and other constants are substituted to (8.18)–(8.23), completing the power loss calculation.

The finite state machine controls the execution sequence for other hardware subunits in the SM hardware unit and receives the control signals from the FSM

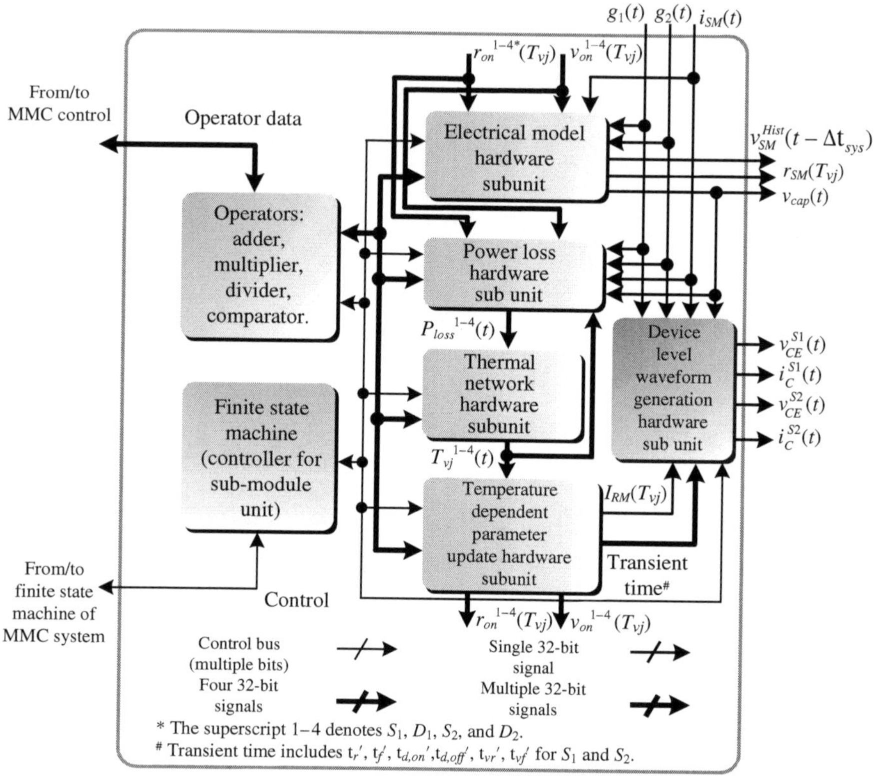

Figure 8.6 Submodule hardware unit emulation on FPGA.

of the entire MMC emulation hardware. Figure 8.7 shows the periodic state chart for the SM hardware unit, which begins from S_1, the start of a new emulation time-step. After the gate signals are received, the Stage 1 of the SM hardware unit is activated by first calculating $v_{SM}^{Hist}(t-\Delta t_{sys})$ and $r_{SM}(T_{vj})$, which are then used for system network calculation. After $i_{SM}(t)$ is received from the system network solution, Stage 2 starts from S_4 to S_7, corresponding to the four hardware subunits in the middle column of Figure 8.6: electrical model, power loss, thermal network, temperature-dependent parameter update hardware subunits. When all the calculations are finished, the FSM return to S_0 to wait for the arrival of next time-step.

Because of the sequential relation among the hardware subunits except for the device-level waveform generation hardware subunit, they share the same set of floating-point operators with pipelined structure and multiplexed input registers in this design, including one adder (3 clocks), one multiplier (3 clocks), one divider (9 clocks), and one comparator (2 clocks). The calculation times for the subunits

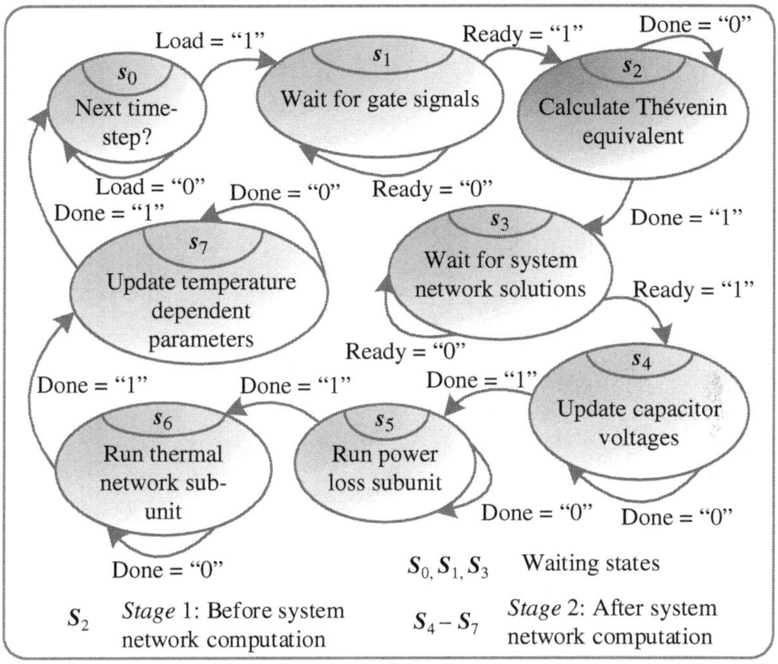

Figure 8.7 Finite state machine for SM hardware unit.

Table 8.3 Latencies of various subunits in the SM hardware unit.

Subunit	Total computation time with one set of operators (clock cycles)
Stage 1 electrical models	27
Stage 2 electrical models	18
Power loss	38
Thermal network	72
Parameters update	18
SM hardware unit total clock cycles	173
Percentage in 1000 clock cycles (%)	17.3

with one set of operators and the percentage of total computation time for the SM hardware unit are listed in Table 8.3. Since the FPGA runs at the clock frequency of 100 MHz, and system-level time-step is 10 μs, there are 1000 clock cycles available for a complete computation for one system-level time-step. Using more operators can decrease the delay significantly due to the parallelism existing in those hardware subunits, which requires more FPGA resources.

The device-level waveform generation hardware subunit generates the voltage and current waveforms of the upper and lower IGBTs, $v_{CE}^{S1}(t)$, $i_C^{S1}(t)$, $v_{CE}^{S2}(t)$, and $i_C^{S2}(t)$, with linearized transients continuously at the time-step of 10 ns, which means the output waveforms will update at each FPGA clock cycle. The received voltage and current will be first converted into fixed-point format, since fixed-point adders are much faster than those of floating-point. The fixed-point operators will add or substrate a specific value to the voltage and current during a specific time period determined by temperature dependent transient times.

8.3.4 MMC System Hardware Emulation

The MMC system hardware emulation and state chart on FPGA are shown in Figures 8.8 and 8.9, respectively. The finite state machine controls the entire emulation flow by interacting with the state machines of other hardware units through the control bus. It activates different hardware units according to the sequence

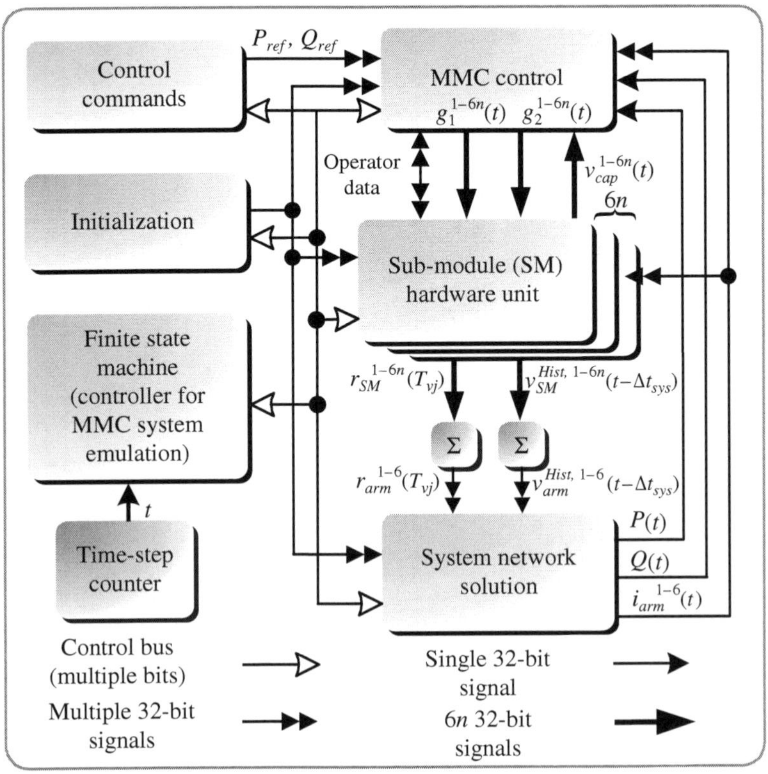

Figure 8.8 MMC system hardware emulation.

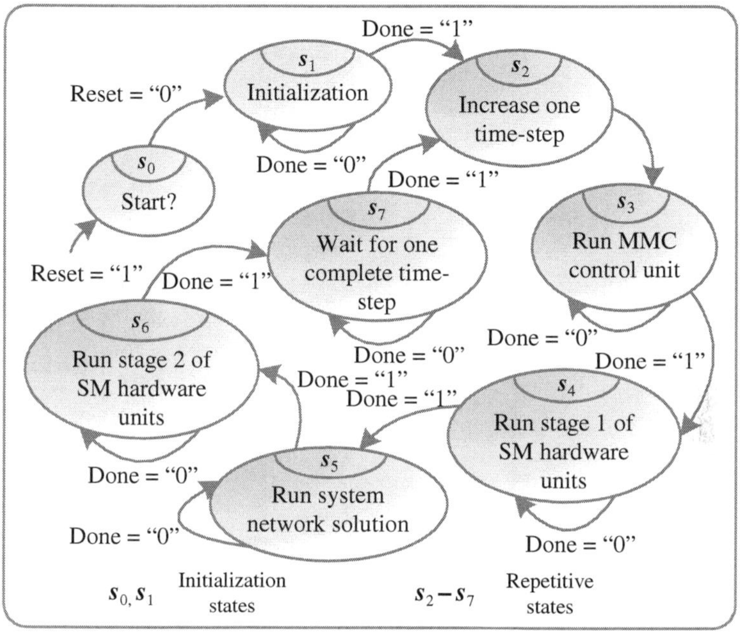

Figure 8.9 Finite state machine for the MMC system.

shown in the state chart. Following a definitive reset order, the FPGA-based emulator goes into the initialization state S_1 through S_0. During the initialization, some constants which will be periodically used in later computation, such as α_i in (8.26), are calculated, and the matrix simplification requested by system network solution is also done. Then the time-step counter starts to work, providing the exact time t of the real-time emulation to the FSM, which is used to determine the arrival moment of next time-step in S_2.

In S_3, the MMC control unit receives the active and reactive commands from the control commands unit as well as the data from the SM hardware units and the system network solution unit, and then generates the gate signals to SM hardware units. In practice, the MMC control unit and the control commands unit are the devices under test during the real-time emulation. In this work, they are integrated onto the FPGA board in the demonstration design. Since the pulse width modulation (PWM) method based on phase-shift control (PSC) strategy generates drive signals for each SM, the process requires $6n$ sets of operators to ensure full parallelism. In this design, the MMC control unit and the SM hardware units share the same sets of operators, indicated by the bidirectional operator data bus. The detailed internal structure and algorithm of MMC control unit and SM hardware units are presented in Sections 8.3.1–8.3.4. State S_4 involves the Stage

1 of SM hardware units and the summation for $r_{SM}^{1-6n}(T_{vj})$ and $v_{SM}^{Hist,1-6n}(t - \Delta t_{sys})$. For clarity, two summation units are drawn in Figure 8.8. However, they are multiplexed in the actual design to save resources without affecting the speed, since the calculation of $v_{SM}^{Hist,1-6n}(t - \Delta t_{sys})$ involves more steps and are completed later than $r_{SM}^{1-6n}(T_{vj})$. The system network solution unit receives the interface components $r_{arm}(T_{vj})$ and $v_{arm}^{Hist}(t - \Delta t_{sys})$ for the six arms from Stage 1 of SM hardware units and then solves the nodal equations for the entire circuit topology in S_5. The outcome $i_{arm}^{1-6}(t)$ are then used to complete the Stage 2 of the SM hardware units in S_6, including the power loss and thermal network calculation. When the moment of next time-step arrives, the finite state machine changes from S_7 to S_2, starting a new emulation cycle.

8.3.5 Real-Time Emulation Results

8.3.5.1 Test Circuit and Hardware Resource Utilization

Two case studies for MMC systems: a single-phase five-level MMC and a three-phase nine-level MMC, were emulated on the Xilinx® Virtex-7® XC7VX485T FPGA (Table 1.2) at the clock frequency of 100 MHz to validate both system-level and device-level results. The three-phase test case has the same topology of Figure 8.1, while the single-phase test case is connected to a passive load instead of the grid. The circuit parameters for the two cases are given in Appendix A.7. The emulation results were captured by a four-channel oscilloscope connected to the 16-bit 4-channel digital-to-analog converter, which received the data from the FPGA. The major resource utilization summary is presented in Table 8.4, and the percentage data are indicated for Virtex-7 XC7VX485T and XC7V2000T FPGAs, respectively. Since the designs only use distributed registers as memory units, the BRAM utilization is 0. The 17-level and 25-level MMCs were synthesized on the Xilinx® Virtex-7® XC7V2000T FPGA [12], which has more logic resources (4× more logic cells and CLB slices, and 2.6× more distributed RAM) than the XC7V485T FPGA used for demonstration. Figure 8.10 shows the exact number of operators, including adders, multipliers, and dividers, utilized for MMC emulation hardware with different levels. The FPGA designs which only contain SM hardware units and summation units by the electrothermal model and the Thévenin equivalence method with two-state resistor model for IGBT modules were synthesized on XC7V2000T for resource comparison purpose. The same FPGA design methodology was applied for both cases, which is using one set of operators for each SM. When using two-state resistor model for IGBT modules, only one adder and one multiplier are required for each SM. Sharing the same set of operators among multiple SM hardware units reduces the resource consumption; however, can increase the time delay, which is not adopted in this work. The maximum SM number accommodated on XC7V2000T device for the

Table 8.4 Hardware resource utilization for three-phase MMC.

Device	Levels	SM numbers	Registers	LUTs	DSP slices
XC7VX485T	5	24	180 447(28%)	157 735(51%)	130(4%)
	9	48	315 931(52%)	294 554(97%)	236(8%)
XC7V2000T	17	96	686 767(28%)	639 887(52%)	420(19%)
	25	144	1 020 845(41%)	991 004(81%)	612(28%)

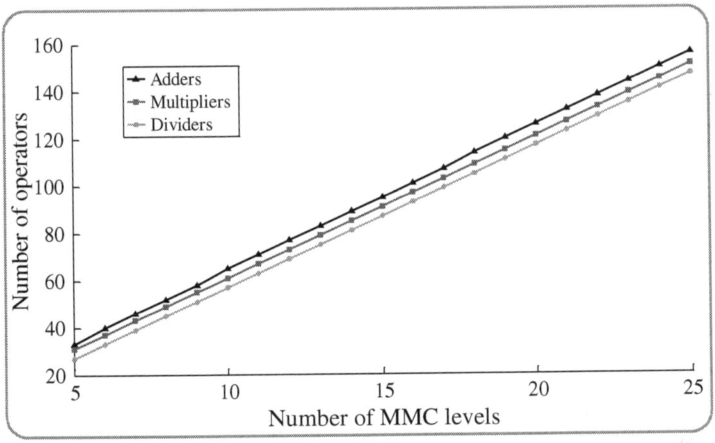

Figure 8.10 Operator count for the MMC system with increasing number of levels.

electrothermal model and two-state resistor model are 276 and 1104, respectively. Under normal operation, the power losses and junction temperatures are in the same level for different SMs, verified by Table 8.5, therefore, emulating all SMs by the electro-thermal model is not necessary with limited FPGA resources. A combination of using the above two methods can greatly reduce the resource consumption. With these optimization and simplification schemes, the FPGA design can be very flexible to meet the resource usage and time delay requirements.

Moreover, according to Table 8.4 and Figure 8.10, the resource usage and the operator numbers for the electrothermal model increase almost linearly with the number of MMC levels, which indicates that with more advanced FPGA devices, such as the Xilinx® UltraScale series, or using multiple FPGA boards, MMCs with larger number of levels can be configured and emulated quite efficiently.

The off-line simulation tool SaberRD® was used to validate the real-time emulated results for the single-phase five-level MMC test case. Electrothermal behavioral IGBT and power diode model with parameters extracted from the datasheet were employed in the SaberRD® model with a variable time-step strategy.

Table 8.5 Average power dissipation and junction temperature of IGBTs and diodes in steady-state operation.

		S_1		D_1		S_2		D_2	
Arm	**Submodule**	P_{loss} (W)	T_{vj} (°C)	P_{loss} (W)	T_{vj} (°C)	P_{loss} (W)	T_{vj} (°C)	P_{loss} (W)	T_{vj} (°C)
Pos. arm	SM(1)	256.3	72.44	243.2	83.83	1425.1	130.28	51.1	64.81
	SM(2)	255.9	72.40	242.9	83.78	1424.2	130.23	50.1	64.68
	SM(3)	255.1	72.31	239.5	83.39	1424.3	130.19	49.1	64.53
	SM(4)	257.0	72.48	242.3	83.75	1425.8	130.33	51.5	64.86
Neg. arm	SM(1)	255.4	72.31	239.6	83.39	1424.7	130.20	49.1	64.54
	SM(2)	257.1	72.49	242.7	83.75	1426.3	130.33	51.5	64.86
	SM(3)	256.3	72.44	243.3	83.83	1425.0	130.30	51.1	64.81
	SM(4)	256.3	72.41	243.0	83.78	1424.3	130.24	50.3	64.70

8.3.5.2 Results and Comparison for Single-Phase Five-Level MMC

Figure 8.11a–e present the system-level results including the converter output voltage v_o, the load current i_L, the converter arm current i_p and i_n, and the capacitor voltage of all SMs in positive and negative converter arm $v_{cap,p}$ and $v_{cap,n}$. The FFT analysis was applied to v_o. Despite the line frequency, harmonics around 2 kHz are notable in both hardware emulation and SaberRD® results, which is expected, since the switching frequency is 500 Hz and each converter arm has four SMs. Figures 8.11f–i to 8.13 present the device-level results for the IGBTs and diodes in the first SM of the positive converter arm. Figure 8.11f,h shows the averaged power losses of S_1 and S_2 in each system-level time-step (P^{S1}_{loss} and P^{S2}_{loss}). The conduction power losses P^{S1}_{cond} and P^{S2}_{cond} are particular shown in Figure 8.11g,i, where switching power losses are removed for clarity. The junction temperature of S_1, S_2, D_1, and D_2 (T^{S1}_{vj}, T^{D1}_{vj}, T^{S2}_{vj} and T^{D2}_{vj}) in first five seconds and the zoomed waveforms during 0.5 second are shown in Figure 8.12a,b, respectively. Figure 8.13a,b shows the switching transients of S_2 during IGBT turn-off and turn-on (v_{CE} and i_C). Table 8.6 shows the averaged switching and conduction power losses between 0.175 and 0.225 second from SaberRD® and FPGA hardware emulation. The apparent errors exist for device-level results, since the solver and IGBT module model used in the emulation and SaberRD® are very different. The datasheet does not provide adequate data for switching power losses when the current through the IGBT module is low. The switching power losses from curve fitting in hardware emulation are higher than those from the off-line simulation results from SaberRD®, which can be observed from Figure 8.11e,f. This explains the largest power loss error occurs in the reverse recovery loss of D_2 in Table 8.6, where the smallest current goes through. The IGBT switching transient parameters are extracted based

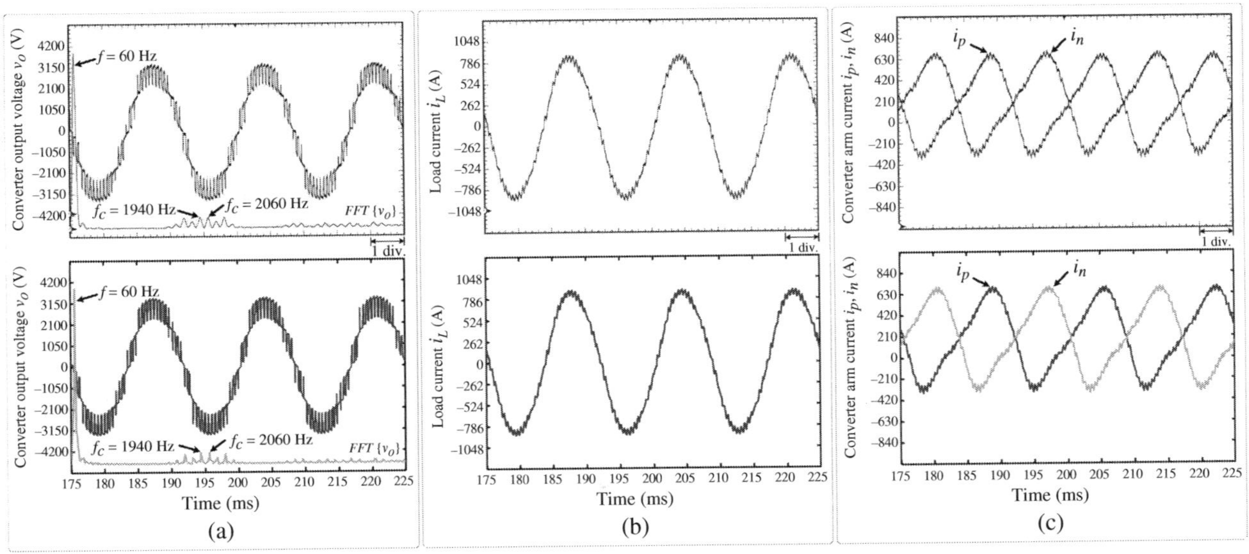

Figure 8.11 System-level and device-level power loss results for single-phase five-level MMC system from real-time hardware emulation (top in (a)–(i)) and off-line simulation by SaberRD® (bottom in (a)–(i)) at 500 Hz switching frequency. Scale: (a)–(i) x-axis: 5.0 ms/div.

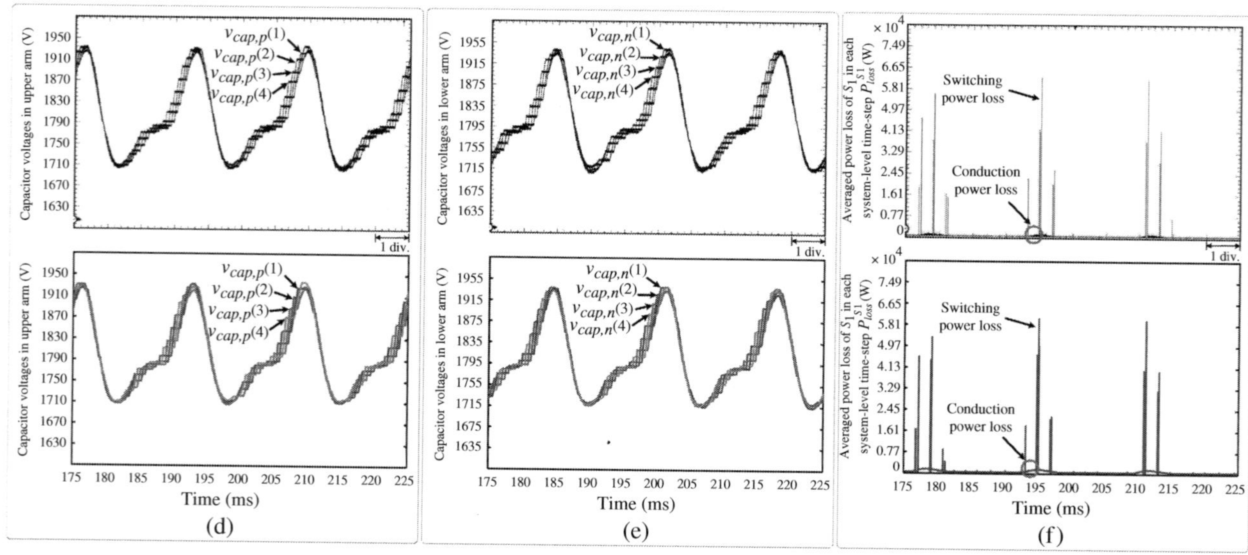

Figure 8.11 *(Continued)*

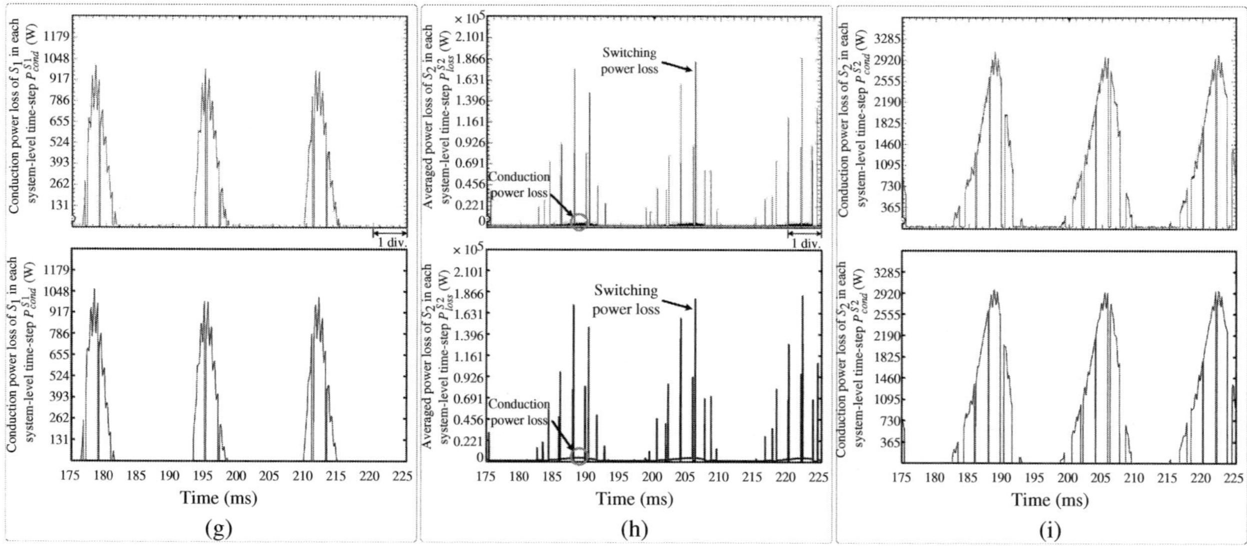

Figure 8.11 (*Continued*)

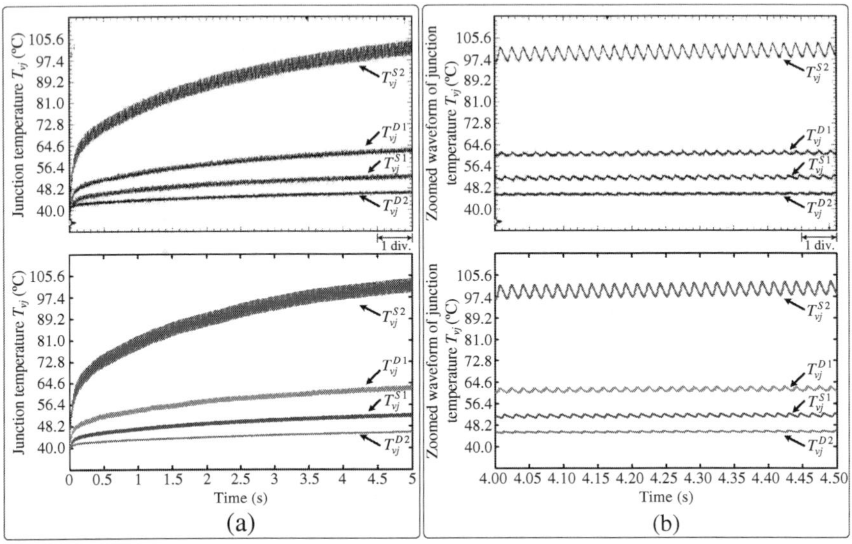

Figure 8.12 Junction temperature for single-phase five-level MMC system from real-time hardware emulation (top sub-figure) and off-line simulation by SaberRD® (bottom sub-figure). Scale: (a) x-axis: 0.5 s/div, (b) x-axis: 0.05 s/div.

on the power losses and the current rise and fall time under rated cases. The linear interpolation scheme used to estimate the nonlinear curves for other non-rated cases induced the errors. From the perspective of power loss matching, the transient time of nonlinear SaberRD® waveforms are slightly longer than the linear curves from hardware emulation. The diode reverse recovery process greatly expedites the turn-on process of the IGBT, therefore, t_r of IGBT from both the real-time and off-line simulation shown in Figure 8.13b is almost half the value in the datasheet, which is 0.4 μs. Since the datasheet does not provide the detailed information for the drive circuit, they are simply modeled as the ideal voltage sources in series with the gate resistors in SaberRD®'s model, which caused the differences for turn-on and turn-off delay times. Despite all of the above, the hardware emulation and SaberRD® results are quite close.

Table 8.5 also shows the average power dissipation and junction temperature of IGBTs and diodes for all SMs under steady-state operation. These data are collected from the real-time emulation after 369 seconds, which is more than eight times longer than the largest thermal time constant (45 seconds). The power losses and junction temperatures of the corresponding devices in different SMs are very close to each other, which verifies the advantage of even usage of SMs for PSC–PWM control scheme.

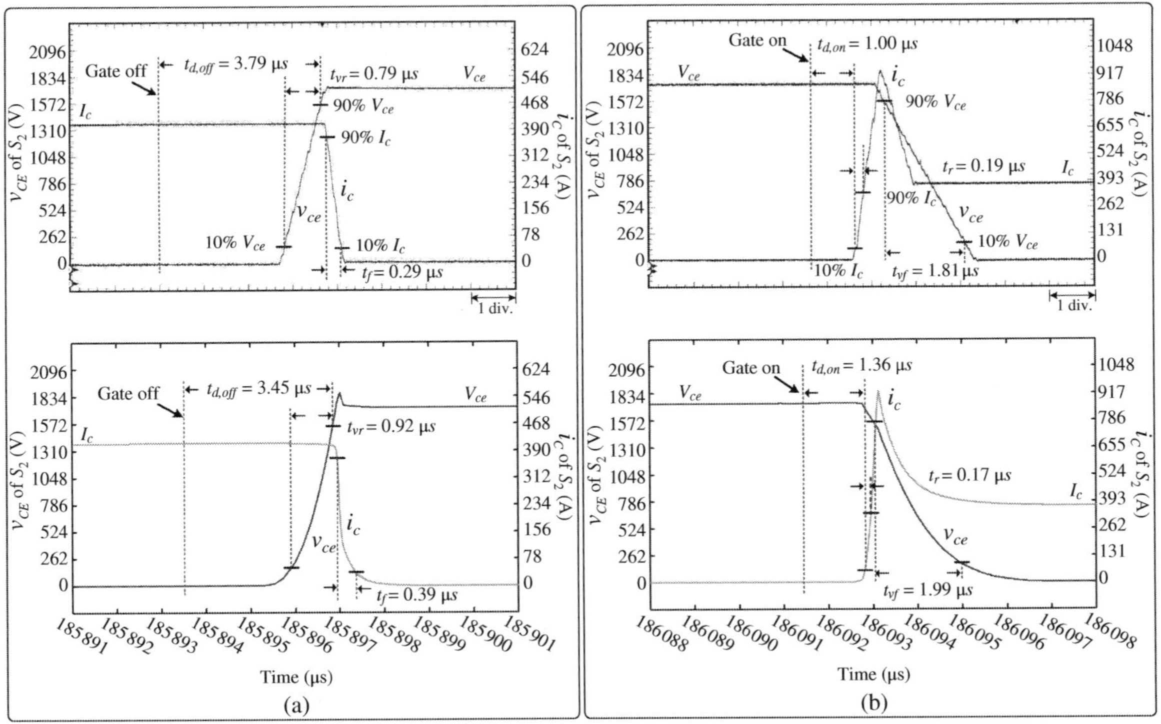

Figure 8.13 Device-level switching waveforms for IGBTs in single-phase five-level MMC system from real-time hardware emulation (top sub-figure) and off-line simulation by SaberRD® (bottom sub-figure). Scale: (a) and (b) *x*-axis: 1 μs/div.

Table 8.6 Comparison of power dissipations of IGBTs and diodes between SaberRD® and real-time emulation.

Power dissipation	SaberRD® (W)	FPGA (W)	Error (%)
P_{on}^{S1}	60.6	61.2	1.0
P_{off}^{S1}	41.7	42.2	1.2
P_{cond}^{S1}	137.3	134.4	2.1
P_{rr}^{D1}	72.2	74.3	2.9
P_{cond}^{D1}	136.3	130.3	4.4
P_{on}^{S2}	293.7	280.1	4.6
P_{off}^{S2}	168.1	163.6	2.7
P_{cond}^{S2}	722.0	723.5	0.2
P_{rr}^{D2}	24.7	26.8	8.5
P_{cond}^{D2}	13.5	14.0	3.7

8.3.5.3 Results for Three-Phase Nine-Level MMC

The emulated three-phase nine-level MMC is the largest circuit which can be configured in the Virtex-7® VC707 board with XC7VX485T FPGA whose resources are listed in Table 1.2. Figure 8.14a,b shows the converter output phase voltages (v_o^a, v_o^b, and v_o^c) during 50ms and the zoomed waveforms during 20 ms, while Figure 8.14c shows the output line current (i_a, i_b, and i_c). Figure 8.14d shows the active and reactive power tracking performance, with the rated power of 7 MW, and Figure 8.14e shows the junction temperature transients of the switching devices in the first positive arm SM of phase A during the same time period. Figure 8.14f shows the steady-state junction temperatures with 1.0 p.u. active power and −0.2 p.u. reactive power after 369 seconds.

8.4 Virtual-Line-Partitioned Device-Level Models

While MMC arm aggregation is enabled by the linear curve-fitting IGBT/diode switch model in Section 8.3, it is not always feasible in device-level simulation of the modularized converter, e.g. the existence of transconductance in the nonlinear behavioral model deprives it of the discrete time domain equivalent circuit, not to mention merging the elements to yield the Norton equivalent circuit. Consequently, the many submodules connecting to each other introduce plenty of meshes and nodes, making direct solution of the converter impractical. Therefore, a more general method applicable to all MMCs regardless of the power

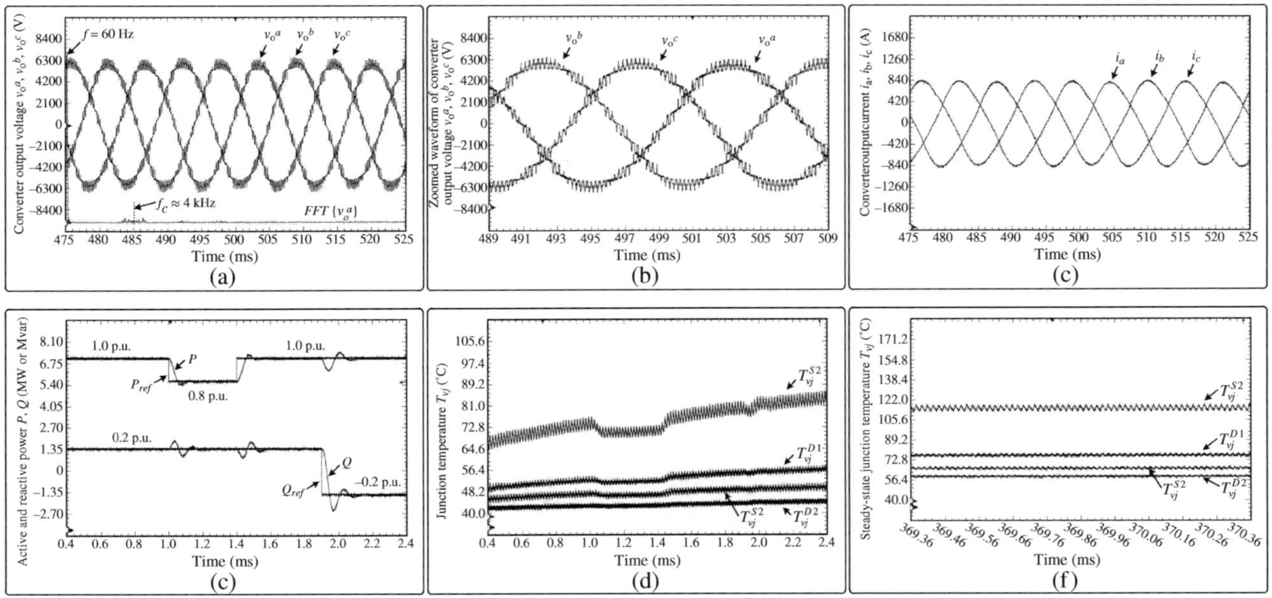

Figure 8.14 System-level and device-level results for three-phase nine-level MMC system from real-time hardware emulation (oscilloscope) at 500 Hz switching frequency. Scale: (a) *x*-axis: 5.0 ms/div, (b) *x*-axis: 2.0 ms/div, (c) *x*-axis: 5.0 ms/div, (d) and (e) *x*-axis: 0.2 s/div, (f) *x*-axis: 0.1 s/div.

semiconductor switch model is introduced in this section. Circuit partitioning is an effective method in reducing the corresponding matrix equation's dimension. Based on the fundamental principle that the section to be split should have a stiff voltage or current, and the fact that the complexity of MMC model is caused by its power semiconductor switches in the SM, the MMC arm turns out to be the ideal partitioning interface to create a group of independent subcircuits. Therefore, the original large admittance matrix for the MMC is split into a number of smaller matrices and parallel computation can be achieved on the FPGA to accelerate hardware-in-the-loop (HIL) emulation.

8.4.1 TLM-Link Partitioning

As the three-phase MMC is symmetrical, it is reasonable to carry out analysis based on one phase. Figure 8.15a shows the process of splitting a large MMC network consisting of a considerable number of nodes and meshes into several structurally independent, electrically related subcircuits. The arm inductor is first divided into $(N+1)$ parts which are redistributed so that a new inductor δL is connected in series with each submodule to constitute a two-port network, and consequently, the remaining inductance for the arm inductor is $L_{u,d} - N \cdot \delta L$. Then, these new inductors are replaced by transmission line modeling (TLM) links, discretization of which leads to the separation of submodules from the rest of the converter (MMC main circuit), enabling the replacement of the originally large impedance or admittance matrix by a number of submatrices with smaller dimensions, which, if processed in parallel on the FPGA, would be much more time and resource efficient.

The selection of the value of δL, which decides the characteristic impedance Z_0 and vice versa, plays a significant role on the emulation results. The principle, as stated, is an appropriate value of δL should ideally keep the arm inductor current identical to before its introduction [163]. Nevertheless, neighboring values resulting in a tiny change might also be allowed depending on the requirement of simulation accuracy. Thus, the optimum value can be selected from its range by running Matlab simulation of the MMC and comparing the current changes. It shows that the final value of δL is negligible compared with the arm inductance so that the latter can still be deemed as $L_{u,d}$, e.g. around 0.1%.

For the MMC main circuit, where the Norton equivalent circuit part of TLM link's hybrid model locates, by merging all the Norton circuits in the upper and lower arms, respectively, there is actually only one node since the potentials at all other three nodes are known, as shown in Figure 8.15b, the overall schematic for MMC. Then, the nodal voltage equation at nth time-step can be derived by applying KCL:

$$ {}_{n}\mathbf{V}_{s} = \mathbf{G}^{-1} \cdot {}_{n}\mathbf{I}, \tag{8.28} $$

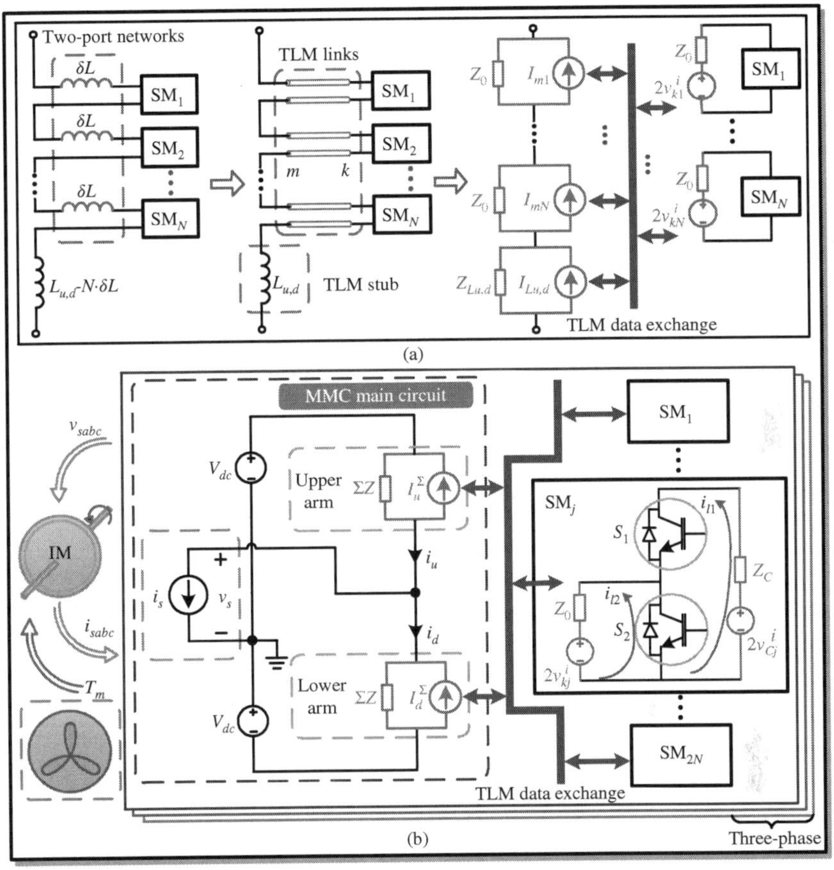

Figure 8.15 TLM-based model for $(N + 1)$-level MMC: (a) MMC partitioning approach, and (b) discretized schematic for the overall system.

where these 1×1 matrices are

$$\mathbf{G} = \left[\frac{2}{N \cdot Z_0 + Z_{Lu,d}}\right], \tag{8.29}$$

$${}_n\mathbf{I} = \left[\frac{{}_nJ_u^{\Sigma} - {}_nJ_d^{\Sigma}}{\Sigma Z} - {}_ni_s\right] = \left[2 \cdot \frac{\sum_{j=N+1}^{2N} {}_nv_{mj}^i + {}_nv_{Ld}^i - \left(\sum_{j=1}^{N} {}_nv_{mj}^i + {}_nv_{Lu}^i\right)}{N \cdot Z_0 + Z_{Lu,d}} - {}_ni_s\right]. \tag{8.30}$$

In (8.30), ΣZ, ${}_nI_u^{\Sigma}$, and ${}_nI_d^{\Sigma}$ are the impedance and current sources of the Norton equivalent circuitw voltage vector ${}_n\mathbf{V}_s$ is numerically equal to the stator voltage ${}_nv_s$. It should also be pointed out that all variables keep constant for a whole time-step. Prior to calculating reflected pulses for the next time-step, the upper and lower arm currents should be updated based on the obtained nodal voltages, as briefly expressed by

$$\begin{bmatrix} {}_ni_u \\ {}_ni_d \end{bmatrix} = \begin{bmatrix} \frac{1}{\Sigma Z} & \frac{-1}{\Sigma Z} \\ \frac{1}{\Sigma Z} & \frac{1}{\Sigma Z} \end{bmatrix} \cdot \begin{bmatrix} V_{dc} \\ {}_nv_s \end{bmatrix} + \begin{bmatrix} -{}_nI_u^{\Sigma} \\ -{}_nI_d^{\Sigma} \end{bmatrix}. \tag{8.31}$$

The 2N sub-circuits containing the Thévenin equivalent circuit part of the TLM link are identical. When the CFM is adopted for the IGBT/diode, all SMs have two meshes, so according to KVL, the mesh current equations under steady state can be written uniformly as

$$\begin{bmatrix} {}_ni_{l1} \\ {}_ni_{l2} \end{bmatrix} = \begin{bmatrix} Z_C + r_1 + r_2 & -r_2 \\ -r_2 & r_2 + Z_0 \end{bmatrix}^{-1} \begin{bmatrix} 2 \cdot {}_nv_{Cj}^i \\ -2 \cdot {}_nv_{kj}^i \end{bmatrix}, \tag{8.32}$$

where ${}_ni_{l1}$ and ${}_ni_{l2}$ are mesh currents, Z_C is the characteristic impedance of DC capacitor. On the other hand, when the transient stage takes place, the collector current can be expressed as a definitive percentage of the arm current, leading to

$${}_ni_{C1,2} = {}_nk \cdot |{}_ni_{u,d}|, \tag{8.33}$$

from which it is convenient to calculate the mesh currents by

$${}_ni_{l1} = {}_ni_{C1}, \tag{8.34}$$

$${}_ni_{l2} = {}_ni_{C1} - {}_ni_{C2}. \tag{8.35}$$

Then, the voltage across each Thévenin equivalent circuit, regardless of the stage, can be calculated by

$$\begin{bmatrix} {}_nv_{Cj} \\ {}_nv_{kj} \end{bmatrix} = \begin{bmatrix} -Z_C & 0 \\ 0 & Z_0 \end{bmatrix} \cdot \begin{bmatrix} {}_ni_{l1} \\ {}_ni_{l2} \end{bmatrix} + \begin{bmatrix} 2{}_nv_{Cj}^i \\ 2{}_nv_{kj}^i \end{bmatrix}. \tag{8.36}$$

Thus, calculation of reflected pulses can be carried out by substituting the acquired terminal voltages, and the time-step ends with updating the incident pulses.

8.4.2 Hardware Design on FPGA

In this section, the MMC with curve-fitting model is applied to drive an induction machine (IM). The hardware design on FPGA is first carried out, followed by real-time emulation results demonstration, analysis, and validation.

Table 8.7 Hardware utilization of the MMC-IM system.

FPGA	System	LUT	FF	DSP	f_{max} (MHz)
	MMC$_5$(3ph)	233 K (76.74%)	134 K (22.08%)	966 (34.50%)	116
XC7V-	MMC$_7$(1ph)	114 K (37.50%)	64 K (10.57%)	490 (17.50%)	115
X485T	MMC$_1$1(1ph)	168 K (55.38%)	109 K (18.03%)	901 (32.18%)	116
	MMC$_5$-IM	249 K (82.06%)	143 K (23.62%)	1155 (41.25%)	115
	MMC$_5$(3ph)	233 K (19.08%)	134 K (5.49%)	966 (44.73%)	125
XC7V-	MMC$_7$(3ph)	350 K (28.67%)	194 K (7.94%)	1530 (70.83%)	121
2000T	MMC$_1$1(1ph)	168 K (13.80%)	111 K (4.53%)	901 (41.71%)	121
	MMC$_5$-IM	250 K (20.51%)	143 K (5.87%)	1155 (53.47%)	125

8.4.2.1 Hardware Platform

The hardware design of the MMC-IM system was carried out on the Xilinx® XC7VX485T FPGA, and Table 8.7 lists an estimation of hardware utilization when different levels of MMCs are implemented on two types of FPGA devices, and the maximum operational frequency f_{max} of each design is also shown. A higher operational frequency gives a larger speed margin for a certain time-step, but the chip power dissipation increases along with it; on the contrary, a lower frequency leads to less power dissipation but the design may fail to attain real-time execution. Therefore, a trade-off is made and the operational frequency of 100 MHz is chosen, with the corresponding clock period of the FPGA T_{clk} as 10 ns.

The hardware resources of XC7VX485T are sufficient for running a single-phase 11-level MMC but falls short of driving the induction machine with even seven-level MMC due to a lack of LUTs. As can be seen from the table, the demand for one phase accounts for 37.50% and will exceed the total available resources if the size triples. This can be avoided if the design is deployed to another FPGA device with abundant LUTs like the XC7V2000T, although it has fewer DSPs for implementing the three-phase 11-level MMC, as shown in the same table.

8.4.2.2 Controller Emulation

For the MMC-IM system, the controller is twofold, referred to as the MMC inner control and induction machine outer control, respectively. The former is in charge of SM DC capacitor voltages, and the latter regulates the induction machine's angular velocity. Detailed control algorithms for MMC and the induction machine have been separately developed [164, 165], and the overall controller combing both is shown in Figure 8.16. Three-phase stator currents i_{sa}, i_{sb} and i_{sc} as well as ω_m and its reference ω_m^* are the inputs for the outer controller, which produces three-phase modulation signals v_{abc}^* and sends them to the inner controller as its inputs. Then the three-phase MMC inner controller generates driving pulses to control the switches.

As can be seen, regardless of what the conditions of surrounding devices such as the induction machine are, the hardware latency of the outer controller is restricted in a small range between 379 and 382 T_{clk}, while the latency of the inner controller is a logarithmic function of the number of submodules in a leg due to the averaging of DC capacitor voltages, and the hardware delay is

$$L_{inner} = (T_{adder} \cdot \lceil \log_2(2N) \rceil + 40) \cdot T_{clk}, \tag{8.37}$$

where T_{adder} is the latency of the adder, which takes four clock cycles for single precision numbers, and the rounding function is equivalent to setting N to its nearest even number times of 4. Hence, for the five-level MMC that has eight submodules, the controller latency is 52 T_{clk}, or 520 ns time delay, slightly over the time-step of 500 ns for the MMC circuit, and for the seven-level MMC this delay increases to 560 ns, so using the same time-step would hinder achieving real-time. The solution is to utilize multiple time-steps for these two subsystems emulated in parallel: 1 μs for the MMC inner controller and 4 μs for the IM outer controller. Theoretically, with this time-step setting, the inner controller is able to deal with MMCs with thousands of levels for real-time HIL emulation purpose, but in reality, the number of voltage levels is restricted by hardware resources.

8.4.2.3 MMC Emulation on FPGA

Table 8.8 is a summary of the latencies of each hardware module in the five-level MMC-IM system and the emulation time-steps for these subsystems. Based on the update frequency of variables, the whole system is dispatched to three layers, each satisfying the following criterion that ensures real-time:

$$T_{clk} \cdot max\{L_1^i, L_2^i, \dots, L_n^i\} \leq \Delta t_i, \tag{8.38}$$

where $L_1^i, L_2^i, \dots, L_n^i$ are the latencies of hardware modules that the ith layer with the time-step Δt_i contains.

In order to run the five-level MMC HIL emulation in real-time, the time-step for *Layer 1* should be close to 370 ns, if the induction machine is not taken into account. According to the device datasheet, this minimum time-step is approximately the rise/fall time of the selected Infineon® IGBT FZ400R33KL2C_B5 (V_{CES} = 3300 V, I_C = 400 A) when its gate resistor is 10 Ω. This means that under these circumstances, a maximum of two values can be caught during rise/fall

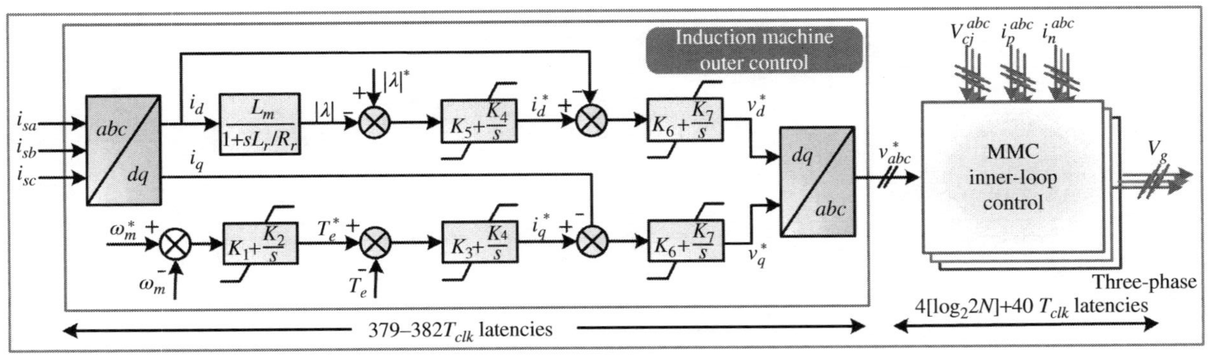

Figure 8.16 Control algorithm for the MMC-IM system.

Table 8.8 Latencies of different hardware modules in the five-level MMC-IM system.

Hardware module	Maximum latency	Time-step (μs)	Layer
MMC main circuit	$37T_{clk}$		
Induction machine	$41T_{clk}$	$\Delta t_1 = 0.5$	*Layer 1*
Submodule	$37T_{clk}$		
MMC controller	$52T_{clk}$	$\Delta t_2 = 1.0$	*Layer 2*
IM controller	$382T_{clk}$	$\Delta t_3 = 4.0$	*Layer 3*

process and that section of the switching curve is straightened. On the other hand, the transient process is not limited to the aforementioned region, and there are other sections of the curves that distribute beyond it; thus, the time-step can be set a little larger to 500 ns and the voltage and current waveforms can be represented by piecewise linearized lines, one of which contains the rise/fall process.

Table 8.8 also shows that the induction machine has the largest latency in *Layer 1*. However, when the number of submodules increases, as the only part whose latency is affected, the MMC main circuit latency begins to overtake the IM as the dominant factor to determine real-time operation. The latency incremental for $(M + 1)$-level MMC main circuit can be deduced from its $(N + 1)$-level counterpart by

$$\Delta t_{N\rightarrow M} = \left(T_{adder} \cdot \left\lceil \log_2 \frac{M+1}{2^{\lceil \log_2 (N+1) \rceil}} \right\rceil \right) \cdot T_{clk}. \tag{8.39}$$

Thus, the maximum number of levels that can achieve for real-time HIL emulation with a 500 ns time-step is 64, when the latency of the MMC main circuit reaches 49 T_{clk}.

The hardware structure and signal flow routes for the MMC-IM system are drawn in Figure 8.17, where *j*th submodule structure can be seen out of the total $2N$ submodules. There are two levels of parallelism in the design: layers with different time-steps run simultaneously, and all hardware modules within a certain layer are also in parallel. In *Layer 1*, after each time-step Δt_1, the MMC main circuit exchanges TLM link information with the submodules and updates the three-phase voltages for the induction machine, from which stator currents are received. Then there are information exchanges between the layers. Data going to the IM outer controller will not take effect unless an entire time-step Δt_3 ends and produces the three-phase modulation waves for the inner one. For *Layer 2*, since Δt_2 is between two other time-steps, the values of modulation waves are kept constant for $\frac{\Delta t_3}{\Delta t_2}$ cycles and the DC capacitor voltages $\mathbf{V}_C$ and arm currents from *Layer 1* can participate in the control only when a new time-step begins.

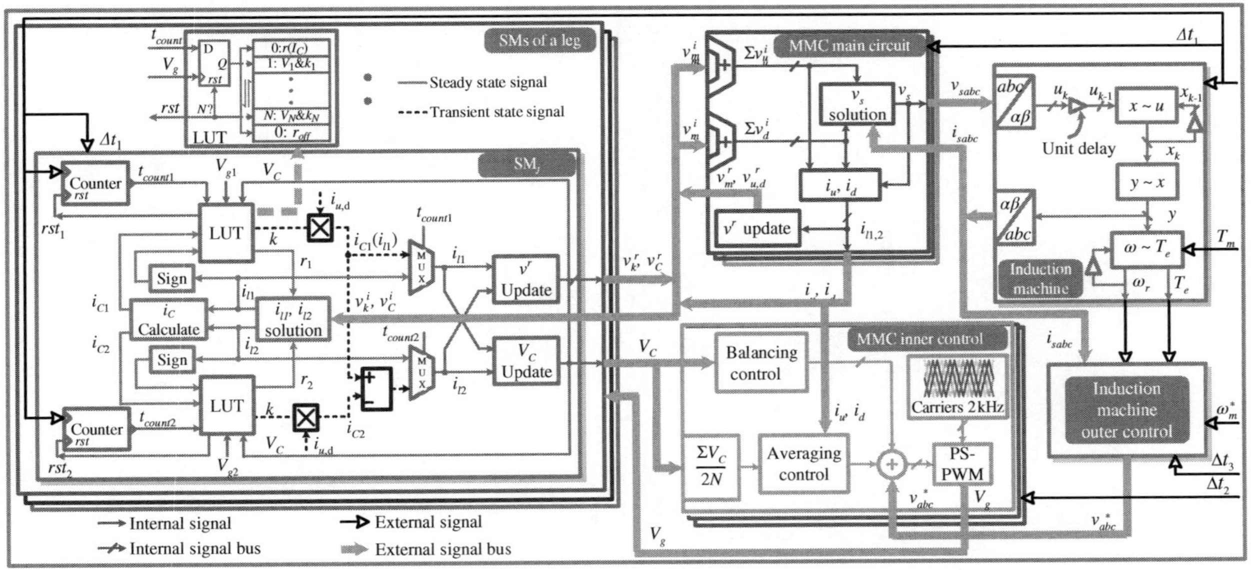

Figure 8.17 Hardware structure and signal flow diagram for the FPGA emulation of the MMC-IM system.

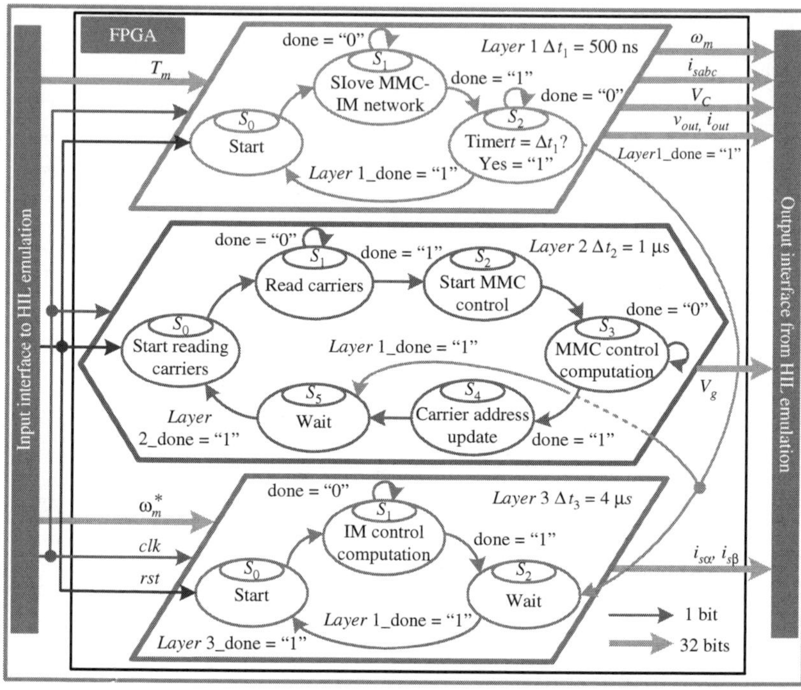

Figure 8.18 Finite state machine of the overall MMC-IM system for hardware emulation.

One of the benefits with such a hardware design is that all external and internal signals as well as the hardware other than LUTs in the submodules will not change if a new piecewise linear switch model is established to replace the original one. Even if a more complex switch model such as physics-based model is introduced, the only alteration occurs within submodules, thus there is no necessity to redesign the hardware for other parts.

With regard to the specific structure of each hardware module, their corresponding functions are written in C/C++ in Xilinx® Vivado HLS®. In this hardware design there are totally five types of function blocks: the induction machine, the MMC main circuit and submodules as well as the two controllers. Each is coded as an independent function in a separate program, whose inputs and outputs include all external signals of that block. Meanwhile, detailed mathematical as well as logic operations within a function block, such as those in Figure 8.17, are represented by the programming language in a pipelined fashion. Although Vivado HLS® also has a pipeline directive option, which could further increase the maximum operational frequency of the designs, it was not used because the frequency improvement is at the cost of more hardware resource utilization and 100 MHz was deemed sufficient to ensure real-time execution. By running C synthesis of the completed code and the exporting RTL operation that follows, an IP core, the hardware module corresponding to the function block, is generated. However, these modules are yet to be linked with each other. This is realized by very high-speed integrated circuit hardware description language (VHDL) coding in the form of signal exchange that takes place at the end of every time-step, and so is the finite state machine that achieves the multilayer design and decides the time sequence of each module.

Figure 8.18 shows the relationship between different layers and how they cooperate to execute the entire MMC-IM system by finite state machine. It should be pointed out that the maximum latency in each layer is smaller than corresponding time-step, meaning that the MMC-IM system will proceed *faster* than real-time. Therefore, a timer is introduced in *Layer 1* to achieve exact real-time, when it counts to Δt_1, that value is reset and the calculation for the next time-step begins. The command is also sent to the other two layers to enable their respective FSMs to enter a new stage, if they are already waiting. In *Layer 2*, the values of carriers are needed before the control starts, and near the end of each time-step, the carrier addresses are updated so that in the next time-step new values can be referred to. For the last layer, the operation is similar to the first layer, other than the fact that shifting to state S_0 is controlled by the command from the latter. When the reset order is issued, the states in all three layers begin to circulate and the HIL emulation of the MMC-IM system is ongoing. Thus, by giving proper speed and torque orders through the input interface, the status of the overall system can be observed via the output interface.

8.4.3 Real-Time Emulation Results

8.4.3.1 MMC

In this part, functions of different levels of MMCs are tested with R–L load. In the test, the DC line voltage is maintained at 900 V, meaning that when the number of levels increases, the DC capacitor voltages will decline accordingly. However, the values of other circuit components such as the arm inductance will not be changed, as shown in Appendix A.7, and the switching frequency is 2000 Hz. To validate the results from HIL emulation, SaberRD® simulations are also carried out with a maximum time-step 500 ns to ensure transient processes are recorded. The IGBT and diode models employed in simulations are *igbt1_3* and *dp1*.

In Figure 8.19 specific system-level performances of the five-level MMC and its seven-level counterpart are shown. Figure 8.19a,d are the 60 Hz, single-phase output voltages of the five-level and seven-level converter, respectively. As can be observed, the voltage waveform in the latter has two more levels than the former, but their peak values are virtually the same, both close to 430 V, and high

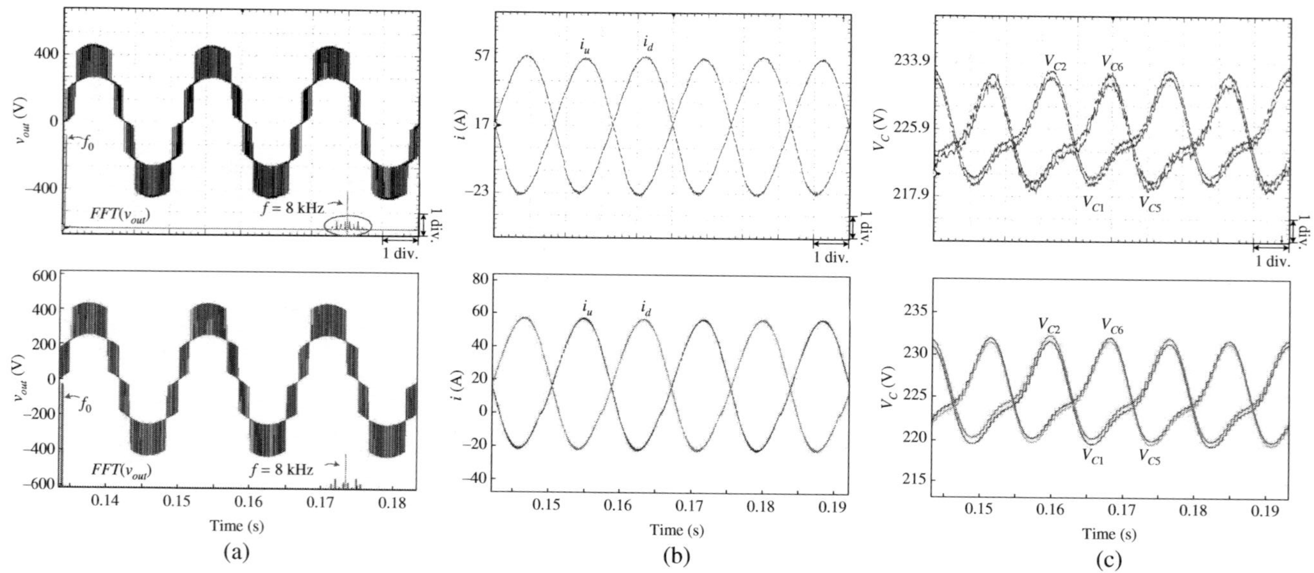

Figure 8.19 Comparison of performances of five-level ((a), (b) and (c)) and seven-level ((d), (e), and (f)) MMC between real-time HIL emulation (top) and SaberRD® (bottom). (a) five-level MMC output voltage, (b) arm currents, (c) DC voltage ripples of submodules in upper and lower arms, (d) seven-level MMC output voltage, and (e), (f) DC voltage ripples of submodules. Oscilloscope axes settings: (a), (d) *x*-axis 5 ms/div, *y*-axis 133.34 V/div (v_{out}) and 66.67 V/div (FFT); (b) *x*-axis 5 ms/div, *y*-axis 13.333 A/div; (c), (e) and (f) *x*-axis 5 ms/div, *y*-axis 2.667 V/div.

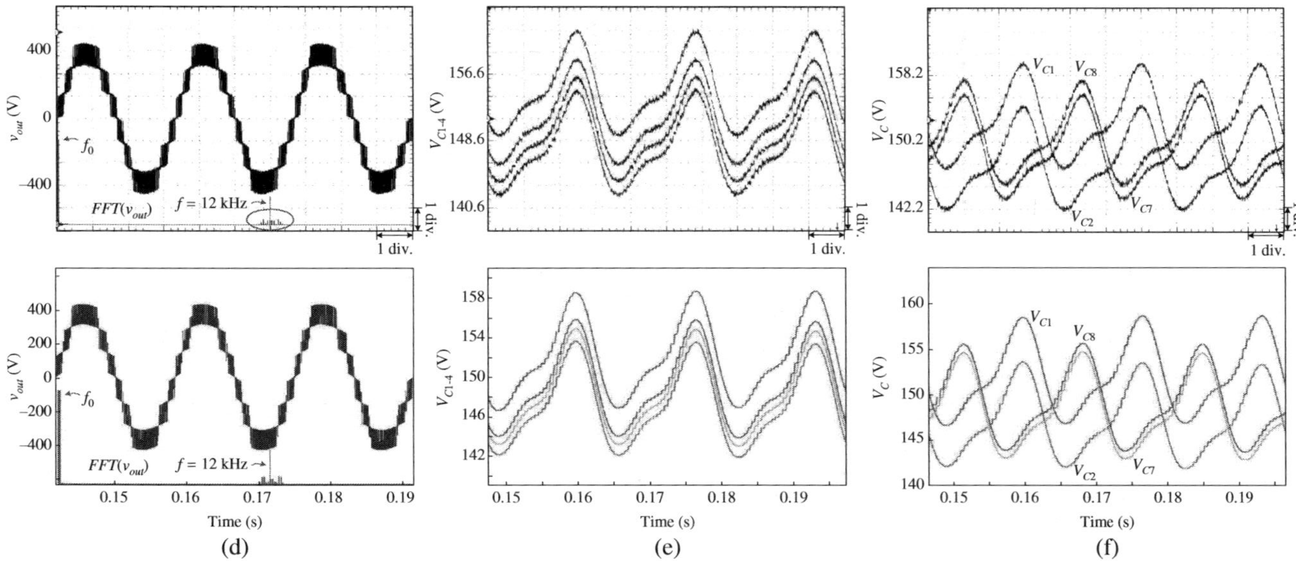

Figure 8.19 (*Continued*)

symmetry is also observed. Moreover, voltage spectral analysis is carried out by the oscilloscope, which demonstrates that for the five-level MMC, its output voltage harmonics mainly distribute around 8 kHz – four times higher than the switching frequency, while for the seven-level converter, the major harmonics center around 12 kHz. This phenomenon agrees with the theory that for $(N+1)$-level MMC, the effective switching frequency is N times higher. The results are verified by SaberRD® simulations as they give identical waveforms. Figure 8.19b demonstrates the upper and lower arm currents of the five-level converter, the results from oscilloscope and simulation agree with each other quite well in both waveshape and values. Figure 8.19c shows the DC voltage ripples of the submodules in upper and lower arms for the five-level converter. These values fluctuate around the reference of 225 V, indicating the inner controller is working properly. The peak-valley difference is estimated to be around 13.3 V from the oscilloscope and simulation. In Figure 8.19e,f, some DC capacitor voltages of seven-level MMC are shown and compared. The former indicates that for submodules in the same arm, the rising/declining trends of DC voltage ripples are same, while the latter shows the trend in the opposite arm is totally in contrary. The average values of these DC voltages, as can be read from these figures, are about 150 V since the number of submodules in an arm increases to six while DC line voltage is kept constant.

Figure 8.20 gives the switching process and power losses in the five-level MMC and the shape of these waveforms in seven-level MMC are almost the same and are therefore not shown. Figure 8.20a,b are the transient IGBT voltage and current waveforms during the turning on and off processes. After exerting a positive driving pulse on the gate and a period of turn-on delay lasting for 1 μs, the voltage begins to drop and a current surge can be observed from both HIL emulation and SaberRD® simulation. Then the current gradually stabilizes and the voltage finally remains slightly above zero due to the conduction resistance. The rise time is defined as the time interval between 10% and 90% of collector current under steady state, which is around 0.33 μs, slightly below 0.4 μs provided in the datasheet. When the driving pulse disappears, the turning off process takes place after a turn-off delay of approximately 4 μs; it is an opposite process during which v_{CE} rises to DC capacitor voltage and collector current goes to zero, but the fall time has a similar definition to rise time and its value is near 0.42 μs, a little larger than the datasheet value of 0.35 μs. In Figure 8.20c, diode reverse recovery process is shown. As can be seen after plunging to peak value, which virtually has the same amplitude as the steady-state current, the reverse current begins to decay to zero and voltage over the diode climbs to DC capacitor voltage. It is observable that the current tail in the SaberRD® simulation is a little longer, but since the value of the final stage is extremely small it is forced to zero in the diode model and that will not cause a significant error when calculating power loss. Meanwhile, the forward voltage of diode is also nonzero attributing to the exponential static

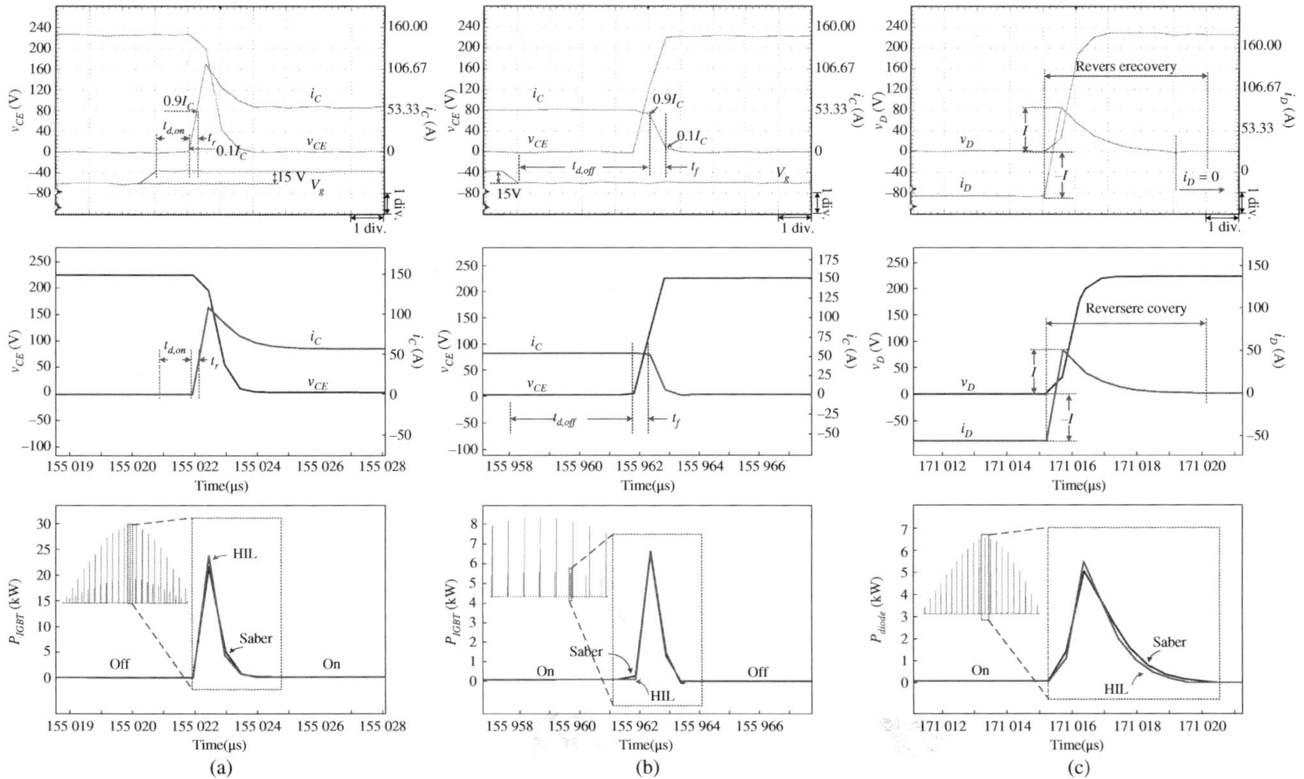

Figure 8.20 Details of switching processes and power losses of IGBT or diode from HIL emulation (top) and SaberRD® simulation (bottom). (a) IGBT turning on, (b) IGBT turning off, and (c) diode reverse recovery. Oscilloscope axes settings: x-axis 1 μs/div, y-axis 40 V/div and 26.67 A/div.

I–V characteristics. The power loss corresponding to each process is also shown, a high degree of consistency between HIL emulation and SaberRD® simulation is observed.

To validate the effectiveness and convenience of the circuit partitioning method in achieving real-time, the seven-level MMC is expanded to 11-level and emulated execution on the FPGA. Figure 8.21a is the 11-level output voltage and the load current from HIL emulation, compared with those of five- and seven-level MMC. The voltage quality is higher and as anticipated that the voltage spectral analysis yields an array of harmonics around 20 kHz, but they are almost negligible. The root mean square value of fundamental component is same to those of other two,

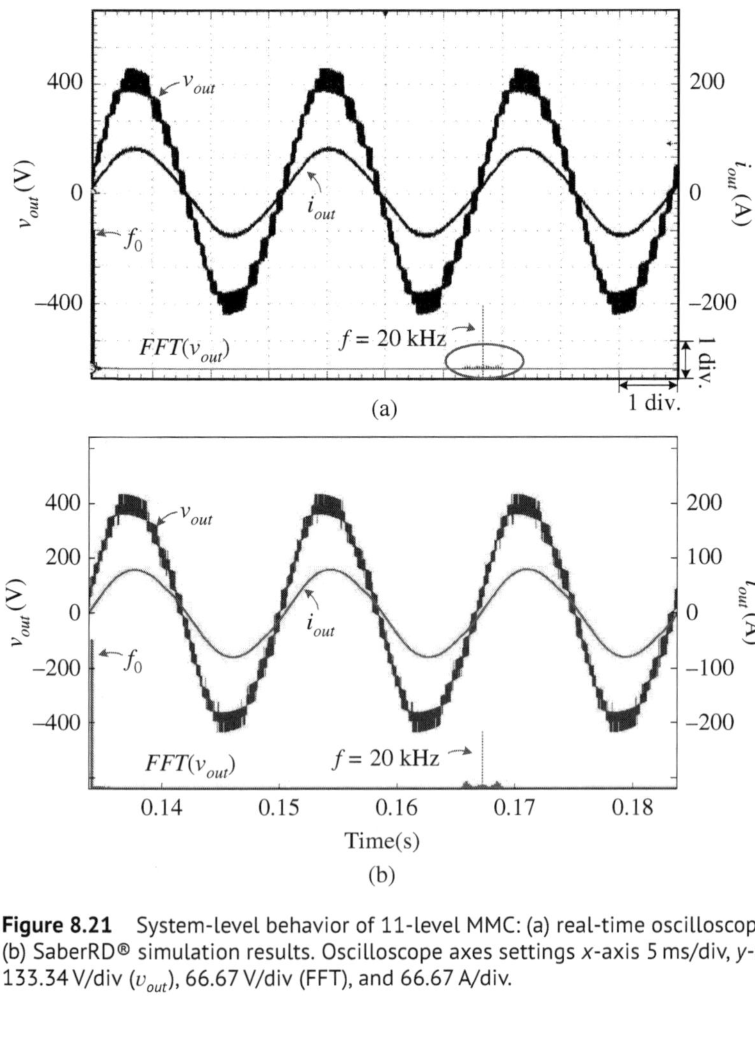

Figure 8.21 System-level behavior of 11-level MMC: (a) real-time oscilloscope results; (b) SaberRD® simulation results. Oscilloscope axes settings x-axis 5 ms/div, y-axis 133.34 V/div (v_{out}), 66.67 V/div (FFT), and 66.67 A/div.

all about 280 V. The output current, due to filtering effect of inductors, is sinusoidal and it reaches a peak value of 80 A, agrees with its theoretical value. The results from SaberRD® are also shown in Figure 8.21b for comparison, which indicates the hardware implementation of MMC is correct.

Table 8.9 lists the time each switching process takes. The IGBT turn-on delay from HIL emulation is exactly what was provided in the datasheet, while its turn-off delay and diode reverse recovery time are both rounded to integers because the HIL emulation time-step is 500 ns. The errors for IGBT rise and fall time are relatively large, because their values are smaller than the time-step and, consequently, both processes are located on straightened lines and affected by the slopes.

For the upper switch in a submodule, the maximum current flows through the antiparallel diode while for the lower switch, the maximum current emerges in the IGBT; thus, their power losses are important. Table 8.10 shows the energy consumption of IGBT and diode during the transient process and steady-state when the current reaches largest in five- and seven-level MMC, where the errors are in their absolute forms to avoid negative values. As can be seen from Figure 8.20, the maximum steady-state current for both IGBT/diode pairs is about 60 A.

Table 8.9 Switching times of IGBT and diode.

Time	Description	HIL (μs)	Datasheet/ SaberRD® (μs)
$t_{IGBT}^{d,on}$	Turn-on delay	1.00	1.00
t_{IGBT}^{r}	Rise time	0.33	0.40
$t_{IGBT}^{d,off}$	Turn-off delay	4.00	3.90
t_{IGBT}^{f}	Fall time	0.42	0.35
t_{diode}^{rr}	Reverse recovery time	5.00	4.80

Table 8.10 Energy consumption validation of IGBT and diode model in MMC.

MMC	5L-MMC		7L-MMC	
Energy (mJ)	HIL/SaberRD®	Error (%)	HIL/SaberRD®	Error (%)
$E_{IGBT}^{turn-on}$	14.01/13.43	4.32	7.93/8.33	4.80
$E_{IGBT}^{turn-off}$	6.38/6.77	5.76	5.41/5.43	0.37
$E_{IGBT}^{conduct}$	2.59/2.58	0.39	2.40/2.39	0.42
E_{diode}^{rr}	9.28/9.56	2.90	4.69/5.03	6.76
$E_{diode}^{conduct}$	1.75/1.76	0.57	1.69/1.64	3.05

The steady-state power losses are quite accurate because the static V–I characteristics are provided in the datasheet, whereas the transient waveforms are obtained by curve-fitting, and therefore, the error is a bit larger, but still they are precise and can be referred to when designing the MMC as well as the cooling system. Moreover, with the increase of output voltage level, the power consumed by switches decreases along with voltage and current stresses. Generally, the HIL system is able to offer accurate power losses of both steady-state and transient stages in the MMC despite the variation of its voltage level and the load. It is more convenient compared with measuring power losses by setting up an experimental MMC prototype whose excitations as well as the loads should be adjusted repeatedly in order to provide the switches with the same electromagnetic environment. In addition, although knowing the steady-state current from simulation of conventional MMC models with ideal switches enables direct acquisition of steady-state power loss from the device datasheet, estimating the transient portion based on the turn-on and turn-off energy losses provided by datasheet is less accurate, since they were obtained in an experimental setup with distinct testing conditions.

8.4.3.2 Induction Machine Driven by Five-Level MMC

The speed of the induction machine can only be regulated by five-level MMC when the emulation was done on the XC7VX485T FPGA. As shown in Table 8.7, the LUT is not enough for the other two MMCs to extend to three phases.

Figure 8.22a shows the regulation of the mechanical angular velocity by real-time HIL emulation. The initial speed reference is 160 rad/s, so the machine starts and the velocity goes up to the reference value in about one second. Meanwhile, a large stator current can be observed in all three phases and only phase A is shown since they are symmetrical. After one second, the actual speed is very close to the reference and the machine operates under steady-state with stator currents reduce significantly to around 20 A in amplitude. Then at $t_1 = 3$ seconds, ω_m^* plummets to −160 rad/s, meaning that the rotation direction is reversed so that the positive speed slows down to zero and later increases in the opposite direction until it reaches the reference value, which sees a slight increase at $t_2 = 6$ seconds to −80 rad/s. Consequently, the real speed follows and the machine quickly enters steady state. Between $t_3 = 8$ seconds and $t_4 = 9$ seconds, a pulse of 100 N·m is applied to the torque, following this change is a temporary rise of stator current, but the impact it has on the angular velocity is negligible. As can be seen throughout the whole period, a large angular velocity leads to a higher current frequency demonstrated by the density of the waveform. For comparison, Matlab/Simulink simulation is carried out, and corresponding system-level performance is shown in Figure 8.22b, which proves that both controllers are functioning normally and the design theory is correct.

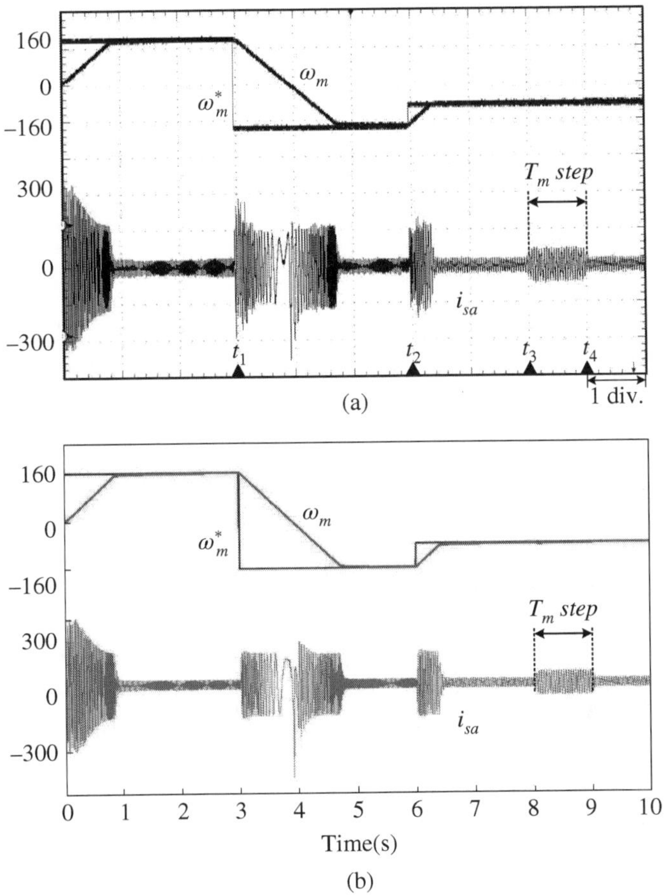

Figure 8.22 Regulation of induction machine speed by five-level MMC: (a) real-time oscilloscope results, and (b) off-line simulation results. Oscilloscope x-axis: 1 s/div.

The starting of the induction machine with different values of torques was also tested. In Figure 8.23a, the locus of stator currents in α–β frame are drawn for the starting period when the mechanical angular velocity climbs up from 0 to 160 rad/s without any load. A momentary current surge at the vertical axis is observed immediately after starting, indicated by curve A. Then, as can be seen from curve B, the current steadily reduces from 300 to 150 A, and following a sudden decline shown by curve C, the current finally stabilizes around the region D.

The loci of stator currents for three torques under steady-state are shown in Figure 8.23b. As expected it shows that a larger torque yields a circle with greater

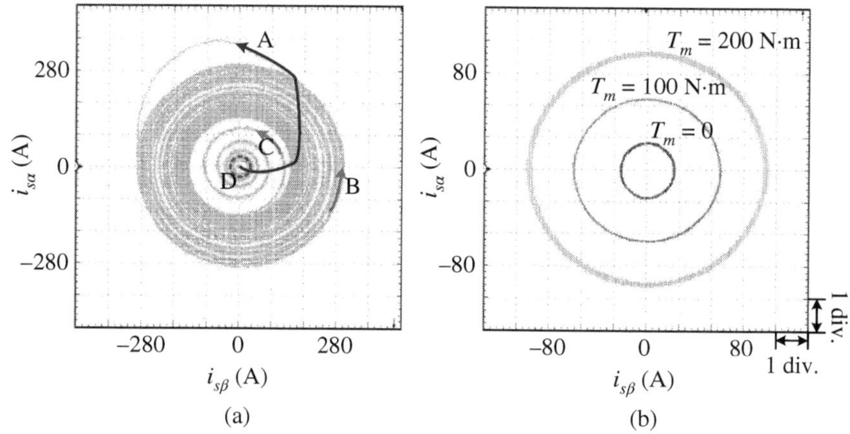

Figure 8.23 Real-time oscilloscope results of stator current in α–β frame under (a) starting period, and (b) steady-state with $T_m = 0$, 100 and 200 N·m, respectively. Oscilloscope x- and y-axis settings: (a) 93.34 A/div; (b) 26.67 A/div.

radius. Other information such as the relation between the duration of transient process and torque is also available. When $\omega_m^* = 160$ rad/s, it takes 0.52, 0.65, 0.86, 1.31, and 2.68 seconds for the machine to reach 95% of ω_m^* when the torques are −200, −100, 0, 100, and 200 N·m, respectively, indicating a larger torque leads to a longer time to approach steady state, while the reverse is true for $\omega_m^* = -160$ rad/s.

8.5 MMC Partitioned by Coupled Voltage–Current Sources

8.5.1 *V–I* Coupling

The coupled voltage–current sources (V–I) can be inserted between the arm and submodule in an exact manner, where the TLM-link is applied [166]. However, the V–I coupling does not require the additional characteristic impedance that TLM-link has, as shown in Figure 8.24. The current source is placed on the submodule side since it contains nonlinear IGBT/diode models solved by nodal equations. While on the left side, the MMC main circuit can either be solved by mesh current or nodal voltage equations – in the latter case, the coupled voltage sources need to be converted to current sources first.

The partitioning method induces a unit delay to both sides. At the instant $t - \Delta t$, $i_u(t - \Delta t)$ and $i_d(t - \Delta t)$ are obtained by solving the matrix equation corresponding to the left circuit, and they are sent to the submodules. Then, the time instant

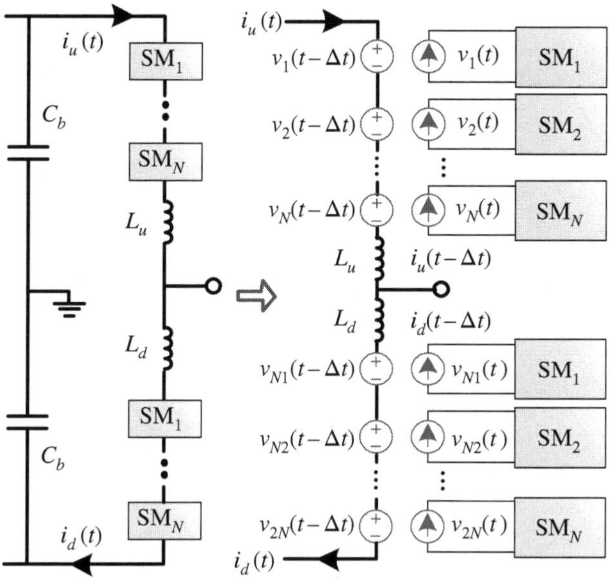

Figure 8.24 MMC partitioning by V–I coupling.

t begins. On the submodule side, based on the SM current it just received, its port voltage $v_k(t)$ can be derived. Thus, v_k is one time-step ahead of i_u and i_d on the SM side, while the reverse is the case for the MMC arm. Nevertheless, under the circumstance that the circuit computation frequency is much higher than that of the arm current, i_u and i_d can be deemed as constants in two neighboring time-steps. Consequently, the impact of unit-delay methods, e.g. V–I coupling and TLM-link, on simulation accuracy is negligible.

The nodal voltage equation on the SM side is determined by the IGBT/diode type it uses, with an k-node IGBT/diode model, the SM generally has the following admittance matrix:

$$\mathbf{G}_{SM} = \begin{bmatrix} G_C & \mathbf{0}_{1\times(2k-3)} \\ \mathbf{0}_{(2k-3)\times 1} & \mathbf{0}_{(2k-3)\times(2k-3)} \end{bmatrix} + \begin{bmatrix} m\cdot\mathbf{G}_{\mathbf{S}k\times k} & \mathbf{0}_{(k-2)\times(k-2)} \\ \mathbf{0}_{(k-2)\times(k-2)} & \mathbf{0}_{(k-2)\times(k-2)} \end{bmatrix}$$
$$+ \begin{bmatrix} \mathbf{0}_{(k-1)\times(k-1)} & \mathbf{0}_{(k-1)\times(k-1)} \\ \mathbf{0}_{(k-1)\times(k-1)} & m\cdot\begin{bmatrix} G_{S11} & \cdots & G_{S1(k-1)} \\ \vdots & \ddots & \vdots \\ G_{S(k-1)1} & \cdots & G_{S(k-1)(k-1)} \end{bmatrix} \end{bmatrix}, \tag{8.40}$$

where the element G_C represents the conductance of SM capacitor; the second matrix lists all elements in the k-node upper switch, while its lower counterpart is placed in the third matrix, which only contains $(k-1)\times(k-1)$ elements after the kth node is naturally grounded. Similarly, the current contribution vector can be expressed by

$$\mathbf{J}_{SM} = \left[I_{hisC}\ \ 0\ \ \cdots\ \ 0\ \ J_s|_{(k)}\ \ 0\ \ \cdots\right] + m\cdot\left[J_{S1}\ \ J_{S2}\ \ \cdots\ \ J_{Sk}\ \ 0\ \ 0\ \ \cdots\right] + m\cdot\left[0\ \ 0\ \ \cdots\ \ J_{S1}|_{(k)}\ \ J_{S2}\ \ \cdots\ \ J_{S(k-1)}\right], \tag{8.41}$$

where I_{hisC} and J_s are the capacitor's history current and arm current, respectively. The SM nodal voltages are subsequently obtained by

$$\mathbf{U}_{SM} = \mathbf{G}_{SM}^{-1}\cdot\mathbf{J}_{SM}. \tag{8.42}$$

On the main circuit side, the arms have a fixed form, e.g. the Thévenin equivalent circuit is

$$Z_{arm} = Z_{Lu,d} + r_{arm}, \tag{8.43}$$

$$U_{arm} = 2v^i_{Lu,d} + \sum_{k=1}^{N} v_k(t-\Delta t), \tag{8.44}$$

where $Z_{Lu,d}$ and $v^i_{Lu,d}$ constitute the TLM-stub model of the inductor, and r_{arm} is its parasitic resistance.

8.5.2 Hardware Emulation Case of NBM-Based MMC

8.5.2.1 Power Converter HIL Emulation

A medium-voltage DC (MVDC) system is implemented on the FPGA for demonstration of the nonlinear behavioral IGBT and diode models involved in system-level emulation, as shown in Figure 8.25a. The station controller is

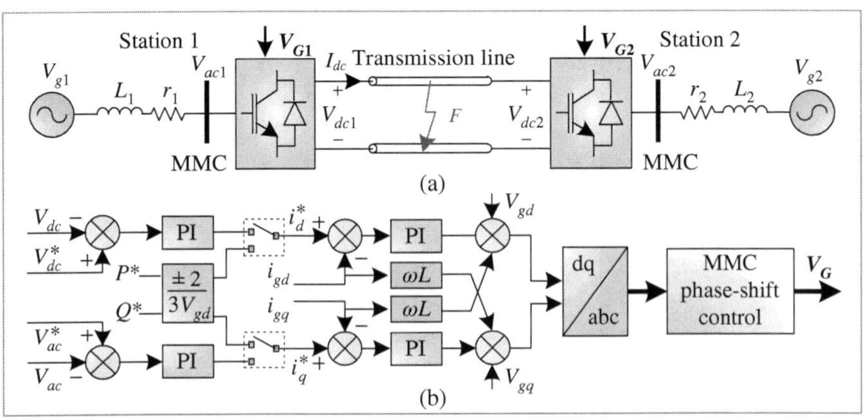

Figure 8.25 MMC-based MVDC system: (a) system configuration, and (b) station control scheme.

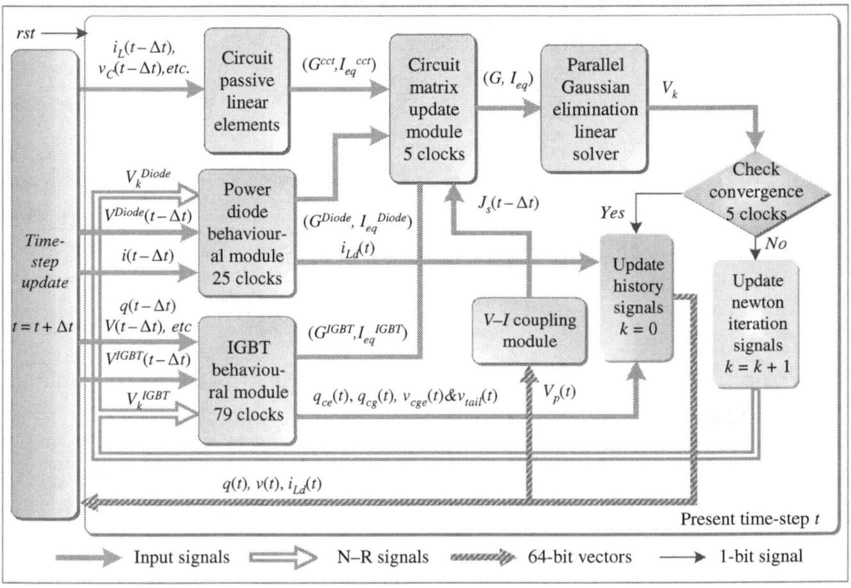

Figure 8.26 Hardware architecture and its signal flow routes for the MMC submodule with nonlinear behavioral switch models.

shown in Figure 8.25b. The inverter is set to control the DC line voltage, while the rectifier is in charge of instantaneous power regulation.

In Figure 8.26, the iterative HIL emulation process of the MMC submodule containing nonlinear behavioral IGBT and diode models is depicted. It should be noted that the *V–I* coupling module is designed specifically for the converter part with circuit partitioning. The hardware parallelism indicates that the two IGBT and diode pairs in a submodule can be processed simultaneously and thus does not require additional time despite an increase in the number of devices. The outputs, namely, the admittance and history currents, are sent to the *Circuit Matrix Update Module* to form the matrix equation which is solved by the following hardware modules that constitute the Newton–Raphson iteration. The emergence of convergent results indicates the beginning of a new time-step following the update of history signals.

8.5.2.2 HIL Emulation Results and Validation

To showcase the versatility of nonlinear behavioral models, HIL emulation results from device-level to system-level captured by the Tektronix DPO 7054 Digital Phosphor Oscilloscope are validated by off-line simulation tools running under 64-bit Windows® 7 Enterprise SP1 operating system with 3.40 GHz Intel® Core™i7 CPU and 8.00 GB of RAM. The parameters of employed IGBT and power diode modules are listed in Tables A.4 and A.5, and SaberRD®, which has experimentally verified IGBT/diode device-level models are used for validation.

8.5.2.3 Islanded MMC Performance

The MMC topology in Figure 8.1 is used as an inverter with DC-link voltage $V_{dc} = 3$ kV and AC side inductive load 5 Ω–6 mH for demonstrating the performance of nonlinear behavioral IGBT and diode models. In device-level simulation, the selection of a switch type should consider the device's capacity. The BSM300GA160D IGBT (1600 V/400 A) is suitable for this DC voltage rating and thus is chosen. The frequency of switches and AC output are 2.0 kHz and 60 Hz, respectively.

Due to the nonlinearities in a submodule, a minimum of five N–R iterations are needed for convergent results, and each iteration has a latency of 209 clock cycles. The HIL emulation time-step is set as 200 ns and FPGA clock frequency is 100 MHz. Table 8.11 summarizes the time some EMT simulators and the HIL system need to conduct the computation of a number of circuits for a 100 ms period. To achieve high fidelity, multiple switches are considered. The time SaberRD® needs to complete simulation of simple circuits, e.g. a single diode and IGBT, is acceptable, and the hardware speedup is medium. However, it rises dramatically along with the circuit scale and the number of parallel switches, as it is witnessed that SP1 – the HIL system's speedup over SaberRD® – could reach 65 times for a three-phase five-level MMC, while it is 275 for 11-level MMC. Meanwhile, compared with PSCAD/EMTDC® running under a much larger time-step of 20 μs, the HIL system's speedup SP2 indicates that it has a similar, or even faster simulation speed in three-phase MMC cases. Thus, it can be inferred that with higher-voltage levels, more converters, or parallel devices, the speedup becomes more significant because the MMC latency keeps the same.

Table 8.11 Simulation execution times from EMT simulators and HIL systems.

		Execution time (s)			
Tool	m	**Diode**	**IGBT**	**5L-MMC**	**11L-MMC**
	$m_1 = 1$	2.96	4.2	340	715
SaberRD®	$m_2 = 2$	4.15	6.5	528	1060
	$m_3 = 3$	5.10	8.6	620	1430
PSCAD®	1	0.3	0.3	4.5	17.5
HIL system	$m_{1,2,3}$	0.68	2.2	5.2	5.2
Speedup SP1	$m_{1,2,3}$	4.3/6.1/7.5	1.9/2.9/3.9	65/101/119	137/204/275
Speedup SP2	$m_{1,2,3}$	0.44	0.13	0.87	3.37

The oscilloscope results in Figure 8.27a–c show starting of the five-level MMC. Slightly irregular in the first two cycles, the output voltage later stabilizes with an evident level of 5. DC capacitor overcharge is observed in all submodules, with those in the lower arm having larger amplitudes to around 1200 V, but finally, all of them manage to maintain around 750 V, as shown in Figure 8.27b, indicating proper functioning of the controller. Figure 8.27c shows two arm currents, the opposite phase relation explains the submodule capacitor voltages in the upper and lower arms reach their peaks alternately. Moreover, a momentary current surge at the beginning explains the overcharge in DC capacitors. The impact of the number of behavioral SMs in an MMC is also tested by setting all of them nonlinear, and the results are given in the middle, which are verified by SaberRD® using the same configuration in the bottom. The ideal switch model leads to some minor differences in the output voltage around the third cycle; other than that, its outcomes are virtually the same to the other two rows, indicating that the partitioned MMC arm structure retains high fidelity.

From the perspective of a real converter design, switching dead-time is always set to protect switches in a submodule, and the gate driver circuit also affects their safe operation. Figure 8.28a shows the turn-on waveforms of an IGBT without dead-time, and a gate voltage $V_G = +15\,\text{V}/0\,\text{V}$ exerted on the device via a gate resistance of 10 Ω. A collector current surge up to over 1200 A appears due to overlapped conduction of the two complimentary switches and consequently, the energy stored in the DC capacitor discharges dramatically through that path. To avoid the hazardous current that may damage the switches, as well as to demonstrate the versatility of the behavioral model, different gate driving conditions are set. As depicted in Figure 8.28b,c, the current surge, caused by diode reverse recovery, witnesses a remarkable mitigation to about two times the amplitude of the steady-state current by simply setting a sufficient dead-time to 5 μs. Reducing the off-state gate voltage V_G^{off} would lose the requirement on dead-time, as demonstrated in Figure 8.28d,e. By setting a 2 μs dead-time and $V_G^{off} = 0$, a current surge up to 1000 A can still be observed. In contrast, it disappears when $V_G^{off} = -10\,\text{V}$. Figure 8.28f are the overview of switching waveforms of the upper and lower IGBT-diode pairs in a submodule. During Stage 1, the arm current is positive and consequently, the upper diode conducts to charge the DC capacitor, as can be noticed from the rising envelopes of v_{CE1} and v_{CE2}. Reverse recovery accompanies the diode operation, and correspondingly, current overshoot is induced to the lower IGBT. At Stage 2, the arm current becomes negative so the upper IGBT is ordered to turn on repeatedly, and the lower diode acts in concert to discharge energy stored in the DC capacitor. These device-level results prove that the nonlinear behavioral model has a high versatility to variations of electromagnetic environment since its switching waveforms can change accordingly along with external circuits without any adjustment on its parameters once they are obtained;

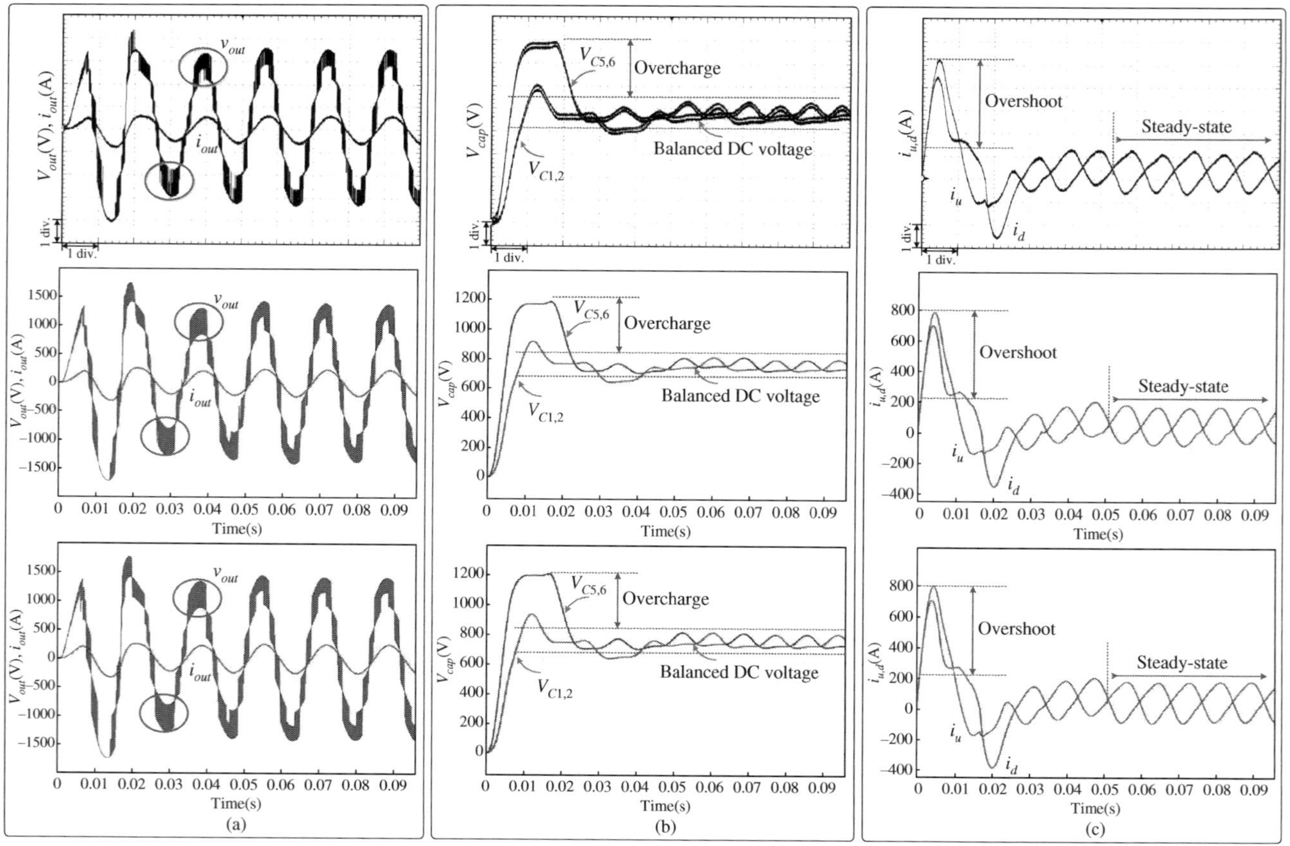

Figure 8.27 System-level performance of MMC with nonlinear behavioral models (top, middle) and in SaberRD® simulation (bottom): (a) Output voltage, (b) capacitor voltages, and (c) arm currents. Oscilloscope *y*-axis: (a) 396 V(A)/div., (b) 155 V/div., (c) 155 A/div.; *x*-axis: 50 ms/div.

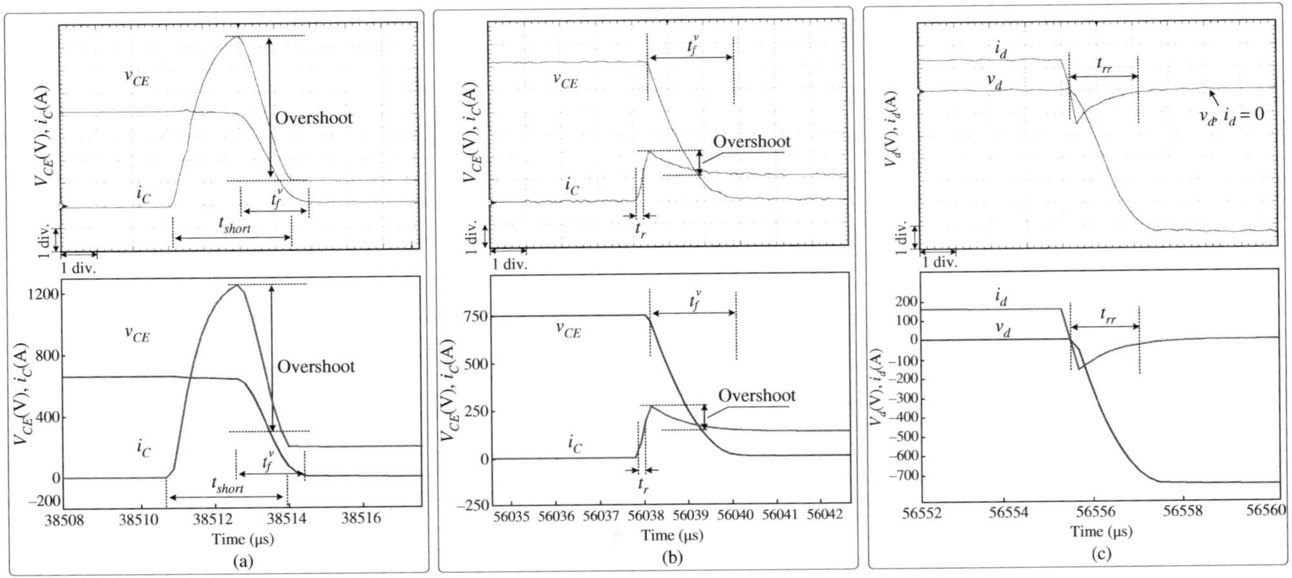

Figure 8.28 Performance of MMC with nonlinear behavioral models from HIL emulation (top) and SaberRD® simulation (bottom): (a) IGBT turn-on without dead-time, (b) and (c) switching transients with 5 μs dead-time, (d) and (e) switching transients with 2 μs dead-time, and (f) operation of complimentary switches in a SM from HIL emulation. Oscilloscope *y*-axis: (a) 156 V(A)/div., (b)–(e) 130 V(A)/div., (f) 255 V(A)/div.; *x*-axis: (a)–(e) 5 μs/div., (f) 10 ms/div.

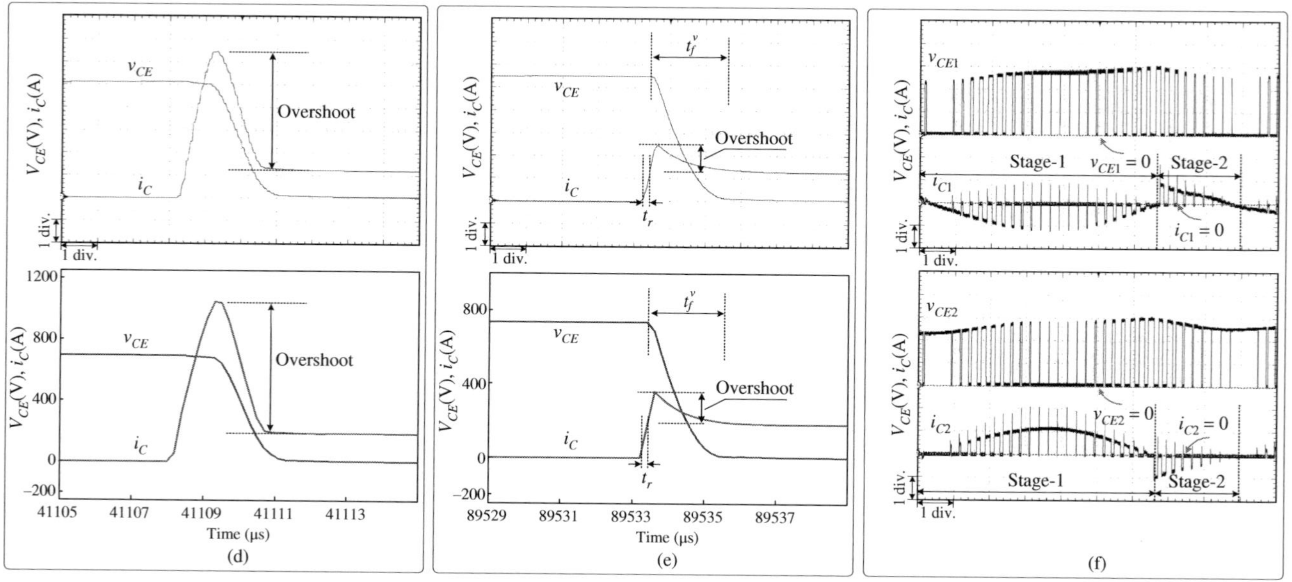

Figure 8.28 *(Continued)*

Table 8.12 Validation of IGBT and power diode nonlinear behavioral models by SaberRD®.

	SaberRD®	**FPGA**	**Error (%)**
	Transient time (ns)		
t_{rr}^{Diode}	2080	1970	5.2
t_r^{IGBT}	200	205	2.5
t_f^{IGBT}	640	600	6.6
	Energy consumption (mJ)		
E_{rr}^{Diode}	3.71	3.51	5.4
E_{cond}^{Diode}	7.26	7.27	0.2
E_r^{IGBT}	18.18	18.11	0.4
E_f^{IGBT}	103.48	99.96	3.4
E_{cond}^{IGBT}	6.58	6.58	<0.1

on the contrary, the ideal switch model and the averaged value model do not have transients. It is also impractical to enable the curve-fitting model to have that capability because potentially there could be numerous switching cases, and selection of an appropriate case is difficult. Moreover, it is also restricted by the availability of hardware resources when implemented on the FPGA. In Table 8.12, some static and dynamic features of IGBT and diode models are validated by SaberRD® simulation. It shows that the reverse recovery time of diode lasts up to 2 μs, much longer than IGBT's turn-on and -off period, which are around 200 and 640 ns, respectively. The conduction energy consumption distinguished by subscript *cond* is measured when the collector current reaches its maximum, i.e. 300 A. The error with respect to SaberRD® is negligible because essentially, it is a comparison of the static I–V characteristics, which is easy to model. The transient energy dissipation covers the overall switching period, i.e. from the time prior to the process to the switch's re-entry into steady state. Thus, the energy consumption E_{Tr} is calculated over the duration of switching period:

$$E_{Tr} = \int_0^{T_{Tr}} (v \cdot i)dt. \tag{8.45}$$

In HIL simulation, the Trapezoidal method is applied to the above equation, leading to

$$E_{Tr} = \frac{\sum_{i=1}^{N_{Tr}} (v_i \cdot i_i + v_{i+1} \cdot i_{i+1})\Delta t}{2}, \tag{8.46}$$

where the entire duration T_{Tr} is divided into $N_{Tr} = T_{Tr}/\Delta t$ intervals. Though the mathematical model for the switching transients is more complex, the energy

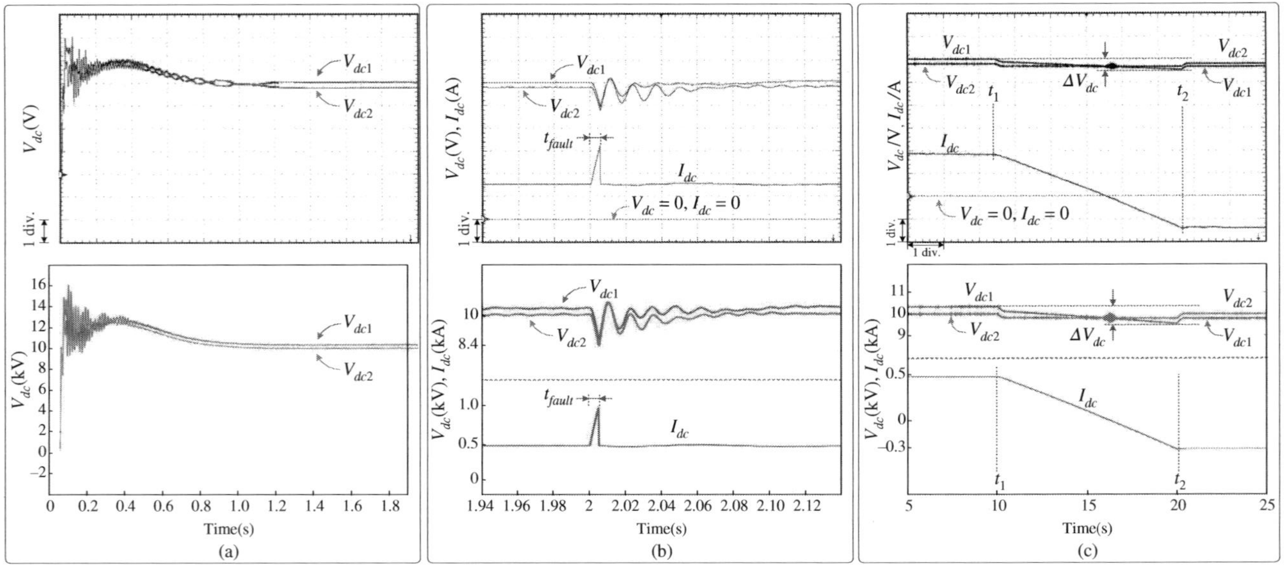

Figure 8.29 MVDC System-level performance from HIL emulation (top) and PSCAD/EMTDC® (bottom): (a) system start, (b) line-to-line fault response, and (c) power reversal. Oscilloscope y-axis: (a) 2.58 kV/div., (b) 1.73 kV/div., 272 A/div., (c) 1.72 kV/div., 246 A/div.; x-axis: (a) 1 s/div., (b) 100 ms/div., (c) 10 s/div.

loss from HIL emulation is still precise, with diode reverse recovery energy consumption having the largest error of 5.4% and IGBT turn-off loss next to it, at 3.4%. Moreover, the numerical results indicate that transient power losses are much higher, underlining the importance of device-level nonlinear switch models for evaluation of the safe operation of a converter.

8.5.2.4 MMC–MVDC Performance

To enable a higher DC voltage with the same five-level MMC configuration, IGBTs with a larger capacity, such as the 5SNA 2000K450300 StakPak IGBT Module (4500 V/2000 A), should be used. HIL emulation of a 10 kV/0.8 kA MVDC system is conducted while results validation relied on PSCAD/EMTDC® as SaberRD® is unable to simulate such a large system for a long period. In Figure 8.29a, system start is conducted, after a few oscillations at the beginning, the DC voltages stabilize at around 1s, with the rectifier station slightly over 10 kV. At $t = 2$ seconds, pole-to-pole fault lasting 5 ms is occurring to the center of the transmission line, as Figure 8.29b depicts, the DC voltages fall immediately, and the transmission line sees a large current, from initial 500 A to approximately 1 kA. In Figure 8.29c, power reversal is carried out. The power reference in the rectifier station is ordered to ramp down from −5 to 3 MW in a time interval of 10 seconds, and consequently, the DC line current I_{dc} declines from approximately +500 to −300 A. Therefore, before $t_1 = 10$ seconds, the energy is transferred to the inverter side, and the DC voltage at rectifier station V_{dc1} is slightly higher than V_{dc2} at inverter station to ensure energy flow. Then, I_{dc} starts ramping down, accompanied by a minor decrease of DC voltages at both terminals. At $t_2 = 20$ seconds, the process is ceased and noticeably, the numerical relationship between the two DC voltages has also reversed. These results prove that the decoupled hardware modules of the nonlinear behavioral switch models can be effectively employed for system-level studies when the fully iterative solution provides the same results as a transient simulation tool PSCAD/EMTDC® performing a noniterative solution using ideal switch models, particularly when an obvious speed advantage is witnessed, i.e. it takes around 752 seconds for the latter tool to simulate a 10 seconds interval with a much larger time-step of 20 μs, while the HIL system only requires 520s even though its time-step is 100 times smaller.

8.6 Clamped Double Submodule MMC

A major challenge of implementing the DC grid is its strict demand of safe operation under contingencies, especially the DC side fault, which can lead to a complete malfunction of the multiterminal system and cause severe damage to the converter stations. As the half-bridge SM is too simple to block the DC fault current, this section focuses on the clamp double submodule (CDSM) topology, which has a complex structure compared with half-bridge and full-bridge SMs, making the device-level modeling challenging.

The traditional Thévenin equivalence method is not suitable for a complex topology such as the CDSM because the calculation can grow cubically with the increase of internal nodes. The surrogate network and equivalent circuit methods [167, 168], employed by RTDS® and OPAL-RT® real-time simulators, respectively, use a finite number of circuit operation states based on IGBT gate signals and arm current direction, and therefore, do not consider and present detailed temperature-dependent static and dynamic characteristics. However, in a realistic environment, multiple factors have a great influence on the performance of power electronics devices. Consideration of multiphysics domains, which are the electromagnetic and thermal networks in this case, can effectively reflect the interactive phenomenon and increase the overall simulation accuracy.

With complex converter topologies, control and protection schemes applied in the modern power system, using analytical calculation to estimate the power losses and the thermal information is difficult and inaccurate, especially during system contingencies. It is necessary to use accurate EMT waveform to calculate the detailed device-level data, which are of great value for the system designers and operators. This section introduces a device-level electrothermal model of CDSM MMC for real-time emulation, which utilizes temperature-dependent electrical parameters, and presents the power losses, the junction temperatures, and the switching transients of the individual IGBTs and diodes during normal operation and fault conditions. The IGBT and diode characteristics are based on the piecewise polynomial scheme. Compared with the conventional modeling scheme for CDSM MMC, the presented modeling scheme has the following advantages:

1. increasing the accuracy of the system-level and device-level simulation;
2. providing the thermal data during fault condition, which is the critical information to evaluate protection schemes and parameters;
3. providing significant indicators for system designer to evaluate the thermal performance, which can be useful for IGBT module selection, heat sinker design, etc.;
4. helpful for converter control system designer to evaluate the overall energy efficiency and balance the thermal performance of all individual IGBT modules and to adjust valve-level control scheme and control parameters.

This section describes CDSM MMC modeling schemes and explains the parallelism from algorithm perspective, which includes device-level, SM-level, and converter-level modeling schemes [169]. The 32-bit floating point precision is used for the emulation system, except for the switching transient waveform generation, which uses fixed-point data and can update the values on every FPGA clock cycle (10 ns). PSCAD/EMTDC® and SaberRD® are employed to validate both system-level and device-level emulation results captured on the oscilloscope for steady-state operation and DC fault transient.

8.6.1 Operation Principles of CDSM

The MMC is composed of multiple SMs in each converter arm as shown in Figure 8.30a. The capacitors in submodules can be either inserted or bypassed by controlling the gate signals of IGBTs. Therefore, multiple voltage levels can be generated forming the low-harmonic sinusoidal waveforms at the AC side. Inductors are connected in series with the SMs to suppress the circulating current.

Breaking DC over-current is a major challenge for the HVDC transmission system. If the capacitor voltage in the SM can be negatively inserted into the converter arm, the DC fault current can be decreased and limited rapidly due to the reversed DC side voltage at the converter. The half-bridge SM, the full-bridge SM (FBSM), and the clamp double SM are the most popular ones shown in Figure 8.30b,c,d. The half-bridge topology is the simplest one, although it cannot reverse the polarity of the inserted capacitor voltage. The full-bridge submodule has the fault-tolerant capability with the cost of doubling the power electronics switches. In addition, the FBSM provides the capability of eventually changing the DC voltage rating and polarity, although it is normally not required for an HVDC system. As the name indicates, the CDSM contains two half-bridge SMs and a clamping circuit

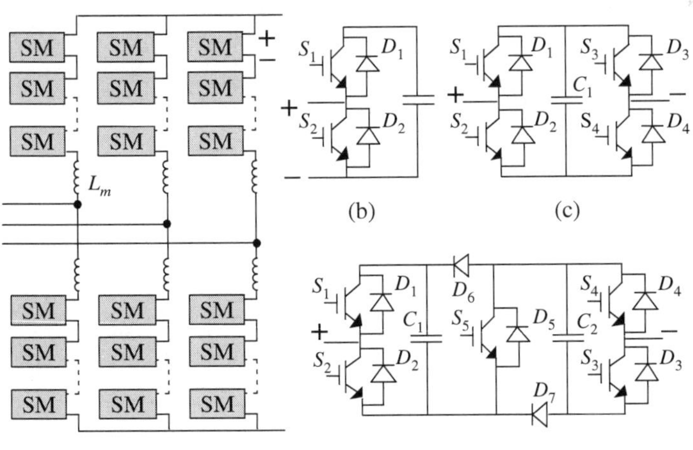

Figure 8.30 MMC and SM topologies: (a) modular multilevel converter, (b) half-bridge submodule, (c) full-bridge submodule, and (d) clamp double submodule.

composed of two additional diodes and one IGBT module. One CDSM is equivalent to two FBSM providing the same voltage rating and levels; therefore, the CDSM MMC utilizes fewer switches than FBSM MMC.

Figure 8.31 presents some cases of the normal operation mode and the protection mode during faults. In normal operation, the IGBT gate signal of S_5 is always on as shown in Figure 8.31a,b. The existence of Diodes D_6 and D_7 can clamp the capacitor voltage to positive values and can prevent internal loop current. Thus, the two half-bridge structures are connected in series and can be controlled independently. The IGBT modules of the half-bridge structures are numbered in such a way that if the odd number switches are turned on, the corresponding capacitor

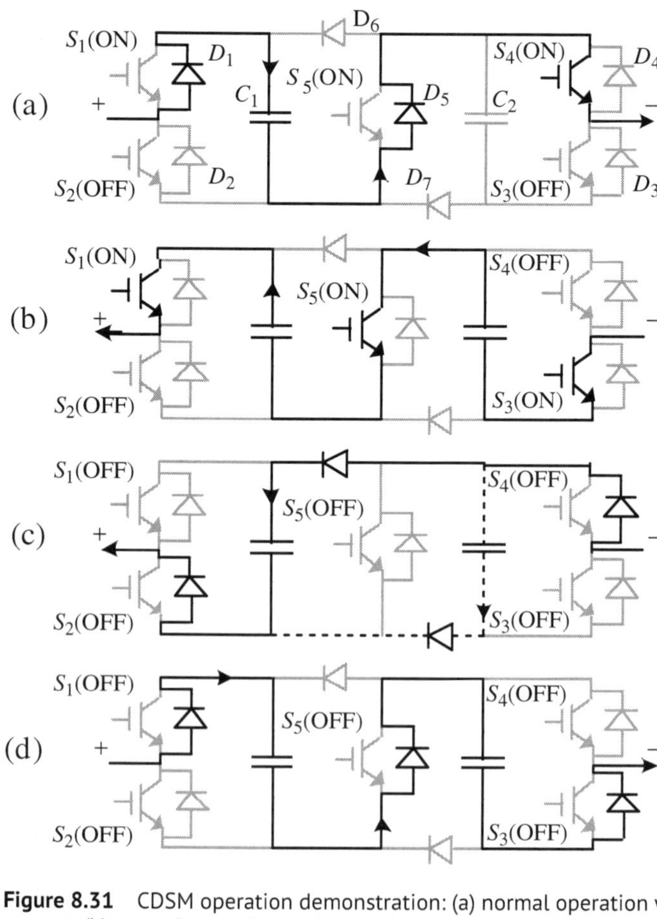

Figure 8.31 CDSM operation demonstration: (a) normal operation with positive arm current, (b) normal operation with negative arm current, (c) protection mode with negative arm current, and (d) protection mode with positive arm current.

Table 8.13 Operation mode of each half-bridge structure: Case I (S_5 turned-on combined with any SM current direction) and Case II (S_5 turned-off combined with positive SM current).

Mode	S_1/S_3	S_2/S_4	Current direction	Inserted voltage
Insert	1	0	+/−	$+V_{c1}/+V_{c2}$
Bypass	0	1	+/−	0
Block	0	0	+	$+V_{c1}/+V_{c2}$
	0	0	−	0
Fault	1	1	±	0

1 (0) means turned-on (turned-off), and + (−) means current flows from SM positive side to negative side (negative side to positive side).

is inserted in the converter arm. By doing so, the control scheme of HBSM MMC is compatible to the CDSM MMC during normal operation. When a fault is detected, S_5 and other gate signals will be turned off. The capacitor voltages are inserted into the arm limiting the fault current with opposing polarity. In Figure 8.31c, current flows from the negative side to the positive side with $-V_c$ inserted, which will go through capacitors C_1, C_2 or both in parallel based on the comparison of the capacitor voltages. In Figure 8.31d, current flows from the positive side to the negative side with $+2V_c$ inserted. Table 8.13 lists the operation modes of each half-bridge structure for Case I when S_5 is on combined with any SM current direction and Case II when S_5 is off combined with positive SM current, while Table 8.14 lists the modes of CDSM for Case III when S_5 is off combined with negative SM current.

8.6.2 Device-Level Modeling Scheme

Device-level modeling refers to the generation of the circuit model of individual IGBT modules and diodes, which are the voltage source v_{sw} (subscript *sw* indicates the IGBT module or diode) in series with the resistor r_{sw} as shown in Figure 8.32a. In this work, the detailed device-level electrothermal model of the IGBT modules for CDSM MMC is built using the information obtained from the manufacture's datasheet for the Infineon® FZ400R33KL2C_B5 IGBT module. The algorithm contains the temperature-dependent electrical interface parameter calculation, the power loss calculation, the thermal network calculation, and the device-level linearized transient waveform calculation.

8.6.2.1 Temperature-Dependent Electrical Interface Parameter Calculation

The electrical interface of the IGBT or diode is composed of the voltage source v_{sw} in series with a resistance r_{sw}. The temperature-dependent IGBT and diode output

Table 8.14 Operation mode of CDSM: Case III (S_5 turned-off combined with negative SM current).

Mode	S_1	S_2	S_3	S_4	V_{c1} vs. V_{c2}	Inserted voltage
Positive Insert	1	0	1	0	$V_{c1} < V_{c2}$ $V_{c1} < V_{c2}$ $V_{c1} = V_{c2}$	$+V_{c2}$ $+V_{c1}$ $+V_{c1}/+V_{c2}$
Negative Insert	0	1/0	0	1/0	$V_{c1} < V_{c2}$ $V_{c1} < V_{c2}$ $V_{c1} = V_{c2}$	$-V_{c1}$ $-V_{c2}$ $-V_{c1}/-V_{c2}$
Bypass	0 1	1/0 0	1 0	0 1/0	Any	0
Fault	1/0 1	1/0 1	1 1/0	1 1/0	Any	0 or V_{c1} 0 or V_{c2}

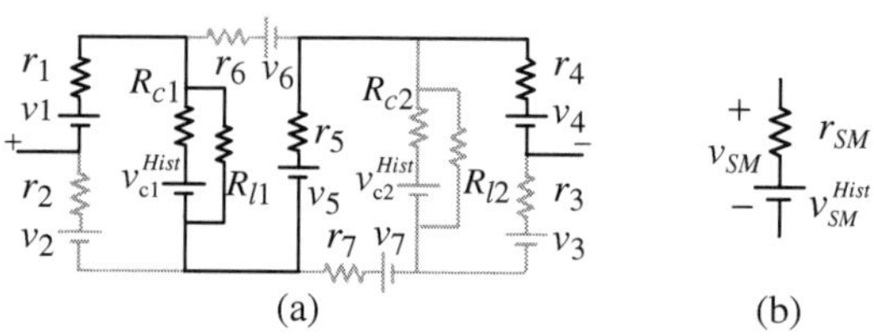

Figure 8.32 CDSM models: (a) complete circuit model with active route and (b) simplified SM interface model.

characteristics can be obtained from the datasheet, as shown in Figure 8.33a. The piece-wise polynomial curve fitting scheme is applied, with the following equations:

$$v(i) = \sum_{i=0}^{n} a_n i^n, \quad r_{sw} = \frac{\mathbf{d}v}{\mathbf{d}i} = \sum_{i=0}^{n-1} b_n i^n, \quad v_{sw} = v(i) - r_{sw} i, \tag{8.47}$$

where v and i are the voltage and current across the device; a_n and b_n are the correspondent polynomial coefficients. In this work, the curve is divided into three sections: the threshold section, the nonlinear section, and the linear section. In the threshold section, since the current is very small and has negligible influence on the electrical and thermal performance, the device is simply modeled as a large fixed resistor. In the nonlinear section, a third- or a fourth-order polynomial function is used, while the first-order fitting is applied when current is large in the linear section as shown in Figure 8.33a. Since the datasheet only provides the characteristics at $T_1 = 25$ °C and $T_2 = 125$ °C, linear interpolation is applied to calculate the values at arbitrary temperatures, which can also be used for other

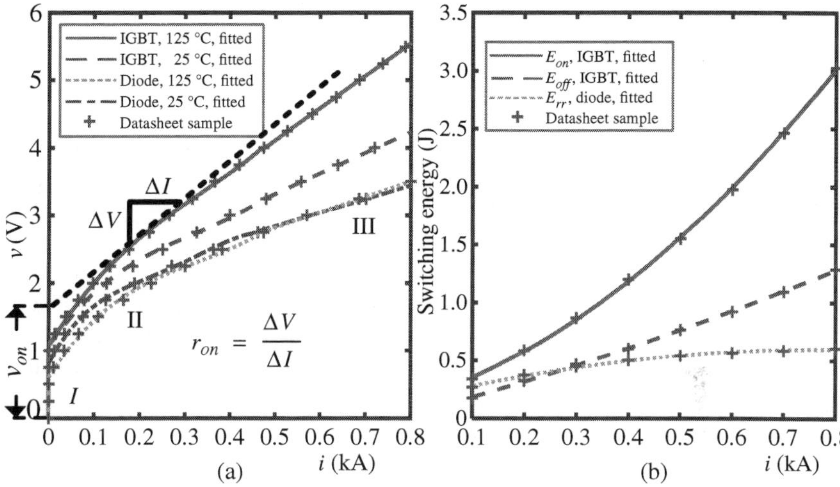

Figure 8.33 Datasheet sample and fitted curves of the Infineon® FZ400R33KL2C_B5 IGBT and diode characteristics: (a) output characteristics of IGBT and diode at 125 and 25 °C and (b) switching energy losses of IGBT turn-on, IGBT turn-off, and diode reverse recovery processes at 125 °C.

temperature-dependent variables given in datasheet, such as switching energy, IGBT current rise, fall time. Taking the example of r_{sw}, the interpolation equation is given as follows:

$$r_{sw}(T_{vj}) = \frac{T_{vj} - T_2}{T_2 - T_1}\left(r_{sw}^{T_2} - r_{sw}^{T_1}\right) + r_{sw}^{T_2}, \tag{8.48}$$

where T_{vj} is the junction temperature of the corresponding device.

The switching energies at 25°C and 125°C are available in the datasheet, and an example is given in Figure 8.33b. It is assumed to be proportional to the voltage between the device when turned off, and therefore, the fitting method in (8.21)–(8.23) can still be applied.

When using equivalent circuit method for SM-level modeling, only a portion of the devices are turned on in a certain operation case. Therefore, not all device-level model parameters are required for the calculation of SM interface. However, if the thermal performances of all power electronic devices in the SM are required to be monitored, the conditions and variables for all devices shall be calculated and stored.

8.6.2.2 Device-Level Linearized Transient Waveform Calculation

The piecewise linear IGBT curve-fitting model is used in this work for low hardware resource utilization as well as a low latency. The temperature-dependent rise and fall times of the IGBT are obtained from the datasheet. The slopes of those transients are assumed to remain constant; therefore, the rise time and fall time of the current and voltage waveform are proportional to the values of the turn-on current and turn-off voltage of the device, respectively. Since the datasheet only provides the rise time and fall time of the current waveform, the corresponding parameters of the voltage waveform are estimated by the switching energy losses.

8.6.3 SM-Level Modeling Scheme

The task of SM-level modeling is to obtain the simplified SM model composed of the voltage source and resistor. The computation effort of determining Thévenin equivalent is similar to solving the matrix equation of the SM circuit model. The number of effective nodes is calculated as the total nodes of the circuit topology minus one, which is assumed to be the ground node. The HBSM only contains two effective nodes, while the CDSM topology contains five effective nodes. The computation effort increases cubically with the matrix size to be solved, which is the same as the number of effective nodes. Therefore, it can be computationally expensive to use Thévenin equivalence method to calculate for all the CDSMs.

This work uses multiple equivalent circuits for different switching combinations, which is essentially the same as the surrogate network method, and the equivalent circuit method. With a certain switching combination, current flows into some determined branches. The leakage current of the IGBT module can be as small as several milliamperes, and is therefore neglected. Under most operating conditions, there is only one current path between the terminal nodes. Figure 8.32a shows the CDSM circuit model, and the black path indicates the active current route, when S_1 and S_4 are turned on. The capacitor is modeled as the impedance R_c in series with the history voltage source v_c^{Hist} using Trapezoidal rule, and a leakage resistance R_l is also included.

The SM-level model can be obtained by summing the circuit model elements of the devices in the active path to a voltage source in series with a resistor, as shown in Figure 8.32b. The determination of the circuit topology is based on the gate signals of the IGBTs, the current direction and the relation between the two capacitor voltages as listed in Tables 8.13 and 8.14. Occasionally, symmetrical current paths may happen, which can be easily combined by adding the conductances of the two paths.

8.6.4 Converter-Level Modeling Scheme

The MMC has a large number of switching devices and relatively complex circuit topology, which is a major challenge for its modeling and simulation. To accelerate the computation speed, the size of the system matrix shall be minimized. In this work, equivalent voltage sources and impedances are used for all SMs. Using the

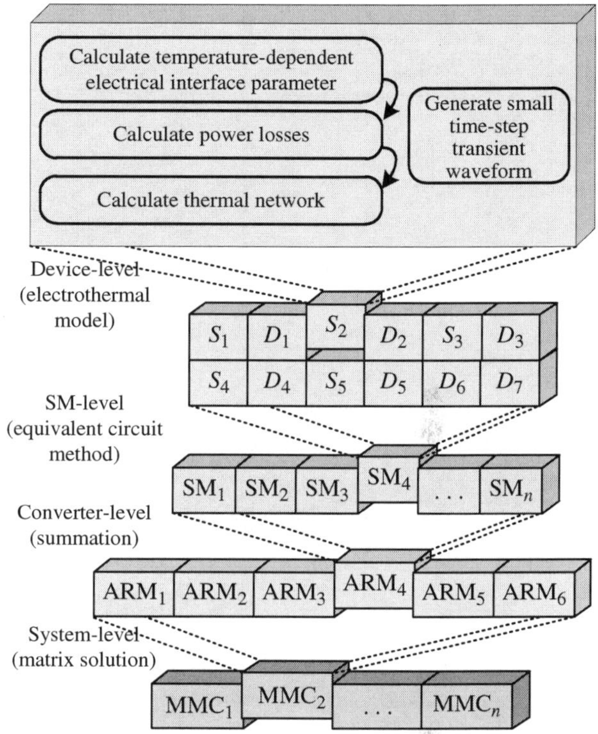

Figure 8.34 Illustration of hierarchical MMC modeling algorithm.

arm current of the last time-step, the simplified SM model can be summed to a single voltage source to interface with the external circuit. Therefore, the nodes of the interface elements for the converter arms are substantially decreased, and all SMs can be calculated in parallel.

Figure 8.34 illustrates the major algorithm of the electrothermal CDSM MMC model. The complete EMT simulation of MMC system is composed of device-level, SM-level, converter-level, and system-level calculation. The sequential relation exists between different levels, which means only when the interface elements of the fundamental level are obtained, the results of the next level can be accomplished. The calculation among the units of the same level can be parallelized well. When expanding each level, the amount of the paralleled device-level units is substantial and requires large computation capability. The parallelism also exists in each calculation step of the device-level modeling scheme.

8.6.5 Case Study and Hardware Implementation

A three-terminal DC system is used as the case study as shown in Figure 8.35a to illustrate the effectiveness of the parallel modeling schemes. The multiterminal

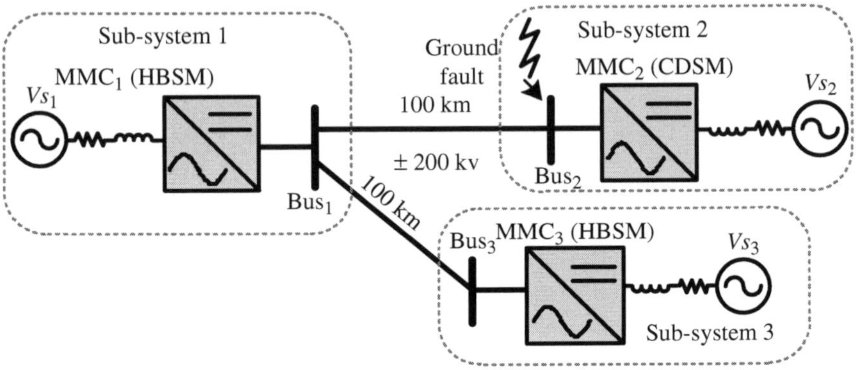

(a)

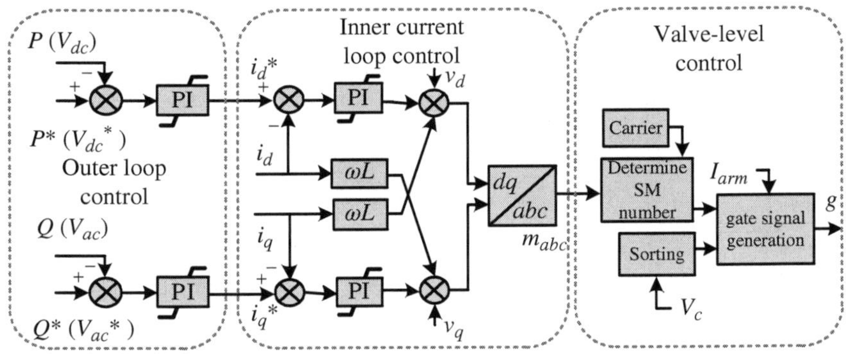

(b)

Figure 8.35 (a) Circuit topology of MMC-based three-terminal MTDC system and (b) control diagram of MMC. (The signals with the superscript * are reference signals.)

DC (MTDC) system is composed of two HBSM MMC stations and one CDSM MMC station connected by two 100 km ± 200 kV DC transmission lines. The major parameters of the case study are listed in Appendix A.7. A ground fault of both poles at Bus_2 is applied to the test condition. Since this work is focused on the modeling of CDSM MMC, the electrothermal model is only used for MMC_2, while MMC_1 and MMC_3 are using equivalent circuit models. All MMCs have 17 levels, which means that each MMC arm has 16 HBSMs or 8 CDSMs. To provide sufficient current rating and voltage rating, four FZ400R33KL2C_B5 IGBT modules or diode are connected in parallel, and 14 such paralleled structures are connected in series. The transmission lines are modeled using distributed line or TWM, and the system simulation time-step is 20 μs.

8.6.5.1 Design Partition

The computation of the MTDC system is allocated to the four application processing unit (APU) cores and programmable logic resources of the Zynq UltraScale+ MPSoC system. The major challenge of the partition process is to minimize the communication latency between different computation units. In other words, the exchanged data between different units shall be minimized. It is known that the programmable logic is more suited for highly paralleled tasks, which are the MMC SMs related computation in this work. To save the PL resource, other computations are located in PS part, though some of them could be accelerated in PL.

With the above consideration, the system design is first partitioned into the three subsystems separated by the distributed line model. Besides the process control signals for synchronizing purpose, the exchanged data during periodic calculation is only the history current of the transmission line model. The system matrix can also be divided into three smaller submatrices. One APU core is responsible for the calculation of each subsystem, and the fourth core is used for the system-level control calculation which generates the modulation signals for valve-level control. The valve-level control requires the SM capacitor voltage sorting process and generates gate signals for all SMs, which needs a large amount of data exchange from the SM calculation units. Therefore, the valve-level control is located in PL section to avoid the long latency between PS and PL communication.

Figure 8.36 shows the major computation processes of all applied units in Zynq UltraScale+ MPSoC. APU 1 is the master core that controls the simulation processes. At the beginning, it executes the hardware initialization, which includes waking up other cores, setting system counter, and initializing the PL module. Then it begins the initialization of the application, which are the calculation of the circuit element parameters, and the variable initialization. The initialized data is sent to other cores during the receiving process.

Then the periodic calculation begins, APU 1–3 can execute simultaneously for the correspondent subsystems. They first exchange the transmission line data, and then calculate the variables for other components, such as the transmission lines, arm inductances, voltage sources. In the meantime, the valve-level control, HBSM MMC, CDSM MMC modules of the subsystems on PL will start their respective calculations. With all variables calculated, the system matrix equations are calculated and the history data of the transmission lines and the inductances are then updated. APU 1 will first wait for the accomplishment of other cores and then wait until the next time-step comes in real-time before entering to the next loop. The system-level control can have its independent time-step and can transfer its system-level control data with a fixed delay. It uses classical *dq*-axis decomposition and PI controllers for the DC voltage and power outer loop control and the inner

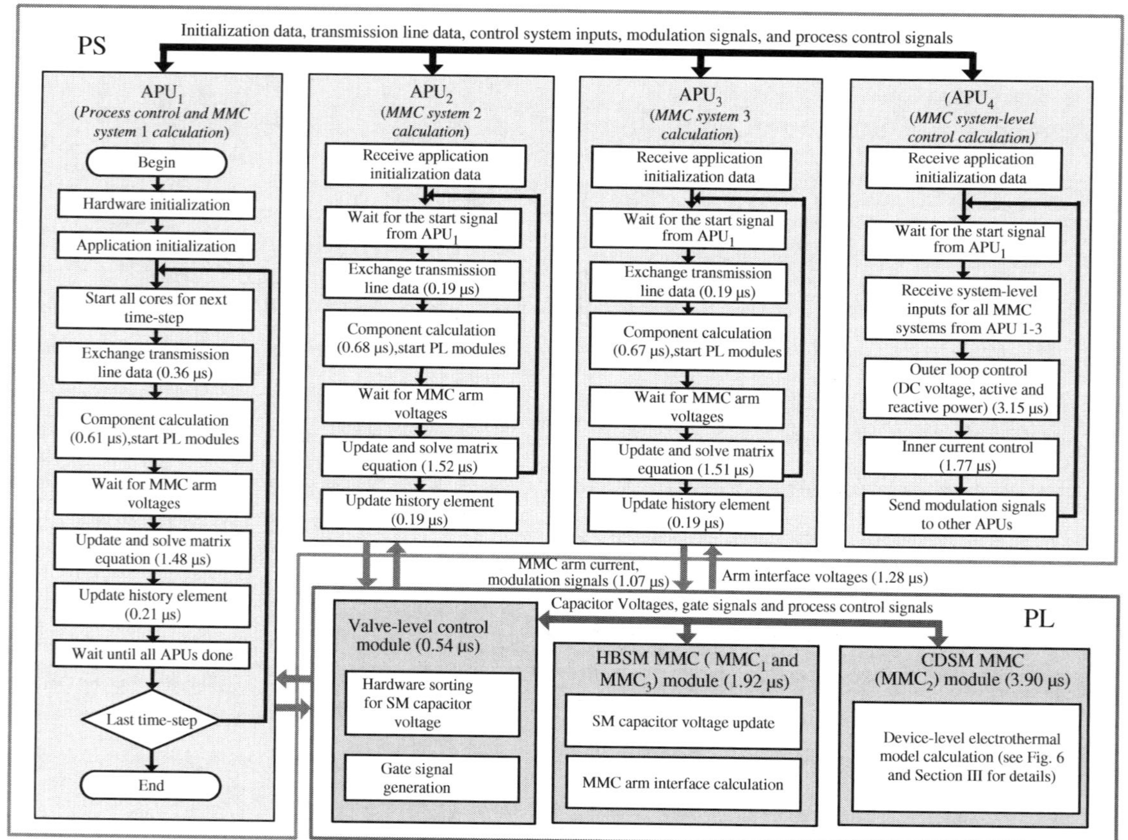

Figure 8.36 Design of the MTDC emulation system on MPSoC.

current loop control as shown in Figure 8.35b [170, 171]. In this work, the control on APU 4 uses the same time-step as the simulation time-step, and transfers its control variables during the component calculation process of other cores.

On the PL, the valve-level control module generates the gate signals for the HBSM MMC and CDSM MMC module using capacitor voltage sorting-based control scheme. Sorting n capacitor voltages in one converter arm is implemented with full $\frac{n(n-1)}{2}$ times comparison, which can be conducted in parallel on hardware. By adding the corresponding one-bit comparison results, the sequence is generated. Compared with sequential software sorting, the algorithm based on hardware is substantially faster which uses two FPGA clock cycles (20 ns) for a SM in the arm. It is noted that software sorting can also use $\frac{n(n-1)}{2}$ times comparison in the worst-case scenario, which must be considered for real-time application. In HBSM MMC module, the major tasks are the SM capacitor voltage update and MMC arm interface calculation. The interface calculation for SMs in one arm is conducted in parallel. The interface voltage sources of SMs are summed to generate the arm interface, and then sent back to corresponding APU cores. The general process of CDSM MMC is similar to the more complex process of the SM interface generation, where the device-level electrothermal model is applied as described in the device-level modeling part.

8.6.5.2 Latency and Resource Consumption

The latencies of the major processes during the periodic calculation are shown in Figure 8.36. For instance, the calculation time for the component calculation takes 0.61 μs and the calculation of the matrix equation takes 1.48 μs on APU 1. The calculation time on different APU cores has slight variance due to the minor difference of the included components. For instance, MMC system 2 contains the two-state resistance connecting to the ground for short-circuit test purpose, and MMC system 1 contains two outgoing transmission lines. Although the computation can be fully paralleled on FPGA resources as long as the algorithm allows, it may not be necessary and can consume much hardware resource. Multiplexing and pipelining are used for the resource optimization. The valve-level control module for one arm takes 30 ns, and it takes 540 ns for 18 arms in three MMCs by multiplexing. The HBSM MMC module includes three sets of the paralleled arm calculation units, and the CDSM MMC module has two sets of arm calculation units. The resource consumption for the modules are listed in Table 8.15, which is relevant to the allocation of corresponding arm-level units. In other words, the resource consumption can be lower with a higher cost of the latency by using fewer paralleled units. Another strategy to reduce the resource consumption and increase the MMC level number is to apply a hybrid modeling scheme by using both electrothermal model and simple equivalent circuit model in the same MMC, which will be elaborated in Chapter 9.

Table 8.15 Resource consumption of modules on PL.

Module	LUT	FF	DSP
Valve-level control module	24 635 (9.0%)	8 234 (1.5%)	0 (0%)
HBSM MMC module	13 896 (5.1%)	12 390 (2.3%)	135 (5.4%)
CDSM MMC module	213 104 (77.7%)	160 800 (29.3%)	970 (38.5%)
Available	274 080 (100%)	548 160 (100%)	2 520 (100%)

8.6.6 Real-Time Emulation Results and Analysis

The emulation system hardware is set up as presented in Chapter 1. A host computer downloads the programming files including the bitstream to the Xilinx® ZCU102 MPSoC board [172] through the USB cable and the on-board JTAG chip. On the Zynq device, the CPU cores are running at 1.2 GHz, while the FPGA is running at 100 MHz. The real-time results are transferred to the digital-to-analog converter through the FMC connector and the FMC-DAC adapter. The analog results are then collected on the oscilloscope through SMA cables.

8.6.6.1 Steady-State Results

Figure 8.37 presents the CDSM MMC AC voltages, AC currents, and the first SM capacitor voltages of the upper and lower arm in phase-A. The left-side oscilloscope results are compared with the right-side PSCAD/EMTDC® results. The small control error can be accumulated and can lead to different conditions of SMs, especially for the voltage sorting-based algorithm. Therefore, the switching patterns for individual SMs cannot exactly match between the real-time results and PSCAD/EMTDC® results, which explains that the overall performance matches quite well, however, the harmonics can have small differences. Figure 8.38 presents the turn-on and turn-off switching transients of S_1. The results are compared with SaberRD®, which uses detailed datasheet-based device-level behavior model for IGBT and diode. The current rise time t_r, current fall time t_f, voltage rise time t_{Vr}, voltage fall time t_{Vf} are measured, with the definition of 10–90% change of the steady-state I_c and V_{ce} as shown in Figure 8.38. The overshoot current during the turn-on process is generated due to the reverse-recovery current of the turn-off of D_2. The linearized waveforms can give a good estimation of the switching transients based on the datasheet.

8.6.6.2 DC Power Flow Control

This section presents the test results of the power flow control of the MTDC system as shown in Figure 8.39. Among the three MMC stations, MMC_1 maintains the

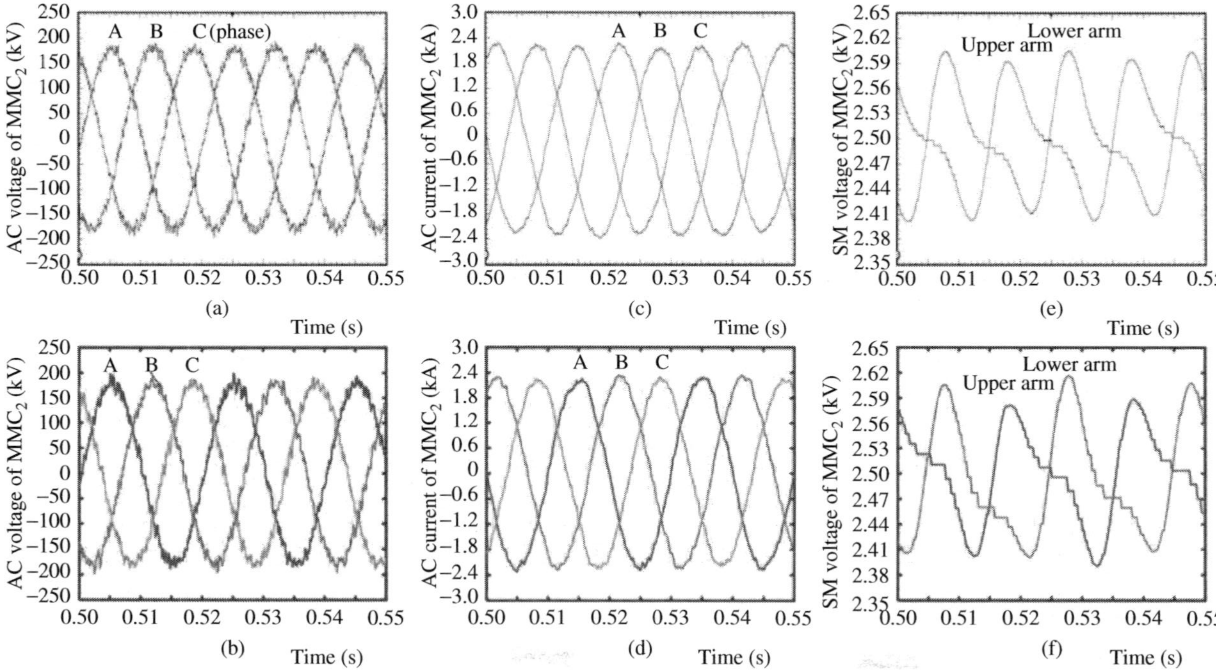

Figure 8.37 Steady state results: (a) (b) AC voltages, (c) (d) AC currents, and (e) (f) SM voltages. (Top sub-figures are real-time results, and bottom sub-figures are offline PSCAD/EMTDC® results.)

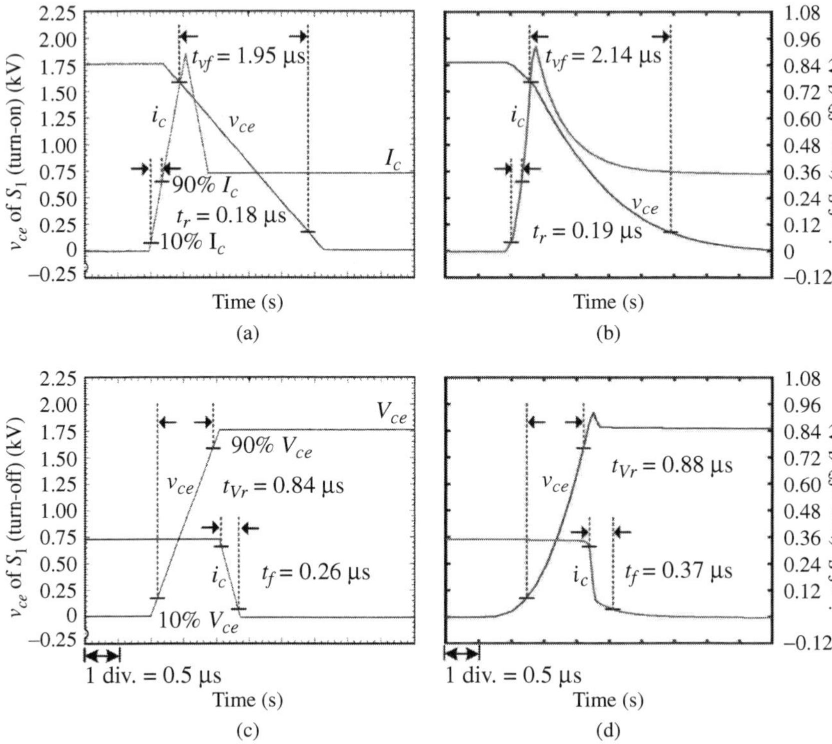

Figure 8.38 Voltage and current switching transient waveforms of S_1: (a) (b) turn-on process and (c) (d) turn-off process. (Left sub-figures are real-time results, and the right sub-figures are offline SaberRD® results.)

DC voltage of the MTDC grid, while MMC_2 and MMC_3 control the power flow. All the MMCs can adjust the reactive power of the corresponding AC systems independently. The reference active power of MMC_2 changes from −600 to 400 MW from 1.2 to 1.5 seconds of the simulation run, and the reference active power of MMC_3 changes from −400 to 200 MW from 2.0 to 2.3 seconds of the simulation run. The positive direction of the active power is defined as the direction from the DC side to the AC side of the converter. The reactive power of the converter is all controlled to be 0 Mvar. As shown in Figure 8.39a, the reference power tracking performances of MMC_2 and MMC_3 is satisfactory, and the power flow of MMC_1 can eventually reach steady-state with the disturbance due to the regulation of DC voltage. Figure 8.39b shows the change and variation of DC current and DC voltage of MMC_1. Figure 8.39c shows the power loss of S_1 (MMC_2), which first decreases to almost 0 kW and then rises to a steady condition during the power flow reverse.

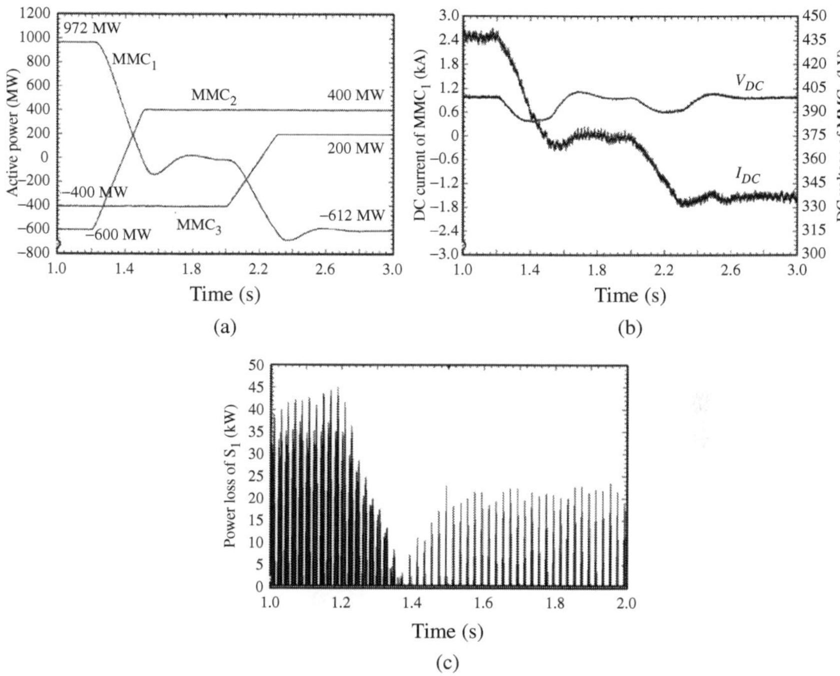

Figure 8.39 Real-time results: (a) Active power of MMC_1, MMC_2 and MMC_3, (b) DC current and DC voltage of MMC_1, and (c) power loss of S_1.

8.6.6.3 DC Fault Transient Results

A DC ground fault is applied to the case study system at 0.8 second of the simulation run. Once the instantaneous over-current is detected which is ±10 kA in this work, the protection mode is activated and S_5 is turned off. The capacitor voltages are then inserted into the arm with the reversed polarity to suppress the fault current. Once the arm current reaches 0, the breaker in the converter arm is then opened. Figure 8.40 shows the comparison of the arm current fault transients of the upper and lower arms between the real-time results on the oscilloscope and the PSCAD/EMTDC® results. It is observed that the rate changes during the decrease of the current. It happens because when the other arm current in the same converter leg reaches 0, the effective circuit topology is changed. Figure 8.41 presents the power losses of S_1 and S_5 in the first CDSM of phase-A upper arm. The conduction power losses of the switches are zoomed between 0.775 and 0.795 second. The switching power is not exactly instantaneous power, but the average power during every simulation time-step. The power losses during the fault transient are substantially larger than the losses during steady-state operation. The composition of the power losses between S_1 and S_5 is very different due to the different

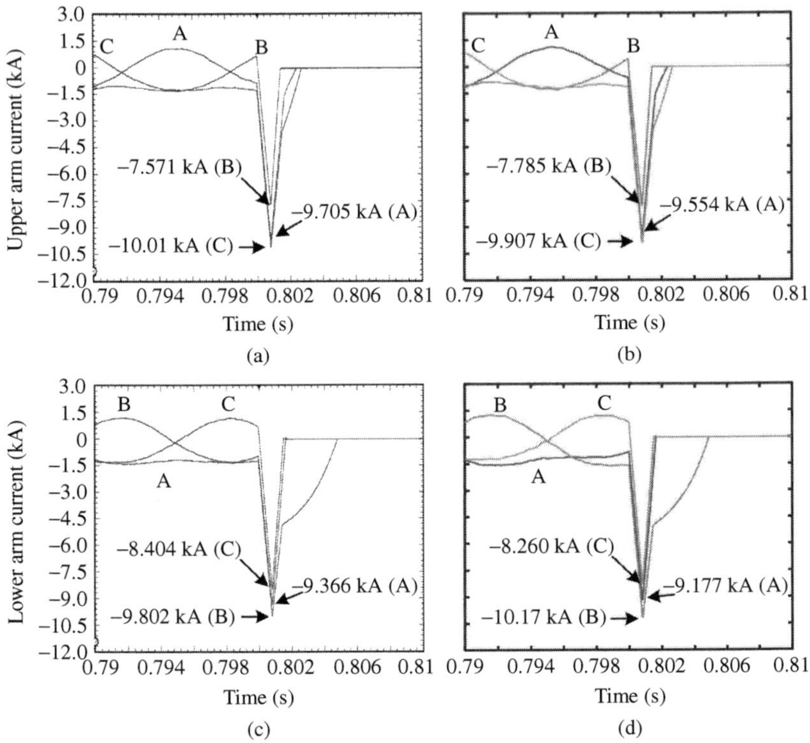

Figure 8.40 Arm current transients: (a) (b) upper arm currents, (c) (d) lower arm currents. (Left sub-figures are real-time results, and the right sub-figures are offline PSCAD/EMTDC® results.)

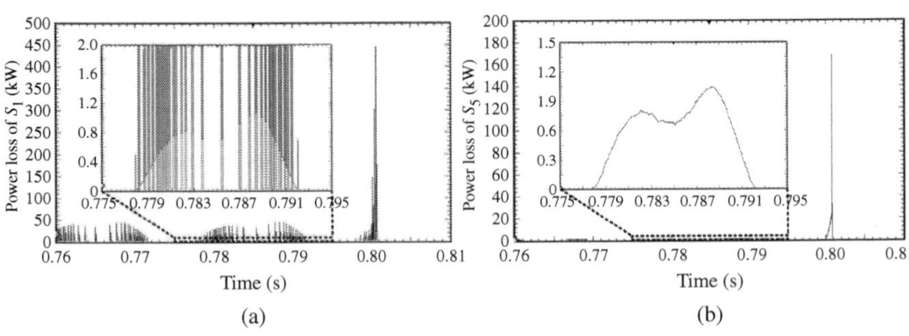

Figure 8.41 Real-time power losses: (a) power loss of S_1 and (b) power loss of S_5.

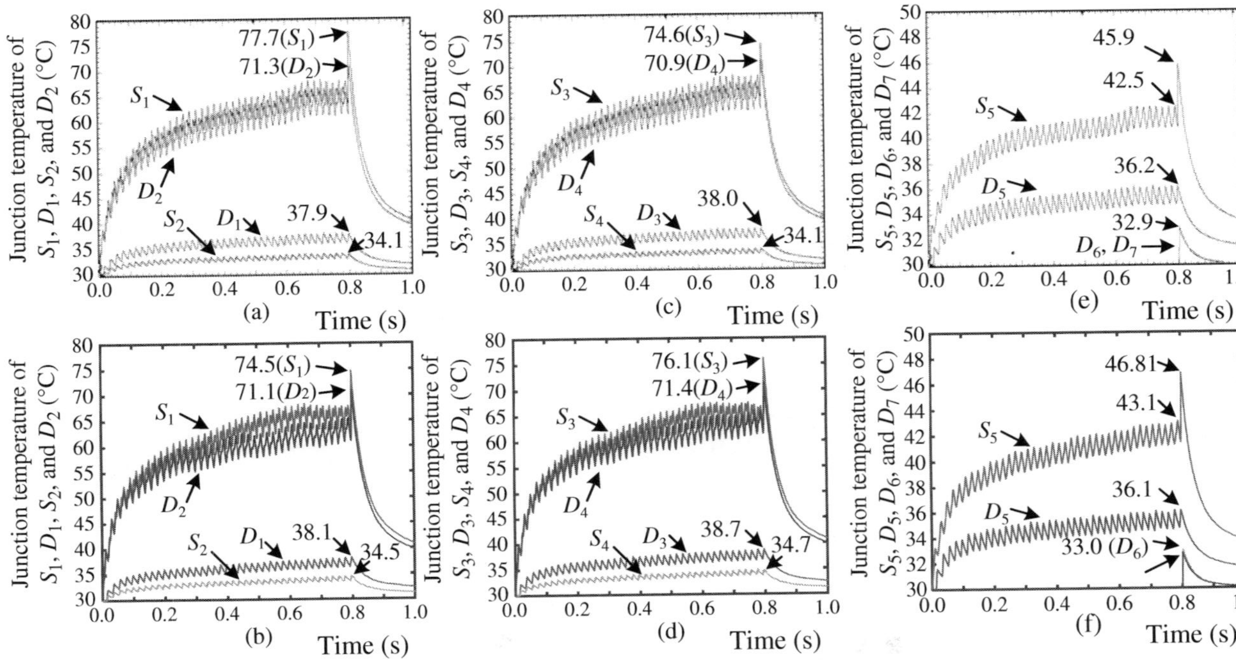

Figure 8.42 Junction temperatures: (a) (b) S_1, D_1, S_2, and D_2, (c) (d) S_3, D_3, S_4, and D_4, and (e) (f) S_5, D_5, S_6, and D_7. (Top sub-figures are real-time results, and bottom sub-figures are offline SaberRD® results.)

switching pattern of the devices. S_5 is always turned on unless the protection mode is active. Therefore, during normal operation, there are only conduction power losses. Figure 8.42 shows the junction temperatures of all devices in the first CDSM in phase-A upper arm, which are compared with SaberRD® results. Due to the difference of the switching pattern, the existence of the MMC loop current, the junction temperatures can be quite different for the devices though they may have same electrothermal characteristics. When the fault happens, the junction temperatures of the active devices first rise sharply due to the abnormally high power losses, then decrease to a lower value since the converter is shut down. Due to the assumption that the paralleled devices are mounted on the same heatsink, the junction temperatures of D_6 and D_7 are not influenced by other devices and do not change until the fault mode is active. When S_5 is turned off, the two capacitor voltages are balanced very quickly, and D_6 and D_7 evenly share the current as long as they have same characteristics. The introduced method assumes that the voltages will be instantaneously balanced, otherwise numerical oscillation may occur with EMT type algorithm. It explains that the junction temperatures of D_6 and D_7 are exactly same in oscilloscope results, while there is a minor difference in the SaberRD® results. Due to the difference of the modeling schemes and solution methods between the hardware emulation and SaberRD®, a small amount of error is expected.

8.7 Summary

This chapter emphasizes on detailed device-level electro-thermal modeling of the modular multilevel converter as an interface between the AC and DC grids for hardware acceleration or real-time simulation for converter design evaluation under both normal operation and contingencies for specific control and protection schemes that otherwise would not be achievable by the ideal switch model which are adopted by prevalent commercial offline simulation tools. Two major circuit reconfiguration strategies are presented, i.e. the merging of an MMC arm and its partitioning using either the lossless TLM link or coupled voltage-current sources. It was demonstrated that when the curve-fitting model is applied, the submodule can be deemed as a Thévenin circuit regardless of its complexity, while splitting submodules from the arms is the most effective method when high-order nonlinear iterative models are included for better versatility.

Circuit partitioning enables the coexistence of separated nonlinear submodules and the linear MMC circuit even though they have distinct time-steps, and from a mathematical point of view, it achieved the decomposition of a large matrix equation corresponding to the integral MMC circuit by a set of smaller equations, which when solved in parallel, significantly accelerated computational speed while avoids numerical divergence. Meanwhile, the method offered a new perspective for hardware design in which the overall system is represented by several hardware modules and any change specific to one of them has no impact on others; thus scalability and modularity could be attained, just as in a real MMC system.

Regarding the performance of device-level models in system-level simulation, the nonlinear behavioral model is more adaptive to electromagnetic environment variation than its curve-fitting counterpart, and consequently, can be applied to various power converters to obtain their precise behavior for thorough circuit design evaluation. The consistency between HIL emulation results and those from off-line simulation tools indicated that device-level modeling has a wide application prospect ranging from nanoseconds behavior observation to system-level performance preview.

9

DC-DC Converters

9.1 Overview

A general DC grid could contain a number of subsystems with different voltage ratings. Like the power transformers being pivotal in the AC transmission and distribution network, the formation of an extensive DC grid relies on its own voltage transformation equipment, the DC-DC converters. The solid-state transformer (SST), which achieves voltage conversion by medium- or high-frequency isolation in conjunction with power converters, is a prevalent DC-DC converter type in high-voltage scenarios, and it is able to fully control the voltages and currents of its terminals and, therefore, offers a new approach to regulate the power flow in the DC grid.

The modular multi-level converter (MMC) has prompted the development of high voltage direct current (HVDC) technology for various applications, including the SST, which enabled the connection of a number of HVDC stations to form a multi-terminal direct current (MTDC) system [173, 174], in addition to other benefits such as fault isolation, and power flow control [175–177]. The MMC-based SST is composed of an inverter, a rectifier, and a transformer, which is designed to be physically isolated from the AC grid and can, therefore, operate at a medium frequency to make the SST less bulky [178]. However, such a benefit is accompanied by a corresponding rise of switching frequency and higher switch power loss.

Real-time simulation of HVDC system plays a pivotal role in validating control and protection strategies as well as providing a platform for system performance study [179–181]. Simple converter models such as the averaged model can be employed for gaining speed advantage while the accuracy is compromised [182]. Modeling the power converter specifically using the ideal switch model with a fixed low on-state resistance R_{on} and a high off-state resistance R_{off} is sufficient for system-level studies which can show effects such as switch transition and

Real-Time Electromagnetic Transient Simulation of AC-DC Networks, First Edition.
Venkata Dinavahi and Ning Lin.

harmonics [183, 184]. However, the preference for high-power density by increasing the switching frequency and reducing the converter volume underscores the necessity of acquiring switching power loss for converter design evaluation. Consequently, it is better to include device-level switch models so long as they do not hinder achieving real-time execution. The inability of the ideal model in providing any switching transients precludes it from being employed for electro-magnetic-thermal simulations.

Meanwhile, efforts have been expended in developing control algorithms to reduce insulated gate bipolar transistor (IGBT) thermal stresses [185], and with switching details, determination of an appropriate algorithm prior to setting up a real converter becomes feasible. Other detailed IGBT models were developed to provide insight into accurate switching waveforms and are therefore frequently referred to for power converter design assessment [133, 186]. Their main drawbacks are low computation efficiency due to the solution of device physics and the tendency of numerical divergence even when the converter scale is small. The curve-fitting model is, therefore, favored in large circuit simulation for electro-thermal analysis, where the transient values are programmed in advance or stored in a look-up table (LUT) [187, 188], yet to achieve a maximum versatility of transient waveforms to electromagnetic surroundings, the dynamic curve-fitting electrothermal IGBT model described in Chapter 7 is adopted so that more device-level information can be revealed. A hybrid arm is constructed by combing the dynamic curve-fitting model (DCFM) with ideal submodules, which possesses merits such as lower hardware resource requirement and faster execution speed when hardware-in-the-loop (HIL) emulation is performed on the field programmable gate array (FPGA).

With a primary focus on real-time hardware emulation of a device-level MMC-based SST, this chapter also narrates modeling and hardware design of a four-quadrant Buck–Boost converter, as it is fully capable of carrying out system-level simulation of a DC grid. The two DC-DC converters are shown in Figure 9.1.

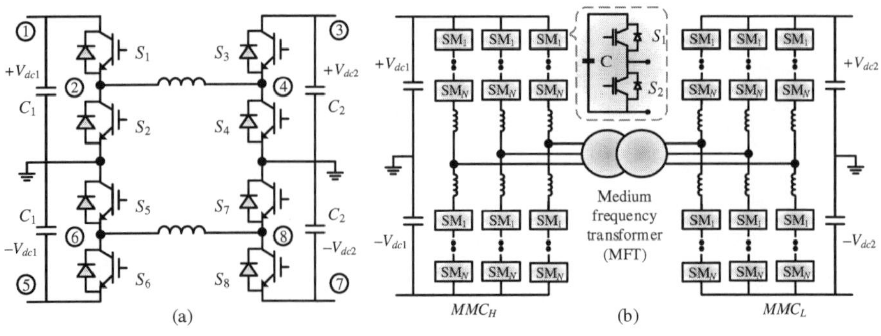

Figure 9.1 DC-DC converters: (a) nonisolated four-quadrant Buck–Boost converter, and (b) detailed MMC-based solid-state transformer.

9.2 Buck–Boost Converter

9.2.1 System-Level Modeling

When the simulation of a large-scale DC grid is conducted for purposes such as power flow study, the necessity of employing a fully detailed SST becomes low and therefore, simplification is justified. In such scenarios, basic nonisolated DC-DC converters with Type 3 power semiconductor switch model or Type 5 average model are sufficient.

Figure 9.1a shows the configuration of a neutral-grounded Buck–Boost converter. With the IGBT being represented by Type 3 switch model, the overall circuit contains eight nodes excluding the ground, and its symmetry indicates the positive pole can be decoupled from its negative counterpart in nodal voltage solution, or mathematically, the admittance matrix is a block diagonal matrix and can thus be decomposed. As a result, it is reasonable to carry out modeling and hardware design of an arbitrary pole. With power semiconductor switch resistances determined by the controller, the matrix equation takes the form of

$$\begin{bmatrix} U_1(t) \\ U_2(t) \\ U_3(t) \\ U_4(t) \end{bmatrix} = \begin{bmatrix} G_{S_1} + G_{C_1} & -G_{S_1} & 0 & 0 \\ -G_{S_1} & \sum_1^2 G_{S_i} + G_L & 0 & -G_L \\ 0 & 0 & G_{S_3} + G_{C_2} & -G_{S_3} \\ 0 & -G_L & -G_{S_3} & \sum_3^4 G_{S_i} + G_L \end{bmatrix}^{-1} \cdot \begin{bmatrix} 2v^i_{C_1}(t) \cdot G_{C_1} \\ 2v^i_L(t) \cdot G_L \\ 2v^i_{C_2}(t) \cdot G_{C_2} \\ -2v^i_L(t) \cdot G_L \end{bmatrix}, \tag{9.1}$$

where G_{C_1}, G_{C_2}, and G_L are conductance of two capacitors and the inductor, respectively, $v^i_{C_{1,2}}$ and v^i_L are their transmission line modeling (TLM) model incident pulses, and $G_{S_{1-4}}$ is the IGBT conductance.

The Buck–Boost converter control can simply be realized by the classic proportion-integration controller, as given in Figure 9.2. The integration part of PI controller can mathematically be expressed in the following formula in digital simulation:

$$s(t) = K_I e(t) + s(t - \Delta t), \tag{9.2}$$

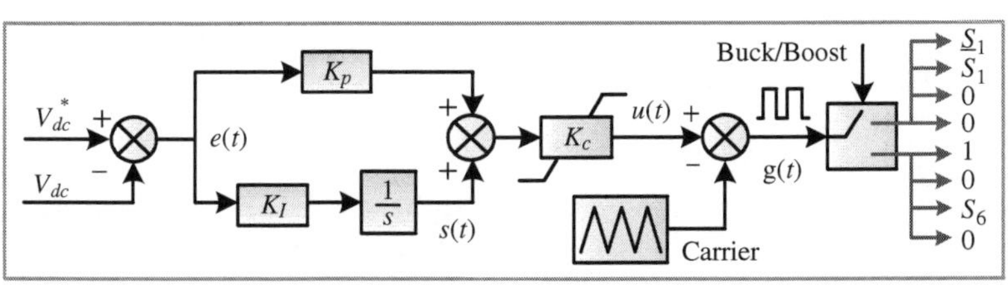

Figure 9.2 Buck–Boost converter controller.

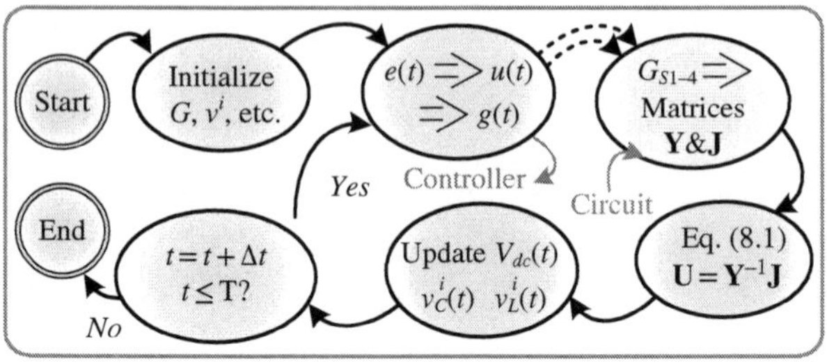

Figure 9.3 Buck–Boost converter EMT simulation process.

where $s(t-\Delta t)$ is the output of that section in the previous time step, and then the output of PI section can be expressed as

$$u(t) = K_p e(t) + s(t). \tag{9.3}$$

Knowing the circuit matrix equation and the controller, the electromagnetic transient (EMT) simulation process of the DC-DC converter can be briefly depicted by the flowchart in Figure 9.3. The circuit parameters and controller variables need to be initialized once the simulation starts. The controller and circuit are relatively independent, and therefore, they can proceed concurrently on parallel hardware architectures, or the controller can be computed first to generate IGBT gate signals which subsequently ascertain the circuit parameters. Following the establishment of admittance matrix **Y** and history current vector **J**, the nodal voltages can be derived by solving the matrix equation (9.1). Before the simulation steps into the next cycle or ends, the DC voltage and history items should be updated.

9.2.2 Hardware Implementation

Based on the description, the Buck–Boost converter can be divided into two modules, namely the circuit and controller when hardware implementation is carried out using the high-level synthesis tool Vivado HLS®. The two parts are first written as C/C++ functions, and then are converted into hardware modules, as demonstrated in Figure 9.4. The gate signals are generated by the controller and thus are the inputs of the circuit module, and the output DC voltage as the control target is in turn fed back. In addition to interblock exchange, both parts also have updates from their own history items due to the PI controller and reactive components, and each of these signals could either be taken as two separate input and output variables, or just one variable identified by its address in the function's declaration. As shown in the figure, these internal variables signals are identified by their addresses, for example, once the iteration count reaches its maximum four times, the nodal voltage vector is ascertained and used to calculate incident pulses v_C^i and v_L^i of reactive components, which are sent to the vector **J** without assigning the value to a new signal.

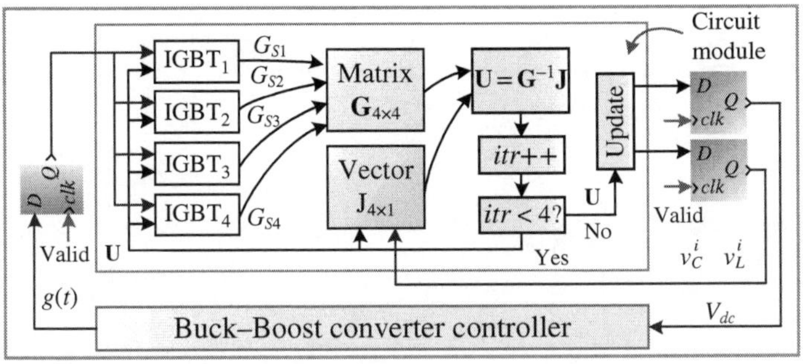

Figure 9.4 Buck–Boost converter top-level hardware implementation in Vivado HLS®.

Hardware parallelism is realized in the Buck–Boost converter emulation. The four IGBTs are computed simultaneously, and so are formation of matrix **G** and vector **J**. The circuit and controller are also enabled to be concurrent following the automatic register-transfer level (RTL) synthesis of D latches. It indicates that at the moment, a hardware module or a component inside it completes calculation, its outputs turn valid and therefore are stored in the D latches. When the next time-step begins, these stored values can be read by the other module. In this case, the circuit and controller read their own inputs and start calculation when an emulation cycle starts, the controller has a lower latency and completes first, but it waits until the circuit part finishes. Prior to the next emulation cycle, the outputs of the two modules have already been stored in the D latches, and therefore, they are available for the next time-step.

9.3 Solid-State Transformer Modeling

In detailed DC grids simulation that integrates power electronic device-level models into large-scale systems, a practical DC-DC converter, such as the front-to-front MMC-based SST shown in Figure 9.1b, is required to derive the electrothermal information for equipment operation evaluation. Figure 9.5 shows a typical way that the SST is integrated into a multiterminal HVDC system, where MMC_1 is set as the rectifier station, and MMC_2 and MMC_3 are inverter stations. The SST comprises of two MMCs, denoted as MMC_H and MMC_L, and one medium-frequency transformer (MFT).

The complexity of MMC model determines the speed of DC grid EMT simulation. Among a number of choices, the ideal switch-based detailed equivalent model (DEM) has been proven by EMT tools and is prevalent for achieving faster simulation speed compared with traditional models [160]. However, it could not describe the exact on/off state of an IGBT with a freewheeling diode, let alone a

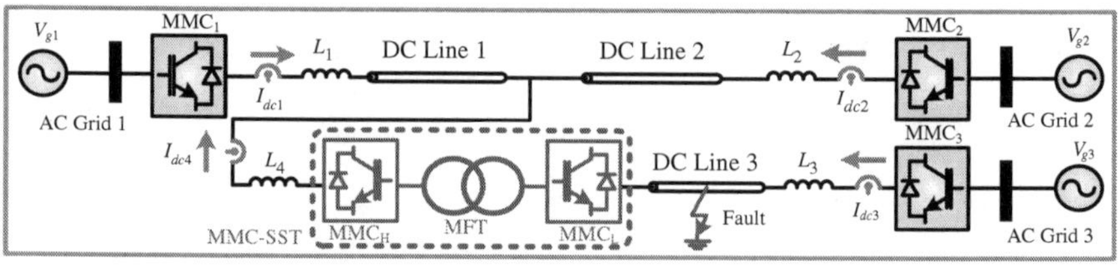

Figure 9.5 MMC-based DC-DC converter for MTDC system.

lack of device nonlinearities. On the contrary, high-order nonlinear switch models are highly inefficient for CPU simulation and usually require large amount hardware resource for FPGA implementation due to the iterative nature of the solution. Thus, for real-time emulation purpose, a hybrid MMC arm model which features device details and computational efficiency is adopted.

9.3.1 MMC Arm Models

9.3.1.1 TLM-Stub Model (TLM-S)

Based on an MMC submodule's performance, another detailed model based on the Thévenin or Norton equivalent circuit can be derived. According to Figure 8.2, when an arbitrary submodule numbered k is under on-state, the capacitor is being charged through the upper switch which is instantaneously a small resistance, and if the submodule is off, the capacitor is short-circuited and consequently the equivalent circuit is purely a small resistor, as shown in Figure 9.6a,b. Thus, an on-state resistance is always in the conducting path during operation. The existence of SM capacitor can be determined by the gate signal of the upper switch, deemed as a binary, i.e. $V_{gk} = 1$ for on-state and $V_{gk} = 0$ for off-state. For the blocked state, simply ordering $V_{gk} = 0$ and $R_2 = R_{off}$ omits the free-wheeling diode effect, just as the DEM. To enable correct SM ON/OFF mode of the blocked state, the gate signal is determined by the direction of the arm current i_{SM}: if it flows into the SM through node a, which is defined as the positive direction, then $V_{gk} = 1$; otherwise, $V_{gk} = 0$. This criterion leads to two equivalent circuits similar to Figure 9.6a,b. For the

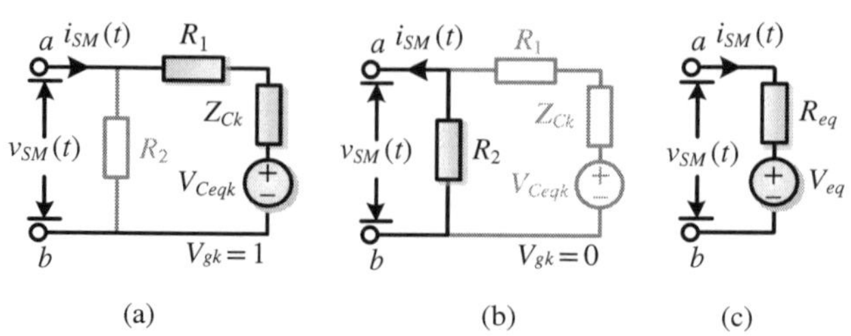

Figure 9.6 MMC TLM-stub model: (a) SM on-state/blocked state, (b) SM off-state/blocked state, and (c) generic representation.

former state, the submodule impedance equals to $R_{on} + Z_{Ck}$, while this value is R_{on} for the latter. Correspondingly, the capacitor voltage v_{Ck} alternates between $i_{SM} \cdot Z_{Ck} + V_{Ceqk}$ and V_{Ceqk}. Nevertheless, to simulate the high-impedance mode of the blocked state when both diodes are off, v_{SM} and v_C are required to judge whether the upper diode should be turned on. In the TLM-stub theory [189], a capacitor is equivalent to a section of lossless line with its far end an open-circuit and a round-trip time the simulation time-step. Thus, its Thévenin equivalent circuit takes the form of (2.55), where the backward voltage $v^-(t + x/v)$ is termed as the incident pulse $v^i_C(t)$, and the characteristic impedance expressed as $Z_C = \Delta t/(2C)$. The submodule capacitor voltage, alongside its incident pulse $v^i_{Ck}(t)$ being updated each time-step, can then be written as

$$v_{Ck}(t) = V_{gk}(t) \cdot i_{SM}(t) \cdot Z_{Ck} + 2v^i_{Ck}(t), \tag{9.4}$$

$$v^i_{Ck}(t + \Delta t) = v_{Ck}(t) - v^i_{Ck}(t). \tag{9.5}$$

The Thévenin equivalent circuit of an SM can be obtained as in Figure 9.6c, in which

$$R_{eq}(t) = R_{SM} + V_{gk}(t) \cdot Z_{Ck}, \tag{9.6}$$

$$V_{eq}(t) = 2V_{gk}(t) \cdot v^i_{Ck}(t), \tag{9.7}$$

where R_{SM} equals to R_{on} for all states except high-impedance mode when it should be R_{off}.

Then, for an MMC arm containing N submodules, the Thévenin equivalent circuit can be expressed as

$$V_{arm\_eq}(t) = V_{Leq}(t) + \sum_{k=1}^{N} V_{eqk}(t) = V_{Leq}(t) + 2\sum_{k=1}^{N} \left(V_{gk}(t) \cdot v^i_{Ck}(t)\right), \tag{9.8}$$

$$R_{arm\_eq} = Z_{Lu,d} + \sum_{k=1}^{N} R_{eq}(t) = Z_{Lu,d} + NR_{SM} + Z_{Ck}\sum_{k=1}^{N} V_{gk}(t). \tag{9.9}$$

Similar to the detailed equivalent circuit model, the above two equations can also be converted into the form of Norton equivalent circuit when the nodal voltages are to be solved.

9.3.1.2 Nonlinear Switch-Based Model (NSM)

When the operation frequency of MFT increases for a higher power density, the switching frequency should follow, incurring larger IGBT power loss and higher junction temperature. However, as introduced in Chapter 7, MMC models based on the ideal switch such as Type 3 and Type 4 are incapable of electrothermal calculation. Thus, the datasheet-driven DCFM involving switching transients is adopted, along with submodule partitioning schemes narrated in Chapter 8.

9.3.1.3 Hybrid Arm Model

The introduction of nonlinear switch model to the arm leads to a more complicated MMC network, for which the submodule merging approach is not instantly feasible. Voltage-current source coupling is used to separate submodules from MMC arms, as shown in Figure 9.7, where voltage sources are placed on the linear arm side. The hybrid arm can contain a flexible number of split submodules while the rest adopt TLM-S for efficient computation and less hardware utilization when deployed to FPGA. Then, the arm's Thévenin equivalent circuit can be expressed as

$$v_{arm}(t) = i_{SM}(t) \cdot Z_{Lu,d} + 2v_L^i(t) + \sum_{k=1}^{n} v_{SMk}(t-\Delta t) + i_{SM}(t) \cdot \sum_{k=n+1}^{N} R_{eqk}(t) + \sum_{k=n+1}^{N} V_{eqk}(t-\Delta t), \tag{9.10}$$

where $Z_{Lu,d}$ and $2v_L^i$ is the TLM stub model for an arm inductor. v_{SMk} is the voltage coupling of kth submodule, and the number $n \in [1, N-1]$.

On the nonlinear submodule side, the computation approach relies on switch state. Under steady state, both switches of the SM are taken as resistors, then

$$\begin{bmatrix} v_{Ck}(t) \\ v_{SMk}(t) \end{bmatrix} = \begin{bmatrix} Z_{Ck}^{-1} + R_1^{-1} & -R_1^{-1} \\ -R_1^{-1} & R_1^{-1} + R_2^{-1} \end{bmatrix} \cdot \begin{bmatrix} v_{Ck}^i(t) \cdot Z_{Ck}^{-1} \\ i_{SMk}(t-\Delta t) \end{bmatrix}, \tag{9.11}$$

where $v_{SMk}(t)$ is the voltage to be sent to the opposite side. The SM blocked state is a special steady-state, where R_1 and R_2 are determined by the arm current: a positive i_{SM} indicates that the upper diode is on and R_1 is small, otherwise, R_2 has a small resistance. During transient state, the IGBT is taken as a controlled current source, and the current in the complementary switch, defined as flowing from collector to emitter is given as follows:

$$i_C'(t) = i_C(t) \pm i_{SM}(t-\Delta t). \tag{9.12}$$

Lastly, knowing the branch currents enables the calculation of circuit other variables, such as $v_{Ck}(t)$ and $v_{SMk}(t)$.

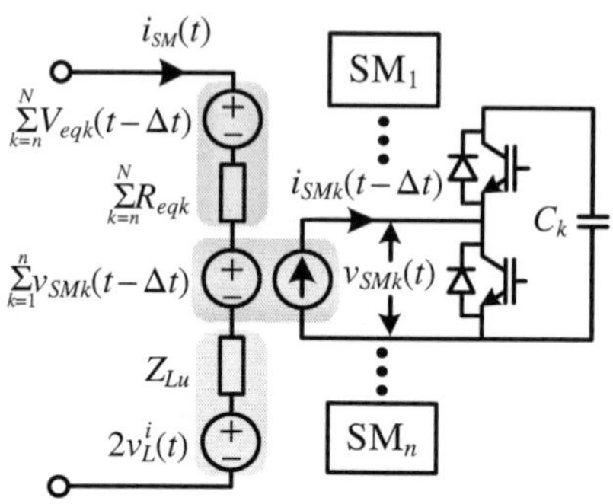

Figure 9.7 MMC hybrid arm model with $V-I$ couplings.

The partitioning method induces a unit delay to both sides. At the instant $t - \Delta t$, $i_{SM}(t - \Delta t)$ is obtained by solving the matrix equation corresponding to the left circuit, and it is sent to the SM. Then, instant t begins. In the submodule, based on $i_{SM}(t - \Delta t)$ it just received, $v_{SMk}(t)$ can be derived. Thus, v_{SMk} is one time-step ahead of i_{SM} on the SM side, while the reverse is the case for the MMC arm. Nevertheless, the fact that the circuit computation frequency is much higher than that of the arm current means i_{SM} can be deemed as a constant current source in two neighboring time-steps and its impact on simulation accuracy is negligible.

9.3.2 Three-Phase Saturable Transformer Model

The medium-frequency transformer is an essential part to the SST. The lumped admittance-based model introduced in Chapter 3 is adopted for lower resource utilization and smaller latency when deployed on to the FPGA board. The matrix equation for a three-phase transformer can be rewritten as follows:

$$\mathbf{v} = \mathbf{Ri} + \mathbf{L} \cdot \frac{d\mathbf{i}}{dt}, \tag{9.13}$$

where $\mathbf{R}$, $\mathbf{L}$ are the matrices of winding resistances and inductance, respectively. The discretized form for EMT computation after using Trapezoidal Rule is

$$\mathbf{i}(t) = \mathbf{Gv}(t) + \mathbf{I}_{his}(t - \Delta t), \tag{9.14}$$

where the iterative history current takes the form of

$$\mathbf{I}_{his}(t) = 2(\mathbf{G} - \mathbf{GRG})\mathbf{v}(t) + (\mathbf{I} - 2\mathbf{GR})\mathbf{I}_{his}(t - \Delta t), \tag{9.15}$$

$$\mathbf{G} = \left[\mathbf{I} + \frac{\Delta t}{2}\mathbf{L}^{-1}\mathbf{R}\right]^{-1} \frac{\Delta t}{2}\mathbf{L}^{-1}. \tag{9.16}$$

The nonlinearity caused by saturation is superimposed on the nodal voltage vector solved from the linear network,

$$\mathbf{v}_F = \mathbf{v}_0 - \mathbf{R}_{Thev}\mathbf{i}_{comp}, \tag{9.17}$$

where $\mathbf{v}_0$ and $\mathbf{v}_F$ are linear and final nodal voltage vectors, $\mathbf{R}_{Thev}$ is the transformer's Thévenin equivalent resistance matrix, and $\mathbf{i}_{comp}$, the compensation current injected into the linear network, is obtained from the saturation curve by Newton–Raphson iteration.

9.3.3 SST EMT Model

Figure 9.8 is the basic three-phase SST circuit configuration for EMT simulation. In each phase, the upper and lower arms of an MMC are connected to a common MFT terminal, and the Norton equivalent circuit is adopted since (9.14) is based on nodal voltage. The four branches on both sides have the same form of admittance

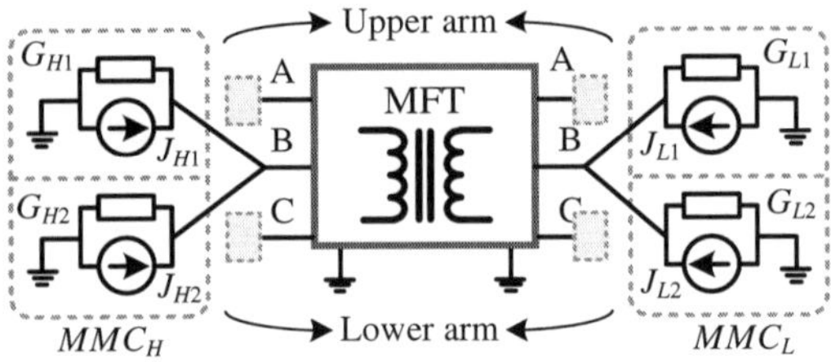

Figure 9.8 Three-phase SST EMT model.

and current contribution. Taking the upper arm for instance, its conductance and current contribution can be computed as follows:

$$G_{H1/L1} = \left(Z_{Lu} + \sum_{k=n+1}^{N} R_{eq}(t) \right)^{-1}, \tag{9.18}$$

$$J_{H1/L1} = G_{H1/L1} \left(V_{dcu} - 2v_L^i - \sum_{k=1}^{n} v_{SMk} - \sum_{k=n+1}^{N} V_{eqk} \right), \tag{9.19}$$

where the DC voltage V_{dcu} corresponds to the upper arm. These elements are then combined with inherent elements of the MFT so that the nodal voltage equation of the circuit in Figure 9.8 can be solved.

9.3.4 SST HIL Emulation

The MMC DEM and TLM-S are implemented on the Xilinx® Virtex®-7 VC707 XC7VX485T FPGA whose resources are listed in Table 1.2. As summarized in Table 9.1, TLM-S enables faster simulation by CPU and smaller hardware latency on FPGA, making it more suitable for real-time simulation, particularly when a small time-step is required. For instance, DEM is nearly 30% slower than the TLM-S in simulating a 51-level MMC using a 20-μs time-step when it is run by 64-bit Windows® 7 Enterprise SP1 operating system on the 3.40 GHz Intel®

Table 9.1 MMC model simulation speed comparison.

	5-s simulation duration			Latency	
MMC-level	**DEM**	**TLM-S**	**Speedup**	**DEM**	**TLM-S**
5-L	7.8 s	6.8 s	1.15	76 T_{clk}	64 T_{clk}
11-L	9.2 s	7.7 s	1.19	92 T_{clk}	80 T_{clk}
51-L	17.8 s	13.6 s	1.31	108 T_{clk}	96 T_{clk}
101-L	28.8 s	21.1 s	1.36	116 T_{clk}	104 T_{clk}
501-L	115.0 s	80.1 s	1.44	132 T_{clk}	120 T_{clk}
1001-L	222.0 s	153.4 s	1.45	140 T_{clk}	128 T_{clk}

Core™i7 CPU and 8.00 GB RAM, while the accuracy of the two models is the same. Moreover, the resource utilization of DEM is much higher than TLM-S, for example, Xilinx® Vivado HLS® estimates that for an 11-level MMC, one DEM controller takes around 14% of LUT while it is only 4% for TLM-S.

Hardware design of MMC_H is taken as an example because centering around the MFT, the topology of SST is symmetrical. Vivado HLS® which enables C/C++ functions to be synthesized into hardware modules is employed to shorten the design cycle. Signals with the same attribute are grouped in an array, rather than being taken individually, for a significant reduction in hardware resource utilization when C synthesis is conducted. The drawback is that the latency will increase slightly along with the array size. Therefore, such hardware design strategy is mainly adopted in controllers which tolerate a large latency. The unroll directive option provided by the design tool is also chosen for achieving parallelism of signals in the same array.

Table 9.2 gives hardware design specifics. Around 15% of LUT and digital signal processing (processor) (DSP) are required by the one-phase 55-level MMC and its controller. With a FPGA clock frequency of 100 MHz, the time-step for phase-shift

Table 9.2 MMC hardware design specifications.

Module latency.

Module	**Description**	**Latency**	**Time-step**
NSM	Nonlinear SM	35 T_{clk}	500 ns
MMCP	1-phase MMC	95 T_{clk}	15 μs
MFT	Transformer	205 T_{clk}	15 μs
DQT	*abc-dq*	78 T_{clk}	15 μs
OLC	Outer loop control	110 T_{clk}	15 μs
PLL	Phase reference	8 T_{clk}	15 μs
PSC	Phase-shift control	1429 T_{clk}	15 μs
THM	Thermal network	61 T_{clk}	15 μs

Hardware resource utilization

Resources	**55L-TLM-S (1p)**	**NSM (1SM)**
LUT	45 685 (15.05%)	5 848 (1.93%)
LUTRAM	47 (0.04%)	49 (0.04%)
FF	38 749 (6.38%)	3 140 (0.52%)
BRAM	31.50 (3.06%)	0 (0%)
DSP	426 (15.21%)	32 (1.14%)
BUFG	4 (12.5%)	4 (12.5%)

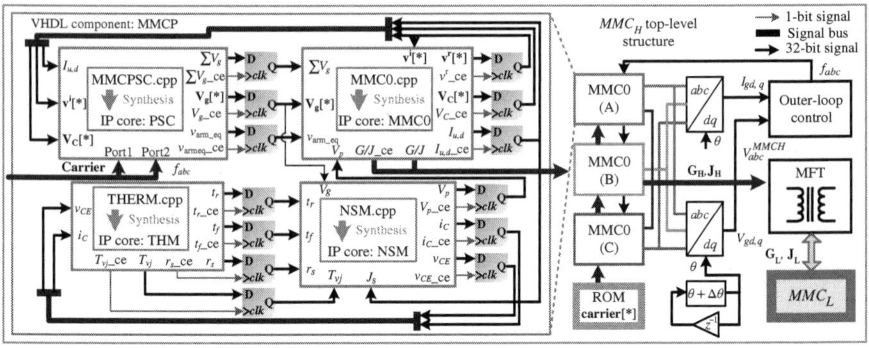

Figure 9.9 Top-level hardware structure of MMC_H.

control (PSC) should be no less than 13.75 μs according to its 1375 T_{clk} latency, where T_{clk} = 10 ns, while it is flexible for the remaining modules other than NSM, so that all of them can be set at 15 μs. The parallelism of the 55-level MMC is purposely weakened to save hardware resources and, consequently, the latency of TLM-S increases to 95 T_{clk}. Meanwhile, the time step for NSM is 500 ns to ensure the accuracy of switching transients.

Figure 9.9 depicts the general hardware structure of MMC_H, where the input and output ports of MMC and its controller – grouped as one component – are specifically shown, while other functional blocks in the top-level are represented by their simplified forms. As can be seen, manipulating gate-level logics is avoided with Vivado HLS®, and only the input and output ports of a component are required during hardware design. Since the hybrid arm is used in the simulation, one MMC contains mainly four blocks, i.e. PSC, MMC linear part (MMC0), MMC nonlinear submodule (NSM), and the thermal network (THM). Among them, the THM is independent of other modules to shorten the hardware latency: the NSM sends v_{CE} and i_C to THM whenever it completes calculation, and the newest values take effect when the THM starts a new computation cycle; similarly, the THM calculates $t_{r,f}$ and sends them immediately to the NSM. Meanwhile, the six arms of an MMC are calculated concurrently, and the outputs G and J are sent to the MFT module, where the nodal voltages are sought.

Connecting these modules either by wire or through D latches is achieved by very high-speed integrated circuit hardware description language (VHDL) in Vivado®, where the top-level finite state machine defining the operation sequence in Figure 9.10 is realized. Once hardware emulation starts, two independent loops run simultaneously. Loop 1 contains solely NSM and repeats every Δt_1 = 500 ns, while all remaining modules constitute Loop 2 that has a period of Δt_2 = 15 μs. Thus, this multiple time-stepping scheme avoids compromising switching transients by other hardware modules that must have a large time-step. The carrier

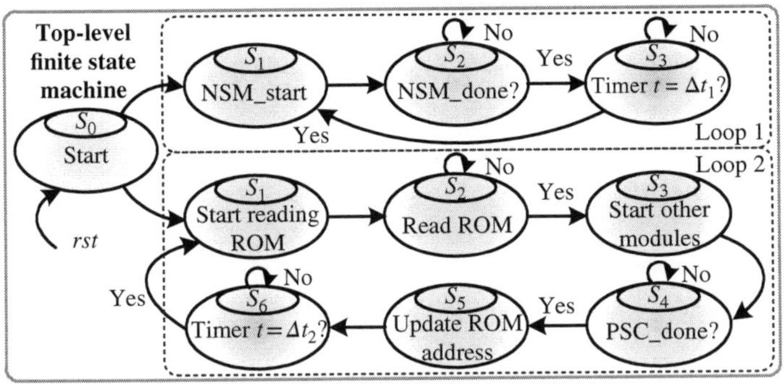

Figure 9.10 SST top-level finite state machine.

waveforms are stored in ROM so that MMC PSC can operate properly. Table 9.2 shows that for each loop, the time-step is larger than the maximum latency, so two timers are set: once an exact time-step runs out, a new calculation cycle will begin.

Figure 9.11 shows the outer-loop controllers of the SST. MMC_L regulates active and reactive power, while its counterpart is in charge of the MFT AC voltage on the primary side. The control scheme is carried out in *d*-*q* frame and is largely the same to other voltage-source converters, except an additional MMC inner loop employing PSC is adopted. The angle θ for Inverse Park's Transformation is the reference which determines the MFT operation frequency. Internal variables with superscripts *H* and *L* correspond to the primary and secondary sides of the transformer, respectively.

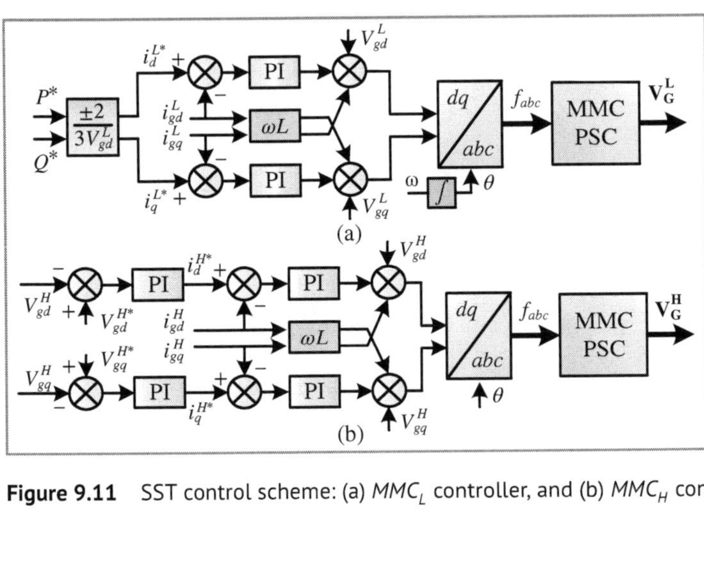

Figure 9.11 SST control scheme: (a) MMC_L controller, and (b) MMC_H controller.

9.3.5 SST Real-Time HIL Emulation Results

9.3.5.1 Device-Level Behavior

A high degree of agreement between simulation and experimental data indicates that the IGBT dynamic curve-fitting model qualifies for MMC simulation to give design guidance. In Figure 9.12 and thereafter, simulations are conducted based on the MTDC system in Figure 9.5, whose parameters are listed in Appendix A.8. Stipulating that the switching frequency is kept 10 times higher than that of MFT, MMC_H lower IGBT operation status in delivering 200 MW to MMC_3 is shown in Figure 9.12. The power loss waveforms at two switching frequencies are given in Figure 9.12a, which indicates that with a higher density of power pulses, the latter has a more significant impact on junction temperature. Figure 9.12b verifies this viewpoint: with switching frequencies of 600 Hz, 1800 Hz, and 3000 Hz, the junction temperatures center around 45 °C, 75 °C, and 103 °C, respectively. The fluctuations in the temperature are caused by the alternating turn-on and turn-off processes, and consequently, the higher the switching frequency, the denser the ripples appear. With respect to safe operation, it can be inferred that measures such as using external cooling

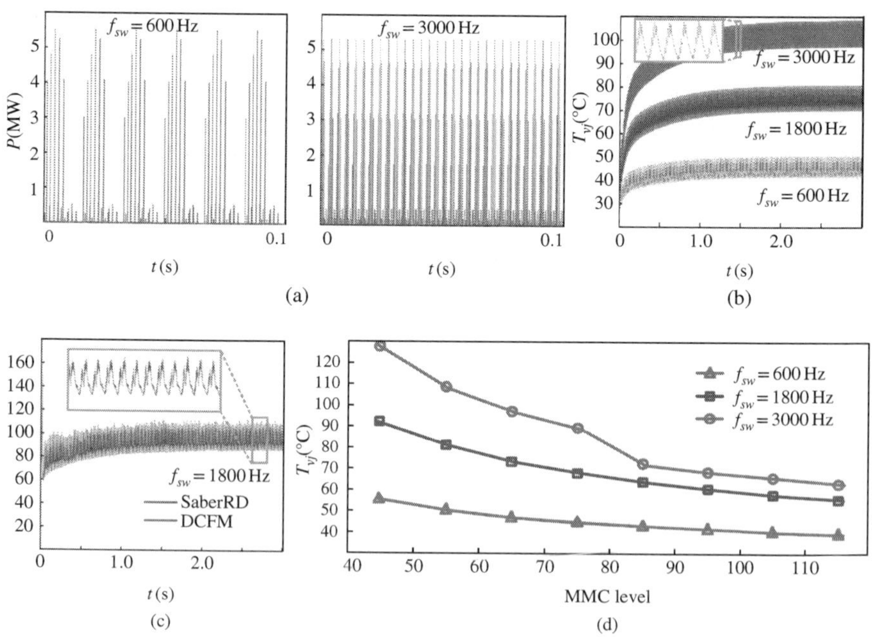

Figure 9.12 MMC_H lower IGBT operation status: (a) power loss waveforms for 55L-MMC, (b) junction temperature waveforms for 55L-MMC, (c) 5L-MMC SaberRD® validation, and (d) relation between T_{vj} and MMC level.

apparatus and increasing the MMC level are required when $f_{sw} = 3000$ Hz, while natural cooling is sufficient for steady-state operation with $f_{sw} = 600$ Hz. For further validation, the transferred power is reduced to 20 MW, and SaberRD® is used to simulate a five-level MMC considering numerical divergence will occur if the number of level is higher. The results in Figure 9.12c demonstrate that the junction temperatures from DCFM and the simulation tool's own IGBT model are largely the same, meaning that the DCFM is as accurate as commercial simulation tools. Figure 9.12d gives the relation between MMC voltage levels and maximum IGBT junction temperatures, which demonstrates that by increasing the MMC voltage level, a dramatic junction temperature drop can be achieved if $f_{sw} = 3000$ Hz, while the improvement is not significant when f_{sw} is 600 Hz.

9.3.5.2 Converter Performance

The MMC_H control target V_{gd}^{H*} is 90 kV, and the MFT frequency for Figure 9.13a,b are set to be 60 and 180 Hz, respectively. The real-time results from the oscilloscope show that the MFT primary voltages v_{pri} are exactly the control objects. Consequently, on the secondary side, the value is halved to about 45 kV. The SM capacitor voltages are also given in Figure 9.13c, which indicates increasing the MFT frequency leads to smaller sizes of transformers as well as arm inductors and capacitors. Under 60 Hz, SM capacitor voltage ripples for both MMC_H and MMC_L are still larger than those under 180 Hz even though its SM capacitance and arm inductance are two to four times larger, as shown in Appendix A.8. As expected, the introduction of more accurate DCFM causes some trivial differences with respect to PSCAD/EMTDC® results, which are based on Type 3 semiconductor switch model; however, the average values and variations of these signals are

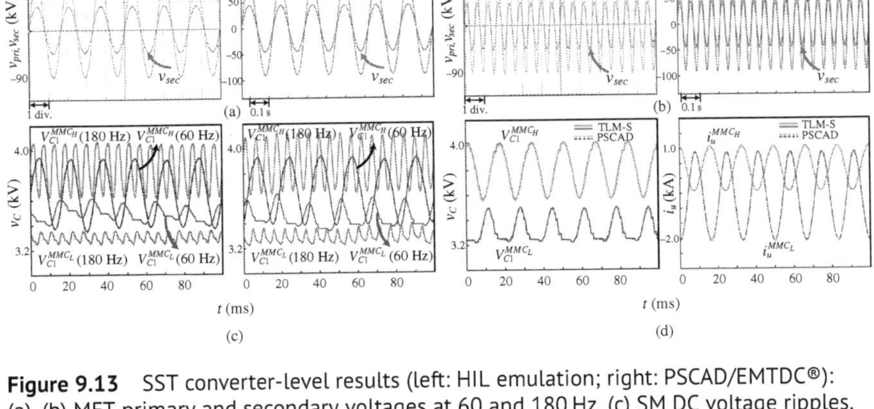

Figure 9.13 SST converter-level results (left: HIL emulation; right: PSCAD/EMTDC®): (a), (b) MFT primary and secondary voltages at 60 and 180 Hz, (c) SM DC voltage ripples, and (d) ideal MMC models comparison. Oscilloscope horizontal axes setting: 10 ms/div.

similar. Figure 9.13d supports this viewpoint, the arm currents and SM DC voltages are compared between TLM-S and PSCAD/EMTDC®, it can be seen that these two types of ideal MMC models fit well, implying the adoption of more detailed device-level IGBT models leads to a higher accuracy.

9.3.5.3 System Tests

Some tests demonstrating the function of SST are carried out as a further validation of the MMC model constituted by hybrid arms. In Figure 9.14, power reversal is conducted, and all power flowing to DC yard is defined as positive. Initially, the power delivered to Station-2 and Station-3 are 300 and 100 MW, respectively, thus I_{dc2} and I_{dc4} are approximately 1.5 and 0.5 kA, and I_{dc3} maintains around two times that of I_{dc4}. At $t_1 = 2$ seconds, the power order in MMC_L begins to ramp from −100 to 100 MW, i.e. Station-3 is diverted to a rectifier station, and consequently, Station-2 receives 500 MW power from the other two stations and I_{dc2} finally stabilizes at 2.5 kA. It shows that the DC currents are a clear indicator of power variation, because the DC voltage at Station-2 U_{dc2} is precisely controlled at 200 kV. During the reversal process, the MFT currents undergo ramping while its voltages keep constant due to MMC_H's control, as demonstrated by v_{sec}. Another notable feature of SST is fault isolation, which is shown in Figure 9.15. Immediately after $t_0 = 1$ second when the fault on DC Line-3 is detected, both MMC_H and

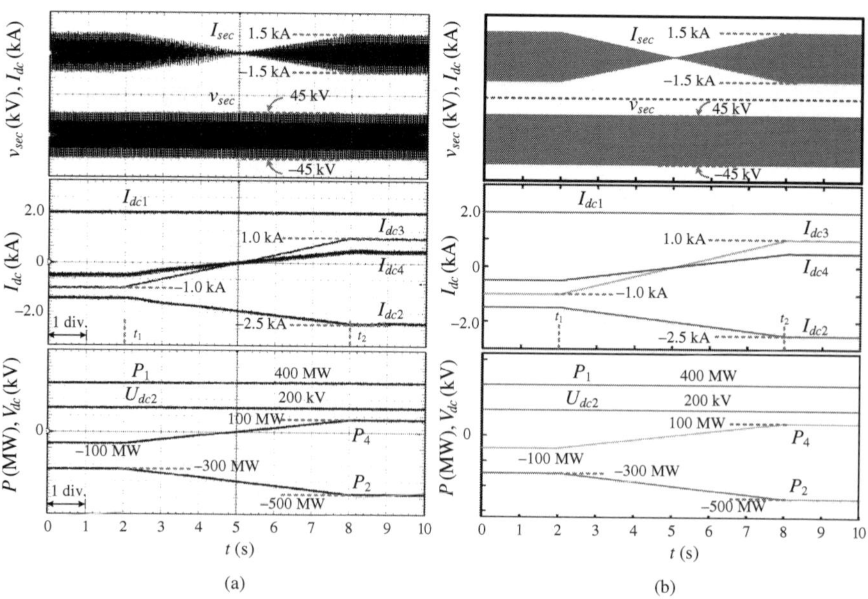

Figure 9.14 MTDC system power reversal from HIL emulation (a) and PSCAD/EMTDC® (b). Oscilloscope horizontal axes setting: 1 s/div.

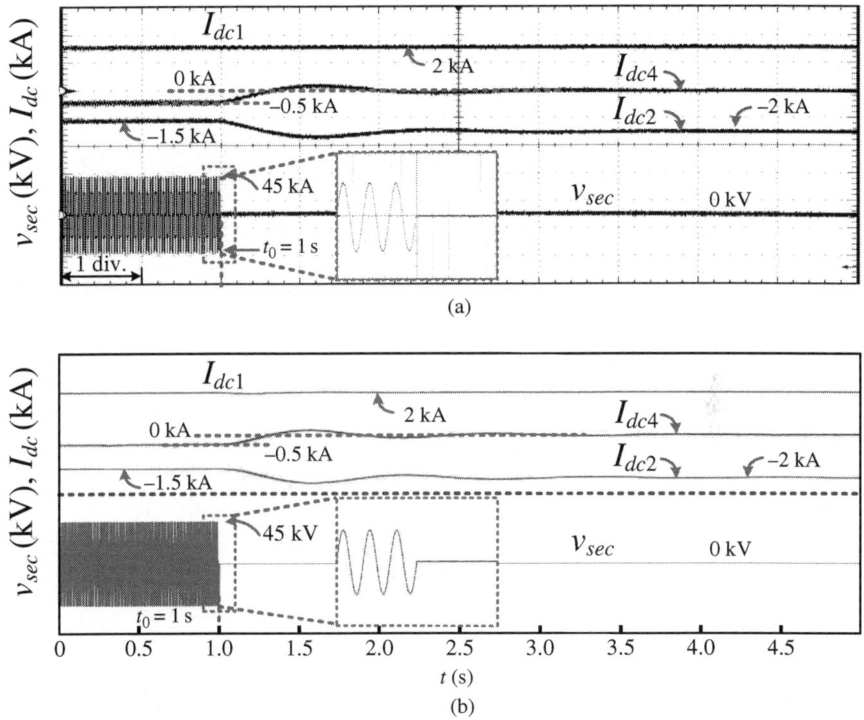

Figure 9.15 SST fault isolation test waveforms from HIL emulation (a) and PSCAD/EMTDC® (b). Oscilloscope horizontal axes setting: 0.5 s/div.

MMC_L are ordered to block their driving pulses. As a result, the voltage on both sides of the MFT vanish, indicating that the SST has fault isolation capability. Meanwhile, the power from Station-1 is diverted solely to Station-2 because I_{dc2} has the same amplitude to I_{dc1} and I_{dc4} reduces to 0. Corresponding results from PSCAD/EMTDC® confirm these statements, indicating the hybrid-arm-based MMC model can be used for MTDC grid studies.

Since the SM blocked state cannot be explicitly shown due to the SST's fault isolation capability, it is proven by applying the improved MMC model and the original TLM-S to MMC_1 in Figure 9.5 where a 1 Ω line-to-ground fault is imposed on DC Line 1 right after inductor L_1, and corresponding PSCAD/EMTDC® simulations are conducted for validation, as given in Figure 9.16. Both models produce the same correct results until $t = 0.1$ second when an obvious bifurcation emerges. The improved MMC model yields result identical to that of the traditional model with each IGBT having a free-wheeling diode in PSCAD/EMTDC®. The AC voltage under this scenario is still being rectified by the diodes so V_{dc1} is about 30 kV, and the fault current stabilizes at around 30 kA. Moreover, in the controller, i_{gd}

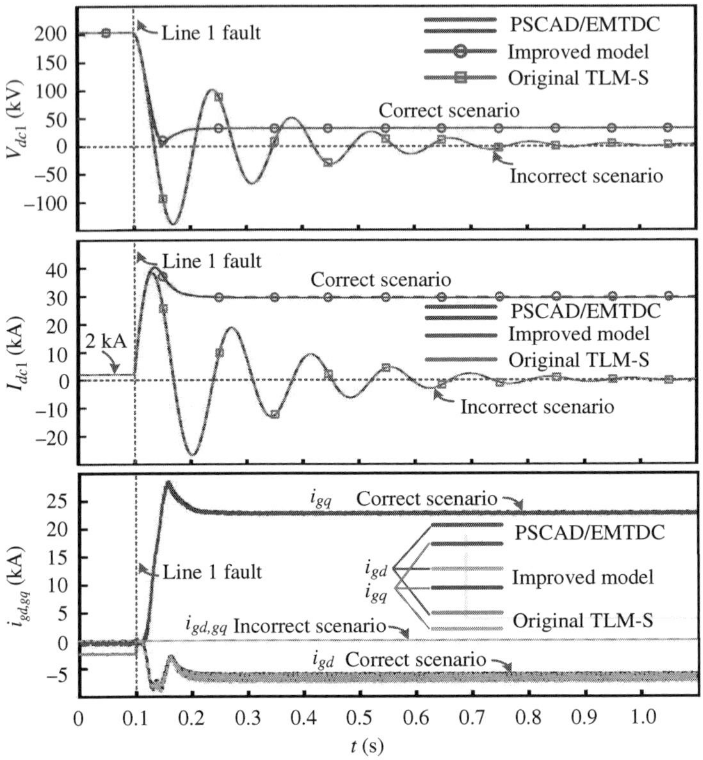

Figure 9.16 MMC_1 blocked state test results.

and i_{gq} deviate from the control target, and their nonzero values prove energy flow between the AC and DC grids. However, with the original TLM-S that fully blocks the MMC, V_{dc1} finally stabilizes at 0 after periods of oscillation, and similar behavior can be observed with the DC current. Meanwhile, i_{gd} and i_{gq} are zero, meaning that no current is flowing from the AC side to the DC side. These incorrect results are identical to those in PSCAD/EMTDC® when a large resistor is inserted between the MMC and DC yard, proving that the original TLM-S has a high-impedance blocked state.

9.4 Summary

This chapter introduced two typical DC-DC converters suitable for DC grid hardware emulation. When system-level results such as power flow are the main concern, simple converter configurations such as buck, boost, and buck–boost topologies can be used for lower hardware resource utilization and smaller latency. On the other hand, when the simulation integrates devices with systems, fully detailed converters such as the MMC-based SST is mandatory.

An MMC hybrid arm model is presented for real-time HIL emulation of the DC grid. The dynamic curve-fitting model provides guidance on converter design by revealing the device's electro-magnetic-thermal information such as power loss and junction temperature under various electromagnetic environments, which facilitates determination of the number of MMC submodules, the MFT's operation frequency, and adoption of external cooling apparatus. Meanwhile, the MMC TLM-stub model alleviates hardware resource burden and shows its speed advantage in both CPU simulation and HIL emulation on FPGA in conjunction with other complex switch models. Circuit partitioning enables the coexistence of separated nonlinear submodules and the linear MMC circuit even though they have distinct time-steps, and with smaller matrix dimension, a significant speedup can be attained in addition to avoiding numerical divergence. Other than the SST, both models can be applied individually or jointly by constituting the hybrid model to MMCs in various applications.

10

DC Circuit Breakers

10.1 Overview

Recent rapid development of DC grids formed by linking multiple converter stations raises the concern of safe operation under various contingencies, especially line faults which are more challenging to interrupt than in an AC grid as the DC current will instantly witness a dramatic surge and a persistently unidirectional injection into the fault location. With an inherent quick response, power-electronics-based hybrid DC circuit breakers are a promising solution to containing the fault whose hazardous impact may spread to other parts of the DC grid or even its neighboring AC grid and consequently cause malfunctioning of devices in the whole power system. Mainly based on the insulated gate bipolar transistor (IGBT) or thyristor, a variety of high voltage direct current (HVDC) circuit breakers have been proposed, and they all share the similarity of composing dozens or even hundreds of power semiconductor switches in series and parallel to withstand the common high voltage and large current, respectively. In this chapter, modeling and hardware implementation of two prevalent DC circuit breakers, the IGBT-based hybrid HVDC breaker (HHB) and the ultra-fast mechatronic circuit breaker (UFMCB) employing thyristors are carried out.

Determined by the types of power semiconductor switch models summarized in Chapter 7, modeling of a circuit breaker can be conducted from two perspectives: system-level and device-level. A precise model should be a full-scale model so that internal details can be investigated in addition to the overall performance for system preview, although taking consideration hundreds of power semiconductor switches even in the form of Type-3 switch model would lead to an extremely heavy computational burden as well as to a high field programmable gate array (FPGA) resource utilization caused by a massive array. The inclusion of device-level models can further improve simulation results by providing more details. Like the predominant two-state switch model, the curve-fitting model (CFM) is also a two-node switch model, whose on-state resistance is

Real-Time Electromagnetic Transient Simulation of AC-DC Networks, First Edition.
Venkata Dinavahi and Ning Lin.

obtained from the static I–V characteristics, and the dynamic waveforms acquired directly from experimental measurement or indirectly from the device datasheet to achieve fast circuit computation. Thus, the calculation of steady state and transient power losses are more accurate. Its major shortcoming is that the stored values for transient waveforms need to be altered repeatedly along with the variation of electromagnetic environment in order to keep accuracy. On the other hand, the high-order nonlinear behavioral model (NBM) is widely used in off-line device-level tools such as SaberRD® to provide every detail of the circuit accurately. Another merit is the versatility, take the IGBT NBM for example, it is deemed to be able to represent a real IGBT under most conditions without changing its parameters, while the drawback of inefficient numerical computation is also noticed since the nonlinear model contains multiple nodes solved usually by many iterations of the Newton–Raphson method, making it prone to numerical divergence and sensitive to initial conditions.

It is, in any sense, meaningful to include an integrated HVDC circuit breaker model as a fundamental component in the libraries of various electro-magnetic transient (EMT) tools. Hitherto, precise models suitable for fast simulation or capable of real-time hardware-in-the-loop (HIL) execution are yet to be developed. Despite the fact that hundreds of IGBTs in the main path of the HHB are constructed in series and parallel – and so are thyristors in the ultra-fast mechatronic circuit breaker (UFMCB) – for the purpose of fast simulation, the prevalent modeling work has focused on a scaled-down model, i.e. the number of IGBTs in the main path is only one or two for unidirectional and bidirectional HHBs, respectively, which is far less than that of a real DC breaker. Such a simplification is reasonable and could provide good results for system-level grid studies, as the main interest is to validate the protection concepts [190, 191]. In the meantime, the IGBTs and their auxiliary circuits such as the snubber are also omitted to keep a minimum number of meshes or nodes, which consequently avoid matrices of large dimensions in the simulation process that would take an extraordinarily long time for processors like the central processing unit (CPU) and FPGA to compute. Nevertheless, the models fall short of providing guidance on the circuit breaker design and the state of components within the breaker for operation monitoring.

In contrast, a full-scale breaker model contains the exact number of power semiconductor switches and other devices so that more system-level details can be shown [192–194]. However, the subsequent hundreds or even more circuit nodes bring the same computational challenge as the adoption of complex models yields a corresponding large matrix equation. Therefore, it is impractical for HIL platforms to test the control and protection strategies of a multi-terminal direct current (MTDC) system in real-time due to extremely slow speed and high hardware resource utilization. Moreover, integrating device-level models into a full-scale circuit configuration in an effort to achieve the maximum possible fidelity induces the heaviest computational burden despite it enables to evaluate device-level behavior, such as switching transients, power loss, as well as the junction temperature, which in turn affects the performance of the power semiconductor itself.

In this chapter, two distinct topological simplification schemes are introduced and applied to the HHB and UFMCB, respectively, since they differ significantly in configurations. The HHB as a prevalent DC breaker and consequently the main focus of this chapter is first specified, i.e. three types of its full-scale models – classified according to IGBT models defined in Chapter 7 – are adopted for efficient real-time HIL emulation of HVDC grids on FPGAs as well as for achieving fast off-line electromagnetic transient simulation. Type-3 model utilizing two state switch model (TSSM) is the most straightforward model, and the Type-2 model based on the curve-fitting technique is a further improvement to realize two goals simultaneously: it can be used in real-time HIL emulation, and device-level phenomena are included to enable the model to provide more details. The second-order NBM-based DC breaker is also introduced as the most accurate one and is categorized as the Type-1 model. As an essential part of device-level modeling, the electro-thermal network is included to enable the acquisition of operation statuses such as IGBT power loss and junction temperature, and consequently, the HHB design including the selection of IGBT type and its number can be evaluated. To reduce FPGA hardware resource utilization caused by a large number of IGBTs, circuit partitioning is first applied and based on that, one of the subcircuits is used to represent all other identical ones.

In the UFMCB modeling, in contrast, merging elements with the same attribute, rather than partitioning, is carried out due to an absence of topological homogeneity. Similar to the IGBT, the nonlinear silicon controlled rectifier (SCR) is modeled with static I–V characteristics, the transient features, as well as the electrothermal dynamics. Then, circuit dimension reduction is applied to alleviate the computational burden so as to attain real-time implementation for HIL testing of the multiterminal DC grid.

10.2 HHB in MTDC System

10.2.1 MTDC Test System Schematic

Figure 10.1 shows a three-terminal HVDC system which demonstrates the way that HHBs are installed. The configurations of the three converter stations are symmetrical. STN_1 (REC) is set as the rectifier station, while the other two, denoted as STN_2 (INV1) and STN_3 (INV2), are inverter stations. The former is in charge

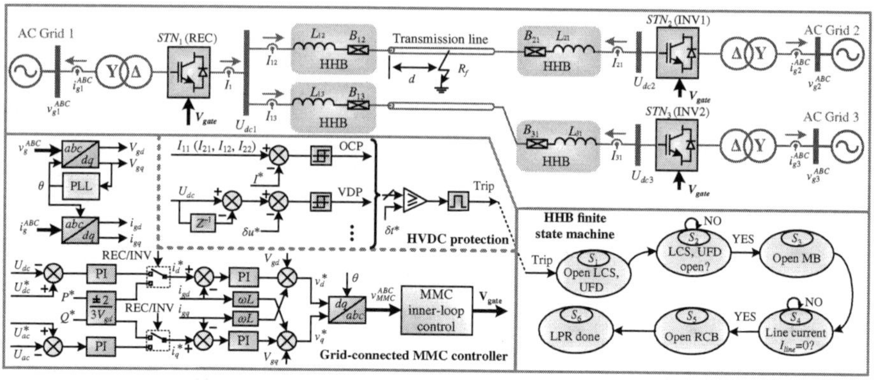

Figure 10.1 Schematic of a three-terminal monopole HVDC system installed with hybrid HVDC breakers and its control and protection concepts.

of power while the latter controls individual DC bus voltage. I_1, I_{12}, and I_{13} are rectifier side DC currents, and I_{21} as well as I_{31} represent inverter side DC currents. L_{12}, L_{13}, L_{21}, and L_{31} are current limiting inductors in DC yards, which, together with symbols B_{12}, B_{13}, B_{21}, and B_{31}, constitute HHBs. Line faults with resistance R_f can be simulated on both transmission lines linking inverters station with the rectifier station.

As modular multilevel converters (MMCs) are gaining popularity and presumed to be dominant in future MTDC projects, the HHB models are inserted in the DC yards of such a system. Considering the main focus is on the performance of HHBs and since a proper modification of discounting AC side reactance and resistance into DC side enables the MMC averaged value model to predict system-level behaviors when DC line fault occurs [136, 195], it is adopted to achieve low computational burden. Furthermore, the installation of an appropriately designed HHB guarantees that the DC fault current from an AVM-based MTDC system to be similar to that of a detailed equivalent model based on a period longer than the HHB protection time.

The control scheme of the MMC is also shown in Figure 10.1, which demonstrates that the strategies for the rectifier and inverter are largely the same, other than that the control objective is selected according to the state of the converter station. Meanwhile, the scheme based on *d*-*q* frame is identical to that of other grid-connected voltage sourced converters, except the modulation signals v_{MMC}^{ABC} are sent to an additional inner-loop controller, which, in this case, adopts a phase-shift strategy to generate driving pulses denoted by the vector V_{gate}. In AVM, the output voltage of a submodule is determined by the state of its upper switch. Thus, the combined output voltage of submodules in one arm of a

$(N+1)$-level MMC can be uniformly calculated by

$$v_{u,d} = \sum_{1}^{N} V_{g,i}^{upper} \times \frac{U_{dc}}{N}, \tag{10.1}$$

where $V_{g,i}^{upper}$ is a binary number indicating on/off state of the upper switch in the ith submodule by 1 and 0, respectively, and U_{dc} denotes the converter side DC line voltage.

10.2.2 DC Line Protection

The DC line protection (LPR) concepts for MMC-HVDC systems share a great similarity with line-commutated converter-based HVDC, which means that a variety of criteria, such as voltage derivative protection (VDP), under-voltage protection (UVP), and over current protection (OCP), can also be applied to judge line faults. The difference lies in the fact that isolating the faulty section is mainly achieved by HHBs on both inverter and rectifier sides. In Figure 10.1, two popular protection concepts for HHB testing are shown.

10.2.2.1 Voltage Derivative Protection

VDP has a fast reaction to line faults. The principle is when the DC line contacts ground via a small resistance R_f, the voltage drops instantly from hundreds of kilovolts to close to zero or a negative value. Thus, the voltage change rate DUDT is extremely large, which is calculated by

$$DUDT = \frac{dU_{dc}(t)}{dt} = \frac{U_{dc}(t) - U_{dc}(t-\delta t)}{\delta t}, \tag{10.2}$$

where δt is the digital sampling rate and, consequently, $U_{dc}(t-\delta t)$ indicates DC line voltage of the previous sampling. The protection threshold δu^* should be far larger than the DUDT value under the steady-state condition and during the start or stop of the converter. Even so, to avoid maloperation, a width comparison section is introduced: if DUDT keeps larger than δu^* for a preset time δt^*, a trip order will be issued to activate HHB protection process.

10.2.2.2 Over Current Protection

OCP has a relatively slower response to line faults compared with VDP and consequently, it has a higher requirement on the breaking capability of an HHB. Nevertheless, it is still useful in protecting electrical facilities and can be used as a backup. The principle consists of the following: when the line current rises beyond the setting, a tripping pulse with a predefined width will be issued which will be followed by HHB operation sequence.

10.3 Proactive Hybrid HVDC Breaker

Figure 10.2a is the circuit configuration of a unidirectional HHB, which, as the real equipment does, contains six essential parts: current limiting inductor L, residual current breaker (RCB), ultrafast disconnector (UFD), load commutation switch (LCS), metal oxide varistor (MOV), and main breaker (MB) with the snubber circuit [196, 197]. Under normal conduction, the LCS accounts for the majority of energy consumption. On the contrary, when DC line protection is triggered, the power loss of LCS is negligible compared with that caused by MB as well as the MOV which absorbs most of the energy stored in the energy transmission corridor [198], including the current limiting inductor. Thus, an accurate device-level IGBT model is necessary for MB so that the switching power loss can be calculated for HHB design evaluation, while for LCS, its steady-state power loss is more concerned.

As part of line protection concepts, the operation sequence of the HVDC breaker is shown in the right corner of Figure 10.1. After receiving the trip order, the LCS gate signals are immediately retrieved and the UFD is commanded to open, which takes around 2 ms to complete. The MB gate voltages should vanish as soon as the previous actions are confirmed. The protection procedure ends with the opening of the RCB when the line current declines to zero so as to protect the varistor from overheating.

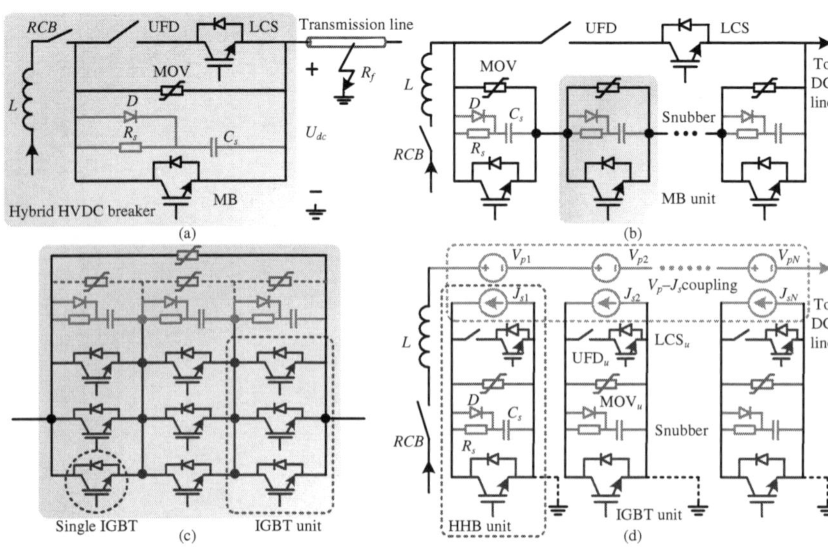

Figure 10.2 Models of unidirectional HHB for EMT simulation: (a) circuit configuration, (b) conventional full-scale model, (c) 3 × 3 IGBT array, and (d) decomposition of HHB full-scale model using v–i coupling.

10.3.1 HHB EMT Model

The design theory and operation principle of HHB have been illustrated in detail under the assumption that all IGBTs in the MB chain are synchronized, which is reasonable and also implies that all internal nodes are well balanced. The scaled-down model can be derived by representing each HHB part as one component, and the equivalent circuit is identical to that in Figure 10.2a. The MB and LCS are taken as two-state resistors, with on-state and off-state resistances R_{on} and R_{off}, respectively, and the snubber circuit is rarely included. Thus, the total node number is 3 as UFD and LCS can be merged into one resistor so that the internal node between them is eliminated. For further simplification, the UFD–LCS branch is merged with the MB branch and this constitutes the simplest HHB model for EMT simulation. When a fault occurs on the transmission line, this model can give an approximate performance of the HHB. However, as stated above, it is unable to provide further information about the circuit breaker and may give inaccurate results. In Figure 10.2b, the full-scale HHB model is depicted, which has the exact configuration as that of a real one, and two types of commonly used snubber circuits for the circuit breaker are employed, i.e. RC and RCD, the latter is shown in the subfigures.

Depending on the requirement of HHB capacity, an IGBT symbol for both LCS and MB in the full model may actually consist of only one or a number of such devices that are organized in a $N \times N$ array. As indicated in Figure 10.2c, a 3×3 array of 5SNA 2000K450300 StarkPak IGBT Module ($V_{CE} = 4500$ V, $I_C = 2000$ A) [199] is able to endure a DC current over 6000 A. However, modeling the circuit breaker as it actually appears would result in a large system admittance matrix due to hundreds of IGBTs in the MB branch, and their snubber circuits that yield a similar number of nodes, making the original HHB model highly time- and resource-inefficient for commercial off-line as well as real-time EMT simulation tools to solve. Therefore, the voltage–current source coupling method is applied for circuit partitioning to achieve speedup in simulation, as shown in Figure 10.2d, the V_p–J_s coupling enables all HHB units to be physically separated but electrically linked to the power transmission corridor. The UFD and LCS are equally divided to have the same number of HHB units, and so are their resistances. It should be stressed that one IGBT unit in Figure 10.2d denotes an IGBT array, i.e. three IGBTs in parallel, and the MOV in MB unit is also equally divided and denoted by MOV_u since the three IGBT units sharing it has been forcibly disassembled.

Figure 10.3a shows the power transfer corridor, the configurations on both sides of the transmission line are symmetrical and, therefore, only the rectifier station side is shown. The current source I_{dc} and capacitor C_e represent the DC part of the MMC (MMC-DC). By applying transmission line theory, the equivalent circuit for EMT simulation can be obtained, as shown in Figure 10.3b, where C_{Ce} is represented by its transmission line modeling (TLM) link model

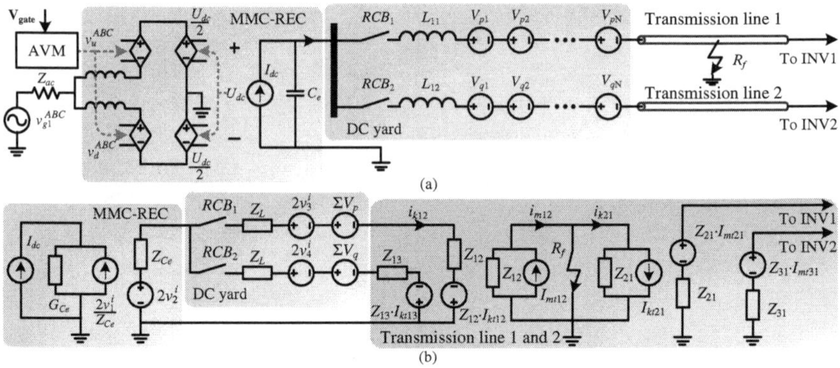

Figure 10.3 HVDC power transfer corridor with HHB separated: (a) equivalent circuit topology, and (b) EMT simulation model.

for circuit partitioning, and both transmission lines employ the traveling wave model (TWM), which adopts a hybrid Thévenin–Norton structure, leading to a number of one-node circuits whose calculation becomes very convenient. The only exception is the subcircuit, where the DC yard is located, whose matrix dimension is 2, as expressed below:

$$\mathbf{Z} = \begin{bmatrix} Z_{Ce} + RCB_2 + Z_L + Z_{13} & -RCB_2 - Z_L - Z_{13} \\ -RCB_2 - Z_L - Z_{13} & \sum RCB + 2Z_L + \sum Z_{1i} \end{bmatrix}, \tag{10.3}$$

$$\mathbf{U} = \begin{bmatrix} 2v_2^i - 2v_4^i - \sum V_q + Z_{13} \cdot I_{kt13} \\ 2v_4^i + \sum V_q - Z_{13} \cdot I_{kt13} - 2v_3^i - \sum V_p + Z_{12} \cdot I_{kt12} \end{bmatrix}, \tag{10.4}$$

where v_2^i, v_3^i, and v_4^i are incident pulses of the TLM link and stub models of capacitor and inductors, respectively. Z_{Ce} and Z_L are characteristic impedance, Z is the transmission line's characteristic impedance, $\sum V_{p,q}$ means voltage sources coupled with HHB units. Thus, the DC yard is linked to MMC-DC by incident pulses, while it connects to the transmission line by the coupling between current sources I_{kt} and I_{mt}. Noticing that line faults are simulated on transmission line 1, a special line section independent from DC yard is constructed, while the model of transmission line 2 has only two sections.

The advantage of such a partitioning method is that HHBs in the power transmission path will not introduce any additional mesh; thus, mesh currents, rather than nodal voltages, are taken as variables, making the solution of its corresponding matrix equation fast.

10.3.2 Varistor Model

The varistor is modeled as a nonlinear resistor, whose value plummets when a current surge is witnessed. The rating of the virtual varistor unit MOV_u can be determined from a real one, and since the current flowing through them are the same, their distinction lies in the voltage rating. For a 3 × 3 IGBT array, the voltage rating of MOV_u should be reduced by two-thirds, and the I–V relation is expressed by

$$i_v = \left(\frac{v_v}{k_v \cdot V_{ref}}\right)^{\alpha_v} \cdot I_{ref}, \tag{10.5}$$

where k_v, α_v are coefficients, V_{ref} denotes protection voltage and I_{ref} is the corresponding current, and v_v and i_v are varistor's voltage and current, respectively.

Based on (10.5), the Norton equivalent model of the nonlinear varistor can be derived using partial derivatives, as

$$G_v = \frac{\partial i_v}{\partial v_v} = \frac{\alpha_v \cdot I_{ref}}{k_v \cdot V_{ref}} \cdot \left(\frac{v_v}{k_v \cdot V_{ref}}\right)^{\alpha_v - 1}, \tag{10.6}$$

$$I_{veq} = i_v - G_v \cdot v_v. \tag{10.7}$$

Since (10.6) and (10.7) are nonlinear equations, Newton–Raphson iteration is mandatory to obtain correct results. However, Matlab off-line simulation of the HHB showed that calculation of the transient stage requires over 20 iterations, which is too many and would significantly prolong the computational time. Thus, piecewise linearization is used to divide its entire I–V curve into 11 segments, which are proven to be sufficient to retain a high accuracy by off-line simulation, with each having the form of

$$i_v = G_v \cdot v_v + I_0, \tag{10.8}$$

where I_0 is a constant in the segment, and the conductance G_v is linearly interpolated by the terminal values as

$$G_v = (I_{n+1} - I_n)(V_{n+1} - V_n)^{-1}, \tag{10.9}$$

where (V_{n+1}, I_{n+1}), and (V_n, I_n) represent the two terminals of a piecewise linear segment of the I–V characteristics. Then, the equivalent current of its companion model can be obtained using (10.7).

The time HHB takes to block the DC current since line fault occurs consists of two parts: the reaction time of the UFD, known as breaking time, which takes about $\Delta t_1 = 2$ ms, and the fault clearance time Δt_2 that is decided by the inherent I–V characteristics of the varistor. After a line-to-ground fault occurs, the DC line current increases and reaches breaking current at Δt_1

$$I_{f,max} = \frac{U_{dc}}{r} \cdot \left(1 - e^{-\frac{\Delta t_1}{\tau}}\right) + I_{dc} \cdot e^{-\frac{\Delta t_1}{\tau}}, \tag{10.10}$$

where r is fault path resistance and to facilitate calculation it is deemed as R_f, and $\tau = L/r$ is the time constant. To quench that amount of current within Δt_2, the protection voltage V_{ref} should meet the following criterion:

$$V_{ref} = \frac{I_{f,max} \cdot L}{\Delta t_2} + U_{dc}, \tag{10.11}$$

which indicates that the MOV's protection voltage, reached during the fault clearance period, should be set higher than U_{dc}.

10.3.3 General HHB Unit Model

Three types of IGBT models are used to meet the different simulation requirements of accuracy and speed. For real-time HIL emulation that aims to acquire system-level performance, the Type-3 model is a good choice as the simulation speed of two-state switch is fast and utilization of FPGA resources is low. In the case that both high simulation speed and specifics of HHB are demanded, the Type-2 model with curve-fitting IGBT is preferred. When the generality of IGBT model is prioritized, the Type-1 model which adopts the NBM becomes the best alternative.

With regard to the first type, the parallel IGBTs in HHB unit can ultimately be replaced by an ideal two-state switch, with a small on-state resistance R_{on} and a large off-state resistance R_{off}. For the second type, the IGBT is normally deemed as a gate voltage controlled, time-varying current source. Therefore, both models have only two nodes and do not introduce any additional node to the HHB unit. As for the NBM which supposedly has N nodes, it adds an extra $N - 2$ nodes to the originally four-node unit. Therefore, the total number of nodes reaches $N + 2$, among which one is considered the virtual ground, as shown in Figure 10.2d. To achieve real-time with a time-step in the scale of hundreds of nanoseconds, the smallest matrix dimension is favored. Thus, the internal nodes in the UFD–LCS branch and the snubber circuit are merged. For the former, it is naturally eliminated by taking the branch as one resistor, while for the latter, the diode is dominant when the capacitor C_s is quickly charged during the turn-off stage of MB, and when it turns on, R_s decides the rate of discharging. Based on this, the RCD snubber is equivalent to RC snubber, whose discretized model, by applying Backward Euler integration method, can be written as

$$R_{eq} = R_{sD} + \frac{\Delta t}{C_s}, \tag{10.12}$$

$$I_h(t) = \frac{(R_{sD} - R_{eq})}{R_{eq}^2} \cdot v_s(t) + \frac{R_{sD}}{R_{eq}} \cdot I_h(t - \Delta t), \tag{10.13}$$

where R_{sD} is the equivalent resistance of R_s-D pair, $v_s(t)$ is the voltage over the snubber, I_h, which is iterative, represents the history current of C_s and Δt is the simulation time-step.

Noticing that all HHB units are identical, it is not necessary to model all of them. Instead, an arbitrary unit is selected for conducting the modeling work, bringing in a significant speedup for off-line simulation, as well as a great reduction in hardware resource utilization when the HHB model is deployed on FPGA. Furthermore, the parallel IGBTs in an IGBT unit are identical and synchronized, indicating they can be represented by one of them to avoid additional nodes caused by the rest [200]. Then, the matrix equation for the N-node HHB unit can be generally written as

$$\mathbf{U_{HHB}} = \begin{bmatrix} G_{11} & mG^I_{12} & \dots & mG^I_{1N} \\ mG^I_{21} & mG^I_{22} & \dots & mG^I_{2N} \\ \vdots & \vdots & \ddots & \vdots \\ mG^I_{N1} & mG^I_{N2} & \dots & mG^I_{NN} \end{bmatrix}^{-1} \begin{bmatrix} J_1 \\ mJ^I_2 \\ \vdots \\ mJ^I_N \end{bmatrix}, \tag{10.14}$$

where elements G_{11} and J_1 take the form of

$$G_{11} = G_v + R^{-1}_{eq} + (UFD + LCS)^{-1} + mG^I_{11}, \tag{10.15}$$

$$J_1 = J_s(t) - I_{veq}(t) - I_h(t) + mJ^I_1. \tag{10.16}$$

In (10.14), m is the number of IGBTs in parallel. The elements from IGBT, which are distinguished by superscript I, are multiplied with m since these parallel IGBTs are identical.

10.3.4 Two-Node IGBT Models

Both TSSM and CFM have only two nodes. As one of the most popular models in EMT simulation for its simplicity, TSSM realizes the function of an IGBT by shifting between R_{on} and R_{off} when it is commanded to turn on and off, respectively, as shown in Figure 10.4a. Thus, the switching transient cannot be shown by this model and the on-state voltage and current are not sufficiently accurate for power loss calculation, let alone thermal analysis. However, in this case, (10.14) is one-dimensional since only one node with unknown voltage is left in HHB unit, and it can easily be obtained by solving the following algebraic equation:

$$U(t) = \frac{\sum J(t)}{\sum G} = \frac{J_s(t) - I_{veq}(t) - I_h(t)}{\dfrac{1}{UFD + LCS} + \dfrac{1}{R_{eq}} + G_v + G_{MB}}, \tag{10.17}$$

where G_{MB} is the conductance of the IGBT unit and, consequently, is a reciprocal of its resistance.

CFM overcomes the aforementioned shortcomings, as its on-state resistance can be obtained from the IGBT static characteristics and the switching features can

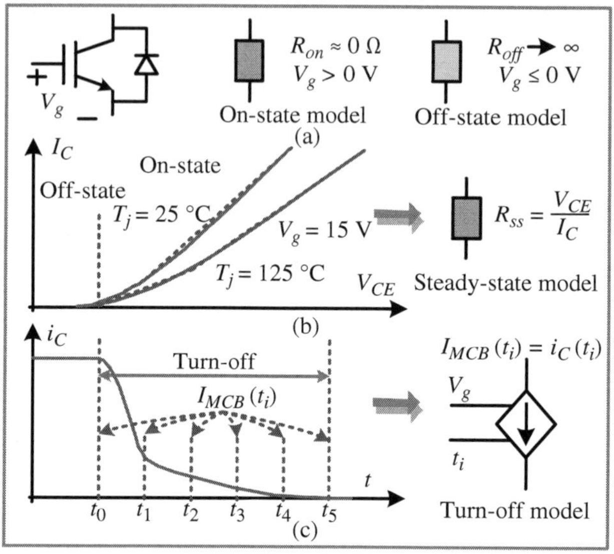

Figure 10.4 (a) IGBT two-state switch model, (b) steady-state representation of IGBT curve-fitting model, and (c) controlled current source for the turn-off of IGBT curve-fitting model.

be preset in the program. As shown in Figure 10.4b, the V–I relation of IGBT is embodied by a piecewise linear resistor that can be expressed by a set of functions taking the form of

$$R_{ss,i} = \frac{V_{CE}}{I_C} = \frac{1}{k_i(T_j)} + \frac{b_i(T_j)}{k_i(T_j) \cdot I_C}, \tag{10.18}$$

where subscript i means the ith linear segment, k_i and b_i are linear functions of junction temperature.

While its turn-off voltage shape is prone to change, the robustness of the shape of IGBT turn-off current becomes critical to establishing its curve-fitting model for the transient stage [201]. These current values, measured at different times by either experiment or simulation of commercial software and stored in a look-up table (LUT), are programmed as a time-controlled current source so that its value after a given number of time-steps can be accessed, as depicted in Figure 10.4c. Although during steady state (10.17) is still applicable to CFM-based HHB, its model for turn-off stage distinguishes itself from TSSM, with the nodal voltage expressed by

$$U(t) = \frac{\sum J(t)}{\sum G} = \frac{J_s(t) - I_{veq}(t) - I_h(t) - I_{MB}(t)}{\dfrac{1}{UFD + LCS} + \dfrac{1}{R_{eq}} + G_v}, \tag{10.19}$$

where I_{MB} is the programmed current source representing IGBT. A combination of (10.17) and (10.19) makes CFM a little more complex than TSSM because the way nodal voltage should be solved is dependent on the operating conditions of the switch. Thus, as a status indicator t is introduced in the hardware design to decide which of the two equations should be used.

10.3.5 IGBT Low-Order Nonlinear Behavioral Model

One prominent merit brought by the aforementioned two-node models is efficient computation. However, they both have limitations, i.e. the TSSM is incapable of showing switching details and power calculation is not accurate, and the CFM lacks versatility since its transient current waveform will not change along with the electromagnetic environment, and consequently, the curve should be amended. There are models that have generality while at the same time provide details of a switch, such as the fourth-order NBM. The main disadvantage is the complexity and relatively slow computational speed. To facilitate HIL emulation as well as simulation of circuits comprised of such models, it is simplified in a way that maintains its accuracy.

10.3.5.1 IGBT Fourth-Order Behavioral Model

In Chapter 7, the full NBM of IGBT is introduced, as also shown in Figure 10.5a, which can mainly be categorized as the metal-oxide semiconductor field-effect transistor (MOSFET) behavior represented by voltage-controlled current source i_{mos} and inter-electrode capacitors C_{ce}, C_{ge} and C_{cg}, tail current i_{tail} that controlled by v_{tail}, the voltage over C_{tail} and R_{tail}, and a piecewise linear diode D that sets the minimum on-state collector-emitter voltage drop.

As can be seen, the full model contains five nodes, which means any circuit containing it will yield at least a 4×4 admittance matrix, and solving its corresponding equation requires multiple Newton–Raphson iterations and more often than not it is prone to divergence. Thus, model simplification is carried out to improve its computational speed as well as robustness to divergence.

10.3.5.2 Parameters Extraction

Different IGBT types are distinguished by the parameters which can be extracted from the device datasheet using the IGBT tool in the off-line simulation tool SaberRD®. Similar to the CFM, the parameters of IGBT NBM can be categorized as the static set and the dynamic set reflecting individual characteristics. The former mainly concentrates on i_{mos}, while the latter is applied to the remaining components.

It should be pointed out that the curves and data in the device datasheet are experimentally measured, which means that the linearities and nonlinearities,

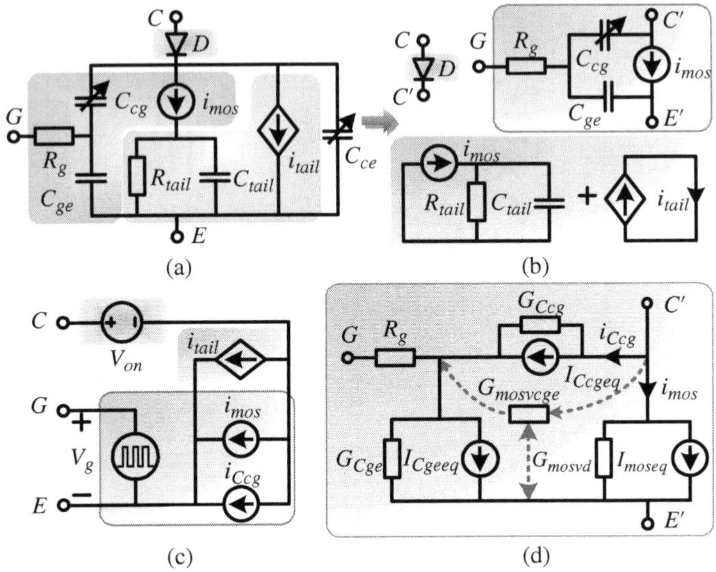

Figure 10.5 (a) IGBT fourth-order nonlinear behavioral model, (b) IGBT second-order nonlinear behavioral model, (c) generic representation of IGBT behavioral model, and (d) linearized discrete-time equivalent model for EMT simulation.

including the nonlinear nature of IGBT capacitances, are fully considered and can be reflected by the NBM. A number of curves from the device datasheet, including typical on-state characteristics, typical transfer characteristics, output characteristics are imported into the IGBT tool for extracting static parameters, such as a, b, x, y, z, V_t, and R_g. In the meantime, dynamic features, such as the relationship between typical capacitances and collector-emitter voltage, turn-on time and turn-off time are used for obtaining the remaining parameters to ensure that transient characteristics are sufficiently reflected and properly modeled as well. Specific procedures for parameter extraction are provided by SaberRD® [142].

10.3.5.3 Sensitivity Analysis

As can be seen from the IGBT NBM, each node links to several branches, making the element G^I_{ij} in the admittance matrix a sum of individual admittances. Thus, when calculating the matrix, a considerable amount of time will be spent on addition and subtraction operations. Based on Jacobian sensitivity analysis, the matrix can further be simplified. To accomplish that goal, the weakly coupled items, which can be identified by putting the IGBT into a test circuit, have to be distinguished from those that are dominant.

At an arbitrary node that connects to N branches, if the conductance or transconductance of kth branch is negligible at any time compared with the sum of the rest, that is,

$$\frac{\partial i_k}{\partial f_k(v_1, v_2, \ldots)} \ll \sum_{j=1}^{j=N} \frac{\partial i_j}{\partial f_j(v_1, v_2, \ldots)}. \tag{10.20}$$

Then, that item can be removed from the admittance matrix for fewer algebraic operation times. The analysis outcome showed that G_{tailvd}, $G_{tailvcge}$, and G_{Cce} can be omitted. Similarly, sensitivity analysis of $\mathbf{J}^I$ leads to a removal of I_{taileq}.

10.3.5.4 Model Parallelization

Noticing that the IGBT behavior can be largely categorized into two types, i.e. the basic MOSFET behavior determined by v_{Cge} and v_d, and the tail current phenomenon that can be deemed as an augment part solely dependent on v_{tail} according to sensitivity analysis, parallelization of the full behavioral model can be achieved. The former mainly includes components such as i_{mos}, C_{cg}, C_{ge}, and R_g, while the latter is a combination of R_{tail}, C_{tail}, and i_{tail}. Thus, the overall model can be deemed as a superposition of both behaviors, and consequently, it is possible to detach these two parts from each other so as to reduce the number of nodes in each part. Since i_{tail} can be deemed as a current- and voltage-controlled current source that its own terminal voltage has no impact on, there is no necessity to physically connect it to other circuit components. Therefore, the tail current itself constitutes an independent circuit. With regard to the parallel R_{tail}–C_{tail} combination, the voltage across them is so small compared with that of i_{mos}, that their existence has negligible influence on the MOSFET behavior. Thus, it can also be detached. Then the superposition model of IGBT can be derived as a collection of several submodels, as shown in Figure 10.5b. The second-order MOSFET subcircuit is the only part that participates in circuit nodal voltage calculation, while the other three are used for result correction, as shown in Figure 10.5c the general representation of IGBT behavioral model. Hence, the collector current is calculated as

$$i_C = i_{mos} + i_{Ccg} + i_{tail}, \tag{10.21}$$

where i_{mos} and i_{Ccg} are obtained through solving the circuit in which the MOSFET part locates and i_{tail} is calculated directly by (7.23). Similarly, the device's voltage can be deemed as a summation of V_{on} and v_d. In Figure 10.5d, the discretized and linearized circuit for MOSFET part which contains merely one transconductance $G_{mosvcge}$ is shown, where the arrow with a dashed line indicates that $G_{mosvcge}$ is related to the node it points to. Based on that (10.22) and (10.23) can be obtained, and the dimension of the matrix equation reduces to 3.

Such improvements of the full model leads to multifold benefits: the new model is as precise as its original counterpart, and the number of N–R iterations reduces substantially due to fewer nonlinearities as well as smaller matrix dimension; meanwhile, the maximum time-step to compute the model is also increased so that simulation can run much faster.

$$\mathbf{G}^I = \begin{bmatrix} G_{mosvd} + G_{Ccg} & G_{mosvcge} - G_{Ccg} & -G_{mosvcge} - G_{mosvd} \\ -G_{Ccg} & G_{Cge} + G_{Ccg} + R_g^{-1} & -G_{Cge} - R_g^{-1} \\ -G_{mosvd} & -G_{cge} - R_g^{-1} - G_{mosvcge} & G_{Cge} + G_{mosvd} + G_{mosvcge} + R_g^{-1} \end{bmatrix}, \tag{10.22}$$

$$\mathbf{J}^I = \left[-I_{moseq} - I_{Ccgeq}, I_{Ccgeq} + \frac{V_g}{Rg} - I_{Cgeeq}, I_{Cgeeq} + I_{moseq} - \frac{V_g}{Rg} \right]. \tag{10.23}$$

10.3.6 Electro-Thermal Network

The heat induced by IGBT power loss will raise the junction temperature which in turn affects the HHB performance. Meanwhile, determination of the size of IGBT array in LCS and MB unit also relies heavily on the junction temperature. Thus, the inherent electrothermal network established in Figure 7.9 is included as part of an accurate IGBT model. The cooling system is not included in the MB for the reason that the selected IGBT type should have sufficient capacity to withstand the DC line current momentarily, while the adoption of a cooling system in LCS can be determined by calculating the IGBT junction temperature using the electrothermal network. It should be pointed out that this network is suitable for all the three HHB models, and only one detailed electrothermal network corresponding to the selected IGBT is established due to the fact that all IGBTs including their electrothermal networks are identical. Therefore, their operation status such as the junction temperature is immediately known when computation of the selected HHB unit is completed.

As illustrated in Chapter 9, linear functions taking the form of (7.39) are applied to reflect the impact of junction temperature on IGBT performance. The coefficients of the thermal network are listed in Table A.6.

10.3.7 HHB Hardware Implementation on FPGA

The hardware design of the HHB integrated into the MTDC system is targeted onto the Virtex-7® FPGA XC7VX485TFFG 1761-2 whose resources are listed in Table 1.2. To achieve a pipelined structure, the overall system is divided into a number of subcircuits, and hardware modules are designed specifically for each one of them, such as three types of HHBs and the thermal network, transmission

Table 10.1 Latencies and hardware resource utilization of principal hardware modules in the three-terminal HVDC system.

Module	Description	Latency	LUT	FF	DSP
ABCDQ	*abc-dq*	77–78	4723 (1.56%)	3044 (0.50%)	34 (1.21%)
DQABC	*dq-abc*	75–76	4628 (1.53%)	3041 (0.50%)	34 (1.21%)
LPR	Protection	4	665 (0.22%)	308 (0.05%)	2 (0.07%)
MMC	MMC Ctrl	38	6362 (2.10%)	4323 (0.72%)	35 (1.25%)
REC/INV	Rec/Inv control	35	867 (0.29%)	697 (0.11%)	10 (0.36%)
DCYARD	DC yard	43	3235 (1.07%)	1984 (0.33%)	18 (0.64%)
TL	Line fault	31	1646 (0.54%)	1135 (0.19%)	10 (0.36%)
MMC-AC	AC part	23	3588 (1.18%)	2021 (0.33%)	15 (0.54%)
MMC-DC	DC part	13	486 (0.16%)	409 (0.07%)	8 (0.29%)
ITAIL	Tail current	26	1291 (0.43%)	737 (0.12%)	5 (0.18%)
THERM	Thermal network	31	2893 (0.95%)	1779 (0.29%)	15 (0.54%)
HHB-1	Type-1	67–125	14471 (4.77%)	6502 (1.07%)	106 (3.79%)
HHB-2	Type-2	38	5032 (1.66%)	2662 (0.44%)	26 (0.93%)
HHB-3	Type-3	40	5431 (1.79%)	2233 (0.37%)	23 (0.82%)

line with fault stimulus, the controllers for rectifier as well as the inverter and their DC yards, MMC inner loop controller, and its specific circuits. Like previous chapters, Vivado HLS® is employed to enable C/C++ coding of a subcircuit in the form of a function whose input and output variables are turned into corresponding hardware module's physical ports after being synthesized and exported as an IP core. Table 10.1 shows individual latencies obtained from Vivado HLS® synthesis as well as FPGA resource utilization of some key modules of the design.

It demonstrates that the longest hardware delay in the MTDC system employing either HHB-2 or HHB-3 can be attributed to the *abc-dq* transform module, which takes up to 78 clock cycles. However, it is not the factor that determines whether the design can attain real-time execution; instead, the HHB module is decisive because at least one N–R iteration is required in calculating the transients that take place after activating line protection. Therefore, the actual latencies for HHB-2 and HHB-3 are doubled to around 80 T_{clk} considering a few intervals are inserted between two calculations. Hence, to accomplish the goal of real-time, the time-step should be larger than that value. In contrast, the NBM-based HHB, which, according to Table 10.1, has the largest maximum latency of 125 T_{clk} among all of the components and is the determinant of HIL emulation speed. The varying latency makes output results last for different periods at different stages, leading to distorted waveforms. To avoid that, a timer is included to unify the actual latency of Type-1 model to a fixed 125 T_{clk}. Meanwhile, each type of HHB module has a very low percentage of resource utilization compared with the power converter. And since only one HHB unit containing one IGBT model is necessary to be designed into a hardware module, the resource utilization for HHB-1 is quite low, let alone the other two much simpler types. The Type-2 model normally requires more resources than the Type-3 model, but after transferring the calculation of (10.18) to the electro-thermal network, it has a similar scale to the latter and its latency is also reduced from over 50 T_{clk} to 38 T_{clk}.

Figure 10.6 depicts the pipelined hardware structure of a portion of the MTDC system as well as signal exchange routes, in which all hardware modules sealed in blocks achieve parallelism. Those modules related to MMC AC part and its control are represented by the module *Grid-connected MMC*, which receives reactive and active power or DC voltage orders and generates AC side current and voltage information in *dq* frame for the DC part of MMC, where the transmitted power is obtained and converted to DC current and voltage. Then, the incident pulse v_2^i is calculated and sent to the rectifier DC yard so as to obtain two mesh currents, based on which, other variables, such as DC line current can also be acquired. The module for Type-1 circuit breaker is shown as an example in the figure, while

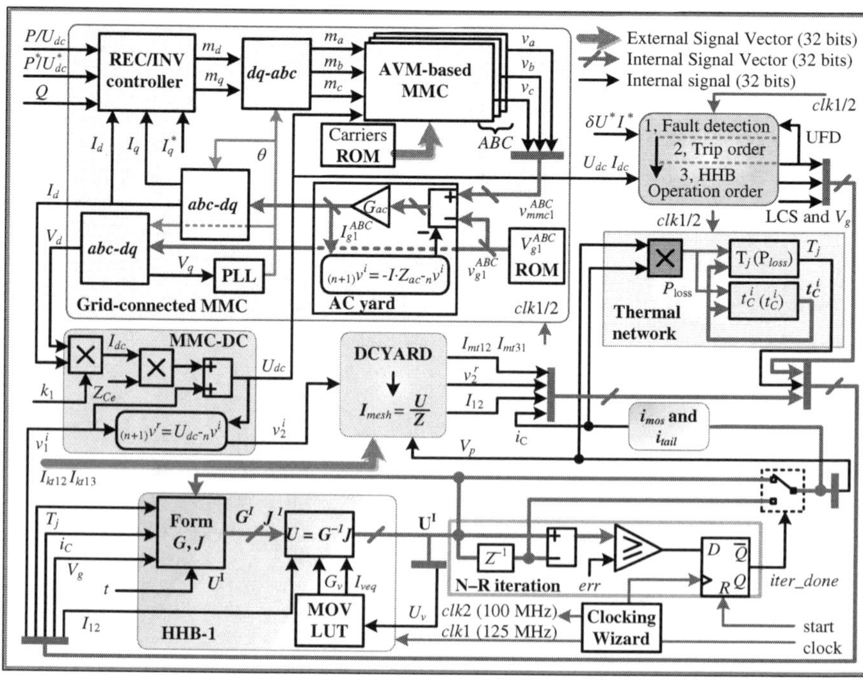

Figure 10.6 Hardware design of HHB integrated with MMC in a pipelined structure and signal flow routes.

the other two types have the same ports. The DC line current acts as excitation and based on the status of the IGBT, the nodal voltages can be calculated. The signal t is introduced specifically for Type-2 HHB, for indication of operation status and consequently, the transient current can be ascertained in the LUT. The voltage and current obtained directly or indirectly are delivered to the electrothermal network so that the power loss and the junction temperature can be obtained, which are in turn sent to the circuit breaker module to update IGBT parameters for the next time-step. The virtual block *N–R Iteration* is not a hardware module designed by Vivado HLS®. In fact, it is realized by very high-speed integrated circuit hardware description language (VHDL) coding in Vivado®, where all modules are arranged and connected to each other to form the top-level module.

A proper operation sequence for hardware modules is coordinated by a top-level state machine, as shown in Figure 10.7 for MTDC system with Type-2 and Type-3 HHBs. With regard to Type-1, the only difference is the criterion in State S_5 should change to whether the calculation of HHB-1 has been completed. The overall system starts to operate under the command *rst* that is generated by pressing the reset button on FPGA board. External signals such as three-phase AC grid voltages and carriers for MMC modulation are stored in ROM, and those data are accessed prior to the operation of all hardware modules. State S_5 takes 43 T_{clk} since the start order is given, which ensures at the end of that state, HHB-2 and

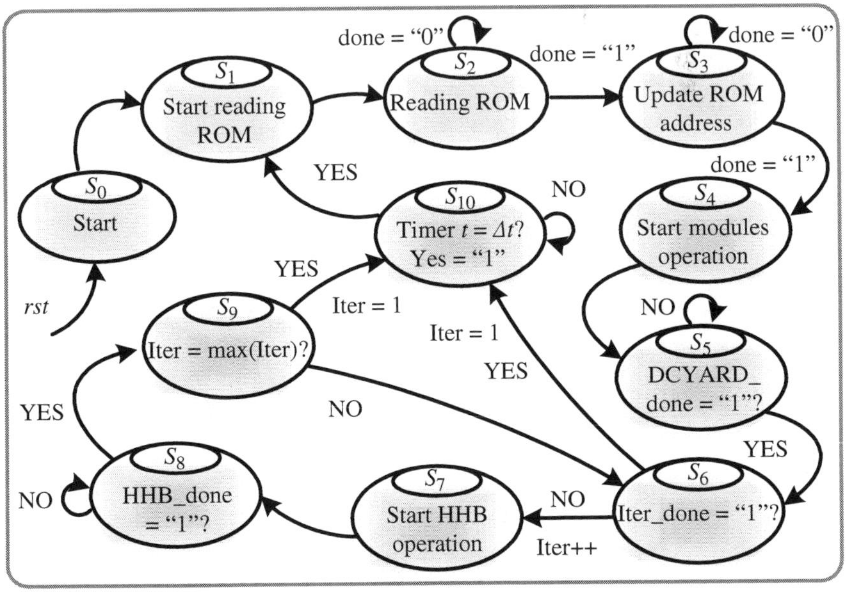

Figure 10.7 Top-level state machine for coordinated operation of MTDC hardware modules.

HHB-3 have already completed their first computation and have been waiting to enter a new phase. Then, if the results converge, those finished modules will keep idle in State S_{10} until one time-step runs out, and by the time Park's transformation and its reverse have also been finished. On the contrary, if the results are not convergent, only the nonlinear HHB module will be executed again until it converges or the maximum number of iterations has been reached. For Type-2 and Type-3 HHBs, the maximum iteration number can be set to two while for NBM-based HHB, three iterations are required. The selection of operating frequency is a trade-off between time-step and FPGA capability. For the first two types of models, the frequency is set to 100 MHz, which means $T_{clk} = 10$ ns and consequently, the minimum time to wait to synchronize to real-time in State S_{10} is approximately 100 ns. Since the nominal latency of Type-1 model is unified to 125 T_{clk}, the frequency is chosen as 125 MHz so that $T_{clk} = 8$ ns and the design will execute three times slower than real-time.

10.3.8 HHB HIL Emulation Results

The functions of HHB in guaranteeing normal operation of healthy power transmission corridors and isolating the faulty section are tested by HIL emulation of the three-terminal HVDC system in which the center of transmission line 1 is subjected to short-circuit while line 2 keeps operating, as indicated in Figure 10.1. The reaction of the overall system as well as the performance of some of its components – especially the DC circuit breaker – is investigated and validated by comparison with results from industry-standard transient simulation tools PSCAD/EMTDC® and SaberRD®, respectively. The MMC model established in such tools is AVM-based, since they are particularly inefficient or even unable to compute a circuit with nearly a thousand nodes if the detailed switching function model is applied to the 17-level MMC adopted in the simulation together with two conventional full-scale HHBs. Specific parameters of the MTDC system are given in Appendix A.9 which shows that the combined protection voltage of all MOVs is 340 kV, and the number of HHB units is 100, which means that the protection voltage of MOV_u is 3.4 kV. It can be estimated by applying (10.11) that the fault clearance time is close to 4 ms considering the actual MOV voltage during that period is a little lower than its protection voltage.

10.3.8.1 Device-Level Performance

The device-level behavior of the HHB mainly includes the voltage and current waveforms of its internal components as well as IGBT's junction temperature. SaberRD® is chosen for results validation since it provides detailed nonlinear behavioral IGBT models with the thermal network, such as the chosen *igbt1_3x* model.

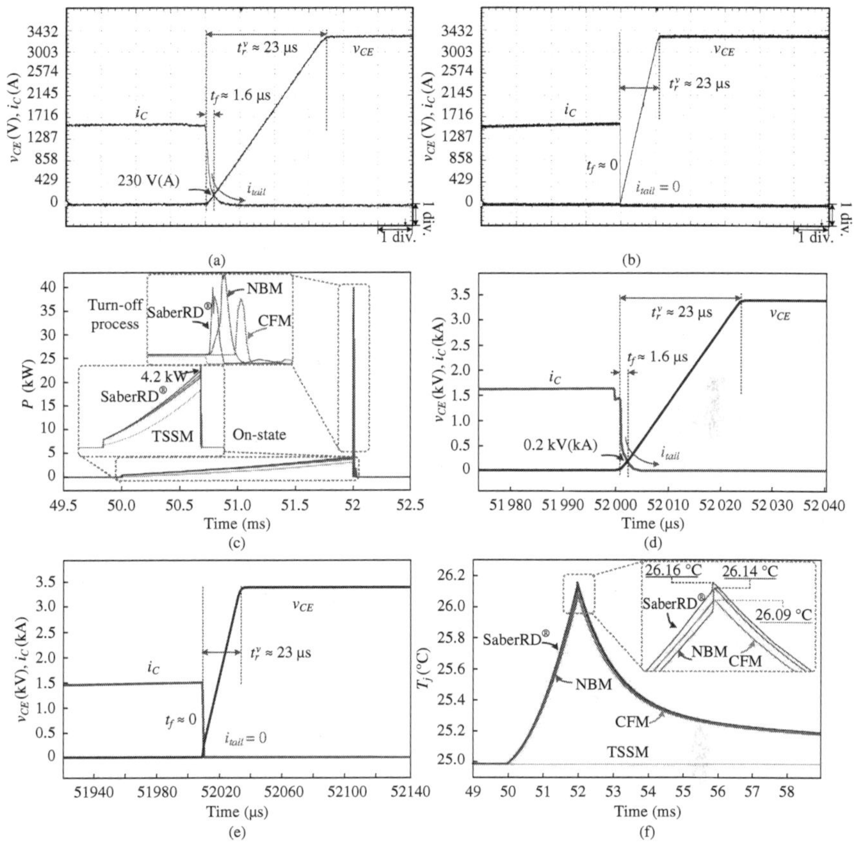

Figure 10.8 Turn-off performance of HHB with RCD snubber circuit: (a) MB NBM model of IGBT from HIL emulation, (b) MB TSSM model of IGBT from HIL emulation, (c) single MB IGBT power loss, (d) MB NBM model of IGBT from SaberRD®, (e) MB TSSM model of IGBT from PSCAD/EMTDC®, and (f) MB IGBT junction temperature. Oscilloscope horizontal axes settings: 20 μs/div.

Figure 10.8 shows the transient waveforms of an MB IGBT with RCD snubber during turn-off process. The HIL emulation results of the nonlinear behavioral IGBT model is shown in Figure 10.8a, which indicates that during the 23 μs period, v_{CE} slowly rises to around the varistor's protection voltage because the diode is under conduction state and consequently the RCD snubber circuit is equivalent to a capacitor that is being gradually charged. Even though the device still turns off within about 1.6 μs, an obvious tail current can also be observed. The v_{CE} curve of CFM is not shown since the result is identical, while the i_c waveform slightly differs with turn-off time of 2.5 μs to fit with the RC snubber case. Figure 10.8d demonstrates the same process by SaberRD®, which validates the correctness of the second-order NBM as well as the partitioning approach. It should be pointed out that variables shown in the oscilloscope are annotated based on the actual time the process needs to complete, and therefore, voltage rise time t_r^v and IGBT turn-off time t_f are divided by a factor of 3, which is the time of speed that the HIL emulation runs slower than real-time. In Figure 10.8b,e, the turn-off process of TSSM IGBT is shown to illustrate the usefulness of complex IGBT models. Although the voltage rise process is exactly the same as that of NBM IGBT, the current waveform is straightened. As a consequence, the power loss during the transient stage is much smaller, while in the NBM IGBT case, the power loss soars to 37 kW, as shown in Figure 10.8c. The power loss at on-state of CFM and NBM are almost the same, with the former having a little closer result to that of SaberRD®. However, the TSSM with estimated on-state resistance of IGBT is incapable of describing the power loss accurately. The junction temperature variation is shown in Figure 10.8f. Since a complete loop in the electrothermal network cannot be formed in TSSM, its temperature remains slightly above 25 °C. On the contrary, the IGBT junction temperature in the other two models rises shortly after 50 ms because of the power dissipation induced by the DC line current transferred from UFD–LCS branch to the MB branch by forcing the LCS to turn off immediately after the protection sequence is activated by the line fault which occurs at 50 ms. As can be seen, both NBM and CFM approaches lead to curves that almost agree with the one from SaberRD® simulation, meaning that HHBs employing these two models are sufficient for HIL emulation. Moreover, in this case, the latter shows its advantage by running in real-time.

In comparison, Figure 10.9 shows voltage and current waveforms of HHBs with RC snubber. Figure 10.9a,b are turn-off waveforms of the MB with NBM and CFM IGBT respectively, and part (d) shows the simulated v_{CE}–i_C curves from SaberRD®, which validates the device-level models and partitioning approach. The IGBT turn-off process becomes a little longer and MB terminal voltage rises more quickly, from approximately 23 μs of RCD snubber to around 3.6 μs. The two current curves for RCD and RC snubber cases of CFM IGBT are the same: both have a turn-off time of 2.5 μs, proving that it is incapable of adjusting to variations in the electromagnetic environment unless the LUT is modified. With RCD snubber, v_{CE} keeps low during IGBT turn-off process and consequently, the power loss is small, while in RC snubber case, the rise of v_{CE} occurs simultaneously with the fall of i_C, and they cross at about 0.8 kV(kA), which indicates the power loss is much larger, as shown in Figure 10.9c. The on-state power loss is virtually identical to that of MB with RCD snubber, and thus the junction temperature in both cases is approximately 26.1 °C at the beginning of 52 ms before the IGBT turn-off process; nevertheless, its transient power loss reaches over 10 times higher to about 500 kW, leading to an instantaneous junction temperature jump to around 26.8 °C – an increment of 0.7 °C. As an obvious comparison, the temperature jump

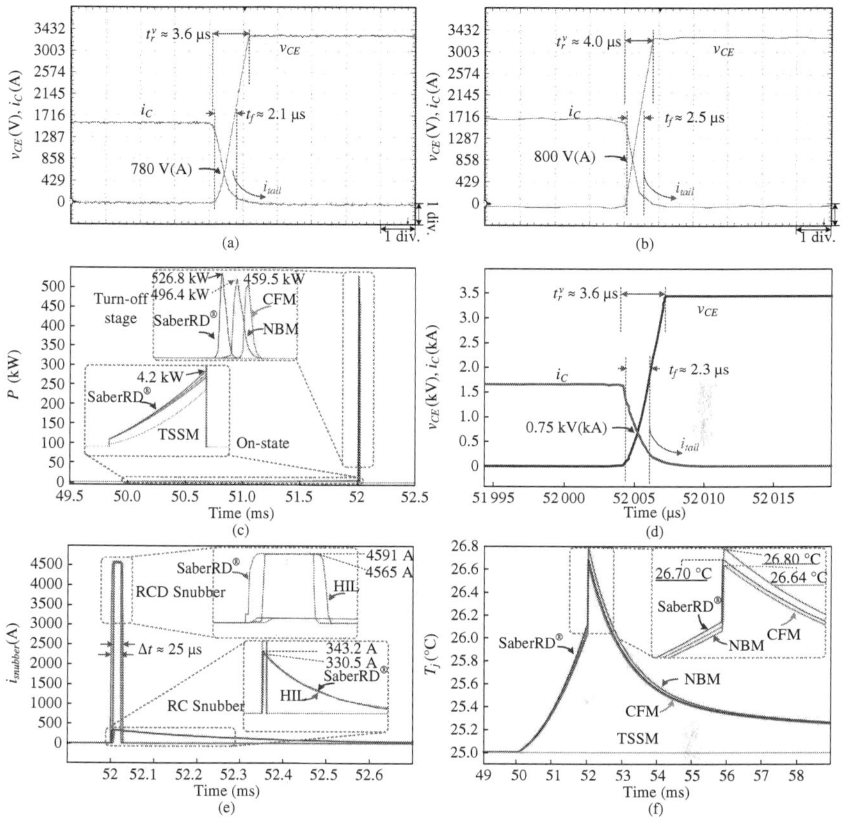

Figure 10.9 Turn-off performance of HHB with RC snubber circuit: (a) MB NBM model of IGBT from HIL emulation, (b) MB CFM model of IGBT from HIL emulation, (c) single MB IGBT power loss, (d) MB NBM IGBT from SaberRD®, (e) comparison between RCD and RC snubber, and (f) MB IGBT junction temperature. Oscilloscope horizontal axes settings: (a) 10μs/div, (b) 5μs/div.

by IGBT turn-off behavior in the RCD snubber case is much smaller, estimated at 0.08 °C, indicating the effectiveness of RCD snubber in reducing IGBT switching loss. Thus, the low junction temperature verifies the statement that no cooling system is required for the IGBT stacks in the MB branch, and the close agreement with results from SaberRD® in Figure 10.9c,f again prove the accuracy of the HHB models. In Figure 10.9e, the snubber currents are compared, all three models generate almost the same waveforms and, therefore, they are represented by the Type-1 model. It shows the credibility of HIL emulation which leads to results similar to that of SaberRD® in both RCD and RC snubber cases. For the RCD snubber, combined with Figure 10.8a, the MB and varistor operating process can

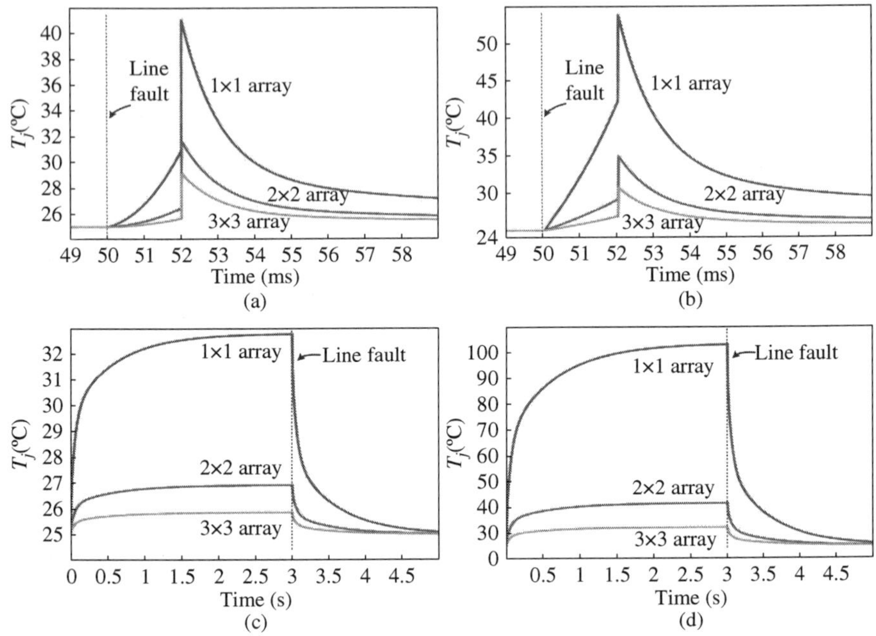

Figure 10.10 Junction temperature variation during operation: (a) MB IGBT under DC current 1 kA, (b) MB IGBT under DC current 4 kA, (c) LCS IGBT under DC current 1 kA, and (d) LCS IGBT under DC current 4 kA.

be derived. After receiving block order from line protection, the MB turns off. In the meantime, the DC fault current diverts to the snubber and after its voltage, as well as the varistor's voltage, increases to the protection voltage it again diverts to the varistor where it gradually vanishes. For RC snubber, as soon as MB turns off, R_s endures the protection voltage because the voltage over C_s is very low due to a slow charging rate limited by the resistor. Therefore, the snubber current is much smaller, but due to an early establishment of a voltage around 3300 V, the power loss of MB is high.

The above two cases show that with a proper snubber circuit and three IGBTs in parallel, the junction temperature rise is negligible. However, such a benefit is accompanied by adopting extra IGBTs. With the help of electrothermal network, evaluation of an appropriate size for IGBT array becomes feasible, and CFM is employed in the emulation. In Figure 10.10, line protection tests of two HVDC systems with steady-state DC currents 1 and 4 kA are conducted to show the significance of electro-thermal network in guiding HHB design. Figure 10.10a,b are MB IGBT junction temperature variations, which indicate that the chosen IGBT type has enough capacity to construct an MB unit with 1 × 1 IGBT array to protect a transmission line with even 4 kA steady-state DC current since the maximum temperature is only about 55 °C. The LCS IGBT temperature variation is given in Figure 10.10c,d. The temperature steadily rises after the HHB starts normal operation and at the entry into steady state at $t = 3$ seconds, a line fault is simulated. As can be observed, a single-IGBT LCS is enough to accommodate 1 kA, while when the steady-state DC current increases to 4 kA, the junction temperature could rise beyond 100 °C with self-cooling, indicating the margin from safe operation is too small and consequently other types of IGBT arrays, such as 2×2, or an external cooling system, are required. In the meantime, a comparison between MB and LCS junction temperature variation validates the theory that switching transients modeling is particularly important for MB IGBTs while it is negligible for LCS, whose static characteristics dominates junction temperature rise, meaning that even the steady-state part of CFM is sufficient to satisfy simulation requirements.

Figure 10.11 shows the overall performance of different types of HHB models. Figure 10.11a,b are the results from HIL emulation in which Type-1 and Type-3 models are employed, respectively. The shapes of these voltages and currents in both figures are virtually the same, which indicates that the breaking time is 2 ms and the fault clearance time is approximately 4 ms. The results of Type-2 breaker are omitted since they are identical. During the breaking time, the DC fault current is equally divided among three paralleled IGBTs so that each accounts for one-third of the total. The main difference between these two figures is their emulation speed. For Type-1 model, it runs three times slower than real-time, thus Δt_1 and Δt_2 by horizontal-axis are 6 ms and 12 ms, respectively, which are discounted into 2 ms and 4 ms. On the contrary, the other model is executed in real-time. In Figure 10.11c, the simulation results from PSCAD/EMTDC® are given, which have exactly the same shapes and values as the HIL emulation results.

In Figure 10.12, the importance of developing a full-scale HHB model for HVDC system performance prediction is demonstrated. It would be misleading that the snubber parameters such as those in Appendix A.9 are applied to a scaled-down model since it produces some unexpected oscillations in v_{hcb}, the circuit breaker's voltage, as well as line current that a full detailed model will not cause, as shown in Figure 10.12a, which gives the results from real-time HIL emulation and PSCAD/EMTDC® simulation. Therefore, evaluating the behavior of an HVDC system will be inaccurate. On the other hand, the parameters that enable the scaled-down model to produce device-level waveforms as shown in Figure 10.11 are probably inappropriate for the full model, as can be observed from both HIL emulation and PSCAD/EMTDC® simulation in Figure 10.12b that some oscillations are introduced to the breaker's voltage and line current when the snubber resistance is altered to 200 Ω.

Table 10.2 summarizes the energy consumed by the three main parts of HHB under two snubber circuits after line protection is activated. We can see that regardless of the snubber circuit, MOV absorbs the majority of remaining energy,

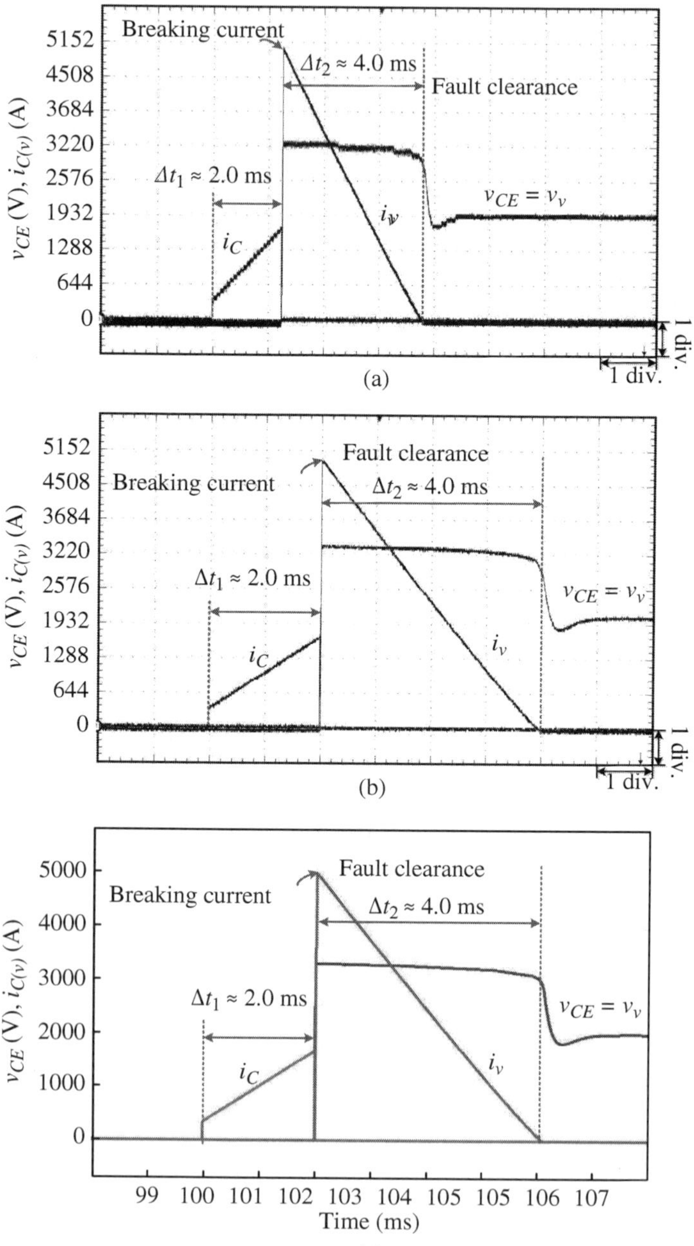

Figure 10.11 Varistor voltage and current during protection: (a) HIL emulation of Type-1 HHB model, (b) HIL emulation of Type-3 HHB model, and (c) PSCAD/EMTDC® simulation results. Oscilloscope horizontal axes settings: (a) 5 ms/div, (b) 1 ms/div.

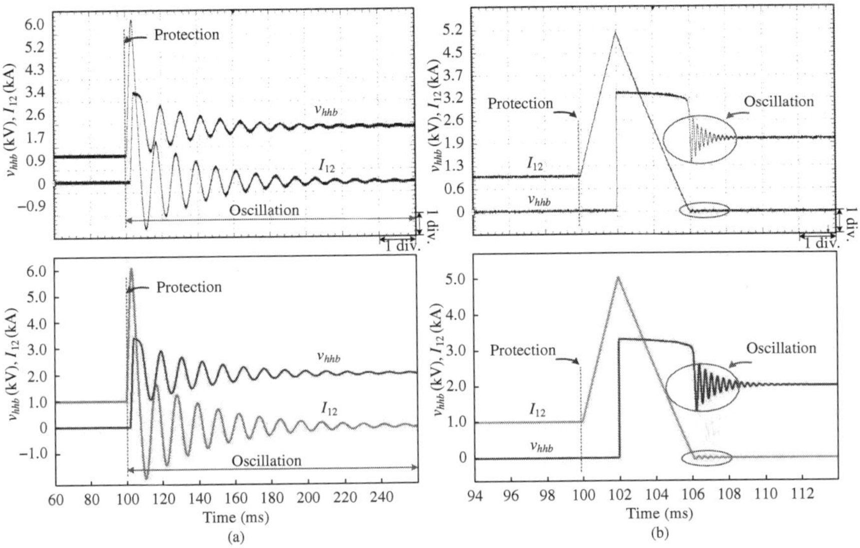

Figure 10.12 Varistor voltage and line current during protection from HIL emulation (top) and PSCAD/EMTDC® simulation (bottom): (a) scaled-down model, (b) full-scale model. Oscilloscope horizontal axes settings: (a) 20 ms/div, (b) 2 ms/div.

Table 10.2 Energy consumed by different HHB components.

Snubber type	**RC snubber/RCD snubber**		
Models	**MOV**	**Snubber**	**MB**
Type-1 (NBM)	3.231 MJ/3.254 MJ	23.9 kJ/8072.0 J	627.7 J/444.5 J
Type-2 (CFM)	3.071 MJ/3.111 MJ	24.0 kJ/7866.0 J	622.9 J/428.8 J
Type-3 (TSSM)	3.075 MJ/3.114 MJ	24.0 kJ/7866.9 J	288.0 J/288.0 J
SaberRD®	3.193 MJ/3.233 MJ	23.1 kJ/8258.3 J	633.8 J/442.1 J

slightly over 3 MJ in both cases. However, the amount of energy dissipated by the other two parts is heavily dependent on the type of snubber. Energy absorbed by the RCD snubber is about one-third of the RC circuit. On the other hand, all three types of circuit breakers yield a close energy consumption for MOV and snubber, and there is disagreement in the energy consumption of the MB path. The Type-3 model shows the least energy because the turn-off process of the TSSM is inaccurate. The Type-2 and Type-1 models have similar energy consumption, and as can be observed, the one based on the second-order NBM has closer results to SaberRD®, where the fourth-order IGBT model *igbt1_3x* is used, while for Type-2 model to attain more precise power consumption under RCD snubber case, the current curve of CFM should be adjusted. Moreover, it can be observed from the table that different IGBT models can cause minor differences in energy consumed by the MOV and snubber, which underlines the importance of precise switch models.

10.3.8.2 System-Level Performance

All three models are applicable to the MTDC system as they produce the same system-level results, so the Type-2 model is used for real-time purpose, and the waveforms are validated by PSCAD/EMTDC® simulation, as shown in Figure 10.13.

Before line fault, all converter side DC voltages are maintained at around 200 kV, with the rectifier having a small margin over the other two to ensure power transfer. Immediately after Line 1 contacts the ground, the DC voltages at all stations sag, and after the faulty section is isolated by HHBs on both sides within around 6 ms, the DC voltages are gradually restored since they are controlled by one of the inverters connecting to the rectifier, as shown in Figure 10.13a.

Figure 10.13b is the converter side DC currents during the same period. As can be seen, during breaking time, the line fault leads to current surges in the Rectifier side as well as Inverter-1 side, which sees a polarity inversion from −1 kA to over 2 kA as the fault forces the inverter, along with the rectifier, to provide energy to the ground. As a consequence, the energy received by Inverter-2 reduces, but with a much slower speed since energy is also stored in the 200 km-long path, including two current limiting inductors. After the fault is isolated by two HHBs, currents flowing from Inverter-1 and the Rectifier to the ground are interrupted, and therefore, the 2 kA rectifier current diverts to Inverter-2. In less than 100 ms, the current stabilizes at 2 kA as it is still controlled by the Rectifier. Figure 10.13c shows the power delivered or consumed by the three stations, and we can see that during the fault, the rectifier station can provide as much as 1 GW power to the ground, but after completing the protection process, the power is restored and Inverter-2 receives all that amount of energy. As a comparison, results from the scaled-down HHB model with the same snubber parameters are also shown. Figure 10.13d indicates the voltages are less affected by the simplification of the model. However, the current waveforms have a remarkable difference, with high-frequency oscillations lasting up to 100 ms and the DC fault currents in Line 1 are not quenched immediately. As a consequence, the power transmission is also unstable during that period, with the Rectifier power reducing to zero momentarily and multiple energy exchanges between Inverter-1 and the rest of the system.

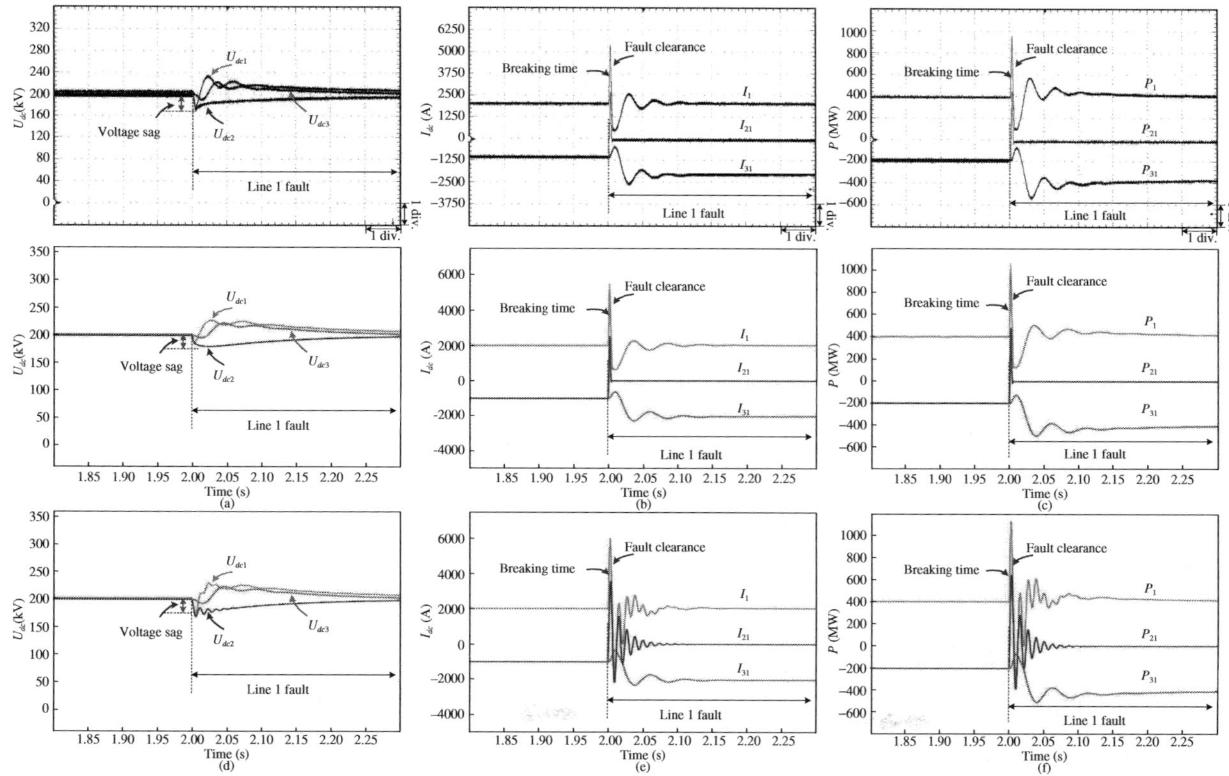

Figure 10.13 System-level performance of the MTDC system during long-term line fault with detailed device-level and scaled-down HHB models from HIL emulation (top) and PSCAD/EMTDC® simulation (middle and bottom). (a)(b)(c) Converter side DC voltages, currents and active powers with detailed HHB models, (d)(e)(f) Converter side DC voltages, currents, and active powers with the scaled-down HHB model. Oscilloscope horizontal axes settings: (a)(b)(c) 50 ms/div.

10.4 Ultrafast Mechatronic Circuit Breaker

In addition to the above ABB's hybrid circuit breaker, Alstom Grid has also developed its own prototype which employs SCR that features a larger capacity [202, 203], and its design process was also specified [204]. The actual installation of the ultrafast mechatronic circuit breaker is illustrated in Figure 10.14a, where the direct integration into the DC transmission path induces too many nodes – including those from the HVDC power transmission path – that would cause a heavy computational burden. Similar to the previous case, the *V*–*I* coupling mechanism is also applicable regarding splitting the DC circuit breaker from the transmission path for node number reduction, given in Figure 10.14b, for the same reason.

It should be noticed in the figure that a single thyristor symbol, i.e. SCR_1, SCR_{11}, SCR_{12}, and SCR_2, in the UFMCB is actually a chain of dozens or even over a hundred thyristors. Thus, it is obvious that although the circuit breaker has been separated from the external system, its detailed model still contains – depending on the thyristor model used in EMT simulation – hundreds or even thousands of nodes, making instant circuit solution impractical. Meanwhile, creating a fundamental unit as in the IGBT-based HHB is also infeasible, due to its highly irregular configuration in the auxiliary branch. In addition to circuit splitting, merging the cascaded thyristors becomes the other option for node number reduction. The details are narrated in this section [205].

10.4.1 Nonlinear Device-Level Thyristor Model

10.4.1.1 Basic Device Characteristics

The ideal thyristor model categorized as Type-3 is a two-state resistor whose on- and off-state is controlled by the thyristor logic, which it turns on after a positive pulse is exerted on the gate of the device whose anode–cathode voltage v_{AK} meanwhile is greater than the threshold V_f; when the current i_{AK} vanishes, the thyristor turns off if v_{AK} keeps below 0 over a period longer than the turn-off time T_q.

In a device-level thyristor model, the on-state *I*–*V* characteristics should be considered, along with some switching transients. Figure 10.15a shows the nonlinear behavioral thyristor model that shares some similarities with the reverse-recovery diode model. The additional SCR logic controls the proper state of the device by regulating the resistance of r_s: a large resistance for the off-state while a fixed small value is set when the thyristor is turned on.

The nonlinear diode symboled by *NLD* reflects the static on-state characteristics between v_{AK} and the current i_{AK}. It is approximated by the following analytical function:

$$v_{AK} = A + B \cdot i_{AK} + C \cdot \sqrt{i_{AK}} + D \cdot \ln(i_{AK} + 1), \tag{10.24}$$

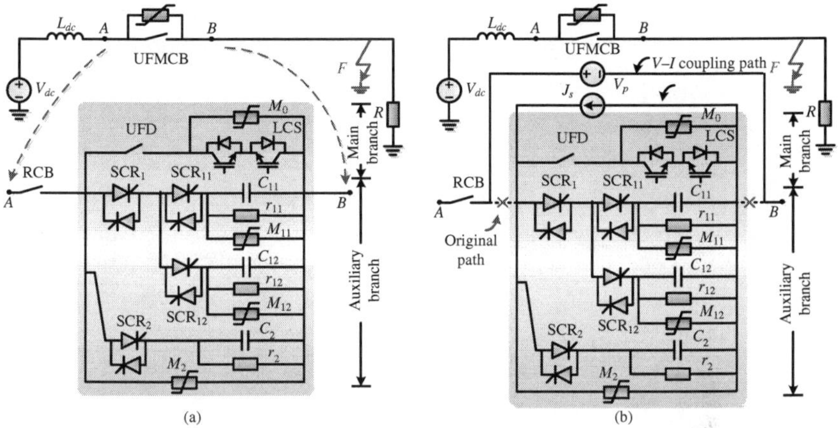

Figure 10.14 Ultrafast mechatronic circuit breaker: (a) actual installation in the transmission path, and (b) decoupled model.

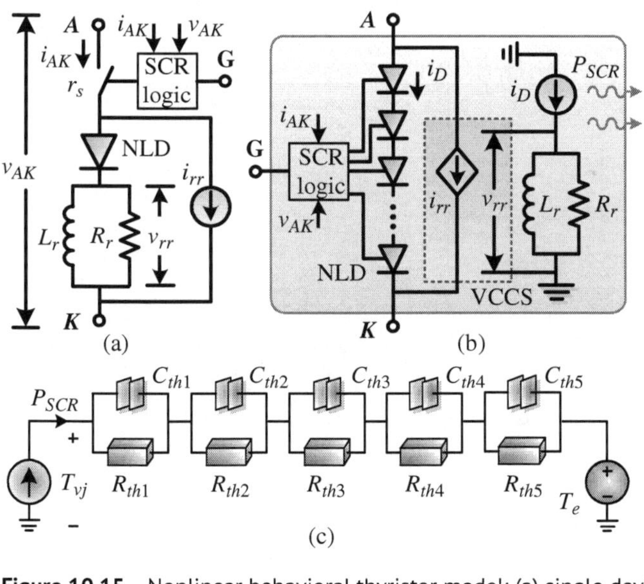

Figure 10.15 Nonlinear behavioral thyristor model: (a) single device, and (b) scalable model for cascaded thyristors with (c) cascaded R–C network for thermal impedance.

where A, B, C, and D are temperature-dependent coefficients, which are either provided by the device manufacturer's datasheet or can be estimated according to available I–V curves normally under 25 and 125 °C. Therefore, the four coefficients as a function of temperature can be linearly interpolated. For digital simulation based on nodal equations, (10.24) is usually discretized by taking partial derivatives so the companion circuit becomes available. However, for simplicity as well as achieving a low hardware latency, *NLD* is taken as a pure nonlinear resistor with its conductance a function of the current, i.e.

$$G_{AK}(i_{AK}) = \frac{i_{AK}}{v_{AK}(i_{AK})}. \tag{10.25}$$

The SCR logic block distinguishes a thyristor from the diode. It turns on after a positive pulse is exerted on the gate and a v_{AK} greater than the forward voltage drop V_{To}; it turns off when the current i_{AK} vanishes and v_{AK} keeps under 0 over a period longer than the turn-off time t_q. Correspondingly, when the thyristor is under conduction, the controlled switch r_s exhibits a low impedance, otherwise, it approaches infinity. Thus, in the model, *NLD* is enabled to be only in charge of on-state characteristics.

Nevertheless, like the diode, a thyristor also has the reverse recovery phenomenon, which is why the current source i_{rr} is retained to yield proper reverse direction current whose value is dependent on the voltage over R_r–L_r pair. The current source i_{rr} is dependent on the voltage over R_r–L_r pair, which has the following relation:

$$L_r = \frac{R_r}{\ln 10}\left(t_{rr} - I_{rrm}\left(\frac{dI_r}{dt}\right)^{-1}\right), \tag{10.26}$$

where t_{rr} and I_{rrm} are the reverse recovery time and peak current, respectively, I_r is the reverse recovery current.

10.4.1.2 Scalable Cascaded Thyristor Model

The aforementioned nonlinear behavioral thyristor model contains four nodes. Different from the ideal switch, a device-level model has the mandatory requirement that its voltage and current should not exceed the actual ratings. Therefore, it is always organized in series or parallel to withstand high voltages or large currents, resulting in inefficient circuit solution and a deficiency of simulation hardware/software resource.

A computationally efficient model for cascaded thyristors is presented in Figure 10.15b. The R_r–L_r pair is separated since their values are too small to affect the static feature, and the *NLD* of each thyristor in series is merged with SCR logic module to eliminate the internal nodes, resulting in the following equivalent conductance of N_{SCR} cascaded thyristors:

$$G_{AK}^{N_{SCR}}(i_{AK}) = \frac{G_{AK}(i_{AK})}{N_{SCR}}. \tag{10.27}$$

The *NLD* branch, along with i_{rr}, constitutes the basic part that participates in external circuit solution. Following the acquisition of the diode current i_D, the reverse recovery current is calculated directly using

$$i_{rr} = K \cdot (G_{Lr} + G_r) \cdot (i_D - I_{Leq}), \tag{10.28}$$

where G_{Lr} along with I_{Leq} constitutes the inductor's companion circuit. As can be seen, no amplification is required for the reverse recovery current since all serial thyristors share the same current. Regarding the SCR junction temperature, it is calculated based on the thermal network in Figure 10.15c using the method presented in (7.36).

The SCR logic in Figure 10.16 was tested by the circuit in Figure 10.17a where ABB fast thyristor 5STF 28H2060 was modeled, and ANSYS Simplorer® was deployed for device-level model validation. To show an obvious thyristor reverse recovery phenomenon, the voltage source v_s is set square as ±200 V/60 Hz, while the thyristor gate voltage has a frequency of 100 Hz, and their phase relationship is given on the top. When positive gate voltage P_1 arrives, the thyristor turns

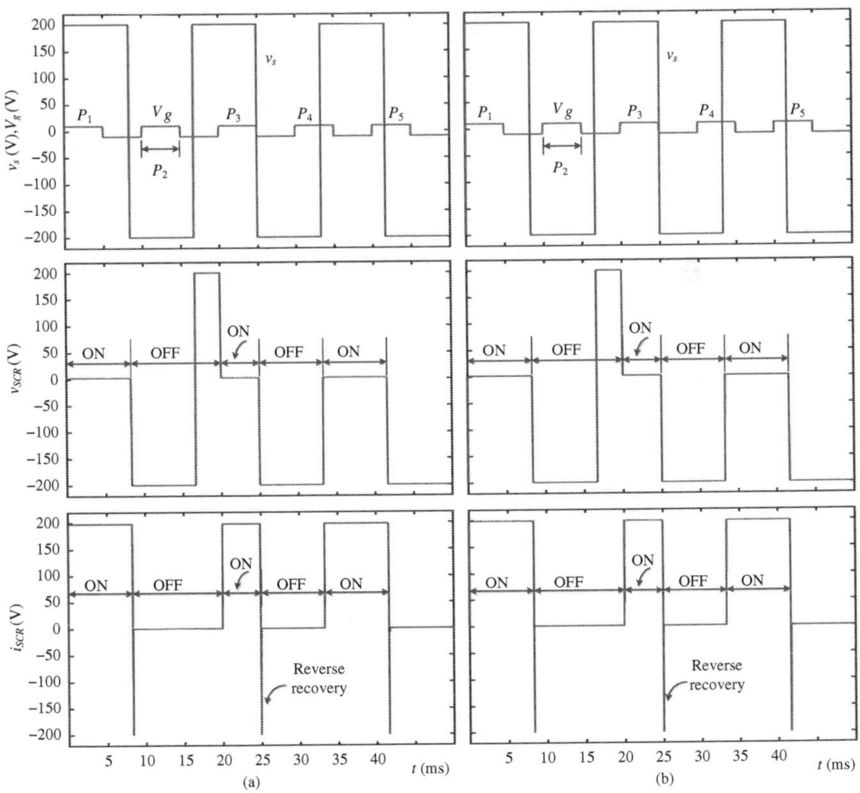

Figure 10.16 SCR logic validation (top: v_s and V_g; middle: thyristor voltage; bottom: thyristor current): (a) scalable model, and (b) ANSYS/Simplorer®.

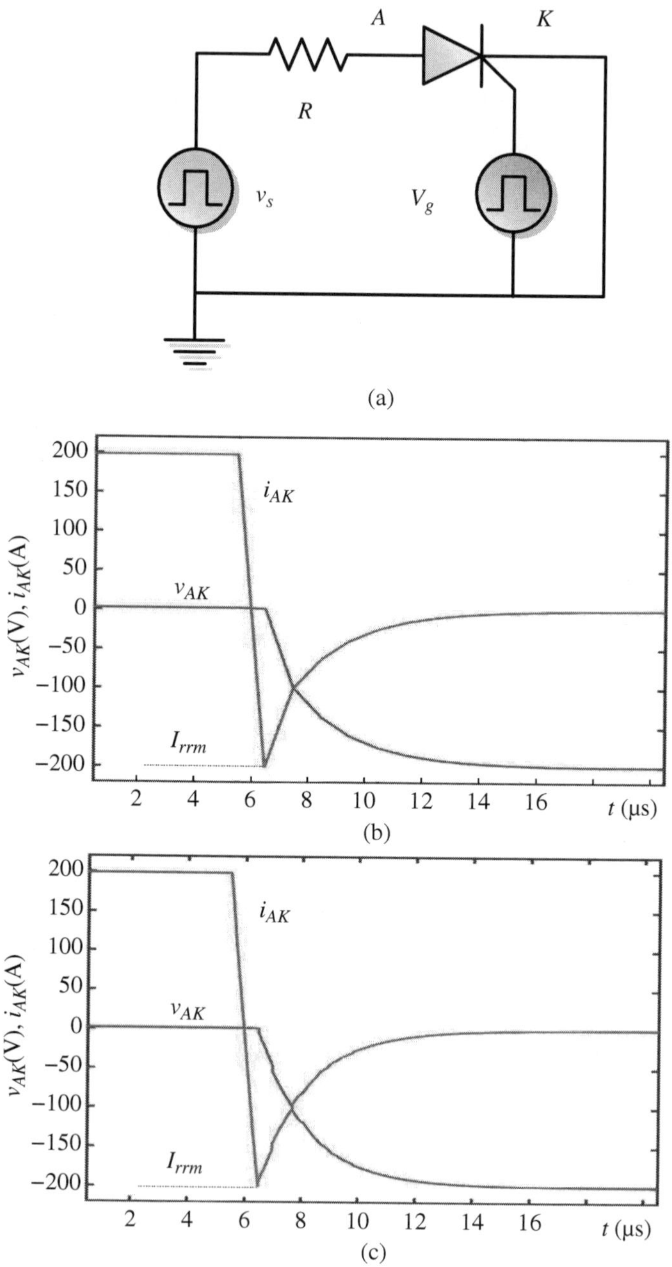

Figure 10.17 (a) Basic thyristor test circuit. SCR reverse recovery: (b) scalable model, and (c) ANSYS/Simplorer®.

on since it has already been forward-biased. In contrast, when P_2 appears, the thyristor is not turned on because of reverse-biased state, and the off-state lasts until the arrival of P_3 even though the biased condition is changed prior to that. Another noticeable feature of the SCR logic is the turn-off process during P_5, the thyristor turns off when it is reverse-biased even though the gate voltage is still in effect.

The device-level modeling was validated by the reverse recovery process in Figure 10.17b,c, which was obtained from the same test circuit. While the current is ascending from I_{rrm} to 0, v_{AK} is steadily approaching −200 V in a process lasting around 10 μs. The results are very close to that of ANSYS Simplorer®, indicating that the dynamic part was also correctly modeled.

10.4.2 UFMCB Modeling

The bidirectional UFMCB given in Figure 10.14 is symmetric since all unidirectional components have their counterparts in the opposite direction. In EMT simulation of the HVDC grid, the source V_{dc} is replaced by an MMC, and therefore, a heavy computational burden is imposed by the device-level UFMCB and the MMC if not simplified. In this work, the MMC's fully detailed model instead of the average value model is preferred since the latter is proved to be less accurate regarding DC fault. Meanwhile, the UFMCB model is anyhow a major source of the computational burden since it contains a large number of nodes and nonlinearities that as a consequence require an iterative calculation process to be completed in a small time-step. Thus, as an independent apparatus, the UFMCB is separated from the DC yard by a pair of coupled voltage-current source. The impact of the UFMCB on DC grid is realized by the voltage source V_p, which means when the DC yard operates without any circuit breaker, its configuration remains the same by setting V_p constant zero. Meanwhile, although all SCRs are connected in an antiparallel manner to accommodate bidirectional power flow, no additional node is induced in the EMT model.

In the main branch, the principle of the DC breaker decides that the LCS always turns on and off under a low voltage clamped by the auxiliary branch, meaning its primary power loss is induced by normal conduction. Therefore, the switching transients can be ignored and the LCS is taken as a current-dependent resistor, whose conductance G_{LCS}, according to the static I–V characteristics in the manufacturer's datasheet, can be calculated by the terminal voltage divided by current, as in (10.25). The IGBT nonlinear I–V characteristics is piecewise linearized with each taking the form of

$$i_C = k_1(T_{vj}) \cdot v_{CE} + k_2(T_{vj}), \tag{10.29}$$

where k_1 and k_2 are temperature-dependent and, consequently, linearly interpolated based on two temperatures, i.e. 25 and 125 °C. The two IGBTs with opposite directions in LCS can be further merged and eventually taken as one resistor.

To lower the voltage and current stresses, each IGBT symbol in the LCS is a 3×3 array of the ABB 5SNA 0650J450300 HiPak modules with inherent cooling mechanism. The analytical function for its transient thermal impedance shares the same form to (7.34), other than there are only three cascaded R–C pairs and the parameters can be found in the datasheet [206].

Like in the HHB, the metal-oxide varistors denoted as M_0, M_{11}, M_{12}, and M_2 are specifically installed to protect the circuit breaker's components from over-voltage. Its model, as introduced, can eventually be represented by the Norton equivalent circuit G_v and I_{veq}.

The UFD, on the other hand, is usually taken as an ideal switch with fixed on- and off-state conductances. As further work to HHB, the arc phenomenon is considered by employing the Mayr arc model, which is [207]

$$\frac{dlng}{dt} = \frac{1}{\tau_M}\left(\frac{ui}{P_M} - 1\right), \tag{10.30}$$

where g is the arc conductance, u and i are the arc voltage and current, and τ_M as well as P_M represents the time constant and cooling power constant, respectively. Taking the logarithm of arc conductance lng as a variable, the above transcendental equation can be rewritten in its integrated form as

$$lng = lng_0 + \frac{1}{\tau_M}\int\left(\frac{i^2}{gP_M} - 1\right)dt, \tag{10.31}$$

and eventually discretized as

$$g(t+\Delta t) = g_0 \cdot \exp\left(\frac{b}{\tau_M}\left(Z(t) + \left(\frac{i^2}{g(t)P_M} - 1\right)\Delta t\right)\right), \tag{10.32}$$

where g_0 is the initial conductance, b is a binary denoting the open or closed status of the UFD, and the iterative $Z(t)$ is

$$Z(t) = Z(t-\Delta t) + \left(\frac{i^2}{g(t-\Delta t)P_M} - 1\right)\Delta t. \tag{10.33}$$

For simplicity, the UFD conductance is denoted by G_{UFD}, and whichever model is adopted, its existence induces in the MB an internal node, which can be eliminated by merging with the LCS. Then, the companion model, in the form of Norton equivalent circuit with G_{MB} and I_{MBeq} the conductance and equivalent current, respectively, is expressed by

$$G_{MB} = \frac{G_{UFD} \cdot (G_{LCS} + G_{v0})}{G_{UFD} + G_{LCS} + G_{v0}}, \tag{10.34}$$

$$I_{MBeq} = \frac{G_{MB}}{G_{LCS} + G_{v0}} I_{veq}. \tag{10.35}$$

With the breaking period restricted by a relatively slow UFD reaction, the UFMCB operation time is mainly determined by the protection voltage of M_2 as a higher rating curtails the duration, and a larger DC current leads to a longer operation period. Their exact relationship can be ascertained with a real-time emulator as part of the design guide.

10.4.3 Relaxed Scalar Newton–Raphson (RSNR)

The decoupled UFMCB model introduced above still has a dimension of 5 in its admittance matrix and current vector, which are subjected to the Newton–Raphson iteration since it contains a number of nonlinearities contributed by the thyristor and MOV. The nodal matrix equation takes the form of

$$\mathbf{U}^k = \left(\mathbf{G}^{(-1)}\right)^k \cdot \mathbf{J}^k, \tag{10.36}$$

where the subscription k is the iteration number. Following the acquisition of the kth iteration nodal voltage vector $\mathbf{U}^k$, the admittance matrix $\mathbf{G}$ and the current vector $\mathbf{J}$ are updated for the next calculation until the results are convergent, i.e.

$$\left|\frac{\mathbf{U}^{k+1} - \mathbf{U}^k}{\mathbf{U}^{k+1}}\right| \leq \epsilon, \tag{10.37}$$

where ϵ is the threshold. To solve (10.36) with a high dimension and varying admittance matrix, the Gaussian Elimination should be applied repeatedly, which is quite inefficient and consequently not used for the 5-node UFMCB.

A relaxed scalar Newton-Raphson (RSNR) method that decomposes the circuit and consequently avoids calculating the matrix equation is presented here. At the kth Newton–Raphson iteration, the general matrix equation (10.36) can be expanded and sorted out as

$$\begin{bmatrix} G_{11} & \cdots & G_{1i} & \cdots & G_{1N} \\ G_{21} & \cdots & G_{2i} & \cdots & G_{2N} \\ \vdots & \cdots & \vdots & \cdots & \vdots \\ \underline{G_{i1}} & \cdots & \underline{G_{ii}} & \cdots & \underline{G_{iN}} \\ \vdots & \cdots & \vdots & \cdots & \vdots \\ G_{N1} & \cdots & G_{Ni} & \cdots & G_{NN} \end{bmatrix}^k \begin{bmatrix} U_1 \\ U_2 \\ \vdots \\ \underline{U_i} \\ \vdots \\ U_N \end{bmatrix}^k = \begin{bmatrix} J_{eq1} \\ J_{eq2} \\ \vdots \\ \underline{J_{eqi}} \\ \vdots \\ J_{eqN} \end{bmatrix}^k. \tag{10.38}$$

Therefore, at an arbitrary node i, its voltage can be extracted from the ith row which is underlined in (10.38) as

$$U_i^k = (G_{ii}^k)^{-1} \cdot \left(J_{eqi}^k - \sum_{j=1, j\neq i}^{N} G_{ij}^k U_j^k \right). \tag{10.39}$$

To solve the ith nodal voltage U_i^k, voltages at other nodes U_j^k are yet to be known. However, the nodal voltage at kth iteration can be expressed by the previous calculation, i.e.

$$U_i^k = (G_{ii}^k)^{-1} \cdot \left(J_{eqi}^k - \sum_{j=1, j \neq i}^{N} G_{ij}^k \left(U_j^{k-1} + \xi_j \right) \right), \tag{10.40}$$

where ξ_j denotes the error between two successive iterations at the jth node. It means that the solution of an arbitrary node can be realized without solving the large matrix equation as in (10.38) which is time-consuming. Compared with the Jacobi method for linear systems, the elements J_{eqi}, G_{ii}, and G_{ij} are possibly nonlinear in RSNR, and therefore, their values should also be updated in every iteration. With the ongoing repetitive process updating nodal voltages, ξ_j decreases until it is smaller than ϵ when the results are deemed convergent.

Figure 10.18 describes the RSNR method by an equivalent circuit. The node n_0 to be solved is neighbored by a number of other nodes, denoted as n_1, n_2, etc., and the branches between two nodes are represented by the Norton equivalent circuit J_{eqi} and G_i $(i = 1, 2, \ldots)$. The potentials at other nodes U_1, U_2, etc., are represented by voltage sources to form a complete circuit. Then, the nodal voltage to be solved can be derived by

$$U_0 = \left(\sum_{i=1}^{N} J_{eqi} + \sum_{i=1}^{N} G_i U_i \right) \left(\sum_{i=1}^{N} G_i \right)^{-1}. \tag{10.41}$$

Thus, the partitioned five-node UFMCB evolved into five subcircuits as in Figure 10.19, or from a mathematical point of view, the five-dimensional (5-D) matrix equation is split into five single equations that can be solved in a more efficient parallel manner. For instance, Node 1 is neighbored by Nodes 2 and 5, and therefore, the impact of other components on these two nodes are converted to voltage sources U_2 and U_5, while for Node 5 solution, the potential U_1 is assumed available.

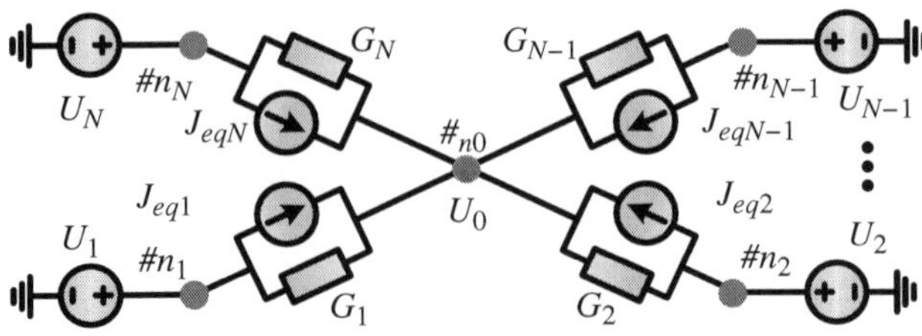

Figure 10.18 Equivalent circuit of the RSNR method.

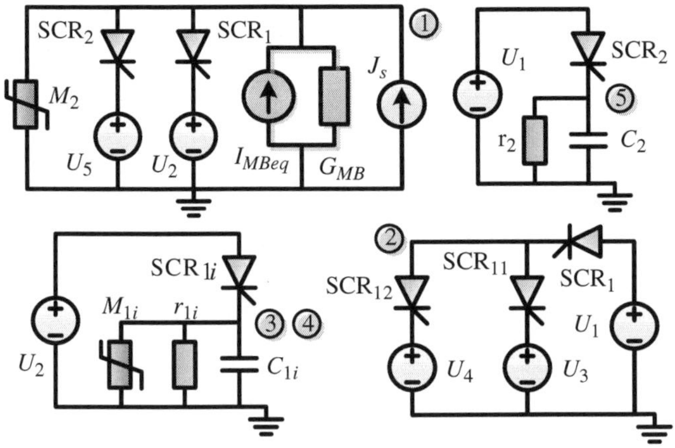

Figure 10.19 UFMCB nodal domain decomposition.

10.4.4 UFMCB Hardware Design

The UFMCB model has five main parts which are designed into corresponding hardware modules by the high-level synthesis tool Vivado HLS® enabling the design by C/C++ coding. In Table 10.3, the five types of hardware modules are listed, i.e. the SCR, the MB, the MOV, the solver, and the signal update function (Update), when the UFMCB hardware design is deployed onto the Xilinx® UltraScale+ XCVU9P FPGA. Due to the introduction of the scalable thyristor model, the hardware resource utilization rate is low, less than 3% for all three types. Meanwhile, for the 5-D nodal matrix equation, the original solver using Gaussian Elimination requires more resources than the RSNR solver, whereas the main improvement is on the latency that is a key factor for real-time execution. A time-step around 8 μs required by *Solver*1 would compromise the resolution, especially considering that device-level results are needed; in contrast, the RSNR enables a four times speedup since the time-step can be set to 2 μs. The fact that the MOV latency remains 0 due to hardware parallelism indicates that the selected time-step will not change even if the MOV piecewise linearization yields more segments.

The synthesized UFMCB hardware architecture is given in Figure 10.20. The four SCR modules receive their individual gate signals from the protection device and then give the calculated conductances to the solver module. Meanwhile, the MOVs and the module *MB* conduct the companion model calculation, and the outcomes are also sent to the solver where the nodal voltages are computed in parallel. It is noticed that the outputs from SCR, MOV, and *MB* are not instantly

Table 10.3 UFMCB parts hardware design summary.

Module	LUT	FF	DSP48	Latency
GauElim (Solver1)	20 580	11 606	68	242
RSNR (Solver2)	2 149	1 776	31	49
SCR (×4)	2 496	1 366	18	10
MB	1 205	494	9	12–25
Update	2 043	1 605	26	5
MOV (×4)	2 375	0	0	0
UFMCB (Solver1)	28 699 (2.43%)	15 071 (0.64%)	121 (1.77%)	**756**
UFMCB (Solver2)	10 268 (0.87%)	5 241 (0.22%)	84 (1.23%)	**177**
Available	1 182 240	2 364 480	6 840	—

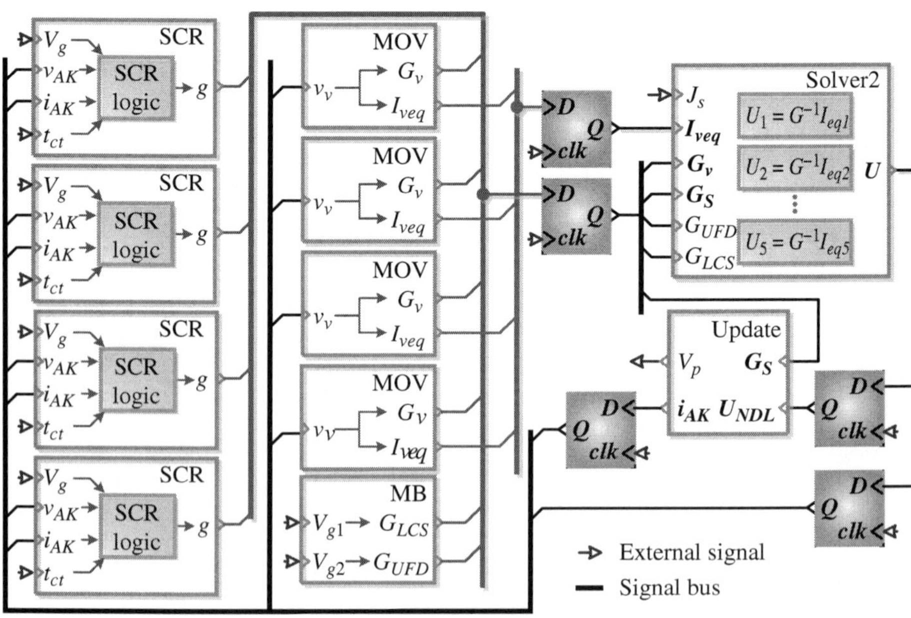

Figure 10.20 Synthesized hardware architecture of the UFMCB.

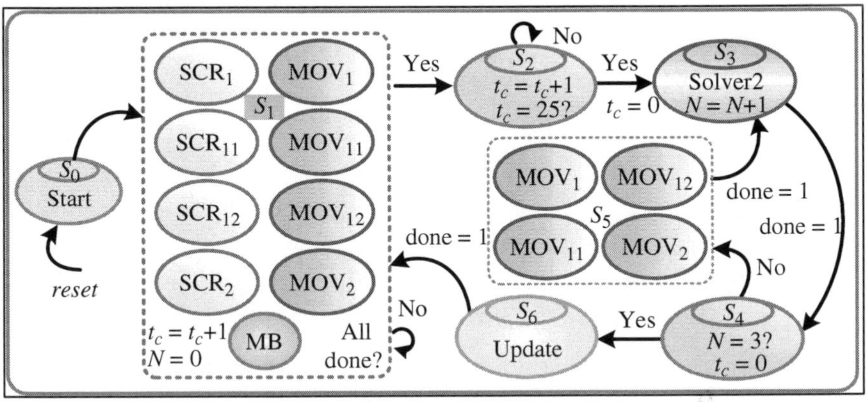

Figure 10.21 UFMCB top-level finite state machine.

sent to the solver, instead, the D-latch whose clock port is fed by the data valid indicator is always between the outputs and inputs of two hardware modules, and this is realized in the top-level formation using VHDL in Vivado®. Once correct nodal voltages are obtained from *Solver2*, they are sent to the *Update* module for calculating the thyristor currents and the voltage coupling V_p. As can be seen, a closed-loop is formed, and the inputs of a module are acquired from its upstream counterparts; however, parallelism is available for most of the blocks, and whenever the output data is valid, it is sent to the downstream modules.

A corresponding finite state machine (FSM) is designed to coordinate those modules since the involvement of nonlinear elements forces the adoption of Newton–Raphson iteration. According to Figure 10.21, repetitive operations of the MOV and *Solver*2 modules are possibly needed. In the beginning, the reset order is issued and the emulation starts from state S_1 when *MB* and all SCR and MOV modules are running. Completion of these modules does not indicate an immediate start of circuit solution; instead, *Solver*2 begins until 25 clock cycles counted since S_1 has passed due to the variable latency of the MB whose maximum latency is 25 clock cycles after taking its arc phenomenon into account. Following the nodal voltage solution is the convergence check, and because a maximum of three iterations is needed, the calculation in each time-step is repeated three times to ensure the results will not be distorted. As can be seen, only the MOV modules are involved in N–R iteration before entering to the nodal voltage solution module for gathering convergent results in the time domain. Hence, according to the FSM, a simulation time-step involves three N–R iterations and the total latency is 177, as specified in Table 10.3.

Therefore, the Xilinx® UltraScale+ FPGA providing a virtual primary power system can work as a HIL emulator under either the open-loop mode as introduced

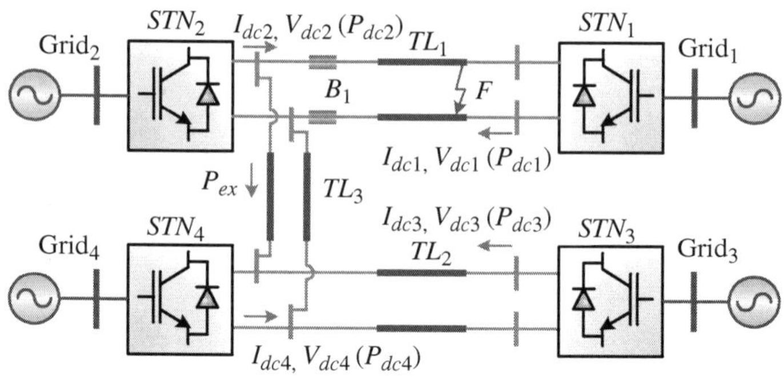

Figure 10.22 Four-terminal HVDC grid based on mixed MMC topologies and UFMCB.

here for DC circuit breaker design evaluation and system study or the closed-loop mode by interconnecting external control and protection devices for real-time testing.

10.4.5 UFMCB Real-Time Tests and Validation

10.4.5.1 Four-Terminal DC Grid Test Case

Figure 10.22 shows the four-terminal DC grid, where the rectifier stations STN_2 and STN_4 employ HBSM-MMCs, while the inverter stations STN_1 and STN_3 are clamp double submodule (CDSM)-based MMCs, all of which are symmetric monopole with the DC neutral point grounded in a fashion as in Figure 8.1. Therefore, the UFMCBs should be deployed to the two rectifier stations, while their counterparts do not require it for the fault isolation capability; since the fault is simulated on line TL_1, only the UFMCB B_1 is installed at STN_2. Once the pole-to-pole fault F takes places on that line, all IGBTs of STN_1 are ordered to turn off, while on the rectifier side, only the breaker B_1 is ordered to operate so that power transmission from STN_2 to the remaining healthy system continues.

The real-time HIL emulation setup provides both device- and system-level information, which is validated by ANSYS/Simplorer® and PSCAD/EMTDC®, respectively, as the former tool contains electro-thermal models of power electronics, while the latter is popular for grid-level study.

10.4.5.2 UFMCB Design Evaluation by HIL System

The DC circuit breaker hardware design is validated by tests conducted in the system given in Figure 10.14, where a 120 kV DC voltage is applied to an 80 Ω load via the UFMCB and an 100 mH inductor L_{dc} which has a parasitic resistance of 1 Ω, and each SCR symbol contains 60 serial 5STF 28H2060 thyristors. At t = 100 ms, a line-to-ground fault with a resistance of 1 Ω takes place, and the protection is triggered 50 μs later. Figure 10.23a gives the UFMCB's terminal

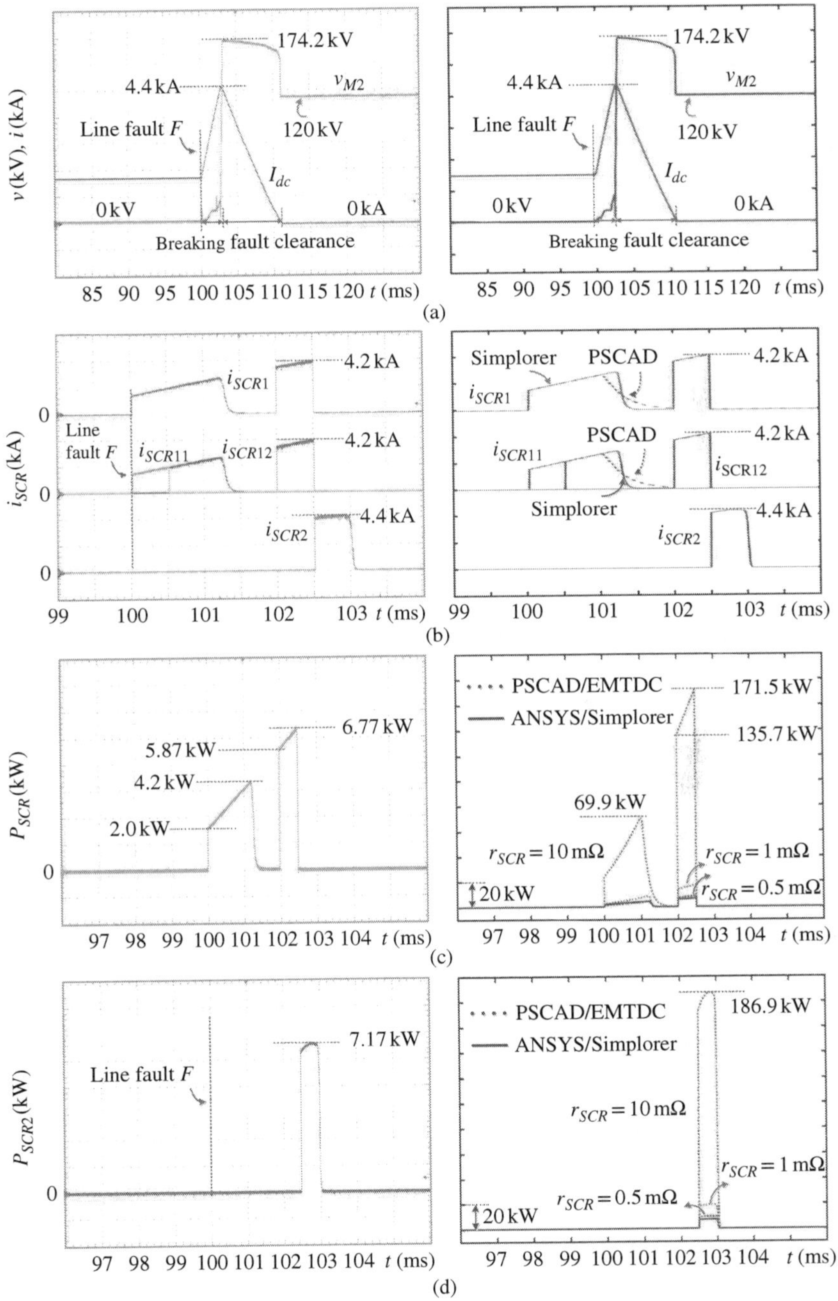

Figure 10.23 UFMCB real-time emulation oscilloscope results (left) validated by ANSYS/Simplorer® and PSCAD/EMTDC® (right): (a) protection waveforms, (b) SCR currents, (c) SCR_1 power consumption, and (d) SCR_2 power consumption.

voltage and current. The current rises in an approximately linear fashion during the breaking period, followed by an opposite trend in fault clearance period when it is eventually quenched to 0 by the MOV M_2 which maintains a protection voltage v_{M2} of around 170 kV. The insights peculiar to a device-level SCR model are given and compared with the ideal model in PSCAD/EMTDC®. It is noticed in Figure 10.23b that the currents of the thyristor model are identical to those from ANSYS/Simplorer®, while PSCAD/EMTDC® leads to some differences in two of them even though the waveforms are in system-level. In Figure 10.23c,d, the power losses of SCR_1 and SCR_2 from different tools are compared to demonstrate the importance of a device-level model. The HIL emulation results give definitive values at different time instants, while for PSCAD/EMTDC®, a small on-state resistance should be given to the ideal thyristor before simulation, and its various values yield distinct power losses that will be misleading if adopted for the design evaluation. The bidirectional current breaking capability of the UFMCB is also confirmed in the test, and all the results under reverse operation are absolutely identical to those shown above.

Based on the power loss, the selection of an appropriate thyristor type can be conducted by judging the junction temperature. With 5STF 28H2060 which has a peak nonrepetitive surge current I_{TSM} of 46.5 kA, the highest junction temperature appears in SCR_1, and it is merely around 27 °C as shown in Figure 10.24a, meaning the capacity of the thyristor is not fully utilized. The 5STF 05D2425 thyristor with I_{TSM} = 7.0 kA and a lower cost yields the maximum temperature at 41 °C in Figure 10.24b. Therefore, the latter type is more suitable while an adequate

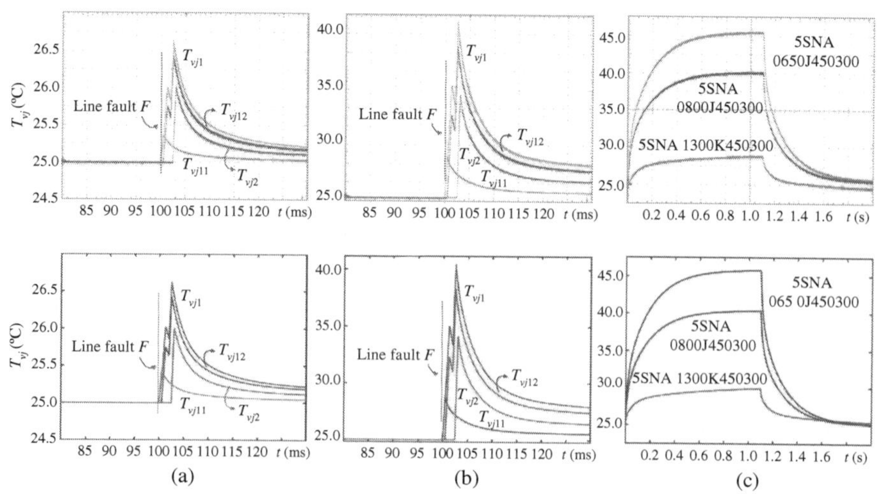

Figure 10.24 UFMCB power switch junction temperatures during protection from real-time HIL emulation (top) and ANSYS/Simplorer® (bottom). (a) 5STF 28H2060, (b) 5STF 05D2425, and (c) 3 IGBT types for LCS.

capacity margin is maintained. The 5SNA 0650J450300 IGBT is selected for the LCS according to Figure 10.24c, because even with a smaller capacity than its counterparts, the junction temperature during normal operation will not exceed 50 °C. The above results are validated by identical waveforms on the right side from ANSYS/Simplorer®, which is among popular device-level EMT simulation tools referred to for power electronic apparatus design guide before the equipment is galvanized and tested.

In Figure 10.25, the two thyristor types are compared regarding the power loss and junction temperature when the UFMCB is closed to establish normal power transmission. Only SCR_1 and SCR_{11} conduct during the process and, therefore, their device-level information is revealed. Although the 5STF 05D2425 with a smaller capacity yields a higher junction temperature due to larger power loss, it is only 30 °C – lower than the opening process. Therefore, this thyristor type is preferred over 5STF 28H2060 for constructing the UFMCB for applications with a similar voltage and power ratings to the tested DC system, and the HIL platform can always be referred to when the system changes.

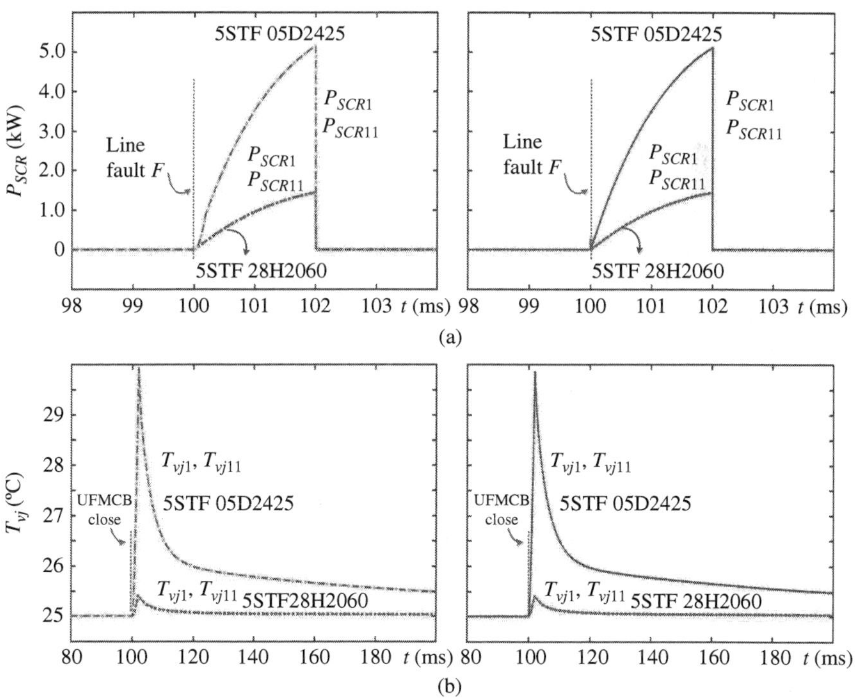

Figure 10.25 UFMCB thyristor device-level results during closing from real-time HIL emulation (left) and ANSYS/Simplorer® (right): (a) power loss, and (b) junction temperature.

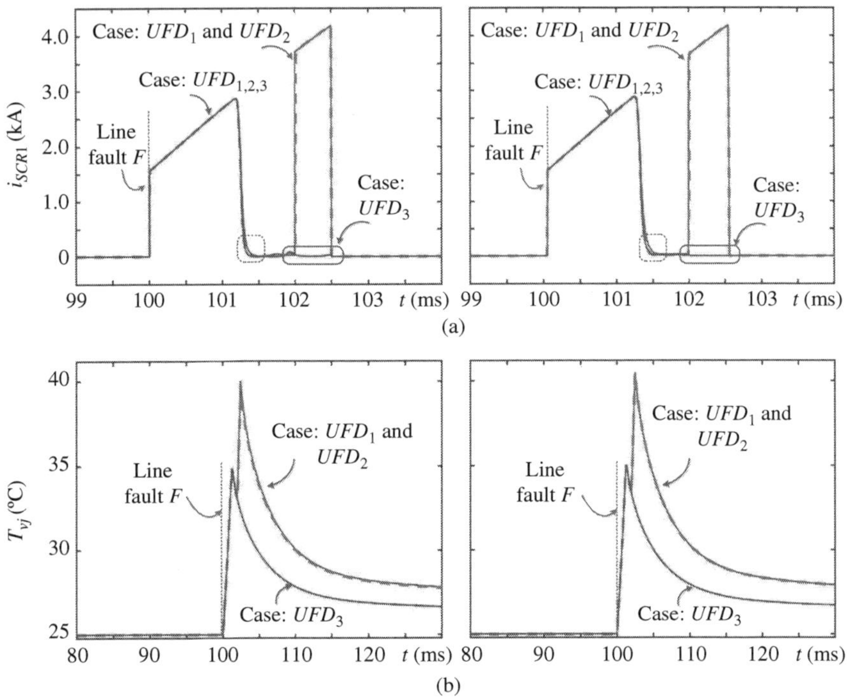

Figure 10.26 Impact of UFD arc on UFMCB thyristors from real-time HIL emulation (left) and ANSYS/Simplorer® (right): (a) SCR currents, and (b) junction temperature.

The impact of three sets of UFD parameters on the DC circuit breaker, mainly the thyristors SCR_1 and SCR_{12}, are demonstrated in Figure 10.26. The currents of SCR_1 are given as an example in Figure 10.26a. UFD_2 which has a cooling power constant of 1000 kW leads to the SCR current i_{SCR1} virtually identical to that of the ideal switch model UFD_1; in contrast, when the UFD_3 has a longer arc duration, i.e. the cooling power constant reduces to 100 kW, it can be seen that between 102 and 102.5 ms, the current i_{SCR1} is 0 as it is diverted to the UFD-M_0 path in the main branch. The junction temperatures of SCR_1 in Figure 10.26b are a direct reflection of the current: the thyristor junction temperature rises to over 40 °C when the UFD model is ideal and $P = 1000$ kW, while it is only 35 °C if $P = 100$ kW.

10.4.5.3 UFMCB in HVDC Grid

The four-terminal DC grid in Figure 10.22 is taken as the testbench for studying the system's performance under the joint protection scheme of the self-protective CDSM-MMC and the external apparatus UFMCB. As given in Figure 10.27, initially, all DC currents at the four terminals are around ±2 kA, and their DC voltages

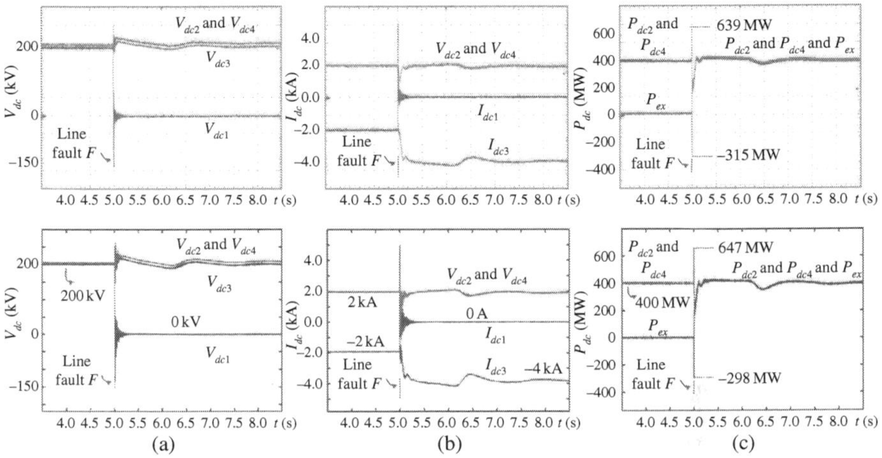

Figure 10.27 Real-time MTDC grid DC line fault emulation results (top) and PSCAD/EMTDC® validation (bottom): (a) DC voltages, (b) DC currents, and (c) power.

are approximately 200 kV, with the rectifier side having a slight margin, meaning each rectifier is delivering 400 MW power solely to the corresponding inverter, which accounts for no power flow between the two HVDC links as $P_{ex} = 0$. At $t = 5$ seconds, the pole-to-pole fault takes place near STN_1, followed by protections from both the UFMCB and the station itself. Due to CDSM-MMC's blocking capability, the DC voltage on STN_1 side V_{dc1} plummets to 0, and so is its current I_{dc1} with a short period of oscillations. On the other hand, the prompt response of the UFMCB exempts STN_2 from blocking and its power P_{dc2} is able to restore immediately. Consequently, P_{ex} increases to 400 MW and the power STN_3 received is doubled to about 800 MW, as it can be seen that its DC current I_{dc3} rises from 2 kA to around 4 kA while its DC voltage V_{dc3} is still 200 kV.

In Figure 10.28, the DC grid protection performance of the CDSM-MMC and UFMCB are given. In Figure 10.28a, after the permanent fault, all IGBT gate signals in the submodule are retrieved, and the SM capacitor is slightly charged and maintained around 4 kV for the 51-level MMC even though at the DC side its voltage V_{dc1} has become 0, and due to the high impedance of the SMs, the arm current drops to 0. As an important property of DC circuit breakers, the UFMCB reclosing function is tested in Figure 10.28b. After the fault, the DC current keeps rising when it is commutating from one branch to another in the circuit breaker and decreases to 0 after all SCRs are turned off so M_2 maintains a voltage of nearly 290 kV – higher than the rated 200 kV DC yard voltage – to extinguish the current; a reclosure is attempted 160 ms later to restore the power transmission, but since the fault still exists, another interruption is witnessed.

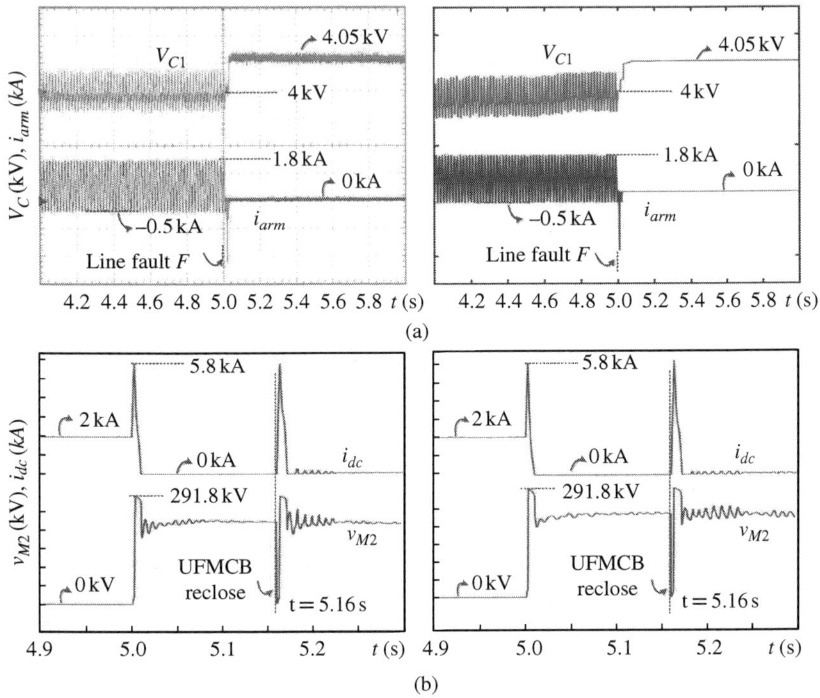

Figure 10.28 Real-time HIL emulation results (left) and PSCAD/EMTDC® validation (right) of protections initiated by (a) CDSM-MMC, and (b) UFMCB with reclosure.

10.5 Summary

This chapter began with a proposal of three full-scale ABB's HHB models for the purpose of accurate real-time HIL emulation as well as electrothermal transient simulation under the circumstance that their conventional counterpart is highly burdensome and resource-consuming. The approach that partitions the HHB full model into a number of fundamental and identical subcircuits results in a remarkable reduction of the dimension of corresponding matrix equations and therefore can be referred to for the modeling of other power electronic apparatus. On the other hand, FPGA hardware resource utilization declined drastically by at least two orders of magnitude compared with the conventional full-scale model, and the burden of computing HHB models is virtually the same as that of the scaled-down model. Therefore, it substantially accelerates the computational speed and is feasible for HIL execution to validate the control and protection strategies of an MTDC system. Meanwhile, as a pivotal part of HHB, three types of IGBT models were adopted to give a variety of guidance in the breaker design process. It is demonstrated by comparison with the scaled-down model that all three models are capable of verifying whether a selection of parameters is reasonable by investigating the impacts of the designed HHB on the overall system that the latter is unable to achieve. And particularly, the curving-fitting model and improved nonlinear behavioral IGBT model are able to provide extra information unavailable in system-level simulation studies of HVDC circuit breakers, such as IGBT power loss and subsequently its junction temperature, which is meaningful in the determination of an appropriate IGBT type and the size of its array for LCS and MB, as well as evaluation of the cooling condition so that a lower cost incurs. HIL emulation results demonstrated that the CFM-based HHB model can be executed in real-time, while the NBM HHB model provides better versatility. Meanwhile, it showed that precise IGBT models with switching transients are needed for IGBT type selection, while a simple steady-state model is sufficient for LCS IGBT. Moreover, the simplified IGBT NBM is computationally more efficient and robust against numerical divergence, so it can be applied for the simulation of other power converters.

Meanwhile, the device-merging method was also tested in modeling the ultrafast mechatronic circuit breaker. A thyristor with reverse recovery phenomenon was presented, and to reduce the node number when they are cascaded, an equivalent circuit was derived. The merging method leads to a significant reduction in hardware resource utilization when the design is deployed on FPGA. Like the IGBT model, the modeling methodology of taking a device-level thyristor as a combination of basic characteristics and augment parts can be applied to other power semiconductor switches for efficient computing.

At the end of each case, the nonlinear power semiconductor switch-based DC breaker was tested in a multiterminal DC grid. The hardware emulation results, as well as the speed, demonstrated that system-level simulation involving device-level details was feasible, and consequently more comprehensive and accurate information was accessible that could be referred to during the prototype design stage to lower the cost.

11

Large-Scale AC and DC Networks

11.1 Overview

High-voltage direct current transmission has been playing a significant role in delivering electricity from one area to another since its first commercial debut in 1950s. The breakthrough in semiconductor industry enabled the insulated-gate bipolar transistor (IGBT) to be able to withstand a higher voltage while increasing the capacity, and consequently, transmission technology has also witnessed a great leap. The two-level converter was initially used in voltagesource converter (VSC)-based high voltage direct current (HVDC) with limitations such as high switching loss, strict demand for synchronous gate driving, and abundant harmonics. Therefore, the neutral-point-clamped converter was later adopted. Modular multi-level converter (MMC)-HVDC, first commercially operated in the Trans Bay Cable project [208], is now the most prevalent HVDC type. One common drawback of the aforementioned VSC types, compared with a line-commutated converter, is a lack of DC fault block capability. Therefore, in an endeavor to minimize the adverse impact brought by a DC line fault, a number of new components were proposed and tested, and their hardware design and real-time simulation have been described in Chapters 7–9. For example, instead of feeding energy into the fault, some MMC variants such as full-bridge submodule and the clamp double submodule (CDSM) charge their submodule capacitors once the converter block order is implemented. The DC-DC converter prevents the fault location from affecting the other side with its inherent blocking capability. Compared with them, installing hybrid HVDC breakers (HHBs) in the converter station is more proactive and therefore, it restricts the impact to the minimum possible by isolating only the faulty line when an appropriate fault detection algorithm is adopted.

With sufficient participation of these novel power electronic equipment to ensure high transmission security and reliability, DC grids are being constructed by linking multiple HVDC terminals, and driven by high demand for diversified energy supplies, their expansion in scale is expected in applications such as

Real-Time Electromagnetic Transient Simulation of AC-DC Networks, First Edition.
Venkata Dinavahi and Ning Lin.

offshore wind farm interconnection. The very spontaneous formation of multi-terminal direct current (MTDC) systems leads to the coexistence of AC and DC networks that are interactive, making the power electronics-integrated power system more complex than a single grid.

Real-time electromagnetic transient (EMT) simulators are required not only to reproduce detailed high-frequency transients but also to be able to simulate large-scale power system transients due to the increasing complexity of modern power systems. But in practical large-scale real-time EMT simulators accuracy and computational efficiency are conflicting requirements that have ramifications in terms of simulator hardware and cost. A realistic reproduction of high-frequency transients requires detailed modeling of power system components and a small simulation time-step. Thus, large AC–DC system size entails excessive computational burden. To lower computational burden using simplified modeling or a large time-step would lower fidelity of the simulation. Inevitably, to meet real-time constraints and to accommodate large system sizes, a compromise is needed in component model complexity to curtail simulator cost at the expense of accuracy. In such cases, simplified modeling of system components is tolerated. Simplified models include pi or traveling wave representation for lines and cables, Thévenin equivalent representation for machines, and switched piecewise linear approximation for nonlinear elements. In this chapter, a functional decomposition method is presented that is ideally suited for paralleled and pipelined emulation of large-scale power systems on multiple field programmable gate arrays (FPGAs). Implementation details are given, along with performance analysis such as hardware resource consumption and latencies. Both large-scale AC and DC grids are emulated in hardware, and the real-time emulation results are validated against off-line simulation. The developed emulation system can significantly enhance the accuracy and efficiency of hardware-in-the-loop (HIL) test for integrated AC–DC grid planning, operation, and control.

With the expansion of smart grids worldwide, modeling the interaction between power systems and communication networks becomes paramount and has created a new challenge of cosimulating the two domains. Existing cosimulation methods mostly concentrate on the off-line software-level interface design to synchronize messages between the simulators of both domains. Instead of simulating in software with a large latency, a real-time coemulation (RTCE) framework on FPGA-multiprocessor system-on-chip (MPSoC) based hardware architecture is presented for a more practical emulation of real-world cyber-physical systems (CPS).

Fast dynamic simulation is required in energy control centers for contingency analysis and to determine security indices of the grid. Toward the end of the chapter, a faster-than-real-time hybrid dynamic-EMT emulation of AC–DC networks on multiple FPGAs is presented. To gain high acceleration, a flexible time-stepping–based dynamic emulation is applied to the AC grid with detailed generator modeling; meanwhile, EMT emulation revealing the transient performance of the DC grid provided an insight into the impact on its AC counterpart. Case studies involving large-scale AC systems interfaced with multiterminal DC grids are emulated to demonstrate the efficacy of the proposed approach on FPGAs.

11.2 Spatial Decomposition and Parallelism

11.2.1 Functional Decomposition for Large-Scale Real-Time Emulation

Parallel processing is extensively used in existing real-time simulators to execute large-scale system models cooperatively on multiple sequential processors. This parallel processing relies on the fundamental premise that a power system can be decomposed into smaller subsystems (Figure 11.1) due to the natural travel time delay of the transmission line or cable which provides the decoupling necessary for the subsystem calculations to occur without timing conflict; thus, one or more subsystems can be assigned to multiple sequential processors which share the computational workload, subject to the condition that the simulation time-step is less than the travel time on the link lines. However, this method has some drawbacks:

1. If there is no real transmission line or cable connecting any two subsystems or if two neighboring subsystems are tightly coupled, fictitious lines or cables with travel time are necessary to partition the network. Such artificial lines introduce errors in the frequency response of the simulated transient; although the errors can be compensated or minimized, the location and the length of the link lines still need to be carefully chosen to maintain accuracy.
2. After executing their respective calculation, the sequential processors need to exchange subsystem data with each other at the cost of extra communication latency within the time-step which limits the achievable bandwidth of the real-time simulator.
3. The partitioning scheme to divide the original system into smaller subsystems is arbitrary at best, usually based on the experience and the specific requirements of the user, albeit some simulators have an automatic partitioning method.
4. Based on a given network topology, it is quite possible to find a large subsystem which cannot be decomposed further leading to uneven computational workload for the processors. In such cases, the overall computational bandwidth of the real-time simulator is limited by the speed of the processor to which the largest subsystem is assigned.

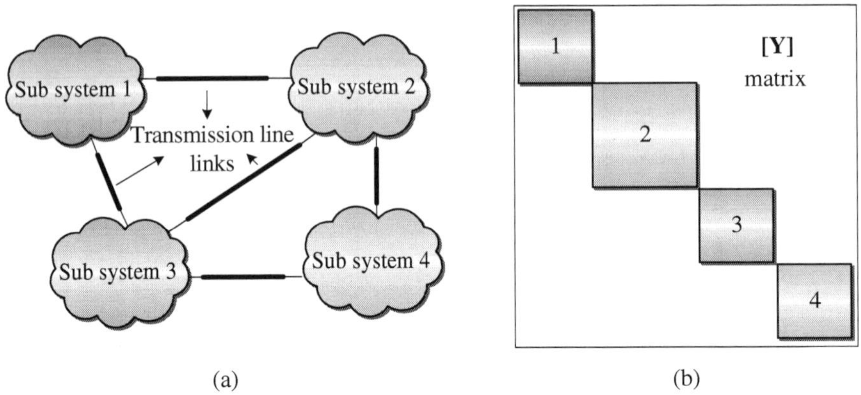

Figure 11.1 (a) Partitioning of a large power system into subsystems interconnected with transmission lines, and (b) block-diagonal system admittance matrix.

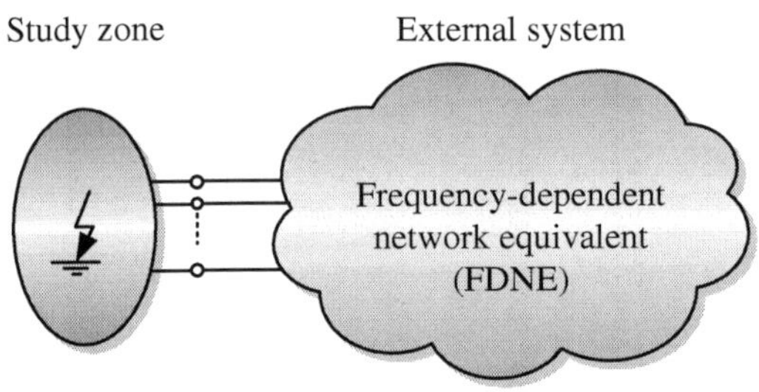

Figure 11.2 Decomposition of a system into study zone and a frequency-dependent equivalent for the external system.

The traditional method for accommodating large network sizes on limited simulator hardware is to use a frequency-dependent network equivalent (FDNE) where the original system is divided into a study zone (modeled in detail) and an external system (modeled as lumped RLCG groups derived from approximating the frequency response of the external system at the interfacing port), as shown in Figure 11.2. While network equivalents are still used in some off-line and real-time simulators [23], the prevailing requirement is to represent the original system in full detail without using simplified equivalents.

Digital hardware emulation of large-scale networks has the potential to alleviate all of the above compromises and uncertainties inherent in large-scale real-time EMT simulation, using the FPGA as the core simulator hardware. A *functional decomposition* method is introduced for allocating the system model calculation to the multi-FPGA hardware resource [209]. This method of power system decomposition relies on systematically clustering the model calculation of system components on individual FPGAs based on the component functionality independent of the network topology. As seen in Figure 11.3, a generic power system composed of lines, generators, nonlinear elements, loads, buses, circuit breakers,

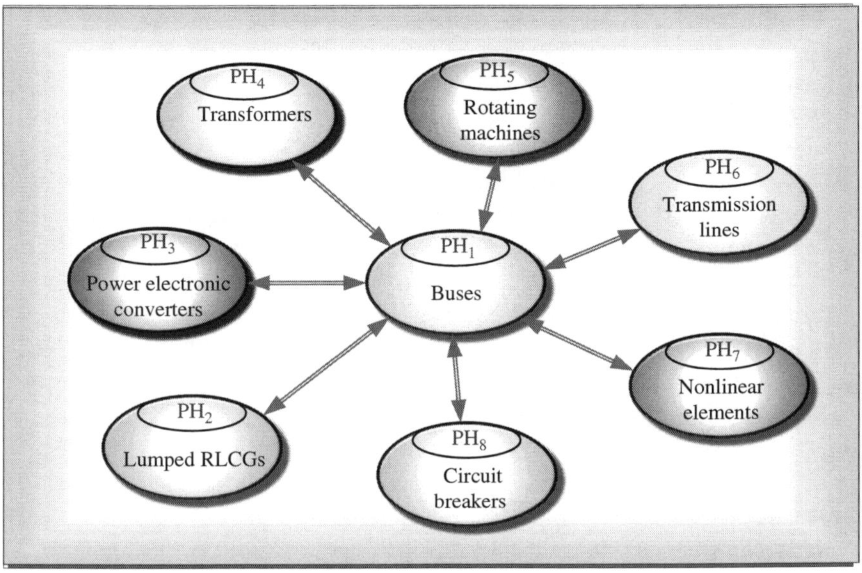

Figure 11.3 Functional decomposition of a power system for hardware emulation.

etc., can be mapped into a multi-FPGA architecture such that each processing hardware (PH) is responsible for the model calculation of one type of system component. For example, all transmission lines are simulated in the specified PH. Conventional network partitioning using transmission line links is no longer required to simulate a large-scale system. Alternately, the models for each type of system component are calculated simultaneously in their own PHs. As shown later, functional decomposition lends itself to the full exploitation of pipelining and hardware parallelism inherent in an FPGA. It does not require artificial lines or cables of fixed latency to be inserted, rendering the hardware emulation true to the original system. It removes the uncertainty related to partition boundaries, while allowing detailed modeling of all system components with an even distribution of computational workload. Since system components of the same type are clustered together, this method naturally allows multirate simulation where multiple time-steps can be chosen for various functional types to account for model complexity or to increase computational efficiency.

11.2.2 Hardware Module Parallelism

The hardware modules corresponding to the system functional components are implemented in the 65 nm Stratix III EP3SL340 FPGA from Altera whose logic resources include 270 400 combinational adaptive look-up table (ALUT) (equivalent 340 000 logic elements), 16 272 kb block memory, 576 18-bit multipliers, 12 phase locked loops (PLLs) and maximum 1104 user input/output (I/O) pins. The emulated modules include the *RLCG* module for lumped RLCG elements and transformers, the universal line model (*ULM*) module for transmission lines and cables, the universal machine (*UM*) module for machines, the *NR* module for nonlinear elements, the *Switch* module for circuit breakers, and the *Network Solver* module is realized solving the network equation. In addition, a *Control* module is implemented to coordinate the calculations of all the other modules. Since each module is hardware-independent with respect to the other, the parallelism between various functional modules is achieved naturally. This means that the processing in all modules can be carried out simultaneously. This is *system-level* parallelism existing in the real-time EMT emulator. Parallelism also exists within each of the functional modules which have been shown in previous chapters. For example, in the *ULM* module (Chapter 2), the calculations at both sending-end and receiving-end is conducted fully in parallel. For multiphase transmission lines, the calculation in each phase is also executed simultaneously. This is *module-level* parallelism.

Table 11.1 lists the main logic resources utilized by each functional module in the design. As can be seen, the *ULM* module utilizes the most logic resources. To emulate a large-scale power system network in real-time, hardware parallelism and pipelining need to be exploited on a larger-scale which can be realized by using multiple hardware functional modules; however, as seen from Table 11.1, this is hard to achieve on a single Stratix III FPGA due to the limitations of its logic resources. Multiple FPGAs had to be employed to enable massive parallelism and pipelining for large-scale systems.

Table 11.1 Hardware resources utilized by individual system functional modules.

Functional module	Combinational ALUTs	DSP 18-bit multipliers	Memory (Kbits)
RLCG	3 701 (1.36%)	8 (1.39%)	123.16 (0.76%)
ULM	44 610 (16.49%)	96 (16.67%)	707.27 (4.35%)
UM	24 679 (9.12%)	88 (15.27%)	870.74 (5.35%)
NR	11 729 (4.33%)	16 (2.78%)	217.39 (1.33%)
Switch	124 (0.05%)	0 (0%)	0 (0%)
N/W Solver	2 151 (0.79%)	4 (0.69%)	43.8 (0.27%)
Control	200(0.07%)	0 (0%)	0 (0%)

11.3 Multi-FPGA Hardware Design for Real-Time EMT Emulation

Based on the hardware resource utilization of the individual functional modules, the number of modules of a specific type to replicate and the number of pipelined components per module are decided in the multi-FPGA design. From the EMT user viewpoint, there are three main variables for real-time simulation: (i) the simulation time-step Δt, (ii) number of nodes in the modeled system, and (iii) the number of FPGAs employed. These variables influence specific aspects of the real-time EMT simulation: the time-step Δt influences the maximum frequency of the simulated transient; the number of nodes influences the size or scale of the system simulated; and the number of FPGAs influences the hardware cost. Accordingly, there are three questions a hardware designer is faced with

1. For a given system size and hardware configuration, what is the minimum time-step Δt_{min} achievable for real-time EMT simulation?
2. For a given system size and a specified time-step Δt, what is the minimum number of FPGAs required for real-time simulation?
3. For a fixed hardware configuration and a specified time-step Δt, what is the largest power system network that can be simulated in real time?

The hardware designs presented in this section attempt to reveal the answers to these questions.

The DN7020k10 multi-FPGA board from the Dini Group® is utilized to realize these designs. This board is populated with 10 Altera Stratix III EP3SL340 FPGAs arranged in a 2 × 5 matrix as shown in Figure 11.4. Taken together $FPGA_0$-$FPGA_9$ provide 3 380 000 equivalent logic elements, 162 720 memory Kbits, and 5760 18-bit multipliers. Ample pin connections are also provided between adjacent FPGAs; each pair of adjacent FPGAs share a maximum of 220 pins for high-speed bidirectional data transfer. A 48-pin bus (*MainBus*) is connected to all FPGAs. The FPGA interconnects are either single-ended or low-voltage differential signaling (LVDS). The multi-FPGA board has programmable clock synthesizers (2 kHz–710 MHz), and global clock networks that reach every FPGA. The hardware designs are performed in VHDL on the host PC using the Altera Quartus II environment. The configuration files (bitstreams) are downloaded from the host PC into the individual FPGAs using the Joint Test Action Group (JTAG) interface through a USB cable. Fast configuration using Compact FLASH is also possible when the debugging is finished. A 125 MSPS DAC card was connected through a high-speed connector to the multi-FPGA board to enable the capture of real-time results on the oscilloscope.

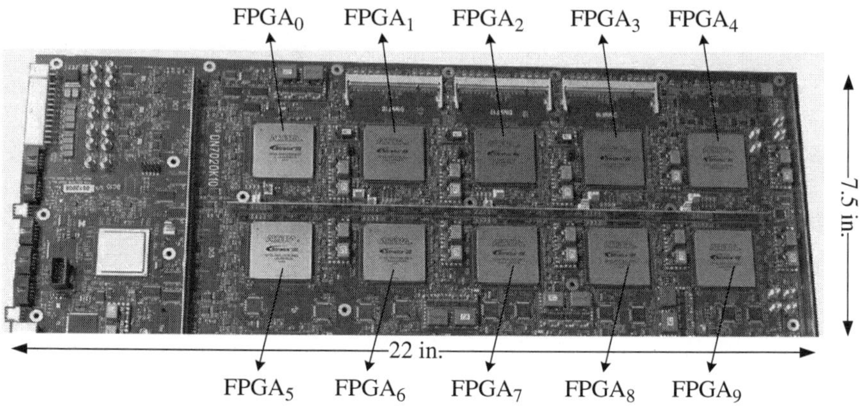

Figure 11.4 Multi-FPGA prototyping board.

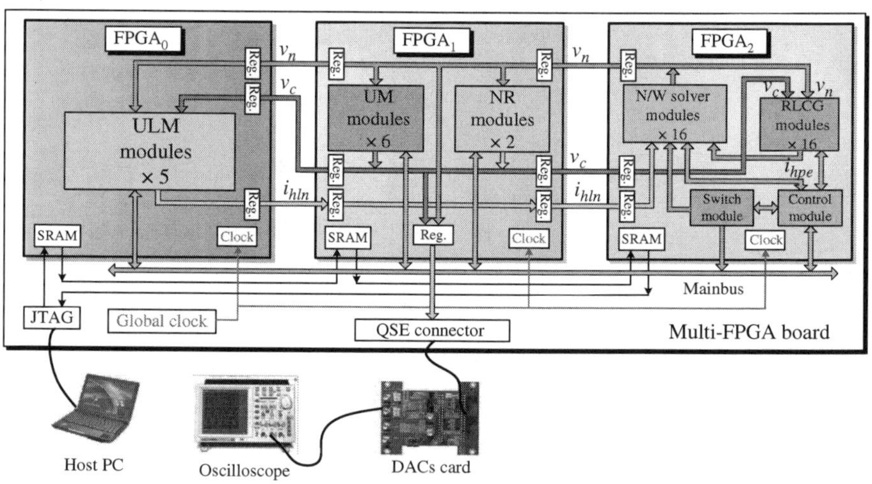

Figure 11.5 3-FPGA hardware architecture for real-time EMT simulation in Case I.

11.3.1 Case I: 3-FPGA Hardware Design

This case study shows the design details of a 3-FPGA functionally decomposed real-time EMT simulator. As can be seen in Figure 11.5, $FPGA_0$ is fully employed to realize 5 *ULM* hardware modules; $FPGA_1$ is used to implement 6 *UM* and 2 *NR* modules, whereas $FPGA_2$ implements other modules which includes 16 *Network Solver*, 8 *RLCG*, 1 *Switch*, and 1 *Control* modules. This arrangement of the various modules into the 3 FPGAs minimizes the interconnected signals between FPGAs. This is important because on the one hand, the limited FPGA pins cannot support massively parallel data input/output, while on the other hand, on a multi-FPGA board only neighboring FPGAs are usually interconnected and support high-speed data transfer, for example, the 225 MHz data transfer for single-ended signals on adjacent FPGAs on the DN2070k10 board.

In the 3-FPGA hardware configuration, the transferred signals include the node voltages $\boldsymbol{v}_n$ which are calculated in the *Network Solver* modules in $FPGA_2$ and are sent to $FPGA_1$ and $FPGA_0$; the compensated voltages $\boldsymbol{v}_c$ which are calculated in the *UM* and *NR* modules in $FPGA_1$ and are sent to $FPGA_0$ and $FPGA_2$; and the line history currents $\boldsymbol{i}_{hln}$ which are calculated in the *ULM* modules in $FPGA_0$ and are sent to $FPGA_2$ via $FPGA_1$. Buffer registers are inserted for signal input and output between the 3 FPGAs. The *MainBus* is used by the *Control* module in $FPGA_2$ to send control signals and receive acknowledge signals. The *Switch* module in $FPGA_2$ also uses it to send switch status signals. The logic resource utilization for this 3-FPGA design is shown in Table 11.2. As can be seen, the hardware space for $FPGA_0$ and $FPGA_1$ are filled to capacity with modules of the specific functional components with little room for further expansion, while $FPGA_2$ still has leftover capacity. Although some resources may appear to be underutilized, the resource that reaches its limit first determines the maximum capacity for any specific FPGA, e.g. DSP block in the $FPGA_1$. Table 11.2 also shows the f_{max} of each FPGA. The f_{max} is the maximum clock frequency that the designed hardware can operate at, and it is calculated based on the longest signal path latency in the design. It is obvious that the higher the resource utilization of the FPGA, the lower is the f_{max}, i.e. $f_{max0} < f_{max1} < f_{max2}$. Furthermore, although multirate simulation is clearly available in various different modules, for simplicity, a 100 MHz clock frequency is chosen for all FPGAs in this design.

Table 11.2 Resource utilization for the 3-FPGA hardware design.

Resource	$FPGA_0$	$FPGA_1$	$FPGA_2$
Logic utilization (%)	96	80	35
–Combinational ALUTs (%)	84	64	27
–Dedicated logic registers (%)	44	37	14
DSP block 18-bit elements (%)	83	97	22
Block memory bits (%)	22	35	17
PLLs (%)	8	8	8
Pins (%)	13	32	52
f_{max} (MHz)	100.4	136.9	170.8

To test the 3-FPGA hardware design a sample power system shown in Figure 4.14 is modeled. It is a modified version of the IEEE 39-bus New England test system. This system consists of 35 three-phase transmission lines modeled using the ULM; 10 three-phase generators modeled using the UM model; 19 three-phase loads modeled using RLCG elements, 11 three-phase transformers modeled using equivalent RLCG elements, and 3 series compensation capacitors, resulting in a total of 99 lumped RLCG elements; and 3 three-phase nonlinear surge arresters which protect 3 series compensation capacitors. The total number of nodes in the network is 126.

The spatiotemporal design workflow to model this system in real time on the 3-FPGA hardware design is shown in Figure 11.6. As can be seen, the 35 lines are allocated to the 5 *ULM* modules with 7 lines pipelined through each module. The 10 generators are allocated to the 5 *UM* modules with 2 generators pipelined per module and with one module empty. The 99 RLCG elements are allocated into 8 *RLCG* modules (12 RLCGs each for the first 5 modules and 13 RLCGs each for the remaining 3 modules). The 3 surge arresters are allocated to the 2 *NR* modules (2 in the first module and 1 in the second). The 126 node voltages are solved in 16 *Network Solver* modules (8 nodes each in first 14 modules and 7 nodes each in the other 2 modules). Figure 11.6 also shows the constitution of a simulation time-step in the 3-FPGA real-time EMT simulator. One simulation time-step Δt has three periods: *Period I* in which the node voltages $\boldsymbol{v}_n$ (without taking into account the electric machines and nonlinear elements) are solved in *Network Solver* module; *Period II* in which the *UM* and *NR* modules start to solve their respective equations and the compensated voltages $\boldsymbol{v}_c$ are calculated; meanwhile, the *ULM* modules compute part of their convolutions; and *Period III* in which the *RLCG*, *UM*, and *ULM* modules update their history terms. The minimum time-step Δt achieved for modeling the system in Figure 4.14 in the 3-FPGA hardware design is 11.55 μs, with *Period I* taking the minimum execution time of 0.64 μs for the *Network Solver* modules, and *Period II* taking the maximum execution time of 6.21 μs for the *NR* modules for an average three iterations for the nonlinear solution. In *Period II*, the execution time is determined by the *NR* module in $FPGA_1$, while in *Period III* the execution time is determined by the *ULM* module in $FPGA_0$.

A three-phase-to-ground fault at Bus 2 which occurs at $t = 0.05$ s is emulated in the 3-FPGA hardware design. Figure 11.7a shows the three-phase voltages waveforms at Bus 1 captured from a real-time oscilloscope connected to the DAC card. The voltage transient lasts about two cycles and its peak value falls from 20 to 15 kV. Identical behavior can be observed from Figure 11.7b which shows the off-line PSCAD/EMTDC® simulation.

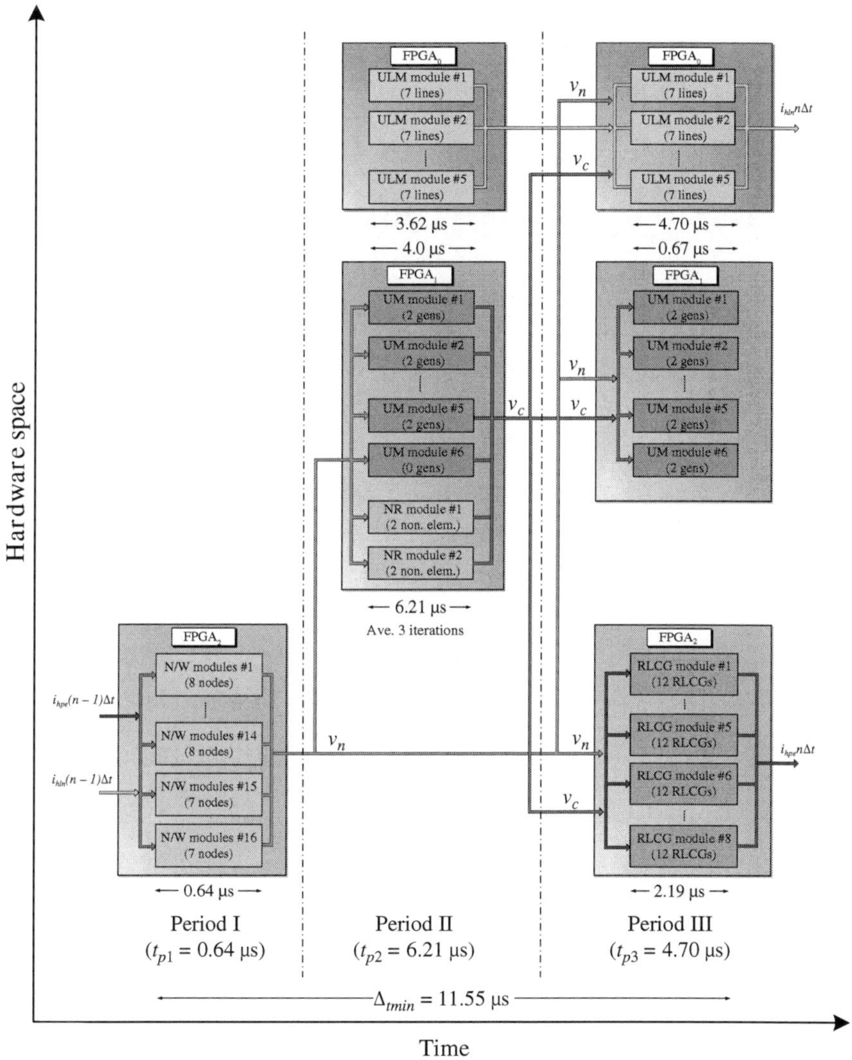

Figure 11.6 Spatiotemporal design workflow for the 3-FPGA real-time EMT emulator for Case I.

11.3.2 Case II: 10-FPGA Hardware Design

This case study utilized all the 10 FPGAs to exploit maximum hardware space on the multi-FPGA board. The 10-FPGA hardware architecture is shown in Figure 11.8. $FPGA_0$, $FPGA_1$, and $FPGA_2$ remain the same as in Case I. $FPGA_3$

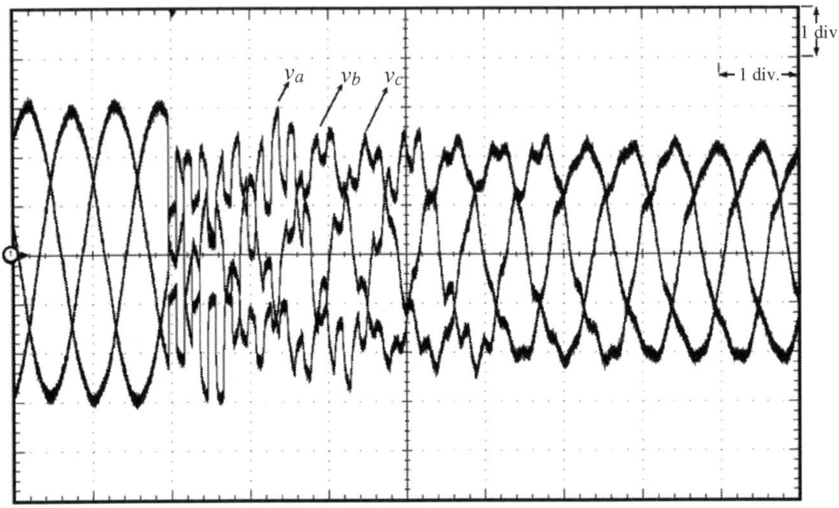

(a)

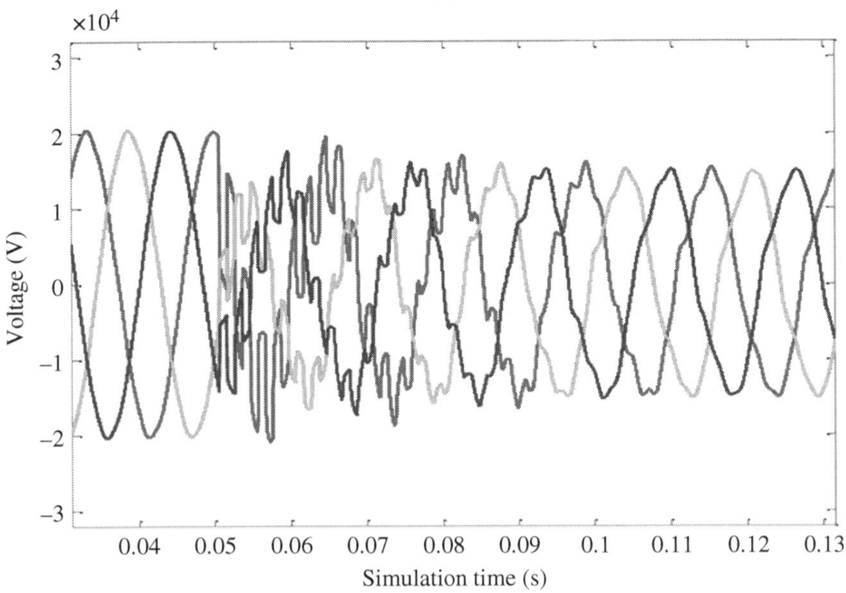

(b)

Figure 11.7 (a) Real-time oscilloscope traces, and (b) off-line simulation results from PSCAD/EMTDC® of Bus 1 voltages for a three-phase fault at Bus 2, Scale: x-axis: 1 div. = 10 ms, y-axis: 1 div. = 6.8 kV.

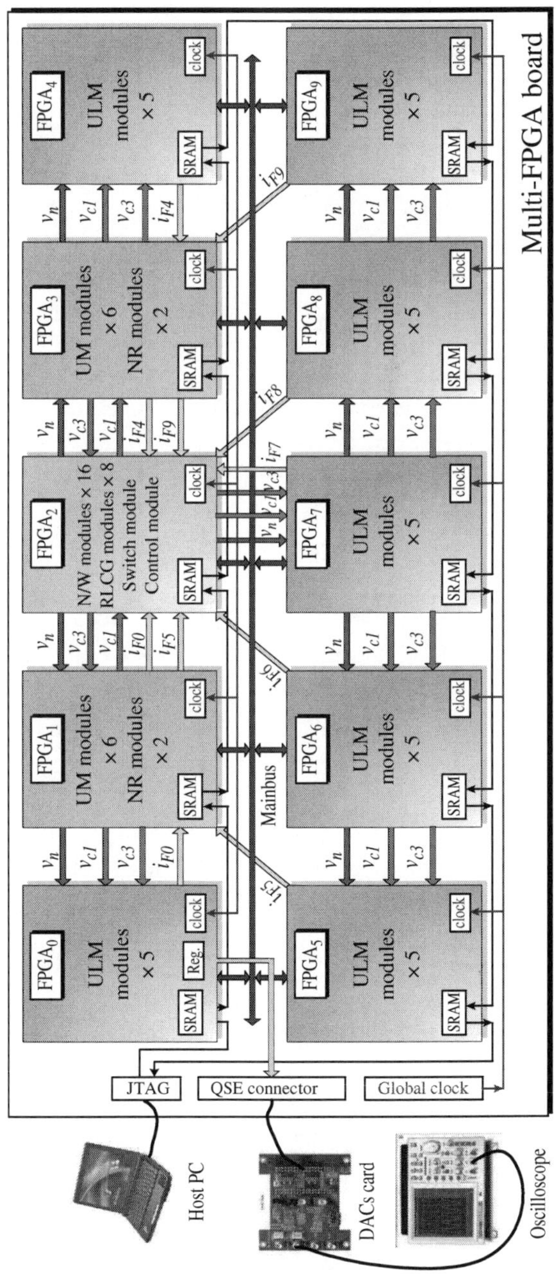

Figure 11.8 10-FPGA hardware architecture for real-time EMT simulation in Case II.

is employed for realizing the *UM* and *NR* modules as FPGA_1. FPGA_4 through FPGA_9 are employed for realizing *ULM* modules. In Figure 11.8, $\boldsymbol{v}_{c1}$, $\boldsymbol{v}_{c3}$ denote the compensated voltages computed in FPGA_1 and FPGA_3, respectively, and $\boldsymbol{i}_{Fk}$ $\{k = 0, 4, 5, 6, 7, 8, 9\}$ are history current vectors calculated in FPGA_k.

The number of any specific functional modules implemented in this multi-FPGA architecture depends on three criteria: demand, logic utilization, and module execution time. The demand is determined by the number of each type of functional component in the power system. For example, there are many more transmission lines than generators in a power system; thus the number of the *ULM* modules is expected to be more than that of the *UM* modules. The logic utilization of modules (Table 11.1) determined the possible number of modules implemented on any FPGA. The module execution time affects the time-step directly. More parallel modules are required for those components that have longer execution times. The detailed timing analysis for each functional module is discussed in Section 11.3.3.

A large-scale power system shown in Figure 11.9 is used to test the 10-FPGA real-time EMT simulator. This system is constructed by replicating the Case Study I system 10 times and interconnecting using transmission lines. The augmented system consists of

- 376 three-phase transmission lines modeled using the ULM;
- 100 three-phase generators modeled using the UM model;
- 190 three-phase loads modeled using RLCG elements;
- 110 three-phase transformers modeled using equivalent RLCG elements with a total 990 lumped RLCG elements;
- 30 three-phase nonlinear surge arresters which protect 30 series compensation capacitors.

The number of network nodes in this system is 1260.

The allocation of all components into this 10-FPGA design is shown in Figure 11.10. The overall clock frequency driving all FPGAs for the 10-FPGA hardware design is 100 MHz. The achieved minimum time-step Δt in the 10-FPGA hardware design is 36.12 μs.

Figure 11.11a shows the terminal currents of generator G_9 during the three-phase-to-ground fault at Bus 2, where the transients can be seen clearly. Again, the detailed agreement between real-time and off-line simulations can be observed in Figure 11.11.

11.3.3 Performance and Scalability of the Real-Time EMT Emulator

To answer the three questions posed earlier, first a detailed analysis of the time-step is in order. As shown in Figures 11.6 and 11.10, the simulation time-step

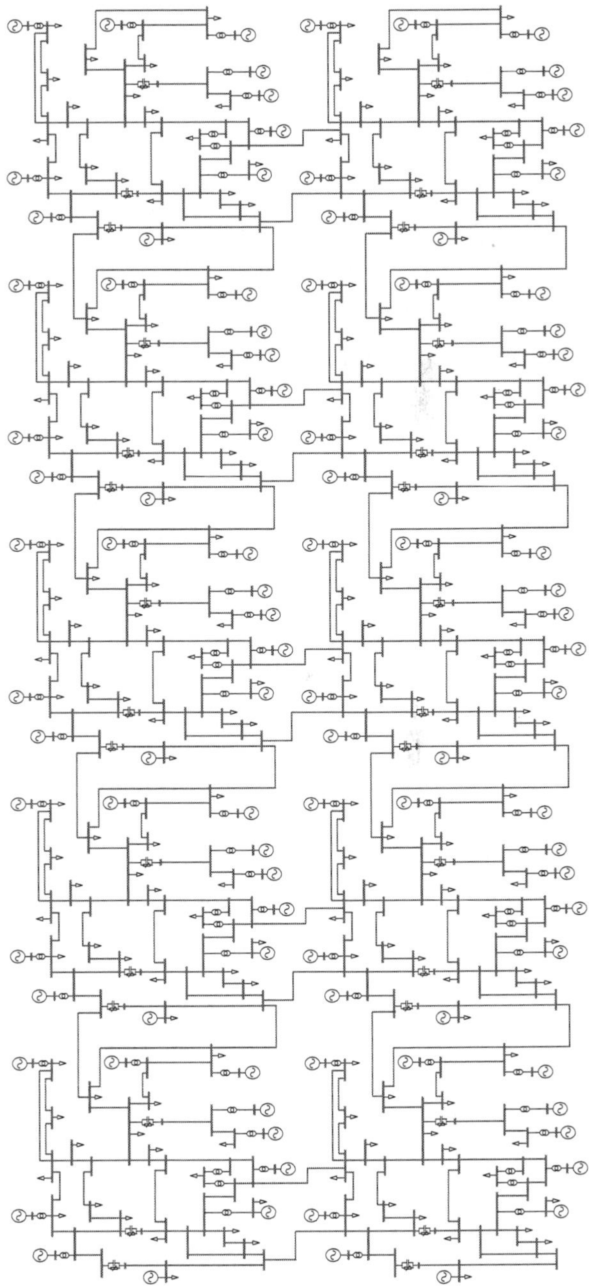

Figure 11.9 Single-line diagram of the power system modeled in Case II.

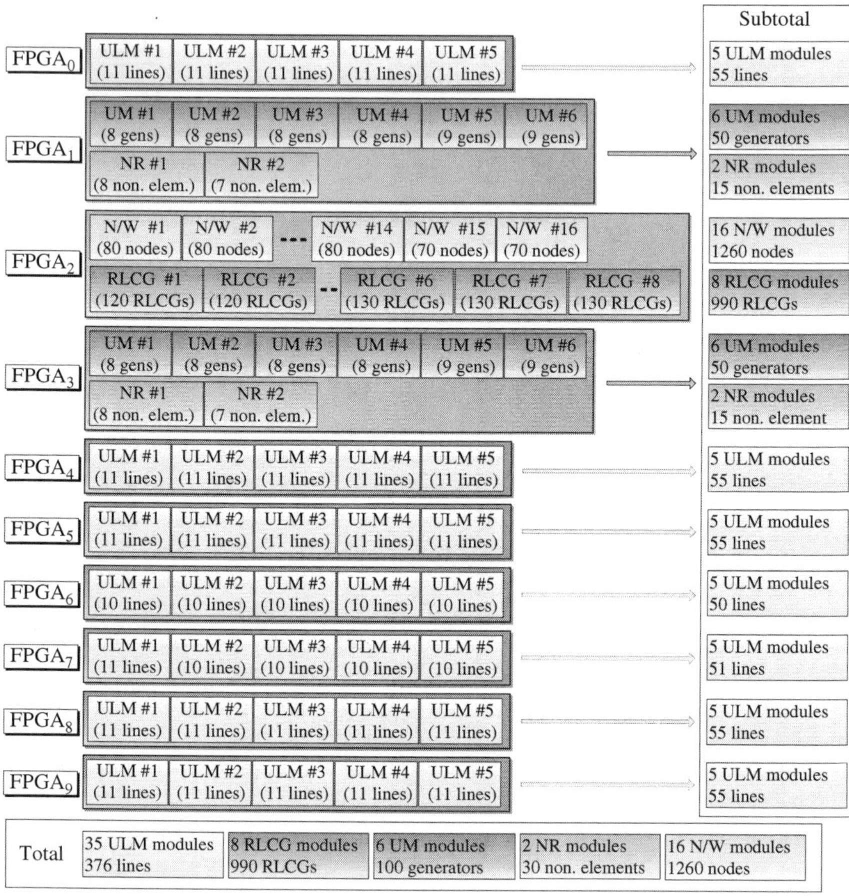

Figure 11.10 Allocation of components of Case II in the 10-FPGA design.

Δt can be determined as

$$\Delta t = t_{p1} + t_{p2} + t_{p3}, \tag{11.1}$$

where t_{p1}, t_{p2}, and t_{p3} are the execution times in *Period I*, *Period II*, and *Period III*, respectively, which are calculated as

$$t_{p1} = t_{NW}, \tag{11.2a}$$

$$t_{p2} = max\{t_{ULM}, t_{UM}, t_{NR}\}_{p2}, \tag{11.2b}$$

$$t_{p3} = max\{t_{ULM}, t_{UM}, t_{RLCG}\}_{p3}, \tag{11.2c}$$

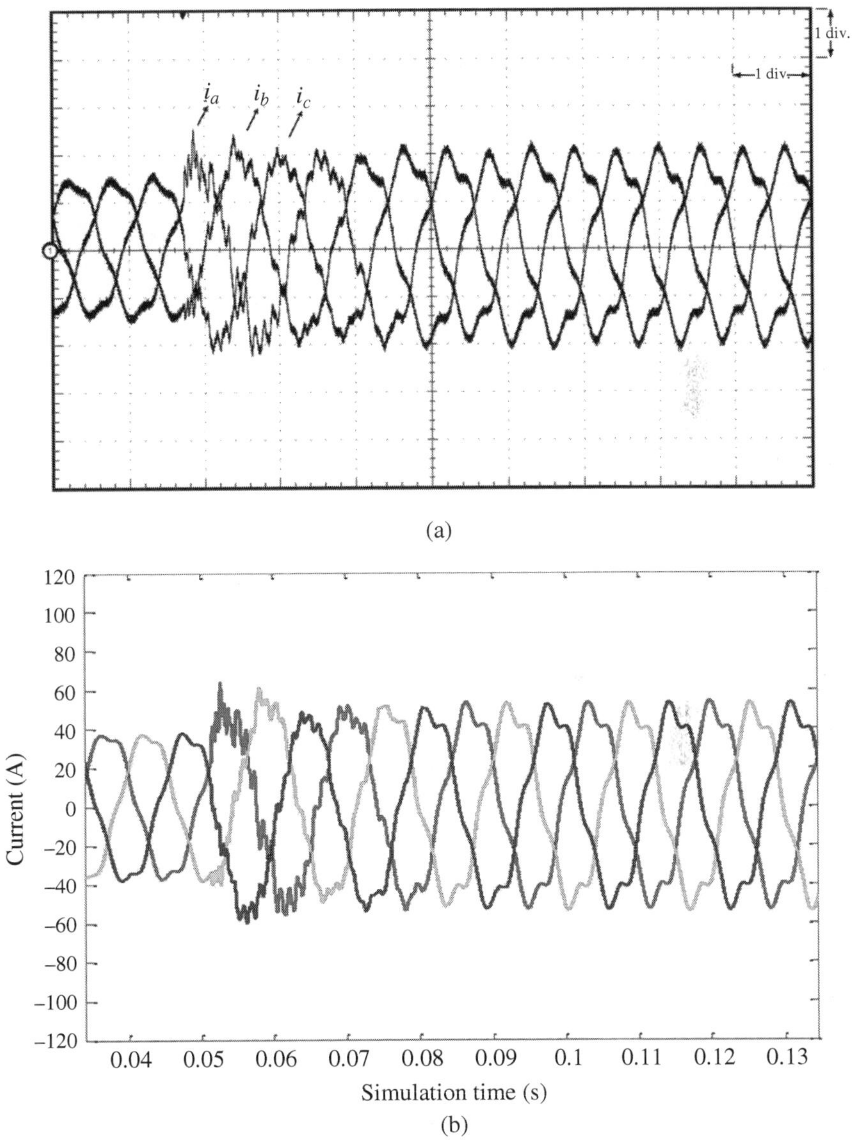

Figure 11.11 (a) Real-time oscilloscope traces, (b) off-line simulation results from PSCAD/EMTDC®, of generator G_9 terminal currents during a three-phase fault at Bus 2, Scale: x-axis: 1 div. = 10 ms, y-axis: 1 div. = 25 A.

where t_{NW}, t_{ULM}, t_{UM}, t_{NR}, t_{RLCG} are execution times for *Network Solver* module, *ULM* module, *UM* module, *NR* module, and *RLCG* module, respectively. The execution time for each module is calculated according to the total hardware latency of the module and pipelined components fed into the module. For example, t_{ULM} in *Period II* can be determined as

$$t_{ULM} = (34 + 8 * N_{line}) * T_f, \tag{11.3}$$

where the number "34" denotes the total hardware latency in terms of clock cycles; N_{line} is the number of lines pipelined into the module, and T_f is the working clock period of FPGA (10 ns in this design). The number "8" is related to the assumption that each transmission line has a 9th-order ($N_p = 9$ in (2.119)) fitted rational functions for the characteristic admittance matrix and total 13th-order ($N_{p,1} = 4$ and $N_{p,2} = 9$ in (2.120)) of fitted rational functions in 2 ($N_g = 2$ in (2.120)) propagation modes for the propagation function matrix. The execution time of the *ULM* module in *Period II* and *Period III* with respect to the number of lines pipelined through the module is plotted in Figure 11.12b. Similarly, Figure 11.12a shows t_{RLCG} with respect to the number of RLCG elements; Figure 11.12c shows t_{UM} with respect to the number of machines; Figure 11.12d shows t_{NR} with respect to the number of nonlinear elements; and Figure 11.12e shows t_{NW} with respect to the number of network nodes.

The performance of the proposed multi-FPGA hardware design is investigated by the minimum time-step with respect to the simulated system size. The Case I test system (Figure 4.14) is used as a reference system to determine the number of RLCG elements, transmission lines, machines, and nonlinear elements with respect to the system size quantified in terms of the number of network nodes.

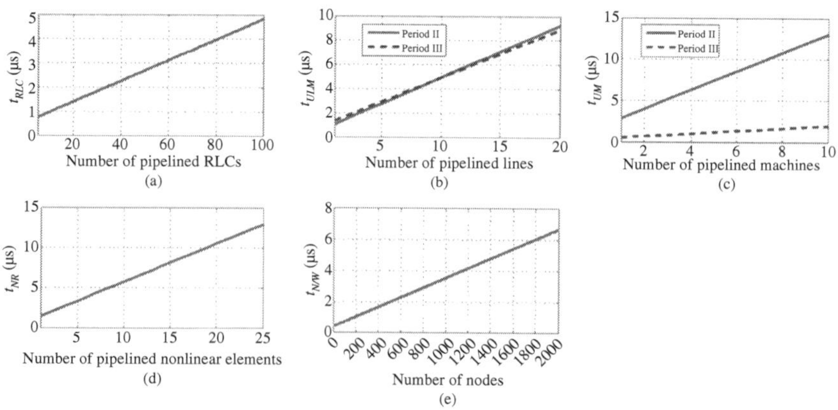

Figure 11.12 (a)–(e) Execution time of each functional module with respect to the number of pipelined elements per module.

Based on this information, the minimum time-step is calculated for a given size of power system, as shown in Figure 11.13. As can be seen in this figure, the minimum time-step of 11.55 μs and 36.12 μs are achieved for Case Study I for a system of 126 nodes in the 3-FPGA design and Case Study II for a system of 1260 nodes in the 10-FPGA design, respectively. Figure 11.13a also shows the largest system size that can be simulated with a specified time-step. For example, with a 50 μs time-step using detailed modeling a system of 850 nodes can be simulated on the 3-FPGA design, while a system of 1860 nodes can be simulated on the 10-FPGA hardware design. Note that with simplified models much larger systems can be simulated with the same time-step.

The number of FPGAs required to carry out real-time EMT simulation varies with the specific time-step and the given system size. Figure 11.13b shows the surface plot of the variation of number of FPGAs required with respect to system size and the minimum time-step achieved. This plot shows the possible combinations of these three variables, and the two case studies presented before represent two extremities of this 3-D surface. Clearly, the variation of the minimum time-step is nonlinear with respect to the system size and the number of FPGAs used. The slope of Δt vs. system size curve tends to decrease as the number of FPGAs increases. In general, for a fixed large-scale system size employing more FPGAs would allow to achieve a lower time-step for real-time EMT simulation but with diminishing returns. Nevertheless, the achieved time-step would be so small as to allay any concerns related to detailed system modeling. Meanwhile, larger and faster FPGAs would allow more parallel functional modules to be emulated and would also raise the maximum frequency (f_{max}) of the design, which will ultimately lead to further reduction of time-step.

11.4 CIGRÉ DC Grid Hybrid Modeling Methodology

A hybrid MPSoC-FPGA platform using Xilinx® Zynq Ultrascale+ XCZU9EG and Virtex Ultrascale+ XCVU9P devices (Table 1.2) for the realization of CIGRÉ DC grid real-time emulation is introduced in the following three sections. For the purpose of providing both accurate and detailed results of components in the local study zone, and the global interactive results of other components in the DC grid with relatively low hardware resource and design cost, a feasible hybrid modeling scheme is utilized for representing various MMCs in the DC grid. Device-level electrothermal model, equivalent circuit model, and the average value model are used for representing different components and zones in the emulated system. The complete system is fully decomposed to 20 subsystems to achieve high parallelism and modularity.

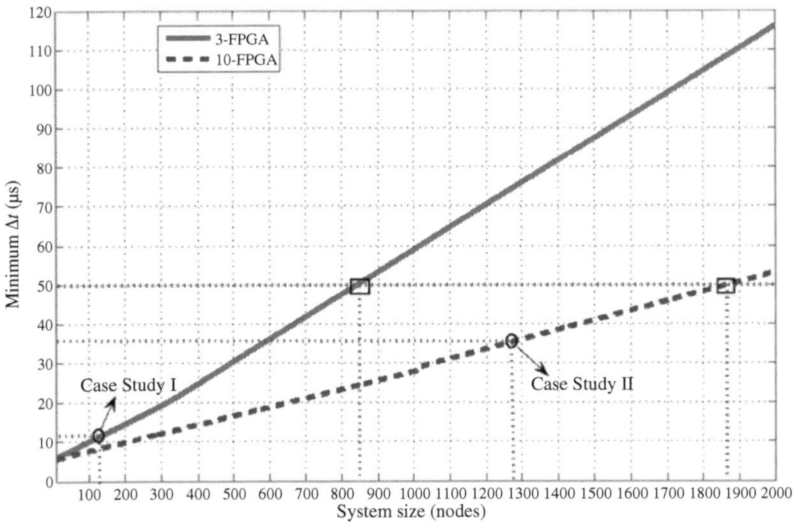

(a)

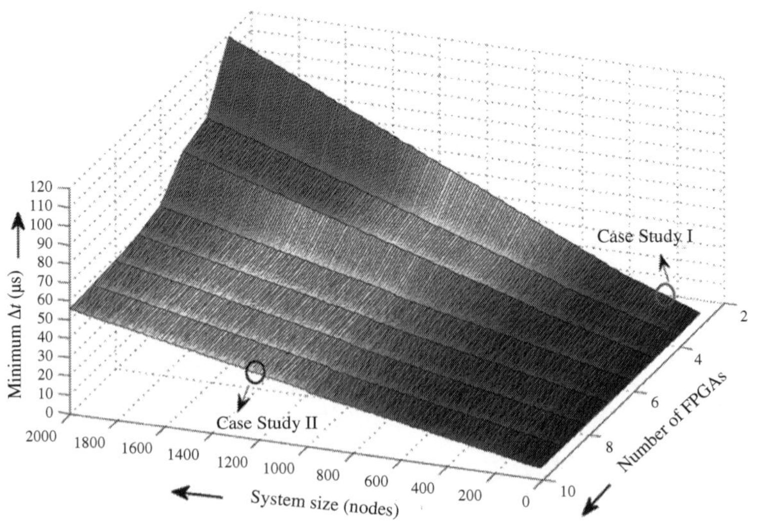

(b)

Figure 11.13 (a) Minimum time-step achieved in the 3-FPGA and 10-FPGA hardware designs for real-time EMT simulation, and (b) variation of number of FPGAs with system size and the time-step.

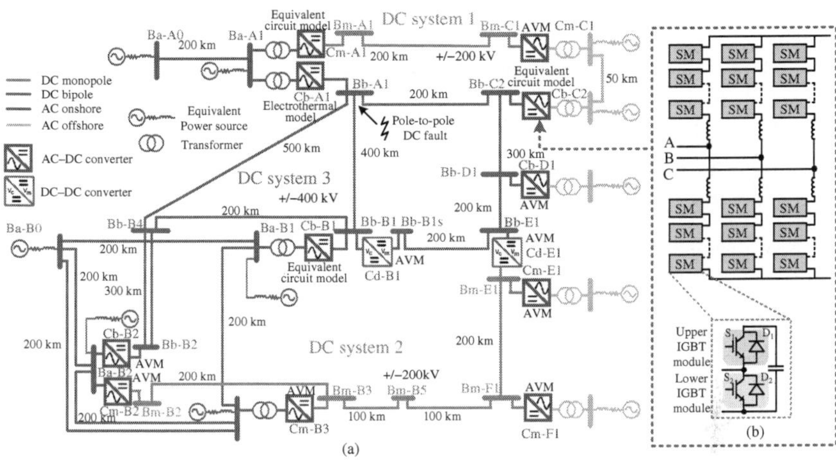

Figure 11.14 CIGRÉ DC grid test system: (a) topology, and (b) HBSM-MMC converter stations.

11.4.1 Network Topology

The CIGRÉ B4 DC grid [210] is shown in Figure 11.14a and its parameters are given in Appendix A.10. It can be divided into three subsystems based on the voltage rating and location. DC System 3 which has the highest $\pm$400 kV bipolar voltage has a meshed configuration with the DC-DC converter Cd-B1 providing an extra degree of freedom to control the loop power flow. The $\pm$200 kV DC System 1 shares the same AC grids with DC System 3 on both sides, while DC System 2 with a rated voltage of $\pm$200 kV borders DC System 3 on multiple buses, including the direct connection by the DC-DC converter Cd-E1. The HBSM-MMC, shown in Figure 11.14b, is used for all AC–DC converters.

11.4.2 Control Scheme

The control diagram of the MMC is shown in Figure 11.15, which is composed of the outer loop control, inner current loop control, and valve-level control. As narrated in previous chapters, the converter has two control objectives in both d- and q-axes, and the mode is determined by the role of converter station and the robustness of the AC system it connects to. In this chapter, Converters Cm-A1, Cb-A1, Cm-B2 connecting to a strong on-shore AC grid are set to regulate the DC voltage, while other AC–DC converters primarily control the active power flow. The reactive power flow can be regulated independently for all converters. Capacitor-sorting based control scheme is used for the valve-level control of MMC [143]. It is noted that the computation effort of valve-level control is

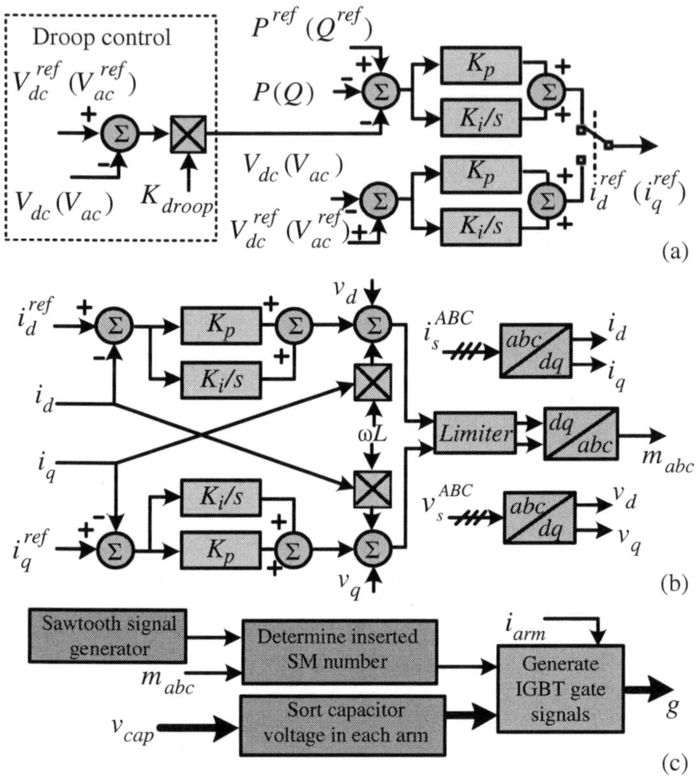

Figure 11.15 Control diagram of MMC: (a) outer loop control for power and DC voltage, (b) inner current loop control, and (c) MMC valve-level control.

dependent on the SM number in an arm, and can be significantly larger than the system-level control including the outer and inner loop controls.

11.4.3 Hybrid Modeling Methodology

A major challenge of the DC grid modeling and real-time EMT simulation lies in the implementation of all converters with limited hardware resources. Although the FPGA board can provide high parallelism, the realization of all converters using detailed models can consume a huge amount of resources, which exceeds the capacity of a single board. Therefore, hybrid modeling strategy is applied to reduce the resource consumption and maintain the accuracy of the study zone and the complete size of the system. Among various modeling methods, device-level electrothermal model, equivalent circuit model, and average value model, which cover a wide range of modeling complexity are adopted for various locations in the

DC grid. The detailed modeling scheme is used at the local element or zone which is of most interest and can be the closest converter to the fault location for transient analysis; the equivalent circuit modeling scheme is applied for the converters surrounding the local element; the average value model is used for the converters in remote locations. In Figure 11.14, the modeling schemes adopted for various converters are annotated. Such strategy is also applicable for other elements, such as the transmission lines, etc.

11.4.3.1 Device-Level Electrothermal Model

Device-level electrothermal model is the most complex method adopted in this chapter, which can provide the power losses, the junction temperatures, and device-level switching transients of any single IGBT or diode. The parameters of the electrical model interface obtained from manufacture's data-sheet [162] are dependent on the values of the junction temperature, which matches the actual physical phenomena of power electronic devices. Using such a modeling scheme in the real-time simulation can provide the comprehensive on-line evaluation of the power converter for efficiency and security under normal condition and fault transients with direct thermal indicators.

A schematic process of the electrothermal model is illustrated in Figure 11.16. During each system-level time-step, the system-level results, such as current waveforms and switching signals, from the electromagnetic transient simulation are used to calculate the conduction and switching power losses. Then, the power losses of individual switches are imported as equivalent current sources into a thermal network, which is composed of multiple cascaded thermal impedance of various devices in the same heatsink. The calculated junction temperatures are used to update the electrical parameters of the interface, which is used for the next time-step calculation. The device-level calculation requires the temperature-dependent parameters, such as the rise-time and fall-time of the voltage and current waveforms, and the system-level results. When implemented on hardware, the device-level waveform generation module can update at a selected FPGA clock frequency of 100 MHz.

11.4.3.2 Equivalent Circuit Model

Since a high-level MMC contains many SMs that can generate a large number of system matrix nodes, the conventional EMT simulation can be extremely time-consuming and is not able to execute in real-time. Equivalent or surrogate circuit is thus commonly used for the simplification of a submodule model. Taking the example of upper IGBT module, based on its temperature-dependent characteristics shown in Figure 11.17a and the operating conditions determined by the gate signal and the arm current direction, the resistor r_1 and voltage source v_1 in Figure 11.17b can be calculated using piecewise polynomial functions. For

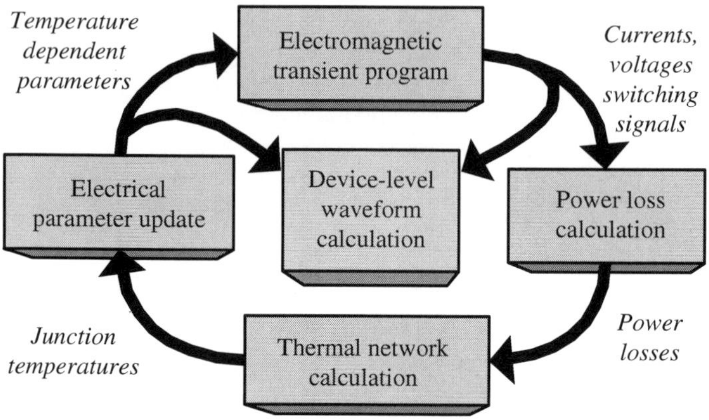

Figure 11.16 Procedures of device-level electrothermal model.

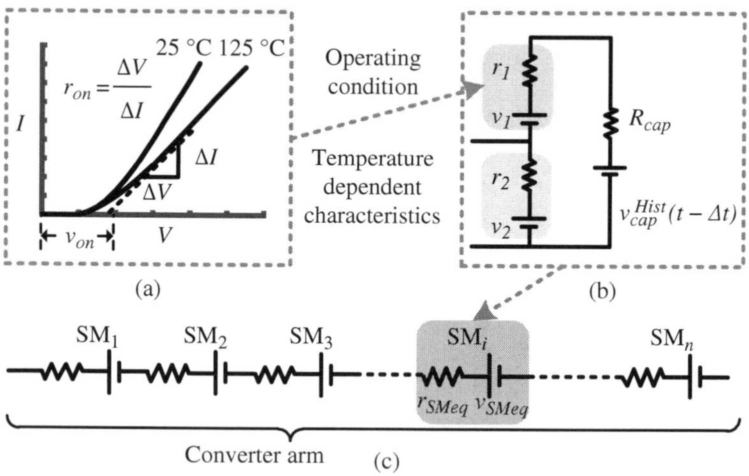

Figure 11.17 (a) Temperature-dependent IGBT module output characteristics,(b) circuit model of half-bridge SM, and (c) simplified SM model.

conventional equivalent circuit modeling scheme, r_1 and v_1 only have two fixed values to represent the on-state and off-state, respectively. The simplified SMs shown in Figure 11.17c allow to be merged with each other and, therefore, leads to a substantial electrical nodes reduction. The equivalent circuit modeling scheme accurately calculates the SM capacitor voltages and verifies the fundamental function of the complete control algorithm, in addition to being applied to combine the device-level electrothermal model of IGBT modules to the interface of MMC model to ensure real-time performance.

11.4.3.3 Average Value Model

The AVM is the simplest EMT modeling scheme for an MMC. The voltage waveform with fundamental frequency is generated at the AC side of a converter, and the DC side current is generated conserving the transferred power. The valve-level control cannot be verified with this model, since all the SM capacitor voltages are assumed to be balanced at all time. The average value model uses the controlled voltage and current sources with one time-step latency, which provides the matrix separation capability between the AC and DC side. Such a feature is beneficial for system decomposition to reduce the matrix size, while may also bring numerical consequences during faults.

11.4.3.4 Transmission Line Model

A hybrid modeling methodology is also applied to the transmission line by using both ULM and the TWM described in Section 2.6. The ULM fully considers the frequency-dependent characteristics of the transmission lines, while the time-domain convolution required for such model consumes relatively larger computation and memory resources. On the other hand, TWM models the traveling wave with constant parameters and therefore is simpler. ULM is applied to the lines in the study zone connecting the interested local elements, while the TWM is used for short lines or remote locations in AC system. The PI section model as another prevalent representation is excluded in the DC grid modeling due to its inaccuracy and the absence of latency feature for network decomposition.

In this chapter, Converter Cb-A1 is chosen as the component of interest; and the detailed electrothermal model is applied to all IGBT modules of this converter. The nearby three converters, i.e. Cb-B1, Cb-C2, and Cm-A1, utilize the equivalent circuit model. The modeling schemes used for various converters are summarized in Table 11.3, where each of the above converters has 65 voltage levels or 384 modules. 10 IGBT modules are connected in parallel and 7 in series to provide sufficient voltage and current rating for the converter. The transmission lines connecting between these converters utilize the ULM. The rest of the converters and transmission lines use AVM and BLM, respectively. A uniform modeling scheme is used for the power sources and transformers in the grid. In fact, hybrid modeling schemes could also be extended to other elements, such as a detailed off-shore wind power plant.

11.4.4 Real-Time MPSoC-FPGA Based DC Grid Emulator

11.4.4.1 System Decomposition

In addition to the rapid development of IC technology, the feasibility of conducting real-time EMT simulation of large-scale power systems relies on an appropriate decomposition or relaxation, which aims to divide the complete system into

Table 11.3 Modeling schemes for converter stations.

Modeling scheme	Converter station
Electrothermal model	Cb-A1
Equivalent circuit model	Cb-B1, Cb-C2, Cm-A1
Average value model	Cb-D1, Cb-B2, Cm-B2, Cm-B3, Cm-C10, Cm-E1, Cm-F1, Cd-B1, Cd-E1

multiple smaller subsystems. As a result, the large matrix is correspondingly decomposed into multiple small submatrices, which can be calculated in parallel. Time-step delay accompanies the decomposition scheme which can be realized by means of

1. using distributed line models, such as TWM, FDLM, or ULM, for the widely existing transmission lines in the CIGRÉ DC grid;
2. using the inherent delay property of the average value model of the converters.

Regarding the use of disturbed parameter line models, the line must be sufficiently long, which makes the traveling time at least longer than the system simulation time-step, which is typically 20 μs. The traveling time of some specific transmission lines shall also be longer than the communication latencies between various hardware computing units. Such requirements can be generally met for a transmission network, while may be an issue for a smaller system, such as a microgrid. It can be solved by inserting short distributed line models for decomposition. As previously explained, average value model of MMC can also provide system decomposition with the sacrifice of accuracy and numerical stability.

During the periodical calculation by the hardware, the EMT simulation includes three major steps, which are the electrical and control element calculation, data exchange between subsystems, and matrix equation formation and solution as shown in Figure 11.18a. Except for the exchange of history values of transmission line models and the voltage and current values of MMC average value model, the calculations among the subsystems are independent and can thus be parallelized.

11.4.4.2 Hardware Resource Allocation and Task Partitioning

The multicore CPU is often used for the real-time EMT simulation. Admittedly, the communication latencies among the CPU cores are small, but frequent and large-amount of data exchange between cores is a heavy burden and is seldom recommended. For such a reason, the complete calculation of a subsystem often resides in a single thread or core illustrated in Figure 11.18a. Since a CPU core is a sequential calculation device, the computing capability limits the number of subsystems that can be processed by a single core. When complex topologies and

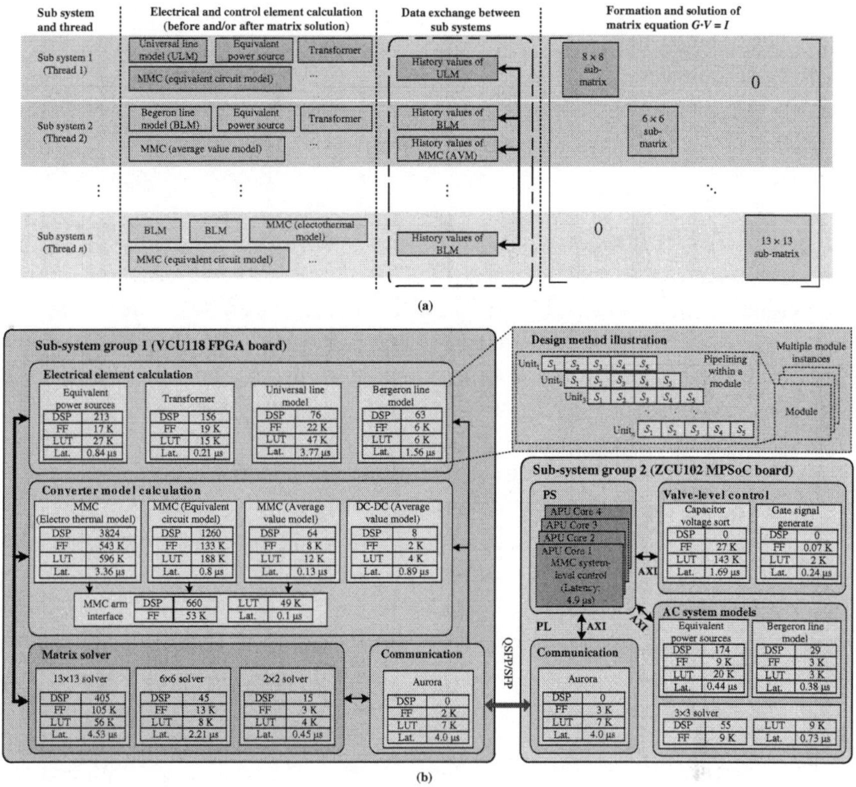

Figure 11.18 (a) System decomposition and conventional processor-based partition scheme, (b) MPSoC-FPGA based partition and implementation details.

components exist, it is necessary to use multiple cores to improve parallelism and subsequently ensure real-time performance, at the cost of an increase in the inter-core communication burden.

The utilization of FPGA resources can effectively eliminate the above dilemma, since various modules are connected directly by appropriate routing design without the complex mechanism for generalized communication. In fact, the module-level connections have no fundamental differences from the internal connections for various logic resources within an FPGA module. Instead of partitioning the system by subsystems used in the conventional CPU-based simulator, it is also feasible to partition the system by different functions. The internal design of the modules uses pipelined structures to optimize the resource utilization. One or multiple module instances are applied to different components based on the timing requirements. Such processes are automatically done by high-level synthesis (HLS) tools with the input of C code and directives enforcing specific resource and latency requirements. Generally, resource and latency optimization cannot be achieved simultaneously, which may require multiple design iterations to achieve a proper tradeoff.

The resources on one FPGA or MPSoC board may not be sufficient for the emulator of a large and detailed system. Multiple boards can be interconnected to provide enormous computation capability. The board-level partition scheme is similar to the subsystem based one, due to the relatively high communication cost. A subsystem group containing a large number of subsystems is assigned to a board. In order to present the versatility of the proposed CIGRÉ DC grid emulator, which can conduct control device HIL test, power equipment HIL test, and the multi-board expansion, the MMC control of all 11 converters and the calculation of two AC buses Ba-A0 and Ba-B0 alone with the corresponding equivalent power source and solver modules reside in the MPSoC board.

11.4.4.3 Design and Implementation

Figure 11.18b shows the hardware implementation of the CIGRÉ DC grid emulator on the hybrid MPSoC-FPGA platform. The resource consumption and the latency of major components or functions are also specified. When the emulation starts, the initialized coefficients and variables are loaded to all the modules. Then calculations of electrical elements and converters begin simultaneously. When all the values of the matrix equations are updated, the solver modules run concurrently. The communication modules exchange the data between the FPGA board and the MPSoC board through Aurora IP core, which handles the data conversion for serial transceivers. The transferred data includes the emulation control signals, MMC control input and output data, transmission line data, and emulation results chosen for observation. The MPSoC board follows the control signals from the FPGA board for synchronization and accomplishes tasks of MMC control and the calculation of two AC buses. When the exact time of the next time-step reaches, the periodical calculation starts again.

There are 20 submatrices existing in the DC grid emulator with sizes varying from 2×2 to 13×13. Four types of solvers are implemented to calculate all of them to achieve the optimized resource and latency performance as shown in Figure 11.18b. For example, a 9×9 matrix equation can be solved by a 13×13 solver, while the matrix equation with smaller size such as 6×6 can use another smaller matrix solver.

The implementation of the MMC controller takes the advantages of MPSoC devices. The relatively complex and sequential calculations of the system-level control of the 11 MMC converters are completed in the four application processing unit (APU) cores. The system-level control schemes and parameters can also be conveniently modified in the APU cores for HIL test. The valve-level control required by Converters Cb-A1, Cb-B1, Cb-C2, and Cm-A1, includes the sorting of all the SM capacitors in an arm and the gate signal generation of all IGBT modules, which are computationally demanding and thus assigned on PL for acceleration.

There are various design flexibilities regarding the latency and resource consumption, when using the same devices or other devices, such as Xilinx® XCVU13P, which has 46% more logic blocks and 80% more DSP slices compared with Xilinx® XCVU9P FPGA. In the design DC grid emulator, the calculation modules for converters on the FPGA board wait for the MMC control signals from the MPSoC board within the same time-step. Sometimes there could be a delay for the MMC control signals, and the converter calculation modules can use the control signals of the previous time-step. In this way, the simulation time-step can be further reduced. In terms of resource consumption, the number of SMs using electrothermal model within an MMC model, the number of MMC converters using electrothermal model or equivalent circuit model, and the MMC level numbers can be adjusted based on the available resources and emulation requirements.

The VCU118 FPGA board is connected to the ZCU102 board through quad small form-factor pluggable (QSFP) to 4 × SFP cable. The programmable logics of both boards run at 100 MHz. The USB-JTAG cables of the two boards are connected to the host computer for downloading the configuration files. A digital-analog converter board is used to connect the FPGA mezzanine card of the VCU118 board and the oscilloscope to capture the real-time results from the emulator.

11.4.5 Real-Time Emulation Results and Validation

Various studies and tests along with substantial results can be accomplished with the proposed real-time CIGRÉ DC grid emulator. The emulator uses 20 μs as the time-step for the system-level calculation and 10 ns as the time-step for the device-level transient waveform update. This section presents some of the system-level and device-level real-time results under both normal operation and DC fault validated with commercial software PSCAD/EMTDC® and SaberRD®, respectively. The complete CIGRÉ DC grid is modeled in PSCAD/EMTDC® with the same hybrid modeling scheme as the real-time emulator, except that the electrothermal model is not used. However, the complete DC grid could not be implemented in SaberRD® due to the numerical stability and convergence issue for the complex DC grid system, and the external system-level waveforms are imported to the precisely modeled submodule for device-level results verification.

11.4.5.1 Steady-State Operation

Figure 11.19 shows the system-level steady-state operation results, including the AC voltages of Converters Cb-A1 and Cm-C1, and the SM capacitor voltages of the first upper arm SM of Converter Cb-A1, which are compared

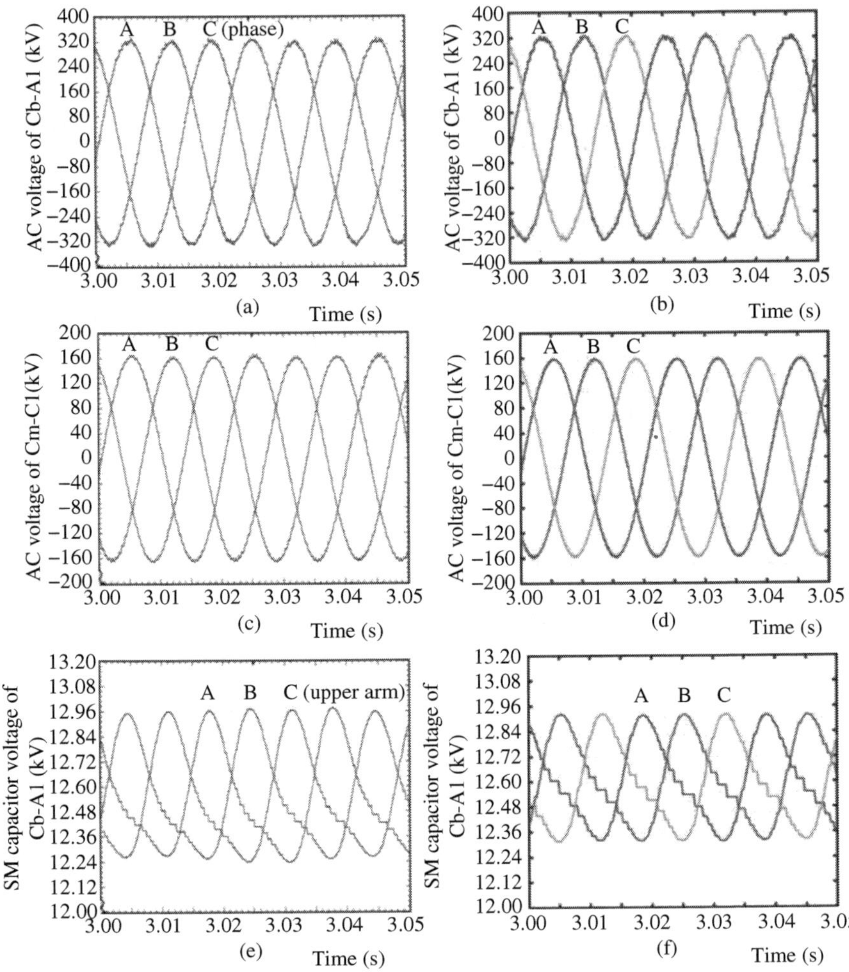

Figure 11.19 System-level steady-state results: (a) (d) AC voltages of Converter Cb-A1, (b) (e) AC voltages of Converter Cm-C1, and (c) (f) the first upper arm SM capacitor voltages of Converter Cb-A1. ((a) (b) (c) are real-time results; (d) (e) (f) are the offline PSCAD/EMTDC® results.)

with PSCAD/EMTDC®. Since the MMCs have 65 levels, the AC voltage of the Converter Cb-A1 is close to sinusoidal with few harmonics. These harmonics can also be observed for Converter Cm-C1 modeled with average value model since the other Converter Cm-A1 uses the equivalent circuit model and can affect the waveforms of Converter Cm-C1, which is in the same DC system. The SM capacitor voltages of the real-time results and that of PSCAD/EMTDC® have some differences, due to different modeling schemes. The IGBT model used in PSCAD/EMTDC® simulation is the two-state resistor model with a small resistance representing on-state and a large resistance representing the off-state. While the electrothermal model used for Converter Cb-A1 is based on the detailed and accurate temperature-dependent nonlinear output characteristics of the IGBT module obtained from the manufacturer's datasheet for FZ400R33KL2C_B5 IGBT module.

The device-level results of IGBTs and diodes are presented in Figure 11.20 with the steady-state initial condition. SaberRD® is utilized for the verification, which uses the semiconductor switch's dynamic thermal. The final values and dynamical changing patterns of the junction temperatures for S_1, S_2, D_1, and D_2 are consistent between the real-time results and SaberRD® results shown in Figure 11.20a–d. The linearized voltage and current waveforms can give a good estimation for the switching transients with small amount of additional computation effort shown in Figure 11.20e–h.

11.4.5.2 Power Flow Command Change

Figure 11.21 shows the results during power flow command change. The power generation from Converter Cb-C2 changes from 600 to 0 MW from 3.5 to 4.5 seconds of the simulation run representing the scenario of cutting-off an off-shore power plant. Figure 11.21a,b shows the active power tracking performance of Converter Cb-C2. Since Converter Cb-A1 is responsible for DC voltage regulation, the power flow increases responding to the power flow change of Converter Cb-C2, which influences the corresponding power losses of the IGBT as shown in Figure 11.21c,d.

11.4.5.3 DC Fault

A DC ground fault of both poles is applied to the Bus Bb-A1 at 1.0 second of the simulation run with the steady-state initial condition. The DC voltages of Converter Cb-A1, Cm-B3, and Cm-C1 are shown in Figure 11.22a–f, which are located in DC System 3, DC System 2, and DC System 1. The real-time results are accurate compared with PSCAD/EMTDC®. Although Converter Cm-B3 is far away from the fault location, the DC voltage is seriously affected by the fault. It is because the DC System 3 and DC System 2 are directly connected by the DC-DC converter without galvanic isolation. On the other hand, the fault effect is small for DC System 1 due to the galvanic isolation and the support from a relatively strong AC system. Figure 11.22g–j presents active powers of Converters Cb-D1 and Cm-F1 during the DC fault. Figure 11.22k,l shows the junction temperatures of $\mathbf{S}_1$ and $\mathbf{D}_2$, where the fault current flows through. The characteristics and parameters of high temperatures and high currents are estimated from the normal operation data from the datasheet and the IGBT and diode parameter fitting tools provided by SaberRD®. It

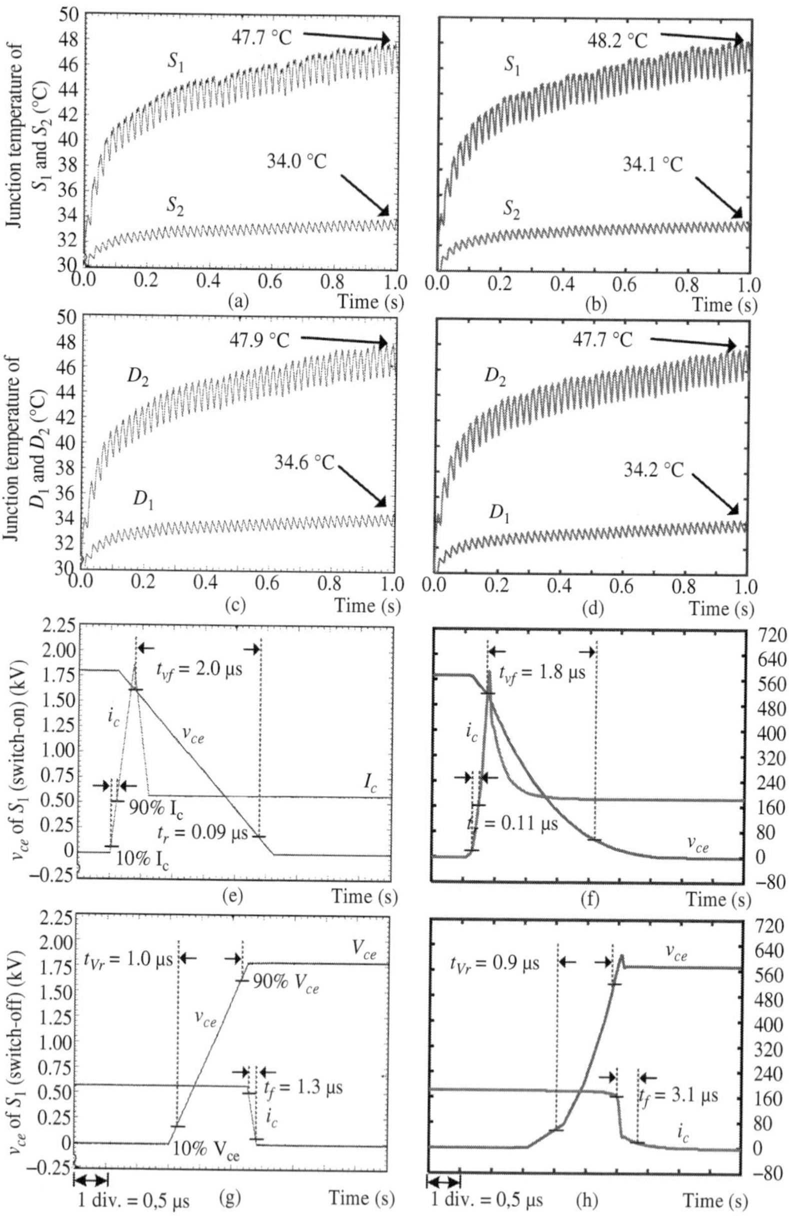

Figure 11.20 Device-level steady-state results of the first SM in Converter Cb-A1 upper arm phase A: (a) (b) junction temperatures of IGBTs S_1 and S_2, (c) (d) junction temperatures of Diodes D_1 and D_2, (e) (f) voltage v_{ce} and current i_c of S_1 during switching-on transient, (g) (h) voltage v_{ce} and current i_c of S_1 during switching-off transient, ((a) (c) (e) (g) are the real-time results; (b) (d) (f) (h) are the offline SaberRD® results.)

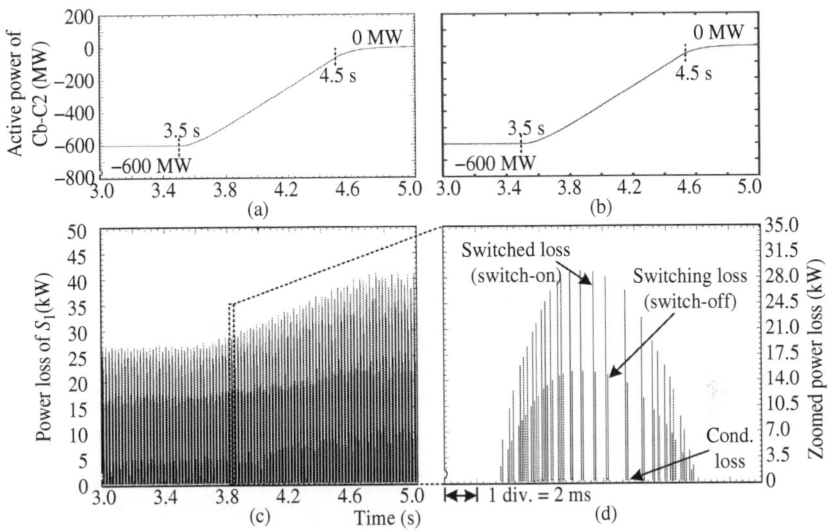

Figure 11.21 Results during power flow command change: (a) (b) active power of Converter Cb-C2, (c) power loss of IGBT S_1 from the first SM in Converter Cb-A1 upper arm phase A, and (d) zoomed power loss. ((a)(c)(d) are the real-time results; (b) is the offline PSCAD/EMTDC® result.)

is observed that the temperatures increase significantly within few milliseconds. The times for the devices reaching 150 °C since the begin of the fault are measured and compared with SaberRD®. When temperatures exceed 150 °C, the devices are no longer safe for operation. It is noted that such results may not be accurate with extreme high temperatures, where the devices could be damaged and the characteristics are fundamentally changed in practical situation. The electrothermal model can still give a rough estimation of the junction temperatures during the faults to evaluate the potential damage to the devices and to verify the control and protection algorithms for these devices.

11.5 Real-Time Co-Emulation Framework for Cyber-Physical Systems

With the innovations in electric power grid and information and communication technologies (ICT), traditional power systems are evolving into complex CPS, which increasingly leverage the capabilities of data communication and computation to enhance the flexibility of power transfer in the smart grid. One major concern in smart grid research is to simulate the entire behavior of the system to adequately evaluate the influence of the interplay between digital world and

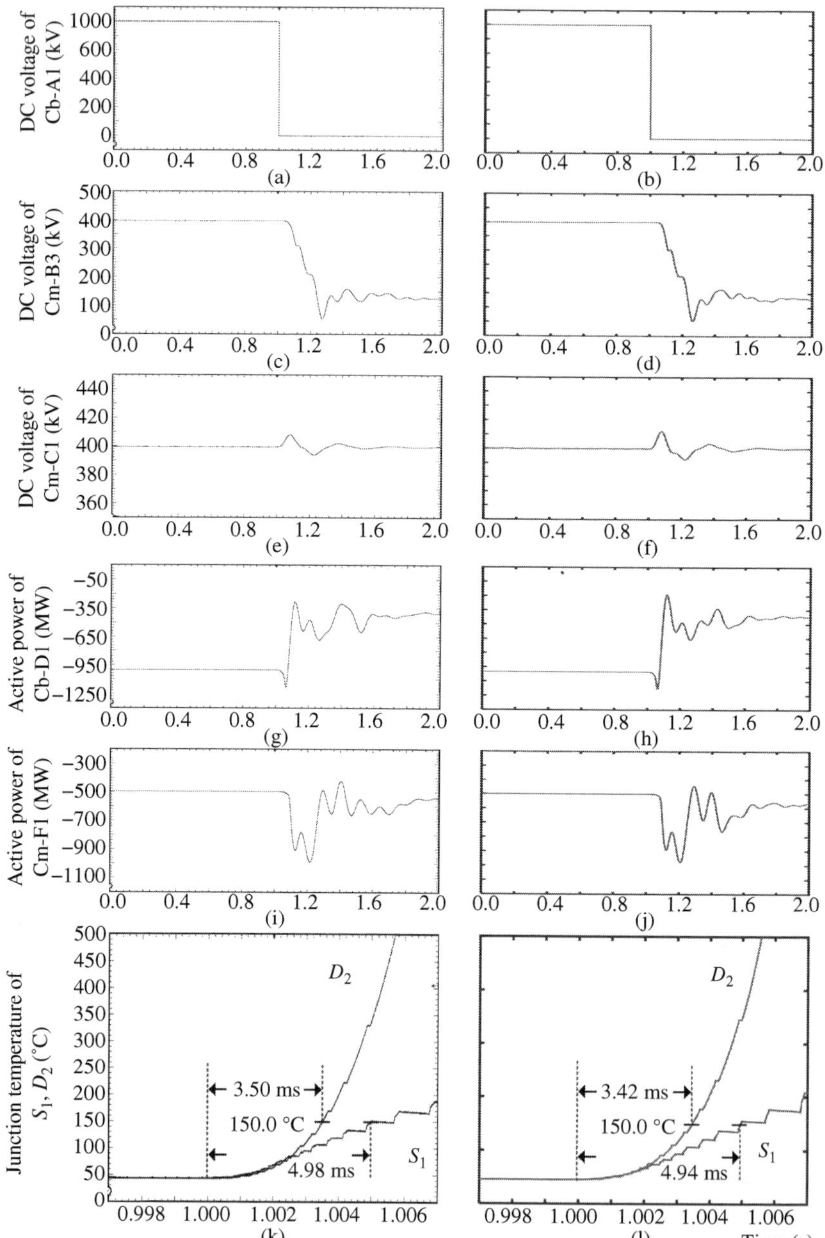

Figure 11.22 Results during DC fault: (a) (b) DC voltage of Converter CB-A1, (c) (d) DC voltage of Converter Cm-B3, (e) (f) DC voltage of Converter Cm-C1, (g) (h) active power of Converter Cb-D1, (i) (j) active power of Converter Cm-F1, and (k) (l) junction temperatures of $\mathbf{S}_1$ and $\mathbf{D}_2$. ((a) (c) (e) (g) (i) (k) are the real-time results; (b) (d) (f) (h) (j) are the offline PSCAD/EMTDC® results; (l) is the offline SaberRD® result.)

physical equipment. However, since the power system and communication network have distinctly different working principles and simulation tools, simulating the entire behavior of the two domains (so-called "co-simulation") is quite challenging [211, 212]. Power system EMT simulation is a *discrete-time* process while the communication network simulation is a *discrete-event* process.

Various co-simulation frameworks have been proposed in the recent past since the first interface for the PSCAD/EMTDC® off-line simulator to integrate an agent-based distributed application was developed [213]. Most of these works are not to design a complete simulator that could finish all required functions in one software package but attempt to interface two existing simulators of each domain, because there are already various mature power system and communication network simulators available. Unfortunately, there does not exist universally accepted interfaces for data exchange between simulators of the two domains due to their different working principles or even operation systems. Thus, the main concern of these co-simulator frameworks [214–216] is to properly handle data exchange and synchronization for related events in both domains at runtime. Most co-simulators mainly distinguish from each other by the type of power system/communication network simulators used, the methods applied for synchronization, and the application scenarios. However, the performance of the software-based simulators is relatively low compared with actual power and communication network devices even without taking the data exchange and synchronization time between two simulators into account. With respect to EMT the simulation process will be extremely slow due to the small time-step size and massive synchronization requirements. It is, therefore, difficult to practically simulate and test the adequacy of manufactured protection and control equipment responding to potential damage created by transient events in real-time.

In this section, a RTCE framework on FPGA-MPSoC–based hardware architecture is presented for a more practical emulation of real-world CPS.

11.5.1 Communication Network Simulation and Co-Simulation

Different from power systems, the elements in communication networks are modeled not based on physical principles but based on their *functionality*, and thus there are no differential equations or matrix equations to be solved. Instead, the components in communication networks simply receive, modify, and send data packets, and thus the process of each network element is always modeled into a sequence of *discrete events* that occur unevenly in time, where each event represents an operation of packet processing or transmission. In this context, the communication network simulator refers to a program running on software or hardware to simulate the behavior of a user-defined network topology, and each node in the simulator executes the packet processing procedure similar to real-world equipment.

The basic network component in the cyber-physical power system is the sampling and reporting device, or a smart meter, which is responsible to measure the electrical values of the power system and report the measurements to the system controller. The phasor measurement unit (PMU) [217] is a commonly used basic network component that is deployed in each single process bus or substation model to measure the electrical quantities; and after sampling the data values, it computes the corresponding phasor values and then reports them to the phasor data concentrator (PDC), which is responsible to monitor a power area and collects the measurements from the PMUs in the area. In fact, the main function of the communication network is to provide a two-way communication between smart meters, data concentrators, and controllers to modify the related circuit parameters according to the service requirements.

Over the past, many communication network simulators have been adopted in co-simulation research, of which the representative examples are NS-2/NS-3, OMNeT++, OPNET, etc. NS-2/3 (from network simulator) is a series of discrete-event network simulators that are primarily used in academic research. The network protocols such as (Transmission Control Protocol) TCP/(User Datagram Protocol) UDP, routing algorithms, multicast schemes are supported in this simulator, thus the traditional network function can be well evaluated. For example, both NS-2 and NS-3 were used in [218, 219] for communication network modeling in the co-simulation. OMNeT++ is an open-source discrete event based simulator for communication networks, which was used to model smart grid in [220, 221]. Different from NS-2/3 and OMNeT++, OPNET is a more powerful discrete event network simulator mainly for commercial use, enabling abundant network models such as wired networks, wireless networks, and ad-hoc networks. OPNET offers a visual high-level user interface to access the C/C++ based blocks that are used for different models and functions, which also enables users to customize network applications in multiple co-simulation approaches [222, 223].

Co-simulation refers to simulating the power system operations while taking the communication network layer into consideration, so that the impact of interaction between the two domains can be evaluated and the entire system features can be revealed. However, power system simulation is time-continuous and the simulation proceeds step-by-step in discrete time, while the communication network simulation is event-driven where the events are unevenly distributed in time. Thus, the main challenge in co-simulation is how to interface the two domain simulators with significantly different operating principles.

The interaction between the two domains includes the sampling and reporting operations. The sampling operation refers to measuring and digitizing the values of the concerned electrical quantities, which is also called *domain-level synchronization*, since it is used to exchange the data of the two domain simulators. The domain-level synchronization does not involve the network packet encapsulation, which is usually performed periodically at a constant sampling rate for a stable measurement. The reporting operation refers to reporting the measured electrical values from a smart meter to a power area data concentrator, and from a data concentrator to the system controller. It is called the *application-level synchronization*, since it is based on the network data packet and is handled within the communication network domain for the application-level purpose. The application-level synchronization contains a two-stage reporting. The representative approaches for synchronization in each level are the periodic and on-demand methods: when applying the periodic synchronization strategy, the reporting only happens at each synchronization point; for the on-demand synchronization, the data-exchange is only initiated if an event of interest is detected, and under normal conditions the corresponding measurement values are regarded as unchanged. An example is shown in Figure 11.23, it can be seen that the domain-level synchronization is performed at a constant sampling rate; while the two stages of the application-level synchronization are performed using the periodic and on-demand strategies, respectively.

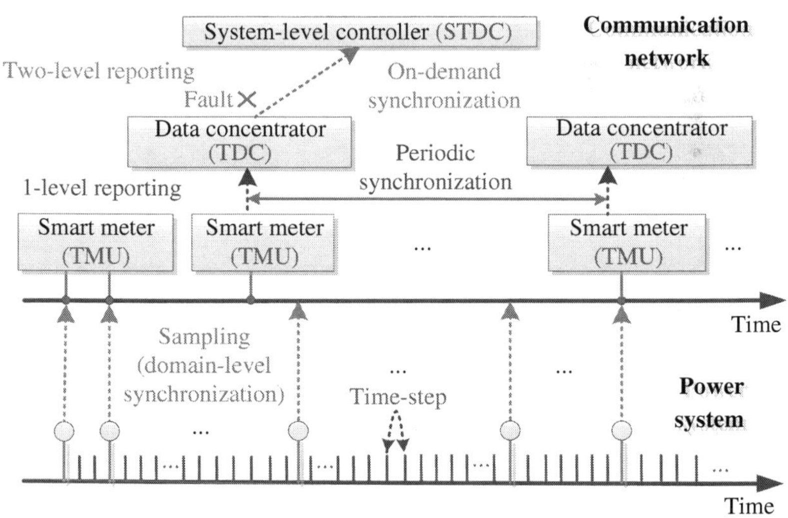

Figure 11.23 Example of interaction between power system and communication network simulation.

11.5.2 Real-Time Co-Emulation Framework

Based on the co-simulation background, most leading research in this area is either built on existing simulators or the development of a middleware to interface simulators from the two domains; thus, their performance is dramatically limited by the software environment and achieving fast simulation for large-scale cyber-physical power systems becomes difficult. An alternative strategy that combines both programmability and high-speed computation is hardware-based co-emulation [224]. Furthermore, the real-time power system EMT simulation has already been extensively carried on the FPGA as seen from preceding chapters, and communication network functions such as Ethernet, TCP/IP are also available as vendor specific IP cores or soft-codes built on FPGA. The MPSoC [20] integrates the FPGA-based programmable hardware logic and the ARM® based multicore processor systems within one platform, which enable complex embedded applications to be developed. This motivates the case for an integrated FPGA-MPSoC based power-communication hardware co-emulation.

The FPGA has a large amount of hardware programmable logic resources that can be customized by the engineer to achieve specific functions; secondly, there are rich I/O transceivers in FPGA that enable several prototyping boards to be interconnected to extend the compute capability for real-time power system EMT emulation or for emulating real communication network functions. In addition, the processors integrated on the FPGA chip (such as Microblaze processor in Xilinx® FPGAs) or the additional ARM® based APU processors in the MPSoC board provide the ability to develop monitor and control systems for the cyber-physical system.

11.5.2.1 RTCE Hardware Architecture

The top-level architecture of RTCE framework is shown in Figure 11.24, wherein the power system EMT emulation and the communication network emulation are integrated on each board. In power system domain, each board emulates a subsystem of the entire system, and emulation results are exchanged with connected boards in each time-step; and in communication network domain, each board emulates the information networking between the buses located in its emulated area. Different from the commonly used PMU and PDC concepts, in this work, the concepts of transient measurement unit (TMU) and transient data concentrator (TDC) are applied, since the measurement data is obtained from the instantaneous transient level EMT power system simulation. The TDC is responsible to monitor the emulated subsystem and collect the measurements from the TMUs; TDC does not have a global view and control applications of the whole power system, thus the super transient data concentrator (STDC) is also modeled to receive messages from TDCs and compute the centralized control strategies when detecting abnormal conditions. The corresponding TMU, TDC, and STDC components are marked in Figure 11.23.

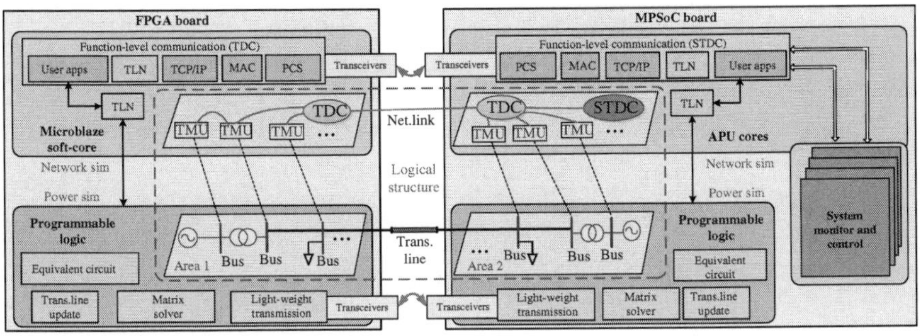

Figure 11.24 Demonstration of real-time co-emulation (RTCE) architecture on FPGA-MPSoC platform.

Power System EMT Emulation For large power systems, it is difficult to emulate the whole circuit topology in one board due to the requirements of huge amount of hardware logic resources and memory. Thus, a common approach is to decompose the system into several subsystems using the inherent latency of the widely distributed transmission lines as seen in Section11.2, and these decomposed subsystems can be allocated to different boards for parallel processing. After the results are obtained in each time-step, they should be exchanged among adjacent subsystems using fast data exchange path, as shown in Figure 11.24. It is sufficient to use light-weight communication such as the Xilinx® Aurora protocol for fast data exchange. In EMT emulation, each subsystem can be abstracted into an equivalent circuit along with a matrix equation to be solved, and transmission line data should also be calculated and exchanged between the adjacent subsystems. Thus, the power subsystem emulation can be divided into three parts: update of equivalent circuit parameters, matrix equation solution, and update of transmission line data, as illustrated in Figure 11.24.

Communication Network Emulation Since each FPGA/MPSoC board has limited I/O resources and soft-processors, it can only operate as one network device. However, the network nodes (TMUs) in a smart grid can even have the same number as the buses, so it is impossible to emulate all of the network nodes with bit-level details in hardware and obtain the correct transmission parameters between each TMU and STDC. Thus, in this work, RTCE emulates the communication network with the two-level model: transmission-level networking (TLN) and function-level networking. The TLN is used to model the end-to-end packet transmission characteristics such as delay and loss, while the function-level networking is used to emulate the TDCs and model the specific network functions or protocols such as Ethernet or TCP/IP. The TLN emulation models individual network components such as routers, hosts, links, and protocols with specific abstractions rather than completely implementing all the details to avoid complex hardware-based networking executions that are not concerned in power system analysis. The TLN models all of the transmission characteristics between TMUs and TDCs and between TDCs and STDC using parameters such as the link losses and delays, which can be obtained through software simulators such as NS-2/3. The function-level networking is carried out directly on the I/O resources of FPGA/MPSoC boards, because the physical coding sublayer (PCS) and media access controller (MAC) functions can be implemented using hardware logic resources, and the Internet protocol or other upper network layer functions (TCP, UDP, etc.) can run in the soft-processors (such as Xilinx® MicroBlaze) to handle the complex control tasks such as routing, connection establishment, and flow control.

As shown in Figure 11.24, each TMU receives the measurement from power system emulation as the sampling operation, and then sends the data packet to the corresponding function-level TDC module via the TLN module as the reporting operation. Then the TDCs send data packets to the function-level STDC module via the real-world communication link if detecting abnormal conditions, but still need to pass the TLN module in the function-level networking to emulate the transmission process between the TDC and STDC since TDCs/STDC are not directly connected in practical network. Using this method, the specific network functionalities can be emulated to observe the interaction between power and communication domains, while the detailed transmission processes between TMUs, TDCs, and the STDC are simplified and abstracted as crucial and practical transmission parameters.

Power-communication Co-emulation In each board, power system EMT emulation and communication network emulation run concurrently. Thus, the data exchange and synchronization between the two domains are simplified: the results of power system emulation in each bus and each time-step can be output to the network emulation module directly without any delays to emulate the sampling process of the TMU in a single process bus. This means, at every hardware clock, the results of the two domains can be exchanged within the FPGA logic. If the sampling rate is assigned practically, this data measuring process can also be done for every specific time space. The TMU in each bus is connected with the TDC via the TLN module to send measurement data packets periodically, and TDCs check the received measurement data to find if there exists abnormal conditions; if some transients are detected, the TDC will send messages to STDC for system control command. This on-demand TDC-to-STDC reporting strategy could reduce the amount of generated data packets and thus reduce the congestion of the communication network. The control command from the STDC that may change the circuit topology or other circuit parameters

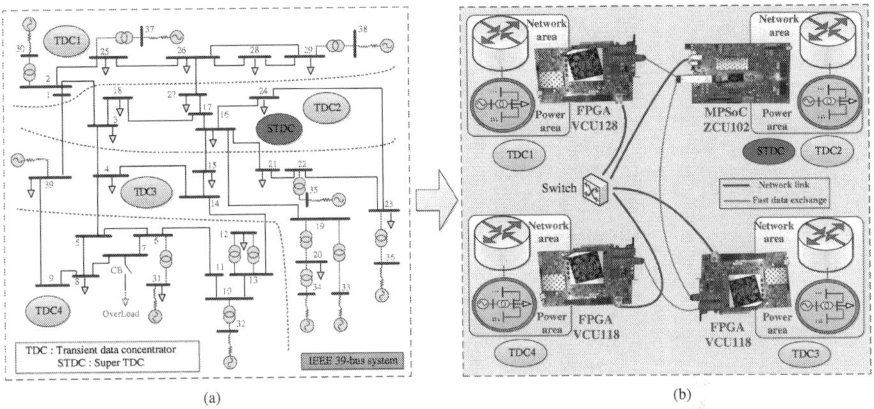

Figure 11.25 Example of co-emulating a cyber-physical system on multiboard hardware platform.

can also be sent to the controllable power devices in the same way. TDCs can run in the soft-processors of FPGA platform, but the system-level monitor and control of STDC should run in the soft-processors of MPSoC platform because the soft-processors in the MPSoC board have more computation resources and a higher clock frequency.

11.5.3 Hardware Implementation of RTCE

The RTCE framework was tested on the IEEE 39-bus system as shown in Figure 11.25a. The system is partitioned into four areas, and TDCs are deployed in each area (at Bus 25, 24, 4, 9, respectively) to accumulate the measurements from each bus and inform the STDC deployed in area 2 (at Bus 24) if detecting abnormal conditions. The hardware RTCE emulator was developed to co-emulate the test system in real-time. As shown in Figure 11.25b, there are two Xilinx® VCU118 FPGA boards, one Xilinx® VCU128 FPGA board and one Xilinx® ZCU102 MPSoC board used in this emulator, which emulate the partitioned four areas concurrently. The specifications of the XCVU9P FPGA on VCU118 board and the XCZU9EG MPSoC on ZCU102 board are given in Table 1.2. The Virtex® Ultrascale+ XCVU37P FPGA on the VCU128 board [225] consists of 2 852 000 programmable logic cells, 340.9 Mb BRAM, and 9024 DSP slices. The communication network links between RJ-45 ports of each board are connected via a switch, while the fast data exchange links for power system emulation are connected on the SFP/SFP+ ports via the 10G optical fiber. The detailed hardware block design in each board is shown in Figure 11.26.

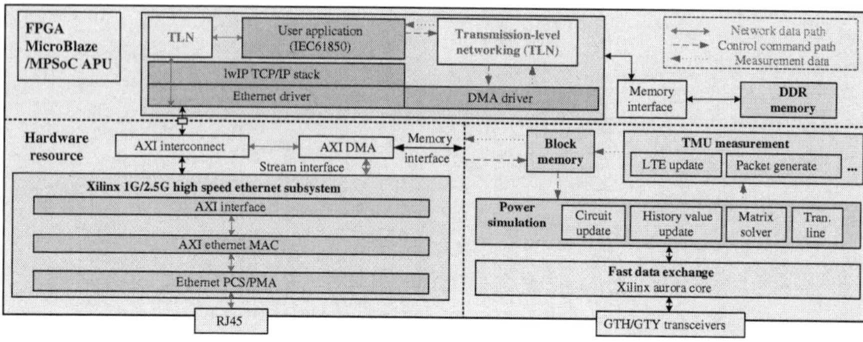

Figure 11.26 Illustration of detailed block design on a single FPGA/MPSoC board.

11.5.3.1 Multi-Board EMT Emulation

In this work, the applied power equipment models (synchronous machine, power transformer, transmission line and loads, etc.) are similar to those of PSCAD/EMTDC® used for verification. For real-time multiboard EMT emulation, the power system is usually partitioned by the long transmission lines. Then the fast data exchange path is required to exchange the transmission line model data among adjacent subsystems. In this work, the lightweight communication module is used: Xilinx® 64-bit Aurora core. Instead of using the complex communication functionalities such as the Ethernet protocol, the simple high-speed Aurora communication core enables direct data transfer through different types of transceivers on FPGA/MPSoC boards with up to 10 Gb/s speed.

To interact with the communication networking emulation, the abnormal condition detection is the prerequisite to determine the data packets generation. State estimator is usually used to predict the next state variable values and be compared with the real value to determine if the system is running in control. For transient-level analysis of this work, the local truncation error (LTE) is used to estimate the perturbation, and the abnormal condition is found once the LTE or the measured current and voltage exceed the predetermined threshold. After the main conductance matrix is solved in each time-step, the state variables of linear equipment are obtained, and then the LTE at the nth time-step is computed for the linear equipment given by (6.7). For the subsystems, where nonlinear equipment exists the iterative approach is involved. The standard method is to first use an explicit method or interpolation polynomial (called the predictor) to compute a candidate value of the state variable to be solved, and then use it as the initial solution to apply Newton's method for the implicit integrator (called the corrector) until convergence is achieved. Then the LTE can be estimated by comparing the initial solution $\boldsymbol{x}_n^0$ and final solution $\boldsymbol{x}_n$:

$$\mathrm{LTE}(t_n) \approx \frac{C_{p+1}}{1 - C_{p+1}} \left(\boldsymbol{x}_n - \boldsymbol{x}_n^0\right). \tag{11.4}$$

11.5.3.2 Communication Protocol and Implementation

In order to practically emulate the communication network, the standardized communication protocol, IEC 61850 [226] was utilized which can be used to manage a large number of communication devices. The existing communication protocols such as the Ethernet, TCP/IP typically define how data bytes are transmitted on the wire, however, they usually do not specify the application layer data organization. The IEC 61850 defines how the application data bits should be arranged and how the created data items/objects and services are mapped to any other existing under layer protocols that can meet the service requirements. In the RTCE framework, all the communication packets are encapsulated based on the IEC 61850 standard.

In the communication network emulation, each bus is aligned with a TMU, which is able to digitize the base measurement quantities at the source and transmit the resulting sample values to the TDC. The measurement data includes voltages, currents, and some status information (LTE). In the RTCE architecture, the data is transferred from power emulation to communication networking emulation in each time-step to perform real-time EMT analysis, which means the sampling rate of TMU is related to the actually applied time-step size. Different from the TMU measurement gathering that does not involve network packet generation, the TDC collects data from the TMUs in its area based on a specific reporting rate. The reporting rate from TMU to TDC was set at 60 Hz. As mentioned before, each TDC is required to compare the LTE, current or voltage with the specific threshold; once the measurement exceeds the threshold, the abnormal condition is detected, and then it is responsible to create a measurement data packet to be sent to the STDC.

Transmission-Level Networking The data packet networking is achieved in two modes: transmission-level and function-level networking. In the transmission-level emulation, the transmission process between TMUs and the TDC and between TDCs, and the STDC is modeled as bandwidth, communication delay, and loss rate. The TLN is implemented in the soft-processor of FPGA/MPSoC boards because the transmission parameters can be flexibly configured in software without modification of the hardware design. As illustrated in Figure 11.26, for example, if the TMU in Bus 6 generates a packet to be sent to the TDC at Bus 9, then the data packet is generated in the hardware module and then is sent to the soft-processor and passes the TLN function; and if the TDC (run as the function-level networking module) at Bus 9 detects an abnormal condition and send a data packet to the STDC at Bus 24, the packet needs to first pass the TLN function in the application layer of TCP/IP stack and then be sent to the lower TCP/IP and MAC layer. The implementation of the TLN function is quite straightforward: based on the Xilinx® timer() function, the TLN function detects the source and destination of the input packet and searches for the corresponding end-to-end transmission delay; then, after the delay outputs the input data packet. The real communication delay is related to the distance and hops between the source and destination, while the loss rate refers to the possibility of an unsuccessful transmission process. In this implementation, the values chosen for the delay and loss are determined based on the testing results from the real network simulator NS-3.

Function-Level Networking For function-level communication network emulation, the high-level view is shown in Figure 11.26. The Xilinx® 1G/2.5G High Speed Ethernet Subsystem core is used to implement the MAC with a PCS based on the hardware logic and I/O resources. The upper network layer functions are implemented in the Lightweight IP (lwIP) stack [227], which is an open source TCP/IP networking stack for embedded systems and is available in the Xilinx® programming tools. The Xilinx® Software Development Kit (SDK) provides lwIP software customized to run on various embedded systems that can be MicroBlaze soft-processor in FPGA chip or the ARM®-based SoC devices. Through the lwIP application programming interface, users can add networking capability to an embedded system. Since the echo server application that can create a TCP connection and sets up a callback on a connection being accepted is already provided by Xilinx® SDK, the IEC 61850 protocol and data encapsulation are implemented based on the existing echo server code to achieve the specific communication patterns of smart grid.

The function-level networking data path is marked as solid lines with bidirectional arrows in Figure 11.26, which is achieved by the stream interface between direct memory access (DMA) module and Ethernet module; the control command path that is used to send control commands from the controller in the soft-processor to the power emulation module is marked with dashed line in Figure 11.26, which is achieved through the memory interface between DMA module and the block memory; the measurement data path that is used to send measurement data from each TMU to the TDC is marked as the dotted line, which is transmitted via the TLN module. Note that the measurement data in the block memory on FPGA can only be passively read from the DMA driver function; that means the reporting operation from TMU to TDC is achieved by the application on the lwIP stack to use the driver function to read the values from the block memory periodically at a reporting rate of 60 Hz.

Table 11.4 Hardware resource consumption of the case study.

Resource	VCU128	ZCU102	VCU118-1	VCU118-2
LUT (%)	40.3	95.2	57.1	55.2
FF (%)	36.1	70.1	50.3	48.9
BRAM (%)	73.3	86.9	77.4	75.5
DSP (%)	35.6	87.2	64.1	62.8
Latency (μs)	10.1	8.7	16.2	13.1

11.5.4 Real-Time Emulation Results and Verification

Based on the implemented RTCE hardware emulator, the test system was emulated to evaluate how the communication failure affected the power system, and how the power system transients impacted the communication network.

11.5.4.1 Processing Delay and Hardware Resource Cost

For the hardware emulation, the FPGA boards (including the Microblaze soft-processor) run at the clock frequency of 100 MHz. According to the system partition and RTCE implementation described above, the hardware resource consumption and maximum processing latency for one time-step are presented in Table 11.4. The VCU118-1/2 boards emulate the power areas 3 and 4, while the VCU128 and ZCU102 boards emulate the areas 1 and 2, respectively. It can be observed that the resource cost is relatively low, which means a larger system can also be accommodated in this multiboard platform.

The processing latency of one time-step includes the latency of power equipment model computation, matrix solution, transmission line updating, history item exchange (including board-to-board fast data exchange), and LTE computation, which indicates the minimum time-step size that can be applied for real-time EMT emulation. In this work, the power areas in the four boards are emulated using the same time-step, 20 μs, which is larger than the latencies of each subsystem and is sufficient for EMT-level analysis.

For the communication network emulation that runs concurrently, since the transmission-level communication behaviors are parameterized by real emulator results and the function-level emulation is just executed on the hardware transceivers for real-world networking, the communication network emulation also runs in real-time. However, the real-time networking performance is actually confined by the transmission-level parameters and function-level hardware capabilities. In addition, the TCP/IP protocol execution and system-level control performance are limited by the computational capabilities of soft-processors in FPGA and APU cores on the MPSoC board. In this work, the tested function-level networking performance is 984 Mbit/s based on the lwIP performance testing program; the transmission delay of each communication link in the NS-3 simulator is set at 1 ms/200 km, which is a classical value used in communication simulation; and the forwarding rate of each network device that impacts the delay of packet processing within a device is set at 100 Mbit/s, which is also a commonly used value.

For the interaction between the two domains, the delay actually refers to the delay of the reporting operation between the FPGA MicroBlaze/MPSoC APUs soft-processor and the block memory on FPGA, since the sampling operation is performed directly on the programmable logic with zero delay. On the FPGA board, the tested throughput of the reading data operation of the driver function is 186 Mbit/s; on the MPSoC board, the test throughput was 214 Mbit/s. Note that the driver read function does not influence the power system simulation, which means, the data exchange delay between the two domains is sufficiently small for real-time interaction.

11.5.4.2 Case Study 1: Over-Current Fault

To study how the power system fault conditions affects the entire cyber-physical system, a sudden load increase case is applied on Bus 7 at emulation time two seconds, which results in a over-current fault. Upon detecting this condition, the STDC will perform the centralized algorithm for protection. The emulation results are shown in Figure 11.27.

As shown in the results, after the fault is initiated, the corresponding message (including the bus voltage, load current, and LTE) is sent to STDC due to the sudden transient-level increase of LTE. Subsequent messages are generated because the load current exceeds the threshold value. And about 50 ms after the fault, the controllable circuit breaker at Bus 7 receives the control command from STDC to open the load circuit for protection. The time delay between the fault and the response message mainly includes the transmission time and computation latency for control command. In this work, the control program was developed only for this case study and thus, it consumes a very low latency. The transmission time consists of the in-board communication (referred to TLN delay between Bus 7 and TDC4 and between TDC4 and STDC) and board-to-board communication latency. Since the board-to-board communication delay is extremely low, it can be omitted compared to the TLN delay that is measured through the software network simulator. The real-time emulation results indicate that the interaction between the two domains is quite fast. Besides, as shown in Figure 11.27b,c, when co-emulating the two domains, the transient-level waveforms can also be captured and output in real-time to observe the details of the value change due to the small time-step applied.

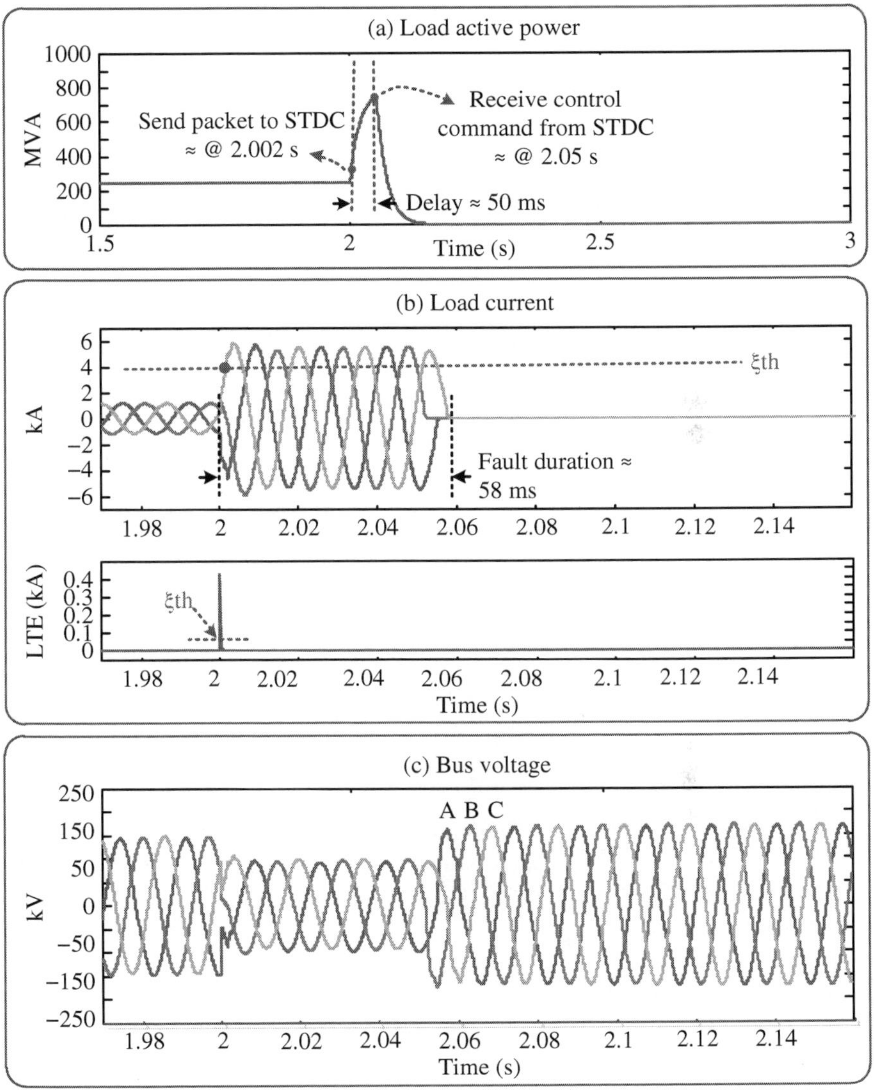

Figure 11.27 Overcurrent fault case study: (a) active power of load at Bus 7; (b) total load current at Bus 7; (c) voltage of Bus 7.

11.5.4.3 Case Study 2: Communication Link Failure

To investigate the effect of network link failure, the link between Bus 8 and Bus 9 that is essential for the communication between TMUs and TDC breaks down at emulation time two seconds, while the over-current fault is also applied at

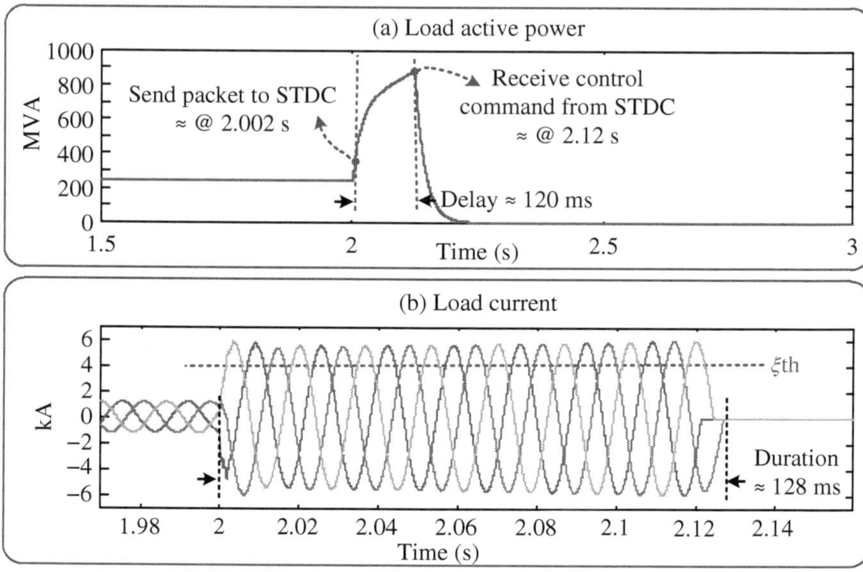

Figure 11.28 Communication link failure case study: (a) active power of load at Bus 7; (b) total load current at Bus 7.

the same time. Since the on-demand synchronization strategy is applied, during normal conditions the influence is small, because the power system is regarded as working normally if there is no message sent from TDCs. However, when the overcurrent fault happens, this impact is essential to the system response. The system controller in STDC is required to respond to this situation according to the messages received from TDCs, however, since the link broke down, re-computing a new route costs a considerable time, and the generated new route has a much longer distance between the TMU at Bus 7 and TDC at Bus 9, which also results in an increased communication delay.

As shown in Figure 11.28, the system is supposed to respond to the over-current fault caused by the sudden load change within a short delay, however, due to the transmission link failure, the system response becomes relatively slow. Since the link failure impacts the transmission-level communication delay, the increased latency is actually obtained from the NS-3 simulator. From Figure 11.28b, it can be observed that the total duration of overcurrent fault raises to 128 ms, which greatly increases the duration of impact.

11.6 Faster-Than-Real-Time Hybrid Dynamic-EMT Emulation of AC–DC Grids

Dynamic simulation of the integrated AC–DC grids plays a crucial role in energy control centers for wide area monitoring, control, and protection. To ensure stability of the grid dynamic security assessment (DSA) is carried out. A complete DSA is composed of five stages [228, 229]: measurements, modeling, computing, visualization, and control. The real-time measurements are captured by the supervisory control and data acquisition (SCADA) for basic state estimation, and then a detailed time-domain simulation is carried out for imminent contingencies. The obtained security indices are delivered to the system operators in the control center. The entire DSA cycle is conducted within 10–30 minutes to ensure that the operators have sufficient time to take remedial actions.

Due to the enormous computational burden in DSA systems and the expansion of power system scale, the hardware resources are stretched. For instance, to realize near real-time simulation of a power transmission system with more than 50 000 nodes and 3000 generators, the DSA is conducted on the supercomputer with more than 24 000 CPU cores for parallel power system simulation [230]. On the other hand, the SCADA system collects data from the field every two to five seconds, which may be sufficient for steady-state tasks such as online power flow or state estimation, however, it falls short of predicting system stability and conducting online DSA. The PMU technology provides a faster refresh rate (50/60 samples per second) which is at least two orders of magnitude faster than SCADA; however, PMUs are expensive and cannot be installed on a wide-area to gather system dynamic situational awareness. Therefore, an integrated hardware platform based on FPGAs is presented in this section for faster-than-real-time/SCADA (FTRT/SCADA) purpose so as to predict system dynamics faster. FTRT dynamic emulation [231, 232] deployed in the energy control center can assess the impact of excessive energy penetration at critical locations, predict destabilizing outages, mitigate power system oscillations, and devise newer control strategies for enhancing system stability and security.

To provide fast, accurate, and time-synchronized data for the complex AC–DC grids, a high-fidelity simulation strategy is also needed. Traditional off-line dynamic simulation tools utilize fixed time-steps which requires more hardware resources and prolongs the simulation when the system scale becomes larger. As various equipment have different dynamic responses, the local equipment-based flexible time-stepping methodology is employed for the FTRT emulation of the hybrid network wherein the AC grid undergoes the dynamic emulation, while for the DC grid, an EMT emulation is carried out. A power-voltage interface is inserted between the AC and DC grids since distinct emulation strategies are utilized, and the EMT-dynamic co-emulation is able to run concurrently on multiple interconnected FPGA boards. Various contingencies, such as three-phase-to-ground fault, generator outage, and sudden load change are emulated to reveal the efficiency of the local equipment based flexible time-stepping algorithm for the co-emulation. Thus, following the occurrence of a disturbance, the FTRT emulator can generate an optimal solution in advance to maintain the stability of the hybrid AC–DC grid.

11.6.1 Flexible Time-Stepping Algorithm for Dynamic Emulation

11.6.1.1 Transient Stability Emulation Methodology

A practical power transmission system includes synchronous generators, buses, transmission lines, as well as various loads. The differential equations describe the dynamics of the synchronous machines, and the remaining components contribute to the algebraic equations which solve the network in conjunction with stator voltages of the generators:

$$\dot{\mathbf{x}} = \mathbf{f}(\mathbf{x}, \mathbf{u}, t), \tag{11.5}$$

$$\mathbf{g}(\mathbf{x}, \mathbf{u}, t) = 0, \tag{11.6}$$

where $\mathbf{x}$ refers to the vector of state variables and $\mathbf{u}$ represents the bus voltages of a typical power system. The initial conditions can be written as follows:

$$\mathbf{x}_0 = \mathbf{x}(t_0). \tag{11.7}$$

The main challenge of emulating the dynamics of a power system is basically solving a set of differential algebraic equations (DAEs) [233]. Due to the nonlinear nature of the equations, the Newton–Raphson (NR) algorithm is required. The explicit numerical integration methods such as Forward Euler, or fourth-order Runge–Kutta (RK4) can be utilized in hardware emulation of the DAEs due to the low resource usage. In a trade-off between emulation efficiency and the accuracy, RK_4 is adopted as the solution for solving the nonlinear DAEs, as given below:

$$\text{RK}_1 = dt \cdot f\left(t_n, x_n\right), \tag{11.8}$$

$$\text{RK}_2 = dt \cdot f\left(t_n + \frac{dt}{2}, x_n + \frac{RK_1}{2}\right), \tag{11.9}$$

$$\text{RK}_3 = dt \cdot f\left(t_n + \frac{dt}{2}, x_n + \frac{RK_2}{2}\right), \tag{11.10}$$

$$\text{RK}_4 = dt \cdot f\left(t_n + dt, x_n + RK_3\right), \tag{11.11}$$

$$x_{n+1} = x_n + \frac{1}{6}(\text{RK}_1 + 2\text{RK}_2 + 2\text{RK}_3 + \text{RK}_4), \tag{11.12}$$

where x_n refers to the system state at the nth time-step, and dt is the emulation step.

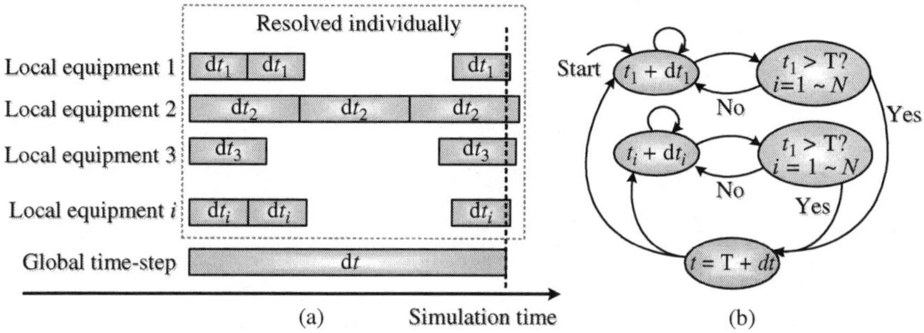

Figure 11.29 Flexible time-stepping for hybrid AC–DC grid emulation: (a) principle of the algorithm, (b) time instant of synchronization.

11.6.1.2 Local Equipment Based Flexible Time-stepping

According to the sensitivity and stability of the various components in the power transmission system, different time-stepping control strategies can be applied, which is beneficial to reduce the computational burden in low sensitivity parts. While the utilization of various time-stepping strategies may lead to an asynchronous time instant at the interface, which may lead to an unstable emulation. Therefore, a hybrid time-step control algorithm should be generated. For instance, the output mechanical torque (T_m) of the governor control system in the synchronous generator has a low sensitivity to time-step variation; in other words, a relatively large time-step will not influence the stability of the governor system. The event-based variable time-stepping method (based on rate of change of voltage or rotor angle) which provides time-steps ranging from 5 to 10 ms is employed in the generator control system as shown in Figure 11.29a. To ensure the accuracy of the mechanical and electrical parts, the flexible time-stepping strategy should be adopted, which is usually executed in smaller time-steps after a disturbance occurs.

The local equipment-based flexible time-stepping algorithm contributes a localized time-step dt_i produced in each component. Once a disparity emerges among them, the emulation goes forward only when the time instants of all variable time-stepping systems exceed the global time-step which is determined by the largest time-step of the subsystems. In the proposed emulation platform, the generator control systems are always computed with the largest time-step, which is therefore chosen as the global time-step. The synchronization mechanism is as follows: the control systems proceed at a much lower sample frequency, and they enter the next time-step only when the time instants of all local equipment reach beyond their current value; otherwise, all the components compute individually,

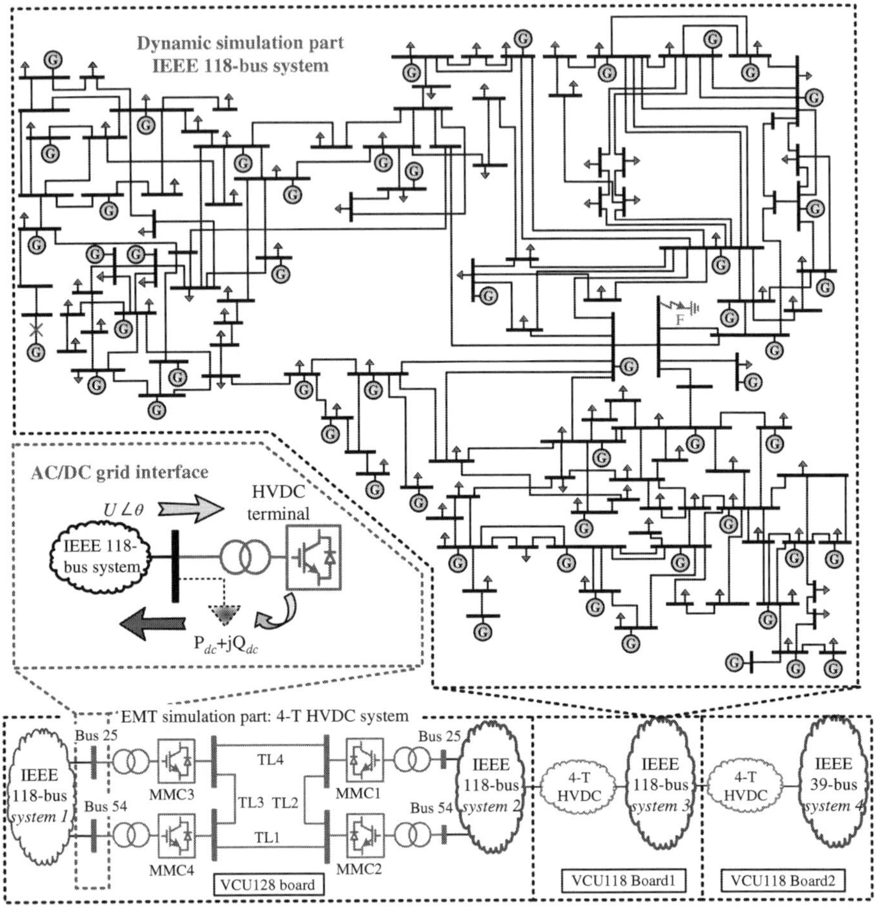

Figure 11.30 Integrated AC/DC grid for dynamic-EMT co-emulation.

while the global system waits for them to finish. The time instant of synchronization of the local equipment-based flexible time-stepping algorithm is illustrated in Figure 11.29b.

11.6.2 AC–DC Grid Component Modeling

11.6.2.1 AC–DC Grid Interface

Figure 11.30 shows a typical integrated AC/DC grid, where the IEEE 118-bus test systems [234] are connected with four-terminal HVDC system at Bus 25 and Bus 54. *MMC*3 and *MMC*4 act as rectifier stations, while *MMC*1 and *MMC*2 are treated as inverter stations. The IEEE 118-bus systems undergoing the transient stability analysis have a flexible time-step ranging from 1 to 10 ms. On the other hand, to reveal the dynamic process of the four-terminal HVDC, the EMT emulation is adopted for the DC grid, which has a fixed time-step of 200 μs. The local equipment-based flexible time-stepping is applied for establishing an integrated co-emulation time-step scheme. Due to the distinct emulation strategies of the AC and DC grid, a proper interface should be designed.

In this case, the converter stations are utilized for delivering power among IEEE 118-bus systems. The converters can be treated as time-varying $P+jQ$ loads, which means both P and Q values are updated in every time-step to reveal the dynamic process of the converter stations. At the point of common coupling (PCC), the power injections of MMC stations can be calculated in (11.13) and introduced in the admittance matrix of the AC network.

$$Y_{Load} = \frac{(P_{Load} + j \cdot Q_{Load})}{V_{Bus}^2}. \tag{11.13}$$

The instantaneous voltages represented by a combination of amplitude U and phase angle θ at PCC are treated as the input of the EMT emulation. Hence, the mechanism of the interface strategy is as follows: the EMT emulation of MMC stations provide the time-varying $P+jQ$ loads to the AC transmission system at the coupling bus, and the updated $P+jQ$ loads act as the input of the AC grid for calculating the admittance matrix and subsequently solve the differential equations of synchronous generators and network equations in parallel. The resulting coupling bus phase voltages $U\angle\theta$ are in turn delivered to the DC grid. Meanwhile, step control should also be considered due to various time-steps in AC and DC grid, and the mechanism is similar to that of the flexible time-stepping algorithm. For instance, the DC and AC grids adopt a fixed time-step of 200 μs and 1 ms during and after a disturbance, respectively. Therefore, each AC grid computation is accompanied by five independent DC grid calculations for synchronization of time instants in both grids.

In the EMT-dynamic co-emulation of the hybrid AC–DC grid, the bus voltage angle can be directly obtained from the solution of the AC network equations, and therefore, the PLL is not used in the converter controller. Due to parallelism in the DC grid, the PLL has no impact on the emulation speed, albeit the hardware resource utilization increases slightly, as shown in Table 11.5. To emulate the integrated AC/DC grid with high accuracy and low hardware resource burden, and also due to the fact that the PLL can trace the PCC phase angle instantly, it is reasonable to adopt the calculated phase angle.

11.6.2.2 AC Grid Modeling

The main components in the AC grid are synchronous generators and the network. The synchronous generator model contains the mechanical part, electrical

Table 11.5 Specifics of major AC/DC grid hardware modules.

Module	BRAM	DSP	FF	LUT	Latency
MMCCNT	0	54	5 733	11 085	$85T_{clk}$
MMCAVM	16	208	8 292	19 277	$90T_{clk}$
DCGrid	0	17	2 209	2 868	$73T_{clk}$
PLL	0	8	892	1 357	$18T_{clk}$
YMatrix	12	1045	123 388	126 767	$1470T_{clk}$
RK4	0	36	6 970	7 520	$83T_{clk}$
FTS	0	12	521	780	$18T_{clk}$
Network	16	534	43 928	57 664	$269T_{clk}$
Governor	0	17	3 783	4 598	$29T_{clk}$
Update	0	38	4 951	6 875	$33T_{clk}$
Total	44	4758	753 450	848 458	$1873T_{clk}$
	1.02%	69.56%	31.87%	71.76%	—
XCVU9P FPGA	4320	6840	2 364 480	1 182 240	—

part, and excitation system. The transmission lines, compensators, as well as loads, contribute to the network.

Synchronous Machine Model The differential equation (11.5) represents the dynamics of a synchronous generator which has nine state variables [235]. Among the ninth-order DAEs, the mechanical equations which refer to the derivative of rotor angle and angular velocity are given below:

$$\dot{\delta} = \omega_R \cdot \Delta\omega(t), \tag{11.14}$$

$$\Delta\dot{\omega}(t) = \frac{1}{2H}[T_e(t) + T_m(t) - D \cdot \Delta\omega(t)]. \tag{11.15}$$

The rotor electrical circuit equations adopt two windings on the d-axis and two damping windings on the q-axis, which can be expressed as

$$\dot{\psi}_{fd}(t) = \omega_R \cdot [e_{fd}(t) - R_{fd} i_{fd}(t)], \tag{11.16}$$

$$\dot{\psi}_{1d}(t) = -\omega_R \cdot R_{1d} i_{1d}(t), \tag{11.17}$$

$$\dot{\psi}_{1q}(t) = -\omega_R \cdot R_{1q} i_{1q}(t), \tag{11.18}$$

$$\dot{\psi}_{2q}(t) = -\omega_R \cdot R_{2q} i_{2q}(t). \tag{11.19}$$

Meanwhile, the control system of the synchronous generator includes an AVR and PSS, and the differential equations have the following expression:

$$\dot{v}_1(t) = \frac{1}{T_R} \cdot [v_t(t) - v_1(t)], \tag{11.20}$$

$$\dot{v}_2(t) = K_{stab} \cdot \Delta\dot{\omega}(t) - \frac{1}{T_\omega} v_2(t), \tag{11.21}$$

$$\dot{v}_3(t) = \frac{1}{T_2} \cdot [T_1 \dot{v}_2(t) + v_2(t) - v_3(t)]. \tag{11.22}$$

Therefore, the vector $\mathbf{x}$ in (11.5) is a 9-D unit for a single generator, which can be formulated as follows:

$$\mathbf{x} = [\delta, \Delta\omega, \psi_{fd}, \psi_{1d}, \psi_{1q}, \psi_{2q}, v_1, v_2, v_3], \tag{11.23}$$

where the definition of constant parameters in (11.14)–(11.22), such as ω_R, H, D, R_{fd}, R_{1d}, R_{1q}, R_{2q}, T_R, K_{stab}, T_ω, T_1, and T_2 can be found in [235].

AC Network Model The lines, loads, and fixed compensators can be expressed as (11.13). The transmission lines are represented by a lumped π equivalent circuits. The DC system taken as loads at the AC–DC coupling points will be updated in every time-step. Therefore, all the components of the AC–DC network can be included in a single admittance matrix. Sharing the same admittance matrix, the generators are solved together with the AC network in the following equation:

$$\begin{bmatrix} \mathbf{I}_m \\ \mathbf{I}_r \end{bmatrix} = \begin{bmatrix} \mathbf{Y}_{mm} & \mathbf{Y}_{mr} \\ \mathbf{Y}_{rm} & \mathbf{Y}_{rr} \end{bmatrix} \begin{bmatrix} \mathbf{V}_m \\ \mathbf{V}_r \end{bmatrix}, \tag{11.24}$$

where m, r represent the number of synchronous generator nodes and the number of remaining nodes, respectively. Since the current injections remain zero on the nongenerator buses, i.e. $\mathbf{I}_r = [\mathbf{0}]$, the voltages of the remaining buses can be solved directly.

11.6.2.3 DC Grid Modeling

Modular Multilevel Converters Figure 11.31a shows the common half-bridge submodule based MMC which is treated as the converter station in an HVDC grid. The operating principle is as follows: once the upper IGBT is switched-on and the lower IGBT is switched-off the capacitor will charge or discharge depending on the direction of the arm current i_{arm}. On the contrary, when the upper IGBT is switched off and the lower IGBT is switched-on, the capacitor is bypassed. Therefore, the submodule voltage V_{SM} can be expressed as the following:

$$V_{SM} = \int \left(\frac{i_{arm}}{C}\right) dt + i_{arm} \cdot r_{on}, \ (V_{g1} = 1), \tag{11.25}$$

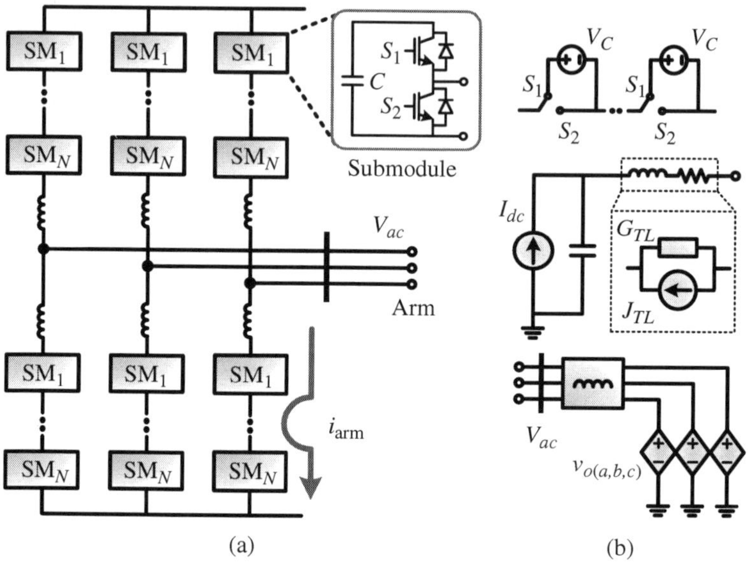

Figure 11.31 Modular multilevel converter model: (a) three-phase topology, (b) average value model.

where V_{g1} is a binary denoting the gate signal of the upper switch S_1, if $V_{g1} = 1$, the upper IGBT is switched-on, and r_{on} represents the on-state resistance of the switch.

MMC Average Value Model Under steady-state, assuming that the MMC is internally balanced, then each submodule can be treated as a voltage source as shown in Figure 11.31b. The equivalent voltage source can be expressed as

$$\begin{cases} v_{sm(i)}(t) = \frac{V_{dc}}{N}, \ (V_{g(i)}(t) \geq V_{th}), \\ v_{sm(i)}(t) = 0, \ (V_{g(i)}(t) < V_{th}), \end{cases} \tag{11.26}$$

where $V_{g(i)}(t)$ refers to a time-varying gate signal of the upper IGBT, V_{th} represents the threshold voltage, and V_{dc} is the sum of all capacitor voltages of an arm. The arm voltage is given as

$$v_{arm} = \sum_{i=1}^{N} v_{sm(i)} + L_{arm} \frac{di_{arm}}{dt}, \tag{11.27}$$

where L_{arm} represents the arm inductance.

Compared with the detailed MMC model, AVM has the feature of lower latency under hardware implementation. Based on the aforementioned interface strategy,

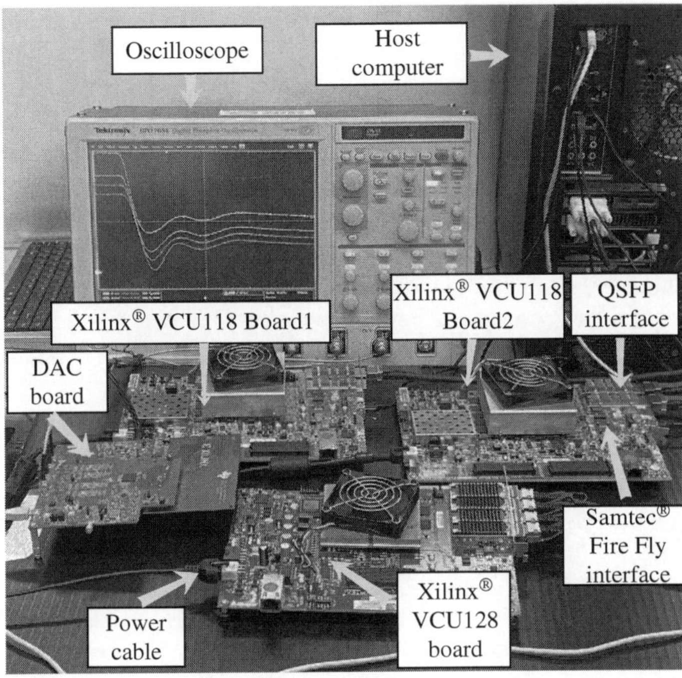

Figure 11.32 Hardware setup for FTRT dynamic-EMT emulation.

regardless of the influence of the MMC model, there is no significant difference in power injection from using the AVM or a detailed model. However, the detailed MMC model requires more hardware resources, as well as the detailed model is more sensitive to the time-step.

11.6.3 FTRT Emulation on FPGAs

The hardware emulation of the hybrid AC/DC grids was conducted on Xilinx Virtex® UltraScale+™ FPGAs. The integrated hardware platform includes three FPGA boards: two VCU118 evaluation boards featuring the XCVU9P FPGAs and a VCU128 board [225] with XCVU37P FPGA. Two IEEE 118-bus systems connecting with one 4-T HVDC system are accommodated on the VCU128 board. The remaining parts are emulated on the two VCU118 boards, as shown in Figure 11.30. The mechanism of the hardware design is as follows: the Xilinx® Vivado® package enables designing hardware modules by coding in C/C++ in its HLS tool to shorten the design cycle. The components or functions which consist of the integrated AC–DC grid in Figure 11.30 are transformed into a hardware module in HLS that can be imported into Vivado® to form the top-level using the hardware description language VHDL. Figure 11.32 shows the hardware configuration for the FTRT emulation.

In the integrated AC–DC grid, each subsystem and function can be designed as a reconfigurable hardware module, and the HIL emulation [236] is achieved following proper connection of these modules and deployment onto the FPGA boards. For the FTRT emulation, the initial conditions and the functions are downloaded from the host computer via the JTAG interface. The current operating conditions could be emulated on the FTRT platform based on the signals it receives. Each module reserves one or several input data channels for receiving real-time data. If a disturbance occurs, the received current operating conditions will be sent to the relative subsystems, such as the dynamic power injection data will be delivered to *Ymatrix* module and the output mechanical torque of steam turbine will be calculated in the module which represent the synchronous generator. With the required real-time data of the grid and the emulation model downloaded from the host computer, the FTRT predictive regulation control can be realized. The QSFP cables are used to interconnect the boards and the data communication among the FPGA boards utilizes Xilinx® Aurora protocol.

As per the AC–DC grid interface strategy, the VCU118 Board1 sends time-varying P and Q values calculated from the four-terminal HVDC system to the VCU128 Board, and the instantaneous phase voltages are in turn delivered to VCU118 Board1 via the bidirectional QSFP interface, as given in Figure 11.32. From each board's viewpoint, the P and Q values and the phase voltages are real-time measurements. In addition, the constantly updated P and Q values act as the inputs of the AC grid for calculating the admittance matrix, which can be treated as changing the network topology or dynamic conditions in every time-step.

Table 11.5 gives the hardware resources and the latencies of major modules on the VCU118 Board1, where MMCCNT refers to the control system of MMC stations, MMCAVM represents the functions of MMC average value model, and DCGrid denotes the network equations of the DC grid. The modules in the 4-T HVDC system are calculated in parallel. The latency of the DC grid is 90 T_{clk}, meaning with an EMT emulation time-step of 200 μs, the FTRT ratio is over 222 times under an FPGA clock cycle of 10 ns. Meanwhile, the *YMatrix*, *RK4*, *Network*, *Governor*, and *Update* denote the components in IEEE 118-bus system, where the proposed local equipment-based flexible time-stepping algorithm is applied on the AC grids. The performances of the proposed flexible time-stepping are relative to the contingencies. For instance, a severe fault may lead to a long-term oscillation until the system retains to steady-state, and the proposed algorithm will operate in a small time-step for a long period. Therefore, the total time-steps are different under various disturbances. Table 11.6 gives the FTRT ratio, and the total time-steps in the AC system under various conditions. Although a relatively small

Table 11.6 FTRT execution under various contingencies.

Contingencies	Hardware execution time of FTRT emulation (ms)	Emulation time span (s)	FTRT ratio
Load change	274.56	30	109.23
Three-phase fault	271.17	30	110.63
Gen. outage	296.18	30	101.29

time-step is adopted during or after the fault, a minimum FTRT ratio of 101 can be achieved by utilizing the proposed flexible time-stepping method. If the EMT emulation time-step reduced to 50 μs, the FTRT emulation can still be realized. However, the FTRT ratio could be highly dependent on the DC circuits, only 27 times faster than SCADA/real-time can be reached.

It is noticed that the hardware resources such as *DSP* and *LUT* are nearly full. The relationship between hardware resources and the number of synchronous generators is shown in Figure 11.33, where the synchronous machines are modeled in a set of ninth order DAEs, and each generator has a governor system. The governor is mainly responsible for providing mechanical power P_m (the same as T_m in per unit) to the synchronous generator. Meanwhile, the control of mechanical power is regulated by the steam turbine, which is a relatively slow process. The input of T_m in *RK4* is provided by the results of *Governor* from previous time-step, and thereby *RK4* and *Governor* modules can be calculated in parallel. Taking the IEEE 118-bus system as an example, the DSP utilization can be calculated as $(36 + 17) \times 54 + 1045 + 534 + 38 = 4479$, where the 54 refers to the number of generators. The DSP utilization of *RK4* and *Governor* modules are proportional to the synchronous machines. The remaining modules are solved in series in a specific time-step, such as *YMatrix* and *Network*. The hardware resources of those modules are directly related to the bus numbers, which increases along with grid nodes. Therefore, the hardware resources are nearly proportional to the synchronous machines, and a single VCU118 FPGA board is able to accommodate around 70 synchronous generators in parallel.

11.6.4 FTRT Emulation Results and Validation

Various contingencies of the integrated AC/DC grid in Figure 11.30 are emulated on the FTRT platform, and the results are validated by the off-line transient stability simulation tool TSAT® in the DSAToolsTM suite.

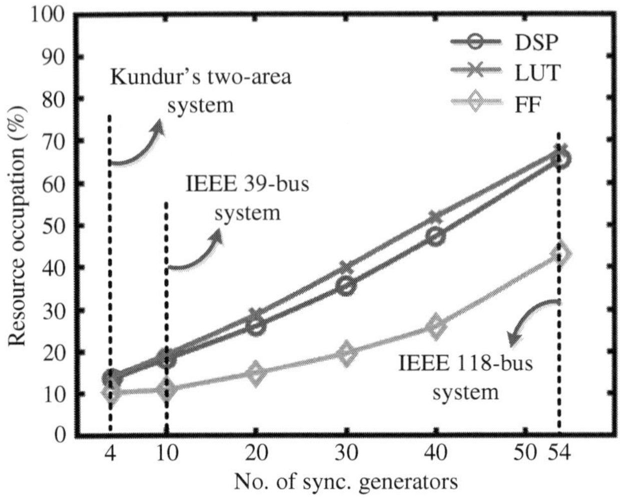

Figure 11.33 Relationship between hardware resources and synchronous generators.

11.6.4.1 Three-Phase-to-Ground Fault

At the time of five seconds, a three-phase-to-ground fault occurs at Bus 68 in *System 1*, as shown in Figure 11.30. The impacts after the disturbance are severe to the AC system, including the generators' rotor angles, bus voltages, and the frequencies, as shown in Figure 11.34. The synchronous generators lose synchronism after the ground fault as the relative rotor angles rise significantly as shown in Figure 11.34a,b. Meanwhile, in Figure 11.34c,d, the bus voltage of G_5 decreases to 0.67 p.u. in less than 0.1 second. The frequency of $G_1$0 plummets to 59.2 Hz which is beyond the threshold of $\pm 1\%$ as shown in Figure 11.34e,f. At $t_2 = 5.1$ seconds, the three-phase-to-ground fault is cleared, and the system return to the steady-state in about five seconds as given in Figure 11.34.

In a real power transmission system, a more than 100 ms ground fault is unwanted in a power control center. The long-term ground fault may lead to subsynchronous resonance in a series compensated transmission system and subsequent generator damage. Therefore, the flexible time-stepping–based FTRT emulation is applied to the control center to predict or mitigate the oscillations after a severe disturbance. The zoomed-in plots in Figure 11.34d and Figure 11.37a demonstrate that the time-steps shrink right after the bus voltages drop. At steady state, the time-step is 10 ms, once the fault occurs, the time-step reduces to the lower limit to maintain accuracy.

11.6.4.2 Generator Outage and Sudden Load Change

Figure 11.35 shows the results of the rotor angles, generator voltages, and the frequencies before and after the generator G_5 outage. The rotor angles of the generators decrease significantly after t_1, especially in G_5, which is shown in Figure 11.35a,b. Meanwhile, the generator voltages decrease to 0.83 p.u. in less than 0.2 second as shown in Figure 11.35c,d. Due to the robustness in a large power transmission system, the generator voltages and rotor angles may transfer to a new steady state. The main impact of the generator outage contingency is the divisions on the frequencies. Figure 11.35e,f present the frequencies drop to 59.7 Hz, which may cause the system to operate in a long-term unstable state.

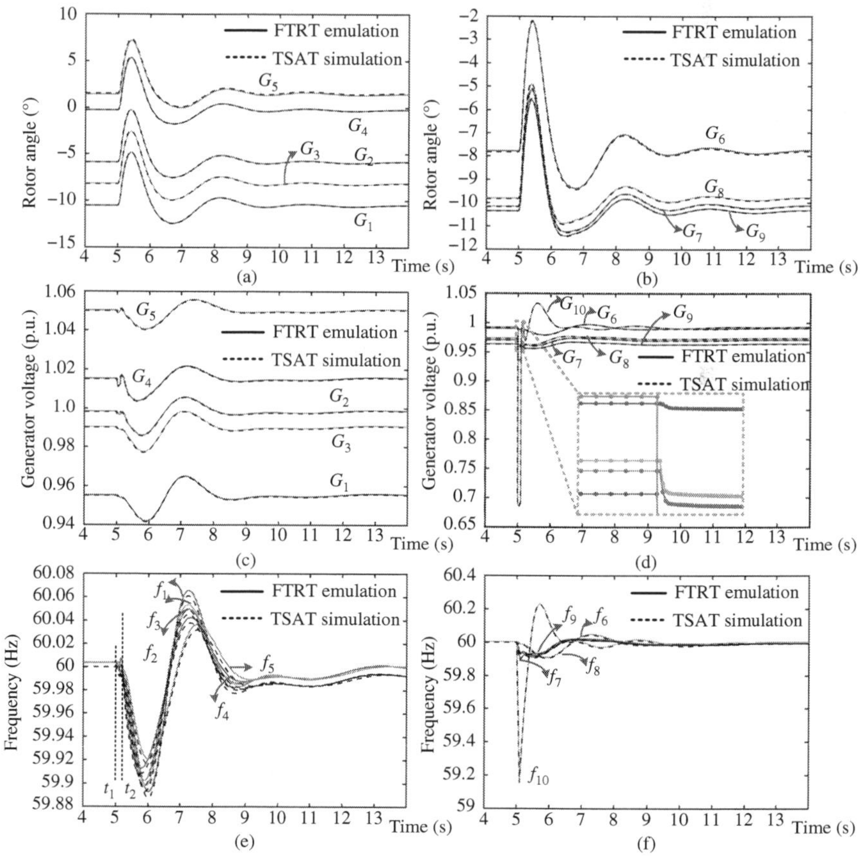

Figure 11.34 FTRT emulation using flexible time-stepping algorithm for preview of generator behaviors under AC system ground fault: (a) generator relative rotor angles (G_1-G_5), (b) generator relative rotor angles (G_6-G_9), (c) generator output voltages (G_1-G_5), (d) generator output voltages (G_6-G_10), (e) frequencies (G_1-G_5), and (f) frequencies (G_6-G_10).

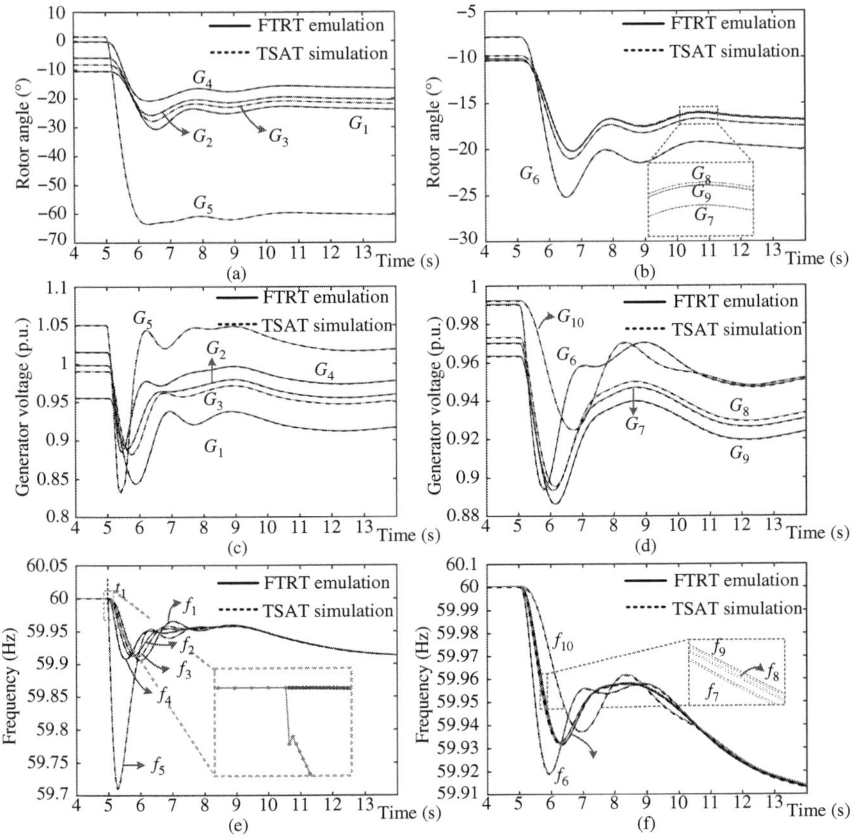

Figure 11.35 FTRT emulation using flexible time-stepping algorithm for preview of AC–DC grid behavior under generator outage: (a) generator relative rotor angles (G_1–G_5), (b) generator relative rotor angles (G_6–G_9), (c) generator output voltages (G_1–G_5), (d) generator output voltages (G_6–G_10), (e) frequencies (G_1–G_5), and (f) frequencies (G_6–G_10).

The zoomed-in plots in Figure 11.35e and Figure 11.37b illustrate the proposed flexible time-stepping algorithm can significantly change the time-steps after disturbance occurs to ensure the emulation accuracy. Meanwhile, a relatively large time-step is applied to accelerate the emulation.

The sudden load change disturbance happens at the load Buses 25 and 54 in IEEE 118-bus *System 1*. At $t_1 = 5.0$ seconds, a load of 200 MW and 100 MW are removed from Buses 25 and 54, respectively, causing the disturbance of the AC system which cannot be restored even the loads are recovered at $t_2 = 10$ seconds to its original capacity of 277 and 184 MW. The load change factor is relatively small in the large power transmission system, resulting in the slight changes in rotor

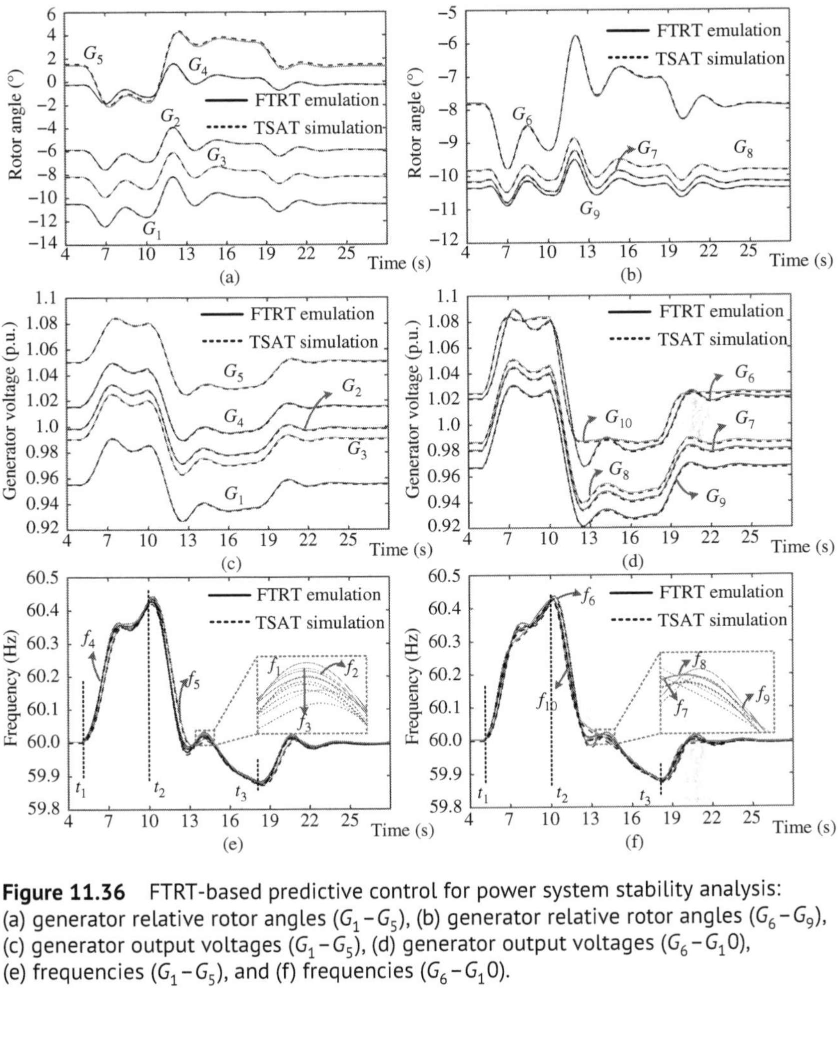

Figure 11.36 FTRT-based predictive control for power system stability analysis: (a) generator relative rotor angles ($G_1 - G_5$), (b) generator relative rotor angles ($G_6 - G_9$), (c) generator output voltages ($G_1 - G_5$), (d) generator output voltages ($G_6 - G_10$), (e) frequencies ($G_1 - G_5$), and (f) frequencies ($G_6 - G_10$).

angles and generator output voltages, as shown in Figure 11.36a–d. However, the main impacts of the load change contingency are the frequencies, which reach the maximum allowed threshold of 1% as shown in Figure 11.36e–f. The control center should select a proper strategy to mitigate the impacts after the load change. At $t_2 = 10.0$ seconds, *MMC*3 and *MMC*4 deliver extra 100 MW and 50 MW active power to Buses 25 and 54, respectively, from IEEE 118-bus *System 1* to *System 2*. At $t_3 = 18.0$ seconds, the extra power injection is removed, and the whole system is therefore restored stable gradually in about seven seconds. As a result, the introduction of the HVDC system improves the stability of the integrated AC/DC

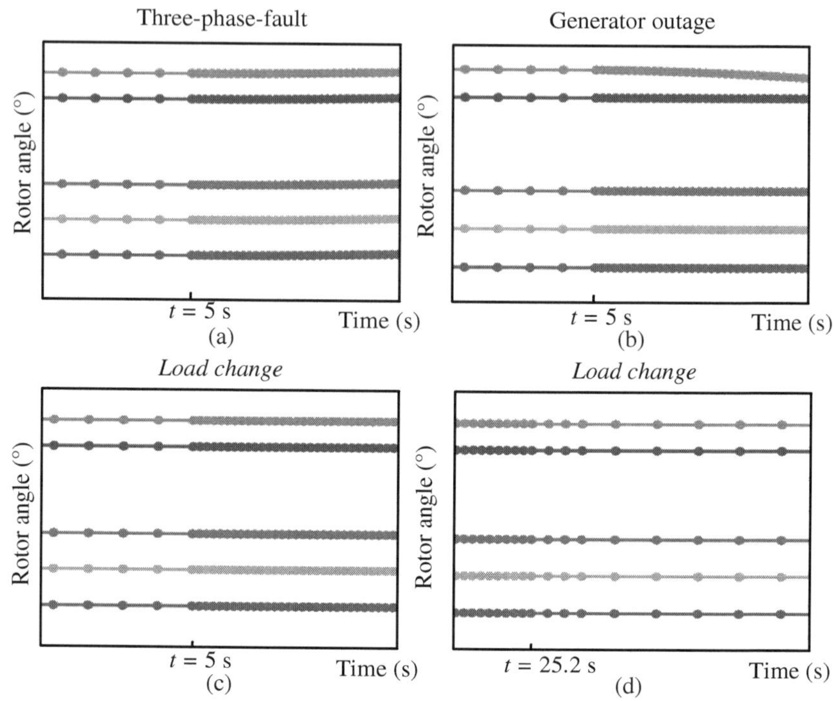

Figure 11.37 Zoomed-in plots of rotor angles for three contingencies: (a) three-phase-to-ground fault (G_1-G_5), (b) generator outage (G_1-G_5), (c), and (d) sudden load change (G_1-G_5).

grid by injecting active power into the AC grid or absorbing power from AC system. The zoomed-in plots in Figure 11.37c demonstrates that the proposed flexible time-stepping reduced the time-step to the lower limit for higher accuracy, and the time-step increases gradually when the system restores to steady-state after the power injection, as shown in Figure 11.37d.

11.7 Summary

Hardware emulation on FPGAs makes it possible to overcome many of the limitations related to the real-time EMT simulation of large-scale AC and DC networks by allowing the user to achieve such level of detail that is seldom achieved on traditional sequential processors. Facilitated by transmission delay, spatial partitioning resulting in numerous smaller subsystems particularly caters to the hardware parallelism of the FPGA board, and interconnected multiboard architectures.

This chapter first presented a functional decomposition methodology for hardware emulation on multiple FPGAs of power systems using detailed modeling of the individual components. The functional decomposition method circumvents the need for artificial lines to divide the system, reduces subsystem communication latencies, and is independent of a specific partitioning scheme. Furthermore, this method is ideally suited for achieving full parallelism and pipelining in the implemented system modules on the FPGAs. Although a single time-step is chosen to illustrate this method, multirate simulation with multiple time-steps for various functional components is clearly possible. Two case studies involving a three-phase 42-bus and a 420-bus power systems are presented to analyze the performance of the proposed hardware design with detailed modeling of the power system components. The achieved minimum time-steps for these systems are 11.55 μs and 36.12 μs on the 3-FPGA and the 10-FPGA hardware designs, respectively, on a clock frequency of 100 MHz. Smaller time-steps and even larger system sizes can be easily achieved as the design is fully scalable and extensible for an FPGA-cluster.

As an extended application of FPGA boards, the design and implementation of a hybrid real-time DC grid emulator on FPGA-MPSoC platform. It demonstrated that the hybrid modeling scheme and detailed hardware implementation of the real-time EMT emulation system of the benchmark CIGRÉ DC grid test system are entirely flexible and extensible. PSCAD/EMTDC® and SaberRD® were used to validate the results of system-level and device-level hardware emulation results. The improvement of the modeling schemes and implementation techniques can enlarge the functionality and the scope of EMT studies and provide comprehensive analyses of the AC/DC grid. Every new generation of FPGA technology features devices with a significant increase in logic resources and clock speed, as well as lower inter-board communication latency; consequently the scale of integrated AC/DC networks that could be deployed onto the hardware emulator should also increase to satisfy even the most demanding real-time EMT emulation.

For the analysis of CPS, this chapter also presented a RTCE framework on the FPGA-MPSoC hardware architecture. This approach is advantageous over the cumbersome software-based co-simulation of power and communication networks. The power system EMT emulation is carried out on programmable hardware units so that transients can be captured in real-time; and the communication network emulation is modeled in abstraction for transmission-level characteristics and directly conducted on real-world physical ports to form the specific networking functions, which can emulate the communication networking in real-time. The power system and communication network emulation modules are embedded within each FPGA and MPSoC board so that the data exchange and synchronization between the two domains are sufficiently fast for real-time interaction. The IEC-61850 communication model was applied, and the

IEEE 39-bus test system was implemented and emulated on FPGA-MPSoC platform. The hardware resource costs and processing latency were evaluated. System monitoring and control applications are executed within the soft processors of FPGA-MPSoC platform so that the system control could be programmed flexibly to respond to the physical system events such as faults or communication link failures. This RTCE framework can be extended to large-scale AC–DC networks with integrated renewable energy resources [227] for the implementation of ICT-enabled global control schemes on heterogeneous hardware.

Finally, a faster-than-real-time hybrid EMT-dynamic emulation of AC–DC grid is presented. The local equipment based flexible time-stepping algorithm has the ability to modify the time-steps ranging from 1 to 10 ms, which facilitated the FTRT emulation of a complex power grid on the FPGAs. The flexible scheme can instantly find the appropriate time-step and it reduces the computational burden by utilizing large time-steps under steady-state and small time-steps during or after the disturbance. On the other hand, the parallel architecture of the hardware platform makes the transient stability emulation more efficient than traditional CPU based simulators. With the minimum FTRT ratio of 101, the energy control center has sufficient time to assess the dynamic security, predict power system responses, and select an optimal solution to maintain the system stability. The results of the hybrid EMT-dynamic emulation are highly matched with the off-line dynamic simulation tool TSAT®. When more system contingencies need to be considered, the FTRT platform can be applied to accelerate contingencies screening and ranking in DSA systems.

Bibliography

1 Alfke, P., Bolsens, I., Carter, B. et al. (2011). It's an FPGA! *IEEE Solid-State Circuits Magazine* 3 (4): 15–20.

2 Fawcett, B. (1996). FPGAs as reconfigurable processing elements. *IEEE Circuits and Devices Magazine* 12 (2): 8–10.

3 Trimberger, S.M.S. (2018). Three ages of FPGAs: a retrospective on the first thirty years of FPGA technology. *IEEE Solid-State Circuits Magazine* 10 (2): 16–29.

4 Dini, M. (2018). The practical applications and uses of FPGAs: insights from historical, current, and future perspectives. *IEEE Solid-State Circuits Magazine* 10 (2): 36–45.

5 Xilinx (2016). *Unleash the Unparalleled Power and Flexibility of Zynq UltraScale+ MPSoCs*. Xilinx, Inc.

6 Xilinx (2019). *'Versal: The First Adaptive Compute Acceleration Platform (ACAP)*. Xilinx, Inc.

7 Caulfield, A.M., Chung, E.S. and Putnam, A. et al. (2017). Configurable Clouds. *IEEE Micro* 37 (3): 52–61.

8 Kilts, S. (2007). *Advanced FPGA Design: Architecture, Implementation, and Optimization*. Wiley-IEEE Press.

9 Farooq, U., Marrakchi, Z., and Mehrez, H. (2012). *Tree-Based Heterogeneous FPGA Architectures*. New York: Springer.

10 Xilinx (2014). *VC707 Evaluation Board User Guide (UG885)*. San Jose, CA: Xilinx Inc.

11 Xilinx (2018). *VCU118 Evaluation Board User Guide (UG1224)*. San Jose, CA: Xilinx Inc.

12 Xilinx (2018). *7 Series FPGAs Data Sheet: Overview*. Xilinx, Inc.

13 Xilinx (2018). *UltraScale Architecture and Product Data Sheet: Overview*. Xilinx, Inc.

14 Xilinx (2016). *7 Series FPGAs Configurable Logic Block*. Xilinx, Inc.

Real-Time Electromagnetic Transient Simulation of AC-DC Networks, First Edition.
Venkata Dinavahi and Ning Lin.

15 Xilinx (2017). *UltraScale Architecture Configurable Logic Block User Guide*. Xilinx, Inc.

16 Xilinx (2016). *7 Series FPGAs Memory Resources*. Xilinx, Inc.

17 Xilinx (2019). *UltraScale Architecture Memory Resources User Guide*. Xilinx, Inc.

18 Xilinx (2018). *7 Series DSP48E1 Slice User Guide*. Xilinx, Inc.

19 Xilinx (2018). *UltraScale Architecture DSP Slice User Guide*. Xilinx, Inc.

20 Xilinx (2016–2018). *Zynq UltraScale+ MPSoCs Product Tables and Product Selection Guide*. Xilinx, Inc.

21 Xilinx (2019). *ZCU102 Evaluation Board User Guide*. Xilinx, Inc.

22 Xilinx (2017). *Aurora 64B/66B LogiCORE IP Product Guide*, PG074 (v11.2). Xilinx, Inc.

23 Nie, X., Chen, Y., and Dinavahi, V. (2007). Real-time transient simulation based on a robust two-layer network equivalent. *IEEE Transactions on Power Systems* 22 (4): 1771–1781.

24 Chen, Y. and Dinavahi, V. (2014). Hardware emulation building blocks for large-scale real-time simulation of power grids. *IEEE Transactions on Industrial Informatics* 10 (1): 373–381.

25 Gear, C.W. (1971). *Numerical Initial Value Problems in Ordinary Differential Equations*. Prentice-Hall Inc.

26 Dommel, H.W. (1969). Digital computer solution of electromagnetic transients in single and multiphase networks. *IEEE Transactions on Power Apparatus and Systems* 88: 388–395.

27 Dommel, H.W. (1984). *EMTP Theory Book*. Portland, OR: Bonneville Power Administration.

28 Greenwood, A. (1991). *Electrical Transients in Power Systems*, 2e. New York: Wiley Interscience.

29 Bewley, L.V. (1951). *Travelling Waves on Transmission Systems*, 2e. New York: Wiley.

30 Budner, A. (1970). Introduction of frequency-dependent line parameters into an electromagnetics transients program. *IEEE Transactions on Power Apparatus and Systems* PAS-89: 88–97.

31 Meyer, W.S. and Dommel, H.W. (1974). Numerical modeling of frequency-dependent transmission line parameters in an electromagnetic transients program. *IEEE Transactions on Power Apparatus and Systems* PAS-99: 1401–1409.

32 Semlyen, A. and Dabuleanu, A. (1975). Fast and accurate switching transient calculations on transmission lines with ground return using recursive convolutions. *IEEE Transactions on Power Apparatus and Systems* PAS-94 (2): 561–571.

33 Marti, J.R. (1982). Accurate modeling of frequency-dependent transmission lines in electromagnetic transient simulations. *IEEE Transactions on Power Apparatus and Systems* PAS-101 (1): 147–155.

34 Marti, L. (1988). Simulation of transients in underground cables with frequency-dependent modal transformation matrices. *IEEE Transactions on Power Delivery* 3 (3): 1099–1110.

35 Noda, T., Nagaoka, N., and Ametani, A. (1996). Phase domain modeling of frequency-dependent transmission lines by means of an ARMA model. *IEEE Transactions on Power Delivery* 11 (1): 401–411.

36 Nguyen, H.V., Dommel, H.W., and Marti, J.R. (1997). Direct phase-domain modeling of frequency-dependent overhead transmission lines. *IEEE Transactions on Power Delivery* 12 (3): 1335–1342.

37 Morched, A., Gustavsen, B., and Tartibi, M. (1999). A universal model for accurate calculation of electromagnetic transients on overhead lines and underground cables. *IEEE Transactions on Power Delivery* 14 (3): 1032–1038.

38 Bergeron, L. (1961). *Water Hammer in Hydraulics and Wave Surges in Electricity*. New York: Wiley.

39 Hedman, D.E. (1965). Propagation on overhead transmission lines: I- Theory of modal analysis and II- Earth conduction effects and practical results. *IEEE Transactions on Power Apparatus and Systems* 84 (3): 200–211.

40 Chen, Y. and Dinavahi, V. (2009). FPGA-based real-time EMTP. *IEEE Transactions on Power Delivery* 24 (2): 892–902.

41 Gustavsen, B. and Semlyen, A. (1998). Simulation of transmission line transients using vector fitting and modal decomposition. *IEEE Transactions on Power Delivery* 13 (2): 605–614.

42 Chua, L.O. (1971). Efficient computer algorithms for piecewise linear analysis of resistive nonlinear networks. *IEEE Transactions on Circuit Theory* 18 (1): 73–85.

43 Dommel, H.W. (1971). Nonlinear and time-varying elements in digital simulation of electromagnetic transients. *IEEE Transactions on Power Apparatus and Systems* 90 (4): 2561–2567.

44 Chen, Y. and Dinavahi, V. (2011). An iterative real-time nonlinear electromagnetic transient solver on FPGA. *IEEE Transactions on Industrial Electronics* 58 (6): 2547–2555.

45 Lachowicz, S. and Pfleiderer, H. (2008). Fast evaluation of the square root and other nonlinear functions in FPGA. *Proceedings of 4th IEEE International Symposium on Electronic Design Teat and Applications*, Hongkong, China, pp. 1–4.

46 Boullis, N., Mencer, O., Luk, W., and Styles, H. (2001). Pipelined function evaluation on FPGAs. *Proceedings of 9th IEEE Field-Programmable Custom Computing Machines*, pp. 1–3.

47 Martinez, J.A. and Mork, B.A. (2007). Transformer modeling for low and midfrequency transients - a review. *IEEE Transactions on Power Delivery* 22 (2): 1235–1246.

48 Brandwajn, V., Dommel, H.W., and Dommel, I. (1982). Matrix representation of three-phase n-winding transformers for steady-state and transient studies. *IEEE Transactions on Power Apparatus and Systems* PAS-101 (6): 1369–1378.

49 Narang, A. and Brierley, R.H. (1994). Topology based magnetic model for steady-state and transient studies for three-phase core type transformers. *IEEE Transactions on Power Systems* 9 (3): 1337–1349.

50 Mayergoyz, I.D. (1991). *Mathematical Models of Hysteresis*. Berlin: Springer-Verlag.

51 Naidu, S.R. (1990). Simulation of the hysteresis phenomenon using Preisach's theory. *Proceedings of the Institution of Electrical Engineers A* 137 (2): 73–79.

52 Avila-Rosales, J. and Semlyen, A. (1985). Iron core modeling for electrical transients. *IEEE Transactions on Power Apparatus and Systems* PAS-104 (11): 3189–3194.

53 Tarasiewicz, E.J., Morched, A.S., Narang, A., and Dick, E.P. (1993). Frequency dependent eddy current models for nonlinear iron cores. *IEEE Transactions on Power Systems* 8 (2): 588–597.

54 Liu, J. and Dinavahi, V. (2014). A real-time nonlinear hysteretic power transformer transient model on FPGA. *IEEE Transactions on Industrial Electronics* 61 (7): 3587–3597.

55 Sen, K.K. and Sen, M.L. (2003). Introducing the family of "Sen" transformer: a set of power flow controlling transformers. *IEEE Transactions on Power Delivery* 18 (1): 149–157.

56 Sen, K.K. and Sen, M.L. (2003). Comparision of the "Sen" transformer with the unified power flow controller. *IEEE Transactions on Power Delivery* 18 (4): 1523–1533.

57 Enright, W., Nayak, O.B., Irwin, G.D., and Arrillaga, J. (1997). An electromagnetic transients model of multi-limb transformers using normalized core concept. *Proceedings of the International Conference on Power Systems Transients*, Seattle, WA, June 1997, pp. 93–98.

58 Liu, J. and Dinavahi, V. (2016). Detailed magnetic equivalent circuit based real-time nonlinear power transformer model on FPGA for electromagnetic transient studies. *IEEE Transactions on Industrial Electronics* 63 (2): 1191–1202.

59 Faruque, M.O. and Dinavahi, V. (2007). A tap-changing algorithm for the implementation of "Sen" transformer. *IEEE Transactions on Power Delivery* 22 (3): 1750–1757.

60 Asghari, B., Faruque, M.O., and Dinavahi, V. (2008). Detailed real-time transient model of the "Sen" transformer. *IEEE Transactions on Power Delivery* 23 (3): 1513–1521.

61 Hayt, W.H. and Buck, J.A. (2001). *Engineering electromagnetics*. McGraw Hill.

62 Liu, J. and Dinavahi, V. (2016). Nonlinear magnetic equivalent circuit based real-time Sen transformer electro- magnetic transient model on FPGA for HIL emulation. *IEEE Transactions on Power Delivery* 31 (6): 2483–2493.

63 Liu, P. and Dinavahi, V. (2018). Real-time finite element simulation of electromagnetic transients of transformer on FPGA. *IEEE Transactions on Power Delivery* 33 (4): 1991–2001.

64 Zienkiewicz, O.C., Taylor, R.L., and Zhu, J.Z. (2005). *The Finite Element Method*, 6e. Amsterdam, London: Elsevier Butterworth-Heinemann.

65 Fuzi, J. (2003). Analytical approximation of Preisach distribution functions. *IEEE Transactions on Magnetics* 39 (3): 1357–1360.

66 Johns, P.B. and Beurle, R.L. (1971). Numerical solution of 2-dimensional scattering problems using a transmission-line matrix. *Proceedings of the Institution of Electrical Engineers* 118 (9): 1203–1208.

67 Johns, P.B. and O'Brien, M. (1980). Use of the transmission-line modelling (t.l.m.) method to solve non-linear lumped networks. *Radio and Electronic Engineer* 50 (1–2): 59–70.

68 Hoefer, W.J.R. (1985). The transmission-line matrix method - theory and applications. *IEEE Transactions on Microwave Theory and Techniques* 33 (10): 882–893.

69 Hui, S.Y.R. and Christopoulos, C. (1990). Physical science, measurement and instrumentation, management and education. *IEE Proceedings A (Physical Science, Measurement and Instrumentation, Management and Education)* 137 (6): 379–384.

70 Dinavahi, V. (2004). Transient analysis of systems with multiple nonlinear elements using TLM. *IEEE Transactions on Power Systems* 19 (4): 2102–2103.

71 Asghari, B. and Dinavahi, V. (2011). Real-time nonlinear transient simulation based on optimized transmission line modeling. *IEEE Transactions on Power Systems* 26 (2): 699–709.

72 Lobry, J., Trecat, J., and Broche, C. (1996). The transmission line modeling (TLM) method as a new iterative technique in nonlinear 2-D magnetostatics. *IEEE Transactions on Magnetics* 32 (2): 559–566.

73 Deblecker, O., Lobry, J., and Broche, C. (1998). Use of transmission-line modelling method in FEM for solution of nonlinear eddy-current problems. *IEE Proceedings - Science, Measurement and Technology* 145 (1): 31–38.

74 Asghari, B., Dinavahi, V., Rioual, M. et al. (2009). Interfacing techniques for electromagnetic field and circuit simulation programs IEEE task force on interfacing techniques for simulation tools. *IEEE Transactions on Power Delivery* 24 (2): 939–950.

75 Lombard, P. and Meunier, G. (1992). A general method for electric and magnetic coupled problem in 2D and magnetodynamic domain. *IEEE Transactions on Magnetics* 28 (2): 1291–1294.

76 Wang, L., Jatskevich, J., Dinavahi, V. et al. (2010). Methods for interfacing rotating machine models with transient simulation pro- grams. *IEEE Transactions on Power Delivery* 25 (2): 891–903.

77 Lauw, H.K. and Meyer, W.S. (1982). Universal machine modeling for the representation of rotating electric machinery in an electromagnetic transients program. *IEEE Transactions on Power Apparatus and Systems* 101 (6): 1342–1350.

78 Lauw, H.K. (1985). Interfacing for universal multi-machine system modeling in an electromagnetic transients program. *IEEE Transactions on Power Apparatus and Systems* 104 (9): 2367–2373.

79 Chen, Y. and Dinavahi, V. (2012). Digital hardware emulation of universal machine and universal line models for real-time electromagnetic transient simulation. *IEEE Transactions on Industrial Electronics* 59 (2): 1300–1309.

80 Monmasson, E. and Cirstea, M.N. (2007). FPGA design methodology for industrial control systems - a review. *IEEE Transactions on Industrial Electronics* 54 (4): 1824–1842.

81 Krause, P.C., Wasynczuk, O., and Sudhoff, S. (1994). *Analysis of Electric Machinery and Drive Systems*. New York: IEEE Press.

82 Korn, G.A. and Korn, T.M. (1961). *Mathematical Handbook for Scientists and Engineers: Definitions, Theorems, and Formulas for Reference and Review*. McGraw Hill Inc.

83 Tavana, N.R. and Dinavahi, V. (2015). A general framework for FPGA-based real-time emulation of electrical machines for HIL applications. *IEEE Transactions on Industrial Electronics* 62 (4): 2041–2053.

84 Xilinx, Inc. (2012). *Xilinx System Generator*. Xilinx User Guide, October 2012.

85 Sudhoff, S.D., Kuhn, B.T., Corzine, K.A., and Branecky, B.T. (2007). Magnetic equivalent circuit modeling of induction motors. *IEEE Transactions on Energy Conversion* 22 (2): 259–270.

86 Amrhein, M. and Krein, P.T. (2009). 3-D magnetic equivalent circuit framework for modeling electromechanical devices. *IEEE Transactions on Energy Conversion* 24 (2): 397–405.

87 Perho, J. (2002). Reluctance network for analysing induction machines. *Acta Polytechnica Scandinavica / Electrical Engineering Series* 110: 1–147.

88 Derbas, H.W., Williams, J.M., Koenig, A.C., and Pekarek, S.D. (2009). A comparison of nodal- and mesh-based magnetic equivalent circuit models. *IEEE Transactions on Energy Conversion* 24 (2): 388–396.

89 Tavana, N.R. and Dinavahi, V. (2016). Real-time nonlinear magnetic equivalent circuit model of induction machine on FPGA for hardware-in-the-loop simulation. *IEEE Transactions on Energy Conversion* 31 (2): 520–530.

90 Neagoe, C. and Ossart, F. (1994). Analysis of convergence in nonlinear magnetostatic finite elements problems. *IEEE Transactions on Magnetics* 30 (5): 2865–2868.

91 Jandaghi, B. and Dinavahi, V. (2017). Prototyping of nonlinear time-stepped finite element simulation of linear induction machines on parallel reconfigurable hardware. *IEEE Transactions on Industrial Electronics* 64 (10): 7711–7720.

92 Jandaghi, B. and Dinavahi, V. (2018). Real-time FEM computation of nonlinear magnetodynamics of moving structures on FPGA for HIL emulation. *IEEE Transactions on Industrial Electronics* 65 (10): 7709–7718.

93 Horowitz, S.H. and Phadke, A.G. (2008). *Power System Relaying*, 3e. Wiley.

94 ALSTOM (2012). *Network Protection & Automation Guide.*

95 IEEE Power System Relaying Committee (2009). Understanding microprocessor-based technology applied to relaying. *Power System Relaying Committee, Report of Working Group I-01*, January 2009.

96 Anderson, D., Zhao, C., Hauser, C. et al. (2012). A virtual smart grid. *IEEE Power and Energy Magazine* 10 (1): 49–57.

97 Podmore, R. and Robinson, M.R. (2010). The role of simulators for smart grid development. *IEEE Transactions on Smart Grid* 1 (2): 205–212.

98 Qi, H., Wang, X., Tolbert, L.M. et al. (2011). A resilient real-time system design for a secure and reconfigurable power grid. *IEEE Transactions on Smart Grid* 2 (4): 770–781.

99 Yinger, R.J., Venkata, S.S., and Centeno, V.A. (2012). Southern California Edison's advanced distribution protection demonstrations. *IEEE Transactions on Smart Grid* 3: 1012–1019.

100 Borowski, J.F., Hopkinson, K.M., Humphries, J.W., and Borghetti, B.J. (2011). Reputation-based trust for a cooperative agent-based backup protection scheme. *IEEE Transactions on Smart Grid* 2 (2): 287–301.

101 Liu, H., Chen, X., Yu, K., and Hou, Y. (2012). The control and analysis of self-healing urban power grid. *IEEE Transactions on Smart Grid* 3 (3): 1119–1129.

102 Wang, Y. and Dinavahi, V. (2016). Real-time digital multi-function protection system on reconfigurable hard- ware. *IET Generation, Transmission and Distribution* 10 (10): 2295–2305.

103 Volder, J.E. (1959). The CORDIC trigonometric computing technique. *IRE Transactions on Electronic Computers* EC-8 (3): 330–334.

104 Walther, J.S. (1971). A unified algorithm for elementary functions. *AFIPS Spring Joint Computer Conferences* 38: 379–385.

105 Glover, J.D., Sarma, M.S., and Overbye, T.J. (2012). *Power System Analysis and Design*. Cengage Learning.

106 Gu, J.C. and Yu, S.L. (2000). Removal of DC offset in current and voltage signals using a novel Fourier filter algorithm. *IEEE Transactions on Power Delivery* 15 (1): 73–79.

107 Begovic, M.M., Djuric, P.M., Dunlap, S., and Phadke, A.G. (1993). Frequency tracking in power networks in the presence of harmonics. *IEEE Transactions on Power Delivery* 8 (2) 480–486.

108 Golub, G.H. and Van Loan, C.F. (1996). *Matrix Computations*, 3e. The Johns Hopkins University Press.

109 GE Digital Energy. Line Protection with Overcurrent Relays. http://www.gedigitalenergy.com/multilin/notes/artsci/art13.pdf (accessed 20 January 2021).

110 Blackburn, J.L. and Domin, T.J. (2006). *Protective Relaying Principles and Applications*, 3e. Taylor and Francis Group, LLC.

111 Horak, J. (2006). Directional overcurrent relaying (67) concepts. *59th Annual Conference for Protective Relay Engineers*, p. 13.

112 Zimmerman, K. and Costello, D. (2010). Fundamentals and improvements for directional relays. *63rd Annual Conference for Protective Relay Engineers*, March-April 2010, pp. 1–12.

113 Anderson, P.M. (1999). *Power System Protection*. New York: McGraw-Hill.

114 Schweitzer, E.O. III and Roberts, J. (1993). *Distance Relay Element Design*. Schweitzer Engineering Laboratories, Inc.

115 Wang, Y. and Dinavahi, V. (2014). Low-latency distance protective relay on FPGA. *IEEE Transactions on Smart Grid* 5 (2): 896–905.

116 IEEE Std C37.102-2006 (2013). *IEEE guide for AC generator protection*, 21 August 2013.

117 PSpice® (2000). *User's Guide*. Cadence Design Systems, Inc.

118 HSPICE® (2008). *User Guide: Simulation and Analysis, Version B-2008.09*. Synopsys, Inc.

119 Saber® (2004). *User Guide, Version V-2004.06-SP1*. Synopsys, Inc.

120 Shin, K. and Ramanathan, P. (1994). Real-time computing: a new discipline of computer science and engineering. *Proceedings IEEE* 82 (1): 6–24.

121 EMTDC® (2010). *User's Guide: A Comprehensive Resource for EMTDC, Version 4.7*. Manitoba HVDC Research Centre.

122 Soderlind, G. and Wang, L. (2006). Adaptive time-stepping and computational stability. *Journal of Computational and Applied Mathematics* 185 (2): 225–243.

123 Deml, C. and Turkes, P. (1999). Fast simulation technique for power electronic circuits with widely different time constants. *IEEE Transactions on Industry Applications* 35 (3): 657–662.

124 Lauritzen, P.O. and Ma, C.L. (1991). A simple diode model with reverse recovery. *IEEE Transactions on Power Electronics* 6 (2): 188–191.

125 IEEE Std C62.22-2009 (2009). *IEEE guide for the application of metal-oxide surge arresters for alternating-current systems*. 3 July 2009, pp. 1–142.

126 Standler, R.B. (1995). *Protection against lightning electromagnetic impulse - Part I: General principles*, IEC Standard 1312-I, February 1995.

127 Golub, G.H. and Van Loan, C.F. (1996). *Matrix Computations*, 3e. Baltimore, MD: Johns Hopkins University Press.

128 Shen, Z. and Dinavahi, V. (2017). Dynamic variable time-stepping schemes for real-time FPGA-based nonlinear electromagnetic transient emulation. *IEEE Transactions on Industrial Electronics* 64 (5): 4006–4016.

129 Wang, L. and Jatskevich, J. (2006). A voltage-behind-reactance synchronous machine model for the EMTP-type solution. *IEEE Transactions on Power Systems* 21 (4): 1539–1549.

130 Duan, T. and Dinavahi, V. (2020). Adaptive time-stepping universal line and universal machine models for real- time and faster-than-real-time hardware emulation. *IEEE Transactions on Industrial Electronics* 67 (8): 6173–6182.

131 Xilinx (2018). *Aurora 64B/66B LogiCORE IP Product Guide, PG074 (V11.2)*. San Jose, CA: Xilinx Inc.

132 CIGRE (2014). Guide for the development of models for HVDC converters in a HVDC grid. Working Group B4.57.

133 Sheng, K., Williams, B.W., and Finney, S.J. (2000). A review of IGBT models. *IEEE Transactions on Power Electronics* 15 (6): 1250–1266.

134 Dinavahi, V.R., Iravani, M.R., and Bonert, R. (2001). Real-time digital simulation of power electronic apparatus interfaced with digital controllers. *IEEE Transactions on Power Delivery* 16 (4): 775–781.

135 Dinavahi, V., Iravani, R., and Bonert, R. (2004). Design of a real-time digital Simulator for a D-STATCOM system. *IEEE Transactions on Industrial Electronics* 51 (5): 1001–1008.

136 Beddard, A., Sheridan, C.E., Barnes, M., and Green, T.C. (2016). Improved accuracy average value models of modular multilevel converters. *IEEE Transactions on Power Delivery* 31 (5): 2260–2269.

137 Bazzano, G., Cavallaro, D.G., and Greco, G. (2011). An analog behavioral thermal macro-model aimed at representing an elementary portion of a discrete IGBT power device. *2011 17th THERMINIC*, Paris, France, September 2011, pp. 1–6.

138 Hsu, J. and Ngo, K.D.T. (1996). Behavioral modeling of the IGBT using the Hammerstein configuration. *IEEE Transactions on Power Electronics* 11 (6): 746–754.

139 Courtay, A. (1995). MAST power diode and thyristor models including automatic parameter extraction. *SABER User Group Metting*, Brighton, UK, September 1995.

140 Lin, N., Shi, B., and Dinavahi, V. (2018). Non-linear behavioural modelling of device-level transients for complex power electronic converter circuit hardware realisation on FPGA. *IET Power Electronics* 11 (9): 1566–1574.

141 Zhang, M., Courtay, A., and Yang, Z. (2000). An improved behavioral IGBT model and its characterization tool. *Proceedings IEEE Electron Devices Meeting*, Hong Kong, pp. 142–145.

142 Synopsys (2009). *Saber Model Architect Tool User Guide*. Synopsys, Inc.

143 Ma, C.L. and Lauritzen, P.O. (1993). A simple power diode model with forward and reverse recovery. *IEEE Transactions on Power Electronics* 8 (4): 342–346.

144 Hefner, A.R. and Diebolt, D.M. (1994). An experimentally verifed IGBT model implemented in the Saber circuit simulator. *IEEE Transactions on Power Electronics* 9 (5): 532–542.

145 Oziemkiewicz, G.T. (1995). Implementation and development of the NIST IGBT model in a SPICE-based commercial circuit simulator. MSc dissertation. University of Florida.

146 Chua, L.O. and Lin, P. (1975). *Computer-Aided Analysis of Electronic Circuits: Algorithms and Computational Techniques*. Englewood Cliffs, NJ: Prentice-Hall, Inc.

147 Wang, W., Shen, Z., and Dinavahi, V. (2014). Physics-based device-level power electronic circuit hardware emulation on FPGA. *IEEE Transactions on Industrial Informatics* 10 (4): 2166–2179.

148 Lin, N. and Dinavahi, V. (2017). Behavioral device-level modeling of modular multilevel converters in real time for variable-speed drive applications. *IEEE Journal on Emerging and Selected Topics in Power Electronics* 5 (3): 1177–1191.

149 Parma, G.G. and Dinavahi, V. (2007). Real-time digital hardware simulation of power electronics and drives. *IEEE Transactions on Power Delivery* 22 (2): 1235–1246.

150 Myaing, A. and Dinavahi, V. (2011). FPGA-based real-time emulation of power electronic systems with detailed representation of device characteristics. *IEEE Transactions on Industrial Electronics* 58 (1): 358–368.

151 Myaing, A., Faruque, M.O., Dinavahi, V., and Dufour, C. (2012). Comparison of insulated gate bipolar transistor models for FPGA-based real-time simulation of electric drives and application guideline. *IET Power Electronics* 5 (3): 293–303.

152 Lin, N. and Dinavahi, V. (2018). Dynamic electro-magnetic-thermal modeling of MMC-based DC–DC converter for real-time simulation of MTDC grid. *IEEE Transactions on Power Delivery* 33 (3): 1337–1347.

153 Shen, Z. and Dinavahi, V. (2016). Real-time device-level transient electrothermal model for modular multilevel converter on FPGA. *IEEE Transactions on Power Electronics* 31 (9): 6155–6168.

154 Infineon (2018). FZ400R33KL2C_B5 datasheet. https://www.infineon.com/cms/en/product/power/igbt/ (accessed 20 January 2021).

155 ABB Library (2012). ABB applying IGBTs. http://www.abb.com/abblibrary/DownloadCenter/ (accessed 20 January 2021).

156 Liang, T. and Dinavahi, V. (2018). Real-time system-on-chip emulation of electrothermal models for power electronic devices via hammerstein configuration. *IEEE Journal of Emerging and Selected Topics in Power Electronics* 6 (1): 203–218.

157 Liang, T. and Dinavahi, V. (2018). Real-time device-level simulation of MMC-based MVDC traction power system on MPSoC. *IEEE Transactions on Transportation Electrification* 4 (2): 626–641.

158 Huang, Z. and Dinavahi, V. (2019). An efficient hierarchical zonal method for large-scale circuit simulation and its real-time application on more electric aircraft microgrid. *IEEE Transactions on Industrial Electronics* 66 (7): 5778–5786.

159 Liang, T., Huang, Z., and Dinavahi, V. (2020). Adaptive real-time hybrid neural network-based device-level modeling for DC traction HIL application. *IEEE Access* 8: 69543–69556.

160 Gnanarathna, U.N., Gole, A.M., and Jayasinghe, R.P. (2011). Efficient modeling of modular multilevel HVDC converters (MMC) on electromagnetic transient simulation programs. *IEEE Transactions on Power Delivery* 26 (1): 316–324.

161 Rohner, S., Bernet, S., Hiller, M., and Sommer, R. (2010). Modulation, losses, and semiconductor requirements of modular multilevel converters. *IEEE Transactions on Industrial Electronics* 57 (8): 2633–2642.

162 Infineon (2018). Technical information, Infineon IGBT modules FZ400R33KL2C_B5. https://www.infineon.com (accessed 20 January 2021).

163 Fung, K.K. and Hui, S.Y.R. (1996). Fast simulation of multistage power electronic systems with widely separated operating frequencies. *IEEE Transactions on Power Electronics* 11 (3): 405–412.

164 Hagiwara, M. and Akagi, H. (2009). Control and experiment of pulsewidth-modulated modular multilevel converters. *IEEE Transactions on Power Electronics* 24 (7): 1737–1746.

165 Blaschke, F. (1972). The principles of field orientation as applied to the new transvector closed-loop system for rotating field machines. *Siemens Review* 34: 217–220.

166 Lin, N. and Dinavahi, V. (2019). Exact nonlinear micromodeling for fine-grained parallel EMT simulation of MTDC grid interaction with wind farm. *IEEE Transactions on Industrial Electronics* 66 (8): 6427–6436.

167 Maguire, T., Warkentin, B., Chen, Y., and Hasler, J. (2013). Efficient techniques for real time simulation of MMC systems. *Proceedings of the International Conference on Power Systems Transients (IPST'13) in Vancouver, Canada* (18–20 July 2013).

168 Li, W. and Bélanger, J. (2016). An equivalent circuit method for modelling and simulation of modular multilevel converters in real-time HIL test bench. *IEEE Transactions on Power Delivery* 31 (5): 2401–2409.

169 Shen, Z. and Dinavahi, V. (2019). Real-time MPSoC-based electrothermal transient simulation of fault tolerant MMC topology. *IEEE Transactions on Power Delivery* 34 (1): 260–270.

170 Wu, B. (2006). *High-Power Converters and AC Drives*. Hoboken, NJ: Wiley-IEEE Press.

171 Saeedifard, M. and Iravani, R. (2010). Dynamic performance of a modular multilevel back-to-back HVDC system. *IEEE Transactions on Power Delivery* 25 (4): 2903–2912.

172 Xilinx (2017). *ZCU102 Evaluation Board User Guide, UG1182 (v1.3)*. Xilinx, Inc.

173 Gowaid, I.A., Adam, G.P., Massoud, A.M. et al. (2015). Quasi two-level operation of modular multilevel converter for use in a high-power DC transformer with DC fault isolation capability. *IEEE Transactions on Power Electronics* 30 (1): 108–123.

174 Xing, Z., Ruan, X., You, H. et al. (2016). Soft-switching operation of isolated modular DC/DC converters for application in HVDC grids. *IEEE Transactions on Power Electronics* 31 (4): 2753–2766.

175 She, X., Huang, A.Q., and Burgos, R. (2013). Review of solid-state transformer technologies and their application in power distribution systems. *IEEE Journal on Emerging and Selected Topics in Power Electronics* 1 (3): 186–198.

176 Li, R., Xu, L., Yao, L., and Williams, B.W. (2016). Active control of DC fault currents in DC solid-state transformers during ride-through operation of multi-terminal HVDC systems. *IEEE Transactions on Energy Conversion* 31 (4): 1336–1346.

177 Jovcic, D. and Zhang, H. (2017). Dual channel control with DC fault ride through for MMC-based, isolated DC/DC converter. *IEEE Transactions on Power Delivery* 32 (3): 1574–1582.

178 Lüth, T., Merlin, M.M.C., Green, T.C. et al. (2014). High-frequency operation of a DC/AC/DC system for HVDC applications. *IEEE Transactions on Power Electronics* 29 (8): 4107–4115.

179 Kong, F., Hao, Z., Zhang, S., and Zhang, B. (2014). Development of a novel protection device for bipolar HVDC transmission lines. *IEEE Transactions on Power Delivery* 29 (5): 2270–2278.

180 Hassanpoor, A., Roostaei, A., Norrga, S., and Lindgren, M. (2016). Optimization-based cell selection method for grid-connected modular multilevel converters. *IEEE Transactions on Power Electronics* 31 (4): 2780–2790.

181 Zhang, F., Li, W., and Joós, G. (2016). A voltage-level-based model predictive control of modular multilevel converter. *IEEE Transactions on Industrial Electronics* 63 (8): 5301–5312.

182 Karaagac, U., Mahseredjian, J., Cai, L., and Saad, H. (2017). Offshore wind farm modeling accuracy and efficiency in MMC-based multiterminal HVDC connection. *IEEE Transactions on Power Delivery* 32 (2): 617–627.

183 Saad, H., Peralta, J., Dennetiére, S. et al. (2013). Dynamic averaged and simplified models for MMC-based HVDC transmission systems. *IEEE Transactions on Power Delivery* 28 (3): 1723–1730.

184 Ahmed, N., Ängquist, L., Norrga, S. et al. (2014). A computationally efficient continuous model for the modular multilevel converter. *IEEE Journal on Emerging and Selected Topics in Power Electronics* 2 (4): 1139–1148.

185 Wei, L., Kerkman, R.J., Lukaszewski, R.A. et al. (2011). Analysis of IGBT power cycling capabilities used in doubly fed induction generator wind power system. *IEEE Transactions on Industry Applications* 47 (4): 1794–1801.

186 Baliga, B.J. (2013). Analytical modeling of IGBTs: challenges and solutions. *IEEE Transactions on Electron Devices* 60 (2): 535–543.

187 Xu, D., Lu, H., Huang, L. et al. (2002). Power loss and junction temperature analysis of power semiconductor devices. *IEEE Transactions on Industry Applications* 38 (5): 1426–1431.

188 Rajapakse, A.D., Gole, A.M., and Wilson, P.L. (2005). Electromagnetic transients simulation models for accurate representation of switching losses and thermal performance in power electronic systems. *IEEE Transactions on Power Delivery* 20 (1): 319–327.

189 Selhi, H., Christopoulos, C., Howe, A.F., and Hui, S.Y.R. (1996). The application of transmission-line modelling to the simulation of an induction motor drive. *IEEE Transactions on Energy Conversion* 11 (2): 287–297.

190 Ahmed, N., Angquist, L., Mahmood, S. et al. (2015). Efficient modeling of an MMC-based multiterminal DC system employing hybrid HVDC breakers. *IEEE Transactions on Power Delivery* 30 (4): 1792–1801.

191 Lin, W., Jovcic, D., Nguefeu, S., and Saad, H. (2016). Modelling of high-power hybrid DC circuit breaker for grid-level studies. *IET Power Electronics* 9 (2): 237–246.

192 Bucher, M.K. and Franck, C.M. (2016). Fault current interruption in multiterminal HVDC networks. *IEEE Transactions on Power Delivery* 31 (1): 87–95.

193 Cwikowski, O., Barnes, M., Shuttleworth, R., and Chang, B. (2015). Analysis and simulation of the proactive hybrid circuit breaker. *Proceedings IEEE 11th International Conference on Power Electronics and Drive Systems*, Sydney, Australia (9–12 June 2015), pp. 4–11.

194 Martinez, J.A. and Magnusson, J. (2015). EMTP modeling of hybrid HVDC breakers. *Proceedings of IEEE Power & Energy Society General Meeting* (26-30 July 2015), pp. 1–5.

195 Xu, J., Gole, A.M., and Zhao, C. (2015). The use of averaged-value model of modular multilevel converter in DC grid. *IEEE Transactions on Power Delivery* 30 (2): 519–528.

196 Häfner, J. and Jacobson, B. (2011). Proactive hybrid HVDC breakers-A key innovation for reliable HVDC grids. *Proceedings Cigré Symposium*, Bologna, Italy (13-15 September 2011), pp. 1–8.

197 Hassanpoor, A., Häfner, J., and Jacobson, B. (2015). Technical assessment of load commutation switch in hybrid HVDC breaker. *IEEE Transactions on Power Electronics* 30 (10): 5393–5400.

198 Döring, D., Ergin, D., Würflinger, K. et al. (2016). System integration aspects of DC circuit breakers. *IET Power Electronics* 9 (2): 219–227.

199 (2013). ABB 5SNA-2000K450300 StakPak IGBT module, doc. No. 5SYA1431-00, May 2013. http://new.abb.com/semiconductors/stakpak (accessed 20 January 2021).

200 Lin, N. and Dinavahi, V. (2018). Detailed device-level electrothermal modeling of the poactive hybrid HVDC breaker for real-time hardware-in-the-loop simulation of DC grids. *IEEE Transactions on Power Electronics* 33 (2): 1118–1134.

201 Wong, C. (1997). EMTP modeling of IGBT dynamic performance for power dissipation estimation. *IEEE Transactions on Industry Applications* 33 (1): 64–71.

202 Grieshaber, W., Dupraz, J.P., Penache, D.L., and Violleau, L. (2014). Development and test of a 120 kV direct current circuit breaker. *CIGRE 2014*, Paris, France, August 2014, pp. 1–11.

203 Davidson, C.C., Whitehouse, R.S., Barker, C.D. et al. (2015). A new ultra-fast HVDC circuit breaker for meshed DC networks. *Proceedings 11th IET International Conference on AC-DC Power Transmission*, pp. 1–7.

204 Far, A.J. and Jovcic, D. (2018). Design, modeling and control of hybrid DC circuit breaker based on fast thyristors. *IEEE Transactions on Power Delivery* 33 (2): 919–927.

205 Lin, N. and Dinavahi, V. (2020). Electrothermal transient behavioral modeling of thyristor-based ultrafast mechatronic circuit breaker for real-time DC grid emulation. *IEEE Transactions on Industrial Electronics* 67 (3): 1660–1670.

206 ABB 5STF 05D2425, doc. No. TS-TR/239/07b. http://www.milimsys.com/ (20 January 2021).

207 Bizjak, G., Zunko, P., and Povh, D. (1995). Circuit breaker model for digital simulation based on Mayr's and Cassie's differential arc equations. *IEEE Transactions on Power Delivery* 10 (3): 1310–1315.

208 Lesnicar, A. and Marquardt, R. (2003). An innovative modular multi-level converter topology for a wide power range. *IEEE Power Tech Conference*, Bologna, Italy, June 2003.

209 Chen, Y. and Dinavahi, V. (2013). Multi-FPGA digital hardware design for detailed large-scale real-time electro- magnetic transient simulation of power systems. *IET Generation, Transmission and Distribution* 7 (5): 451–463.

210 Vrana, T.K., Yang, Y., Jovcic, D. et al. (2013). The CIGRE B4 DC grid test system. *Electra* 270: 10–19.

211 Mets, K., Ojea, J.A., and Develder, C. (2014). Combining power and communication network simulation for cost-effective smart grid analysis. *IEEE Communications Surveys and Tutorials* 16 (3): 1771–1796.

212 IEEE Task Force on Interfacing Techniques for Simulation Tools, Müller, S.C., and Georg, H. et al. (2018). IEEE Task Force on Interfacing Techniques for Simulation Tools, Interfacing power system and ICT simulators: challenges, state-of-the-art, and case studies *IEEE Transactions on Smart Grid* 9: (1): 14–24.

213 Baran, M., Sreenath, R., and Mahajan, N. (2002). Extending EMTP for simulating agent based distributed applications. *IEEE Power Engineering Review* 22 (2) 52–54.

214 Hopkinson, K., Wang, X., Giovanini, R. et al. (2006). EPOCHS: a platform for agent-based electric power and communication simulation built from commercial off-the-shelf components. *IEEE Transactions on Power Systems* 21 (2): 548–558.

215 Lin, H., Veda, S., Shukla, S. et al. (2012). GECO: Global event-driven co-simulation framework for interconnected power system and communication networks. *IEEE Transactions on Smart Grid* 3 (3): 1444–1456.

216 Georg, H., Müller, S.C., Rehtanz, C., and Wietfeld, C. (2014). Analyzing cyberphysical energy systems: the INSPIRE co-simulation of power and ICT systems using HLA. *IEEE Transactions on Industrial Informatics* 10 (4): 2364–2373.

217 IEEE Std C37.118.1 (2011). *IEEE Standard for synchrophasor measurements for power systems*. New York, NY: IEEE Power & Energy Society.

218 Bergmann, J., Glomb, C., Goandtz, J. et al. (2010). Scalability of smart grid protocols: protocols and their simulative evaluation for massively distributed DERs. *Proceedings of the 1st IEEE International Conference on Smart Grid Communications (SmartGridComm'10)*, October 2010, pp. 131–136.

219 Godfrey, T., Mullen, S., Dugan, R.C. et al. (2010). Modeling smart grid applications with co-simulation. *Proceedings of the 1st IEEE Internaitonal Conference on Smart Grid Communications (SmartGrid-Comm'10)*, pp. 1–6.

220 Nutaro, J. (2011). Designing power system simulators for the smart grid: combining controls, communications, and electro-mechanical dynamics. *Proceedings of IEEE Power and Energy Society General Meeting (PES'11)*, pp. 1–5.

221 Mets, K., Verschueren, T., Develder, C. et al. (2011). Integrated simulation of power and communication networks for smart grid applications. *Proceedings of IEEE 16th International Workshop on Computer Aided Modeling and Design of Communications Links and Networks (CAMAD'11)*, Kyoto, Japan, June 2011, pp. 10–11.

222 Sidhu, T. and Yin, Y. (2007). Modeling and simulation for performance evaluation of IEC 61850-based substation communication systems. *IEEE Transactions on Power Delivery* 22 (3): 1482–1489.

223 Zhu, K., Chenine, M., and Nordstrom, L. (2011). ICT architecture impact on wide area monitoring and control systems' reliability. *IEEE Transactions on Power Delivery* 26 (4): 2801–2808.

224 Duan, T., Huang, Z., and Dinavahi, V. (2021). RTCE: Real-time co-emulation framework for EMT-based power system and communication network on FPGA-MPSoC hardware architecture. *IEEE Transactions on Smart Grid* 1–10 (early access).

225 Xilinx (2018). *VCU128 Evaluation Board User Guide (UG1302)*. San Jose, CA: Xilinx Inc.

226 IEC Technical Report (2013). Communication Networks and Systems for Power Utility Automation - Part 1: Introduction and Overview. *IEC TR 61850-1:2013*.

227 Xilinx Inc. (2014). Xilinx Doc. XAPP1026 (v5.1). LightWeight IP Application Examples. 21 November 2014.

228 Morison, K., Wang, L., and Kundur, P. (2004). Power system security assessment. *IEEE Power and Energy Magazine* 2 (5): 30–39.

229 Wang, L. and Morison, K. (2006). Implementation of online security assessment. *IEEE Power and Energy Magazine* 4 (5): 46–59.

230 Dai, R., Liu, G., and Zhang, X. (2020). Transmission technologies and implementations: building a stronger, smarter power grid in China. *IEEE Power and Energy Magazine* 18 (2): 53–59.

231 Cao, S., Lin, N., and Dinavahi, V. (2020). Faster-than-real-time dynamic simulation of AC/DC grids on reconfigurable hardware. *IEEE Transactions on Power Systems* 35 (2): 1539–1548.

232 Cao, S., Lin, N., and Dinavahi, V. (2020). Mitigation of subsynchronous interaction in hybrid AC/DC grid with renewable energy using faster-than-real-time dynamic simulation. *IEEE Transactions on Power Systems* 36 (1): 670–679.

233 Stott, B. (1979). Power system dynamic response calculations. *Proceedings of IEEE* 67 (2): 219–241.

234 Christie, R. (1993). *Power Systems Test Case Archive*. Seattle, WA: University Washington. https://www.ee.washington.edu/research/pstca/pf118/pg_tca118bus.htm.

235 Kundur, P. (1994). *Power System Stability and Control*, New York: McGraw-Hill.

236 Cao, S., Lin, N., and Dinavahi, V. (2021). Flexible time-stepping dynamic emulation of AC/DC grid for faster-than-SCADA applications,. *IEEE Transactions on Power Systems* 36 (3): 2674–2683.

237 Duan, T., Cheng, T., and Dinavahi, V. (2020). Heterogeneous real-time co-emulator for communication-enabled global control of AC/DC grids integrated with renewable energy. *IEEE Open Journal of Industrial Electronics Society* 1: 261–270.

Appendix A

Parameters for Case Studies

This appendix provides the relevant parameters for all the case studies presented in the chapters.

A.1 Chapter 2

A.1.1 Case in Section 2.7

Loads: 1200 Ω, 500 mH; 2150 Ω, 380 mH; 250 Ω, 25 mH; 350 Ω, 60 mH; 250 Ω, 25 mH; 420 Ω, 30 mH; 200 Ω, 130 mH; 650 Ω, 250 mH; $C_1 = 5\,\mu F$, $C_2 = 20\,\mu F$.
G1–G4: $1.03\angle 20.2^\circ$, 1.2 Ω, 38.98 mH; $1.01\angle 10.5^\circ$, 1.1 Ω, 45.52 mH; $1.03\angle -6.8^\circ$, 0.9 Ω, 38.98 mH; $1.01\angle -17.0^\circ$, 0.8 Ω, 35.23 mH.
T1–T4: 1.5 Ω, 23.4 mH; 0.8 Ω, 29.5 mH; 1.6 Ω, 23.4 mH; 0.6 Ω, 20.8 mH.

A.1.2 Cases in Section 2.8

Case 1: $V_s = 20$ kV peak; $R_s = 5\,\Omega$, $L_s = 24.1$ mH; line mode +: 1.273e−5 Ω/m, 9.337e−4 mH/m, 1.274e−5 μF/m; line mode 0: 3.864e−5 Ω/m, 4.126e−3 mH/m, 7.751e−6 μF/m; length 100 km; $C = 100\,\mu F$; $R_{L1} = 8\,\Omega$, $L_{L1} = 4000$ mH, $R_{L2} = 270\,\Omega$, $R_f = 0.1\,m\Omega$; $V_{ref} = 8192$ V, $p = 600$ A, $q = 6$.
Case 2: $V_s = 125$ kV peak; $R_s = 1\,\Omega$, $L_s = 15$ mH; $R_{fe} = 200\,M\Omega$, $C_W = 5$ nF, $C_S = 1.25$ nF.

A.2 Chapter 3

A.2.1 Cases in Section 3.2

A.2.1.1 Cases Study I

Inrush current transients: T: Y–Y connection, 250 MVA, 200/400 kV; G: 20 kV; $R = 1\,\Omega$.

Real-Time Electromagnetic Transient Simulation of AC-DC Networks, First Edition.
Venkata Dinavahi and Ning Lin.

Ferroresonance and Eddy current transients: T: Y–Y connection, 250 MVA, 125 kV/500 kV; G: 50 kV; $C_s = 0.05\,\mu F$, $C_p = 0.0125\,\mu F$, $R = 500\,\Omega$.

A.2.1.2 Cases Study II

T_1: Y–Y connection, 20 MVA, 15/35 kV.
T_2: Y–Y connection, 20 MVA, 32/14 kV.
T_3: Y–Y connection, 20 MVA, 32/20 kV.
Line: mode +: 1.156 mH/m, 0.0194 μF/m; mode 0: 5.819 mH/m, 0.0120 μF/m; L_1, L_2, L_3 length: 120 m, 140 m, 180 m.

A.2.2 Cases in Section 3.3

A.2.2.1 Transformer

Geometry: Limb length and cross section area: 7.18 m, 0.454 m^2; yoke length and cross section area: 2.66 m, 0.454 m^2.
MEC model: Material resistivity: $5.2 \times 10^{-7}\,\Omega/m$; lamination thickness: 0.0005 m; $X_{leakage} = 0.25$ mH; turn ratio: 64/26; hysteresis: $a_1 = 0.7900$, $a_2 = 486$, $a_3 = 312$, $b_1 = 0.00186$, $b_2 = 0.011236$, $b_3 = 0.01898$, $c_1 = 0.4271$, $c_2 = 0.3824$, $c_3 = 1.0406$.

A.2.2.2 System

Energization: $V_{base} = 20/\sqrt{3}$ kV, $R_s = 0.507\,\Omega$.
Hysteresis: $V_{base} = 16/\sqrt{3}$ kV, $R_s = 0.507\,\Omega$.
Power-flow control and fault: $V_{base} = 28$ kV, $S_{base} = 260$ MVA; ST: 260 MVA; $G_{s1} = 28$ kV, $G_{s2} = 28\,\text{kV}/20°$ lag; Z_1, Z_2: 0.5 H/19.17 μH, 0.4 H/10.17 μH; load: 10 Ω; Line mode +: 1.556×10^{-7} H/m, 0.0194×10^{-8} F/m; line mode 0: 5.819×10^{-7} H/m, 0.012×10^{-8} F/m; length: $L_1 = 10$ km, $L_2 = 12$ km; tap changer: 8 positions, 0.05 p.u./step, 0.5 s/step.

A.2.3 Cases in Section 3.4

Preisach: $a_0 = a_1 = 0$, $b_0 = b_1 = 3162$, $c = 0.6$.
Transformer: Outer/inner size: 1.3 m × 1.8 m/0.7 m × 1.2 m; coil size: 0.3 m × 0.5 m; turn ratio: 200/1076; $R_p = 0.667\,\Omega$, $R_s = 3.588\,\Omega$, $\sigma^e = 1000$.
System: $V_{ac} = 37.5$ kV, $R_1 = 5\,\Omega$, $L_1 = 2$ mH, $R_{L1} = 200\,\Omega$, $R_{L2} = 100\,\Omega$, $L_{L1} = 10$ mH, $L_{L2} = 5$ mH; Line: 15 km, $Z_{TL} = 200\,\Omega$; 2nd/4th harmonics: 9.38/4.69 kV.

A.3 Chapter 4

A.3.1 UM Case in Section 4.2

Lines: Three conductors, diameter 3.105 cm, resistance 0.0583 Ω/km, and 150 km. The tower geometry is given in Figure A.1.

Synchronous machine: 1000 MVA/22 kV, Y-connected, field current 2494 A, 2 poles, 60 Hz, single-mass, $J = 5.5628\text{e}4\ \text{kg·m}^2$, and $D = 0.678\text{e}4\ \text{N·m·s/rad}$. The winding resistances (Ω) and leakage inductances (H) are listed in the Table A.1.

Loads and transformers: $\text{Load}_{1,2,3}$: 5 Ω/4 H; XT1, XT2, and XT3: 15 mH; and C: 20 μF.

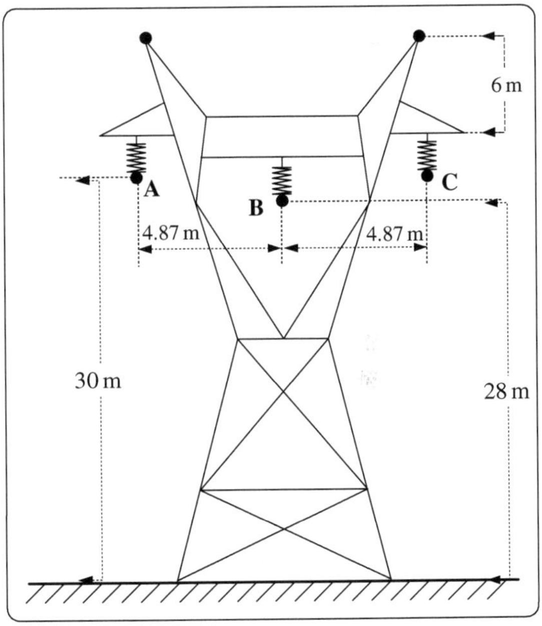

Figure A.1 Tower geometry of transmission lines.

Table A.1 UM parameters.

Parameter	R_d	R_q	R_0	R_f	R_{D1}	R_{Q1}	R_{Q2}
Value	9.68e−4	9.68e−4	9.68e−4	1.1109	3.4985	0.7627	1.2270
Parameter	L_d	L_q	L_0	L_f	L_{D1}	L_{Q1}	L_{Q2}
Value	0.0018	0.0017	2.4136e−4	0.6345	0.5483	1.1811	0.5882
Parameter	M_{df}	M_{dD1}	M_{qQ1}	M_{qQ2}	M_{fD1}	M_{Q1Q2}	
Value	0.0234	0.0234	0.0226	0.0226	0.5480	0.4184	

A.3.2 Cases in Section 4.3

A.3.2.1 State-Space Matrices of Rotating Machines

Based on the following matrix notation, the state-space definitions for various machines are given below.

$$\mathbf{A} = \begin{bmatrix} \mathbf{A}_{11} & \mathbf{A}_{12} \\ \mathbf{A}_{21} & \mathbf{A}_{22} \end{bmatrix}, \quad \mathbf{B} = \begin{bmatrix} \mathbf{B}_{11} & \mathbf{B}_{12} \\ \mathbf{B}_{21} & \mathbf{B}_{22} \end{bmatrix}, \quad \mathbf{C} = \begin{bmatrix} \mathbf{C}_{11} & \mathbf{C}_{12} \\ \mathbf{C}_{21} & \mathbf{C}_{22} \end{bmatrix}.$$

Induction Machine

$$\mathbf{x} = \begin{bmatrix} \lambda_{qs} & \lambda_{ds} & \lambda_{qr} & \lambda_{dr} \end{bmatrix}^T, \tag{A1}$$

$$\mathbf{u} = \begin{bmatrix} v_{qs} & v_{ds} & v_{qr} & v_{dr} \end{bmatrix}^T, \tag{A2}$$

$$\mathbf{y} = \begin{bmatrix} i_{qs} & i_{ds} & i_{qr} & i_{dr} \end{bmatrix}^T, \tag{A3}$$

$$\mathbf{A}_{11} = \begin{bmatrix} \frac{r_s}{L_{ls}}\left(\frac{L_{aq}}{L_{ls}} - 1\right) & -\omega \\ \omega & \frac{r_s}{L_{ls}}\left(\frac{L_{ad}}{L_{ls}} - 1\right) \end{bmatrix},$$

$$\mathbf{A}_{21} = \begin{bmatrix} \frac{r_r}{L_{lr}}\frac{L_{aq}}{L_{ls}} & 0 \\ 0 & \frac{r_r}{L_{lr}}\frac{L_{ad}}{L_{ls}} \end{bmatrix}, \quad \mathbf{A}_{12} = \begin{bmatrix} \frac{r_s}{L_{ls}}\frac{L_{aq}}{L_{lr}} & 0 \\ 0 & \frac{r_s}{L_{ls}}\frac{L_{ad}}{L_{lr}} \end{bmatrix},$$

$$\mathbf{A}_{22} = \begin{bmatrix} \frac{r_r}{L_{lr}}\left(\frac{L_{aq}}{L_{lr}} - 1\right) & \omega_r - \omega \\ \omega - \omega_r & \frac{r_r}{L_{lr}}\left(\frac{L_{ad}}{L_{lr}} - 1\right) \end{bmatrix}, \tag{A4}$$

$$\mathbf{B} = eye(4, 4), \tag{A5}$$

$$\mathbf{C} = \frac{1}{D}\begin{bmatrix} L_{rr} & 0 & -L_M & 0 \\ 0 & L_{rr} & 0 & -L_M \\ -L_M & 0 & L_{ss} & 0 \\ 0 & -L_M & 0 & L_{ss} \end{bmatrix}, \tag{A6}$$

$$\mathbf{D} = zeros(4, 4), \tag{A7}$$

where

$$L_{ss} = L_{ls} + L_m, \quad L_{rr} = L_{lr} + L_m, \quad D = L_{ss}L_{rr} + L_m^2, \tag{A8}$$

$$L_{ad}^{-1} = L_{aq}^{-1} = 1/L_m + 1/L_{ls} + 1/L_{lr}. \tag{A9}$$

Synchronous Machine

$$\mathbf{x} = \begin{bmatrix} \lambda_{qs} & \lambda_{ds} & \lambda_{kq1} & \lambda_{kq2} & \lambda_{fd} & \lambda_{kd} \end{bmatrix}^T, \tag{A10}$$

$$\mathbf{u} = \begin{bmatrix} v_{qs} & v_{ds} & v_{kq1} & v_{kq2} & e_{xfd} & v_{kd} \end{bmatrix}^T, \tag{A11}$$

$$\mathbf{y} = \begin{bmatrix} i_{qs} & i_{ds} & i_{kq1} & i_{kq2} & i_{fd} & i_{kd} \end{bmatrix}^T, \tag{A12}$$

$$\mathbf{A}_{11} = \begin{bmatrix} \frac{r_s}{L_{ls}}\left(\frac{L_{aq}}{L_{ls}} - 1\right) & -\omega_r & \frac{r_s}{L_{ls}}\frac{L_{aq}}{L_{lkq1}} \\ \omega_r & \frac{r_s}{L_{ls}}\left(\frac{L_{ad}}{L_{ls}} - 1\right) & 0 \\ \frac{r_{kq1}}{L_{lkq1}}\frac{L_{aq}}{L_{ls}} & 0 & \frac{r_{kq1}}{L_{lkq1}}\left(\frac{L_{aq}}{L_{lkq1}} - 1\right) \end{bmatrix},$$

$$\mathbf{A}_{21} = \begin{bmatrix} \frac{r_{kq2}}{L_{lkq2}}\frac{L_{aq}}{L_{ls}} & 0 & \frac{r_{kq2}}{L_{lkq2}}\frac{L_{aq}}{L_{lkq1}} \\ 0 & \frac{r_{fd}}{L_{lfd}}\frac{L_{ad}}{L_{ls}} & 0 \\ 0 & \frac{r_{kd}}{L_{lkd}}\frac{L_{ad}}{L_{ls}} & 0 \end{bmatrix},$$

$$\mathbf{A}_{12} = \begin{bmatrix} \frac{r_s}{L_{ls}}\frac{L_{aq}}{L_{lkq2}} & 0 & 0 \\ 0 & \frac{r_s}{L_{ls}}\frac{L_{ad}}{L_{lfd}} & \frac{r_s}{L_{ls}}\frac{L_{ad}}{L_{lkd}} \\ \frac{r_{kq1}}{L_{lkq1}}\frac{L_{aq}}{L_{lkq2}} & 0 & 0 \end{bmatrix},$$

$$\mathbf{A}_{22} = \begin{bmatrix} \frac{r_{kq2}}{L_{lkq2}}\left(\frac{L_{aq}}{L_{lkq2}} - 1\right) & 0 & 0 \\ 0 & \frac{r_{fd}}{L_{lfd}}\left(\frac{L_{ad}}{L_{lfd}} - 1\right) & \frac{r_{fd}}{L_{lfd}}\frac{L_{ad}}{L_{lkd}} \\ 0 & \frac{r_{kd}}{L_{lkd}}\frac{L_{ad}}{L_{lfd}} & \frac{r_{kd}}{L_{lkd}}\left(\frac{L_{ad}}{L_{lkd}} - 1\right) \end{bmatrix}, \tag{A13}$$

$$\mathbf{B}_{11} = eye(3,3), \quad \mathbf{B}_{12} = \mathbf{B}_{21} = zeros(3,3), \quad \mathbf{B}_{22} = \begin{bmatrix} 1 & 0 & 0 \\ 0 & \frac{r_{fd}}{L_{md}} & 0 \\ 0 & 0 & 1 \end{bmatrix}, \tag{A14}$$

$$\mathbf{C}_{11}=\begin{bmatrix}\dfrac{(L_{kq1}L_{kq2}-L_{mq}^2)}{Dq} & 0 & \dfrac{(-L_{mq}L_{kq2}+L_{mq}^2)}{Dq}\\ 0 & \dfrac{(L_{fd}L_{kd}-L_{md}^2)}{Dd} & 0\\ \dfrac{(L_{mq}L_{kq2}-L_{mq}^2)}{Dq} & 0 & \dfrac{(-L_{mq}L_{kq2}+L_{mq}^2)}{Dq}\end{bmatrix},$$

$$\mathbf{C}_{21}=\begin{bmatrix}\dfrac{(L_{mq}L_{kq1}-L_{mq}^2)}{Dq} & 0 & \dfrac{(L_qL_{mq}+L_{mq}^2)}{Dq}\\ 0 & \dfrac{(L_{md}L_{kd}-L_{md}^2)}{Dd} & 0\\ 0 & \dfrac{(L_{md}L_{fd}-L_{md}^2)}{Dd} & 0\end{bmatrix},$$

$$\mathbf{C}_{12}=\begin{bmatrix}\dfrac{(-L_{mq}L_{kq1}+L_{mq}^2)}{Dq} & 0 & 0\\ 0 & \dfrac{(-L_{md}L_{kd}+L_{md}^2)}{Dd} & \dfrac{(-L_{md}L_{fd}+L_{md}^2)}{Dd}\\ \dfrac{(L_qL_{mq}+L_{mq}^2)}{Dq} & 0 & 0\end{bmatrix},$$

$$\mathbf{C}_{22}=\begin{bmatrix}\dfrac{(-L_qL_{kq1}+L_{mq}^2)}{Dq} & 0 & 0\\ 0 & \dfrac{(-L_dL_{kd}+L_{md}^2)}{Dd} & \dfrac{(L_dL_{md}-L_{md}^2)}{Dd}\\ 0 & \dfrac{(L_dL_{md}-L_{md}^2)}{Dd} & \dfrac{(-L_dL_{fd}-L_{md}^2)}{Dd}\end{bmatrix}, \tag{A15}$$

$$\mathbf{D}=zeros(6,6), \tag{A16}$$

where

$$L_q=L_{ls}+L_{mq},\quad L_d=L_{ls}+L_{md},\quad L_{kq1}=L_{lkq1}+L_{mq}, \tag{A17}$$

$$L_{kq2}=L_{lkq2}+L_{mq},\quad L_{fd}=L_{lfd}+L_{md},\quad L_{kd}=L_{lkd}+L_{md}, \tag{A18}$$

$$D_q=L_{mq}^2\left(L_q-2L_{mq}+L_{kq1}+L_{kq2}\right)-L_qL_{kq1}L_{kq2}, \tag{A19}$$

$$D_d=L_{md}^2\left(L_d-2L_{md}+L_{fd}+L_{kd}\right)-L_dL_{fd}L_{kd}, \tag{A20}$$

$$L_{aq}^{-1}=1/L_{mq}+1/L_{ls}+1/L_{lkq1}+1/L_{lkq2}, \tag{A21}$$

$$L_{ad}^{-1}=1/L_{md}+1/L_{ls}+1/L_{lfd}+1/L_{lkd}. \tag{A22}$$

Line Start Permanent Magnet Synchronous Motor

$$\mathbf{x} = \begin{bmatrix} \lambda_q & \lambda_d & \lambda_{kq} & \lambda_{kd} \end{bmatrix}^T, \tag{A23}$$

$$\mathbf{u} = \begin{bmatrix} v_q & v_d + r_s \frac{L_{md}}{L_{ls}} i_m & v_{kq} & v_{kd} + r_{kd} \frac{L_{md}}{L_{lkd}} i_m \end{bmatrix}^T, \tag{A24}$$

$$\mathbf{y} = \begin{bmatrix} i_q & i_d & i_{kq} & i_{kd} \end{bmatrix}^T, \tag{A25}$$

$$\mathbf{A}_{11} = \begin{bmatrix} \frac{r_s}{L_{ls}} \left(\frac{L_{aq}}{L_{ls}} - 1 \right) & -\omega_r \\ \omega_r & \frac{r_s}{L_{ls}} \left(\frac{L_{ad}}{L_{ls}} - 1 \right) \end{bmatrix},$$

$$\mathbf{A}_{21} = \begin{bmatrix} \frac{r_{kq}}{L_{lkq}} \frac{L_{aq}}{L_{ls}} & 0 \\ 0 & \frac{r_r}{L_{lr}} \frac{L_{ad}}{L_{ls}} \end{bmatrix}, \quad \mathbf{A}_{12} = \begin{bmatrix} \frac{r_s}{L_{ls}} \frac{L_{aq}}{L_{lkq}} & 0 \\ 0 & \frac{r_s}{L_{ls}} \frac{L_{ad}}{L_{lkd}} \end{bmatrix},$$

$$\mathbf{A}_{22} = \begin{bmatrix} \frac{r_{kq}}{L_{lkq}} \left(\frac{L_{aq}}{L_{lkq}} - 1 \right) & 0 \\ 0 & \frac{r_{kd}}{L_{lkd}} \left(\frac{L_{ad}}{L_{lkd}} - 1 \right) \end{bmatrix}, \tag{A26}$$

$$\mathbf{B} = eye(4, 4), \tag{A27}$$

$$\mathbf{C} = \begin{bmatrix} -L_{kq}/D_q & 0 & L_{mq}/D_q & 0 \\ 0 & -L_{kd}/D_d & 0 & -L_{md}/D_d \\ L_{mq}/D_q & 0 & -L_q/D_q & 0 \\ 0 & L_{md}/D_d & 0 & -L_d/D_d \end{bmatrix}, \tag{A28}$$

$$\mathbf{D} = zeros(4, 4), \tag{A29}$$

where

$$L_q = L_{ls} + L_{mq}, \quad D_q = L_{mq}^2 - L_{kq} L_q, \tag{A30}$$

$$L_d = L_{ls} + L_{md}, \quad D_d = L_{md}^2 - L_{kd} L_d, \tag{A31}$$

$$L_{aq}^{-1} = 1/L_{mq} + 1/L_{ls} + 1/L_{lkq}, \quad L_{ad}^{-1} = 1/L_{md} + 1/L_{ls} + 1/L_{lkd}. \tag{A32}$$

DC Machine

$$\mathbf{x} = \begin{bmatrix} \lambda_f & \lambda_a \end{bmatrix}^T, \quad \mathbf{u} = \begin{bmatrix} v_f & v_a \end{bmatrix}^T, \quad \mathbf{y} = \begin{bmatrix} i_f & i_a \end{bmatrix}^T, \tag{A33}$$

$$\mathbf{A} = \begin{bmatrix} -\dfrac{r_f}{L_{ff}} & 0 \\ -\omega_r \dfrac{r_{af}}{L_{ff}} & -\dfrac{r_a}{L_{aa}} \end{bmatrix}, \quad \mathbf{B} = \begin{bmatrix} 1 & 0 \\ 0 & 1 \end{bmatrix}, \tag{A34}$$

$$\mathbf{C} = \begin{bmatrix} 1/L_{ff} & 0 \\ 0 & 1/L_{aa} \end{bmatrix}, \quad \mathbf{D} = zeros(2, 2). \tag{A35}$$

A.3.2.2 Parameters of Rotating Machines

Case I: Induction motor, 3 hp, 230 V, $r_s = 0.5\,\Omega$, $r_r = 0.51\,\Omega$, $l_{ls} = 4\,\text{mH}$, $l_{lr} = 4\,\text{mH}$, $l_m = 89.4\,\text{mH}$.

Case II: Synchronous generator, 26 kV/802.5 MVA, $r_s = 0.0048\,\Omega$, $r_{fd} = 0.58043\,\Omega$, $r_{kd} = 0.0203\,\Omega$, $r_{kq1} = 0.0727\,\Omega$, $L_{ls} = 0.57031\,\text{mH}$, $L_{md} = 4.2\,\text{mH}$, $L_{mq} = 3.8\,\text{mH}$, $L_{fd} = 0.040759\,\text{mH}$, $L_{kd} = 0.27852\,\text{mH}$, $L_{kq1} = 0.16605\,\text{mH}$.

Case III: Line start permanent magnet synchronous motor, 4 hp, 230 V, $r_s = 0.017\,\Omega$, $r_{kq} = 0.108\,\Omega$, $l_{kd} = 0.054\,\Omega$, l_{ls}=0.172 mH, $l_{lkq} = 0.350\,\text{mH}$, $l_{lkd} = 0.350\,\text{mH}$, $l_{mq} = 2.7\,\text{mH}$, $l_{md} = 1.3\,\text{mH}$, $i_m = 1.6203\,\text{A}$.

Case IV: DC motor, 240 V, $r_a = 0.6\,\Omega$, $r_f = 240\,\Omega$, $L_{ff} = 120\,\text{mH}$, $L_{af} = 1.8\,\text{H}$, $l_{aa} = 0.012\,\text{H}$.

A.3.3 MEC Case in Section 4.4

Induction machine: The parameters of squirrel cage, Y-connection induction machine are given in Table A.2.

Iron core permeability: $\mu(H) = \mu(-H)$, where

$$\mu = \begin{cases} 0.183, & H = 0, \\ a_6H^6 + a_5H^5 + a_4H^4 + a_3H^3 + a_2H^2 + a_1H + a_0, & H > 0, \end{cases} \tag{A.1}$$

and H can be found in Table A.3.

A.4 Chapter 5

Case I: Source parameters: $Z_s = 9.2 + j52\,\Omega$, $E_{s1} = 230\angle\,0°$ kV, $E_{s2} = 230\angle\,20°$ kV. Transmission line impedance (Ω/km): $Z_0 = 0.363 + j1.326$, $Z_1 = 0.0357 + j0.5078$, $Z_2 = 0.0357 + j0.5078$.

Case II: G_1, G_2, G_3: 120 MVA, 13.8 kV; Loads: $R_1 = 10\,\Omega$, $L_1 = 14\,\text{mH}$, $R_2 = 1\,\Omega$.

Table A.2 Specification of induction machine.

Parameter	Value	Parameter	Value
Stator outer diameter	195.38 mm	Rotor tooth face width	12.7 mm
Rotor outer diameter	114.9 mm	Stator tooth flange thickness	1.2 mm
Stator inner diameter	115.54 mm	Rotor tooth flange thickness	1.8 mm
Rotor inner diameter	36.5 mm	Mechanical air-gap length	0.31 mm
Stator slot depth	21.1 mm	Number of stator slots	36
Stator tooth width	5.7 mm	Number of rotor slots	28
Rotor slot depth	22.1 mm	Rated voltage	230 V
Rotor tooth width	6.2 mm	Frequency	60 Hz
Motor length	107.95 mm	Number of poles	4
Stator tooth face width	7.4 mm	Rotor inertia	0.025 Kg·m^2

Table A.3 Coefficients of 11-segment sextic polynomial function.

Condition	a_6	a_5	a_4	a_3	a_2	a_1	a_0
$0 < H \leq 6$	8.09e−6	−1.80e−4	0.0017	−0.0084	0.027	−0.0669	0.1832
$6 < H \leq 12$	1.54e−8	−1.00e−6	2.82e−5	−4.50e−4	0.0045	−0.0299	0.1543
$12 < H \leq 20$	5.32e−10	−6.11e−8	3.02e−6	−8.42e−5	0.0015	−0.0162	0.1279
$20 < H \leq$ 1e2	3.73e−13	−1.29e−10	2.23e−8	−2.09e−6	1.15e−4	−0.0038	0.0765
1e2$< H \leq$ 1e3	2.99e−19	−1.12e−15	1.71e−12	−1.36e−9	6.05e−7	−1.51e−4	0.0205
1e3$< H \leq$ 8e3	1.60e−25	−4.93e−21	6.19e−17	−4.08e−13	1.52e−9	−3.18e−6	0.0036
8e3$< H \leq$ 3e4	3.87e−30	−5.09e−25	2.80e−20	−8.33e−16	1.45e−11	−1.48e−7	8.45e−4
3e4$< H \leq$15e4	1.18e−34	−7.32e−29	1.88e−23	−2.57e−18	2.03e−13	−9.28e−9	2.32e−4
15e4$< H \leq$ 5e5	1.17e−38	−2.65e−32	2.50e−26	−1.28e−20	3.83e−15	−6.74e−10	6.60e−5
5e5$< H \leq$ 2e6	1.24e−42	−1.07e−35	3.86e−29	−7.48e−23	8.45e−17	−5.56e−11	2.01e−5
2e6$< H$	0	0	0	0	0	0	1.72e−6

A.5 Chapter 6

A.5.1 Cases in Section 6.2

Case 1: $V_s = 10\,\text{V}/5\,\text{kHz}$, $R_s = 1\,\text{m}\Omega$, $R_L = 1\,\Omega$, $I_s = 1\,\text{pA}$, $\tau = 10\,\mu\text{s}$, $T_m = 5\,\mu\text{s}$, $V_T = 25.9\,\text{mV}$.

Case 2: $P_{bases} = 200$ MW, $f = 60$ Hz; $V_s = 120$ kV, $R_s = 1\,\Omega$; Transformer: 115/230 kV, $L_{leakage} = 0.16$ p.u., copper loss 0.004 p.u.; L_1 and L_2: 50 km; Line mode +: 0.9337 mH/km, 12.74 nF/km; Line mode 0: 4.1264 mH/km, 7.751 nF/km; L_L/R_L: 2 mH/265 Ω; Surge arrester: [(0 kV, 0 kA), (263.82 kV, 0.001 kA), (317.19 kV, 0.1 kA), (362.42 kV, 2.8 kA), (429.89 kV, 200 kA)].

A.5.2 Cases in Section 6.3

$P_{base} = 100$ MVA, $V_{base} = 230$ kV, $f = 60$ Hz; Transformers: 230/230 kV, $L_{leakage} = 0.2$ p.u., copper loss: 0.004 p.u.

A.6 Chapter 7

See Tables A.4 and A.5.

Chopper circuit: $R = 1\,\Omega$, $L = 100\,\mu$H, $R_{gs} = 35\,\Omega$, $V_{dc} = 500$ V.
Buck converter: $R = 5\,\Omega$, $L = 700\,\mu$H, $R_g = 100\,\Omega$, $V_{dc} = 100$ V, $D = 0.5$.

Table A.4 Typical diode nonlinear behavioral model parameters.

r_{on}	r_{off}	V_{on}	I_{Fo}	I_{rrm}	t_{rr}	K	L	R_L	$\frac{dI_r}{dt}$
10 mΩ	100 kΩ	0.7 V	10 A	10 A	2 μs	98 830	10 pH	12.79 μΩ	50 M

Table A.5 Siemens® BSM300GA160D IGBT parameters provided by SaberRD®.

Behavioral IGBT model parameters
$r_{off} = 10^9\,\Omega$, $g_{off} = 10^{-12}$ S, $g_{on} = 10^6$ S, $R_g = 5\,\Omega$, $vce1 = 4.8$ V, $vge1 = 9$ V, $ic1 = 225$ A, $vce2 = 1.8$ V, $vge2 = 7$ V, $ic2 = 20$ A, $vce3 = 4$ V, $vge3 = 17$ V, $ic3 = 400$ A, $V_t = 6.3$ V, $V_{on} = 0.8$ V, $vce4 = 10$ V, $vce5 = 4$ V, $vce6 = 800$ V, $vge4 = 10$ V, $vge5 = 20$ V, $itrat = 20$, $crss1 = 30$ nF, $crss2 = 1.6$ nF, $coss1 = 42$ nF, $coss2 = 5$ nF, $q_1 = 400$ nC, $q_2 = 2000$ nC, $q_3 = 3500$ nC, $\tau = 10$ μs, $M = 0.5$, $R_{tail} = 1$ μΩ, $C_{tail} = 10$ F, $a_1 = 0.0217$, $a_3 = 91.705$, $b_1 = 0.00395$, $b_3 = 3.z221$, $x = 0.973$, $y = 1.428$, $z = 0.369$, $icsat3 = 1.789$ kA, $cceo = 12$ nF, $ccgo = 110$ nF, $cgeo = 40$ nF, $vceo = 0.873$ V, $vcgo = 0.0189$ V

A.7 Chapter 8

A.7.1 Equivalenced Device-Level Model in Section 8.3

1-phase 5-level MMC: $L_m = 2\,\text{mH}$, $C_{SM} = 4\,\text{mF}$, $R_s = 3.6\,\Omega$, $L_s = 1\,\text{mH}$, $V_{dc} = 7.2\,\text{kV}$.
3-phase 9-level MMC: $L_m = 5\,\text{mH}$, $C_{SM} = 4\,\text{mF}$, $R_s = 7.4\,\Omega$, $L_s = 2.5\,\text{mH}$, $V_s(L-L) = 7.2\,\text{kV}$, $V_{dc} = 14.4\,\text{kV}$.

A.7.2 MMC-IM Case in Section 8.4

MMC: $V_{dc} = 900\,\text{V}$, $L_{u,d} = 1\,\text{mH}$, $Z_{Load} = 5\,\Omega + 2\,\text{mH}$, $C_{SM} = 6\,\text{mF}$.
Induction machine: $L_s = 35.5\,\text{mH}$, $L_r = 35.5\,\text{mH}$, $L_m = 34.7\,\text{mH}$, $R_s = 0.087\,\Omega$, $R_r = 0.228\,\Omega$, $J = 1.662\,\text{Kg·m}^2$, pole number $P = 4$.

A.7.3 MVDC Case in Section 8.5

$V_{dc} = 10\,\text{kV}$, $P_{dc} = 8\,\text{MW}$, $C_{SM} = 1\,\text{mF}$, $L_{u,d} = 20\,\text{mF}$, $V_{g1,2} = 5.5\,\text{kV}/60\,\text{Hz}$, $r_{1,2} = 0.4\,\Omega$, $L_{1,2} = 1\,\text{mH}$.

A.7.4 MTDC Case in Section 8.6

$V_s = 220\,\text{kV}/50\,\text{Hz}$, $R_s = 0.6\,\Omega$, $L_s = 30\,\text{mH}$, $L_m = 18.9\,\text{mH}$, $C_{1,2} = 2\,\text{mF}$, $V_{dc} = \pm 200\,\text{kV}$, $D = 100\,\text{km}$, line surge impedance $Z_0 = 394\,\Omega$.

A.8 Chapter 9

IGBT static model:

1. $I_C > 1000\,\text{A}$: $k_1 = -4.428\,T_{vj} + 1567$, $b_1 = -4.263\,T_{vj} + 1975.5$;
2. $I_C \in (500{,}1000]\,\text{A}$: $k_2 = -2.684\,T_{vj} + 1113.1$, $b_2 = -1.867\,T_{vj} + 1107.4$;
3. $I_C \in (300, 500]\,\text{A}$: $k_3 = -2.185\,T_{vj} + 881$, $b_3 = -1.588\,T_{vj} + 772.7$;
4. $I_C \in (200, 300]\,\text{A}$: $k_4 = -1.692\,T_{vj} + 709$, $b_4 = -1.179\,T_{vj} + 562.8$;
5. $I_C \in (0, 200]\,\text{A}$: $k_5 = 200$, $b_5 = 0$;
6. $I_C < 0\,\text{A}$: $k_6 = 10^{-6}$, $b_6 = 0$.

IGBT turn-on coefficients: Segment 1. $k_0 = 0$, $k_1 = 0$, $k_2 = 1$, $k_3 = 0$, $b_0 = 3375$, $b_1 = 1$, $b_2 = -1833.3$, $b_3 = -1.6$;
Segment 2. $k_0 = 0$, $k_1 = 0$, $k_2 = 0$, $k_3 = 0$, $b_0 = 5$, $b_1 = 1$, $b_2 = 0$, $b_3 = 0.24$.

IGBT turn-off coefficients: Segment 1. $k_0 = 0$, $k_1 = 0$, $k_2 = 0$, $k_3 = 0$, $b_0 = 1748.3$, $b_1 = 1$, $b_2 = 33.33$, $b_3 = -0.6867$;
Segment 2. $k_0 = 0$, $k_1 = 0$, $k_2 = 0.1$, $k_3 = 0$, $b_0 = 2048.3$, $b_1 = 1$, $b_2 = -200$, $b_3 = -0.6867$;
Segment 3. $k_0 = 0$, $k_1 = 0$, $k_2 = 0$, $k_3 = 0$, $b_0 = 1420$, $b_1 = 1$, $b_2 = 0$, $b_3 = -0.49$.
IGBT thermal network: $R_1 = 1.601$ K/kW, $R_2 = 1.765$ K/kW, $R_3 = 0.358$ K/kW, $R_4 = 0.328$ K/kW, $C_1 = 0.362898$ kJ/K, $C_2 = 0.033428$ kJ/K, $C_3 = 0.01676$ kJ/K, $C_4 = 0.003049$ kJ/K.
MTDC system: $P_{rate} = 400$ MW, $V_{dc1,2} = 200$ kV, $V_{dc3} = 100$ kV, $L_{1-4} = 100$ mH.
SST under 300/180/60 Hz: $C_{SM}^{MMC_H} = 3/12/20$ mF, $C_{SM}^{MMC_L} = 3/5/10$ mF, $L_{u,d} = 10/15/50$ mH; Y–Y MFT, 600 MVar, 110/55 kV; MMC_H 55-level, MMC_L 31-level.
Transmission line: 100 km, $r = 0.01$ Ω/km, $l = 0.1$ mH/km, $C = 0.2$ μF/km.
Converter Stations: 5-level, $L_{u,d} = 20$ mH, $r_{u,d} = 0.1$ Ω, $C_{SM}^{MMC_{1-3}} = 10$ mF, $V_{g1,2} = 134$ kV, $V_{g3} = 67$ kV.

A.9 Chapter 10

A.9.1 HHB Case

Three-terminal DC system: $Z_{ac} = 0.1 + j11.3$ Ω, $C_e = 500$ μF, $P_{rec} = 400$ MW, $U_{dc1,2} = 200$ kV, $I_{rec}^{rated} = 2$ kA.
Transmission line: $r_0 = 0.012$ Ω/km, $l_0 = 0.106$ mH/km, $c_0 = 0.296$ μF/km, $D = 200$ km.
HHB: $R_s = 10$ Ω, $C_s = 30$ μF, $V_{MOV}^{ref} = 340$ kV, $I_{ref} = 2$ kA, $N_{hcb} = 100$ (see Table A.6).

A.9.2 UFMCB Case

Four-terminal DC grid: MMC level 51, $C_{SM} = 5$ mF, $L_{u,d} = 50$ mH, $L_{dc} = 100$ mH, $C_{b1,2} = 100$ μF, $V_g = 135$ kV/60 Hz, $V_{dc} = \pm 100$ kV.
UFMCB: $r_{11} = 50$ Ω, $r_{12} = 100$ Ω, $r_2 = 4$ kΩ, $C_{11} = 300$ μF, $C_{12} = 114$ μF, $C_2 = 7.8$ μF, $V_{M_0} = 18.33$ kV, $V_{M_{11}} = 11.67$ kV, $V_{M_{12}} = 133.33$ kV, $V_{M_2} = 300$ kV.
UFD arc model: $\tau_M = 0.6$ μs, $P_M = 100$–1000 kW.

Table A.6 IGBT parameters as a function of junction temperature.

Parameter	Coefficient *k*	Coefficient *p*
V_t	−0.012221	8.018885
a	38.3699×10^{-6}	0.004176
b	-0.7738×10^{-6}	464.9903×10^{-6}
x	−0.0013681	1.353853
y	-852.9×10^{-6}	1.475723
z	-982.22×10^{-6}	1.062776

A.10 Chapter 11

A.10.1 CIGRÉ B4 DC Grid Test System

System voltage: Onshore AC grid 380 kV; offshore AC grid 145 kV; DCS1, DCS2 ±200 kV; DCS3 ±400 kV.

AC bus (type/generation (MW)/load (MW)): Ba-A0 (Slack/-/-), Ba-A1 (PQ/2000/1000), Ba-B0 (Slack/-/-), Ba-B1 (PQ/1000/2200), Ba-B2 (PQ/1000/2300), Ba-B3 (PQ/1000/1900), Ba-C1 (PQ/500/0), Ba-C2 (PQ/500/0), Ba-D1 (PQ/1000/0), Ba-E1 (PQ/0/100), Ba-F1 (PQ/500/0).

Converter station power rating: Cm-A1, Cm-C1, Cm-B2, Cm-F1 800 MW; Cm-E1 200 MW; Cm-B3 1200 MW; Cb-A1, Cb-B1, Cb-B2 2 × 1200 MW; Cb-C2 2 × 400 MW; Cb-D1 2 × 800 MW.

Index

a

Adams-Bashforth (A-B) method 161
Adaptive computing acceleration platform (ACAP) 2
Adaptive time-step control 245
Adaptive time-stepping (ATS) schemes
  diode full-bridge circuit 224–225
  DVDT/DIDT method 221
  in EMT simulation 220
  FPGA implementation 229–234
  iteration count method 221
  LTE method 220–222
  nonlinear element solution method 219–220
  oscilloscope 234–236
  power transmission system 225, 227–228
  restrictions and measurements for real-time emulation
    conservative time-step criteria 222–223
    foreknown transient information 223
    limiting the range of time-steps 222
    precalculating constants for different time-steps 223
    smaller maximum iteration count number 223
  scalability 234–236
  ULM (*see* universal line model (ULM))
  UM model (*see* Universal machine (UM) model)
Adaptive/variable time-step 217
ADC *see* Analog-to-digital converter (ADC)
Address generation unit 68
Advanced eXtensible Interface (AXI) channels 6
Alarm signal 214
Analog-to-digital conversion delay 218
Analog-to-digital converter (ADC) 194
Anode–emitter voltage 278
ANSYS/Simplorer® 429, 430
Application processing unit (APU) 365
Average value model (AVM) 301, 471
Average value models based on switching functions 253

b

Backward Euler (BE) 220, 257, 304
BE *see* Backward Euler (BE)
Behavioral model 254
Bergeron line model 225
Block RAM (BRAM) 4
Boltzmann constant 278
Breakers
  DC circuit (*see* DC circuit breakers)
Buck–Boost converter
  hardware implementation 380–381
  system-level modeling 379–380

Real-Time Electromagnetic Transient Simulation of AC-DC Networks, First Edition.
Venkata Dinavahi and Ning Lin.

C

Capacitance *C* 22–23
Capacitors' TLM 268
Capacitor switching transients 99
Capacitor units 263
Cauer equivalent circuit 91
CCCS *see* Current-controlled current source (CCCS)
CFM *see* Curve-fitting model (CFM)
Cholesky–conjugate gradient method 188
Chopper circuit hardware emulation
 IGBT/diode NBM 268–270
CIGRÉ DC grid
 AVM 471
 control scheme 467–468
 converter stations, modeling schemes 469, 472
 device-level electrothermal model 469, 470
 equivalent circuit model 469–470
 network topology 467
 overview 465
 real-time MPSoC-FPGA 471–475
 strategy, hybrid modeling 468
 test system 467
 transmission line model 471
Circuit partitioning method 326, 340
Clamped double submodule (CDSM)
 advantages 356
 arm current transients 371–372
 charge 447
 converter-level modeling scheme 362–363
 DC power flow control 368–371
 device-level electrothermal model 356
 device-level modeling
  linearized transient waveform calculation 361–362
  temperature-dependent electrical interface parameter calculation 359–361
 hardware implementation
  design partition 365–367
  latency and resource consumption 367–368
 junction temperatures 373
 operation principles 357–359
 real-time power losses 372
 SM-level modeling scheme 362
 steady-state results 368
 topology 356
Column-based ASMBL architecture 2
Commercial EMT-type simulators 217
Communication 7–9
Communication delay 218
Companion matrix 201, 203
Compensation method 64–65
$Compi_{dq0}$ unit 151, 152
Complete IGBT dynamic electrothermal model 297
Computer-aided design (CAD) tools 2
Configurable logic block (CLB) 3–4
Continuous-time behavioral model 259, 260
Control data memory 232
Control modules 454, 455
Control system 217, 218
Convolution unit
 FDLM 46–47
 frequency-dependent line model (FDLM) 46–47
 universal line model (ULM) 52–53
Coordinate rotation digital computer (CORDIC) 68, 89
Coupled voltage–current sources
 HIL emulation results 347–348
 islanded MMC performance 348–355
 medium-voltage DC system 346–347
 MMC-MVDC performance 355
 V–I coupling 344–346
Current-controlled current source (CCCS) 297
Curve-fitting model (CFM) 254, 255
 IGBT/diode 292–299
Cyber-physical systems (CPS)
 analysis 511
 application-level synchronization 483
 communication link failure 492–494
 communication network simulators 481–482

co-simulation 482
discrete-time process 481
domain-level synchronization 482–483
Ethernet 484
FPGA-MPSoC based power-communication 484
hardware implementation, RTCE 487–490
hardware resource consumption 491
high-speed computation 484
interaction, power system and communication network simulation 483
over-current fault 492, 493
phasor measurement unit 482
power system and communication network 481
processing delay and hardware resource cost 491–492
PSCAD/EMTDC® off-line simulator 481
research, smart grid 479
RTCE
  architecture 485
  communication network emulation 485–486
  communication protocol 489
  function-level networking 490
  hardware block design, single FPGA/MPSoC board 487–488
  multi-board EMT emulation 487–488
  PMU and PDC concepts 484
  power-communication co-emulation 486–487
  power system emt emulation 484–485
  STDC 484
  test, IEEE 39-bus system 487
  transient data concentrator (TDC) 484
  transient measurement unit 484
  transmission-level networking 489–490
TCP/IP 484

d

DAC *see* Digital to analog converter (DAC)
DC circuit breakers
  CFM 397
  development of DC grids 397
  FPGA 397
  full-scale breaker model 398
  HHB
    MTDC system (*see* Hybrid HVDC breaker (HHB))
    proactive 402–425
  HVDC 397
  IGBT 397
  NBM 398
  power-electronics-based 397
  UFMCB (*see* Ultrafast mechatronic circuit breaker (UFMCB))
DC–DC buck converter circuit 285, 286
DC-DC converters
  detailed MMC-based solid-state transformer (*see* Solid-state transformer modeling)
  nonisolated four-quadrant Buck-Boost converter (*see* Buck-Boost converter)
DC line protection
  concepts 401
  over current protection (OCP) 401
  under-voltage protection (UVP) 401
  voltage derivative protection (VDP) 401
DC motor transients
  real-time trace 178
  speed drive applications 177
Detailed equivalent model (DEM) 253, 381
Device-level models 255
Device-level power electronics circuit simulation tools 217
Device-level transient waveforms 293, 294
Device under-test (DUT) 217
Digital signal processing (DSP) 194, 387
  slices 2, 4–6
  software design 215

Digital to analog converter (DAC) 210
Diode full-bridge circuit
  diode model description 224–225
  off-line circuit simulation results 225, 226
Directional overcurrent protection HEBB 205, 206
Discrete Fourier Transform (DFT) with DC offset removal HEBB 198–200
Discrete-time linearized equivalent circuit, power diode 274
Discrete-time linearized IGBT model 279
Discrete-time Norton equivalent 21
  circuit 256
Discretization methods 220
Distance protection HEBB
  hardware implementation 207
  instantaneous-signal-based distance relay element 208–209
  measurement and evaluation 205
  mho/quadrilateral characteristics 207
  phase-to-ground and phase-to-phase impedance 206
  zone protection 207
Distributed parameter traveling wave model 225
Double data rate fourth-generation (DDR4) memory 7
Drive circuit delay 218
DSP *see* Digital signal processing (DSP)
DUT *see* Device under-test (DUT)
DVDT/DIDT method 221
Dynamic curve-fitting model (DCFM) 378

e

Eddy current transients 97–99
Electrical source module 233
Electro-magnetic transient (EMT)
  simulation 143, 217, 253, 301
  simulators 253
  tools 398
  type solvers 253
Electromechanical (EM) relays 193
Electro-thermal network
  HHB performance 412
  IGBT/diode NBM 267–268
Emitter–base capacitance 278
Emitter–base diffusion capacitance 278
Emitter–base junction depletion capacitance 278
Emitter recombination unit 275
EMT *see* Electro-magnetic transient (EMT)
Euler method 183

f

FACTS *see* Flexible AC transmission system (FACTS)
Faster-than-real-time (FTRT) emulation 237, 245–246, 249, 251
  AC grid
    AC network model 501
    synchronous machine model 499–501
  DC grid
    MMC average value model 501–502
    modular multilevel converters 501, 502
  five stage, DSA systems 494
  flexible time-stepping algorithm
    local equipment based 496–497
    transient stability emulation methodology 496
  FPGAs
    execution, various contingencies 504, 505
    generator outage and sudden load change 507–510
    Governor modules 505
    hardware configuration 503
    hardware resources 500, 504–505
    hardware resources and synchronous generators 504, 505
    IEEE 118-bus systems 502
    integrated AC–DC grid 503

predictive control, power system stability analysis 509
three-phase-to-ground fault 506–507, 510
VCU128 board 502, 504
VCU118 evaluation boards 502, 504
integrated AC/DC grid 497–499
measurements 495
ratio 512
SCADA system 495
Fast Fourier transform (FFT) analysis 303
Fault detected signal 199, 212
Fault detection HEBB 203–204
Fault transient emulation 123
FCDFT *see* Full-cycle DFT (FCDFT)
Ferroresonance transients 96–97
Field-circuit coupling 130–132
Field programmability (FP) 194
FPGA design approaches
HDL 159
TPL 159
Field-programmable gate arrays (FPGAs) 194, 255, 301, 378
block RAM 4
communication 7–9
configurable logic block 3–4
control data memory 232
design approaches
HDL 159
TPL 159
digital signal processing slice 4–6
electrical source module 233
FIFO memory receives 232
finite-state machine 229
floating-point implementation, VHDL 162–167
FSM 232–233
FTRT emulation (*see* Faster-than-real-time emulation)
Gauss elimination 231
Gauss–Jordan method 229, 231
hardware architecture 2–3
hardware implementation and data flow 229, 230
hardware resource utilization 233
initialization and loading process 229
initialization module 233
linear element module 233
LU factorization 231
matrix formation 229
matrix solution module 233
multiprocessing system-on-chip architecture 6–7
nonlinear element module 233
number representation 161–162
parallel design architecture 14–15
pipeline design architecture 14
point to on-chip heterogeneous integration 2
programmable array logic (PAL) devices 1
pseudo-code for time-step control module 231–232
real-time emulation flow control 229, 231
real-time I/O control module synchronizes and exchanges data 232
SRAM 1
state latency 233–234
FIFO memory receives 232
Finite element (FE) based numerical model 79
Finite-element method (FEM) 143, 253
Finite state machine (FSM) 232–233, 334
controller 199, 203
nonlinear behavioral IGBT model 263, 264
power diode behavioral model hardware implementation 258–259
power diode hardware module 275
Firm real-time 218
Fitting method 361
Fixed-point implementation, schematic method
design configuration 169
FLPMAS module 168

Fixed-point implementation, schematic method (*contd.*)
  state-space variable calculator 170
  transformation module 168
  Xilinx System Generator® (XSG) 167
Fixed-point number 13
Fixed time-step (FTS) 217
  PSCAD and hardware emulator 248
Flexible AC transmission system (FACTS) 253
Floating-point implementation, VHDL
  electrical machine module 162
  FSM 163
  main control module 162
  matrix-vector multiplication 167
  new_phase 163
  pipelined configuration 164
  source module 162
  SprMxMul 165, 166
  switch module 162
  timer & switch 164, 165
Floating-point multiplier (FPMUL) 55
Floating-point multiply-accumulator (FPMAC) 155
Floating-point multiply-add/ subtract (FLPMAS) 165
Floating-point number 13–14
Flux & Torque unit 151, 152
Forward Euler (FE) 220
Forward recovery unit 275
FP *see* Field programmability (FP)
FPGA mezzanine (FMC) interface 7
FPGAs *see* Field-programmable gate arrays (FPGAs)
Frequency-dependent eddy current model 90–91
Frequency-dependent line model (FDLM) 32
  backward traveling functions 41
  discrete-time equivalent network 45
  forward traveling functions 41
  hardware emulation
    convolution unit 46–47
    update unit 47
  modal transformation matrices 45
  weighting function 42
Frequency-domain formulation 48–49
FSM *see* Finte state machine (FSM)
FTRT *see* Faster-than-real-time (FTRT)
FTS *see* Fixed time-step (FTS)
Full-bridge SM (FBSM) 357
Full-cycle DFT (FCDFT) 198–199
Full detailed models 253
Full physics-based models 253

g

Gauss elimination 231
Gauss–Jordan elimination (GJE) 90, 149
Gauss–Jordan method 229, 231
General HHB unit model 406–407
General purpose IO (GPIO) 7
Governor modules 505
Gram–Schmidt process 201, 203
Graphics processing units (GPUs) 1
Ground fault transients 99–100

h

Half-bridge circuit topology 295
Half-bridge submodule (HBSM) 302
Hammerstein configuration 255
Hard real-time 218
Hardware description language (HDL) 1, 159
Hardware design of HHB
  abc-dq transform module 413
  electro-thermal network 413
  longest hardware delay, MTDC system 413–414
  NBM-based 413
  resource utilization 412–413
  three-terminal HVDC system 413
Hardware emulation
  emulation procedure 185
  finite-state machine 184
  global control 185
  main control module 185
  MatrxAcc submodule 187
  modules 183
  nonlinear evaluation unit 187
  Parallel Gauss–Jordan elimination unit 185–186

Hardware emulation building blocks (HEBBs)
 concept 18
 CORDIC 196–198
 finite-state machine 149
 FPMAC 155
 GJE 149
 input/output signals 148
 linear lumped passive elements 26–27
 numerical integration 18–20
 power grids 18
 PS HEBB 154
 real-time transients 158
 resources 156
 single-line diagram 157
 sources 27–30
 switches 30–31
 transmission lines (*see* Transmission lines)
Hardware-in-the-loop (HIL)
 emulation 2, 378
  FPGA design schemes 14–15
  FPGA experiment 15
  number representation and operations 13–14
  Vivado® high-level synthesis tool 9–11
  Vivado® top-level design 11–12
 simulation 255
 testing 217
Hardware module parallelism
 adaptive look-up table 451–452
 control module 452
 module-level 452
 resources 452
 system-level 452
 ULM 452
 universal machine 452
Hardware multifunction protection system
 architecture 195
 different sampling rates 213
 fault detected signal 212
 power system 213–214
 PRH (*see* Protective relay HEBB (PRH))
 real-time test 211–212
 resource consumption and latencies 209, 210
 SPH 196–203
 Tektronix® 1 GHz DPO4104 210
 test power system 210–211
 trip times, types of faults 212
 Xilinx® Virtex-7® XC7VX485T FPGA 209
Hardware resource utilization
 fixed-point arithmetic operations 173
 floating-point arithmetic operations 174
Hefner's model 292
HHB EMT model 403–404
High-fidelity nonlinear MEC-based ST hardware emulation 109–112
High-level synthesis (HLS) 233
High-performance compute (HPC) 2
High-voltage direct current (HVDC) 377
 grid 302
 power transmission 253
HIL emulation, HHB
 DC circuit breaker 416
 device-level behavior
  energy consumption 421, 423, 424
  full-scale HHB model 421, 423
  junction temperature variation 420–421
  SaberRD® 416, 419
  transient waveforms 417–418
  varistor voltage and line current 421–423
  voltage and current waveforms 416, 418–420
 MMC model 416
 system-level performance 424–425
 three-terminal HVDC system 416
HLS *see* High-level synthesis (HLS)
HSPICE® 217
HVDC circuit breaker model 398
HVDC power transfer corridor with HHB 404

Hybrid HVDC breaker (HHB)
  MTDC test system schematic
    AVM-based 400
    concepts, DC line protection 401
    control scheme of MMC 400–401
    guarantees 400
    MMCs 400
    three-terminal HVDC system installation 399–400
  proactive
    circuit configuration, unidirectional HHB 402
    electro-thermal network 412
    EMT simulation 403–404
    general HHB unit model 406–407
    hardware implementation 412–416
    HIL emulation outcomes 416–425
    IGBT low-order nonlinear behavioral model 409–412
    two-node IGBT models 407–409
    varistor model 404–406
Hysteresis behavior 94–96
Hysteresis transient emulation 118

i

IGBT/diode CFM
  characteristics 292
  DCFM 293, 296–298
  hardware emulation 298–299
  linear static 293–296
  SCFM 292
IGBT/diode NBM
  CFM 255
  chopper circuit hardware emulation 268–270
  electro-thermal network 267–268
  Hammerstein configuration 255
  hardware module architecture 263–265
  macro-model 255
  model formulation 259–263
  multiple parallel devices 265–266
  power diode 256–259
IGBT low-order nonlinear behavioral model
  advantage 409
  disadvantage 409
  fourth-order behavioral model 409, 410
  parallelization 411–412
  parameters extraction 409–410
  sensitivity analysis 410–411
  TSSM 409
IGBTs *see* Insulated gate bipolar transistors (IGBTs)
IGBT transient waveforms 297
Induction machine (IM) 328
  electromagnetic torque 190
  magnetic flux density distribution 189
  N-R method 189
  resource utilization 188
  start-up transient 189
  stator current 190
  2-D nonlinear time-stepping transient 188
Induction motor transients 174
Initialization module 233
Initial value problem (IVP) 18
Inode unit module 84–85
Inrush current transients 94
Instantaneous-signal-based distance relay element 208–209
Insulated gate bipolar transistors (IGBTs) 302, 378
  and diode NBM 255–270
  hardware resources utilization 265
  low-order nonlinear behavioral model 409–412
  vs. power diode nonlinear models 269–270
  two-node 407–409
Interfacing of magnetic and electric circuits 181–182
Interpolation unit
  universal line model (ULM) 54
Inverse matrix calculator (IMC) 111
Iteration count method 221

j

Jacobian matrix 183
Junction capacitance unit 275

k

Katzenelson methods 217
Kirchhoff's current law (KCL) 32
Kirchhoff's voltage law (KVL) 32

l

Large-scale AC and DC networks
 420-bus power systems 510
 CDSM charge 447
 EMT simulators 448
 FPGAs 448
 FTRT emulation 494–507
 functional decomposition 449–451, 507
 hardware module parallelism 451–452
 HHBs 447
 high-voltage direct current transmission 447
 hybrid modeling methodology, CIGRÉ DC grid (*see* CIGRÉ DC grid)
 hybrid modeling scheme 511
 limitations, HVDC 447
 neutral-point-clamped converter 447
 overview 447–449
 real-time co-emulation framework, CPS (*see* Cyber-physical systems (CPS))
 three-phase 42-bus power system 510
Large-scale real-time EMT simulation
 functional decomposition
  conventional network partitioning 451
  digital hardware emulation 450
  drawbacks 449
  external system, FDNE 450
  multi-FPGA architecture 451
  power system for hardware emulation 450, 451
  smaller subsystems 449, 450
 multi-FPGA hardware design (*see* Multi-FPGA design)
Lauritzen's diode model 292
LCC *see* Line commutated converter (LCC)
Linear curve-fitting IGBT/diode switch model 324
Linear element module 233
Linear inductive element 21
Linear interpolation unit 68
Linearized discrete-time equivalent circuit 256, 259, 260
Linear lumped passive elements
 hardware emulation 26–27
 model formulation
  capacitance *C* 22–23
  inductance *L* 20–21
  LC branch 23–24
  resistance *R* 20
  RL branch 23
  RLCG branch 24–26
Linear model formulation 80–82
Linear module hardware design 82–84
Linear multistep (LMS) method 220–221
Linear static CFM
 static characteristics 293, 294
 switching transients 293–296
Line commutated converter (LCC) 253
Line start-permanent magnet synchronous motor (LSPMSM)
 LSPMSM performance 177
 oscilloscope traces 176
Local truncation errors (LTE) method 219–222, 249
Look-up table (LUT) 378
LU-based decomposition (LUD) 111
LU factorization 231
Lumped models 79
Lumped universal machine (UM) model
 $Compi_{dq0}$ unit 151
 Flux & Torque unit 151
 formulation 144–146
 FrmTran unit 150–151
 HEBB 148–149
 power system network 146–148

Lumped universal machine (UM) model (*contd.*)
  speed & angle unit 149–150
  update & compVc unit 151–152

m

Magnetic circuit
  air-gap permeance 181
  magnetic flux prism 180
  rotor and stator sectors 179
  trapezoidal flux prisms 179
Magnetic equivalent circuit (MEC) 143
Magnetomotive force (MMF) 100
Main breaker (MB) 402
MainControl module 58–59
Matrix formation 229
Matrix solution module 233
Matrix vector multiplication (MVM) unit 82
Maximum torque angle (MTA) 205
Medium-frequency transformer (MFT) 381
Medium-voltage DC (MVDC) system 346–347
Metal-oxide semiconductor field-effect transistor (MOSFET) 261
Metal oxide varistor (MOV) 402
Microprocessor-based relay 194
Miller capacitance 261
MMC *see* Modular multilevel converter (MMC)
MMC-based HVDC application 255
Modular multi-level converter (MMC) 254, 377
  clamped double submodule (*see* Clamped double submodule)
  coupled voltage-current sources (*see* Coupled voltage-current sources)
  detailed model 303–304
  EMT simulation 303
  FFT analysis 303
  finite state machine 315
  half-bridge submodule 302
  hardware emulation 314–316
  Norton equivalent circuit 305
  single-phase five-level MMC (*see* Single-phase five-level MMC)
  sub-module
    hardware unit emulation on FPGA 311–314
    power losses calculation method 307–309
    thermal network calculation 309–311
    Thévenin equivalence 306–307
    two-level half-bridge topology 305
  Thévenin equivalence circuit 304–305
  three-phase nine-level MMC (*see* Three-phase nine-level MMC)
  virtual-line-partitioned device-level models (*see* Virtual-line-partitioned device-level models)
Modular multi-level converter (MMC)-HVDC 447
MOSFET *see* Metal oxide semiconductor field-effect transistor (MOSFET)
MPSoC-FPGA based DC grid emulator
  DC fault 477, 479
  design and implementation 473, 474–475
  device-level steady-state outcomes 477, 478
  hardware resource allocation 472–474
  power flow command change 477, 479
  system decomposition 471–472, 473
  system-level steady-state outcomes 475–477
  task partitioning 472–474
MTA *see* Maximum torque angle (MTA)
MTDC system
  pipelined hardware structure 414–415
  top-level state machine 415–416
Multi-FPGA design
  Dini Group® 453
  3-FPGA hardware architecture 454–457

10-FPGA hardware design 457–460
global clock networks 453
LVDS 453
oscilloscope 453
performance and scalability
calculation, minimum time-step 464–466
execution times, functional module 464
simulation time-step 460, 462
specific time-step 465
programmable clock synthesizers 453
prototyping board 453, 454
variables, real-time simulation 453
Multi-processing system-on-chip (MPSoC) 2
architecture 6–7
Multi-terminal direct current (MTDC) system 364, 377

n

NBM *see* Nonlinear behavioral model (NBM)
Network solver
hardware emulation 55
MainControl module 58–59
modules 454, 455
overview 54
paralleled EMT solution algorithm 55–58
real-time emulation case study 59–63
Neutral-point clamped (NPC) converter 301
Newton–Raphson (NR) method 65–67, 217, 219
NLD *see* Nonlinear diode (NLD)
Nonlinear admittance-based real-time transformer model
frequency-dependent eddy current model 90–91
hardware emulation of power transformer 91–94
inode unit module 84–85
linear model formulation 80–82
linear module hardware design 82–84
nonlinear model solution 85–87
hardware design 89–90
Preisach hysteresis model 88–89
real-time emulation case studies
capacitor switching transients 99
eddy current transients 97–99
ferroresonance transients 96–97
ground fault transients 99–100
hysteresis behavior 94–96
inrush current transients 94
Nonlinear behavioral IGBT model
FSM 263, 264
hardware architecture 263, 264
Nonlinear behavioral model (NBM)
IGBT/diode 255–270
static and dynamic device characteristics 255
Nonlinear DCFM 293, 296–298
Nonlinear device-level thyristor model
basic device characteristics
digital simulation 427–428
reverse-recovery diode model 426, 427
reverse recovery phenomenon 428
SCR logic block 428
static on-state 426–427
scalable cascaded thyristor model
ANSYS/Simplorer® 429, 430
inductor's companion circuit 430
mandatory requirement 429
nodes 429
reverse recovery process 429–431
SCR logic validation 429–430
Nonlinear diode (NLD) 256
Nonlinear element module 233
Nonlinear elements
case studies 71–77
compensation method 64–65
hardware emulation of nonlinear solver 67–68
computing vc 71
nonlinear function evaluation 68

Nonlinear elements (*contd.*)
parallel calculation of J and $F(i_{km})$ 68–70
parallel Gauss–Jordan elimination 71
Newton–Raphson method 65–67
Nonlinear element solution method
and dynamic time-stepping schemes 222
NR 219
PNR 220, 222
PWL 219, 222
Nonlinear function evaluation 68
Nonlinear IGBT EMT model 259, 260
Nonlinear MEC model
Euler method 183
hardware emulation (*see* Hardware emulation)
interfacing of magnetic and electric circuits 181–182
Jacobian matrix 183
magnetic circuit 179–181
stator electric circuit 182
time-marching transient simulation 183
Nonlinear model solution 85–87
hardware design 89–90
Preisach hysteresis model 88–89
Nonlinear submodule (NSM) 388
Nonlinear switch-based model (NSM) 383
Norton equivalent model 260
Numerical integration 18–20
Numerical relays 193, 194

o

Off-line validation 171–172
One-step method 220
Oscilloscope 234–236
Over current protection (OCP) 401
Over/under frequency protection HEBB 209
Over/under voltage protection HEBB 205, 206

p

Parallel design architecture, FPGA 14–15
Paralleled EMT solution algorithm 55–58
Parallel Gauss–Jordan elimination 71, 185–186
Park's transformation matrix 243
Phase-shift control (PSC) 315
Physics-based model 254
Physics-based nonlinear IGBT model
analog equivalent circuit 276, 277
BJT 276–278
discretization and linearization 279–281
hardware emulation on FPGA
architecture 281, 282
capacitance units 281, 283
charge limit unit 283
FSM 283
junction limit units 283
MOSFET 284
operation sequence 283
power converter 284–285
stages 283
MOSFET 276, 277
phenomenological circuit 276, 277
Physics-based nonlinear p–i–n diode model
characteristics 271
contact resistance 272
discretization and linearization 272–274
emitter recombination 272
forward recovery 272
junction capacitance 272
physical structure 271
power diode hardware module
architecture 274, 275
finite state machine 275
reverse recovery 271
Piecewise linear IGBT curve-fitting model 361
Piecewise linearization (PWL) method 219, 222
Piecewise N-R (PNR) method 220

Pipeline design architecture, FPGA 14
Polynomial function 296
Power converters 253, 298
Power diode
  device-level modeling and simulation 257
  diode reverse recovery 256
  diode static characteristics 256
  discrete-time Norton equivalent circuit 256
  hardware module architecture 257–259
  linearized discrete-time equivalent circuit 256
  module 263
  in SaberRD® 256, 257
  simplified power diode model 256
  VCCS 256
Power dissipation 291–292
Power electronic switches
  hardware emulation
    device-level transient results 290
    Hefner's model 292
    Lauritzen's diode model 292
    power dissipation 291–292
    SaberRD®'s model 292
    steady-state results, DC–DC converter 286–289, 291
    switching power loss and conduction power loss 292
    switching times *vs.* power dissipation 290
    test circuit 285–286
    transient results, DC–DC converter 286–289
  IGBT/diode CFM 292–299
  IGBT/diode NBM 255–270
  physics-based nonlinear IGBT model 276–285
  physics-based nonlinear p–i–n diode model 271–276
Power-flow control emulation 118–122
Power-flow regulation 118
Power losses calculation method 307–309
Power semiconductor switch, model categorization 254
Power system 194
Power system hardware emulation building block (*PS* HEBB) 154
Power system network, interfacing UM model
  compensation method 148
  orthogonal Park's transformation 147
  trapezoidal rule 146
Power transformers
  hardware emulation 91–94
  overview 79–80
  real-time finite-element model (*see* Real-time finite-element model)
Power transmission system
  arrester model and lightening surge 227, 539–540
  circuit structure 225, 227
  off-line circuit simulation results 227–228
Preisach hysteresis model 88–89
Preisach Hysteretic Unit (PHU) 89
PRH *see* Protective relay HEBB (PRH)
Proactive hybrid HVDC breaker
  circuit configuration, unidirectional HHB 402
  electro-thermal network 412
  EMT simulation 403–404
  general HHB unit model 406–407
  hardware implementation, FPGA 412–416
  IGBT
    low-order nonlinear behavioral model 409–412
    two-node 407–409
  normal conduction 402
  varistor 404–406
Process-reversed model, ULM 240–241
Programmable array logic (PAL) devices 1
Proposed process-reversed ULM model 247, 248
Protection system 217, 218

Protective relay HEBB (PRH)
directional overcurrent protection HEBB 205, 206
distance protection HEBB 205–209
fault detection HEBB 203–204
over/under frequency protection HEBB 209
over/under voltage protection HEBB 205, 206
Protective relays
EM 193
FP 194
hardware multifunction protection system 195–209
high-capacity and high-bandwidth 194
microprocessor-based 194
numerical 193, 194
in power systems 193
SS 193–194
PSCAD/EMTDC® simulation 416, 417, 421–424
PSCAD/EMTDC® software 223, 227, 228, 234, 243, 247, 248, 250, 252
Pseudo-nonlinear method 219
PSpice® 217, 271
Pulse width modulation (PWM) method 315

q
Quad serial peripheral interface (QSPI) flash 7

r
Real-time emulation
accuracy assessment 170–171
induction motor transients 174, 175
LSPMSM 176–177
off-line validation 171–172
oscilloscope traces 153
resource utilization 188
single-line diagram of power system 153
synchronous generator transients 174–176
three-phase-to-ground fault 153
ULM module 153
Real-time finite-element model
case studies
results and validation 137–140
speed-up and scalability 140–141
hardware emulation of finite element model 132–136
magnetodynamic problem formulation
field-circuit coupling 130–132
refined TLM solution 126–130
Real-time hardware high-fidelity nonlinear MEC-based ST model
energization transient and steady-state emulation 113–117
fault transient emulation 123
hysteresis transient emulation 118
power-flow control emulation 118–122
power-flow regulation 118
Real-time simulators 217, 218
Real-world power system 218
Register transfer level (RTL) 1
Relaxed Scalar Newton–Raphson (RSNR)
admittance matrix and current vector 433
arbitrary node 433, 434
equivalent circuit 434
five-node UFMCB 434, 435
Gaussian Elimination 433
Jacobi method, linear systems 434
nodal voltage 434
Residual current breaker (RCB) 402
Reverse Recovery Unit 275, 276
RL branch 23
RLCG branch 24–26
RMS load-flow models 253
Rotating electrical machines
EMT 143
FEM 143
finite-element simulation 191
hardware designs 191
MEC 143
steady-state 191
UM model (*see* Lumped universal machine (UM) model)
Universal machine (UM) model 143

s

SaberRD® software 217, 221, 223, 225, 226, 234, 256, 257, 259, 262, 268, 269–271, 286, 287, 290–292
Scalability 234–236
SCFM *see* Static curve-fitting model (SCFM)
Semiconductor switch model 254–255
Sen transformer (ST) 100
Series inductance 32
Series resistance 32
Shunt capacitance 32
Shunt conductance 32
Siemens® BSM300GA160D IGBT module 262
Signal flow diagram, MMC-IM system 333
Signal processing HEBB (SPH)
  CORDIC HEBB 196–198
  DFT HEBB 198–200
  symmetrical components HEBB 198
  zero-crossing detection HEBB 199–203
Simplified average value models 253
Simplified power diode model 256
Simplified switchable resistances 253
Single instruction multiple data (SIMD) 1
Single-phase distributed parameter loss-less line 33
Single-phase five-level MMC
  comparison of power dissipations 324
  device-level switching waveforms for IGBTs 323
  junction temperature 322
  system-level and device-level power loss results 318–321
Small form-factor pluggable (SFP) interface 7
Small-scale power converter simulation 255
Soft real-time 218
Solid-state (SS) relays 193–194
Solid-state transformer (SST) 377
  converter performance 391–392
  device-level behavior 390–391
  EMT simulation 385–386
  HIL emulation 386–389
  hybrid arm model 384–385
  medium-frequency transformer 381
  nonlinear switch-based model 383
  switch-based detailed equivalent model 381
  system tests 392–394
  three-phase saturable transformer model 385
  TLM-stub model 382–383
Sources, hardware emulation 28–30
Sparse matrix multiplication *(SprMxMul)* 165
Speed & Angle unit
  pipelined computation scheme 150
  rotor speed 149
SPH *see* Signal processing HEBB (SPH)
State-space electrical machine emulation
  FPGA design approach 159–160
  rotating machine models 158
State-space representation of machine models
  Adams-Bashforth (A-B) method 161
  arbitrary velocity 160
  constant parameters 160
Static curve-fitting model (SCFM) 292
Static random-access memory (SRAM) 1
Static synchronous compensator (STATCOM) 253
Subsystem-based ATS scheme 237–238
Supervisory control and data acquisition (SCADA) 495
Surrogate network method 362
Switches, hardware emulation 30–31
Switching transients 293–296
Switch module 454, 455
Symmetrical components HEBB 198
Synchronous generator transients 174–176

t

Tap-selection algorithm 102
10-FPGA hardware design
  allocation of components 460, 462

10-FPGA hardware design (*contd.*)
architecture 457, 459
demand 460
logic utilization 460
module execution time 460
network nodes 460
off-line simulations 460, 463
real-time oscilloscope traces 460, 463
single-line diagram 460, 461
UM and NR modules 460
Textual programming language (TPL) 159
Thermal network (THM) 388
Thévenin equivalence 254, 316
Thévenin equivalent circuit 304
Thévenin voltage source 242
3-FPGA hardware design
architecture 454
configuration 455
control modules 454, 455
data transfer 455
interconnected signals 454
MainBus 455
network solver modules 454, 455
resource utilization 455
RLCG elements 456
spatiotemporal design workflow 456, 457
switch module 454, 455
test 456
three-phase voltages waveforms 456, 458
Three-phase nine-level MMC
average power dissipation 318
hardware resource utilization 316–317
operator numbers 317
system-level and device-level results 324–325
Thyristor-based LCC 253
Time-domain formulation 49–51
TLM *see* Transmission line modeling (TLM)
Topological ST EMT model
high-fidelity nonlinear MEC-based ST model 102–107
iron core hysteresis and eddy currents 107–109
ST operating principle 102
tap-selection algorithm 102
TR *see* Trapezoidal rule (TR)
Traditional ULM model 247, 248
Transformation unit 39
traveling wave model (TWM) 39
Transient network analyzers (TNAs) 17
Transient recovery voltages (TRV) 97
Transmission line modeling (TLM)
capacitors 268
links 326
method 80
Transmission lines
frequency dependent line model 41–45
hardware emulation of FDLM
convolution unit 46–47
update unit 47
parameters 32
traveling waves 32–35
TWM (*see* Traveling wave model (TWM))
ULM (*see* Universal line model (ULM))
Trapezoidal integration rule 306
Trapezoidal method 353
Trapezoidal rule (TR) 18, 220, 257, 272–273, 304
Traveling wave model (TWM) 32, 404
discrete-time model 35
discrete-time Norton equivalent circuit 36
hardware emulation
transformation unit 39
update unit 39–41
modal transformation 36–39
Traveling waves 32–35
TSSM *see* Two-state switch model (TSSM)
2-D configurable logic blocks (CLBs) 2
Two-state switch model (TSSM) 254

u

ULM *see* Universal line model (ULM)
Ultrafast disconnector (UFD) 402

Ultrafast mechatronic circuit breaker (UFMCB)
  decoupled model 427
  direct integration 426
  EMT simulation 426
  hardware design 435–438
  HVDC grid 442–444
  HVDC power transmission 426
  installation 426, 427
  modelling 431–433
  nonlinear device-level thyristor model 426–431
  real-time tests and validation
    ANSYS/Simplorer® 439, 440
    DC circuit breaker hardware design 438
    four-terminal DC grid test case 438
    impact of UFD 441
    outcomes 438, 439
    power loss 440
    power loss and junction temperature 440–441
    PSCAD/EMTDC® 439, 440
    terminal oscilloscope results 438, 440
  RSNR 433–435
  single thyristor symbol 426
UltraScale+ XCVU9P device 243
UM model formulation
  EMTP framework 145
  leakage inductances 145
  mechanical and electrical quantities 146
  winding representations 144
Under-voltage protection (UVP) 401
Unified power flow controller (UPFC) 253
Unit-delay methods 345
Universal line model (ULM)
  electromagnetic behavior 238
  EMT simulation 239
  equivalent circuit 238, 239
  frequency-domain formulation 48–49
  FTRT 237
  hardware emulation
    convolution unit 52–53
    interpolation unit 54
    update x unit 52
  lightning transient 248, 250, 251
  process-reversed model 240–241
  ransmission lines and rotating machines 237
  residue matrix 239
  time-domain formulation 49–51
  time–domain relationship 238–239
  traditional model 240
  transient behavior 248
  validation 247–248
  VCU118 boards (*see* VCU118 boards)
Universal machine (UM) model
  damper windings 242
  equivalent circuit 242, 243
  exciter module 243
  FTRT 237
  lightning transient 248, 250, 251
  minimum time-step size 249
  ransmission lines and rotating machines 237
  stator windings 242
  Thévenin voltage source 242
  transient behavior 248
  VCU118 boards (*see* VCU118 boards)
Update unit 151
  frequency-dependent line model (FDLM) 47
  traveling wave model (TWM) 39–41
UPFC *see* Unified power flow controller (UPFC)

V

Validation, ULM 247–248
Vandermonde matrix 201
Variable time-step control module (VTCM) 284
Varistor model
  companion 405
  Matlab off-line simulation, HHB 405
  nonlinear 405
  protection voltage 406
  rating of the 404–405

VCCS *see* Voltage-controlled current source (VCCS)
VCU118 boards
  adaptive time-step control 245
  FTRT 245–246
  hardware resource utilization 246–247
  latency 246–247
  subsystem allocation 245
  test system and hardware emulation 244, 245
  Virtex® UltraScale+ FPGA 243
VCU118 FPGAboard 475
Very high-speed integrated circuit hardware description language (VHDL) 215, 335, 388
Very large-scale integration (VLSI) 194
VHDL *see* Very high-speed integrated circuit hardware description language (VHDL)
Virtual-line-partitioned device-level models
  hardware design of MMC-IM system
    control algorithm 331
    controller emulation 329–330
    emulation on FPGA 330–331
    finite state machine 334
    hardware resources 329
    signal flow diagram 333
  hardware-in-the-loop emulation 326
  real-time emulation results
    induction machine driven 342–344
    MMC 335–342
  transmission line modeling links 326–328
Vivado HLS® 257, 413
Voltage-controlled current source (VCCS) 256, 261, 297
Voltage derivative protection (VDP) 401
Voltage source converters (VSCs) 253, 301
  average value models 255
  in DC grid 254
VSCs *see* Voltage source converters (VSCs)
VTCM *see* Variable time-step control module (VTCM)

X
Xilinx® 64-bit Aurora core 488
Xilinx ISE® environment 209–210
Xilinx System Generator® (XSG) 167
Xilinx® XCVU9P FPGA 475

Z
Zero-crossing detection HEBB 199–203

Books in the IEEE Press Series on Power Engineering

Series Editor: M E El-Hawary, Dalhousie University, Halifax, Nova Scotia, Canada

The mission of IEEE Press Series on Power Engineering is to publish leading-edge books that cover the broad spectrum of current and forward-looking technologies in this fast-moving area. The series attracts highly acclaimed authors from industry/academia to provide accessible coverage of current and emerging topics in power engineering and allied fields. Our target audience includes the power engineering professional who is interested in enhancing their knowledge and perspective in their areas of interest.

1. *Electric Power Systems: Design and Analysis, Revised Printing*
   Mohamed E. El-Hawary
2. *Power System Stability*
   Edward W. Kimbark
3. *Analysis of Faulted Power Systems*
   Paul M. Anderson
4. *Inspection of Large Synchronous Machines: Checklists, Failure Identification, and Troubleshooting*
   Isidor Kerszenbaum
5. *Electric Power Applications of Fuzzy Systems*
   Mohamed E. El-Hawary
6. *Power System Protection*
   Paul M. Anderson
7. *Subsynchronous Resonance in Power Systems*
   Paul M. Anderson, B.L. Agrawal, J.E. Van Ness
8. *Understanding Power Quality Problems: Voltage Sags and Interruptions*
   Math H. Bollen
9. *Analysis of Electric Machinery*
   Paul C. Krause, Oleg Wasynczuk, and S.D. Sudhoff
10. *Power System Control and Stability, Revised Printing*
    Paul M. Anderson, A.A. Fouad

Real-Time Electromagnetic Transient Simulation of AC-DC Networks, First Edition.
Venkata Dinavahi and Ning Lin.

11. *Principles of Electric Machines with Power Electronic Applications,* Second Edition
Mohamed E. El-Hawary

12. *Pulse Width Modulation for Power Converters: Principles and Practice*
D. Grahame Holmes and Thomas Lipo

13. *Analysis of Electric Machinery and Drive Systems*, Second Edition
Paul C. Krause, Oleg Wasynczuk, and S.D. Sudhoff

14. *Risk Assessment for Power Systems: Models, Methods, and Applications*
Wenyuan Li

15. *Optimization Principles: Practical Applications to the Operations of Markets of the Electric Power Industry*
Narayan S. Rau

16. *Electric Economics: Regulation and Deregulation*
Geoffrey Rothwell and Tomas Gomez

17. *Electric Power Systems: Analysis and Control*
Fabio Saccomanno

18. *Electrical Insulation for Rotating Machines: Design, Evaluation, Aging, Testing, and Repair*
Greg C. Stone, Edward A. Boulter, Ian Culbert, and Hussein Dhirani

19. *Signal Processing of Power Quality Disturbances*
Math H. J. Bollen and Irene Y. H. Gu

20. *Instantaneous Power Theory and Applications to Power Conditioning*
Hirofumi Akagi, Edson H. Watanabe and Mauricio Aredes

21. *Maintaining Mission Critical Systems in a 24/7 Environment*
Peter M. Curtis

22. *Elements of Tidal-Electric Engineering*
Robert H. Clark

23. *Handbook of Large Turbo-Generator Operation and Maintenance,* Second Edition
Geoff Klempner and Isidor Kerszenbaum

24. *Introduction to Electrical Power Systems*
Mohamed E. El-Hawary

25. *Modeling and Control of Fuel Cells: Distributed Generation Applications*
M. Hashem Nehrir and Caisheng Wang

26. *Power Distribution System Reliability: Practical Methods and Applications*
Ali A. Chowdhury and Don O. Koval

27. *Introduction to FACTS Controllers: Theory, Modeling, and Applications*
Kalyan K. Sen, Mey Ling Sen

28. *Economic Market Design and Planning for Electric Power Systems*
James Momoh and Lamine Mili

29. *Operation and Control of Electric Energy Processing Systems*
James Momoh and Lamine Mili

30. *Restructured Electric Power Systems: Analysis of Electricity Markets with Equilibrium Models*
Xiao-Ping Zhang

31. *An Introduction to Wavelet Modulated Inverters*
S.A. Saleh and M.A. Rahman

32. *Control of Electric Machine Drive Systems*
Seung-Ki Sul

33. *Probabilistic Transmission System Planning*
Wenyuan Li

34. *Electricity Power Generation: The Changing Dimensions*
Digambar M. Tagare

35. *Electric Distribution Systems*
Abdelhay A. Sallam and Om P. Malik

36. *Practical Lighting Design with LEDs*
Ron Lenk, Carol Lenk

37. *High Voltage and Electrical Insulation Engineering*
Ravindra Arora and Wolfgang Mosch

38. *Maintaining Mission Critical Systems in a 24/7 Environment, Second Edition*
Peter Curtis

39. *Power Conversion and Control of Wind Energy Systems*
Bin Wu, Yongqiang Lang, Navid Zargari, Samir Kouro

40. *Integration of Distributed Generation in the Power System*
Math H. Bollen, Fainan Hassan

41. *Doubly Fed Induction Machine: Modeling and Control for Wind Energy Generation Applications*
Gonzalo Abad, Jesús López, Miguel Rodrigues, Luis Marroyo, and Grzegorz Iwanski

42. *High Voltage Protection for Telecommunications*
Steven W. Blume

43. *Smart Grid: Fundamentals of Design and Analysis*
James Momoh

44. *Electromechanical Motion Devices, Second Edition*
Paul Krause, Oleg Wasynczuk, Steven Pekarek

45. *Electrical Energy Conversion and Transport: An Interactive Computer-Based Approach, Second Edition*
George G. Karady, Keith E. Holbert

46. *ARC Flash Hazard and Analysis and Mitigation*
J.C. Das

47. *Handbook of Electrical Power System Dynamics: Modeling, Stability, and Control*
Mircea Eremia, Mohammad Shahidehpour

48. *Analysis of Electric Machinery and Drive Systems, Third Edition*
Paul C. Krause, Oleg Wasynczuk, S.D. Sudhoff, Steven D. Pekarek

49. *Extruded Cables for High-Voltage Direct-Current Transmission: Advances in Research and Development*
Giovanni Mazzanti, Massimo Marzinotto

50. *Power Magnetic Devices: A Multi-Objective Design Approach*
S.D. Sudhoff

51. *Risk Assessment of Power Systems: Models, Methods, and Applications, Second Edition*
Wenyuan Li

52. *Practical Power System Operation*
Ebrahim Vaahedi

53. *The Selection Process of Biomass Materials for the Production of Bio-Fuels and Co-Firing*
Najib Altawell

54. *Electrical Insulation for Rotating Machines: Design, Evaluation, Aging, Testing, and Repair, Second Edition*
Greg C. Stone, Ian Culbert, Edward A. Boulter, and Hussein Dhirani

55. *Principles of Electrical Safety*
Peter E. Sutherland

56. *Advanced Power Electronics Converters: PWM Converters Processing AC Voltages*
Euzeli Cipriano dos Santos Jr., Edison Roberto Cabral da Silva

57. *Optimization of Power System Operation, Second Edition*
Jizhong Zhu

58. *Power System Harmonics and Passive Filter Designs*
J.C. Das

59. *Digital Control of High-Frequency Switched-Mode Power Converters*
Luca Corradini, Dragan Maksimoviæ, Paolo Mattavelli, Regan Zane

60. *Industrial Power Distribution, Second Edition*
Ralph E. Fehr, III

61. *HVDC Grids: For Offshore and Supergrid of the Future*
Dirk Van Hertem, Oriol Gomis-Bellmunt, Jun Liang

62. *Advanced Solutions in Power Systems: HVDC, FACTS, and Artificial Intelligence*
Mircea Eremia, Chen-Ching Liu, Abdel-Aty Edris

63. *Operation and Maintenance of Large Turbo-Generators*
Geoff Klempner, Isidor Kerszenbaum

64. *Electrical Energy Conversion and Transport: An Interactive Computer-Based Approach*
George G. Karady, Keith E. Holbert

65. *Modeling and High-Performance Control of Electric Machines*
John Chiasson

66. *Rating of Electric Power Cables in Unfavorable Thermal Environment*
George J. Anders

67. *Electric Power System Basics for the Nonelectrical Professional*
Steven W. Blume

68. *Modern Heuristic Optimization Techniques: Theory and Applications to Power Systems*
Kwang Y. Lee, Mohamed A. El-Sharkawi

69. *Real-Time Stability Assessment in Modern Power System Control Centers*
Savu C. Savulescu

70. *Optimization of Power System Operation*
Jizhong Zhu

71. *Insulators for Icing and Polluted Environments*
Masoud Farzaneh, William A. Chisholm

72. *PID and Predictive Control of Electric Devices and Power Converters Using MATLAB®/Simulink®*
Liuping Wang, Shan Chai, Dae Yoo, Lu Gan, Ki Ng

73. *Power Grid Operation in a Market Environment: Economic Efficiency and Risk Mitigation*
Hong Chen

74. *Electric Power System Basics for Nonelectrical Professional, Second Edition*
Steven W. Blume

75. *Energy Production Systems Engineering*
Thomas Howard Blair

76. *Model Predictive Control of Wind Energy Conversion Systems*
Venkata Yaramasu, Bin Wu

77. *Understanding Symmetrical Components for Power System Modeling*
J.C. Das

78. *High-Power Converters and AC Drives, Second Edition*
Bin Wu, Mehdi Narimani

79. *Current Signature Analysis for Condition Monitoring of Cage Induction Motors: Industrial Application and Case Histories*
William T. Thomson, Ian Culbert

80. *Introduction to Electric Power and Drive Systems*
Paul Krause, Oleg Wasynczuk, Timothy O'Connell, Maher Hasan

81. *Instantaneous Power Theory and Applications to Power Conditioning, Second Edition*
Hirofumi, Edson Hirokazu Watanabe, Mauricio Aredes

82. *Practical Lighting Design with LEDs, Second Edition*
Ron Lenk, Carol Lenk

83. *Introduction to AC Machine Design*
Thomas A. Lipo

84. *Advances in Electric Power and Energy Systems: Load and Price Forecasting*
Mohamed E. El-Hawary

85. *Electricity Markets: Theories and Applications*
Jeremy Lin, Jernando H. Magnago

86. *Multiphysics Simulation by Design for Electrical Machines, Power Electronics and Drives*
Marius Rosu, Ping Zhou, Dingsheng Lin, Dan M. Ionel, Mircea Popescu, Frede Blaabjerg, Vandana Rallabandi, David Staton

87. *Modular Multilevel Converters: Analysis, Control, and Applications*
Sixing Du, Apparao Dekka, Bin Wu, Navid Zargari

88. *Electrical Railway Transportation Systems*
Morris Brenna, Federica Foiadelli, Dario Zaninelli

89. *Energy Processing and Smart Grid*
James A. Momoh

90. *Handbook of Large Turbo-Generator Operation and Maintenance, 3rd Edition*
Geoff Klempner, Isidor Kerszenbaum

91. *Advanced Control of Doubly Fed Induction Generator for Wind Power Systems*
Dehong Xu, Dr. Frede Blaabjerg, Wenjie Chen, Nan Zhu

92. *Electric Distribution Systems, 2nd Edition*
Abdelhay A. Sallam, Om P. Malik

93. *Power Electronics in Renewable Energy Systems and Smart Grid: Technology and Applications*
Bimal K. Bose

94. *Distributed Fiber Optic Sensing and Dynamic Rating of Power Cables*
Sudhakar Cherukupalli, and George J Anders

95. *Power System and Control and Stability, Third Edition*
Vijay Vittal, James D. McCalley, Paul M. Anderson, and A. A. Fouad

96. *Electromechanical Motion Devices: Rotating Magnetic Field-Based Analysis and Online Animations, Third Edition*
Paul Krause, Oleg Wasynczuk, Steven D. Pekarek, and Timothy O'Connell

97. *Applications of Modern Heuristic Optimization Methods in Power and Energy Systems*
Kwang Y. Lee and Zita A. Vale

98. *Handbook of Large Hydro Generators: Operation and Maintenance*
Glenn Mottershead, Stefano Bomben, Isidor Kerszenbaum, and Geoff Klempner

99. *Advances in Electric Power and Energy: Static State Estimation*
Mohamed E. El-hawary

100. *Arc Flash Hazard Analysis and Mitigation, Second Edition*
J.C. Das

101. *Maintaining Mission Critical Systems in a 24/7 Environment, Third Edition*
Peter M. Curtis

102. *Real-Time Electromagnetic Transient Simulation of AC-DC Networks, First Edition*
Venkata Dinavahi and Ning Lin

Printed and bound by CPI Group (UK) Ltd, Croydon, CR0 4YY